The Art and Science of Medical Practice

Dr David H. Dighton

MediCause

PUBLISHER

MediCause

Published in the UK by Medicause, Medical Publisher
115 High Rd., Loughton, Essex IG10 4JA UK
www.daviddighton.com

DEDICATION

I dedicate this book to my parents and to their lifelong friends,
Robert and Lilian Harvey (my 'Uncle' Bob and 'Aunt' Lil).

My mother, Florence, was a person of faith. She spared none of it encouraging me to become the person she knew I could be.

My father, Henry ('Harry'), was selfless in providing all the support I needed to develop beyond his own aspirations.

My 'Uncle' Bob, and 'Aunt' Lil (Lilian), proudly supported and encouraged me without reserve, from the day I was born until the day they died.

With their love and support, I was blessed from birth.

About the Author

Dr. David H. Dighton qualified at the London Hospital Medical College in 1966 with MB and BS (London) degrees. In 1970, after a short time in NHS general practice, he became a British Heart Foundation Fellow in Cardiology at St. George's Hospital Hyde Park Corner, London, working with cardiologists Dr Aubrey Leatham and Dr Alan Harris. In 1973, he became a MRCP(UK), and later became a Lecturer (London University) in Medicine and Cardiology at Charing Cross Hospital, London.

In 1980, the Vrije University Hospital in Amsterdam appointed him as Chef de Clinique (Assistant Professor). Having returned to the UK in 1982, he worked both in his own private medical practice in Loughton, Essex (The Loughton Clinic, was initially established in 1973 as a medical nursing home), and at the Wellington Hospital, London. In 2000, he started a private diagnostic cardiac centre specialising in heart disease prevention and the early detection of heart and artery disease (The Cardiac Centre Loughton).

He retired from practice having been a medical student and doctor for 60 years. His retirement followed conflicts with three medical regulators. He had disagreements with them about how medicine should be practised, and who is most qualified to regulate and supervise it; views he expressed in his book, *The NHS. Our Sick Sacred Cow.* 2023.

BOOKS BY THE SAME AUTHOR

Eat to Your Heart's Content. The diet and lifestyle for a healthy heart. (2003). HeartShield.
ISBN: 0-9551072-0-2

HeartSense. How to look after your Heart. (2006). HeartShield.
ISBN: 0-9551072-1-0

The NHS: Our Sick Sacred Cow: Causes and Cures (2023)
ISBN: 978-1-3999-6027-4 (also an ebook).

How to Become Heart-Smart. A User's Guide to Heart Health and Heart Disease Prevention. (2023 1st Ed./2nd Ed. 2024. ISBN: 978-1-3999-7461-5 (also an ebook).

Who Loses Wins. Winning Weight Loss Battles: A 'Fat Mentality' v A 'Fit Mentality' (2024).
ISBN: 978-1-7385207-1-8 (ebook: 978-1-7385207-2-5)

Doctors, Nurses & Patients. How to Survive Medical Practice. (2024)
ISBN: 978-1-7385207-5-6 (also an ebook, ISBN: 978-1-7385207-6-3)
In Preparation:
Essential Cardiology for Students.
The Life Equation

CONTACT: go to 'www.daviddighton.com' or email: david@daviddighton.com

Poem: The Art of Medicine

Who will hold a troubled hand,
And take guardianship,
Of enough hope and trust,
To settle a troubled mind?

Who will step forth with knowledge enough,
Sufficient to make disease submit;
Halting its progress,
Against the odds?

Who will come forth with fortitude enough,
To assume the care of a life, or a death,
And through trust and confidence,
Encourage a smile to break the grip of fear, pain, and grief?

Who will come forth,
With humility enough,
To step beyond the science,
And embrace the human spirit?

Only those who can, become complete physicians.

Acknowledgements

Tom Rock was a medical student when he inspired me to write this book. He is now Dr. Tom Rock; this book has taken that long to write. This was after he experienced my efforts to combine the science and art of medical practice at the Loughton Clinic, which was for forty-eight years (1973-2021), a private medical and cardiac investigation centre. Tom suggested I should describe how practising the ancient and dying art of medicine, combined with the science of medicine, can benefit more patients.

I owe a debt of gratitude to my colleague and friend, Dr David Baxter. Together we started the Loughton Clinic in 1973. Without him, I might not have created my practice or faced the clinical challenges that enabled me to write this book.

Brief Table of Contents

Detailed Table of Contents	XI
Preface	XVII
Introduction	1
PART ONE	9
1. Briefing Notes for Medical Apprentices	10
2. How Doctors Function	43
3. Doctors as Characters	63
4. About Patients	105
5. Patient Charicatures	130
6. The Doctor-Patient Relationship	167
7. Cultural Difference and Clinical Medicine	180
PART TWO	199
8. Patient History Taking	200
9. Clinical Examination	229
10. Other Consulting Room Matters	242
11. Patient Investigation	250
12. Clinical Process	272
PART THREE	292

13. Clinical Decision Tools	293
14. Clinical Judgement	336
15. Clinical Management	407
PART FOUR	477
16. Basic Clinical Research	478
Epilogue	487
Glossary of Terms	489
Bibliography, Notes & References	492
Further Acknowledgements	505
Copyright	514
Appendix A	515
Appendix B	521
INDEX	523

Detailed Table of Contents

Introduction: Frontiers

PART 1

CLINICAL ENCOUNTERS

Briefing Notes for Medical Apprentices: Medical students; medical knowledge and starting work; the apprentice and the patient; fundamental skills; a talent for medical practice; 'you can't teach it'; who has the talent; specialisation; functional doctors; castaways; the doctors we get; unexpected sources of advice; who should influence doctors?; other important issues for doctors and other medical professionals; the science and art of medicine; personalised medicine: is it something new?; practising medicine: a sovereign duty?; certainty and medical practice; Security and medical practice; social division, culture and medicine; the call of duty; a few pertinent aphorisms; know your acronyms; doctorates and doctors; can we choose our patients?; The cultural challenge; *Vive la différence*; the medical elite; money and medicine; what can be achieved?; NHS v private care; observation, observation, observation; more observations for apprentices; to be or not to be a Good Samaritan; fame and fortune; life planning; is health the absence of disease?; research matters; a private doctor's apprentice; medical practice ain't what it used to be; Computers Ahoy!; Plus ça change; Good intentions; Final thoughts.

How Doctors Function: The Right Stuff?; worthy doctors; stereotyping; nebula thinking; medical colleagues; doctors as eccentrics; the patient's perspective; Two Medical Cultures: attached and detached; managing patients; senior doctors; Old dogs, new tricks?; stages in a doctor's career; how doctors progress; medical management and sex chromosomes; performance skills; Good behaviour, bad behaviour; belief in perfect health; respect for others; some final words of caution.

Doctors as Characters: Types of Doctor; A Compendium of Doctors as Characters; ABCD; Abrasives; Actors; Altruists; the arrogant; Autistic; Bombast; BBC: Baked Bean Counter; Bias aware; Blackmailers; Bores; BPT: Big Picture Thinkers; Bullies; Carriers; Capricious Creatures; Charming; Club Members; Cold fish; Combustible consultants; Completer finishers; Control freaks; Cool dudes; Corporate kids; DAS: Doctors as Scoundrels; Dedicated doctors; Diagnostniks; Dismissers; Disrespectful; DPSWs: Dedicated Public Sector Workers; DOD: Daft Old Doctors; Dork; Egoists; Emotionally unfit; ERP: Emotionally Reactive Person; Favorito/a; Flirts and Flatterers; Frustrated doctors; Fumblers; GAHAD: Good At Hiding All Deficiencies; GCs: Good Guys; Gossips; Grammar School vs Public School; Gunslinger; Haughty; Hero, Champion and Role Model; High-Minded; Hobby Horse Enthusiasts; Hobbyists; Hollow Men; Humorous Fellows; Idol or Idle?; Jealous Creatures; JGGs: Jolly Good Chaps; Know it All; Lazy on Purpose; Legalist; Little Sh*ts; LMF: Lacking Moral Fibre; Lone Wolf; Lucky Buggers and the Matthew Effect; Manipulators; Mean Bastards; Military; Moralist; NAGs: Needs A Good Sh*g; Names; Naturals; NBWs: Nasty Bits of Work; Nice; NPC: Non-Playable Characters; Obsessives and Nit-Pickers; Old Bailey Types; Old Vic Types; OOSOOM: Out of Sight, Out of Mind; OVIBOI: Overly Impressed By his/her Own Importance; OWT: One-Way Traffic; Pedantic Pedagogues; Persona Non Grata (PNG); Phraseologists; PIMP: Power is My Purpose; PITA: Pain In The Arse; PLASMA: Person Lacking Any Sort of Personality; PLOJ: Person Lacking Observable Judgement; PLOP: Person Lacking and Observable Personality; Politicians; Preppies; Pretty People and Charisma; Quants; RA: Risk Averse and Not Risk Averse (NRA); Reliable / Unreliable; Romantics; Ruthless; Scammers; Scary Guys; Scholar; The Selfless; Serious All The Time (SATT); SD: Shifting Dullness; Slobs; Smart; Smoothies; Spanophiliac; Star-Struck; Stern; Stressed Doctors; Structured and Unstructured Thinkers; Stüm(mer); SS: Supercilious and Sanctimonious; Swimmers and Sinkers; Team Players; Thin Skinned. Thick Skinned; Types A & B; USBN: USeless But Nice; Uptight and Bottoned-Up; Visionaries; Wimps; Wise Sage.

About Patients: Desirable patients?; What do patients need to know?; It takes all sorts; patient scenarios; love lost and lonely; pre-patients?; patient demography; Being a patient; Clarity and Confusion; managing patient information; clashes of socio-economic class; victim or patient?; 'Life', the play; TATT: Tired All The Time; nail in the foot syndrome; *L'impasse*; the Games People Play;

Patient Caricatures: Specific Types of Patient: Advantage takers; ALAC: Acts Like a Child; ALBer: Arrogant Little Bleeder; Amateur doctors; Badly behaved; Bias unaware; Bereaved; Blackmailers; BWMM: butter wouldn't melt in their mouth; clever guys; complainers; controllers; cool dudes; corporate kids; cranks?; criminals; dictator; digital beings; doctors as patients; duper's delight; the fearful; interesting people; gamers; GG: Good Guys; Grammar School, Comprehensive, or Public School?; Hopeful patients; Incapable; Influencers; Interrogators; IS: Internet Savvy; Kidnappers and Hijackers; kind and genteel; lazy buggers; Liars; MAE: Mental Age of Eight; Manipulators; Martyrs; MASH: Mad AS a Hatter; MOMA & POPA; Neanderthal; neurotic & TFBF: Too Fussy by Far; ¿No Entiendo?; OK Patients; objective; obligators; PITA: Pain In The Arse; range riders; recalcitrant recidivists; resentful; Rich, Famous and Conceited; SPP: Self-Promotional Protagonist; Selfish and greedy; SELPO: Self-Possessed; SHUBE: Sentient Human Being; Simple Folk; Slobs; Smiling depressives; Spies and intruders; spiritual; stalkers; stand-ups; Superman Superwoman Complex; Takers and users; troublemakers; TYpeA & B; Vanities and Gods; vulnerable patients; Wealthy; Wimps?; WISC: Wolf in sheep's clothing;

The Doctor-Patient Relationship: Patient education; a few basic communication principles; inductive thinking; the Goldilocks Principle; Catastrophe Theory; Chaos Theory; Clustering; Universal Flux; A British Love Affair; Breaking up is hard to do; Bending the Rules; style; Old Hat?; refugees from the health service; friendly requests; No room for ambiguity; the Hippocratic Oath; some classical background; Do no harm; Good Medical Practice The GMC's Guide; Platitudes;

Cultural Difference and Clinical Medicine: People's Culture; The Nature of Culture; Cultural History; Family Culture; Anthropology and Medical Culture; Cultural Dissonance; Culture and Disease; National Culture; Family and Tribe; Humour and Culture; Language and Culture; Gender and Sexuality; Beauty, Fame, Fortune, and Criminal Culture; Religion and Medicine;The Mixing of Cultures; Equality Issues; The Tendency to Stereotype; The Rich, The Poor, and those between; Indirect Cultural Discrimination; Intra-Cultural Perspectives; Medical Practice and Cultural Bias; Culture and Practice Location;

PART 2

DIAGNOSIS, DIAGNOSIS and DIAGNOSIS

Patient History Taking: The Initial Encounter; Quo Vadis? Patient Reception; In the Consulting Room; Pre-History Evaluation; History Taking; Styles of History Taking: The Executive Style; The Focussed Diagnostic Style; The Interply Style; The Discursive Style; Forensic History Taking; Ask Patients What they Think?; Finkelstein's Shame; Specialised Questioning; Language Use; *Gauch ou Droit* ; Past Medical History; Family Medical History; Social History; The Biological Aspects of Social Stress – 'Clinical Energetics'; Therapeutic History; Attitudes to Drugs; Aspects of Therapeutic management; Drug Rep's and Treatment; History Taking – Points not to be missed;

Clinical Examination: Examination and trust; examining room behaviour; diagnosis at first sight; not to be missed; unhelpful physical examination; Examining the heart; The Lungs; The abdomen; Neurological examination; Leg examination.

Other Consulting Room Matters: Doctor-Patient relations; The Stimulus-Response model; Traffic Analysis, Parallel Information, Meta-Data, and Stress; The early detection of disease; Zoning-out and Emotional Reminiscence; History Taking and Computer Analysis;

Investigation: Diagnostic uncertainty; diagnostic accuracy; Interpretation; appropriateness; urgency, corroboration; clinical management; test worthiness; clinical progress; medico-legal investigations; Why Investigate?; some politics of investigation; Investigation changes; Essential Bayes' Theorem; There will always be error; test appropriateness; test interpretation; 'Why am I always tired, doctor?'; 'Ma': Silence and the space between; Daft investigations; Which tests?; New tests; Unnecessary testing; Investigation management; Cardiac Screening Investigations; My Cardiac Prevention Policy; Worthy and Unworthy Investigations; Investigations: Risk and Benefit; Patients Requesting Investigations; Great Expectations; Patients Refusing Investigation; The Pitfalls of Investigation.

Clinical Process: Common and Rare; Clustering and Entropy; Diagnostic Talent and Process; Information, Information, Information; Time and Diagnosis; Evidence, Evidence, Evidence; Life, Death and Diagnosis; Assigning weight to evidence; Presumptive emergency diagnoses; Timing Issues; Steps in Diagnosis; The Diagnostic Value of Surveillance; Difficult and Elusive Diagnoses; Blood Rhubarb Level?; Dynamic Diagnostic Situations; Pre-Diagnosis; Efficiency and Attitude; Diagnostic Intuition; Discussing the Diagnosis; Presenting Your Diagnosis; Diagnosis in the Future; Reliable New Tests Welcome.

PART 3

THE TOOLS OF CLINICAL MANAGEMENT

Clinical Decision Tools:

PART 1: The Fundamental Tools for Clinical Decision Making.
Fundamental Tools for Clinical Decision Making; Statistics; Information Technology; Clinical Reality and Reliability; Clinical Risk; Bayes' Theorem; Negative Diagnosis and Road Testing; Science, Art and Chance; Corporate Algorithm; Medical Computing; Prediction and Diagnosis; Creativity; Can Computers be Creative? Complicated or Complex; Simple Decisions Making: Tossing Coins; Nebula Processing: Choosing Stars; Chaos Theory, the Butterfly Effect and Medical Practice; The Misuse of Statistics and Algorithms; Clinical Statistics and Anecdote; Aspect; Obvious or Just a Probability Statistic?; The Proliferation of Information; Research Considerations; Dimension as a Consideration; Grand Scale, Small Scale; The Power of Anecdote; God Complex; False Logic and the Wrong Questions; In Search of Truth; Computing the Truth; A Place for Academics; Generalisation and Validity; Summary Thoughts on Clinical Numbers;

PART 2: Computer Considerations; Do we control computers, or do the control us?; Artificial or Real Intelligence or Both?; Machines and Algorithms; Algorithms and Computer Programming; Algorithmic Awe; Big Data.

Clinical Judgement: A Few Aphorisms; A Golden Rule for Judgement; A Golden Rule for Diagnostic Judgement; The Process of Judgement; The Consequences of Poor Judgement; Bounded Judgements: of time, measure, and value; Estimation of value; Judgements of compliance; Everyday judgements; Meta-Information; Clinical Reasoning and Judgement; Basic Rules for Clinical Reasoning; Deductive, Inductive, and Abductive Reasoning; How Reliable is Information?; Fundamental Thinking; Other Modes of Thought: Insular, Nebula, Lateral, Reversed, Wrong; Bunker Thinking; Knight's Move Thinking; Best Guess (Abductive) Thinking; Types of Situation: BLOB, NAEFAD; Straightforward; Complicated, Complex and Flummoxing; Two Flummoxing Cases; To Screen or Not to Screen?; Clinical Wisdom; Judgement Bias; Biases Warp Judgement; List of Biases from Ambiguity to Zero Sum, from Anecdotal to Wizard of Oz Bias; The Wrong Questions; Judging Judgement; Referral Decisions; Judgement and Regulatory Compliance; Beware of Acronyms; Poor Judgement; Faulty Judgement; The Error of Presumption and Assumption; Errors of Omission; Weak Reasoning; Faulty Trains of Thought; Taking Advantage; Taking Advice; Good Judgement; Asking Pertinent Clinical Questions; Clinical Evidence; Insufficient Evidence Situations; Published Research Evidence; Inappropriate Evidence Situations; No Evidence Situations; Personalised Medicine: Evidence-Based Medicine; Clinical Judgement, Anecdote and Algorithmic Thinking; Balance of Risk; Questions of Clinical Judgement; Faulty Decisions Making; Falsification; Judgment at the Hand of Mirth; Judgement by Analogy; Exception-Led Judgement; The 'Classic' Case; Exception-Led Defeat; Exception-Led Judgement; Influencing Wise Judgement; Judgement by Committee; Information and Good Judgement; The Weighting Game; The Music of Decision Making; Judgement under Pressure; Time and Judgement; Speculation; Motivation, Thinking and Judgement; Convincing Others; Judgment and Chance; 'Daft' Ideas and Judgements; Measurement Errors and Errors of Judgement; Meta-Data, Meta-Analysis and Meta-Judgement; Experience: Time Spent Weighting; Road Testing; Efficacy; The Dynamics of Illness: Judgment, Timing and Clinical Trajectory; Golden Rule of Timing in Clinical Practice; Useful Action Theories: Catastrophy Theory, Chaos Theory, Game Theory; Evidence for Presumptions; Dogma and the Halo Bias; Observations and Theories; Irrelevant Information; Serendipity; Data collection and Data Addiction; Theory Bias; Talent and Predictive Talent.

Clinical Management: How cardiac Management Has Changed; The Aims of Clinical Management; Clinical Management Styles; Of What Value is Experience in Clinical Management? Our Sovereign Duty: Management Responsibility; Personal Management before Personal Management?; Coping with Personal Stress; Managing Practice Culture; Briefing; Flying Solo and the Art of Time Management; Clinical Titration; Action Timescales; Prognostication; Managing Patient Expectation; Is Fully Informed Consent Attainable?; Delivering News; Doctors as Therapeutic Agents; Hope and Fear; The Effect of Fear; Horses for courses and Trust; Some Therapeutic Management Tips; Therapeutic Homeostasis; Atrial Fibrillation; Angina; Hypertension; Benzodiazepines, Hypertension and Medical Politics; Team Practice and Management; Team Ethos; Ways of Handling Patients; Crucial Dynamics; Clinical Responsibility and Management; Medical Management and Trust; Telephone Management; Risk Management; Other Aspects of Risk; Risk Assessment and Outcome; Fashion and Clinical Management; Fashion Items; The Equality Bias; Medical Hubris; Some Management Aphorisms; Patient Review; Political and Cultural Factors; Urgent Management; Patient Engagement; Some Cultural, Ethnicity and Communication Issues; Preventative Management; Review, Review, Review; Pharmacist Physicians; Out-of-Hours Requests; The Law of Indulgence; Who Actually Wants Your Advice?; Practice Style and Clinical Management; The Proximity Principle; GP Practice; Jungle Medicine; Village Medicine; Machine Dependent Medicine; Outcome and Patient Type; Diffusing Situations; Making Mistakes; Customer Service and Medical Management; Simple Management and Attention to Detail; Entrances and Exits.

PART 4

Basic Clinical Research: Clinical Research. Is it a duty?; Koch's Postulates; Observational Research; The Cardiac Value of Food; Simple 'Experimental' Clinical Medicine; Therapeutic Research Trials; Questionnaires; Researching the Literature and Publishing; Allowing Patients to Take the Risk.

Preface

The abundance of excellent teaching sources means that anyone with the academic ability, can learn the science of medicine. There is an art, however, to how the science of medicine is applied to each patient.

Those with a natural talent for the art of medicine will leave patients feeling they have consulted a sentient being, proficient in the care of others: one with a vocation to help others; one worthy of their trust; one who understands and respects them as an individual; one who has their best interests at heart. In keeping with the pursuance of all arts, only those naturally able to express themselves aesthetically, without a need for rules or guidance, are capable. The same capability, possessed by those with little knowledge and few practical skills, is present in all professions; they are thus independent features. The science and art of medicine remain two sides of the same coin: one of value to all those with a medical need. Throughout the text, I have noted each art, as it arises.

For millennia, the purpose of those privileged enough to attain medical knowledge has been to relieve mental and physical suffering. The gradual extension of these aims has helped to prevent disease and disability, and to lengthen healthy, normally functioning life. The science of medicine has enabled progress in all these areas, while the art of medical practice has determined how such benefits are best delivered for patient satisfaction. The science and art of medical practice represent two sides of the same coin: inseparable but independent.

For millennia, doctors practiced the art of medicine with little science. Reference to the art of medicine is made many times in the original Hippocratic Oath. Now, many doctors practice medical science with little interest in the art of its application. Since almost every patient appreciates the art of medicine, many now seek care from others.

Patients also appreciate the different ways of taking tea. Consider just two:

1. Zoe puts a coin into a machine on a railway station platform. After one minute, she will walk away with a plastic cup of tea made by using bulk purchased tea leaves collected in Darjeeling.

2. Zoe receives a letter. In it, there is an invitation to meet a friend at Claridge's Hotel in London. During afternoon tea, an elegant teapot containing her choice of tea is served, in surroundings

that ensure her pleasure.

Every provider will deliver their service using their own style; each style creating a specific ambience. Machines can deliver impersonal standardised services, so can people with machine-like qualities. More interesting, admirable and often more memorable, are those who bring a personal form of art to their service.

The rise of the machines has allowed the easy regulation and control of the type and amount of tea served on railway platforms. Machines are functional, cheap to operate, readily serviced, and available to everyone. Personal services are different. They require forethought, talent, product excellence, and a desire to serve discerning people in satisfying ways. Such services are costly, not standardised and not available to all. Those in favour of machines may not appreciate the difference.

The idea of becoming a doctor is awe-inspiring for many, and beyond the aspirations of most. When I was twelve-years old, my GP, Dr Julian Hardman, after learning about my intention to study medicine, informed me that 'many come, but few are chosen'.

Most of those with the potential to gain entry to medical school have a limited idea of what a doctor actually does, and what knowledge, skills and aptitudes are required. Unless they come from a medical family, soap operas rather than reality, may colour their conceptions.

This book aims to expose the fantasy and reality of medical practice. One aim is to provide a map and compass for those who wish to make the journey. Another is to provide the clarity needed to become a carer most patients will appreciate.

For those dedicated to caring for others, few professions are more rewarding than medicine, nursing, caring, or paramedical work. All those who save lives can be equally proud of their work. They include those who work for the police and every fire and rescue service. Together with caring for others and preventing illness, most humanitarian functions have until recently, been sacrosanct, with sovereign duties beyond the remit of bureaucracy. Unfortunately, doctors, nurses, carers and paramedics in the UK, now find themselves undervalued in an over-managed, over-regulated, bureaucratic era, with medical executives, regulators and bureaucrats progressively gaining power over them.

Introduction

> *... so he escaped into your arms,*
> *you wonderful sister of Truth and Goodness alike,*
> *you who console the sufferer -*
> *you, Art, you come from above.*

Franz Grillparzer (1827). Funeral Oration for Ludwig van Beethoven

Science ... is the torch which illuminates the world.

Louis Pasteur

While medical science continues to advance, medical apprenticeship will remain lifelong.

Those who gain entry to medical school must prove their scientific orientation, and their understanding of scientific methods. Because of their youth, they may not have lived long enough to know what else is important - how best to deal with people; how to care for them in a pastoral sense, or how to benefit individual patients.

Those who succeed in becoming medical students will later choose a specialty; one which best reflects their interests and skill-set: manual dexterity for surgeons; an interest in physics and engineering for radiologists and cardiologists; mathematics for epidemiologists and humane and inter-personal skills for every physician, psychiatrist and general practitioner.

What is required of those who become doctors? This book details the commitment, the talent, and the knowledge required to step beyond the kudos and the glamour of the job. Unfortunately, the tide has turned for medical educators. Medical students once needed to be broadly educated, and were chosen as doctors, only if they could think for themselves. Now students must show their willingness to comply with an ever-growing corpus of rules, regulations, and guidelines, made by regulators for two reasons. First, the students chosen must believe that the rules are made to be 'in the public interest'; second, to gain control over medical practice. The guidance they define as rules (with legal advice) allows them to judge our actions, and punish doctors 'in the public interest'.

To comply, doctors, nurses, carers and paramedics need training, not education. The educated will use their knowledge and experience to judge for themselves; those who are trained will be taught to adhere to the explicit instructions of faceless corporate beings, without clinical acumen. The corporate bodies regulating medical practice in the UK, now insist that we all are trained. Over the last forty years, the resulting changes in medical culture have adversely affected the quality of UK medical practice.

Before starting the long journey to become a doctor, it is important that students cast off any delusions about how medical professionals function, now the profession is regulated by politicians, corporate directors, managers and lawyers, who know as much about medicine as the average hospital visitor.

Junior doctors are leaving the UK medical profession in significant numbers, while older doctors are choosing to retire early. Why is this? It is an important question for every would-be doctor at school, medical student, nurse and doctor to ask. Although regrettable, one cannot ignore the many important political issues which now distract nurses and doctors from their patient caring role.

Because there are no fixed rules governing any art form, bureaucrats need to push the art of medicine aside in favour of what they can regulate; namely, the science of medicine. The art of medicine involves coming to understand patients as individuals, as members of a family, within the context of their community and culture. Sometimes it may involve getting to know them like a friend or family member. A doctor may then find it easier to decide the most appropriate path for a patient.

It is by practising the art and science of medicine together that trust is more readily awarded. Knowing when to act, and when to retreat based on clinical evidence, is a vital medical art that flourishes only with experience. It probably depends on an inborn potential for wise judgement, since so few attain it.

There are other vital arts for doctors to practice and deploy. They concern getting to know patients and colleagues alike, learning to 'read' them correctly, revealing their hidden agenda to predict their future behaviour. An open doctor-patient relationship, is the *sine qua non,* of effective medical practice, especially when long-term relationships are involved. Part of the art involves understanding non-verbal behaviour, seen during the history taking and clinical examination process. To be guided by knowledge of the patient and by clinical experience, are indisputable keys for making correct clinical diagnoses. In the light of doctors now offering only tele-medicine, grave errors can be expected, except for those dealing with the most elementary and obvious medical conditions.

Clinical decision making, clinical judgement, and the management that follows are information and logic-guided arts, based on the ability to select only the most appropriate facts from a nebula of both interconnected and disconnected bits of information. Nebulae are cosmic places where stars are born and die. In the same way, what I have referred to as mental nebula processing, generates and kills off new ideas by connecting and weighting the relevance of all the information available. Weighting each bit of clinical information appropriately is an intuitive art, made better only by experience. Although not easily specified, neural networks and AI are attempting it using probability calculations. Once developed, AI systems might allow nurses, carers and paramedics to get closer to a correct diagnosis, without knowing how close they are.

Another art involves the interpretation of evidence, and the results of multiple scientific trials. When there are conflicting results, one must decide which are the most justified and reliable, and which are most applicable to a particular patient. There is an art to clinical research, which is mainly focused on designing experiments to reveal the truth and falsity of any theory.

There are open secrets to creating loyal patient relationships, with clear and honest paths of communication that aid accurate diagnosis and sagacious management, in a safe and trusted manner. Not every doctor sees the need to learn them. For those interested in practising both the art and science of medicine, there are secrets to be revealed. Apprentices once learned these, standing next to a learned clinical tutor. Now precedence is being given to avoiding regulators, and completing the often pointless and mind-numbing records and audits required by bureaucracy. All this wastes the time we once had for patients. Is this why every view we get from inside hospitals shows banks of nurses and doctors sitting in front of computer screens rather sitting with their patients?

What does it take for a doctor to become successful in the eyes of patients? What does it take to become a successful member of the medical profession? What does it take for a doctor to satisfy medical regulators, and are all these separate requirements compatible? These are questions that should concern every patient and doctor.

To learn, understand, and put the science of medicine into practice requires intelligence, cognitive and organisational abilities, together with interpersonal and dextrous skills not available to all. To understand the art of medicine requires a completely different, more nuanced set of qualities. Neither applies much to the management of simple medical cases, but in my experience of handling complicated cases, where the diagnosis is not immediately obvious, only applying both to the full will bring a practicable solution to patients.

Only the art and science of medicine practised together can inspire the trust and faith of patients. Doctors, nurses, and all those who care for patients should be able to adapt their approach to the personality, demeanour, education, gender, culture, and status of each patient. A doctor must be able to translate evidence-based medical science effortlessly into words each patient can understand, while sometimes acting as wise friend, guide or guru. She must be able to relate to each patient as a unique sentient being, with their own culture, values, attitudes, mind-set and emotions, and not as an anonymous participant in a scientific research study.

To practise the art of medicine requires patient handling that recognises each patient's personality, intelligence, education, tradition and culture. Practised at best, each patient will leave their doctor, nurse or carer's presence, feeling they have been treated by someone who cares; someone they can trust, with reliable and trustworthy hands. Someone to whom they might entrust their life.

Frontiers

In the rainforest on Pu'u Kukai, Maui, Hawaii, heaven is said to meet Earth.

Interesting things happen at frontiers. Many frontiers exist between doctors and patients, and between doctors and their colleagues. At the frontier between patients and their acquaintances, new attitudes form constantly about doctors, nurses, paramedics and carers, and how they function.

The important frontiers I will consider are:

- Between the science and the art of medicine.

- Between sentient human beings and computers.

- Between anonymous medical corporations like the National Health Service (NHS), the General Medical Council (GMC), the Care Quality Commission (CQC), and the Professional Standards Authority (PSA), all of which are directed by politicians and the bureaucrats and lawyers they command, without interfacing personally with patients or medical practitioners.

- Between the NHS and independent private practice.

- Between the wealthy and the poor (the health divide)

These all affect patient morbidity and mortality. All affect the health benefits and disadvantages experienced by patients. All influence how we practise medicine in the UK.

The practise of medicine is sacrosanct, for one simple reason: it holds the lives of patients in its hands, with no purpose other than to respond effectively to medical need. Many political issues interfere with this purpose. Medical practice staffing, funding, and the metastasised bureaucracy of the NHS, all have affected patient morbidity and mortality. At present, in private practice in the UK, these influences are less apparent, and patients benefit accordingly.

In the UK, a patient's life and death can depend on their wealth. It can depend on whether they can choose to be treated privately, or must accept NHS treatment. Many patients in the UK are blind to this, choose to ignore it, or have no option but to accept the NHS as a sacred cow, beyond criticism, and perfect in every way. The wealthy can choose their doctors and their medical services; NHS patients must put their faith in whatever politicians and medical bureaucrats choose to provide for them. NHS medical bureaucrats, who are mostly anonymous, believe they are qualified to direct the way doctors and nurses practise. In private medicine, patients and experienced doctors and nurses are the ones who decide.

A frontier formed between the new NHS and private medicine in 1948. That frontier remains and separates two distinct medical cultures.

What do students at school, medical students, junior doctors and the public need to know about medical practice, and what might they learn from this book? Here are a few:

1. A doctor's part in a patient's life, and *vice versa*.

2. What doctors think of their colleagues.

3. What doctors think of patients.

4. What patients think of doctors.

5. How patients can rate doctors.

6. What to expect from the science of medicine.

7. What to expect from the art of medicine.

8. The answer to a question posed by many patients in a tight spot: 'What should I do now, doc?'

9. The proper use of statistical evidence to guide clinical practice.

10. The devils in clinical detail, few patients get to consider.

11. Clinical style.

12. How to make useful discoveries that help patients.

13. How Dr. Nobody made useful discoveries in his garden shed.

14. How to make difficult diagnoses and resolve puzzling clinical situations.

15. The various styles of clinical history taking, including forensic history taking.

16. The value of clinical experience.

17. How to think like an experienced physician.

18. When to see patients repeatedly; sometimes, until the diagnostic penny drops.

19. The many aspects of investigation.

20. The most valuable clinical decision tools.

21. How to manage cases at a technical and personal level.

22. Rules and regulations, and their consequences for patients.

23. The sovereign role, rights, duties, and clinical responsibilities of health professionals.

24. The future of medicine in the UK.

25. When to pack your bags and change jobs.

'Literature (like medicine) is not a mere science, to be studied; but an art to be practised.'

Sir Arthur Thomas Quiller-Couch. Preface (November 1915). *On the Art of Writing*. Cambridge University Press 1917. Reproduced by Dodo Press.

One of the latest trends in art is to achieve self-justification by incorporating some science. Art, however, should revel in not being bound at all by the immutable rules of science. Art is free to add something new to reality, while science continues its struggle to understand reality and how it works.

I have written this book for those intending to become a doctor or nurse; those who wished they could become a more effective doctor or nurse, and any lay person interested in medical practice.

The view I describe of medical practice is my own, taken from my own observations as a doctor, but also acting as a 'fly-on-the-wall' for over fifty years. The perspective I present here has resulted from 58-years of personal observation as a medical student, a onetime NHS GP (1968-70), a British Heart Foundation Junior Research Fellow (St. George's Hospital), London University Lecturer in medicine and cardiology (Charing Cross Hospital), Chef de Clinique in Cardiology (Free University Amsterdam), and a medical and cardiac consultant in private practice working at the Loughton Clinic, and Cardiac Centre Loughton, and the Wellington Hospital, London. Although retired, I remain a perpetual medical student and medical apprentice.

This book supports my belief in both the art and science of medicine, the development of clinical experience through apprenticeship, the need for academic clinical research, with enough evidence and experienced-based thinking to underpin the clinical judgement and decision making that is sagacious enough to benefit patients.

I have reflected on my years of diverse post-graduate clinical experience, offering guidance not only to doctors, nurses, and health workers, but also to those beginning to think about dedicating their lives to healthcare. I hope to interest pre-medical students ('A' level students who are considering medicine as a career), medical students, doctors, nurses, care workers, paramedics, alternative practitioners and patients, in what it takes to be worthy of a medical role. Anyone who wants to understand how medical practice works at its most engaging, personal, relevant, efficient and successful level, will find something of interest within.

For almost forty years, nobody paid me but my patients, most of whom returned to my practice repeatedly. I take this as evidence that I must have helped a few!

Let me define the 'art' of medicine. A dictionary will define art as the product of human creativity and imagination. For me, it is the aesthetic, holistic element I tried to apply to every experienced-based and evidence-based clinical judgement used for patients as distinct individuals, rather than as anonymous beings assumed to be 'average'.

When playing the piano, pianists are free to use the style, timing, and emphasis they think best suits the music. Their aesthetic choices determine their performance. Compare this music to that played on programmed pianos. They used to use perforated, pre-programmed rolls as analogue algorithmic programs; electronic keyboards now use MIDI files to reproduce the same music every time, albeit with something missing: human aesthetic touch and variation.

Like every musician, doctors who employ the art of medicine must learn to use timing, emphasis, style, co-ordination, and executive functioning. They must learn the art of interacting with others, together with the use of compassion, empathy, sympathy, patience, understanding, helpfulness, and kindness as tools. The aim must be to benefit each patient and all those important to them. None of these considerations is necessary for the practice of scientific medicine.

Whereas scientists can readily understand humanitarian concepts (although not necessarily their subtleties), few who studied humanities understand much of science. In the chapter entitled 'Doctors', I examine an important duality: doctors whose primary approach to patients is scientific, 'detached' and anonymous, and those whose approach is 'attached' and humanitarian. Both approaches produce results, but in ways not always acceptable to every patient.

Patients need medical science, but once a person becomes a patient, it is usually those doctors and nurses with a talent for the art of medicine who they welcome most. There are reasons for this which apply to every art. Unlike the transitive purposeful process of training to achieve distinct clinical objectives, the art of medicine embraces the intransitive. It requires us to use our sense of being. There need be no particular end-point in mind, other than patient satisfaction.

From a bureaucratic point of view, there is good reason to displace the art of medicine. The art of medicine defies measurement. One cannot control or manage that which one cannot measure, so it will suit those tasked with regulating medical practice to suppress, ignore, or denounce the art of medicine, in favour of scientific medicine. For this political reason, the trend over the last few decades has been for the science of medicine to displace the art of medicine; but is the art of medicine being actively suppressed? Will the lawyers and business managers who regulate the NHS override it, and give credit only to medical science? If they succeed, patients could easily come to regard doctors as anonymous technicians.

More and more, the public is being tacitly invited to accept NHS doctors with an ever more detached corporate attitude, and to accept diagnostic computer interfaces, rather than human contact. Since these trends offer financial savings for governments, they will gain political favour. Medical bureaucrats will have the job of promoting these ideas as reasonable and necessary. None of them apply in private medicine, where a personal, tailored approach is what patients choose to pay for.

NHS practice culture diverged from private practice culture early in its history. It is UK private medical culture that is mostly found in countries other than the UK. When some NHS patients travel to Europe on holiday, they will meet this medical culture for the first time.

The bureaucratic acculturation of the medical profession is now so complete, many NHS doctors and nurses never question it. It is now a sacred cow; heresy to criticise (see my book: '*The NHS. 'Our Sick Sacred Cow'*). From a regulator's point of view, private doctors often see things differently and need to be brought in line.

Doctors might seem to have a free choice in the way they practise, but in the NHS, bureaucracy insists on it being standardised and controlled. Many doctors restrict themselves entirely to the requirements of medical regulators, and a science-based, check-list form of medicine. Those who dare to use the art of medicine in their clinical management, risk allegations from colleagues and many other sources.

In keeping with every other art, medicine holds many intangible mysteries and secrets. Doctors and nurses express them as they use clinical thinking, intuition, prediction, risk assessment and case management, whether applying them to diagnosis, judgement or decision making. Practising medicine requires many other abilities. Every apprentice, regardless of their talent, can assimilate something of their use by observing a clinical master at work. The desire to progress from clumsy amateur to fluent virtuoso should provide them with some motivation. I wrote this book to encourage this progression.

I advocate that medicine, practised at its best, must always combine the art and science of medicine. It will prove much more difficult, however, for those who work for corporations like the NHS, knowing they must comply with enforceable management rules and streams of directives and targets, rather than think for themselves and express their own opinions.

Much of what it takes for a doctor or nurse to become successful in the eyes of patients has remained unchanged for millennia. What has changed, is not all progress. What it takes to become a successful doctor or nurse is to be knowledgeable and proficient, but also affable and engaging enough to gain the acceptance and respect of patients. The next steps are more difficult: to gain a patient's trust, and to become worthy enough to be entrusted with their life. The process starts by selecting students with 'A' Levels thought most suitable for a medical education (usually based on their science examination grades). A background in science is a must, even though many doctors will spend most of their lives dealing with humanitarian issues.

For millennia, doctors have shared their patriarchal guardianship of citizens with the clergy, and a few wealthy philanthropists. The public has traditionally trusted doctors and nurses with their care when sick, but there are questions. Does medical practice engage in an unacceptable form of patriarchy (or matriarchy), and is caring for others now being derided?

Do doctors still have the sovereign right and duty to hold primary responsibility for the sick, or is the bureaucratic grand plan to reduce their role and replace them with staff less qualified, or with medical AI, entirely controlled by corporate managers? If medical bureaucracy in the UK does nothing to benefit patient morbidity and mortality, is it worth paying for, and does it deserve any role other than housekeeping? I challenge them to provide the evidence of benefit they proclaim, and to justify their existence.

Instead of including it here, and mixing too much of politics with medical practice, I have reviewed issues about controlling the medical profession in a separate book, entitled *'The NHS. Our Sick Sacred Cow.'* For those who have chosen a career in medicine and nursing, many political issues need addressing. One can no longer ignore the difficulty patients experience getting to see an NHS GP, or ignore queues of ambulances outside A&E departments, blocked hospital beds, a lack of joined-up social care, patients being treated in corridors, and deaths on waiting lists. One cannot ignore that 'the NHS is now on its knees' (Keir Starmer, Labour Party Leader, September 2022).

Anyone who works in a team must learn to relate to their colleagues and those who use their services. Apart from the job or practising medicine itself, every doctor must understand and appreciate his colleagues, and his patients. It may not be possible for everyone to understand others like Theophrastus, Shakespeare, or Dickens, but making an attempt is obligatory. In Part one, the chapters on 'Doctors' and 'Patients' should help expose each character to be encountered.

Also in Part One, I discuss the doctor-patient relationship as central to the practice of effective medicine; this includes surgery, and all other medical specialties. Depending on their personality and outlook, doctors will practise in a patient-attached way, or in a patient-detached way. I discuss the importance of this distinction. I discuss the doctor-patient relationship and its most important features. Will it remain sacrosanct, and should we block all attempts of regulators to interfere?

Cultural diversity presents every doctor and nurse with communication problems, issues of mutual understanding, and sensitivity to different values, behaviour patterns, beliefs, and traditions. This applies as much to the wealthy and the uneducated poor as it does to ethnicity, given the enormous differences that now exist in the social distribution of health and disease. In Part One, I discuss the clinical significance of wealth and cultural differences to the health divide.

In Part 2, I have reviewed the essential techniques and processes used by experienced, clinically successful doctors when making diagnoses. These range from simple history taking to forensic history taking; from simple examination, to specialised examination and investigation. Those who display a talent for diagnosis and clinical judgement have the advantage, but only those gifted with both a scientific perspective and an innate ability to understand others will excel.

In Part 3, I examine the validity of investigations and decision-making tools. Should we accept the headline results of research papers and meta-analyses, without a thought for the 'devils' in the detail? Some knowledge of statistics, 'the statistical paradox', diagnostic accuracy, and Bayes' Theorem, is essential. All are necessary for correct clinical judgment and the most appropriate patient management.

The last subject is that of clinical research and what every doctor and nurse can bring anew to the care of patients.

PART ONE

CLINICAL ENCOUNTERS

Chapter One

Briefing Notes for Medical Apprentices

'Give every man thine ear, but few thy voice; take each man's censure, but reserve thy judgment.'

Polonius. Hamlet. William Shakespeare

It is a poor student who does not surpass his master.

(Leonardo da Vinci)

All apprentices need to respect, follow, and serve a master. An Arabic proverb teaches that it is the master of a people who serves.

('Sayyedu al-Qumi khademahum.')

Medical apprenticeship is a lifelong process. In this preliminary chapter, I consider some basic facts about a doctor's work; what it involves, and who is suitable to undertake it. Much the same applies to the work of nurses, paramedics, and all those who care for others. Individual suitability for a caring role in medicine is critical to patients, since those they trust can affect their lives. The topics introduced here, are developed in later chapters. One aim is to guide those whose medical work experience is growing and whose practise needs development. This is pertinent to both junior and senior doctors alike.

Medical Students

How much must a student learn in order to practise basic medicine? Medical student freshers might have thought studying for three to four advanced, 'A' levels exams was challenging. They will find that each of the pre-clinical subjects: biochemistry, anatomy, and physiology is equivalent to studying several advanced level subjects. At least, it was when I was at medical school in the 1960s.

Times have changed. Nobody then walked in the street carrying plastic containers of hot coffee; those seen talking to themselves were schizophrenic and not in normal conversation with their friends on a mobile phone. Only 10% of medical students were female. Medical regulators and those who administered medical practice, held a back seat. We thought those who opened doors for women were gentlemen, not anachronistic patronisers.

Those who qualify as doctors, many never need to call on more knowledge than they learned in their first year of qualified medical work. Others who elect to pursue speciality work will need to acquire many new skills. Most of what medical professionals rely on will be intrinsic to their nature: helpfulness, affability, judgement, the assessment of risk, an ability to communicate, intelligence, dexterity, integrity, honesty and dedication.

Some medical specialties allow for a degree of personal detachment. Those who wish to remain anonymous and objective at all times need to consider surgery, anaesthetics, A&E, cardiology, and research work. Those who want patient engagement rather than detachment might prefer general medicine, general practice, and psychiatry. Regardless of how much patient attachment doctors and nurses choose to have, every patient's emotional needs must be considered, but never allowed to cloud clinical judgement.

Occam had it that the truth is always simple. So, here are three examples of simple clinical truth:
Patients like to be cared for by:

- those who inspire their confidence;

- those who clearly know what they are doing;

- those who clearly have their best interests at heart.

Medical Knowledge and Starting Work

'Experience... the one thing you can't buy, borrow, or pretend you have.'

Johnny Nelson. Former Cruiserweight World Champion Boxer.

Although all medical students who pass their final examinations will have digested reams of information, they may struggle to know its relevance on their first day of work as a doctor. Nurses will seem to know more, and new doctors should be gracious enough to learn everything they can from them.

A newly qualified doctor on her first day at work may feel daunted by the responsibility they have. Their feelings can range between elation to fear. 'How is it', she might ask, 'that I seem to know so little that is useful?' The answer is that experience defines the usefulness of knowledge. Within every practical domain, meta-data is vital. You may know to prescribe aspirin for pain, but which patients are unsuitable (given their age, history of peptic ulcers, and bleeding tendency)? Recognising one's inadequacy is an essential step to towards progress.

A newly qualified doctor stands before a steep wall of obligatory learning. A few will be overwhelmed and decide not to continue; there is so much to daunt the novice. Dealing with sickness, blood, death, grief, fear, and suffering is not for everyone. Some will take to the job like a duckling to water, but only time will tell which ducklings will swim, and which will drown.

At this early stage, doctors will soon find where their talents lie. The talents possessed by some of their colleagues will be outstanding. How they handle and manage patients, their diagnostic ability, their facility with hands-on tasks such as minor surgery, taking blood, setting up iv infusions, resuscitation, and laryngoscopy, all will be obvious.

Compare being a doctor or nurse with being a cook. Imagine you have a barrel of apples (lots of bits of medical information about a patient), and are required to choose the most suitable apples for making a pie. The process requires knowledge and discernment. Some factors are crucial: the other ingredients needed (meta information), how long to cook the pie and at what temperature (knowing natural history); how to judge the resulting product and how best to serve it to those who choose it (forming a management plan suitable for each patient). Many talents are involved. Only experience can shape the usefulness of each piece of knowledge, and its clinical significance. One must learn to recognise, discard or archive unnecessary bits of information.

When first a doctor or nurse experiences medical work, the most useful knowledge can seem mundane: how to manage insomnia, sore throats, coughs, colds, backache, and muscle pains. Beyond the hills and far, far away, in the land of tertiary referral centres, it can be the minutiae of the latest research results that are most applicable to managing those with Addison's disease, craniopharyngiomas, channelopathies, or congenital heart block.

Ascribing the correct relevance to each piece of knowledge should come with experience. Clinical judgement critically depends on the correct weight of significance being attributed to each clinical feature. One will gain meaningful experience from seeing many cases through to the end, and recognising the correct relevance of each symptom, sign and investigation finding. Pathologists have all the answers, so clinico-pathological conferences centred on those who have died, help us evaluate every symptom, sign, and investigation result. Seen in retrospect, many mistakes and many instances of correctly ascribed relevance, will assume their proper perspective. Expertise develops from this heuristic learning process.

Unfortunately, much of the information learned at medical school will be of no immediate use. In a long career, however, every bit of information can assume importance; especially one's fundamental knowledge of anatomy, physiology, and pathology. Like reading, writing and arithmetic, the basics never devalue.

Every medical professional will progressively generate their own map of information. The task before apprentices is to populate their map with points of knowledge and its clinical relevance. Together with the facts we are certain of, will be suggestions, inferences, theories, pieces of evidence and guesses, the perspective of which may need to be altered daily. An experienced physician, surgeon, nurse or paramedic, will have learned the relevance of each piece of clinical information and its contribution to a wide range of final diagnoses. After a time, the map will not need to change much; we will have learned when to magnify clinical significance, and when to down-rate it. Every piece of feedback requires us to edit our information map. The closer to failure we get, the more important our editing.

Every doctor, nurse and paramedic will have plenty of guidelines and rules to follow. Doctors need to remember that rules and guidelines can never reflect the variance of every clinical situation, since many are unpredictable. Unfortunately, medical regulators, with their sparse knowledge of medicine, need hard and fast rules to judge clinical performance. It is beyond their remit to acknowledge the relevance of individual patient diversity, or the invariable need each patient has for an individual approach. The uncertainties and risks all doctors face daily are beyond their appreciation, and without clinical experience they cannot qualify to judge it. More of this important topic later.

A crucial factor for professional survival is the ability to cope with the stress of work.

When sorrows come, they come not single spies but in battalions!

Claudius. Hamlet. William Shakespeare.

Our patients expect us to cope with such battalions, and to have more wisdom and understanding than they can glean from their friends, family, and the internet.

The Apprentice and the Patient

Medical apprentices need to be aware of how medical practice functions. At the centre of it all is the relationship between a doctor (nurse, carer, paramedic) and her patient. Encounters with medical practice can often disappoint those patients looking for the same efficiency they find in some commercial enterprises; without efficiency, those businesses would not have survived in the marketplace. Patients who are successful business owners themselves, often struggle to understand why they have to wait weeks for NHS appointments and investigations. Some apprentices will have a talent for operational efficiency, others will need to learn it through training.

Fundamental Skills

All medical apprentices must engage with fundamental medical skills, many of which may appear simple. History taking and physical examination can range from simple (most common) to complex (many variables, with various inter-connections). It can take decades for some doctors to develop the skills required for complex history taking, given that it is guided more by art, talent and experience, than by rote. There is a German word that almost describes the process – *'fingerspritzengefüln'* (literally, 'finger tips feeling', or following your finger); a process which alters in response to every change a patient makes.

With simple clinical presentations, such as minor injuries, medical screening and insurance medical examinations, tick-box processes can suffice. There are many useful frameworks to follow which will ensure completeness (see my general medical framework in Appendix A). In some research, simple but standard questionnaires provide the consistency required for data collection and standardisation. Recording a patients information is not always simple.

The history taking process changes with each doctor and patient. It involves asking appropriate questions, establishing rapport, and an approach that aims to uncover diagnostic clues. Examining and then investigating patients comes next. Each has a specific purpose: to verify the diagnostic possibilities suggested by the history.

Talking to patients can produce more relevant diagnoses than any other intervention, so doctors need to be good at it. History taking also creates the basis for a fruitful doctor-patient relationship; one which will allow a doctor to understand a patient and his personal circumstances. The knowledge derived from a thorough history can be more useful than some high-tech investigations, many of which are now revered by both patients and doctors alike.

Managerial demands related to patient through-put and target achievement, have now seriously eroded the time needed for history taking. Bureaucrats concerned with patient handling efficiency may soon demand history-taking by algorithmic, data gleaning, neural network programs. Their job is to give thought to the savings made in time and money, without regard for any depersonalisation of medical practice that may result.

A Talent for Medical Practice

What talents are required for medical practice? Medical apprentices must ask themselves if they have the intelligence, aptitude, talent, and personality required.

The doctors we get are the consequence of medical school student selection criteria. All medical school interview committees work with a disadvantage. They cannot hope to deduce how candidates will react to patients, or *vice versa*; at least, not from standard resumé (CV) information they have before them.

Those medical student candidates who have been to recognised schools, and have achieved the highest grades in science examinations, will most impress medical interview committees. Those with the smartest parents will have sent them to work in a leper colony, just to impress the selection board. Medical school selection committees need to honour a candidate's specific talents (as listed on their CV), but as important, would be to observe them helping patients; their talent for people-handling should then be obvious. A fly on the wall might do better. It will take five to six years to prove the worthiness of their selection decisions.

For many years, I took aspiring 'A' Level students for work experience at my clinic. It soon became obvious who had the ability and talent for the job, and who did not. After observing a student talking to a patient, one could answer some important questions: Are they calm, confident, and in control? How easily do they achieve rapport, and how quickly? Are they genuine and perspicacious in their use of empathy and sympathy? Are they helpful, or a hindrance? Where lies their focus? Can they resolve tasks without getting flustered or confused? Are they by-standers by nature, or quick to get involved? Do they always need instructions, or are they pro-active? Do they manage their time well and get things done? Can they function efficiently, meaningfully, and in an executive manner? Finally, there is one crucial question to answer: should one trust them with patients and be able to rely on them as a colleague?

As time progresses, doctors develop their own style. Of paramount importance will be how they relate to others, regardless of their educational, social, and cultural background. Discernment, prioritisation, and judgement in complicated and sometimes confusing situations, are all critical pre-requisites for clinical success. A facility to learn from experience is essential for both clinical safety and career progression.

In deciding from whom to learn, doctors must not only make verbal enquires about how good a particular surgeon is; they must watch him at work. Like a professional concert pianist, his talent will be obvious from watching his fingers. His ability to make judgments and handle patients is another matter. If he is talented, he can afford to be light-hearted. He might tell jokes or be witty while performing complex procedures. Every member of his team will know if he has surgical talent, or a lack of it. One cannot assume that the team leader is always the most talented; just as often, it can be his SHO, or registrar.

Above all, doctors need an ability to manage the processes of illness. Apart from a facility for judgement, the business of medicine requires executive functioning, and an ability to think on many levels at once (binary, multifactorial, nebula thinking). A keen sense of timing is critical. There are parallels in other professions. Chefs, for instance, will cook many items, each for differing lengths of time and at different rates, yet be able to serve them all at the same time. A pilot must do much the same thing. To land an airplane safely, he needs a talent for integration. He must focus on the important parameters of flight, be calm under pressure, and never let irrelevant factors become distracting. Many branches of medicine require nothing of the sort, but in ITU and emergency medicine, surgery, and anaesthetics, they are all *sine qua non*.

You Can't Teach It!

In 1969, Neil Armstrong landed the Apollo 11 Lunar Module 'Eagle' on the moon. He was the first to do it and got it right the first time. Many other very talented pilots accepted him as the best astronaut. He could fly 'by the seat of his pants'; no fuss, no bother, no conscious calculation.

No simulator could have prepared Armstrong for what happened on that first lunar landing. On final descent to the moon's surface, he noticed that the automatic landing system was guiding Eagle to a boulder-strewn crater floor, the size of a football field. He took manual control and skimmed over the crater to land on the flat surface beyond. Eagle had only 30 seconds of fuel left on touchdown.

It will embarrass him for me to mention it, but my son Nicholas had an obvious talent for flying, even as a young teenager. He could drive a tractor, a combined harvester, and fly a light aircraft at an age before he could drive a car on public roads. Thanks to the Chicken family at Stapleford Airfield in Essex, the pilot examiner who granted him his PPL (Private Pilots' Licence) told John Chicken, he was, 'a top rate chap.' His obvious talent meant he was to achieve his ambition to become a commercial pilot. He is now a captain with EasyJet. He recently landed in a storm at Gatwick airport, when nine out of ten others diverted to other airports rather than try to land.

Talent can incite envy, even though natural talent and beauty are 'given', divine attributes, with no work is needed to acquire them. Those who possess natural talents can perform effortlessly and will easily draw the attention of admirers. In flying, as in medicine, talent alone is not enough. Moral fibre (a category used by the RAF until recently), energy, passion, wisdom, judgement, and the confidence to take responsibility are other essential ingredients.

Many experienced pilots say the same thing about talent: 'You can't teach it.' I agree with them. When you see a master at work in any field, their talent is obvious and seen early in their training. Those lucky enough to have viewed Monty Don in his garden, Michel Roux in his kitchen, Ronny O'Sullivan at the snooker table, Jimi Hendrix playing a guitar, Milt Jackson on the vibraphone, Stéphane Grappelli playing a violin, and Mary Berry making a cake, will have witnessed effortless co-ordination and control.

There are many clumsy doctors and nurses. Meeting them can help one appreciate the talented ones. How is it possible for inept, clumsy people to succeed in any field? The answer is simple. Many are good at exams, and adept at influencing others.

Who has the talent?

To handle complex networks of inter-related facts, and be able to plan clear decision pathways that benefit patients, requires talent. One must consider the clinical and personal circumstances of each patient. Clarity of thought, an understanding of all the facts, and an ability to handle them efficiently (executive functioning), are all required. Other factors are desirable:

- Case-handling of intelligent, experienced and resourceful patients, exceeding expectation.

- A talent for learning from mistakes and successes (heuristic learning).

Those who need to calculate their every move will struggle to appreciate why some have no need for it. What seems like instinctive action is the mark of a master. With few exceptions, calculated action is for beginners.

Soon after picking up a golf club for the first time, my friend Kenny could swing a club and hit a golf ball a fair distance. With only a few practice swings, he had gauged its weight and inertia, and performed like somebody who had played for years. He did this, making no conscious effort, measurement, or calculation. He could play football and play a guitar with equal ease.

When I tried to beat my father when playing snooker, I tried to use geometry to calculate the angles. It didn't work. He knew 'instinctively' how to play. His skill was not only to 'pot' the object ball, but to leave the white ball in a position that made his next shot easy. World Snooker Champion Ronnie O'Sullivan, once agreed with me. Calculation is not what it's about.

Josh Cohen, in his book '*Not Working*' (Granta, 2018, p214), describes tennis champion Roger Federer, and the essence of his talent: 'The best tennis brings an unexpected sense to the phrase 'not working'; Federer's prodigious athletic prowess is the counterpart of his psychic and fleshy lightness, his miraculous freedom from the physical and mental loads the rest of us are made to lug around so gracelessly.'

There are a few doctors, nurses and paramedics with the same lightness of touch. Their effortless capability is impressive. Every apprentice should try to find and emulate such people. They possess something to learn from.

Specialisation

There was a time when hospital specialisation attracted only the brightest graduates (those capable of passing the MRCP, FRCS, and FRCOG examinations). Others went into general practice, although those who failed to become a specialist, had few other options. General practice has developed and is now regarded as a specialty, although patients and their problems have not changed much. To encourage doctors into their ranks, the Royal College of General Practice was formed in 1952. Since then, the number of professorial chairs in general practice has multiplied, even though most of the daily medical work of a UK GP has remained unchanged (except for added reams of bureaucratic paperwork), Their work is only occasionally affected by advancing technical science. What has change is their need for bureaucratic compliance (possibly 20% of the workload), and the increased leisure time they now have available (no weekend work or night visits).

Specialist training opportunities have significantly improved since my early years. In my day, one had to be lucky enough (well-connected enough actually), to get an attachment to a 'teaching hospital' specialist. You then had to impress your supervisor, since he would later be called to sit in judgement on you, and act as your referee for promotion. Insecure supervisors (usually stuck on the progress ladder), were less likely to promote their juniors.

Those doctors who think that private practice will help them attain financial security must consider general surgery or gynaecology. Doing something practical – an operation, an echocardiogram, or an ECG, is respected most by patients. Private doctors who do not undertake investigations risk being asked, 'Why have you charged me so much? All you did was sit and talk to me!'

A patient once asked an ENT surgeon why he charged £1000 for a 5-minute tonsillectomy operation. He replied: 'I charged £1 for the operation, and £999 for knowing where to cut!'

Functional Doctors

Like defining a safe airline pilot, the most exacting test for a doctor is to watch how he handles the unexpected. We can program robots to follow rules unerringly, but it takes creativity and imagination to function in an emergency; especially when no established rules exist (or there is no time to find the book of rules). Under these circumstances, both doctors and pilots must be able to 'fly by the seat of their pants' (relying on experience and natural talent). The thought of this will give many medical bureaucrats nightmares. They dream of removing all risk; a delusion fostered by their ill-acquaintance with clinical reality.

Out of interest, the word bureaucrat derives from the French word *bureau* for 'desk', and the Greek word κράτος (kratos) for 'a rule'. Most sit at their desks, dreaming up rules for others to follow.

A sudden bird strike disabled Capt. Chesley Sullenberger's airplane (US Airways flight 1549). Flying geese put a stop to both jet engines. 'We'll be in the Hudson!' is all he told air traffic control. He had no time for discussion with controllers. Nobody died, but that didn't stop regulators and those seeking compensation, accusing him of poor decision making.

Castaways

In my later career, I had to survive in practice with no direct support from colleagues, or the medical profession at large. I cast myself asunder into private practice, and had to help patients well enough for them to want to return, despite competition from a 'free' NHS service. To pass this test, an extensive knowledge of medical science is obligatory, and a facility for the art of medicine essential.

I would like to propose a theoretical competence test for doctors; call it a castaway test, analogous to the challenge Native American Indian braves once had to endure. They were required to go forth and survive alone. Only after surviving could they return to the tribe and be eligible for leadership.

Before 1948, such a test was the rite of passage for all doctors; they had to stand on their own two feet. They had to attract and be responsible for their own patients. In many parts of the world, this remains the case. Most doctors cannot shelter beneath a protective financial umbrella, like that provided by a corporate NHS. They must assume personal responsibility for all they do. While I do not advocate the removal of present day corporate protection, it was self-survival that once defined a doctor's personal aptitude.

My patients were always my alpha and omega. I owed them my professional survival; a result of their loyalty. As an independent doctor, I had the liberty to choose my colleagues: those whose performance I considered good enough for my patients. In the distant past, GPs referred their patients only to the local specialists they knew; mainly those working at the nearest NHS hospital. Nowadays, it must be difficult for them to judge the quality of the clinical teams they refer their patients to. In private medicine, patients often do their own research and form their own referral preferences. When asked, I would never refer a patient to any colleague without knowing how good they were at their work.

Because it was my passion, medical practice became my way of life. I lost none of my individuality to corporate management or standardisation, having to comply with standardised rules and regulations. Corporate loyalty, however, offers many advantages. It will provide security and make a doctor's financial survival certain. I understand why many doctors choose to shelter within it, even if they have to suffer bureaucratic inteference.

Other powerful forces exist for doctors and nurses to contend with. One of them is the in-fighting that is rife in all professions, the NHS and all academic departments included. It is not unknown for academics, worried about competition, and a strong need to publish (or be damned), to use their intelligence and cunning to manipulate others. Naïve apprentices will need to acknowledge their *modus operandi* and acquaint themselves with the edicts of Shang Yang, Han Fei, and Machiavelli. They must be able to spot those who would use them for selfish purposes; I escaped them and chose not to study them further. Those with an obsequious nature might enjoy the association. It takes all sorts, as I explain in the chapter on 'Doctors'.

The Doctors We Get

Some doctors aim to attain a position where both they and their patients are comfortable with one another. In this state, patients will usually forward their true thoughts and innermost feelings. Some will come to trust doctors, nurses and paramedics with their life and welfare.

Some of my colleagues seemed not to want to treat people, only medical cases. They were happier with a case number than a patient's first name. It would have been anathema for them to regard personal patient information as interesting or relevant. They seemed to place objectivity and efficient patient handling above humanity.

Some doctors know all about the ingredients necessary to make a cake, but have no wish (or talent) to bake one. They have a place, working in specialties that demand technical expertise.

Although not a choice any doctor or nurse has to make, would they choose loyalty to their patients over loyalty to their colleagues? For a completely independent private doctor like me, there was only one answer. I was indebted to my patients for supporting me. I never had to follow the NHS paradigm of specialists consulting only with patients referred to them by their GPs. There were never reasons for any NHS GP to refer patients to me (as an independent private cardiologist). While indebted to my consultant colleagues for the service they provided my patients, few helped to sustain my practice. My loyalty to them never arose. The relationship I had with my patients was unconventional, but it was direct, transparent, and without intermediaries. Because of the growing number of independent private doctors, I predict the same independent path I created for myself, will now become more common.

Some doctors fail patients. In trying to understand why, many of my patients came to appreciate the value of a meaningful patient-doctor relationship. Most patients have minor problems and have little need for a meaningful patient-doctor relationship; their mission is to get 'fixed' a.s.a.p., by any available doctor. Any doctor trained in mechanical processing, and the performance art of a parrot, can provide an efficient, pre-programmed answer to minor medical problems. For many doctors, there is little satisfaction in it, except for some pride in their patient throughput numbers, and the financial reward it may bring. This type of practice is not the focus of this book.

To be apprenticed to a first class doctor is the best way for a junior doctor to become accomplished. The apprentice must seek to gain clinical wisdom, but what defines this 'wisdom'? Most will know it when they hear it, although a few will always be unreceptive to it. The clarity it brings to a clinical situation is one measure of clinical wisdom. Knowledge distilled by experience and intelligence is essential, and nebula processing is involved (knowing what to include and what to be ignored from the cloud of information available). In more complicated cases, medical knowledge alone is not usually enough to achieve patient satisfaction. Anyone capable of getting a medical degree can learn enough to practise medicine, but to practise at a level which satisfies both discerning colleagues and patients, requires clinical wisdom.

Science is organised knowledge. Wisdom is organised life.

Immanuel Kant (1724 – 1804)

The medical profession is replete with brilliant guys, good at science. Not all can practise both the art and science of medicine. Even if they attain high status, there are some you would not want to treat your mother. Some conform to the Peter Principle: they have risen to their level of incompetence. They rely a lot on their juniors, some of whom may be more talented. Fortunately, some academic doctors will attain a status high enough to remove them from the clinical arena. Secreted in research institutions, many hide from the maddening crowd. Some have autistic traits, a notable absence of empathy, and little ability to communicate. Some will feel that without patients, work would be more enjoyable.

Among sports people, there are clearly different levels of skill and talent. The same applies to doctors and nurses but is not so apparent. We can categorise footballing talent, but can we compare it to clinical talent?

Let's say that a class 4 football talent allows one to kick a ball, and with a class 3 talent there is an ability to play tricks with the ball; class 2 players will be good enough to turn professional, and first-class players will be the very best professional players. A first-class footballer will have the complete game at his or her feet. They can assess dynamic action at a glance, and know how best to score goals.

Equivalent classes exist for doctors and nurses. This is no trivial matter for patients who need to trust the doctors and nurses engaged with them. We can categorise doctors and nurses using well-known psychological criteria:

- Cognitive: medical knowledge, communication skills, intelligence, judgement, risk assessment, and a facility for nebula and lateral thinking.

- Affective: emotional intelligence and emotional connectivity.

- Conative: practical issues such as commitment, focus, executive and managerial functioning, prioritisation, the effective use of time, and a talent for practical procedures.

By my definition, class 4 doctors will have only basic competence. Some are inept at communicating with patients and unsure when answering questions. They may see their job as a technical management exercise, and not as a caring one. They may dislike personal involvement. Like a postal worker, they may wish to limit themselves to delivering messages. Appropriately, they will often leave any clinical responsibility to others. Many will know to keep a low profile.

Class 3 doctors go further and are competent. They will see discussions with patients as useful. Although this suggests involvement, their primary aim is to comply with the guidelines, rules, and regulations, regardless of any impact on patient care. Class 4 & 3 doctors and nurses can function well believing that illness is a patient's problem.

Class 2 doctors are highly competent and have higher-level diagnostic and management skills. They are capable of sensitive communication and caring. The management plans they make, are in response to individual need. They will readily get involved personally and communicate enough to enable intelligent decisions. While keeping their distance, they can use themselves in patient management. They are realistic about what they can achieve. Unfortunately, they may never feel that the buck stops with them. They are diligent followers of instructions and guidelines. With experience, many doctors and nurses will achieve this level of clinical proficiency.

Class 1 doctors and nurses have achieved mastery. They are capable of clinical wisdom and are among the few. Most will have mastered not only the science and art of medicine, but every necessary practical procedure. Many will have invented novel ways of doing things. They have the gift of clarity and strive to keep clinical matters simple. All they do seems effortless. This level of performance requires an above average ability in each of the three psychological domains detailed before. They will succeed in diagnosis when others fail. Their knowledge and commitment to patients drives their successful results. Their commitment is to do whatever they can to help patients. Those willing to take the step beyond being highly competent, will use themselves not only as a source of information and action, but as the driver of patient management.

A first class doctor or nurse is unquestionably in control of everything they do. Patients readily trust them with their lives. Patients are right to have faith in them, whether they are doctors, nurses, paramedics or alternative therapists. They readily take the transference of responsibility, and assume the duty of care for their patients whenever required; especially when inexperienced or confused patients need clear guidance. They will treat every patient as if any consequences were personal, with no trace of obsequiousness, misplaced empathy or sympathy. Their aim in life is to focus on helping others, unselfishly promoting the maximum health potential of each patient. Bureaucrats may find them difficult, uncompromising, and independent-minded. They dislike those capable of thinking for themselves and prefer sheep to shepherds.

The first-class operator can add inspiration, novelty, and boldness to the list of his superior abilities, knowledge, and know-how; traits that go beyond check-list, algorithmic functioning. They will often inspire the advancement of knowledge.

Patients with problems that are easily 'fixed' during a brief consultation need only doctors from classes 3 and 4 (those who are competent, or have only basic competence). Those with difficult to resolve clinical conditions need doctors who are highly competent. The most complicated cases, and those other doctors have failed to help, need first-class doctors.

When my gynaecological friend and colleague for 50-years, Mr. Sunit Ghatak FRCOG retired, I gave a brief speech about our long-term collaboration. His family, GP friends and medical colleagues formed the audience at his retirement party. I told them I was proud to report my experience of working with him. During the time we worked together, I never found one error in his clinical judgement. Very few of my patients developed a surgical complication while under his care. His operating skills were dextrous, efficient, exact, and effortless. He was skilled enough to operate and converse happily with colleagues at the same time.

Not only could he remember the names of his patients, he never forgot the names of the babies he delivered. Our mutual patients remember him as 'simply the best'. His inter-personal skills were effortless, despite many cultural differences. All my patients saw him as warm and empathetic, knowledgeable, and reliably skilful.

Although at the top of his game, he was always modest about his abilities. He proved to be an epitome of surgical mastery, and a doyen of the art of medicine. A first-class person, and a master surgeon.

Unexpected Sources of Advice

Shortly after being given a place at the London hospital Medical School, I received a succinct lesson in 'Lifemanship' from my uncle, Charlie Dighton. I was working as an office assistant at our family business in Glass Street, Bethnal Green, when he told me that my training would teach me all about the tools of my trade, but not much about the broader humanitarian agenda. He said, 'My success (as a businessperson), has been based on creating allies. Those who have the knack (the art) of handling people, and have a genuine devotion to service, delivered with integrity, generosity, and a personal sense of involvement, will usually succeed. Those who pursue a selfish agenda may become rich, and even achieve high professional status, but few of them will satisfy their customers. It is being genuine that counts – it differentiates the best from the rest and quickly distinguishes those most appreciated by others.'

In 1944, the acquisition of clinical knowledge, gained from the attachment of medical students to clinical masters ended, even though it was the only way they could achieve mastery of the art of medicine.

The Goodenough Committee's recommendation was that 'medical student education should provide the student with a university education, on broad and liberal lines' (BMJ. July 22, 1944). Apprenticeship alone . . . 'needed to be replaced by a system of training in principles, problem-solving, and habits of learning which would equip them for a lifetime of practise in a continually changing scene – rather than to train them to become safe house-officers on graduation.' The proper scientific basis of medicine was to be taught, and the needs of the forthcoming NHS were to be met.

Taken from Prof. Clifford Wilson's obituary, The Independent.
(Dr. R. D. Cohen: 19.11.1997).

Prof. Clifford Wilson was one of my teachers. He stood for service and commitment, and fought for ideas which were to become the accepted norm after the publication of the Royal Commission Report on Medical Education (1965-68). The London Hospital Medical College had already adopted these principles while I was there (1961-66), although I only later realised their significance. As a student at 'the London', I had unknowingly benefitted from Prof. Wilson's influence.

I remember Prof. Wilson telling my student group,

'Never forget this case (polycaethemia rubra vera - PV) . . . you may never see another one!'

The words of Prof. Clifford Wilson on a teaching ward round at the London Hospital c. 1964.

I found my first case of 'PV' in 1978, and then two further cases a few years after, despite the long odds (two or three cases per 100,000 occur each year in the UK).

Who Should Influence Doctors?

We are capable of it at any age, but the older a doctor gets, the more daft his ideas may seem.

Some will embrace the metaphysical, and may seem (to their juniors) to have 'lost the plot'. They may jettison their usual balanced approach, and invest their capital (intellectual or otherwise) in one fixed idea, rather than spreading their bets. One obvious reason for this is the forthrightness of age; something that gathers momentum when there is nothing left to lose.

Dr. Peter Nixon at Charing Cross Hospital thought that stress, and only stress, caused heart attacks (at least, that is all he would admit to). A very senior colleague of mine, Mr. Lang-Stevenson, a general surgeon at Whipp's Cross Hospital, was the first I came across (in the late 1960s) to believe that stress caused cancer. A medical school contemporary of mine, Dr. S. B., has long believed that the mercury amalgam in teeth causes several syndromes (the result of mercury poisoning). These theses are based on anecdotal evidence gathered over decades. I questioned these theories by asking for their evidence, but failed to get satisfactory answers. Proof, they might have thought, was a tiresome matter, best left to researchers. The onus of proof, however, always rests on those who develop novel theories, not with others. Some become tired of making their case to others, and resign themselves to posthumous recognition. Retirement can exaggerate the process; a lack of active criticism from colleagues will sometimes leave complacency and daft ideas unchecked.

James Watson, of Watson, Crick and Wilkins fame, expounded a sensible research principle. In his book, 'The Double Helix', he suggests one should not bother finding things out for oneself when it is easier to ask those who already know the answer. Maurice Wilkins showed Rosalind Franklin's crucial X-ray crystallography data on DNA to Watson (photograph 51 of the hydrated B form of DNA). This left her vital input on the structure of DNA unacknowledged. Watson, Crick and Wilkins (without Franklin) received a Nobel Prize in 1962. Nobel Prizes are never awarded posthumously, and Franklin was dead by then. An important point here is that the Nobel citation reads: 'for their discoveries concerning the molecular structure of nucleic acids and its significance for information transfer in living material.'

One can only speculate how Rosalind Franklin felt, knowing that Watson had been given her data. She developed ovarian cancer in 1956 and died in 1958, aged 38. What were the aetiological factors? Her genetics (Ashkenazi BRCA genes), X-ray exposure, stress or serendipity? Watson and Crick announced their discovery of the double helix structure of DNA in 1953. Did the lack of acknowledgement upset her and cause resentment, and could such upset have played some part in her demise? Was my senior surgical colleague Lang Stevenson right to suggest that stress can cause cancer? The other question it poses is whether one should trust colleagues with your information and ideas? In the chapter on doctors, I have made some personal observations on this topic.

Other Important Issues for Doctors and other Medical Professionals

The Science and Art of Medicine

One can learn the scientific foundations of medicine in a lecture theatre, library or on-line, but to become an effective doctor, worthy of a patient's trust, the application of science needs the art of medicine.

*It is only with the heart that one can see rightly;
what is essential is invisible to the eye.*

Le Petit Prince (1943). Antoine de Saint-Exupéry. (Reynal & Hitchcock).

Character and personality determine the sort of doctor students will become, while success with patients often depends on special knowledge and skills. Since there is little difference in the established medical facts we learn, only knowledge and ignorance, will define how able we are.

There are some generalities about learning medicine to consider. Science is a study written precisely in black and white, employing objective measures and strict rules of evaluation. Art is a discipline written loosely in colour, with few rules and as many styles as there are artists.

In the 18th-century, a young painter suggested to the artist Jean-Baptiste Chardin that he painted with colours. Chardin reproved him, saying: 'I do not. I use colours, and paint with feeling!'

The science of medicine has given us many powerful tools, each of which needs to be applied in a way that is appropriate to each patient's clinical condition. Handling their wishes, emotional responses, understanding, and aspirations for the future will require us to use the art of medicine. Clinical success and eventual patient satisfaction require both.

Personalised Medicine. Is it Something New?

I have relegated the science of medicine to the background in this book, because it is now so easily accessed and learned from hundreds of sources. By comparison, learning the arts of medical practice come best from a master. There are doctors who regard the art of medicine as a step too far; too personal and unnecessary. They are, of course, free to develop their own style without restriction, and to adopt their own particular doctor-patient relationship style. When the science of medicine is of limited benefit, what will they do next?

Both Hippocrates and Richard Asher ('Talking Sense', Pitman, 1972) provided examples of the art of medicine, but it is the experience of each apprentice observing his senior colleagues that will be of most value. Students will need some luck to find a doctor who professes to integrate the art and science of medicine.

Ideally, apprentices should seek a doctor versed in the Five Wisdoms of Buddhism (In Tibetan: □□□□□□□□□□, 'yeshe nga'):

- One with an open, accepting, and peaceful disposition (rather than an apathetic, self-interested denier).

- One who has clarity about the way things are (rather than one whose views are self-opinionated and self-righteous).

- One who is resourceful, fulfilled, and generous with his/her time (rather than puffed-up and selfish).

- One who is compassionate and caring (rather than self-possessed, and avaricious).

- One who is capable and productive in helping others (rather than manipulative, politically oriented, and power hungry).

Practising Medicine: a Sovereign Duty?

Before taking medicine as a career, there are some political questions for students should address. Medical practice now involves interacting with bureaucrats, few of whom understand medical practice. Politics and bureaucracy deserve no place in any treatise about medical practice, but because many doctors now work in fear of them, I must mention their possible pernicious influences. Medical bureaucrats now claim that without the regulations they impose, patients will be at risk from doctors and nurses. Controlling these risks, is how they justify their existence.

Those intending to become a medical apprentice, should research the lives of doctors or nurses. The knowledge and skills they gain, come with a duty to use the knowledge to benefit the lives of others, whoever they may be, and whatever the dictates of bureaucracy. In western cultures, life and its preservation usually take precedence over all other issues, so when you intervene medically you undertake a sovereign duty with ancient origins; a duty that has primacy over all others. (For the religious, the care of souls by their appointed clergy might reign supreme instead).

All doctors and nurses are now aware of being watched while performing their duties. Bureaucrats stand ready to make accusations against doctors and nurses whenever there is any suspicion of non-compliance. Consider proceeding with a medical career, only if you can tolerate this situation, or are capable of changing it.

Certainty and Medical Practice

Because the scientific results we use every day have been peer reviewed, we can mostly accept them without further thought. Peer reviewed, corroborated science, can improve the reliability and certainty of medical practice, and certainty is what patients seek. To assume the validity of research results for each patient, however, is a common error.

The art of medicine comes with little or no certainty, simply because all the affairs of man are inherently unpredictable. The anxious and obsessive among us might find this too challenging. Practising the art of medicine comes with no guarantees. Because its *raison d'être* involves the management of the unpredictable, the art of medicine comes with few guidelines, rules, or formulae. One can only learn it through iterative processes, and master it only with repeated experiences. If a tiger is to stalk its prey successfully, it must respond to circumstances as they change, perceiving every subtle change in its prey's behaviour. Unless a hunter remains composed, alert and responsive, he will risk becoming the prey himself. Many non-medical pursuits require similar responsiveness: test-piloting, tight-rope walking, surfing, playing jazz, stockbroking, and rally driving. They all require alertness, sensitivity to change, and rapid responses in a complex environment. Whereas all apply to practising the art of medicine, almost none apply to practising the science of medicine.

Since the art of medicine is not formulaic, those with an Asperger trait (and other autistic variants), many of whom are mathematically minded, are likely to reject all but evidence-based, scientifically justified medical practice. I wonder whether their concern is to satisfy themselves or patients?

Security and Medical Practice

My pleasure and security as a private doctor came from the loyal patients who consulted me repeatedly over several decades. I was a single-handed, single-minded, private doctor for 48-years; the only practicable role for a non-group oriented, independent thinker with a strong desire to control his own destiny.

I had worked for decades in what was a bureaucrat-free zone. I was self-employed and free to set my own standards of practise (aspiring to what were once called 'teaching hospital standards'). I had every need to acknowledge my elders and colleagues, and no inclination to kowtow to bureaucratic administration. My working conditions differed significantly from those of the NHS. I decided my patient's clinical priorities and planned their clinical management with mutual agreement. All doctors had the same degree of clinical freedom once, but throughout my career, I watched the freedom my NHS colleagues had, slowly evaporate. It wasn't until I had my first CQC inspection that I realised, this was also to be my fate.

Social Division, Culture and Medicine

A glib cliché, maybe, but true in practice: 'doctors get the patients they deserve (and *vice versa*)'. Doctors will endear those who share the same style, attitude, and level of executive functioning. A doctor as a round plug, should avoid working in square holes. Doctors should give thought to their preferred patient demographic. They need to ask: 'with whom do I prefer to engage, if work is to be a fulfilling, rather than a frustrating experience?'

My first colleagues were white Caucasian, Christian and Jewish, with a few Irish and Asian doctors in the mix. The nurses I first encountered were white, British, or Irish. During my long working life, the UK slowly changed into the cosmopolitan, multilingual, multicultural society it is today. The change has required doctors and nurses to adapt and become culturally sensitive. Except for a few places in the UK, all doctors and nurses now need to acquaint themselves with various races, religions, and cultures, representing different customs, ways of living and different values and standards. Whatever the differences might be, the focus must remain the same: patients first.

The Call to Duty

I regard it as my duty to pass on my experience, and anything valuable I have learned about medical practice. Many senior doctors share this sentiment and actively engage in it.

Those patients who bring joy, pleasure, amusement, and satisfaction to medical practice balance the dismay brought by a few others. After a little experience, some doctors may feel ill-equipped to deal with patients and their personal problems, and will choose instead, to devote themselves to anonymous medical disciplines like scientific laboratory work, pathology, epidemiology, unconscious patients (anaesthetics), or those often too sick to communicate (ITU).

All customers bring us happiness. Some when they arrive; some when they leave.

From 'The Saracen Head' pub (The 'Sarry'). Glasgow. Frequented by Billy Connolly.

Doctors will meet many patients who are understanding, modest, delightful, and a pleasure to deal with. They will add satisfaction to an already worthwhile job. Doctors must also accept the challenge of dealing equally with the selfish, arrogant, greedy, deceitful, and jaw-droppingly uncouth. Our job is to serve patients, regardless of their traits. Doctors must learn to grin and bear other people while keeping their chin up and reminding themselves (sometimes hourly) what their purpose was in becoming a doctor. If a doctor ever comes to lose compassion and devotion to others, and no longer finds satisfaction in dealing with them, they should find another job. The situation will not change, unless they change their patient demography.

To hold an untarnished view of the public would be to ignore their drive to survive, and the selfishness they need to succeed. To cling to a glowing view of humankind is the reserve of saints, dedicated do-gooders, some philanthropists, and the deluded; among them are the privileged, like those bureaucrats and politicians, some of whom only rarely need to deal with the public face-to-face. Doctors, nurses, care workers, paramedics, and police officers, who deal with the daily realities of human behaviour, cannot afford their delusions.

A Few Pertinent Aphorisms

To see the true nature of people, look for the small things they do; like simple acts of kindness. It is frequently possible to make a big difference to a patient's well-being with one minor change. Every opportunity to help the well-being of others is a gift to one's self-esteem. If you can accept one gift with humility, you are ready to pursue all other opportunities selflessly.

I have one strong bias. My belief is that those who practise medicine and nursing selflessly are especially valuable beings. They contrast with those whose primary aim is to gain control over others, to accumulate wealth and power, and to achieve status. These ambitions may be less worthy, but are part of human nature, and nothing new.

> '... they in whom is implanted a passion for honour and praise, these are they who differ most from the beasts of the field.'
>
> Xenophon's dialogue, *Ieron (Hiero)*, 4th century BC (Chapter 7, section 3).

For most doctors and nurses, experiencing the good that practising medicine does for patients, brings much satisfaction. The value doctors and nurses bring to any society that primarily respects wealth, power and celebrity, is far less tangible. How NHS medical workers and carers are now treated and valued in the UK, illustrates the point.

Know Your Acronyms

Acronyms and jargon are commonplace in medical practice. All doctors and patients have to deal with them, and may feel foolish if they have to ask their meaning. Doctors need to collect all the common clinical acronyms together (CPR, CRP, BMI, etc.), especially those used by their team (in cardiology: TAVI, VKA's, NOAC, FFR, etc.), as well as important research trial acronyms (like the COURAGE, RAMIT and ISCHEMIA).

Doctorates and Doctors

Academic medicine needs scientists with doctorate (research) degrees. Scientists define a 'doctor' as one who holds a research doctorate degree, not one who holds a bachelor medical degree. It is possible to have both, but most are happy with the distinction. The tools required to gain a doctorate degree, like designing experiments, research methodology, and intellectual rigour, are also important to every physician and surgeon. Along with the art of medicine, knowledge of scientific methods, and an acquaintance with statistical analysis, are essential for all those who wish to evaluate the advances in understanding that might bring benefit to patients.

There is a considerable career advantage for doctors who pursue medical research and gain a doctorate degree (usually a PhD or MD). Post-graduate degrees impress most experienced physicians and surgeons, whether or not their efforts will benefit medical practice. The kudos of a higher degree can overshadow the need patients have for the art of medicine. Because a doctorate degree will favour the appointment of doctors capable of scientific, detached thinking, not all patients will acknowledge an advantage.

Can We Choose our Patients?

Doctors cannot choose their patients directly. Instead, they choose them indirectly, by deciding where they want to work, each with a specific demography. The geographical area in which doctors work, and their status within the profession, act as filters. Professors do not work in non-university status hospitals, and get called upon to advise in the most interesting and challenging cases. Giving their advice, rather than their involvement, may satisfy them. This is changing. It will not be long before consultants, and professors of medicine, have to act as junior doctors. This is likely to happen for two reasons. First, it will save money, and second, controlling managers and regulators, with no knowledge of medicine, may not know how to value clinical experience and expertise.

Socio-economic class distinctions are important. The poorer and more deprived the patient demographic, the more a doctor will encounter coronary heart disease and cancer. This applied also to those who suffered most from COVID-19.

The poor and less well educated present a greater challenge when trying to convey understanding, but coping with these challenges is rewarding. Finding interesting cases, and the chance to meet fascinating people are among the many benefits of practising medicine, wherever one works.

Some doctors choose to treat only those patients who can benefit from their involvement. If they are to agree with any medical advice, patients should understand enough of the details to give their fully informed consent.

Doctors must learn to handle any discord, or misunderstanding, with an appropriate level of mutual respect and sensitivity. Some patients who enjoy basking in arrogance and ignorance will be trusting to luck.

Different societies exhibit various forms of behaviour; some more stressed than others. Few patients (or doctors and nurses), however, have enough experience, wisdom, or common-sense to avoid personal stress and the self-harm it can bring. Exposing themselves and their families to avoidable problems, some will drink too much alcohol, drive while drunk, not wear safety-belts, and smoke in cars when their children are on board. Others take out unserviceable loans, with companies who will charge them over 1000% APR. Some spend more than they can afford. Others will climb mountains (metaphorical and real) with no proper safety equipment or prior training.

Many assume that physical attractiveness alone bodes well for a compatible relationship; divorce rates suggest otherwise. Our biological drive to procreate with the beautiful, powerful, and strong, has a lot to answer for. The Delphic edict: 'know thyself' remains the best unheeded advice.

Once the consequences of self-harm become apparent, those with too little foresight may wonder why they are suffering. Those who wish to take no blame, may blame others. Many expect doctors to help them, and to help sustain their self-harming behaviour. The sad news is that doctors are unlikely to help them improve their weaknesses and liability to medical problems (even for those who see this as their role). Too many negative experiences of this sort can later make a doctor wonder what they achieved throughout their career.

The Cultural Challenge

There were few Afro-Caribbean or Asian people in the UK when I was a teenager. My exposure to other cultures started while I was studying for 'A' levels (at what was then, the South West Essex Technical College, Walthamstow). Most of my fellow students were white British, but among the others were Ghanaians, Nigerians, and Gujarati Indians, a few of whom I amused by trying to pronounce the soft and subtle tones of their native languages.

This started my fascination with different cultures and their languages; different traditions, views, values, beliefs, and behaviour patterns. This is now an absurdly sensitive political subject, but all doctors and nurses need to recognise culture difference and cannot assume 'we are all the same' (albeit, more alike than unlike). In fact, to achieve proficiency in the UK, every doctor and nurse must be acquainted with many races, religions and cultures.

Immigrant families need respect for the hard work needed to understand white British culture(s). This will not apply to those with no wish to integrate. No immigrant family escapes having to cope with the differences between their own culture and that found in a new land. The sub-cultures found in every country provide further challenges. Try moving from London to Glasgow, and you will soon appreciate many sub-cultural differences.

We are all the more one because we are many,
For we have made ample room for love in the gap where we are sundered.
Our unlikeness reveals its breadth of beauty radiant with one common life,
Like mountain peaks in the morning sun.

Rabindranath Tagore

Tagore's wisdom helps us all appreciation of our differences. In the medical domain, racial and cultural differences need to be understood and appreciated, not suppressed by political correctness.

Vive la Différence

Doctors need to consider the many levels of cultural difference that exist in the UK and elsewhere. Cultures tend to merge over one or two generations, resulting in a growing number of people who are at least bilingual and bicultural. The gradual disappearance of cultural polarisation will eventually make the medical consideration of cultural factors less important.

The level of education we each have represents another important difference between us. So does wealth and poverty, privilege and deprivation. These are the fertile subjects of literature, whatever its origin. Dealing with these variables creates a cultural challenge for doctors (unless they are pathologists or laboratory workers). Only an intimate acquaintance with members of each group will enable an easy, unguarded rapport. By remaining detached, and limiting themselves to the practise of evidence-based medicine, some doctors believe they have no need for such rapport.

In trying to learn many languages (including a few British regional dialects), I tried to acquaint myself with the details of several distinct cultures. I have much enjoyed learning about them, and how best to relate to them as a white British doctor with a family that stretches back to the Huguenots on my mother's side, and to Anglo-Saxon and Viking Britain on my father's side.

Apart from my personal interest in cultural differences, my aim has always been to understand how best to approach each patient. With growing acquaintance, the individuality of different cultures becomes better distinguished, and much more fascinating. Just as fascinating are the sub-cultures of every nation. In Holland, the culture of the Protestant north differs from the Catholic south; in the UK, one will encounter Scots, Irish, and Welsh; Liverpudlians, 'Brummies' (from Birmingham UK), Geordies (from Newcastle UK), Cockneys (from East London), and many other sub-groups, each with their own shade of outlook, attitude, and value judgement. What you will experience talking to Lancastrians will differ from talking to Yorkshire folk living in Deighton. I now know that my own strident attitude is pure Yorkshire. There are parallels in every country. Any doctor disinclined to recognise the differences, and is neither intrigued nor fascinated by them, will be at a disadvantage when trying to practise the art of medicine.

The Medical Elite

All over the world, there are many elite teams in hospitals serving patients to the highest of worldwide standards. I am lucky to have worked in several. It was a privilege to work alongside those who extended the boundaries of medical knowledge. To have worked with Geoff Davis, who constructed the first implantable cardiac pacemaker; Aubrey Leatham who wrote the first definitive treatise on the physiological relevance of heart sounds; Keith Jefferson, who defined cardiac radiology; Peter Nixon, who had studied the role of stress in heart disease; Michael Davies, who clarified the histopathology and natural history of atherosclerosis in the coronary artery tree, and Graham Leech, who transformed echocardiography into what we now take for granted, was a privilege; a life-shaping one, in fact. To a man, they were quiet, amusing, self-effacing, modest and confident. None of them received a knighthood, or any prize for their work. They all gained something more important - the satisfaction of advancing medical knowledge, and the respect and affection it brought from their colleagues. All doctors and nurses owe a great debt to those who practised medicine before them, advancing clinical knowledge and enriching our capabilities.

'He who cannot remember the past is condemned to repeat it.'

'George' Santayana. Actually, Jorge Agustin Nicolás Ruiz de Santayana y Borrás (1863–1952).

Money and Medicine

Effective medical practice is expensive to provide, and the politics involved in supplying it, contentious in every nation. Whereas doctors have performed valuable work in tents, Nissen huts, and cottage hospitals without spending much money, well-funded 'places of excellence' have not always produced value for money (a political judgment, not an academic medical one). Without a dedication to discovery, inventiveness and talent, clinical excellence cannot happen.

The basic standard of medical work should not vary with money supply, but this is far from the case in the UK. Those who know little about medical priorities, but a lot about political expedience, are the bureaucrats who decide on NHS medical budgets. Those rich enough to use private facilities need never expose themselves to the same restrictions; they can afford better. Unfortunately, they risk exploitation from money-grabbing, patronising medical professionals and their companies, who may regard them as easy prey. They are not easy prey, since most financially successful people well know how to spot a money grabber from afar.

In normal times, national safety and the economy are the primary considerations for politicians, and precede any consideration of health funding. Some wealthy individuals share this outlook. They regard their money as more important than their health; some would sooner risk their life than spend their money. Together with those of limited means, they welcomed the introduction of the NHS in 1948.

In every medical corporation, like the NHS and private insurance companies, cost and medical outcome are the key considerations. It has been said that no government can afford the best of medical standards for all of its citizens; it is always too expensive. So what does it cost to save or improve a life?

The Quality-Adjusted Life-Year (QALY) measures the value of health outcomes. Since health is a function of length of life and quality of life, the QALY attempts to combine their value into one figure. Most decision makers in the United States regard a cost of less than $50,000 to $60,000 per QALY, as reasonably efficient. They thus regard screening for hypertension (costs $27,519 per life-year gained in 40-year-old men), as money well spent.

(See: Calculating QALY's, comparing QALY's ad DALY calculations. Franco Sassi. Health Policy and Planning, Volume 21. Issue 5, September 2006, Pages 402 – 408, htts://doi.org/10.1093/heapol/cz1018).

As a completely independent private doctor, I worked mostly with patients who could afford their own treatment. For me and my patients, QALY's never needed consideration.

There will always be patients rich enough to buy, or extract, special considerations from some doctors. It is an accepted principle, however, that no doctor (nurse, or healthcare worker) should offer special considerations to any patient, based on their wealth or status. In reality, this is far from the case. Medical services should be altruistic, delivering the same high standard to all. It remains the utopian dream of many NHS managers and executives.

If the medical profession does not remain altruistic when dealing with patients, it will no longer qualify for the sovereign, sacrosanct status we have inherited.

Because of the NHS style of clinical management, some frustrated patients seek private care as 'refugees from the NHS' (a quote from my then boss, Dr. Peter Nixon, circa 1975). Since some cannot afford the cost, private doctors in the UK have a moral duty to re-direct legitimate and eligible patients to the best NHS service available. The primary job of every doctor is to facilitate a diagnosis, and to follow the most appropriate management for each patient, regardless of their financial status. Those without means, I had to refer to their NHS GP, or to a NHS consultant working privately. The NHS blocked private doctors referring patients directly to NHS consultants (even if they were British tax-paying citizens).

In London, I had the luxury of working in world-class private hospitals (the Wellington Hospital was my favourite), and the advantage of getting investigations and treatment done rapidly. Very few patients being managed privately die while waiting for test results, or on a waiting list. Unfortunately, both happen with morbid regularity in the NHS.

To redress the NHS situation, Andy Burnham, and the Labour Party, recently made a pledge to get cancers diagnosed in under one week. Personally, I would never let a patient wait over 48-hours for any test result. Thanks to the Doctors Laboratory in Wimpole St., London, I never had to.

Doctors in the NHS have had to put up with under-funding, under-staffing, many delays, and poor practice management standards, with little or no control over the decisions made by medical bureaucrats. These conditions of work in the NHS have driven many consultants to pursue private practice. Similar set-ups have pertained in other nationalised industries in the past, and caused their failure.

'What can we do?', NHS staff might ask when feeling demoralised. Those who are not politically minded, will simply have to take what comes and focus on their job, under almost impossible conditions.

A reasonable excuse for NHS inefficiency, and that of any 'free' service, is overloading. When decisions were made to deal with the COVID-19 pandemic in 2020, this was of primary concern. In the private system, directors of private hospitals dream of being overloaded. The big difference, of course, is that the NHS must take on all-comers. For financial reasons, the private system will never experience this, unless the government privatises the NHS. So why is the NHS not better funded, with more medical staff, in order to perform better? The question needs to be addressed to whoever holds the poison chalice of NHS management in the Ministry for Health. Don't hold your breath while waiting for a cogent answer.

In 1973, I retreated from the NHS for two reasons. First, I no longer wanted untrained medical executives dictating my clinical standards; I insisted on treating patients to teaching hospital standards in the way I wanted. I wanted to give patients as much time and personal consideration as I thought they needed. Second, I did not want to entrust my future to politically oriented, anonymous bureaucrats.

What Can Be Achieved?

'Doctors, cure sometimes, relieve often, comfort always, and prevent hopefully.'
Dr. John Fry.

Enthusiastic junior doctors may have to adjust their sights when considering what they can achieve as a doctor. Lucky doctors will bring about an extra day of life for some patients (allowing them to play one more game of Bingo, or to kiss their grandchild for the first time). Because a sick patient's outlook may differ from that of their doctor, one should not dismiss what might seem to be a trivial achievement; many patients seek nothing more.

So what can doctors achieve? We can certainly put an end to suffering. We can cure most infections, and stop some patients collapsing in the street (from epilepsy, or Stoke-Adams syncope). We can encourage health, and even prevent a premature death from coronary artery disease occasionally (although one cannot so easily prove it). For a few doctors, these are an everyday experience.

NHS v Private Care

The most NHS doctors and nurses can hope for is to achieve the same freedom found by private doctors; especially freedom from time and budget restrictions imposed by bureaucrats, all of which can harm patients.

Politicians and their medical bureaucrats love to wallow in myth. One is that they will provide 'equality of opportunity for all'; another is that 'nothing but the best of medical services will be available to UK citizens'. Most UK citizens know these to be myths.

For medical novices and professionals alike, there are balancing acts to be performed: professionalism with approachability; evidenced-based medicine with the art of medicine, general need (from a statistic viewpoint) with individual need. Because they represent different cultures, the NHS and private sectors balance these options differently.

Some UK private medical insurance providers control their subscribers by allowing referral to only those doctors on their short-list (those grateful for the work, and willing to accept reduced fees). In this way, they limit the choice of their insured patients and reduce company expenditure. This has created two classes of private patient: those rich enough to pay for their own needs, and those under the control of an insurance provider. My patients were mostly of the former kind.

After two decades of professional acceptance by private medical insurance companies, and no complaints, BUPA de-listed me because I was not on the GMC's Specialist Register. To be so registered, I would have needed to exhume some of my former world-famous colleagues as referees. I asked them a rhetorical question (I sometimes enjoy wasting my time). 'How many more cardiac catheterisations, pacemaker implantations, echocardiograms, and exercise ECGs would I need to do before the GMC and BUPA would classify me as a cardiologist?

The question misses the point, of course. We now live in an age of standardised certification. No certificate - no recognition! No exceptions! Corporations train bureaucratic minds to obey, not to think. When they think, the result is likely to be of the binary sort.

I knew, of course, that I could survive without medical insurance companies and their subscribers, so I took the matter no further.

Despite not having succumbed to the questionable benefits of a higher education, many of my patients were successful financially. They had gained something at least as valuable as an 'A' level: street wisdom or nous. As self-made business owners running their own successful companies, many of my patients could spot sincerity and integrity from afar. Few doctors I met (including myself) could match their level of practicable wisdom. These self-sufficient, self-made men and women, had won many of life's battles. Few had observed life from the same privileged vantage point as a doctor.

When managing successful people, giving them short measure is a mistake. If they can afford first class travel, they will not enjoy second-class doctors or second-class doctor-patient relationships. Cost is of secondary importance to many of them; they have the financial freedom to choose whatever medical care they need. They will rarely take what comes, or to take what others choose to give them. For that reason, they are exacting to deal with.

Apart from being an able diagnostician and therapeutic agent, two factors define an acceptable doctor in the eyes of most patients: approachability and availability.

For seventy-six years, the NHS has restricted patients seeking alternative advice. NHS GPs have traditionally regarded their patients as 'their own'; their possessions, not to be seen by other doctors without their permission. This restriction of patient movement between doctors is a UK phenomenon, almost unknown elsewhere in the world. Elsewhere, it is the patient who decides who they consult. The recent advent of 'on-line' consultation services has irreparably broken this UK stranglehold. Soon, patients may want to bypass their GPs altogether, and go directly to private GP providers and specialists, although GP referral remains a prerequisite for medical insurers agreeing to pay private specialist fees.

I quickly came to realise that few NHS doctors would survive successfully in private practice. Few would be able to gain the allegiance of independent-minded, self-sufficient business owners. Can NHS doctors learn independence of spirit, an uncompromising patient-oriented attitude, and a vocational commitment to patients? I doubt it. Even though these are clearly matters of personal outlook, few NHS doctors and nurses practise them. I often referred to them as having an NHS attitude.

You might think that private doctors need to be condescending and obsequious; closer to Uriah Heap than Doc Martin. You would be wrong. My patients were mostly intolerant of pandering and bullshit. Their experiences in life made them wary of both. Their requirements were easy to define: they wanted a knowledgeable, experienced doctor, with time for detailed mutual discussion and rapport; a speedy diagnosis, and management of their case displaying the same efficiency as they expected in their own business operations (usually, regardless of cost).

The elitism of private practice, while distasteful to many, will persist while there are ever-growing advantages for the rich and privileged. My patients were not wrong to demand efficient clinical management. The personalisation of medical service, for those who choose to afford it, explains why some patients favour private medicine. Among those who despise such elitism, are those happier with the anonymous standardisation and the bureaucratic restrictions of the corporate NHS.

There is a principle known to experienced chefs, but not to too many corporate bureaucrats: an unpleasant tasting soup is not made more palatable by adding extra ingredients. The answer is to discard the unpalatable soup, and start again; preferably with a chef experienced in making soup.

To the outsider, private practice might seem patronising. It can be, but acceptable only it is for mutual benefit. My patients and I patronised one another because we shared the same objectives and attitudes about management efficiency. Doctors should try to survive independently before deriding mutual patronisation.

In 2017, the Red Cross identified the existence of a humanitarian crisis within the NHS (Nursingnotes.co.uk. 8/01/2017*)*. They provided ancillary workers to help on hospital wards.

Since there are many functional differences between private practice and the NHS, there is still much for both to learn. Effective executive functioning, choice for patients, long-term patient-doctor relationships, patient access to experienced doctors and nurses, the provision of equipment and services, and clinical management efficiency, are among them. The willingness of a private doctor (traditionally hospital specialists only, but now private GPs) to get more involved, rather than to delegate, can add a lot to patient and doctor satisfaction. In private hospitals, it is the admitting consultant who dictates what happens to patients, not the hospital management. In the private sector, significant waiting times have never existed. Now that the NHS has co-opted some private hospitals to carry the load, they too might have overload to deal with. That might lead some of them to take on a 'take-it-or leave it' attitude, typical of some nationalised services.

Private services are not now as functionally competent as they were.

I recently tried to negotiate the admission of a friend of mine (Ken F.) to one of two central London private hospitals. He had experienced a minor TIA. Both hospitals told me they no longer admitted such patients.

Although I understood why, I questioned why they called their organisation a 'hospital'. It was a weekend, and I also failed to contact anyone offering private neurological services.

Our duty to patients has not changed since Hippocrates: excellent medicine still depends on genuine commitment, with all doctors and nurses dedicated to improve their patients' medical welfare.

Observation, Observation, and Observation

We are all observers, but not all are reliable observers. The creation of ideas likely to benefit others will usually flow from three initial steps (an epizeuxis): observation, observation, and observation. Observation alone is not enough. Also needed are intelligence, imagination and persistence. A lot of work is required to take an eureka moment through to a relevant clinical innovation.

Doctors using their original ideas, rather following established guidelines, risk retribution. Medical bureaucracy can regard them as non-compliant mavericks.

More Observations for Apprentices

Doctors spend a lot of time getting educated and spend the rest of their life among those who think Stonehenge is just 'a bunch of rocks in the middle of nowhere'. (Stonehenge is just that, but there is a bit more to them).

Since I qualified, I have attended hundreds of conferences, and spent years reading in libraries. I can now say, with complete confidence, that I am no longer sure of anything.

Clinical medicine is a playground for some frustrated scientists.

A master is always ready to learn, accept reason, and reject nonsense.

Membership of the 'medical club' was once conditional on the tenets of chivalry: loyalty, fidelity, honesty, and dependability.

'God gave us two ears and one mouth', said Judge Judy, implying that listening is twice as valuable as speaking.

The most adept surgeons are not always Fellows of the Royal College of Surgeons. The late Mr. P. B. Subramanian ('PB'), an orthopaedic registrar at Whipp's Cross Hospital in the 1960s, failed his FRCS five times! He could pin and plate a fractured hip in 20 minutes, skin to skin!

According to the film character Sir Lancelot Spratt, surgeons need 'the eye of a hawk, the heart of a lion, and the hand of a lady'. (*Doctor in the House.* 1954. Rank Organisation).

Experience, learning, passion, and talent are the building blocks of astute judgement.

US President Bill Clinton said, 'When people feel uncertain, they'd rather have someone strong and wrong than weak and right' (The Washington Post.Michael Powell)

The same can apply to medical professionals.

Like poetry, many appreciate the art of medicine without fully understanding it.

'We live in an age when unnecessary things are our only necessities.' Oscar Wilde. He could have been referring to UK medical managerial edicts, although their introduction was long after his time.

Many spend their lives trying to grow personal value in the eyes of others. Better to engage in activities that ignore this. Happy is the person content with her own values.

To Be or Not to Be a Good Samaritan?

Who needs a good Samaritan when there are excellent emergency services (in 1st world countries)? There are potential problems in helping the suffering, or an injured person on the street. A doctor, nurse, or paramedic might make a helpful difference; even save a life. The police might, however, arrest them for molestation. For doctors in the UK, that would bring the GMC into play. That would cause considerable worry, and a lot of wasted time while they exercise their statutory duties and authority ('in the public interest', and to maintain the good standing of the medical profession).

One consequence of becoming a Samaritan is the obligatory need to provide a detailed witness report. That will take hours. Those who intervene medically risk being sued should things go wrong; even though their intention was to be helpful.

Some patients take doctors and nurses for granted, and even take advantage of them. Others will become suspicious of their motives. Generous behaviour can get punished. In Biblical terms, a genuine Samaritan must accept all such negativity; those in need are all the 'children of God' (John 3, 1-2; Galatians 3, 26). Regardless of all the downsides attached to present-day Samaritanism, it is my view that doctors, nurses, carers, and paramedics, with life-saving skills, hold a sovereign duty to help others, regardless of any medico-legal considerations. All medical apprentices must attune their minds the fact that medical practice can be self-sacrificial.

Fame and Fortune

Medical practice in the UK rarely creates wealth for doctors. Doctors can make a living, but only a few become wealthy. That is the preserve of those who 'make' money.

Very few doctors seek fame. One can have famous patients, but fame is unlikely to be theirs. This is the reserve of those vain souls, adept at self-promotion. Some doctors will quickly change roles after qualifying, and become recognised politicians, TV presenters, comedians, and musicians. Did they waste a medical school place?

If it's power a doctor wants, a professorship, and an appointment to the Royal household will help. They might otherwise consider working for the CQC, or GMC; those in the top jobs will automatically leave with a knighthood.

Anyone drawn to academic study should sign up for a BSc course, and follow it with a research degree (MD, or PhD), before attempting a medical apprenticeship. This will impress non-academic colleagues and smooth the path to promotion. If others recognise the merit of your research, you could become eligible for an award (National CEAs), and get extra pay.

Life Planning

A medical career is not all that needs planning.

In 1967, one of the coolest guys I ever met was a gynaecologist registrar at Whipp's Cross Hospital. Dr. J.C, always had a string of beautiful girlfriends in tow. His junior colleagues were envious. I asked him once, why he hadn't married? After a nonchalant puff on his pipe, this landed gentleman from Cork in Ireland, gave me his answer. 'A man should not marry a day before his 40th birthday, and a woman not a day after her 25th!'

I completely failed to heed his wise counsel.

Is Health the Absence of Disease?

Russian journalist, Anna Politkovskaya, said during a BBC interview: 'the job of doctors is to give health to their patients'. Although I understand her sentiment, 'giving health' is hardly what doctors do. Was she ill-informed, or was the translation from her mother tongue (русский язык – Russky yazik) unreliable? Although her English was better than my Russian, there may have been a problem with what she meant to say. There are many problems with translation. A person from one culture cannot always convey their desired meaning to those from a different culture.

I once visited Istanbul to consult with a patient of mine. He had chronic stable angina. He spoke no English, but two of his multilingual sons accompanied him. They spoke eight languages between them. The elder son translated, and said: 'My father would like to ask you a sensitive question, Dr. Dighton, but I fear my translation will be incorrect. It is embarrassing, but he wishes to know if 'domestic' sex is safer for him than 'commercial' sex?'

Health is a matter of personal perception; disease a matter of pathology. In the early or stable phases of a disease, the health of a patient may not be affected. With advanced disease, ill-health becomes obvious, with occasional exceptions. Those about to die from a sudden cardiac infarction, an aneurysmal brain haemorrhage, or a burst aortic aneurysm, may feel perfectly healthy the moment before. If someone feels well, and has enough energy for life, they feel healthy, even when harbouring a chronic disease like atherosclerosis or osteoporosis.

Forever healthy people may have no interest in disease, or what it does to others (pain, shortness of breath, weakness, depression, fear, etc.). They rarely define their state of health as the absence of disease. In fact, they rarely question their health at all (this makes them reluctant to accept medical screening). Since we train doctors to understand disease, their only acquaintance with 'health', may be personal only.

My experience of several generations of the same family, made me favour the idea that we inherit both 'health' and disease, rather than them being the result of lifestyle. Our genotype predicts our phenotype, despite the trend to consider epigenetics important. That it is possible to change our inherited features, may give hope to some, but is too fanciful? Healthy people beget healthy people. Those with many types of physical problem will often have offspring who bear the same burden. That is why delving into a patient's family history is so valuable.

One word of caution. Withdrawing hope is dangerous, even if guided by genetic science. Hopelessness can promote the deterioration of physical disease and mental health.

Sporting families, and those whose definition of health is the ability to walk up a mountain, run a marathon, or canoe for hours, will often share the same enthusiasm with their progeny. Some will inherit the interest. An interest in sport may suit their family culture, with a need to win, typical of most sporting families. As a person disinterested in sport, my reply to the question, 'Where are you most happy?', has dismayed a few sporty people. My answer used to be, 'in a library'. Now that libraries are disappearing, I am happy at home viewing the internet.

The comedian Billy Connolly once commented that . . . in every library (and bookshop) there is an escape tunnel: one that leads to all the knowledge in the world. Represented there, are all the great brains of the world, and their thoughts can be at your fingertips (*'Made in Scotland'*, p42, Billy Connolly. 2018. Random House).

Age is not a reliable predictor of individual health. For medical entities, biological age is a better predictor. There are many healthy octogenarians, and many very unhealthy teenagers. In my experience, healthy athletes usually have a lower biological age than non-athletes (perhaps by pre-selection with healthy families begetting healthy people). Apart from injuries, many older people now keep healthier for longer. My impression of lifelong smokers, drinkers, and drug addicts is that their biological age has been unduly advanced; they often look older than they are and look unhealthy. It does not follow that they complain of ill health, or suffer from disease.

How many smokers know, that only half of them (on average) will reach 65-years of age? I wonder how many care? Not all of them, I'm sure. A small number will, of course, reach 100-years of age. While doctors use research statistics to guide them, averaged results can never apply to every patient (except by chance). This is the statistical paradox. I discuss it later.

For those who want to improve their health, doctors and healthcare workers have many tired clichés to offer: lose weight reduction, diet, exercise regularly, relax and cease smoking and drinking. Many patients have heard them all before, and many find them tiresome. The fatigued, the depressed, and those with limited resources may find such advice unacceptable. Patients can present many valid excuses for not following such advice, like being too busy working to jog, and not rich enough to join a gym or have regular holidays. With limited choices, fewer are likely to achieve job-satisfaction and a boosted self-esteem. Sadly, drinking and smoking may provide their only enjoyment.

An old impoverished friend of mine once said, 'Why on earth would I want to prolong my life? I have known nothing but hardship and stress.' He continued to smoke and follow in his father's footsteps with peripheral vascular disease, leg amputation, severe type-2 diabetes and angina. He had one heart bypass (CABG), and a few coronary and femoral stents. He insisted on chain smoking and eating lots of carbohydrates – they gave him pleasure. Despite his multiple pathologies, he felt healthy for decades. He died recently (2023), 84-years old.

Health and disease can present as separate coins, not just connected sides of the same coin.

Research Matters

Scientifically minded doctors will often find their way into research. They may wish to become scientists first, and physicians or surgeons, second. Hopefully, those ill-equipped to cope with people and their problems will soon realise these as clinical shortcomings. Only a few will become accomplished doctors and research workers.

For amateurs, clinical research is usually a matter of collecting numbers; enough to confirm or refute an anecdotal observation. These days, a lot of organisational research involves pursuing the **BL**indingly **OB**vious. Government organisations, like think-tanks, will do it to justify their existence. There is so much government money available to them, and so many graduates unemployed, that researching and confirming the obvious is now a growth industry. My advice to research-minded physicians is this: avoid BLOB topics and find a research project that challenges current understanding and holds some promise of changing the way we manage patients.

I started my career in 1966, doing ward rounds during visiting times. I did this to involve the patient's family and friends in the assessment, evaluation, and management. I read with dismay, 47-years afterwards, that 'doctor-patient involvement' is now thought invaluable. A survey of patient opinion showed that 96% of patients welcome involvement in their management. *(Importance of clarifying patients' desired role in shared decision making.* Analysis. BMJ (2013). 347: 18 - 19).

I have some questions about researching something this obvious.
- Who would need to question the desire of some patients to be involved in their own management?
- Who would spend money measuring this desire, just to put a percentage figure to it?
- Who would spend lots of research time and money verifying what doctors have known for centuries?

For those engaged in any worthwhile research, there are principles worth knowing: avoid fooling yourself, pursue objectivity, use tried and tested methods, and always get independent verification. To help us decide what is true or false, we can compute confidence limits and percentages, together with 'p', 'r', 't', and χ^2 values. It is our duty to provide as much certainty as we can for patients and those who question research results, especially for those with no practical knowledge of medicine or clinical insight.

A *Private* Doctor's Apprentice?

For those in independent private practice, there are some basic requirements for success. Doctors must not just help themselves, they must provide their patients with a satisfactory medical service: conveniently appointments; rapid correct diagnoses and effective clinical management with no significant delays. The service provided must be more easily accessible and efficient than that on offer from the NHS. They might otherwise fail financially.

An independent doctor's livelihood can depend on his mastery of both the art and science of medicine. Patients must be confident that he knows what he is doing and is worthy of their trust. He will survive financially, only if he can prove this to many patients (many of whom will be successful people, running their own businesses). In order to do this, he will have to draw on many resources. Not only must he be able to establish an easy rapport with patients, he will need to show commitment to their best interests. Private patients do not mind paying; what they mind is a doctor whose only interest is making money.

From the start, the NHS provided a haven for those who found it difficult to establish and run their own independent practices. Many were altruistic and needed support.

The extra income and job satisfaction of private practice attracts many NHS consultants. The pleasure of relinquishing the yolk of corporate NHS control, albeit temporarily, has led many to give up their NHS posts altogether. Several NHS specialists have asked me how I survived without an NHS income. It helped that I was born independent-minded, and quickly found I could stand on my own two feet. I strived to say, 'the buck stops with me'. Primarily, I wanted to be accountable to my patients and staff, and not to any regulatory intermediaries (unless there was something positive to learn from them).

Medical Practice Ain't What it Used to Be!

There is little point to nostalgic reflection, other than learning from mistakes. The past was only good in parts but remains a reserve of tried and tested strategies and methods.

During my career, I witnessed UK medical practice being modelled on business corporations. The NHS introduced the same personnel structure and copied corporate control methods, strategies, rules, and regulations. In corporations, increasing the turnover and multiplying the benefits are the foremost objectives. Most corporations achieve these by minimising time-wasting and expensive human interactions. NHS executives are likely to think they can achieve the same goals by adopting AI driven avatars to replace doctors, nurses and other caring staff.

When I was young, our family butcher and baker, knew what my mother bought. This made my errands to the shops easy. I then watched rapidly growing businesses like Tesco's and Sainsbury's, slowly ease out many small businesses with improved convenience, but far less personal service. Years later, I watched many independent GPs, who had the loyalty and support of their patients, replaced by grouped practices. Both were the conceptions of corporate thinkers, who long ago decided that anonymity was good for business. It may have improved patient throughput and improve financial efficiency, but what did patients think of the service? I'm not sure anyone cared.

Computers Ahoy!

I wrote my first computer programs to run my practice in 1985. The once bomb-proof DOS operating system and programming language I used, gave way to the more complicated (more profitable and fragile) system of Microsoft 'Windows' programming, networks, websites, and internet connections. I watched the birth and growth of the Microsoft empire as programs became accessible to all.

Doctors and administrators are now so dependent on computers, few would dare ask: 'Do we need more computerisation, or less, to achieve patient satisfaction?' Like questioning the NHS as a sacred cow, questioning the need for computers is heresy. Even though we know how unreliable they can be, and how open to piracy they are, we accept them totally, regardless of many inconveniences.

Computerised systems are the equivalent of golden egg laying ducks to corporations. With fewer employees, executives can now grow their businesses and remain in control. Medical bureaucrats quickly learned the advantages. They could store and process the actions of every health professional. Computers allow all corporations, including the NHS, to monitor and audit the actions of their employees. Computer programs can now detect non-compliance and will store the evidence for later use. Internet connections improve communication possibilities, but even after spending billions on IT, the NHS still doesn't have a completely joined up system. This begs an important question. Would the many millions spent on IT in the UK have been better spent directly on patients?

We can already program computers, fronted by avatars, to put medical questions to patients, relate the answers to those they gave before, and weigh them for clinical relevance. If widely introduced, the time saving for doctors could be significant; they could spend more time on the golf course!

There are two reasons for the computerisation of the doctor-patient interface using AI, to be promoted. First, no country in the world can afford personalised medicine for everyone. Second, the desire for corporate control is expanding and evermore intrusive. AI will be adopted if it saves money and aids regulatory control, even if many more high-earning medical executives are necessary.

Should junior doctors and nurses welcome assistant avatars, even though it will be a long time before any of them passes a Türing test, and threatens their jobs? It will take a little more than the rote use of phrases like, 'Hello, Mrs X. How are you today?', and, 'Goodbye, have a nice day', for patients to put their trust in them.

AI processing is impressive. But so are the magic tricks performed by professional magicians, entertaining their audiences with card tricks, and doves drawn from hats. At least AI might help 'fix' patients with minor complaints and be of use to on-line advisers in call centres. AI might even help to filter out time wasters, and those who do not need a doctor. This might free doctors to deal with the patients who need them.

Using pattern recognition, AI computing has proven to be a useful aid for diagnosing rare diseases. Remarkably, they now offer valid interpretations of retinal images, ECGs, X-Rays, MRI, CT, PET scans and other images. As a filter to prevent some human error, they could prove invaluable. Whether we should trust them completely is another matter. To answer who is liable for their misdiagnoses and any clinical consequences, will employ legions of insurers and lawyers.

Will AI free up time for doctors and allow them to spend more time with patients? I doubt it. In corporations, few changes aimed at reducing personal workload have ever been effective; time-wasting processes usually replace other time-wasting processes. Initially couched in jest about Royal Navy procedures, Parkinson's Law (1955), states that 'work expands to fill the time available'.

Plus Ça Change

I gave no thought to it then, but as I entered medical school, the 2nd World War had been over only sixteen years. The wartime dutiful spirit of doctors that pervaded the era, strongly influenced my medical student ethos.

Change is inevitable as time passes, especially since most young people are too restless to handle stability. Not every change is for the better. So what has changed since I started my career?

Take the medical student selection process. Most of my medical student contemporaries were ex-public school, had private incomes, and came from medical families. Only seven of the 70 students in my year of 1961, were women. Writing about Oxford University in 1938, John Betjeman, while discussing the equality of the sexes, suggested that 'only a hermaphrodite can give an impartial judgement.' (*An Oxford University Chest*. 1979. OUP). Times have changed. In 2022, the GMC reported that 'women account for almost two-thirds (64%) of the 2021/2022 medical student intake.' (*The Workforce Report*. GMC. 2022).

My fellow students and I, learned an old-fashioned, chivalric attitude to duty. Perhaps anachronistically, we held honour, honesty, and respect for others in high regard; we thought these were the essential prerequisites for all doctors, taught as we were by those who served in the 2nd World War and provided us with admirable, albeit formal role models. Their disciplined purpose (like their armed forces' contemporaries) was to *'knock students into shape.'*

The old days I refer to are long gone, but my commitment to the standards of the time never changed. It would now be totally unfair, to ask any intending medical student the question, 'If you were not to be paid, would you still want to become a doctor?' It is a measure of how long ago I was at medical school, to say that many of my contemporaries would have answered 'yes' to a question that now holds no validity. Junior doctors in training have to move around from one hospital to another and find accommodation which they can hardly afford on their pay; pay that is neither commensurate with their value, nor comparable to that of NHS executives. They need to strike for better pay, and are planning to. The vocational dedication of doctors is mostly undoubted, but remains an important issue, because it can determine what patients think of the medical profession.

Although financial privilege still abounds, supporting private education and the payment of university fees, family wealth is less prevalent among medical students than it was. What of the old-fashioned chivalric sense of honour and duty? Has that diminished?. Junior doctors and nurses undoubtedly showed their mettle when the COVID-19 epidemic came to test them. But where were the managers and executives? At home, enjoying their large pay-checks, no doubt. For the first time in decades, junior doctors and nurses working without managers, tasted the clinical autonomy I once enjoyed.

As part of my history, many of the nurses I met in my early days of work came from privileged backgrounds. A few at St. George's Hospital, Hyde Park Corner, where I worked in the early 1970s, lived in Grosvenor Square and rode their horses in Hyde Park. Their concern was their patients, not their pay. They were free to pursue their medical career and vocation. One measure of their dedication was that all medical staff stayed on duty until their sick patients improved. This was once a rite of passage to senior medical and nursing posts, but employment laws came to disallow it.

Although the same dedication to patients undoubtedly remains among junior doctors and nurses, they now need an extra focus - to regard corporate culture as important. Managers now proclaim that no patient should have a tired nurse or doctor treating them. They miss the point, as is typical of those who have no medical background. Primarily, patients want competent nurses and doctors dedicated to their vocation; tiredness is a secondary issue.

Good intentions

'Last month I met with Jane Cummings, chief nursing officer at NHS England. I was asked by her to highlight the important work they are doing around improving the quality of care, and more specifically their campaign named the 'six Cs': care, compassion, communication, competence, commitment and courage.'

Bulletin from the President of the RCP: May, 2014.

Have the 'six Cs' ever been in question? After Florence Nightingale, who would dare question them? Is past knowledge that easy to underrate and forget? Today, executives recycle much of has been known for decades, as if it were something new. It serves a useful function: keeping them employed to reinvent wheels. We obviously need to resurrect the wisdom of Hippocrates and Florence Nightingale. This would immediately improve our education and patient management.

Final Thoughts

Foe most doctors, medical apprenticeship is for life, but kept under review.

A strong desire to become a doctor, nurse, paramedic, or carer will keep many going for years, but only the desire to learn more each day will sustain a career for a lifetime. Unfortunately, a diffident willingness to accept bureaucracy in all its overbearing forms is now a major preoccupation. Bureaucrats did nothing more than interfere unnecessarily in my work. I must say, I learned nothing useful from any of them, except how to waste time. Had they come to me wanting to know what I had learned from fifty years of medical practice experience (a practice that generated no complaints from any patient), I would have been more sympathetic to their role as housekeepers. Instead, like all autocratic rule followers, they now think they know most, when actually, they know least.

Here is what I predict will happen next. The 'blame culture' is now in full swing. Ambulance chasing lawyers are increasing in number. There is now a clear direction of travel for what UK medical professions must contend with next. Doctors and nurses, like the police, will shortly have to video every interaction they have with patients and colleagues. They will have to get every patient to verify what they say, and provide a signed statement, witnessed by a third party. This evidence will be required if doctors and nurses are to defend themselves against the growing number of allegations made by patients, co-workers, the CQC, GMC, and PSA. If doctors and nurses refuse to collect the evidence, medical protection insurers might not insure them. The PSA, GMC and CQC could then refuse to regard their medical work as 'safe' and in the public interest. Many doctors would have to be replaced, but by whom?

The battle for medical sovereignty over the dumb expansion of medical bureaucracy in the UK, is gathering pace. What needs to happen is simple. Medical professionals once told medical bureaucrats what they needed to practice, and they carried out our instructions. This state of affairs needs to return before the progressive failings of the NHS can be reversed. Every medical professional must now ask themselves if they are up for the fight. If not, they must act dumb and lie low, emigrate or change profession.

Chapter Two

How Doctors Function

What wisdom can you find that is greater than kindness?

Jean-Jacques Rousseau.

The behaviour of doctors and nurses can result from the balance they strike between their altruistic duty to patients and their selfishness, greed (for money and/or power) and need for co-operation.

Some members of the public see doctors as arrogant, know-it-all, entitled, over-paid, and suffering from a 'God Complex'. The expedient political need for social equality, has exaggerated their views. The medical profession has thus progressively lost status over the last few decades.

In this chapter, I have attempted to describe the functioning of doctors as characters. It is easy to describe how chess pieces move and the rules of the game, but more difficult to learn strategy, and moves that will lead to eventual success. The same difficulty arises when trying to understand patients, nurses and doctors, and why they behave as they do. Since colleagues can influence a doctor's future life and career, there is a distinct advantage to understanding their character, their agenda(s), and why they behave as they do.

Let me start with a warning. If all doctors were honourable, reliable, and trustworthy, there would be no need to ask, 'Can I trust my colleagues with my personal information, with my research results, with my patients when I am off duty, and with privileged information about me?' Unfortunately, only the naïve will not to ask the question.

Most doctors are politically astute, highly motivated, competitive, self-oriented, and hostages to ambition, with agendas of varying transparency. To survive happily in any relationship with them, one must first discover their mission. Only then can one decide what role to play in their game, and what role to allow them in yours.

'For every man has business and desire.'

Hamlet: Act 1, Scene 5.

The Right Stuff?

Society honours those who organise charities, politicians, and those who relentlessly self-promote themselves. Some of them will be colleagues. Money, fame, power, and honours, do not equate to the many other rewards available to an altruist physician; the rewards of practising medicine are far more intimate and personal, but not mutually exclusive.

Knowing how to save lives and improve the health and lifestyle of patients requires a special set of skills, some of which come with a sovereign duty: to use them whenever necessary. Some medical professionals have skills other than clinical ones. Some can improve public health, and some can advance medical knowledge through research.

Unfortunately, some who qualify as a doctor, may find they have made a mistake. They may find that committing themselves to the health and welfare of others is not what they enjoy. As time progresses, it can be more difficult to change track. There are those who may ask, 'With my intelligence and academic ability, how on earth did I make this banal career choice?' In all walks of life, there are plenty of miserable square pegs trying to fit into round holes. They bring no joy to anyone, least of all themselves.

Worthy Doctors

Some doctors I have known have been remarkable, worthy and interesting people, although I have only known a few as close friends. Every doctor will meet hundreds of distinct characters throughout their career, not all of whom will be amusing or enlightening. They range from the funny to the pedantic; from the driven and ambitious to the contented; from lively to dull; from capable to feckless; from utopian in outlook to utilitarian; from sycophantic to self-assured; from mean to generous; from obsessive to carefree; from abrupt and discourteous to affable; from altruistic to frankly commercial. Like all others one will meet in life, medical colleagues can be a source of pain or joy.

In order to recognise worthy doctors, the now anachronistic characteristics of a 'gentleman', and 'gentlewoman', are worth consideration. These characteristics still define worthy colleagues. In his book, *'The Art of Writing'* (Cambridge UP, 1917), Sir Arthur Thomas Quiller-Couch, quotes Newman's definition of a gentleman:

'He is never mean or little in his disputes, never takes unfair advantage, never mistakes personalities or sharp sayings for arguments, or insinuates evil which he dare not say out. If he engages in controversy of any kind, his disciplined intellect preserves him from the blundering discourtesy of better perhaps, less educated minds: who, like blunt weapons, tear and hack instead of cutting clean, who mistake the point in argument, waste their strength on trifles, misconceive their adversary, and leave the question more involved than they found it. He may be right or wrong in his opinion: but he is too clear-sighted to be unjust. He is simple as he is forcible, and as brief as he is decisive.'

For the gentlewomen and gentleman doctors I have met, I would also add a genuine interest in medical knowledge, and a desire to attain clinical wisdom. Clinical wisdom, I define as clinical knowledge distilled by experience into the capacity for dependable judgement.

In performing his (or her) duty, the gentleman (or gentlewoman) doctor, aims to benefit patients in preference to any personal advantage. Despite the practical need to assuage regulators, his patients will always come first. He is gifted with human understanding, whatever the patient's background and culture, and can communicate his thoughts in words that most will appreciate. He will be a faithful and loyal medical colleague, and with wisdom, guide and accompany his patients on their journey through all the key medical decision-making stages. In this, he will employ not only his experience and expertise, but his compassion, kindness, and humility. A devout student of both the art and science of medicine, he is skilled enough to apply each appropriately, even though he may not yet have mastered them completely. He will have clear insight into his medical capabilities; always ready to learn, and to defer to wiser judgement when necessary. He will graciously help any colleague deficient in knowledge or skill whenever asked for guidance.

Only a few doctors I have met were worthy enough to possess the humility expressed by the ten fictional Scandinavian tenets of Jante Law (*Janteloven*). The first is:

'Du smal ikke tro at du er noget.' ('Don't think you are anything special...')

Axel Sandemose (1933). *'A Fugitive Crosses his Tracks.'*

Stereotyping

When describing an individual, we are all liable to make unfounded assumptions. Many will use biased stereotyping, and not only that based on objective, reproducible data. I have often benefitted from holding some stereotypes loosely, while remaining sceptical about how predictive they might be. Would a stranger visiting Scarborough or Moscow benefit from knowing that 'Brits can be argumentative, but are fair-minded', and 'Russians are mostly phlegmatic, and do not make friends easily?' I think they would, knowing both as I do.

Predicting the character of one stranger from a group stereotype is impossible and heretical from a statistical point of view. One cannot reliably use general characteristics to describe any individual, nor can one map the characteristics of an individual onto an entire group. Used without justification, stereotypes can be discriminatory and unfair. If held loosely, with an awareness of all the biases and errors involved, some stereotypes are useful in certain instances.

Why might this be? The answer is statistical. For any set of individuals, a specific characteristic will have a prevalence within the group. In many normally distributed data sets, 60% of all the occurrences of a characteristic will lie close to the average (within one standard deviation of the mean or mode). This suggests that some stereotypes (but never all) could be 60% reliable, used predictively. This is a little better than tossing a coin for prediction (a 50% chance of heads or tails). A stereotype can bias one's initial opinion, but should never allow bigotry. I see stereotyping as a starting point; a step towards the heuristic learning process of getting to know others, and their affiliated group.

Stereotyping becomes bigotry when a person sees no need for confirmatory evidence. For some, it will suit their political purpose. This has given stereotyping a bad name as with misogyny and racism.

When we choose candidates for medical school education, the selection committee will have a stereotype in mind. Candidates must be educated scientifically, be easily able to learn large amounts of information, be objective in outlook, and have a professional attitude to work. Crucially, they must also have a sincere interest in the medical welfare of others.

Eclectic or Nebula Thinking

The mastery of medicine requires much more than binary thought; it requires *nebula thought* applied within a dynamic time-frame. By nebula or eclectic thought, I mean the ability to select the pertinent elements from a constellation of facts, concepts, and thoughts, in order to arrive at a cogent conclusion, be it a diagnosis or clinical management plan. Such conclusions can appear strange, unpredictable and even magical, to those incapable of them and without experience. A complex system like human pathophysiology has many variables interacting together in a difficult to discern way (changes in the weather are similarly complex). Fixed, predictive processing rules using algorithms (with defined boundary conditions), cannot simulate all the inter-reactions. AI now pretends to predict complex functioning, even though local weather forecasting has yet to become reliable for more than fifteen or thirty minutes. Despite the freedom we all have to choose data from a clinical data set, only an experienced clinician will easily recognise the relevant data when forming a conclusion. Because the processes involved are complex, the nebular thinking required to handle it is best described as an art.

> In wit, as nature, what affects our hearts
> Is not the exactness of peculiar parts,
> 'Tis not a lip, or eye, we beauty call,
> But the joint force and full result of all.
>
> Alexander Pope. *An Essay on Criticism* (1711). Part 2: 243-246.

Biological systems and their dysfunctions, like the weather and the economy, are not just complicated (completely described and predictable, like known chemical reactions), they are complex (multifactorial, with many interacting independent variables). Only those comfortable with the unpredictable, and capable of nebula processing, will rise to the challenge of elusive diagnoses and the handling of perplexing clinical cases.

Because it involves a series of personal value judgements, largely based on anecdote, nebula processing is a fuzzy process, an art with science playing a backup, confirmatory role. Eventually, science may catch up with the complex processes that allow human judgement to form decisions that are wise and durable. Neurobiology may eventually come to understand this in terms of 'on' and 'off' biochemical switches, but for now, nebula thinking is an important aspect of the art of medicine. As with all arts, talented masters emerge who seem to make light of it all, able to function without effort or second thought. The absence of formulaic guidance and definitive processing rules to guide them can expose ineffectual decision makers. Doctors, scientists, and entrepreneurs who can only function algorithmically are distinct from those capable of nebula thinking.

The results of nebular thinking can appear unfathomable, but sometimes the distinguishing mark of genius. Paul Dirac, British theoretical physicist extraordinaire, provided a good example. An eclectic person, he had been an engineer and gifted mathematician before getting interested in quantum physics. He drew on the ideas of other leading physicists (Einstein's special relativity, the work of Schrödinger, Planck, Pauli, and Heisenberg), and combined them with his own ideas (quantum field theory, the need for anti-matter, topology, and the square root of vectors). The Dirac equation (1926), resulted from his integrative work. His equation helped the understanding of how relativistic electrons behave, and much more. It led on to the discovery of the positron. His work ranks with that of Newton and Einstein.

Medical Colleagues

Most medical colleagues are helpful, but some are demanding and unhelpful, interested only in using others. Be warned: the ambitious can steal your ideas and pass them off as their own. I have encountered one or two of them, and felt both amused and abused in equal measure. I have a particular weakness: I object to being used. Using people is a common political gambit and second nature to some. In my experience, the more intelligent and ambitious the character, the more likely they are to use others. Some will try to discredit you, if it means that they can grab the prize – first to publish a paper, or get some recognition from their peers.

Ambitious doctors can also be intensely territorial.

Dr. J. was a general hospital physician with an interest in cardiology. He sent a patient with 'fainting' episodes to my clinic for a routine ECG. My colleague Noreen found the patient to have complete heart block (CHB). When she asked me to confirm her finding, the patient became alarmed. 'What's the matter? Have you found something serious?' she asked

I had to decide whether to say nothing and insist that she saw her consultant again with some urgency (she had an appointment for 10-days' time), or inform her of her situation so that she understood the need for urgent action. I chose the latter. I warned her that CHB can cause sudden blackouts (Adams Stokes syncope), and she should not drive her car. We telephoned her consultant to arrange a next-day appointment, but failed to contact him.

He later chastised me for getting involved in his case. After all, he had only referred her only for an ECG. He regarded the matter as his responsibility alone. We agreed to disagree. He sent me no further patients for ECGs, even though our co-operation had well served his patient's best interest. To have honoured Dr. J's absolute right as the only one to communicate with his patient, would have risked a medical catastrophe. He put himself before his patient, and this I disagreed with.

What should a doctor look for in a colleague? Perhaps camaraderie, being fun to work with, helpfulness in discussing clinical problems, and help with practical issues. They are all of benefit. As physicians, we need to collaborate with experts in other disciplines: imaging, laboratory science, pharmacology, surgery, and psychiatry. We need colleagues who genuinely wish to benefit patients.

Some doctors are theoreticians (physicians usually), others have practical skills (surgeons). Doctors mostly view their world from one of these perspectives. A few can do both. Some will instantly see the discrepancy in a theory or argument; others are quick to see practical faults. The ability to recognise something '**BL**indingly **OB**vious' (BLOB), in any discipline, is valuable, but depends on the perceptiveness, knowledge, and experience of the observer.

On the first day of my first house-job, on 1st June 1966, Colin Davis ('Col') my Registrar (a visiting senior surgeon from Australia), briefed me on how we should work together. My recollection of what he said was something like: 'Dave, I'm a cutter. I know very little about medical matters. If it's a surgical problem, call me anytime. If it's a medical problem, it's entirely down to you.'

In retrospect, this was the making of me, although far, far away, from what any junior doctor will now experience in the UK. Without helpful junior colleagues, some confidence, helpful nurses, some clinical insight, an ability to assess risk, and a desire to take responsibility for my actions, tragedy would not have been far over the horizon. The situation worked well for me. To survive, I had to swim after being thrown into the deep end without warning. I soon discovered my capabilities. Happily, both my patients and I survived the experience. Later on I would meet many doctors, even as consultants, who couldn't act or decide anything without the help of others or with a rule book in their hand. They would remain safe, but what of their patients? Being ill and alone as their patient on a desert island, could prove fatal or lack the necessary action. Doctors unable to decide and struck be inertia, can be more of a risk to patients than those capable of action in risky situations.

Doctors as Eccentrics

Eccentricity is a beguiling feature. For doctors, it sometimes requires years of cultivation.

Donald Hunter, author of The Diseases of Occupations *(1955), practically invented industrial and occupational medicine. He was one of my teachers. At one of his lectures in 1963, he stood quietly behind the front laboratory bench, waiting for silence. He then waited a little longer for absolute silence. He raised a fist, and in time with his words, banged the desk in front of him saying, 'Toy – lead – soldiers – must - be banned!' He banged so hard he gave us all a jolt, and imprinted us with an indelible memory.*

A neurologist I knew used to absent himself from half-completed consultations. Without explanation, he would excuse himself and play his tuba in an adjoining room. After a while, he would return as if nothing had happened.

My father told me about a notorious East-end GP, Dr. Jelly, of Lower Clapton Road, London. I once saw him riding his bicycle around Bethnal Green, wearing a top hat and tails. Before the NHS started, he provided cheap consultations for the poor, and regularly took groups of his poorest patients on day trips to Southend-on-Sea.

Few will have ever met a doctor as eccentric as my good friend Dr. Alan Gardner, a GP in Walthamstow, east London. I saw him first in 1967. He was walking down the long corridor of Whipp's Cross Hospital wearing the white coat, typical of all junior doctors. The difference was, he had a Capuchin monkey sitting on his left shoulder. I had to befriend him; a decision that proved to be one of my best.

Alan was having a house party, so I went. What I saw stays with me. Dressed as a Native American, Apache Indian Chief, he opened the door. He wore a full tribal costume and was holding a Winchester rifle. While visiting the US, he had become the blood-brother of an Apache Chief, and said he had the right to wear the costume.

Alan was multi-talented. Not only was he a talented musician (clarinet, saxophone, and guitar) who taught me many simple guitar chords, he was also a house developer, a private firework maker, inaugurator of the first International Conker (horse chestnut) Contest, an architect, member of the Magic Circle, a private brewer who made more beer for personal consumption than anyone else in the UK, a boat builder, sailor and master mariner, an armourer (he made suits of armour, guided by his friend, a master armourer, Terry English), but also a GP in Penrhyn Avenue, Walthamstow, London, E17.

He used to say to those patients he couldn't help, 'It's time you saw a proper doctor.' He would diagnose challenging cases as 'GOK': 'God Only Knows!' He patients, colleagues, and many friends loved him for his light-hearted approach to life, and his directness and honesty in medical matters. Only his close friends were privy to all his eclectic talents and eccentricities.

Sometimes it is only possible to admire from afar: the fantasy becomes richer, the further away one gets. Get too close, and admiration may fade as various peccadilloes and discordant behaviour, comes to light. Not so with Alan Gardner. All of his talents were real and admirable.

My friend, teacher, colleague, and cardiologist, the late Dr. A.M.H. was called regularly to attend the Saudi Arabian Court of King Faisal. After consulting with the King, along with many other cardiologists from around the world, he would receive an attaché case, full of money. He assumed that no tax was due on it since he had earned it outside the UK. He only rarely counted the money, and never enquired about tax liability. It outraged him when he received an estimated tax bill for £500,000 (1970s), from HMRC. He asked me what to do. He had earned the money abroad, so why would any UK tax be due?

He inadvertently bought a house at an auction. A.M.H. became indignant after a sitting tenant refused to leave. He told me, 'I have asked him to leave nicely, but he says he has the legal right to stay.' I had to explain to him that, since he was correct, why not offer him money to leave? That solved the problem.

When his cleaner cut the tassels off of his very expensive Persian rugs (it made them easier to vacuum, she said), he asked me if I knew someone who might sew them back on. Faced with a daughter suffering with tonsillitis, he would ask me what antibiotic to prescribe. A.M.H. was the best invasive cardiologist I have known; he taught me invasive cardiology (vascular catheterisation), including pacemaker implantation and much more. I will always be indebted to him. Solving simple problems eluded him, and I took it as a sign of our friendship that he asked for my advice.

The Patient's Perspective

Patients prefer their doctors to smile and wear conservative clothing, otherwise patients think them aloof and odd. See: *Judging a book by its cover: descriptive survey of patients' preferences for doctors' appearance and mode of address.* Lill, Marianne M.,Wilkinson, T. J.et al. BMJ. (2005). 331 (7531); 1524-1527). Another commonly held view of doctors and other intelligent beings is that they have no common sense. However clever we are, few of us can make wise decisions with emotions in play.

Being a doctor or a nurse is special, but coming to regard oneself as special, is unacceptable and a common cause of conflict with others. If you expect to be treated as special, those who value only fame, money and appearance, may not acknowledge your worth.

Two Medical Cultures: Attached and Detached

In his 1959 'Rede Lecture', C. P. Snow suggested that are two cultures to be found in society: those trained in science, and those trained in the humanities (languages, literature, philosophy, art, history, and the law, etc.) ('*The Two Cultures and the Scientific Revolution*'. Martino Publishing.2013).

We now select medical students because of their scientific culture, and because they are capable of the emotional detachment required by some medical and scientific work.

Some students who achieve the highest grades in their pre-clinical years will take a science degree, a course leading to a BSc. They might then continue to a research degree (PhD or MD). Some may later choose to circumvent patients. This they can do if working at a teaching or specialised postgraduate hospital (with university hospital status). Those with no interest in being detached from patients become GPs, general physicians, and surgeons in non-academic hospitals (once called 'peripheral' hospitals), many of which now hold 'University Hospital' status. Because this is where most routine hospital work happens in the UK, it is where the most useful research into common conditions could be done (if there was enough time and inclination).

An important problem for patients has developed. The era of the general physician has gone; many specialists with their own narrow interests have replaced them. This change has not served patients well. There was always a place for the knowledgeable generalist. This is perhaps another wheel waiting for bureaucratic re-invention. In countries where doctors have to survive financially without a government subsidy, the more knowledgeable a doctor is, the more she can earn.

Medicine, performed as an art, is intuitive, scarcely rule-based, and interpersonally sensitive; this is caring, based on complex nebula thinking and feeling, but with a humanitarian perspective. Medicine is also a science, based on Laplace's scientific determinism, using only evidence-based knowledge and understanding, to form diagnoses and calculated prognoses. Few doctors practise both. As a result, over-specialisation has led to de-skilling, and the over-inclusion of rarer topics (over-specialisation was one of Richard Asher's '*Seven Sins of Medicine.*' Lancet, August,1949).

According to the late poet, John Betjeman (*An Oxford University Chest*, OUP, 1979) reflecting on his years at Oxford, 'Dons (teachers and researchers) are only human . . . creature comfort tempts on one side, ambition on another, pedantry on a third.' Many doctors are no different.

Post-doctoral, academic doctors may share the title 'Dr.' with experienced physicians, but their expertise can be quite different. Their *forte* must be to pursue scientific discovery, while legitimately detaching themselves from the direct needs of patients; they then risk being accused of 'narcissistic scientism' (*Limits to Medicine*, Ivan Illich, 1975).

'. . . . you know what scholars are – inhuman. Ain't you,' said Miss Fagan, turning to her father with sudden ferocity – 'ain't you inhuman?'

'At times, my dear, I am grateful for what little detachment I have achieved.'

Evelyn Waugh. *Decline and Fall*, 1928.

Two distinct behaviour patterns, and attitudes towards patients (thus referred to as separate cultures) exist among doctors and nurses. They significantly influence the doctor-patient relationship, and the patient's experience of individual medical professionals. I define them as:

The Patient-Attached, or Engaged Type. These are perceptive, empathetic, physicians, surgeons, and GPs, capable of making diagnoses and treating diseases, as well as dealing with patients' pastoral needs. They are versatile enough to vary their approach and language to suit each patient. They are thus exponents of the art of medicine. They will give priority to their experience, but never ignore the results of evidence-based, medical science. They are capable of creativity and the unconventional. They are usually adaptable and responsive to every clinical situation. When necessary, they can detach themselves from the need to comply with generic guidelines and rules, in order to manage cases on their individual merits.

The Patient-Detached, Dis-Engaged, Anonymous Type. These are doctors who cherish the scientific method. They rarely wish to engage with a patient's personal problems or psycho-social factors. They are likely to be guided by rigid, evidence-based views, lacking inter-personal versatility. As proud exponents of scientific medicine, they will usually undervalue clinical experience and anecdote. They are most often binary thinkers. Their sanctimony about the contribution of science to medicine is reminiscent of a religious faith. Because they usually insist on evidence before making a clinical decision, they are unhappy with clinical risk, and uncomfortable with the unconventional and unpredictable. Their need to conform is strong, and quickly become unsure when dealing with unusual or indefinite clinical situations. Their intellectual rigidity can render them impotent in clinical situations that need action, especially when the evidence is incomplete.

Most doctors practise in one or other mode; only a few can switch between them. One cannot profess to have mastered medical practice without mastering both.

Consider an exam question given to two 15-year-old schoolchildren, both intent on studying medicine.

Q. A plantation worker eats 9 bananas in 36 minutes. How long will it take for her to eat five bananas?

The student capable of detachment will follow the rules of arithmetic and correctly calculate the answer: 20 minutes.

The 'attached' student asks, 'why would anyone want to eat 9 bananas in 36 minutes?' The question lacks credulity and defies common-sense. She refuses to answer the question, even though the arithmetic poses no problem.

Which pupil might become the better doctor?

I would choose the one capable of the arithmetic, and with enough conviction and courage to refuse to answer the question. Detached doctors will probably choose the compliant student. Both, however, could deserve a place.

An uncritical altruistic desire to help people is reason enough to consider a career in medicine. Those who do not have this desire, yet seek medical work as an interesting academic exercise, can find a place within the medical profession; hopefully, far away from any bedside. Some doctors get away with having no genuine interest in other human beings. Patients may come to respect their knowledge, but not their disinterest in others. Patients may not warm to them, admire them, or co-operate with them, but in straightforward medical situations, their strictly algorithmic medical tools can help fix some patients for a while.

The nebulous, unpredictable nature of human predicaments can make those who choose to be detached from patients and their problems, a little anxious and insecure. They may feel comfortable only when dealing with strictly defined cognitive issues, like those found in physics and mathematics. They can find biological phenomena difficult to handle because of its diversity, its indefiniteness and unpredictability. Attached doctors will be at home with nebula thinking, and the indefiniteness of nature. Artists own this territory. Indeed, they may rejoice in their work, just because it is nebulous, and not bounded by rules and regulations (no doubt, there will be bureaucrats somewhere, trying to certificate art and artists. Their given role will waste time and other people's money).

Strict enforcement of rules on how to draw and paint took place during the long and controlling reign of Louis XIV. Artists had to meet the King's approval and his prescribed standards.

In later life some detached doctors will allow themselves to wrestle with clinical uncertainty and insecurity. As they approach the end of their career and feel more secure, some will attempt to embrace the unknown and the unknowable. Albeit fleeting and uncomfortable, some may loosen their control over their emotions and escape from intellectual imprisonment (like Montaigne who asked, *'Que je sais?'* What do I know?). Some academics appear on popular science TV programs, but must first learn to loosen their ties and jettison their need for strict rules of engagement.

The 'detached' and 'attached' cultures run parallel to an established philosophical dichotomy. How are we to understand the world? The Cartesian view – cogito ergo sum (I think, therefore I am), versus the existential viewpoint: trusting our senses to inform us of the actual world. The philosopher Maurice Merleau-Ponty extolled this view.

Both the law and medicine are open to conflicting conceptions. Plaintiffs and defendants in Courts of Law, seek fairness and justice. Patients seek caring and cure. Lawyers are concerned primarily with the legal process, and not as many assume, with fairness and justice. In parallel, some doctors pursue medical science, not caring. It might shock some patients, but many doctors will state that caring is the primary role of nurses, social workers and carers, not doctors. As professionals, each doctor and nurse must be honest with their patients about their primary purpose. Is it caring, publishing research papers, or both?

Managing Patients

The functionality of doctors depends not only on their knowledge and skill set (cognitive, emotional, behavioural, and practical), but how they select and co-ordinate their interactions with patients. Although vital, current medical school selection processes do not test this functionality.

I will review the essential features as if they were dichotomies. They co-exist in balance, the most appropriate *modus operandi* being chosen somewhere between them (Aristotle's Golden Mean):

1: ***(a) Objectivity*** (an evidence-based outlook). An ability for data analysis, filtering (being able to see the wood for the trees), the weighting of facts for relevance, and evidence-based decision making.

– versus –

(b) Intuitiveness (an experience-based, anecdotal outlook): The ability to use 'common-sense', a sense of timing, a sense of priority, and a facility for risk evaluation.

2: ***(a) An attached demeanour:*** Use of emotional intelligence and sensitivity. An ability to sympathise, empathise, and understand a patient's outlook (and that of his relatives / friends, when meaningfully involved), together with an appreciation of the background and context.

– versus –

(b) A detached demeanour. No aptitude for, or interest in any patient's emotionalissues,and no desire for any social attachment to them.

3: ***(a) Executive Functioning***: Co-ordination of responses; efficiently forming executive plans (getting it all together). Demonstrating dynamic handling ability: acting within the time frame dictated by clinical urgency.

- versus –

(b) Executive Ineptitude: Little evidence of planning or efficient coordinated action.

4: ***(a) Effective Physical Functioning*** – a talent for practical procedures: venesection and arterial punctures, lumbar puncture, surgery and other interventional procedures.

– versus –

(b) No talent for practical procedures. Inept,clumsy,uncoordinated action.

All capable physicians and surgeons have these characteristics. Versatile doctors can shift the balance, as different clinical situations arise. Such versatility is one feature of clinical mastery.

Some doctors handle patients like they would boil an egg. First, they must find an egg. Then they will rummage around to find a saucepan. After a time, they will fill the saucepan and add water, but cannot remember where they put the egg. The egg might get boiled eventually, or it might not.

Do such people become doctors? I am afraid they do, but how? The answer is that executive functioning remains untested during the medical student selection process. The difference between being good at exams, and performing well as a doctor or nurse, needs to be addressed.

Other directives can help to achieve medical mastery. The Delphic dictum 'know thyself' is rather important. The original dictum implied a need for humility, and an acknowledgement of powerlessness, faced with what the Gods had planned. It acknowledged the advantage of gaining insight into one's personal biases and 'hang-ups'. For doctors, another equally important directive is: 'know thy patient'.

Senior Doctors

Back in the 1960s, older doctors could rest on their laurels. Keeping up to date, some found tiresome. One source of frustration for all junior doctors has always been the effort required to teach old dogs new tricks. Overcoming resistance saps energy, and obdurate and inflexible senior colleagues are no exception. Take heed of one blindingly obvious piece of advice to save energy: walk around brick walls rather than blast holes through them.

Old Dogs, New Tricks?

In 1967, I applied for a senior house job in cardiology at Guy's Hospital, London. Like many of my colleagues, I had become an experienced resuscitator as both a junior doctor, and SHO anaesthetist. At Whipp's Cross Hospital, patient survival depended on the ward location. The hospital still has one of the longest corridors in the UK. When a crash call came in from a distant ward (when we juniors were in the centrally placed doctor's mess, or bar), it could take 10 minutes to reach the patient.

One of the older cardiologists on the Guy's interview board asked me to list the most important factors contributing to patient survival following cardiac arrest. My answer displeased him. He thought I was being trivial, and facetious, when I told him that survival was inversely proportional to the distance between the patient and the hospital bar! None of the interview board members appreciated the truth of my statement; they had been too long away from emergency situations themselves; some might never have undertaken resuscitation (CPR).

In 1965 (just before I qualified), Dr. Frank Pantridge in Belfast, invented the portable defibrillator, and so began the life-saving resuscitation of patients in the community. From there, it became accepted world-wide. To everyone involved in resuscitation, it is soon obvious that timing is the critical factor for patient survival (as it is in so many rapidly changing clinical situations).
Bob Hope, the Anglo-American comedian, once said: 'timing is everything.' How right he was.
It was fortuitous that I didn't get the Guy's job. An unwillingness to listen and learn from those with practical, hands-on experience, can be a mistake.

To-morrow, sir, I wrestle for my credit.

W. Shakespeare: Charles. *'As You Like It'* (Act 1, Scene 1).

Stages in a Doctor's Career

The Peter Principle applies to doctors. Many climb the career ladder until they are beyond their level of competence.

Some patients might want to know about the career ladder doctors climb. Each step has typical characteristics.

Medical Student: Usually in awe of teachers who know so much. The primary aim is to gain knowledge and ability.

Houseman (Foundation years, 1 and 2): Hard working, but the lowest form of medical life; trying to do no harm, despite inexperience. Acting as servant and scribe, their job is to make all necessary clinical information available to their decision-making senior colleagues.

SHO (specialty trainees): Must make an impression and get noticed.

Registrar: Thinks he is almost ready to run the show, and will chance his luck in order to impress. Should be thinking of publishing research papers, and writing letters to medical journals.

SpR: Consultant in waiting. May thinks he knows it all, and often will. Used to think consultants were of value, but now not so sure. She should have published some research papers and letters to medical and scientific journals. May resent her consultant being absent when seeing his private patients.

Young Consultant: May be **OV**erly **I**mpressed **B**y his/her **O**wn **I**mportance (OVIBOI, see later definition). He is confident, and ready to sweep aside some old ideas. Likely to be digitally savvy, a techno-cutie, but not yet with enough experience to have distilled knowledge into repeated clinical wisdom. Although he may believe in the paternalistic value of the NHS, he is likely to dismiss private medicine until his partner decides on private education for their children. He may then commit to the NHS/Private, double-standard; one demeanour for NHS patients, another more affable one, for his private patients.

Old Consultant. Old enough to have gained clinical wisdom. He knows how medicine works, and his place within the system. He will know his place as a minor cog in a large corporate medical machine. His experience may soon count for nothing, as managers try to replace him with more compliant younger doctors. He may question what good he has achieved throughout his career. He may tacitly enjoy private practice, but may also despise it; after all, he only ever considered it a means to make extra money.

How Doctors Progress

Surgeons, cardiologists, and intensivists need practical skills; general physicians and GPs need an ability to understand humans as sentient beings. Taken together, these abilities should combine to help form selfless judgements. Some are born with a facility for wise judgement; those born without it will struggle with the multitude of daily judgments and decisions required for safe and effective medical practice.

Doctors need to decide on their career path as soon as possible. They need to ask, 'what would I like to be doing 15-years from now?' With this question answered, they can plan their journey (in pencil, not ink). My advice to the disaffected, is to review their talents and career choice, before it's too late to change. Early on, take advice not only from those you admire but also from those who know you best and have your interests at heart. An uncomfortable principle remains in all walks of life: who you know, and not what you know, will help most to achieve your ambition. Doctors rarely ignore the fact that they are members of an exclusive club, with many attitudes set in stone, rather than fashioned from ice and snow in a temperate climate.

Patients are quick to recognise a doctor uncomfortable with his role; perhaps even before he does. Some patients are smart enough to get the measure of a doctor, even before he opens his mouth. Such patients are often 'animal lovers', with primal animal instincts. Some are HR people from big business, others are small business owners whose success depended on them knowing others. They all share the same ability to 'tune-in' to others. Patients are often better judges of character than doctors, having had a wider exposure to life and its tribulations.

When a doctor or nurse is disinterest or lacks integrity, the will find it difficult to conceal from those with animal instincts. If detected, a doctor or nurse will quickly lose the respect and trust of their patients.

We modern humans have lost some of our natural instincts. Not being in touch with them is one source of anxiety. Polynesians can confidently navigate to distant islands using instinct alone, while modern sailors will feel lost without a navigation device in their hand. Australian aborigines know where in the outback to look for food and water, and how to avoid crocodiles and poisonous spiders. The average European would become lost and in danger in these environments.

There are those whose aim is to get rich from medical practice. It is possible, but they will need to choose their specialty carefully. They should first become a consultant, specialising in surgery. Surgery alone has the potential to make a private doctor rich. Gynaecology, cosmetic surgery and cardiac surgery all have this potential.

Those who love drama should aim for the emergency specialties: A&E, orthopaedics, anaesthetics, neurosurgery, cardiology, and cardiac surgery. Those whose forte is caring should consider general practice and psychiatry. Other doctors can achieve financial stability and be comfortable, but few will attain wealth. Most of the important rewards of medical practice are not financial.

In many countries, doctors can only make a living by treating self-paying patients. That means they need to know a lot about many clinical subjects, some of which are specialties. An American student once told me, 'To make a living in the US, I need to know as much as possible about every specialty. The more I know, the more I can earn.' By learning some cardiology, he intended to steal the thunder of his cardiac consultant associates. Why shouldn't he earn money doing exercise tests and echocardiograms? Such entrepreneurial spirit among UK doctors is rare; from 1948, when the NHS began, there has been little need for it. Whether patients are better served by many generalists working with a few specialists, or a few generalists working with many specialists, is a debate worth having.

In the Royal Navy, the initial ambition of each cadet is known to define the highest rank they will achieve. If a cadet envisages himself as an admiral, there is some chance he might become one. If a cadet sees herself as progressing only to 2nd lieutenant, that will usually be the highest rank she achieves.

An anecdote I once heard may not be true, but it makes a point about political ambition. When the politician, Mr. M.H. was at Eton, a reporter asked him about his future. His answer was, 'It's a tough decision, whether to become Prime Minister or Chancellor of the Exchequer.'

The Good School Guide once had it that the boys at Eton are 'imbued with impending greatness'. Interestingly, this predisposes them to success.

Dr. Richard Asher posed an important question in his book 'Talking Sense'(1972): 'How can one progress to become a royal physician or surgeon?' The question will only concern those ambitious enough to see themselves in the role. Richard Asher's advice was to select a common ailment for study; one below the aspirations of most doctors, like blackheads, warts, in-growing toenails or haemorrhoids. The next step is to research the subject, and get the work published. After this, inaugurate a new journal dedicated to the subject and start a professional society; one that draws all interested parties together. Become President of that society, and you are almost there. If a member of the royal family is unlucky enough to suffer from your specialist interest, be prepared for a call. With luck, the professional society you started will later become a Royal Society. Now you really have arrived. Congratulations. Practise bowing and kneeling, and overcome any fear you have of swords.

With progress in mind, let's change focus from prestigious 'high-flying' doctors, to the 'low-flying' variety. Consider the influence of sex and money on professional progress. Doctors are only human and sex sometimes changes everything. Nurses, secretaries, doctors, physiotherapists, receptionists, all are candidates for mutual attraction at an age when the sap of spring rises constantly; not just in springtime. Apart from performing an essential public service, hospital communities clubs with many for boys and girls seeking affection and lustful satisfaction. Having sex with a colleague can change professional relationships and sometimes team dynamics.

It is not entirely unknown, even in the medical profession, for a personable boy or girl (or some other gender), to sh*g their way to the top. This dynamic will sometimes confirm the Peter Principle.

If you are lucky, your boss may offer you a sailing weekend, or a weekend break at his country house. Team building could the reason; grooming and manipulation could be others. Letting your hair down is one thing; loosening your standards is quite another. Beware of the consequences of falling off the ladder.

Medical Management and Sex Chromosomes

Do female doctors (possessing XX chromosomes), and male doctors (possessing XY chromosomes), differ in clinical acumen, ability, and style? Do patients have a preference? Is there any practicable distinction to be made, now that over 50% of medical graduates are female?

Germaine Grier is my *gurvi* in matters of female insight. In her book, *The Female Eunuch* (1970), she states that a common female need is to be loved; love that needs repeated confirmation. In addition, females can sometimes display hormone-linked behavioural cycles that will challenge male tolerance; especially for those men with little previous exposure to them. Male doctors will soon discover whether their female colleagues regularly display premenstrual (or menopausal), wasp chewing behaviour, or the more amiable pre-ovulatory sort.

Medical professionals and scientists, unlike politicians, need make no excuse for gender-related differences. The effect of chromosomes is biological, not political, and biology has no need for diplomacy or political correctness, even though biology is now being ignored or the subject of discord. Attitudes may change towards gender issues, but biology won't change.

Grier also contends that 'women often refuse to argue logically', while in contests of will, men will often employ sophistry. She writes that 'Male logic can only deal with simple issues: women, because they are passive and condemned to observe and react rather than initiate, are more aware of complexity.' Both the male facility for logic, and the female awareness of complexity, are useful for practising doctors. In my experience, however, the use of clinical logic and successful engagement with clinical complexity (requiring nebula thinking), are gender-neutral talents.

Overt masculinity and femininity are identities lying at either end of the gender spectrum. For clinical encounters with the many diverse types in between, one must consider various sub-types like the masculine female ('butch' with a male outlook) and the effeminate male (with a female outlook). Many variations exist between the two extremes. Considerable care about gender issues is appropriate, because they are important to some individuals. When in doubt, ask each colleague or patient for their gender orientation. Like some patients, a few doctors have also undergone transition operations.

There is a male/female ratio difference in those pursuing various medical specialties. In the US and Canada, 85% of gynaecologists were female. For paediatrics, the female figure was 75%; for psychiatry 57%, and for general practice 58%. Males dominate surgery (59%); emergency medicine (62%), anaesthetics (63%), and radiology (73%). (From: *Report on Residents in US and Canada*, 2015. Association of Medical Colleges).

One factor driving male choice of specialty is camaraderie: 'men like to do a man's job together'. This will put some women off choosing some specialties like orthopaedics. More women than men prefer enduring, rather than transient relationships. Longer term patient-doctor relationships are found in general practice, gynaecology, psychiatry, and paediatrics. If more females than males choose an attached, rather than a detached mode of practise, more females will practise the art of medicine effectively. Since females are equally good at science, proportionately more woman than men should make highly competent, first class doctors.

As far as patient preference is concerned, more Spanish patients chose a female GP when a longer term relationship is appropriate. For emergency work, and one-off consultations, the same study showed no patient preference for gender (Delgado, A., Lopez-Fernandez, L.A., Luna, J de Dios (1993). Medical Care, 795-800.

In my first year at medical school, there were only seven women in our class of seventy students. The argument then was that women have children, and will work for fewer years than men, so why waste valuable medical school places on them? Apart from not promoting equality, there are errors of assumption here. One lies in valuing the length of work more than the quality of work. Another lies in disregarding the different approach female doctors bring to medical practice.

Performance Skills

Poor performers, in all fields of work, are less likely to recognise their performance as poor, and most likely to over-rate themselves. This is the Dunning-Kruger Effect bias. Although not strictly age-related, I have encountered it more in older doctors. Because of the confidence that arises from long experience, fewer may recognise their mistakes. When they do, they may find them more difficult to admit. Demoralised older doctors will more readily admit to the Dunning-Kruger bias, when they are thinking of retirement.

One wizard of TV and radio presenting, Terry Wogan, used to address his BBC Breakfast Show comments to one listener. He was careful not to generalise too much and lose his listener's interest. He knew that many individual listeners disagreed with the majority view and yet he appealed to millions of followers.

Doctors should copy his trick: one eye on the general case; both eyes on the individual case. Terry's was the voice of calm, humorous, Irish wit. Millions, including me, mourned his death (January 2016). It would do no harm for doctors to take something from his light touch and considerate style.

In entertainment, there have been few all-round performers. Imagine someone able to sing, dance, impersonate others, act, compose music, and play multiple musical instruments. You will need to speak to your grandparents, or great-grandparents, to learn about them. Sammy Davis Jnr., and Bruce Forsyth, are the two who come to mind. Are there equivalent doctors who know a lot about every specialty, and can change their performance to suit each case? Was there ever such a complete physician?

Prof. Tony Dornhorst of St. George's, Hyde Park corner, was an all-rounder. He spent much of his time reading specialty medical journals. As a result, he had detailed knowledge of most medical specialties.

His Guardian obituary in 2003, said of him: 'For the most part, he concentrated on teaching - mainly at the bedside - and theorising. Nevertheless, his influence was immense. We learned the importance of honesty, integrity and fairness, of challenging authority - and ourselves - of attention to detail, of the sin of ignoring evidence, of the value of using novel, sometimes multidisciplinary approaches to solve problems, of the importance of listening to, and trusting the evidence of patients, and of recognising the need to keep up to date.'

He once said to my late friend Alan Gardner, 'If you pass your finals, I will eat my hat!' When Alan finally qualified as a doctor, he lacked the temerity to hold Prof. Dornhorst to his promise.

Good Behaviour. Bad Behaviour.

After being admitted to hospital with acute pancreatitis, I found the senior registrar in attendance, to be researching the macrophage. When I casually asked him if he knew what the words 'macro' and 'phage' referred to, he told me that 'macro' meant 'large', and 'phage' was derived from the Greek verb 'to eat'. (Latin is much less useful to physicians than Greek, unless he wishes to read some original texts from the same era as 'De Motu Cordis', by William Harvey).

Actually, 'makria' in Greek means 'far away'; 'megalo' means large. So, I quizzed him on his use of the word 'macrophage', since it implies that such cells 'eat far away'. So far, so good. He accepted this. I then asked him a crucial question. 'Did Metchnikov, who named the cell (Elie Metchnikov, 1880s), think of it as an amoeba-like cell, using pseudopods to eat 'far away'? Or did he know that the cells he named travelled from the bone marrow to far away sites of inflammation?

I detected a little facial pallor. This SpR had just finished his PhD thesis (prior to submission), but had not researched why the cell was named as it was. Since macrophages aggregate at sites of inflammation, the mechanism by which this happens is important. He announced he had better go to the library, post haste. Because I had caused him visible discomfort, this was 'bad behaviour' on my part.

A useful definition of 'good behaviour' is that which leaves all present, feeling content with what they have witnessed. It costs nothing, so why not employ it all the time? It was once the studied preserve of gentlefolk, and a reason for sending children to a British Public School; there, they are more likely to learn the importance of politeness, humility, and respect for others.

I can write with great fondness about my medical school teachers and their behaviour. To a man, they were gentlemen. Some served their country in two world wars but never mentioned their exploits.

On the day of her state funeral (19/09/2022), I reflected on how fortunate I had been to live throughout the reign of H.M. Queen Elizabeth II. In the future, doctors of my generation will become known as Elizabethan. She provided us with unmatched examples of service to others, while maintaining the highest of standards in every domain of life; treating all with graciousness. She first dedicated her life to the service of our nation when 21-years old. Throughout her long life, she left those she met in awe. She had an engaging smile, dignity, and civility. She listened to what people had to say with an intense interest, leaving them feeling forever respected for their value and individuality. Good behaviour is invaluable.

I sometimes posed a rhetorical question to colleagues dealing with my patients. What would you do differently if my patient was HM the Queen? The question prompted thought about what might be best for the patient. I would ask my patients, 'How do you think your case handling would be different if you were HM the Queen?' This opened a question about the quality of the medical services they were receiving.

Belief in Perfect Health

Provide for others through humanity, what we provide for our own family through affection.

Lactantius (240 – 320 AD)

Ivan Illich had it that rich societies seek 'perfect health' as a commodity *(Limits to Medicine: Medical Nemesis - The Expropriation of Health*' Ivan Illich. 1976. Pantheon Books).If we seek hubris by trying to gain this divine quality (an attribute of Greek Gods, not humans) we deserve the nemesis that follows – iatrogenic diseases, emergent new diseases, and a maladapted society seeking to prevent and solve medical problems before they arise.

Many wealthy societies desire perfect health. Governments who promise medical utopia, imply that perfect health would be possible if only doctors did what they were told and accepted the guidance of bureaucrats. Doctors know they cannot achieve it, so by taking responsibility for it, they are easily made political scapegoats.

A parallel case of political, bureaucratic thinking, applies to education. If only teachers tried harder, guided by school regulation and compliance inspectors, every pupil could become 'above average'.

Despite the desire to cure disease, the relief of suffering has brought more success. Genetic engineering should help both. Once there is a prospect of financial benefit from genetic research (the prospect of fewer patients, and less expenditure), investment will follow. This should promote finding the genetic basis for all pathological entities, foster prevention strategies, and provide gene therapies (the latest of which, announced, July 2022, offers a cure for haemophilia). For the 8740 haemophiliacs in the UK, this represented genuine progress, even though not all are suitable to receive it. The gene instructs the liver to produce factor eight (VIII), but costs millions of pounds per treatment. Meanwhile, many benefit much from the art of medicine, at no extra cost whatever.

In the 2006 film, 'A Good Year' (20th Century Fox), Max Skinner's uncle Henry, concludes he is dying. He knows this, he says, because his doctor has ceased talking to him about his health, and now only talks to him about the weather.

Conveying hope and leaving it in place, is a strategic art. Some invested totally in the science of medicine might regard this as a mischievous trick; a false ploy, serving only to protect the patient from reality.

Lafeu said of the King: 'He hath abandon'd his physicians, madam, under whose practices he hath persecuted time with hope; and finds no other advantage in the process but only the losing of hope by time.'

All's Well that Ends Well'. Act 1, Scene 1. William Shakespeare.

Respect for Others

Respect for the knowledge of medical academics and practising professionals has diminished, as the notion of equality has strengthened. Uneducated men now feel confident enough to discuss the law with experienced judges; medicine with long experienced doctors, and the merits of BREXIT with economists and diplomats. Does the man in the street understand how much he doesn't know? A lack of intellectual humility is now a growing challenge. To fully respect others, one must at least understand their intellect, culture, motivations and past influences.

Some Final Words of Caution

All doctors should try to gain proficiency in identifying genuine people, with genuine motivations; especially those with a genuine interest in them and their progress. The disingenuous may be in the minority, but they come in more flavours than Ben & Jerry's ice-cream. All must learn to spot them as soon as possible. One may be eager not to lose potential friends, but there is no loss in losing a friend who is easily lost.

Doctors and patients in the UK now come from many backgrounds. To work happily together, we must attempt to understand why some want money, status, and security, and others are driven by avarice, jealousy, and a need for power. Look for helpfulness, supportiveness, respect and affection for others, for these are the attributes of those who happily pursue medicine as a vocation.

Despite the current fashion for thinking otherwise, most people are incapable of changing their behaviour. Genetic engineering might someday cause leopards to lose their spots, and become pet pussy cats (without a frontal lobotomy), even though natural changes of such magnitude are yet unknown.

Many seeking power over others need no new ploys to gain influence. Listening intently, nodding in agreement, lip-licking, smooth talking, eyelid quivering and gazing intently with limpid-pool eyes, all signal an interest in you as a person. They can also signal the desire to get something from you. If it's your body they want, and the feeling is mutual, that's fine; if it's your ideas, your status, your wallet, or your intellectual property they want, proceed with caution. Those unable to unmask the motivations of others will waste a lot of time, money and energy on false relationships.

> *'Choose your friends carefully, and your enemies even more carefully.'*

Suitable potential friends will have a compatible personality and background. Their need for control, affection, and shared interests is of crucial importance. If you want the truth about them, ignore what they say, and observe their actions and behaviour. The direction their feet take will reveal most, and all you need to know about their true thoughts and feelings. The prerequisite is to understand non-verbal behaviour. If they say they like you and love your ideas, but never find time to meet you, their true real interests lie elsewhere. Those who distance themselves are loudly declaring their lack of interest. The truth can be upsetting, but you will quickly get over it.

During the Second World War, Gordon Welchman, working at Bletchley Park on the German Enigma codes, invented the concept of traffic analysis. While Alan Türing concentrated on deciphering the code, Welchman worked on the significance of who was communicating with whom. If Otto in Calais communicated with Friedrich in Berlin several times on the same day, some important action was afoot in Calais.

When we refer to actions speaking louder than words, or learning from watching how the feet of others move, we are extolling the virtues of traffic analysis. Welchman wrote about this is his book, Hut 6. While working in the US, he was accused of revealing war secrets; they arrested him forty years after the original events took place, and he lost his job. Not long after, he developed cancer and died. Resulting from an equivalent bureaucratic process, Türing had committed suicide almost thirty years before. Together with Alan Türing, their work together shortened the war by years and saved millions of lives. The fate of both men are instances of the insolence of power, and an illustration of just how stupid and inhuman ruling-following, and political expedience can be.

Are humans complicated, or is this illusory? Humans are all simple to understand, but first you must uncover their true motivations and character. Many fear that they are boring and simple, and try to create the impression of being complex and fascinating. Deep and secretive people are rarely complex or fascinating. Many fear exposure of the fact they have nothing of interest to reveal. As Freud pointed out, few of us understand ourselves, and few of us know what drives us (or what should drive us). Few of us really know what we really want from life and why.

As pointed out by the character played by Michael Douglas in the 2009 film, *Solitary Man*, (Millennium Films) when dealing with any other person, try to understand what benefits there are in transacting with them. Worthwhile relationships achieve mutual satisfaction.

Chapter Three

Doctors as Characters

Nothing can replace a good heart and a sensible head.

Yehudi Menuhin.

My description of doctors appeared first in my book 'Doctors, Nurses & Patients (2024). I have reproduced most of it here.

There are some important questions to consider about the character of doctors (and nurses). Of what significance is a doctor's character to the doctor-patient relationship, patient welfare and professional relationships? A respectful, mutually helpful relationship, based on honesty and honourable behaviour is required, but reality can be quite different.

What follows is an incomplete, occasionally overlapping list of the fundamental character traits of the doctors I observed during my five decade medical career. Although most doctors are easy to relate to, are supportive, and present no relationship challenge at all (at least on the surface), I have mentioned only the more challenging types as a forewarning. I have not given complete descriptions, but enough for them to be recognised. It would be of help to know how frequently one might meet these characters, but that would have required me to have undertaken a random survey. Instead, I have mentioned how common or rare I found them to be, without further specification.

I have tried to follow in the footsteps of Theophrastus, associate of Aristotle, who described the various Greek characters he met 2300 years ago. I have not tried to emulate Shakespeare, who brought every variation of human nature before us in his plays, or Dickens, who populated his novels with every conceivable character. Every canon of literature is replete with every known character, and here I present just a few relevant to medical practice.

I have reduced some character descriptions to acronyms. They might amuse and serve as a diplomatic device for communicating with friends and colleagues.

Types of Doctor

Dr. Kate Granger, who died in 2016, was astounded to find when she was ill herself, that some hospital doctors failed to introduce themselves as they approached her hospital bed. She started a campaign: '#HelloMyNameIs'. Her hope was that NHS doctors would consider patients as people, and not just disease entities.

Theophrastus, in the 3rd century BC, wrote his 'Characters', to describe the behaviour of thirty different human types: the dissembler, flatterer, coward, over-zealous, tactless, shameless, newsmonger, mean, stupid, surly, superstitious, thankless, suspicious, disagreeable, exquisite, garrulous, bore, rough, affable, impudent, gross, boorish, penurious, pompous, braggart, oligarch, backbiter, avaricious, the late learner, and the vicious. These types are all still with us among patients and doctors alike, so his descriptions remain valid.

Of particular importance to every doctor is to recognise colleagues; especially those who are helpful, mean, generous, romantic, practical, and down-to-earth. There are also those with their head in the clouds, and those who see themselves as God's gift to medicine and procreation. Some are nerds; some are the life and soul of every party. Unfortunately, we must also consider those whose desire is to be the centre of attention, the selfish, and the self-possessed. It is important to recognise would-be friends, among likely enemies; the knowledgeable, the talented, and those capable of wise counsel. For doctors and nurses, they are all out there to be met as colleagues or patients.

Like St. Paul and St. Thomas Aquinas, those in possession of divine gifts will progress easily. Wisdom, understanding, knowledge, and good counsel are four of St. Paul's seven charismata pertinent to medical professionals.

St. Paul. New Testament Epistles. Corinthians.
The Summa Theologiæ of St. Thomas Aquinas (1225- 1274).

To master a profession, like medicine, will require many talents. According to the actor Matt Damon, the film directors Ridley Scott and Clint Eastwood both have features of mastery which could apply to doctors. They have talent, virtuosity, passion and experience. Although there are doctors with all these attributes and those with none, those able to master both the art and science of medical practice will need many more.

'Every way of a man is right in his own eyes.'

Proverbs 21.2

A Compendium of Doctors as Characters

A wonderful fact to reflect upon, that every human creature is constituted to be that profound secret and mystery to every other.

Charles Dickens. *Tale of Two Cities.*

ABCD

Above and **B**eyond the **C**all of **D**uty. No personal considerations, trivial social arrangements, matters of self-interest, or convenience will divert these dedicated, selfless medical professionals from their work.

In March 2018, the UK experienced Siberian blizzard weather ('Beast from the East'). At this time, Nurse Reg Barker walked ten miles through the snow. He trudged from Crediton to his ward at the Royal Devon and Exeter Hospital to get to work. Reg's action was an example of an ABCD in action.

The Telegraph. 2/3/2018

Abrasives

The typical abrasive doctor asks curt questions and accepts only brief answers. They do not want to waste time on inconsequential matters. They are usually self-absorbed, time-obsessed, or pressured for time. If they ask, 'Have you noticed any chest tightness on exertion?', they will expect a 'Yes' or 'No' answer, not a narrative description. They do not tolerate garrulousness and will cut patients short. They can be arrogant and forceful, irritable and easily irritated, and can give the impression that they would sooner be somewhere else (and possibly, they should). Perhaps they have missed out on the genes determining generosity, charm, empathy, and sympathy.

Doctors and nurses must prepare well if dealing with them. They must be sure of their facts and argue their case with summarised data. Succinct quotes from research papers will help. They should aim for a knock-out punch.

Do not negotiate out of fear and do not fear to negotiate.

John. F. Kennedy

In the 17th century, apothecary Nicholas Culpepper, took the patient's point of view, and came into conflict with the physicians of the Royal College (one of whom was William Harvey). Culpepper said, 'no man deserved to starve to pay a proud, insulting, domineering physician'.

Actors

There are doctors who failed to become a professional film or stage actor. Instead, some found their way onto hospital wards and into GP surgeries.

Their aim is to create an impression, good enough to enhance their career. As a medical student, one of my surgical consultant tutors was then the ageing Mr. Herman Taylor (he invented a gastroscope). He wore a monocle, and so did his much younger senior registrar (later to become a consultant anaesthetist). The monocle was then useful for non-verbal communication. Allowing it to drop, accompanied by a slight elevation of the chin and eyebrows, signalled surprise, error or astonishment. In response to a student answering a question incorrectly, Mr. H.T. could, without saying a word, let him know that he had made an error. The amusing thing was that his young senior registrar would attempt the same act in unison (successfully playing the role of young fogey).

Other actors would wear a kilt, or some other apparel, just to grab some limelight. I never actually saw a doctor wearing laurel leaves and a toga, but I suspect there have been those who gave it thought.

Altruists

Many doctors are altruists. Their dedication and self-sacrifice to medicine and research is admirable. We need more of them.

The Arrogant

'No good work was ever done by humble men.'

G. H. Hardy, in *A Mathematician's Apology'*. (1940).

Arrogance is a common trait among those who pursue power over others. Arrogant medical decision making can be dangerous. Because one can see it in schoolchildren, arrogance is likely to be inherited. The trait is attractive to those who imagine the arrogant to have halos. They might prefer to bask in the shadow of an alpha, dominant person; one who knows his/her own mind; one who seems totally 'in charge', with no doubt about their wisdom and actions. Arrogance can make the insecure feel secure, even though the basis of their arrogance can itself be insecurity. When the arrogant reject sound advice, merely to justify their status, it can prove fatal.

Arrogance, insolence, and pride are nothing new. Aeschylus wrote about insolence, the arrogance of power and catastrophe:

'You see how insolence,
Once opened into flower,
Produces fields ripe with calamity,
And reaps a harvest-home of sorrow.'

Aeschylus (*The Persians*, 441 BC)

No patient should ever die because of ill-informed arrogance. I have seen rule-abiding doctors arrogantly apply regulations to a patient's management, when obviously not in the patient's best interest.

Attached and Detached

This applies most to doctors. Whereas scientists can readily understand humanitarian concepts (although not necessarily the subtleties), few of those who studied humanities understand much of science. There are doctors whose primary approach to patients is scientific, 'detached' and anonymous, and those whose approach is 'attached' and humanitarian. Both approaches produce results, but in ways that satisfy some patients but not others.

Patients need medical science to help them, but it is usually those doctors and nurses with a talent for the art of medicine they will welcome most. There are reasons for this which apply to every art. Unlike the transitive purposeful process of training to achieve distinct clinical objectives, the art of medicine embraces the intransitive. It requires us to use our sense of being. There need be no particular end-point in mind, other than patient satisfaction.

Autistic

Approaching doctors who are on the autistic spectrum can be like accessing a computer avatar; a relationship that lacks true human responsiveness, empathy, or any appreciation of the plight of others (although one can train them to simulate it). Their approach to life is usually objective, mechanical, and algorithmic. They often enjoy categorising and will struggle with all forms of interpersonal contact. They most enjoy being logical and self-oriented. They operate in a diplomacy-free world of their own, making judgements derived from objective analysis, rather than personal considerations. Their relationships and their decision-making processes can seem scripted. The art of medicine eludes them. Seeing no need for human understanding, they will limit their considerations and suggestions to those for which there is good evidence. If they were artists, they might choose to paint in black and white, rather than in colour.

Martin Clunes in the UK, TV series (2004-2009) *'Doc Martin'*, the creation of Dominic Minghella, offers a perfect representation of an autistic doctor.

Bombast

These are self-assured doctors who deride almost anything suggested by others. With an air of 'knowing it all', their aim is to override anyone who dares to offer a different view. They will counter every criticism with a well-rehearsed journal reference, or a quote from an unimpeachable source. During ward rounds, their aim is to make an impression, and to leave no-one in doubt about their cleverness and superior knowledge.

When I was a student, one of my lecturers at 'the London' would surreptitiously research all the cases to be presented at each weekly case conference. We were often astonished by his astute diagnoses. He made successful diagnoses using only one or two seemingly insignificant pieces of clinical information.

I met him many times later in life. Although he remained a clinical bombast, he was a very competent physician.

BBC: **B**aked **B**ean **C**ounter

It is more important to make what is important measurable, than to make what is measurable important.

Simon Jenkins.

Data, rather than patient welfare, can consume the obsessional interest of these doctors and nurses. They prefer calculation to caring; following orders, to deciding for themselves. They may have missed out on their ideal profession – accountancy, or laboratory science. These doctors are at their best, away from the bedside. Many make successful PhD students, and some should undertake yet further doctorate degrees, to keep them off the wards. They are important 'back-room boys', and make valuable members of an academic team.

The bean counter's virtues are analytic. Their analyses can help us to avoid fooling ourselves. Competent doctors will be pleased to have their management assessed *post hoc*, once all the clinical events have played out. Some clinicians will then say, 'I could have told you so.' Such can be the predictive power of long-acquired practical experience. Data-driven bean-counters will often regard anecdote as hearsay, and devoid of scientific value.

Hindsight is an exact science.

How would you spot a BBC if you have never met one before? I recommend the Boulting brothers', 1959 comedy film, *'I'm All Right Jack'* (Charter Film Productions), in which Peter Sellers plays Fred Kite. Fred is a Union official with stereotypical, measured speech; a typical bean counter.

Bean counters are usually serious, obsessional, and self-controlled. The Nazi SS officer Herr Otto Flick, played by Richard Gibson in the original TV series 'Allo, Allo', demonstrated the BBC character perfectly. Herr Flick delivers carefully chosen words precisely. His speech is commanding, delivered as if a metronome had set its pace. keeping any variation of speech rhythm to a minimum.

I think of these doctors and nurses as having a restricted, black and white character, rather than the full Technicolor version. Patients can regard BBCs as unapproachable, unfriendly, and with limited communication skills. Some have a professional problem with intimacy and friendliness, but with a private persona that may differ from their public one.

Because they devote themselves to accuracy, they are ready to criticise, find fault, and correct others who do not accept their analyses. They often ask others to define what they mean.

Many see no need to endear themselves to patients. They are at their happiest in a totally controlled environment (preferably controlled by themselves), like a laboratory or operating theatre. They occasionally escape onto the wards where carers may choose to avoid them. Patients, unfortunately, will not escape them easily.

I have happily collaborated with many BBCs and have occasionally been in BBC mode myself As a research fellow at St. George's Hospital, London, Aubrey Leatham regarded me as a 'back-room boy'. He dubbed me the cardiac department's 'man of fast cars, and slow rhythms'. He had seen me driving my yellow Lotus Elan 2+2 in the 1970s, while researching blackouts.

Mastery of any subject requires an extensive knowledge of both the known and unknown. Experience and perspective will guide every master in any field, but bean counters are likely to view any clinical success derived from experience as 'luck'; they rarely believe in the usefulness of anecdote.

See also: 'Obsessives, and Nit-Pickers.' These present as perfectionist, so called 'anal' sub-groups, with no autistic element.

Bias Unaware People

All decision making is subject to bias. When making any evaluation or decision, one should consider the many cognitive biases outlined by Kahneman and Tversky in the 1970s (see a list in the chapter on Clinical Judgement). Every doctor needs to be aware of their biases and how they might influence their clinical judgement and decision making.

Big Picture Thinkers

There are undoubtedly those doctors who think big. They have their eye only on the top jobs. Like Winston Churchill, many have the premonition that one day, greatness will be theirs.

In research terms, they know what is best to pursue. They will place themselves at the forefront of ground-breaking research. They often see the value of a subject, others are late to recognise. Recognising the growing need for shopping convenience, Jeff Bezos invented Amazon. He also suggested the need for rapid postal delivery services. Both saved the time needed to search retail outlets.

Blackmailers

It is not entirely unknown for doctors, nurses, and other medical professionals to emphasise the negative consequences of not following their wishes. Some accept tacit threat as allowable professional manoeuvring, soft bullying, or gentle coercion; others call it blackmail.

Bores

Usually serious, often with a depleted sense of humour and a limited range of interesting topics, these characters abound in the medical profession.

Bullies

Many newly qualified doctors feel vulnerable. Because their progress can depend on the impression they make, they might not react wisely to bullying. Bullies like to find a weak spot and continue to irritate it. Some senior doctors are intellectual bullies; they will out-quote their juniors in matters of medical science, just to demean them. They will then quickly change the subject.
In a survey of 416 junior doctors (in Europe-wide microbiology and infectious diseases), 22% reported bullying at work. Two-thirds also felt worn out, having had to work longer hours than normal. (Maraolo et al. 2017).

Carriers

In life terms, these doctors are important to identify. They are the metaphorical geese capable of laying golden eggs. They appear to succeed in everything they do. Doctors and nurses should latch on to them (unless one themselves). Doctors should learn from them, perhaps by becoming their apprentice. They will make waves that are worth riding. They are going places.
My partner, Dr. David Baxter, was one such person. An extremely successful, energetic person, he enabled the development of our own private clinic in the 1973. His determination, energy, and ability made our success a foregone conclusion. After we parted, he continued to succeed and developed his own private hospital.
All such people are non-conformists. They are individuals, not groupies; shepherds, not sheep. Aspiring doctors and nurses will need to be compatible with their traits and brave enough to fly alongside them.

Capricious Creatures

Capricious doctors and nurses can find themselves bound into cycles of indecision and unnecessary repetition. One of my bosses used to read my draft research papers many times before publication. If he didn't like it much, he would suggest alterations. After re-writing it, he would make further suggestions which returned it to the original text. One way of handling such people is to do very little. Their minds eventually tire of making half-hearted suggestions. Left alone to think it through, they sometimes realise that their original suggestions were unhelpful.

Charming

Once a most desirable trait. Nurses more frequently possess it. Beware, though! In its false form, it sometimes can combine with utter ruthlessness.

Club Members

In 1972, I was lucky enough to have a mentor for my forthcoming Royal College of Physicians membership examination (MRCP). He pointed out that passing the examination was contingent on being 'the right sort of chap'. Perhaps this applies less now than it once did, but in the 1970s, passing the exam was not just a matter of knowing lots of medical facts. Then, it was as much to do with how doctors presented themselves to the examiners (Fellows of the Royal College of Physicians).

I came before three senior physicians, after reviewing a patient with complicated heart valve disease. They asked me to describe the heart sounds. This I did to the standard expected by Dr. Aubrey Leatham, then a world expert on heart sounds. They listened to me in silence, then asked, 'For whom do you work?' Then, 'How is Aubrey these days? Does he still walk between the Heart Hospital, and St. Georges every day (1.5 miles)? Does he still sail at West Wittering?' They had obviously accepted me as a member of their club.

Cold Fish

Emotionally deprived. Not unlike talking to a dead fish.

Combustible Consultants

Modelled on the cinema character Sir Lancelot Pratt (played by James Robertson Justice in *Doctor at Large*, Rank Film Distributors, 1957), some doctors can explode emotionally when annoyed.

My medical student friend, Leslie Dobson and I, used to attend the London University Wine Tasting Club every week. We rarely limited ourselves to sips of wine. Les became a connoisseur. The day after one such session, neither of us felt too well. Les had to assist a thoracic surgeon, well-known for his irascibility. I remained at a safe distance in the viewing gallery of the operating theatre. Les had turned slightly green and felt nauseated, while trying his best to assist the surgeon.

It was going to be a bad day. While trying to put on his surgical gloves, the surgeon asked Les how he was feeling (having noted the similarity of his facial colour to the green coloured surgical gowns we then wore). 'Not too good', he replied, understating the case. 'Well, pull yourself together, boy!', the surgeon suggested.

Something rare then happened. A nurse gave the surgeon a packet containing two left gloves. This caused him to show his irritation, and to shout a complaint. Nurses scrabbled around, hither and thither, to remedy the situation. They brought him another new packet. He tried on the second pair, only to find the same thing: two left gloves! The temperature in the operating theatre rose sharply. At this moment, Leslie had to admit that he felt ill, faint, nauseous, and about to vomit. The surgeon at once burst into an incandescent rage. A sympathetic nurse whisked Les away to vomit. A sympathetic senior theatre sister did her best to placate the surgeon and the situation. Fortunately, a normal pair of gloves soon arrived and everyone, including the patient, survived the experience.

My much respected, and greatly missed friend Dr. Leslie P. Dobson, served his North Yorkshire community as a GP for most of his working life. He died on the 2nd of March 2020.

Completer Finishers

Effective executive functioning is crucial for both doctors and nurses. Competent doctors can draw all the technical ends together, loose and otherwise, make a diagnosis and correctly decide on the most effective management. Those unable to do this in a co-ordinated way will often revisit the same problem several times. They will see patients repeatedly and sometimes leave them with unanswered, worrying questions. The number of different diagnoses they generate can be a measure of their indecisiveness. The patient, meanwhile, remains in suspense. This is behaviour all doctors and patients should avoid. Doctors should observe efficient, completer-finishers, and learn from them.

Control Freaks

These doctors and nurses are obsessive neurotics. They can be aggressive and uncompromising. King Canute remained unsure of why he wanted to control the tides, but then, not all control freaks seek reasons for what they do. Some are would-be puppeteers, others are shepherds. Either way, their concern is to take the reins and control others. They are usually selfish and want their way, regardless of any cost to others.

There are reasons some doctors want control. Gaining power and recognition are among them. One method they use involves meticulous data collection and categorisation. They understand that information can be a powerful weapon. Some are harmless, like librarians; a few use information for malign purposes.

That information is power is hardly a new idea. The Ancient Egyptians collected personal information to enable taxation. The information in the Doomsday Book served the same purpose. Those people and organisations who collect personal information have many and various motives, some of them Machiavellian.

Cool Dudes

A few cool dudes are doctors and nurses. Love 'em or hate 'em, they remain cool in all circumstances. Study 'Cool.' *The Complete Handbook, by Harry Armfield (1986),* and consider fictional characters like Sean Connery's James Bond, or Winkler's character Fonzie (A.H. Fonzarelli). Few doctors are Fonzie-like; narcissistic peacocks who are constantly preening themselves. Some ooze pheromones, as they delight in themselves. The élite French recognise some as *BCBG: bon chic, bon genre*. Being cool and phlegmatic is mostly inborn, although some will try to learn how. They will not easily escape recognition by any true BCBG. There are questionnaires that rate 'cool', but no cool dude would ever complete one.

Corporate Kids

These guys are at home in a corporate environment, fenced in by what they feel to be comforting rules and regulations. They try not to be controversial, undiplomatic, too political, or anti-establishmentarian. If they keep their heads down, work hard, and do what others ask of them, their careers (and pensions) should flourish. Since the NHS is run as a corporation, the same applies to UK doctors and nurses.

DOD: Daft Old Doctors (or SOS, or Silly Old Sods)

When I first met the very experienced general surgeon, Mr. Douglas Lang-Stevenson (1912 – 1986), at Whipp's Cross Hospital, he was about to retire. He told me that 'stress causes cancer.' It sounded daft. Many thought him past his best. With decades of observation behind him, he deserved more respect. I have little to add, having seen too few cancer cases as a cardiologist. There is a body of controversial evidence that suggest he was partly correct. It remains a worthy topic for research.

Having nothing to lose, the retired and those who are wealthy, can be as daft and despotic as they like. Old doctors, like me, can be cogent pussy cats or dis-inhibited tigers. After decades of experience and being on the brink of dementia, it becomes easier to fool oneself. When old, we are liable to dream up a theory and stick to it, rather than bother to confirm it with extensive research and evidence. We can become less bothered about hard evidence and statistical likelihood, and more comfortable with the richer anecdotal evidence we derive from experience.

Beware, though, some correct ideas did sound daft at first. Prestigious older scientists are liable to the same folly, despite their track record. Beware of adopting the halo bias in their presence. Like pop-singers, they may only be as good as their last hit.

After winning two Nobel Prizes, Linus Pauling proclaimed two theories. First, that vitamin C could prevent cancer, and second, that it would prevent heart attacks. In 1973, he established an institute in Oregon, partly to research his theories. Convincing confirmatory evidence for either hypothesis is still awaited. Theories are cheap, proof is expensive.

Older doctors are not always quick to accept innovation. When I asked a contemporary of mine from medical school to repair my umbilical hernia, he told me he didn't believe in using 'meshes'. He preferred the tried and trusted method of suturing (with the equivalent of football shoelaces). After a quick internet search that revealed a lower recurrence rate after using meshes, I changed surgeons.

Daft ideas are not the sole preserve of the old.

When a patient of mine needed a pacemaker, the young professor of cardiology dealing with him (who was collecting cases for a trial), announced that he did not believe in 'implanting unnecessary bits of metal and plastic'. Was that enough justification for failing to pace a patient with a very slow heart rate and a history of sudden blackouts? Sudden blackouts can kill. Fifty years spent developing the 'metal and plastic' of pacemakers has proven their safety.

PostScript: after one further blackout, which he was lucky to survive, the patient had a pacemaker implanted. That put an end to his symptoms and improved his well-being.

Dedicated Doctors and Nurses

I have encountered many dedicated doctors and nurses in my time and been honoured to work with them. Typically, they will selflessly support you, the system they work for, and their patients. They can be academics, hospital doctors, or GPs. Whatever their position in the profession, they are worthy of admiration and support for their humility and public service.

I would like to pay homage to one such doctor. He was once an NHS GP with a practice in Leytonstone, London. I met him a few times while working at Whipp's Cross Hospital in the late 1960s.

Irishman, Dr. Swift Daly, drove a large Humber Super Snipe in which he conveyed his sick patients to and from the 'Casualty', at Whipp's Cross Hospital (the term 'A&E' had yet to be used). He would sometimes arrive half-drunk with a sick patient on board, leaving his other patients waiting for him at his surgery; patients happy to await his return. He once said to me, 'This patient isn't well. You're a nice young chap (a little blarney goes a long way). I am sure you will find out what's wrong with him. I'll wait outside while you examine him.' Afterwards, he would drive the patient back to his surgery, where he would continue with his other consultations. Those who took no time to get to know him thought he was a bit of a joke.

The police knew him well, and often escorted him to the hospital (at a time before drink/driving laws). One of them told me he drank whisky during his consultations and then threw the empty bottles out into his garden. 'His garden is a sea of whisky bottles', one informed me.

I came to respect Dr. Swift-Daly's clinical acumen and wisdom. He was realistic and honest enough to know what he didn't know. Such characters did a lot for the profession by providing a selfless, dedicated service. For patients, his service transcended any other personal considerations. His patients undoubtedly respected and cared for him. I am confident that thousands of his loyal patients would acknowledge a debt to him.

Had the Care Quality Commission (CQC) then existed, I doubt they would have found his practice compliant. Would they have closed his practice and withdrawn his attentive services from hundreds of patients? Because their focus is now on potential risks, and not what is most valuable to patients, they would undoubtedly have checked many irrelevant boxes, and taken a negative view. Some regard what they do as progress. I doubt respected doctors, like Swift Daly and his patients, would have agreed.

Diagnostniks

The principle fulfilment of these doctors comes from making challenging diagnoses. Some enjoy the accolade of colleagues more than helping patients.

Dismissers

Whatever is suggested, this doctor will reject it, counter it, and dismiss it. Their opinion *is* the only one they value. Other doctors must know their arguments well and counter them only after thorough research. One can ambush them and defeat them by planning the attack. Although they are usually arrogant and best avoided, they may deserve respect for the contributions they have made. Being correct once gives them no right to think they will always be so.

Disrespectful

Most doctors I met during my career were respectful of the ideas of patients, whatever their educational background, socio-economic standing, profession, or culture. Very few doctors were racist, with little interest in the culture or religion of their patients. Meanwhile, UK multi-racial demography marches on and their position no longer tenable.

Disrespect can have intellectual origins. Seeing themselves as custodians of knowledge and intellect, educated, clever people can look down on the poorly educated. Some are sanctimonious and arrogant, and claim intellectual justification for their disrespect. One can easily find published examples of intentional intellectual disrespect. In Richard Dawkins' *Outgrowing God, and The God Delusion*, or Salman Rushdie's *The Satanic Verses*, some disrespect for the beliefs of others is evident. In their pursuit of truth, the use of fiction, logical argument and parody can leave many feeling wounded and angry.

Doctors in high academic office can afford to disrespect the findings of their clinical and research colleagues. Those who disrespect patients will usually try to remove themselves from (the menial task of) having to deal with them.

DAS: Doctors As Scoundrels

You might think less than honourable behaviour to be found only amongst politicians, second-hand car dealers and criminals, not doctors or nurses. You would be wrong. Some doctors will steal the ideas and secrets of others, and are readily capable of skulduggery, cheating, dishonesty, sycophancy, spying, back stabbing, jealousy, avarice, and vindictiveness. I have experienced them all, detecting most of them early enough to avoid the consequences.

I was born into a business family and exposed to 'street wisdom' from an early age. I grew up learning to avoid advantage-takers, but did not expect to find them amongst my professional colleagues. For a life practising medicine, should one really have to study Machiavelli?

A GP I once worked with always did lots of home visits, or so we thought. He registered them every day in our practice work diary and visited patients after 'surgery' hours. We all admired him for making 5 to 10 home visits every day. That was until one of my smarter colleagues checked on him. He telephoned his patients to ask when last a doctor had visited them. Surprise, surprise! The doctor in question had visited very few.

Later on, at the same practice, I found out that we juniors were paying for two extra doctors attached to the practice. Their wages and expenses were being deducted from our salaries, not from those of the senior partners. Their argument, when discovered, was that we juniors were the ones who needed help. My argument was that they helped us all. I won the day by enquiring into their other, less than scrupulous accounting practises, and by asking, 'Are the Inland Revenue aware of what you are doing?'

All very unpleasant. To discover doctors playing dirty tricks is disappointing.

Some doctors are devious, and just as likely to pursue skulduggery for personal gain, as anyone else. When Dorothy walked along the yellow brick road in the Land of Oz, she did so with loyal friends. Many doctors and nurses also progress in the medical profession, accompanied by tigers, lions, and bears and others who may not have their best interests at heart.

It is easy to steal ideas. Since I have always been inventive, I generate more research and other ideas than I have time to develop. For those who lack ideas, it can help to steal some to progress their career. This has happened to me several times.

> 'Where do you get your ideas from, Victoria?
> 'I don't know. But, if I ever find out, I'm going to live there.'
>
> Victoria Wood.

While I was still practising, I had an enterprising (entrepreneurial) group of doctors try to gain access to my practice information. They were asking too many questions, with no obvious intent to collaborate. 'What's in it for them?' I wondered. I guessed they were after my database of patients and my working methods. I had decades in independent private practice information, and held valuable information. The benefits of experience are invaluable and need protecting.

Ideas make money, and for doctors born into a business culture, there is a simple rule - it is easier and cheaper to copy the work of others. What these doctors will not have appreciated is that there is no shortcut to building a lasting private medical practice. Gaining the confidence and trust of smart, successful patients, takes time. The new trend is for doctors to embrace the sunny slopes of capitalism, but for me, they represent the shady side of commercialised medicine.

Not long after visiting me, these doctors copied my prepayment medical scheme (created by me in 1985) and promoted it on their website. This was not my original idea; I modelled it on American HMO schemes. It became much appreciated by my less wealthy patients and continued unchallenged or copied for 35-years.

Another common form of scoundrel is 'the user'. I am sad to say that among the many doctors I have met, there were just as many 'users' as in other walks of life. Doctors and nurses need to be aware of sweet-talking, less capable colleagues, whose aim is to get their work done for them. For many, being helpful is their natural inclination; that means they risk being used. For less sinister characters, it may simply be a matter of continued laziness or convenient delegation that motivates them.

My suggestion is to work only with those creative enough, not to need to steal ideas; they would sooner promote others rather than use them. Some may think this all sounds paranoid. Once exposed to it, they will think differently.

Many of those 'at the top' of every profession have trodden on others on the way up to their prestigious pinnacle of power. Under the Peter Principle, many of them will have arrived at their level of incompetence.

Doctors and nurses must know that power-seekers, thieves, users, and the avaricious, abound in all walks of life. Among them will be their medical colleagues.

DPSWs: Dedicated **P**ublic **S**ector **W**orkers.

Some doctors and nurses work in public services. Some are to be found managing the NHS. Because they decide what services NHS doctors and nurses can give, they may feel they occupy the moral high ground. Some view their work in public services as 'worthy', and medical services as less worthy. Perhaps they haven't noticed that the NHS can no longer cope.

Over the course of my career, I have known many DPSWs to despise the private sector. Some would never acknowledge any benefit from private hospitals or private medical facilities. They must have found it challenging, knowing that captains of industry, company directors, the self-employed, government ministers, and the Royal Family, prefer private medical services. Few are ready to acknowledge why, even when embarrassed by long NHS waiting times, NHS queues for admission, and NHS bed shortages. Their sanctimony and biases might change with their need to off-load more and more NHS 'cold-cases' to private sector hospitals. Private practices now lessen the burden on the NHS, but this too, some choose to ignore.

Believing that public services are superior in every way to private services means that some still voice the well-worn opinion that 'Patients would be mad to pay when the NHS is free.'

Dork

Those with two left feet may be born inept. Some of them become valuable techno-geeks. Awkward, and sometimes socially inadequate, there are plenty of them in the medical profession. Many are endearingly eccentric and make valuable contributions.

Egoists

You will not have to wait long before you meet your first egoist colleague. They have strong opinions, broadcast forcefully. Not all are bigots. Some are 'Jumpers': jumping quickly to a conclusion, and quickly moving on to avoid reasoned discussion. The implication is, 'I know best, so let's not tarry'. Those with the experience and the evidence-base at their fingertips, deserve respect, but never let their ego suppress independent critical thought.

Emotionally Unfit

The intellectual ability needed to pass medical examinations is not a good guide to a doctor's ability to cope with emotionally laden situations. Given problems at home and stress at work, they may not cope at all well with study or work. There were a few among my medical student colleagues, and I encountered a few more as a junior doctor. In the old days, colleagues would have stepped in to help. Not all recovered sufficiently to re-consider continuing their medical career. Prince Harry and Prince William deserve the congratulations they received for exposing long-suppressed mental health issues.

Michael Palin on his 'Full Circle' trip of the world (BBC ONE, 1997) came across the grave of Commander Stokes. Pringle Stokes was captain of HMS Beagle on its first voyage around South America. Inscribed on his memorial cross (12.8.1828) is, 'he died from the effects of the anxieties and hardships incurred while surveying the western shores of Tierra del Fuego.' This led him to commit suicide. Charles Darwin was on the second voyage of the Beagle (1831-1836).

ERP: Emotionally Reactive People

Such doctors and nurses are quick to 'throw their toys out of their cot', when not getting what they want. With weak parents, they could have successfully used this strategy as a child. Some will continue to do so when approaching their competence limit. They are best avoided, except to offer them psychotherapy.

The aetiology of such reactivity is often simple. Some are underdeveloped emotionally. As children, their parents failed to reprimand them, and they carried on behaving as they wished. Parents too weak to resist their tantrums, spoiled them; fearing the loss of their affection. Throwing tantrums once got them what they wanted, so they continued with the behaviour into adulthood. They may not yet have met someone willing to stand up to them, but they will. My advice is for doctors and nurses to recognise them, and move away as fast as possible. Those who choose to remain in their orbit may hope to change them, claiming them to be misunderstood.

Favorites

There are many reasons for a senior doctor to see a junior doctor or nurse as 'the chosen one'; prized just a little more than seems appropriate. The reason for such favouritism is sometimes unclear. The blindness of uncritical admiration and the striking effect of beauty, wealth, and fame are real enough, but mostly based on 'fantasy'. My advice to third party onlookers is, don't to get involved.

There is a quieter, longer-term strategy to consider for colleagues of favoured ones. Play it like Stalin during the Russian Revolution. He took a back seat and waited for the adored ones (Trotsky and Lenin) to fall from grace. He then stepped in and took over.

Flirts and Flatterers

'Flattery will get you everywhere.'

Mae West, Actress

Very few doctors, nurses, and patients are averse to flattery.

Few flatterers are as overt as the eyelash fluttering Miss Piggy, in her quest to seduce Kermit the Muppet (TV series, ABC Studios). Anther amusing example is the girl dragon in the film *Shrek* (Universal Pictures, 2001). She effectively endears Donkey with her loving, dreamy-eyed gaze. In a more subtle way, criminally flirtatious Wanda (Miss Wanda Gershwitz, played by Jamie Lee Curtis in the film *A Fish Called Wanda*), gets her wicked way with a formal British barrister, Archie Leach (Cary Grant's birth name), played by John Cleese. If you want to study flattery, watch these films.

Persuasion is especially effective when an attractive seducer shows an unexpected interest in one who is not so attractive. Some will complement them as beautiful creatures, falsely claiming to have recognised their superior intelligence. The capacity some have to believe compliments, I always found surprising. Flatterers know this as a weakness, and will escalate their compliments to the limit of acceptability (often to the astonishment of onlookers).

'What do you want?', and, 'What's in it for me?', are the key questions for those being flattered to ask. Succumbing to flattery rarely leads to satisfaction. The best thing can be to play flatterers at their own game and come to understand their motives. What is the desired end-point of this transaction, one should ask. If you call their bluff, you might become an adversary; the more attractive or important the flatterer, the less prepared they will be for rejection.

Frustrated Doctors and Nurses

NHS work conditions have angered and frustrated many doctors and nurses. Some have become disheartened and demoralised. They expected to fly like eagles and instead find themselves grounded like turkeys. Unrequited expectation can cause resentment. It can depress mood, reverse a positive outlook, and limit motivation.

Highly intelligent, educated people, can find it frustrating to deal with the lesser educated, some of whom have searched the internet, and want to dictate their treatment requirements to doctors. Such encounters are potential sources of frustration and disappointment.

Fumblers

Whether it is putting up intravenous drips, taking blood, or examining patients, some doctors and nurses are downright awkward, seeming to fumble over every practical task. The doctors among them will not make talented surgeons. Hopefully, few will attempt to become one.

As an anaesthetist for one year, I had the chance to observe many surgeons at work. What surprised me was the variability of their co-ordination and mechanical dexterity.

GAHAD: Good **A**t **H**iding **A**ll **D**eficiencies

In some cultures, avoiding a loss of face has a higher priority than telling the truth. Whatever their culture, there are some who will always try to cover-up their shortcomings. If a doctor or nurse ever injects the wrong drug or wrongly performs a procedure, they must declare it immediately. Using the principle of 'patients first', every medical team member should help to correct the mistake. The patient's interest is never best served by a cover-up. Patients can die, and the perpetrators barred from practice.

A doctor or nurse will need bravery to become a whistle-blower. Up to now, whistle-blowers have paid a high price for their revelations. There is a worrying lesson to be learned here: doctors and nurses should never expect their colleagues, members of management or regulators to support them. Many will want to save themselves from admonishment by disassociating themselves from a whistle-blower or anyone under investigation by the GMC.

Julius Caesar was not the last to be shocked by the betrayal of his friends.

Glory Seekers

Has the need for glory changed over the centuries?

In Xenophon's dialogue *Ieron (Hiero)*, written in the 4th century BC., it suggests that all humans are free to enjoy food, sex, and drink, but only superior beings lust for honour, distinction and glory.

'But they in whom is implanted a passion for honour and praise, these are they who differ most from the beasts of the field, these are accounted men and not mere human beings.'

Ieron. (Chapter 7, section 3). Xenophon.

Hobbes, in *The Leviathan* (1651), states that the attainment of glory (as a means of recognition), with competition (for gain), and diffidence (for protection), are all strong motivating factors. Together they cause quarrels. Competitiveness and a desire for glory are not uncommon traits among doctors and nurses.

For doctors, glory will come only rarely, and be mostly limited to those who publish 'game-changing' research. Some will attain a lesser form of glory in public office, but gaining a knighthood, or peerage, will take a lot of nodding agreement, networking, and avoidance of conflict.

The focus of some doctors is not clinical at all. Their aim is to make their mark as regulators, NHS executives, or as politicians. For them, studying medicine is but a stepping stone.

GGs: Good Guys

Many good people can be found among patients, nurses, and doctors. There are many pleasant, fair, respectful, diplomatic and reasonable doctors and nurses. Many have grace and charm as virtues. They will display 'good behaviour' at all times, and resist corruption by NBW's (Nasty Bits of Work, see later). Good behaviour is easy to define as that which leaves everyone feeling comfortable. Sadly, the opposite is ubiquitous.

'Good Guys' are usually obvious from the first moment you meet them. Time will not change your impression. I have only met a few in my lifetime.

Gossips ('Newsmongers', according to Theophrastus)

Gossips are everywhere in hospitals and medical practices. Some wish to convey only light-hearted hearsay; others can be sinister, hateful, and derogatory. Often the subject is, 'who is screwing whom?' Gossip gains in interest with its level of secrecy.

The Biblical crime of a loose or evil tongue (defamation), as committed by Miriam, can damage three people: the one who propagates it, the person it concerns, and the one who listens. It seems rather harsh to accept that it is worse than idolatry, incest, and murder, but that was the Biblical view (Dr. Jonathon Sacks, former UK Chief Rabbi, *The Power of Praise*, 2018).

Grammar School versus Public School

UK doctors and nurses typically receive education at grammar or public schools. Grammar school alumnae and alumni are typically intelligent and well informed. Their tendency is to be self-righteous. As insecure socialists, many are outspoken, attention seeking, brusque, and promote utopian ideals.

Michael Portillo observed, in a BBC Four programme about Grammar Schools, that Public School boys had more effortless charm and confidence. Because of financial pre-selection, British public school pupils usually have a more secure background. They do not need to display an intense demeanour and mostly show quiet respect for their peers. These result from being banged up in a boarding school for years. Many have the unfortunate tendency of side-lining those from lower socio-economic backgrounds, but they will leave fewer rattled by their behaviour than their Grammar school contemporaries. Because those we find most attractive will usually come from the same social group, our class-ridden UK society remains steadfast.

It may be out-of-date, but for those working in the major professions: medicine, the Church, and the law, public school characteristics remain *de rigueur* for candidate selection in the UK. The last decade has seen moves to make more places in public schools and Oxbridge for the less privileged. In reality, there will be too many levels of conflict for an enjoyable experience to be had by all.

Gunslinger

This type of doctor is rare. A colleague of mine was one. He was both aggressive and uncompromising, and once wrote to a patient telling him he would be 'mad not to comply' with his advice. He followed this with, 'Why die for the sake of having one straightforward operation?' He said he had no time for 'stupid people', and often suggested that such patients should find another doctor. His policy was to 'take no prisoners', and to challenge problems head on; both features helped him to succeed in life.

His private consultant colleagues once approached him about his behaviour and management style while running his own private hospital. They threatened to get him 'struck-off' of the GMC register. His reply was: 'Since I own this hospital, I can employ as many consultants like you as I like. Don't waste my time with idle threats.' They crept away with their tails between their legs. My advice is not to engage with a professional gunslinger without a lot of prior shooting practice. Doctors and nurses need to remember not to box above their weight!

Haughty

Nose in the air, self-possessed, and consumed with their self-importance, some doctors (and to a much lesser extent nurses) treat most others as lesser beings. This is not an uncommon trait.

Hero, Champion and Role Model

Most doctors and nurses will have identified fellow pupils, teachers, or colleagues, who they admire for their knowledge, wisdom, helpfulness, and effectiveness. Hopefully, they will inspire the same reverence from others someday.

I am sad to say that as a lecturer for seven years, I remember only a few of my students. I taught many, however, who remember me. For any teacher to achieve this status is a worthy achievement. Doctors must not expect this to count in their favour, however, if they ever get entangled with the CQC or the GMC. With the statutory power they brandish, some now believe their value to society is greater than that of any doctor or nurse.

High-Minded

'Glory is my object, and that alone.'

Horatio Lord Nelson.

Nelson's sentiment will ring true for a few doctors. There are plenty who crave the Presidency of their Royal College, or a royal appointment. They are usually a little coy about declaring it.

Many authors have known such doctors. George Eliot, author of *Middlemarch* (1871), mentions the ambitious, but tragic, Dr. Lydgate. His ambitions were to pursue research and treat fevers in a novel way, according to the latest French theories. Not all authors have portrayed doctors in a good light. Both Chaucer and Shakespeare saw doctors as money grabbing, and low-minded (not an entirely unfounded view of the occasional doctor to-day).

Hobby Horse Enthusiasts

Beware of fixated doctors. They will try to explain every condition in terms of their particular hobby horse interest. (See 'Spanophilia').

Mr. Bill Kenny. was an anxious hypochondriac. A highly specialised cardiac research team mapped his cardiac amyloid disease. In consultation with him, they described his problem in worrying detail. The studies were of academic interest, but that is not how the patient came to view them. He asked me what I thought.

My echocardiograms had revealed a slight problem, but then he was 85-years-old. My opinion was that his condition was of academic interest only, and unlikely to affect his life expectancy significantly.

Unfortunately, the harm was done. The repeated explanations of research enthusiasts, who found him to be 'an interesting case', emotionally destabilised Bill. He became depressed and withdrawn, the opposite of his more usual happy self.

Enthusiastic, detached researchers, carried away by their interest in scientific detail, can ignore patient sensitivity. They presumably regard the science of medicine as more important to patients than the art of medicine.

Hobbyists

These are doctors whose hobby is medicine. They delight in meeting patients and everything medical fascinates them, especially interesting cases and the latest research. They have ceased to regard the practise of medicine as work, and feel privileged to get paid for pursuing their hobby.

There is a negative aspect to this. They can regard their work as a game, created just for their amusement. While enjoying their hobby, they can make light of how illness may affect patients. They make agreeable companions, and I have always found their enthusiasm for medicine infectious and edifying. We need more of them.

Hollow Men

T.S. Eliot referred to such people in his poem of the same title (1926); their eyes lacking life, spontaneity, and vigour, 'As the hollow men. The stuffed men . . . The hope only of empty men.'

Some doctors and nurses become depressed, others have an insufficient interest in medicine and its practise.

Humorous Fellows

Although many doctors are capable of wit and banter, many see humour as inappropriate and unprofessional when patients are present. Serious professionals have traditionally discouraged humour. As an advanced inter-personal technique for doctors and nurses, I cannot recommend it, even though it was my own USP. It will only attract those patients happy with the approach, and may offend others.

Seriousness does not always connote competence and expertise, and light-heartedness is not always frivolous. Humour can dispel anxiety and fear and can replace the face of impending doom with a smile. Humour can impart a message, but to use it requires talent and practice. Dr. Patch Adams' style provides a noteworthy example, few are able to emulate.

A few doctors have become professional comics; the rest tempt providence when using humour during consultations. Unless doctors know their patient well, it can fall flat and embarrass. It is perfectly possible to be serious (and scientific) on one occasion, and humorous on another. The ability to judge which is appropriate is an art.

'Timing is everything', said Bob Hope, and this means choosing the right moment. The right remark, the witty aside, and a relevant joke, can inspire confidence and breakdown formality. Jokes can ruin a doctor-patient relationship when used facetiously, or when used in poor taste. One problem with humour is that some patients will be too preoccupied to appreciate it. For others, it can divert their introspection and dissolve their fear for a while.

My patient was in hospital, awaiting an urgent hysterectomy. She was losing a lot of blood. As I entered her room she looked pale and frightened. The nursing staff were rushing around, clearly tense and worried. They had imparted their sense of urgency to her and caused her more worry. 'Am I going to die, doctor?' she asked me. Because I knew her well, and knew she had a good sense of humour, I replied, 'Not without my permission you're not!' It made light of a tense, high-risk situation. I followed that with, 'As a private doctor, I cannot afford to lose patients!' That made her smile, and she relaxed a little.

When I next saw her, she had survived a successful emergency hysterectomy. She thanked me for dispelling some of her fear and giving her some confidence. She said, 'When I knew you weren't worried, I relaxed. You helped me believe I was not about to die. Thank-you.'

When a patient happily acknowledges a doctor's humour, it will be obvious from their reaction. Humour shared, can leave both parties with a sense of 'us', rather than the usual 'you (doctor), me (patient)'. It can affirm mutual understanding. The further apart cultural values are between people, the less likely they are to share humorous moments. The risk is that humour will fall flat or offend, leaving both parties feeling alienated.

The late Mike Reid (Cockney actor and comedian), was once my patient. I had the temerity to tell him a joke. While consulting with him, he stood up, and without another word went straight to my receptionist and asked, 'Who is that guy? Is he a doctor, or a comedian? Doesn't he know that I'm the comedian here?'

Joking and making light of situations is typical of Cockney humour. Even the Luftwaffe failed to dampen it during the Blitz of London in the 2nd World War. My parents, grandparents, and all my great-grandparents were Cockneys (born within the sound of Bow bells, in the East End of London), so Mike and I partly shared a culture which made our understanding of one another an easy matter.

Idol or Idle?

Many of us come to idolise, or at least think well, of the teachers who helped us most. They deserve it for giving their pupils invaluable gifts: knowledge, understanding, and sometimes wisdom.

Some doctors and nurses get bored with medicine and become idle. They may have chosen medicine as a career in order to satisfy family aspirations, or simply to gain status and security. Some will know, right from the start, that their career choice was a mistake, others will happily retire early. If demoralised by 'the system' in which they work, some will disengage, become idle and quit.

The demoralisation of many nurses and doctors working in the NHS is now undoubted. Junior doctors are being demoralised by a growing need to comply with managerial interventions and corporate conventions, rather than to be guided by their medical education, clinical experience and judgement. It is now unwise for doctors to think for themselves. Looking over their shoulder will be a manager or regulator waiting to accuse them of non-compliance.

Jealous Creatures

Among both nurses and doctors, professional jealousy is common. It can be the focus of many things – the talent of another doctor, physical attractiveness, popularity, background, financial or class status, and reputation. The ramifications can lead to trouble and waste a lot of time.

JGGs: Jolly Good Guys

These are very pleasant, honest, and amicable doctors and nurses. They make decent friends and are agreeable and sociable to work with. Some wear their superior knowledge and expertise lightly; others seem to know nothing at all. Supportive and pleasant to be around, they can make arduous work bearable. Those with a sense of humour can ease the burden of work with an anecdote or two, even while performing challenging tasks. Those easiest to associate with are those who need not prove their talent. Doctors and nurses should look for amicable companions among those who have humility, and are generous and open enough to share the benefits of their talents.

Friedrich Nietzsche (Aphorisms on Love and Hate) had it that those who would be our friends are of two types: ladder and circle. The ladder types are going onward and upward. They will only be your friend as they pass you by. Others form circles; circles of friends you will keep for a lifetime.

Know It All

These are knowledgeable guys, likely to use information well, but sometimes to embarrass others. They can use their knowledge to shock or impress. Some can be monotonous, egotistical bores, strongly motivated to correct other people. Those who ask their opinion may need to reserve time for a lecture.

Lazy on Purpose

Doctors and nurses are not usually lazy, but should recognise those with divisive laziness. Laziness can be perversely attractive to those inclined to help others with their work. A lack of adroitness, alacrity, or panache will draw some to their rescue. Amid profuse thanks, some will seduce their helper into becoming their servant. They are good at flattery. They know that the more praise they give, the more vulnerable some become. Some doctors will ask others to put up all their iv infusions, and perform all their routine tasks, while they put their feet up. I exaggerate, of course, but not as much as you might think.

The study of gamesmanship (author Stephen Potter) can help some to become successfully lazy. This is the basis of the film 'School for Scoundrels' (Associated British Picture Corp., 1960) with parts played by Terry Thomas, Ian Carmichael, and Alastair Sim. The film depicts the advantages of one-upmanship and gamesmanship. Like much good comedy, it explores serious, real-life relationships and shows how some can get the better of others. Techniques well worth studying for everyone wanting to gain control. Commonly found in medical professionals.

Legalist

The legalist is forever reminding others of the rules, regulations, and guidelines that apply in any situation. They are mostly obsessive and compulsive; baked-bean counters destined for bureaucracy or the chairmanship of committees.

Highly disciplined, dutiful, authoritarian types of nurse and doctor can be strict about adhering to all the accepted rules, as a matter of principle. The strict and punctilious sometimes dislike those who flout the rules, even when acting as needed.

The TV series 'Blackadder Goes Forth' portrays many clashes between an authoritarian and one who is dismissive of the rules. 'Darling' (Captain K. Darling), is a nit-picker, appropriately assigned to unloading and sorting paper-clips. He clashes with 'Blackadder', who cares little for bureaucratic rules, given the probability of his death as a front-line officer in the First World War.

There are military examples of similar clashes. The cunning, non-rule-based Viet-Cong in Vietnam defeated the US military repeatedly. This is partly because the US army is based on following strict rules of engagement. Such adherence helped to cause their demise, but not only in Vietnam. Al-Qaeda in Afghanistan inflicted a similar fate on the US army. General McCrystal, in his book *'Team of Teams'*, states that in the early days of US Army action in Afghanistan, they lacked the versatility of small terrorist groups. They had insufficient independent mindedness to change tactics as circumstances required.

Rules work best where predictability reigns, like running a bank or a railway network. Rule-based functioning cannot deal with exceptions. Since exceptions are the stuff of medical practice, creative behaviour is more often valuable than stereotyped, rule-based behaviour. For this reason, doctors should only obey rules as 'guidelines', never as 'fixed' rules to be followed blindly. Our regulators do not agree. They can only regulate with fixed rules in place, so they commit clinical heresy when they insist that doctors and nurses use guidelines as fixed immutable rules.

Little Sh*ts

These are doctors and nurses who do not mind dropping others in the sh*t, when it suits them. They are often self-possessed, arrogant, and superior. They can be short and dismissive of both patients and those of lower rank.

LMF (Lacking Moral Fibre)

Battle trauma was first recognised late during the First World War. Some military personnel had 'battle fatigue', others had shell shock. Over two hundred soldiers accused of malingering, desertion and dereliction of duty were executed.

The British armed forces did not officially recognise post-traumatic syndrome (PTSD) until after the Falkland's war of 1982. The RAF, in particular, long referred to such cases as 'Lacking Moral Fibre.'

After qualification, some doctors will experience a few situations they find difficult to cope with. As such, they will struggle to make effective front-line interventionists. Given that many of the doctors who educated me served in the Second World War (WW2), the tacit accusation of LMF must have been on their minds when they witnessed junior doctors failing to cope mentally with emergency situations. As a result, they would direct them to specialities like dermatology, general practice, and psychiatry, rather than interventional cardiology, A&E, and surgery. LMF is a completely unacceptable term, but like many others that are politically incorrect, they often contain more than a grain of truth.

Lone Wolf

Many doctors and nurses will work with anti-social, high achieving colleagues, some of whom will have Asperger features. There are doctors who belong in a world of their own; a world they rule alone. Fortunately, they will mostly choose to minimise their contact with colleagues and patients. (see: 'Doctors as Eccentrics'). Some are simply introverted and shy.

Lucky Buggers and the Matthew Effect

'For unto every one that hath shall be given, and shall have abundance: but from him that hath not shall be taken even that which he hath.'

Matthew 25:29. King James' Bible.

A junior doctor colleague of mine once experienced the 'Matthew' effect when on duty one night in 1966.

Junior doctor Tom was a rich Bahamian who shared alternate nights on duty with me. We were both first year housemen. The night he was on duty, he admitted an octogenarian lady with pneumonia. 'I have no family or heirs', she said. 'Would you agree to be my beneficiary?'

Tom eventually inherited 13 terraced houses, and £70,000 in cash (one could then buy a three-bed, terraced house in the area for £3000). His family owned a chain of hotels throughout the Bahamas, and my guess was that he had little need for extra money. Money goes to money, they say. That's life!

Stand around a roulette table for long enough, or attend a horserace meeting or two, and you will see those who seem to win more than they lose. The majority do not win, of course, yet there are those who always fall on their feet, and those who always fall on their head.

Is having bad and good luck an illusion? There is a division of opinion about it. Chance will have it that unusually lucky people (those who beat the odds most of the time) must emerge, if given enough time. I remember one lucky doctor, selected for a prestigious job only because the interview board could not decide between two more suitable candidates.

Amundsen, the polar explorer, did not rely on luck. At the beginning of the 20th century, Roald Amundsen beat Scott to the South Pole. In fact, he was the first to reach both Poles. He wrote: 'Victory awaits him who has everything in order – luck, people call it. Defeat is certain for him who has neglected to take the precautions in time; this they call bad luck.'

A friend of mine deserved his luck. Nigel Nodolsky generously gave a kidney to someone he hardly knew. Because he became a research program subject he was regularly reviewed. A follow-up chest X-ray revealed a small lung lesion, which proved to be early lung cancer. No evidence of spread was found after its removal. This was good luck, but luck very much deserved because of Nigel's selfless generosity.

As an example of bad luck, my patient Oliver O'Rouke came to me with what proved to be a rare tonsillar cancer. Within one year, he returned with tongue cancer. The year after that, I diagnosed his prostate cancer. I regarded him as very unlucky. Because he was a JGG, he made light of it all. Sadly, he died in 2024.

Once I realised the gift of exceptional luck was not mine, I never expected to win anything by chance. I have stuck to what the American military once called: **PPPP**: **P**reparation **P**revents **P**oor **P**erformance, rather than luck (from the TV series M*A*S*H; 20th Century Fox Television). It has served me well.

Manipulators

> For who would bear the whips and scorns of time,
> Th'oppressor's wrong, the proud man's contumely,
> The pangs of despised love, the law's delay,
> The insolence of office . . .

Hamlet. Act 3, Scene 1.

Unfortunately, doctors and nurses can only choose their colleagues once they achieve seniority. Those who are 'manipulators', and 'controllers' by nature, will want to achieve this status as soon as possible. The naïve, the dedicated, and the altruistic don't usually give a damn about status; not until they have tasted the effects of manipulation. Their first reaction may then be: 'I didn't think doctors or nurses could be like that!'

The medical profession is replete with head-strong, arrogant egotists, who would give Niccolò Machiavelli a run for his money. Where survival and success are the issues, cunning and intelligence are both required for successful strategic planning. Cunning and a desire to manipulate others result from a predatory outlook. Both soon come to light in any relationship.

A common strategy used by predators is to keep secrets. The simple rule is: if you want to keep a secret (and avoid manipulation), tell nobody, and I mean nobody.

Some health professionals are guilty of a minor confidence trick. Some osteopaths and dentists now call themselves 'doctors'. Will boosting their supposed status attract business and instil trust?

I once knew an ex-nurse who falsely purported to be a graduate doctor and psychiatrist. He was neither, although he did qualify as a psychiatric nurse. His wife, who came from the Far East, told her family that she had married an eminent doctor. This harmless folie a deux, suited them both.

'We schoolmasters must temper discretion with deceit.'

Dr. Fagan to Pennyfeather. *Decline and Fall* (1928). Evelyn Waugh.

Phillippe was a suave, Gauloises smoking Lebanese doctor. He came to us as an observer in the cardiac department at Charing Cross Hospital, and constantly boasted about his family fortune: their multimillion dollar yacht moored in Monaco's Hercules harbour, and their villa on the Côte d'Azur. He also boasted that he was soon to open a smart medical Clinique, on Rue Cambon, Paris, alongside Chanel. It was to be equipped with the latest state-of-the-art medical equipment. Exaggeration, surely? He failed to manipulate us into thinking well of him.

Many years later, having not seen him for some time, I telephoned him from Amsterdam to say that I would soon be in Paris for a medical conference. He suggested we meet for lunch and view his medical facilities. We met for lunch, and he then showed me his clinic. It must have cost millions to put together. It was busy with patients drawn from the Parisian élites. Not all that seems hyperbole is a rhetorical confidence trick.

Beware of manipulators versed in *gamesmanship*; they can influence the professional progress of doctors and nurses. Both the real and metaphorical chess players among us plan their moves well ahead of time, while non-gamers simply get on with the job of practising medicine. The gamer's short-term aim is to collect 'brownie points' from influential people; undoubtedly something they have done all their life. They will have employed stealth and even underhandedness to advance their career.

The aim of many manipulators is to attain high office. From that lofty position, they will have a strategic advantage over all challengers. They may then exhibit what Hamlet referred to as 'the insolence of office'. My advice is to identify them early, and side-step them quickly. If you don't, you may witness them walking away with prizes gained from manipulating colleagues. Another strategy would be to befriend them and become their second in command.

Doctors and nurses have all met one or two manipulators at school, or during their training. They must be prepared to meet some more when the stakes are higher. Hopefully, no doctor or nurse has to share their job with one. Those who do, might do all their work for them while they take all the credit. It is best not to share confidences with them. They will file these away for later use against you, at a time that best suits their purpose. Learn to keep useful information secret until it benefits you. I would advise all doctors and nurses to keep their best cards up *their* sleeve, or close to *their* chest, and protect them from cunning predators and manipulators.

In order to prepare for such people, there is no better experience than attending a British independent public school. There, one will find those from successful backgrounds, some with family wealth funded on manipulation. There, one can learn their methods, and the earlier in life they are learned, the better. Contrast such people with doctors and nurses who simply love their job, and will do everything they can to help their colleagues and patients.

Controllers, manipulators and users will want to draw doctors and nurses into their circle. One must resist them, however tempting their inducements. Never risk becoming a puppet on their string.

I was once told that I was in line for a professorship ... but there would be a price to pay. My response was immediate. I doubted I could afford the price, whatever it was. The colleague who suggested it dismissed me with a compliment: 'You are too secure, Dighton', he said, realising that I understood his attempted manipulation. The corporate life was not going to be for me.

Mean Bastards

The university medical degrees, MB., BS are not an acronym for **M**ean **B**astard, **B**latently **S**elfish, but they can be! Some doctors help only themselves. There have always been many of them.

Military

Ex-military personnel are easy to spot. They have a disciplined dress-sense, a respectful manner, highly polished shoes, and a noticeable personal bearing. Confident, dependable, and dutiful, they are usually reliable as doctors, nurses or patients.

Moralist

They are forever reminding others of the need for morality. Ethicists will remind us of the rules of engagement.

NAGS: '**N**eeds of **A** **G**ood **SH*G**!'

Rarely uttered; often thought. This can be the most appropriate treatment for those who have become uptight and irritable. Sexual intercourse will usually calm their angst and make them more amenable (see 1980s TV mini-series *Shōgun*, Paramount TV. Original book by James Clavell). Studious types who have spent too long studying (rather than getting out, having a good time, and loosening up) will also benefit. We urgently need double-blind trials to verify the efficacy of this intervention.

Names

What's in a name? Doctors with unforgettable names, especially those that are double-barrelled, once progressed faster than others in the medical profession. If parents wanted their child to get on in the medical profession, they would name their boys Peregrine, Hugh, Sebastian, Tristram or Guy, and their girls Lucretia, Claudia, Tabatha, or Hermione. These names are not only striking to those in authority, but easily remembered. A distinguished name will not guarantee progress through the ranks, but it's a good start. Later on, they will be required to live up to the image their name projects.

Once highly respected were noble foreign surnames. Today, they can be regarded as pretentious. The French aristocratic name, Chalut de Bascoigne, rhymes with the name of one of my contemporaries at 'The London'. Once qualified, he rapidly progressed through the ranks. The vintage Bentley he drove as a student didn't go unnoticed, either.

Naturals

Are there 'naturals'; doctors and nurses born for medical practice? One might regard those born with all the prerequisite talents as 'naturals', but some of them will decide that medicine is not for them. Those who decide to practise medicine will still need preparation and practice.

Besides the gift, you need the drive.

Neil Sedaka.

Those with a natural talent rarely need conscious calculation to do their job. They are quietly confident of their know-how. I have little doubt that innate talent exists, although for those who possess it, the first step is to become aware of it. The next step is to find the drive to use it effectively.

NBW: Nasty Bits of Work

It's not enough for me to win; others must lose.

Gore Vidal.

My son Nicholas, a commercial airline pilot with Easy Jet, recently had a skin naevus inspected by a dermatologist. Flying exposes pilots to more UV light than usual and a high risk of skin cancer. Once the cosmetic surgeon learned for whom he worked, he asked: 'I suppose that means you failed to become an RAF pilot?' Nicholas replied: 'That's an interesting question. I rejected the idea of joining the RAF because I didn't want to bomb innocent people!' (Nicholas is a typical JGG, and the cosmetic surgeon a typical NBW.)

One must approach all rude and challenging people with care. Some of their gratification comes from putting people down. Like angry dogs, they can attack (verbally) without warning. Few have more bite than bark, but one can't assume it. Frustration and hormonal changes can add to their grumpiness. Some are not getting what they want from life, but nothing should excuse their unpleasant behaviour. There can be a bipolar element to their nature: nasty or affable, depending on their mood swings and motives. Unless we choose more affable and predictable characters to befriend, *schadenfreude* becomes likely.

Nice

To meet genuinely nice people is a wonderful thing. Beware, though, being nice is sometimes a manipulative gambit.

NPC

In the computer gaming world, they recognise Non-Playable Characters, or NPCs. Some doctors and nurses, when pressed, will do their best not to get involved. For this reason, GPs can find it difficult to get deserving and urgent medical cases into hospital. Those answering their call can be NPCs and very unhelpful; some will practice their excuses for not accepting patients. In response, some UK GPs will tell their patients to go to A&E, and bypass the admissions system.

Does every hospital ward have one? A person who floats around with a clipboard, hoping not to get involved.(See: Shifting Dullness).

Obsessives and Nit-Pickers

*Whoever thinks a faultless piece to see.
Thinks what ne'er was, nor is, nor e'er shall be.*

Alexander Pope. *An Essay on Criticism* (1711). Part 2: 253-254.

Q. Can you spot a truly obsessional person?
A. Of course. They eat alphabet spaghetti in alphabetical order!

Because medicine is partly driven by medico-legal considerations these days, nit-pickers have come into their own. Obsession is a double-edged sword. It can foster meticulous work or distract attention from the job in hand. Some doctors and nurses may be so keen to get every 'i' dotted, and every 't' crossed, they may not notice a patient deteriorating. Their drive is to complete the task in hand, regardless of relevance. They can often do only one thing at a time. When there is a need for urgent clinical action, their inability to finish what they are doing could put lives at risk.

I have often instructed colleagues to end their telephone call, or stop whatever they were doing (chatting, or looking at a mobile phone screen), in order to focus on the clinical action. The ability to prioritise clinical situations, and drop everything when necessary, is an essential characteristic for emergency doctors, nurses and paramedics. Some obsessive characters will fail the challenge.

Consider what happened on the flight deck of United Flight 173, when flying from New York to Portland, Oregon, on the 28th December 1978. The two pilots flying the DC8 had thousands of hours of flying experience between them.

As they approached Oregon, a false warning light appeared, suggesting failed landing gear deployment. Actually, a damaged sensor light misled the pilots. The flight crew held the airplane in a holding pattern while they discussed the problem. By the time they decided on their readiness to land, with the airport in sight and emergency vehicles standing by, 70 minutes had passed - 5 minutes longer than they had fuel for. Their engines flamed out one by one, and they crash-landed, ploughing through several houses. They skidded to a halt after 500 yards. Most of the passengers lived. There was no fire because they had burned every drop of fuel. Their obsessive checking of lists and options had drawn their attention away from one crucial flight instrument – the fuel gauge!

Exactitude is important in many disciplines, but can be misplaced. There was, for instance, no point in taking an inventory of deck-chairs on the Titanic as she was sinking. It did perhaps help to avert anxiety. Obsessional people make good proof-readers, and safety checkers, but can be so concerned with the detail, they miss the wider picture.

Fear of failure drives many academic achievers; they fear missing an important detail. Many are obsessional about detail. Few patients will warm to a surgeon who feels it necessary to list all 14 known complications of the operation they are about to perform (regardless of the individual probabilities). This behaviour might comply with good medical practice, but it can diminish patient trust and confidence. For some patients, this will represent a doctor's failure to use the art of medicine.

Old Bailey Types

Every conversation is adversarial. They will call for documentary evidence, the definition of any terms used, and justification for everything said. Relating to them can be hard work.

Old Vic Types

Every inter-reaction with them is a drama. Beware of taking a role in their play.

OOSOOM: *Out Of Sight, Out Of Mind*.

When a doctor or nurse asks their colleague for help, it could be the last they see of them. Why would that be? Are they selfish, or just too busy? Either way, they are unreliable and best avoided. Sorry to repeat it, but when trying to understand people, the direction their feet take is usually more revealing than their words.

OVIBOI: OVerly Impressed By his /her Own Importance.

Such people are self-possessed, conceited and vain. There is only one important seat at any table – theirs. Their needs must take priority. Collect lots of evidence prior to any discussion with them and use it to counter their self-centred views. This will not endear them to you, and with luck, it will encourage them to avoid you.

Try this as a zoning-out exercise that no Russian speaker will appreciate. Like the French, they much dislike their language being adulterated. OVIBOI is not a Russian word, but it sounds like one. If it were not an English acronym, it would rhyme with Bolshoi (masculine for 'Big' in Russian, like the famous Moscow theatre). It is linguistic nonsense to suggest that a doctor could be a Bolshoi OVIBOI (if male). The female equivalent (also nonsense) would be a Bolshaya OVIBAYA. A few British people might find this suggestion faintly amusing; no Russian will. Humour is a culturally sensitive issue.

OWT: One Way Traffic.

Although doctors and nurses give of themselves, and try to help others when possible, others will not always reciprocate with the same generosity of spirit. If you find this upsetting, learn to select your associates more carefully.

Pedantic Pedagogues

The original Greek derivation of pedagogue is from 'rules' (*agoge*), applied to 'children' (*paidia*), like the rules of conduct written by a schoolmaster for his pupils. Medically trained pedagogues are in plentiful supply. They are doggedly adherent to rules, regardless of the benefits or downsides. The trait will diminish their ability to make diagnoses, especially with 'difficult' cases. They should consider a job in administration.

Only pedants will enjoy this fact. The Greek phrase 'Oi Polloi' means 'the people' ('Oi', in Greek, means 'the'). If you say 'The Oi Polloi', you will say 'the, the people.' That's pedantic.

Persona Non Grata (PNG)

For the delivery of any message, there will always be one who is the most effective messenger. A child might completely ignore its mother when commanded not to twirl their hair or pick their nose. Their teacher at school will more likely be obeyed after delivering the same message (perhaps because the entire class might overhear). The fact is, some messengers are more effective than others.

It is important to consider who might be the best person to deliver a message to a patient. The most appropriate person might be a relative, a colleague or a friend. In the relationship between messenger and patient, status, age group, culture, and gender are all relevant factors. If the object is to convey understanding and appreciation, the best person for the job might not be a doctor or nurse.

Ignaz Philipp Semmelweis was the Hungarian gynaecologist who discovered that fever after childbirth occurred only after patients were examined by doctors who had just returned from mortuary dissection. Washing their hands prevented it, although not then the accepted practice. He presented his data to colleagues in a form that they found unintelligible; he thus failed to convince them of his observations. In addition, his background was not prestigious enough to command their respect. His important discovery therefore remained unacknowledged.

Florence Nightingale nearly suffered a similar fate, but she presented her Crimean medical data in a novel, understandable format. She invented the first pie-chart to do so. That she was an aristocrat with many vital political and social connections must also have helped her case.

Phraseologists

Some doctors excel at expressing themselves in politically correct, one-upmanship phrases that win them credit with colleagues. Doctors with an extensive stock of 'in vogue' phrases are more likely to be thought competent. Those who use phrases such as 'time-critical', 'fit-for-purpose', 'statistical significance', 'evidence-based', 'quality of life issue', or who use clinical trial acronyms, will make an impression. Those who use these shorthand, headline phrases do not always understand them fully. Some use them as rhetorical devices to convince others of their competence.

When left undefined, the use of an obscure acronym for a research trial might imply that everyone present knows all about the research. Latin phrases will similarly impress. Few will dare ask for a translation. Few wish to display their ignorance.

PIMP: Power Is My Purpose

Plenty of doctors seek power. Committees, steering groups, and political organisations like the British Medical Association, Royal Society of Medicine, and the Royal Colleges attract them. Perhaps they should consider a full-time political career, and aim to become Minister for Health. That would be a first. Nobody has yet seen the merit of appointing a Minister of health with a medical or nursing qualification and the appropriate clinical experience.

PITA: **P**ain **I**n **T**he **A**rse

Some are doctors, a few are nurses. It would be a pain in the arse to describe them further. They come in many annoying and tedious varieties. They commonly induce sly comments and the eye-rolling of bystanders.

Socrates gained a reputation for asking annoying but pertinent questions. He became known as the Gadfly of Athens.

PLASMA: A **P**erson **L**acking **A**ny **S**ort of **M**edical **A**cumen.

PLOJ: A **P**erson **L**acking **O**bservable **J**udgement.

PLOP: A **P**erson **L**acking any **O**bservable **P**ersonality.

Politicians

Some doctors get involved in both local and nationwide politics. They might help more patients through political change than by practising medicine. They will quickly pass you by as they climb the ladder to their intended public office.

Many bureaucrats believe the path to medical utopia must be paved with regulation manuals. I would prefer to see it laid with mosaics depicting nurses and doctors throughout the ages; those freely using their clinical judgement and experience to help patients.

The doctor-patient relationship may not provide power enough for politically minded nurses and doctors. They might, however, come to represent us all in the halls of power. Their political mission might sometimes conflict their personal medical mission.

Preppies

Those who prepare well and practice often. This makes them more likely to succeed.

Pretty People and Charisma

Only a few precious beings have charisma, undoubted beauty, and star quality. Through no fault of their own, they easily incite staring and jaw-dropping interest from others. A queue of obsequious, lip-licking, eyelash fluttering, 'come and take me whenever you want' people, may trail behind them, hoping for one small crumb of recognition.

Actress Mae West prescribed how to get noticed: tell the beautiful they are smart, and the smart how beautiful they are.

Leslie Phillips played the ultimate smooth operator while playing the character Dr. Gaston Grymsdyke. No actor has ever said 'Hello', with more licentious lasciviousness. (See films like 'Doctor in Clover', Rank Organisation, 1966).

Quants

For stock market analysis, quants do the calculations and weigh the risks of trading deals. In medicine, the same types also exist. Many are what used to be called 'back-room boys.' Valuable colleagues, but not always best placed at a patient's bedside.

RA: Risk Averse, and Not Risk Averse (NRA)

A practicable sense of risk is not possible without considerable clinical experience. Gamblers have an advantage. Those who regularly play roulette know the exact odds of winning or losing. With one zero slot, and 36 other numbers on a roulette wheel (some have a zero and a double zero), the chance of the ball dropping into any one slot is 37:1. If you play with these odds long enough, you will get a feel for the likelihood of winning. Those who despise games of chance will be at a disadvantage. They will find it difficult to put the odds of winning or losing into numerical perspective.

Most people prefer predictability to unpredictability. Many, in fact, have a neurotic desire for predictability, and become anxious without it. They may find it difficult to take risks, even though everything we do in life carries a risk. Doctors soon become accustomed to the fact that medical practise is inherently risky, whether performing emergency surgery, or reporting MRI scans. The risk of error and patient suffering is ever present.

For interventional doctors (surgeons and cardiologists), risk aversion can inhibit action; interventional cardiology is not for the circumspect. Procrastination can prove dangerous. There is some advantage for patients who have a risk averse surgeon. They will operate only when they are 100% sure of success. In very dire surgical situations, however, patients can be better off with a surgeon who will try his best, even when the risks are high. In the last-chance saloon, there is nothing to lose and everything to gain.

I once knew a prestigious, older chest surgeon, who would only operate on very low-risk patients. His patients were all fit and had the best indications for surgery. He had a 100% operative success rate, but operated on very few patients. Be careful how you interpret the reputation of physicians and surgeons. Their criteria for selecting patients and their position on the risk-aversion scale are important when choosing to refer to them.

Only non-risk averse doctors will dispense with caution and get on with the job. Although they will save lives, many will think them reckless and cavalier if things go wrong. Around them will be doctors fearful, and frozen to the spot, unable to contemplate any risk at all. They might not see the danger of inactivity. For the most effective doctors, risk-aversion is a variable; one that is altered case by case.

Important factors influence patient risk evaluation: the patient's general fitness (often related to length of illness), their clinical viability (some patients are tough, while others are frail), and likely complications (given specific predispositions like clotting or infection). To assess the risk, doctors need to combine these considerations with the technical difficulty of the procedure and the skill of the operator.

I was working in Casualty in 1967, when a road traffic accident victim arrived. The patient was pale, sweaty, in pain, and semi-conscious. He slipped into unconsciousness as an Egyptian surgeon strode in and called for a scalpel, sterile gloves, i.v. drips in both arms, blood for immediate infusion, and an anaesthetist (me). He demanded an operating theatre to be prepared – straight away! With no time wasted on explanations, he made an upper abdominal incision, inserted a gloved hand into the abdomen, and compressed the patient's spleen. We then rushed several hundred metres to the nearest operating theatre, where he performed an emergency splenectomy. The patient survived.

The experience was inspiring. I knew then I wanted to be involved in this sort of action. I later did similar things in cardiac emergencies. It is an edifying privilege to rescue patients from the jaws of death.

A pathologist once called our cardiac team to attend the hospital autopsy room ASAP. What the pathologist showed us was a pacing wire, previously inserted by me during resuscitation. It had not entered a vein as intended, but the arch of the patient's aorta. I had advanced the catheter into the left ventricle (not the right ventricle as usual via a vein), with no internal haemorrhage at all. The patient paced successfully until his death from renal failure.

The patient had been blue-lighted with a police escort to our cardiac team at St. George's, Hyde Park Corner, London. Too few doctors then knew much about emergency cardiac pacing. The patient's heart had stopped, and he was on the floor of a ward, as I entered the scene. Chest compression was being performed, and must have squeezed his major artery (aorta) up towards his collar bone. I inserted a subclavian pacing electrode effortlessly, and he paced successfully within 20 seconds. No X-ray was available at the location to check the position of the pacing electrode. The dark blood that came back through the entry needle appeared to come from a vein. In retrospect, it must have been dark, de-oxygenated arterial blood.

A word of caution about risk taking. There is a thin line between correct NRA action and recklessness. I never once met a reckless surgeon who could not resist operating. Inertia is the more common problem in emergency situations, with some doctors and nurses frozen to the spot with a patient bleeding to death in front of them. I have classified these as 'BBC's', 'PLASMAs', 'SDs', 'USBs' and 'RAs.' Some doctors have no instinct for rapid assessment and action. They should give serious thought to treating acne as a specialty.

When any bureaucrat gets involved with medical risk assessment, challenge what they can know about it, without ever having faced the sort of situations illustrated above. Challenge their assumptions and their actual experience before allowing them to criticise you.

Reliable / Unreliable

Patients need reliable doctors; those who turn up on time and do what is necessary. Unreliable doctors and nurses are potentially dangerous.

The surgeon I appointed for one of my patients, delayed her abdominal operation for bowel obstruction. It was Friday, and he had plans for the weekend. The delay left this anxious patient worrying about her diagnosis, the surgery she faced and her fate. I replaced him with a surgeon who operated on her a few days later. She was so exhausted by then, her immune system failed to cope. She died of septicaemia post-operatively.

Romantics

Dr. Negativo: 'I hate medicine.'

Dr. Positiva: 'No, you don't. You love it. Loving it is what you hate!'

Instead of objectivity, some doctors prefer a romantic view of their past achievements and what they might achieve in the future. They may come to support romantic, patient oriented ideas, like 'one can be forever young' (if only they had the best cosmetic surgeon), or, 'every cloud has a silver lining' (after consulting the best doctor; taking the best advice, or submitting to the right operation). A few doctors are true romantics, cock-eyed optimists who could have been talented actors.

Romantically inclined doctors can encourage hope, and nobody ever died from that. Problems lie with false hope and false expectation. Doctors must discourage patients from having procedures that are unlikely to improve their situation. In terminal, dire, or inevitable situations, false hope can be hurtful. Only those who know their patients well enough should exaggerate hope. If it improves the patient's composure, it will sometimes improve their prognosis.

There is plenty of evidence to suggest that grief diminishes survival. This is not a new concept.

In 1630, the Company of Parish Clerks in London reported twenty who died of parental 'Grieffe', perhaps because of the death of their child. The death of young children and infants was then commonplace. Two thousand, three hundred and twenty-eight 'Chrisomes' (children less than one month old) and infants, died in that year. Plague killed 1091, while other causes of death were age (662), 'bloody flux' (438), childbirth (157), and stillbirth (423).

A romantic notion, rather than the plain unmitigated truth, will sometimes encourage patients to beat the odds, if only for a while. Using hope appropriately is not an art possessed by all doctors. The art of medicine might dictate that a bare-faced lie is expedient. If I thought a lie would benefit a patient in the short-term (when their prognosis was poor), or in the long-term (if their prognosis was unknown), I never hesitated. Deception is now discouraged as unethical, and poor practice. That will suit all those doctors and administrators with a detached view of medical practice, and those who desire anonymity. They will usually have little talent for the art of medicine.

I well remember talking to one of my best friends, Tony Smith, about what we would do together once he recovered his strength and left the hospice. Even though his prognosis from widespread melanoma was bleak, we made plans. Although we both tacitly accepted that our ideas were romantic, it helped us both cope with the inevitable. I was not his doctor. I had a more important role to play as his best friend.

Tony and I met at school when we were 11-years-old. We sat together in class. On a school holiday when we were both 13-years-old, we found ourselves abandoned late one night in Riccione, Italy. A taxi driver said he could take us no further than the edge of town. We had to walk the rest of the way to the Albergo Alba D'Oro, in Rimini. We started our 12Km. walk in darkness, at 11.30pm. and arrived back around 1.30am. Tony was forever positive, a cheerful boy, never easily disheartened. This was well before the invention of mobile telephones, so we couldn't let our teachers know where we were. They must have been frantic with worry.

Tony's enthusiasm for life, and that walk in particular, stays with me. What a gift to have a wise and joyful companion for life's journey. It naturally fell to me to encourage him while he was taking his last steps. I still treasure our friendship, and the mutual trust and faith we found in one another. I am happy to have helped his composure in his last days, as we reminisced about our happy school-days and made future plans. Encouraging him in a light-hearted way brought him some peace. Our long companionship served us well. Memories of his wise council stay with me and still guide my decisions today. (Alav Ha-Shalom).

Ruthless

Some doctors ruthlessly pursue their career by cultivating the favour of senior colleagues, and bypassing or denigrating those of little use to them. They should be avoided completely; they will leave a trail of dejected souls in their wake.

Scammers

In private medicine, some doctors have one over-riding aim: to make money. Some will escape the tight bureaucratic control of corporations like the NHS, in order to achieve it. Some will indulge in illegal scams or practises, contrary to the GMC's *Good Medical Practice* guide, and not always in any patient's interest.

One innocent scam I came across involved referrals to a private hospital for cardiac CT scans. I had made hundreds of such referrals to other private hospitals, but I had never referred a patient to one particular private hospital. It surprised me to learn that each patient referred had to be seen by their 'in-house' cardiologist. From this, he generated a fee. There is nothing wrong with suggesting a second opinion, but this unnecessary intervention seemed a scam to both me and my patient.

Other quasi-scams have been around for decades. One involves the cross-referral in the private sector to professional friends and colleagues. The effect is for them to get extra income, and for the patient referrer to be granted support in the future. Second opinions, and opinions from those in different specialties, can be highly desirable, but sometimes unnecessary. There can be a minor scam element to the practice, since the patient will have to pay a lot more in extra fees.

Unnecessary (or difficult to justify) follow-up consultations is another quasi-scam some private patients complain about, especially when each follow-up consultation is costly. This is a common, but inevitably limited practice. It is short-sighted because it will disincline patients to seek follow-up assessments, and could diminish a doctor's future long-term potential. That will not concern most private (also NHS based) consultants because few will see the same patient twice.

Private medical insurers have long known that some private practices add unnecessary extra costs to insured patient accounts. Other payors, like oil-rich foreign embassies, care little about the extras. For decades, they have paid unquestioningly for scarcely justifiable investigations, and much longer stays in London private hospitals than necessary (daily visits by doctors also add a lot to their account). Because these practices are only loosely justifiable, they are best referred to as quasi-scams.

Scary Guys

Jack Nicholson, when acting, was often told to 'lose the scary look.'

A few doctors have a 'mad look' in their eyes. Fixed, starring eyes, and exposed conjunctivae are a universal, non-verbal signs of danger. If they are stern and abrupt as well, few will want to approach them. A kindly pussy cat mentality can hide behind a scary face, but few will be brave enough to confirm it. Patients should have feared Dr. Harold Shipman, but they didn't. I believe he was kind and affable, and not scary at all. He apparently had a supportive and unassuming manner which endeared him to many patients. What other doctors thought of him, I do not know.

Scholar

These are intellectual doctors, interested in medical science and scholarship, and little else. They deserve respect. Although, often high-minded, the trait does not exclude a talent for executive functioning, compassion and the art of medicine.

The Selfless Few

Those who can deal selflessly with the baseness of humanity are saints. Typically, they are shy and self-effacing and not at all interested in seeking recognition. In common with those worthy of the Victoria Cross, they feature modesty and no taste for self-promotion. Such people are rare, but a few exist in the medical profession; all healthcare workers are potential candidates. Despite my cynicism about some doctors, the possession of these traits is something for the medical profession to be proud of.

Two doctors won the Victoria Cross twice: Lt. Col. Surgeon Arthur Martin-Leake, and Captain Noel Chavasse.

Captain Charles Upham, a New Zealand soldier, also won it twice but was not a doctor.

Elizabeth W. Harris (1834 – 1917) was the only woman (nurse) to receive (a replica of) the Victoria Cross (1869). Queen Victoria gave special permission for this. It was for her bravery during a cholera outbreak in India. She remains the only woman to be awarded a VC of any description. Only since 1921 have women been eligible to receive a VC. Not one has been won since.

Serious All The Time (SATT)

These doctors and nurses are never light-hearted, and working with them can be hard work. Some appear to be weighed down by the gravity of their status and knowledge. Many share aspects of Asperger's. See also: 'PITA.'

SD: **S**hifting **D**ullness

This applies to those nurses and doctors who float around from one place to another, looking busy, but doing very little. They usually move in the shadows (many holding a clip-board), hoping to avoid work, decision making, or any sort of meaningful involvement. As avoidance behaviour, they may hope not to have their incompetence discovered. Many are simply lazy. Paradoxically, some become well thought of. This is because they have employed the simplest of strategies - minimal offensiveness.

Slobs

Doctors and nurses are not immune to smelling offensive ('body odour'). This will offend both patients and other staff. It will fall to someone brave enough to point it out to them. They might suggest that soap is cheap, or that regular washing is a usual necessity. It is common in nerds and those with a solitary nature. Never overlook the possibility of their deteriorating mental health or psychiatric illness; uncleanliness can result from failing to cope, depression, and psychosis.

If you are unsure of what a slob might be, go to YouTube, and search for Harry Enfield's *The Slobs*, Wayne and Waynetta. I have met only one or two doctors who emulated them: they failed to wash consistently, only rarely changed their clothes, comb their hair or tidy their rooms. Some will ignore piles of unwashed dishes, and in years gone by, never emptied their ashtrays. In the 1960s, many doctors and nurses smoked; it helped them cope with the stress of medical work (one of the few benefits of smoking), while ignoring lung cancer risk (a concept then in its infancy).

Because behaviour provides an accurate guide to the true nature of a person, the disinterest of slobs in cleanliness says a lot about them and their attention to detail. Is there a connection between a tidy home and a tidy mind? The Victorians thought so. They thought 'cleanliness next to Godliness.' Obviously eccentric, absent-minded intellectuals, with their minds on higher things can overlook self-care. Others are just lazy.

During World War 2, Alan Turing, while making ground-breaking advances in computer technology and cipher analysis (requiring the ultimate sort of tidy mind), would often arrive for work at Bletchley Park wearing pyjamas. Unkempt, he may have been, but a genius none-the-less.

Someone I trust most as an expert on human behaviour (C.D.) once told me:

'Employ no-one who comes for an interview wearing dirty shoes. If they don't take care of their shoes, they won't take care of their duties.'

Smart

Shrewd people always do their reconnaissance. They come to know their true situation from every angle, and then correctly position themselves to take advantage. The medical profession is replete with shrewd guys, all trying to feather their nests (professionally and financially). This makes recognition of this trait an important one for the naïve, tender-hearted, weak-willed, trusting, and inexperienced. It is especially important for nerds whose minds are elsewhere.

In Confucianism, smartness alone is not enough; it must combine with 'Ren 口', to form wisdom. Crudely translated, 'Ren' is the virtue of combining understanding with humanity.

Smoothies

Oleaginous, suave and self-assured, they often speak in mellifluous tones, and ooze composure and confidence. They are mostly harmless.

In both films 'Doctor in Clover' (1966. Rank Organisation), and 'Doctor in Love' (1960. Rank Organisation), actor Leslie Phillips plays the archetypal smoothie.

Spanophilia(c)

A term used by Richard Asher *(Talking Sense.* 1972. Pitman). The word is used to describe an undue interest in obscure medical conditions. Some doctors will find any excuse to introduce their much loved diagnostic hobby horse into every discussion (it thus appears to be an obsession). This can be harmless as a hobby or a research interest. It otherwise risks detracting from a balanced view of a patient's clinical picture. Found in researchers, spanophilia can motivate discovery.

One of my 'A' level biology teachers, Dr. W.B. Broughton, was a renowned UK grasshopper specialist. He gave his name to one: Chorthippus broughtensis, one of three English grasshoppers, then named. Dr. Broughton taught me between 1959-61, and because he was searching for a fourth grasshopper, had all his pupils, including me, scouring Epping Forest. We failed to find an undiscovered one. He was a memorable, inspiring teacher.

Star-Struck

Revering fame is not a passing weakness. For many, it fills a need for worship and reverence. I have seen patients fall stage-struck and starry-eyed at the feet of celebrity, although not all are sycophants. A chance meeting with someone famous might be their only chance of glory, if only reflected glory. Although a feeble source of self-esteem, it will enliven their story-telling. As a sign of their devotion, some will perform any task requested by their idol. They will happily demean themselves for one small scrap of recognition, and one faint promise of good-will.

Stern

According to Mark Forsyth (*The Elements of Eloquence*, p4, 2013, Icon Books) 'Stern people dislike rhetoric, and unfortunately, it is stern people who are in charge: solemn fools who believe that truth is more important than beauty.'

Some stern academics get weighed down by the gravity of their status and knowledge.

The incompetence of some stern people will lie hidden. Many fear them and lack the courage to challenge them. Few naturally stern characters have a hidden sense of humour. Some might explain that it was only after suffering too many fools that they adopted a stern outlook.Some then find it expedient to presume that everyone is stupid until proven otherwise. Adopting a stern approach is an excellent way to cut short inane conversations.

The perfect image of a stern doctor is *Doc Martin* (played by Martin Clunes in the UK, TV series (2004-2009), the creation of Dominic Minghella. In reality, Martin is not at all stern; he is just a great actor.

Stern characters often display other unpleasant traits: rudeness, arrogance and incompetence.

Stressed Doctors

'... those restless thoughts which corrode the sweets of life.'

Izaak Walton. *The Compleat Angler* (1653).

The work of doctors and nurses can bring them enough stress to induce panic attacks, anxiety, depression, and an inability to cope. They may have had to work long hours, with sleep deprivation, and had to suffer trouble at home, problems with their finances, and illness in their family. This may have caused them to get tired and exhausted and to look haggard. The face of US President Richard Nixon after Watergate, and that of Tony Blair after Chilcot, (facing the threat of legal action against his actions 13-years before), once provided graphic examples.

After long periods of stress, some get depressed. Others make promises they cannot keep, lose their sense of humour, and their sense of timing and reliability. They may fail to 'get themselves together'. To make matters worse, they will often make more mistakes than usual. They will need to sleep and get help with what has caused their demise. Many take to alcohol, and a few to drugs; neither help in the long-term. They are prone to 'Knight's Move' decision making and action. These are ill-researched, life-changing moves, taken to relieve the pressure and urgency they find unbearable. Some will suddenly resign their job, move house, join the Foreign Legion or go off to help in a leper colony with very little fore-thought. They can be strangely resistant to accept the help they need most. They may see accepting help as an admission of weakness and a loss of face. When recognised, we should advise them, support them and help them whenever possible.

Structured and Unstructured Thinkers

There are those who always need a plan to follow. Without one, they can lose their composure and become anxious and insecure. They don't mind being in a maze, as long as they have a map to go with it. What they abhor is a lack of guidance. Some feel secure only when they are following rules.

Structured thinkers need to work in a strictly defined environment, with accepted rules and regulations to follow. They will get anxious if called upon to use their nous, to think for themselves, or to make judgements and decisions based on insufficient information.

In order to galvanise their people together, both Jewish and Islamic religions, have long provided compendia of rules and practices for every occasion (for Jews, The Shulchan Aruch; Sharia for Moslems). The most devout follow them to the letter.

There are others with no wish to be restricted by rules and regulations; they prefer the freedom to think, to be creative and to invent new things. They enjoy lateral thinking and solving problems, and will enjoy engaging with the mysterious and the unknown. In flying terms, they can 'fly by the seat of their pants', and like the first pilots, will find out what counts as they go along. They know the value of experience and never lose sight of it when thinking. As adaptable, resourceful, and creative creatures, they see few limits to their problem solving.

Stüm(mer)(*Yiddish. Pronounced: 'shtummer.' Derived from 'mute' in German.*)

A taciturn person, with whom conversation is hard work. They are mostly silent, non-responsive types who cannot, or will not, express their feelings. Some are good listeners. Some are shy initially, and will 'open up' later. They become activated only when discussing their favourite topics. Some are autistic. I have always found it strange that stümmers are often credited with superior knowledge.

There can be a sinister side to a patient remaining mute. A person being dominated may be afraid to speak, especially in the presence of their dominator / dominatrix. The history they give may be filtered. Safeguarding issues then arise. The central questions are whether anyone else is in control, and whether that control needs to be challenged. What are the motives, and who is to gain? Never jump to conclusions: those patients with retarded development or senile dementia need well-intentioned guidance.

'SS': Supercilious and Sanctimonious

These are occasional traits found in doctors and nurses. These attitudes may be 'in-born', and for political reasons, not always on display. They are much more likely to be observed while doctors are dealing with their patients than with their colleagues. Some share features with OVIBOI types. There is a cultural element to these traits, associated with wealth, education and security.

Swimmers and Sinkers

After being thrown into deep clinical water, the coping ability of an inexperienced doctor or nurse will soon be clear.

My first surgical registrar was a visiting Australian surgeon, Colin Davis ('Col'). He told me on my first day of work: 'Dave, I'm a cutter. I know very little about medical matters. If it's a medical case, it's all down to you. If it's surgical, call me.'

'Col', handed me the opportunity to find out what I was capable of; a chance very few will ever get these days in the UK, regulations and paternalistic attitudes being as they are. Having found myself in deep water not knowing whether I could swim, I soon learned to swam away and never looked back.

I would argue that if medical schools wanted to choose medical students capable of coping well with medical practice, they might expose them to cases in A&E, before accepting them as medical students. When thrown into fast-running, deep water, only the lucky and the strongest swimmers will survive, and save the lives of others.

I know, all this sounds non-egalitarian and over the top, but reflect on the fact that the practice of medicine IS a serious matter, and IS often dangerous, and courage is necessary to handle it well. While recognising the need for equality of opportunity, lives are at stake, and medical practice is no place for the weak-minded or weak-hearted. There are, however, thousands of jobs in the medical world where shuffling bits of paper is the much less risky primary function.

Team Players

Like the rest of humanity, we can divide doctors and nurses into loners (squash players, golfers, and chess players), and team players (rugby, and football players). Both types have made notable contributions to science, medicine, and medical science.

Sporting doctors who are team players will thrive in multidisciplinary units. When I worked there, the pacing unit at St. George's, Hyde Park Corner, was one such place; although my only claim to team playing was having rowed in an eight for my school and medical college. Although a loner, I much enjoyed working in a small, tightly integrated team of capable professionals. Top-rate collaborators make development work easy.

Unfortunately, NHS bureaucrats with little or no clinical wisdom once considered disbanding one such crack clinical team. In 2016, they tried to close the Paediatric Cardiology Unit at the Royal Brompton Hospital. They doubtless thought it uneconomic. It called into question whether they understood anything of how doctors, nurses, and patients work best together. In December 2017, perhaps having learned something of how well-developed teamwork counts, they reversed their decision.

Giving authority to bureaucrats with insufficient medical knowledge, no experience, and even less ability is surely an insult to stupidity.

How much more untutored intervention do UK doctors and nurses need to suffer from bureaucrats? Read my book *'The NHS. Our Sick Sacred Cow'* to read what I think, and what I think should be done about it.

Thin Skinned. Thick Skinned

Those doctors and nurses impervious to criticism, we call thick-skinned. They have either learned from experience how to counter criticism, or have enough confidence to ignore it.

Ex-US President Donald Trump is reported to have 'A fat ego in a thin skin.' (PennLive, March 2017: Bob Quarteroni). Trump says of himself, 'I have very strong, very thick skin.'

When Trump was president, was he easily hurt by criticism, and did he react in resentful ways? With his experience of running businesses, I think it just as likely that administrators who knew much less about the real world of business than him, caused him endless frustration.

Types A & B

Although 'time urgency' (Type A behaviour) is an important behavioural characteristic of doctors, it is perhaps more important to recognise it in patients. I have deferred describing these traits until the next chapter.

USBN: USeless But Nice.

A few doctors and nurses seem not to know much and are inept at performing practical procedures. They find venesection, lumbar puncture, and establishing i.v. infusions too challenging.

I once worked with a charming, much older junior doctor, who co-opted his colleagues to help him with every practical procedure. His ambition was to become a child psychiatrist. He saw no point in practising procedures he found difficult, that were unpleasant for patients, and were irrelevant to his future. He was wealthy, and to salve his conscience, often had Fortnum and Mason's hampers delivered to our ward. We helped him and forgave his ineptitude.

Uptight and Buttoned-up

Many people are unrevealing by nature and keep their cards close to their chest. Doctors and nurses are no exception. They will want to know everything about their colleagues, but tell them nothing about themselves. Those with an autistic trait are different; they may have little or interest in others and have no need to reveal anything about themselves. Some doctors are shy, others are hoping to gain control. All of them present a communication problem.

Visionaries

Leonardo da Vinci and H.G. Wells were visionaries.

I was lucky to have encountered a few visionary doctors (Dr. Aubrey Leatham, Dr. Peter Nixon, Dr. Alan Gardner, Dr. Paul Kligfield, Dr. Pim de Feijter, Dr. David Baxter, and Dr. Mike Hodges). All cultivated the future, while working in the present. All were determined to make their vision come to pass. Visionaries rarely benefit from consulting with those grounded in the past, or those with no vision of the future. If you have the courage for it, tag along with one for an exciting ride.

Wimps

Yes, there are doctors and nurses who dislike the sight of blood, and who will run in the opposite direction when emergencies occur. They can feel ill, hyperventilate, and collapse at the sight of someone sick or injured. They will soon reveal themselves once they start work. If they don't consider a career in management or psychology, they could spend time as a patient.

Wise Sage

We will all need wise advice at times. We all get stuck while trying to make a diagnosis, or with a practical procedure, and need help. As doctors and nurses progress, it is important for them to identify those who can help most, like those adept at practical procedures. Go to them and ask for help. There is no need to feel inferior or inadequate. By asking a wise, capable person, you help three people: the patient (who will suffer less trauma); yourself (who will have gained knowledge), and the one you asked (who will gain self-esteem).

Chapter Four

About Patients

Patients do not want to be seen as numbers (the patient in bed 15), or diseases (the patient with MS), but as individual sentient beings.

Many patients need help to help themselves.

Patients should choose doctors whose source of gratification and self-esteem comes from improving patients' physical and mental wellbeing.

Communication between patients and doctors are at best when all differences in culture, intelligence, knowledge and behaviour are resolved by mutual respect and a desire for understanding.

Clinical judgements, and the giving of pertinent advice, are essential clinical arts. Understanding disease is merely a matter of science.

I have made an important assumption in this chapter. The patients referred to are conscious, and able to communicate without cognitive impairment. I have not addressed the safeguarding of those with reduced mental capacity, and those who are unconscious.

My description of patients appeared first in my book 'Doctors, Nurses & Patients (2024). For those who have not read it, I have reproduced much of it here.

The characters doctors will meet as patients overlap in type with those described in my compendium of doctors as characters. We populate the same human jungle, and failing to learn about its inhabitants can have serious consequences.

All doctors will have met notable characters at school, and at medical school, but in medical practice they will meet other characters as patients; some they may be ill-acquainted with. Sharing the same educational background will make relationships with colleagues, partly predictable. Relationships with patients are another matter. Doctor-patient relationships are the subject of this chapter.

It has always been a challenge to discover the true nature of others, and to see past the image they choose to portray. In the past, gaining insight into the character of a medical colleague was a relatively easy matter since many shared a similar social background. Today, medical students and patients in the UK, come from more diverse social and cultural backgrounds. This diversity can make mutual insight more difficult to achieve.

By choosing to distance themselves from gaining insight into their patients and their circumstances, some doctors will become less effective clinically. The best clinical judgements are based on the most complete exposure of the pertinent facts about each patient; the perspective of a fly-on-the-wall will help. The further away one gets from the scene of action, the less reliable our judgments become. Exposure, limited to the medical facts alone, puts patient management at risk. To practise medicine holistically when required, needs time spent getting to know each patient. The need for improved corporate efficiency in the overburdened NHS, has meant that consultation times have become shorter. Patient through-put and efficient patient handling are the bureaucratic measures now sought, even though they can oppose the quality of patient care.

The primary aim of medical practice is to reduce patient morbidity and mortality. Some wealthy nations have gone further than the relief of symptoms; they have invested in prevention and the early detection of disease. The statement *'prevention is better than cure'* may be a worn-out cliché, but it is held dear by many patients. Patients have always seen sense in making early diagnoses, even though for many decades, the focus of the NHS was only on overt disease. Over the last five decades, doctors have slowly accepted prevention as a new role. When I first offered medical screening to my patients in 1973 (for early detection and prevention), my colleagues thought it a waste of time, and merely a money making exercise.

For most doctors, the satisfaction derived from helping patients is enough reason for them to continue work. Helping patients to overcome their symptoms is one major source of self-esteem. Participation in an effective clinical team is another. There are other longer-term benefits in private practice, like the accumulation of patients who become friendly supporters.

All jobs done well can provide a sense of worth, but not much compares with taking responsibility for the health and life of others. Apart from doctors, nurses, and paramedics, everyone with patient contact can share in this, whether they be floor cleaners, porters, or those who serve patients' food. Few medical professionals have difficulty justifying their existence; what they cannot justify and most dislike is being taken for granted, side-lined and undervalued.

Desirable Patients?

The best personal relationships require the least effort.

There are a few key patient characteristics which determine the doctor-patient relationship. Foremost among them is educational attainment. Patients vary in the vocabulary and language skills they have for describing their problems. Age, gender, socio-economic class and culture can also contribute. Coincidentally, these are also strong determinants of longevity and morbidity.

The form consultations take can depend on location. I found quite a difference (in the 1970s) between those patients who attended St. George's Hospital at Hyde Park Corner, London SW1, and those who came to St. George's Hospital Tooting, in South London. The inhabitants of these areas may have changed in the last fifty-four years, but distinct contrasts remain (had the hospital at Hyde Park Corner still been open. It closed in 1980). Some doctors prefer a particular patient demographic, and must give due consideration to where they choose to work. This is no trivial matter since demography has significant associations with disease prevalence and outcome.

Gender at birth determines the prevalence of particular diseases, and the doctor-patient relationship needed. I have often found that female patients (with 'XX' chromosomes) are more open; more in touch with their emotions and more brave in medical matters than men. Even when afraid, more women readily agree to necessary medical interventions than men.

Age influences the doctor-patient relationship. Dealing with children is a specific challenge, not every doctor or nurse can handle well. So much so, some children will refuse to see certain doctors. There is no reliable relationship between advancing age and wisdom, but outspokenness becomes more common with age. At my age, I have nothing to lose by expressing my opinions; I am not looking for a job, and have no duty to please others. When dealing with the retired, prepare for comments prefaced by 'When I was a young . . .' or 'Back in the day . . .' What will follow might be of interest, but may lack current relevance.

There is much to understand about the differing cultural attitudes expressed by the rich and the poor. I will comment on their relevance to the doctor-patient relationship later.

Which aspects of a patient's individual personality will bring joy or dismay to those caring for them? For me, it was always how interesting, charming, appreciative, experienced, passionate and intelligent they were that brought joy. It was how demanding, pugnacious, devious, arrogant, or obsessional they were that brought me dismay. It also mattered to me whether they were users, abusers, or seducers, and whether they were honest professionals or dishonest scoundrels. Like all others, doctors and nurses will have their own list of characters they like and dislike. Not all will be of clinical relevance, but they can all affect the patient-doctor relationship.

No doctor or nurse can choose their patients directly, so any question about which patients might be best, is rhetorical. Our calling is to serve all who suffer.

I was once told by a shipping billionaire what talent he needed most to succeed in business. It was a talent for choosing people (implying an ability to understand others).

Mrs. Anita Griggs, Head of Falkner House School, London, can choose the five-year-old girls, most likely to excel as educated, successful women, whatever path they choose to follow in later life. For many decades, the reputation of her school has depended on her ability to judge the potential of young children. She once referred to it as 'a black art'.

In all fields of human endeavour, it is easier to deal with friendly, helpful people. If they are well-informed, intelligent, perceptive, and have insight, so much the better. As far as patients are concerned, it helps if they are willing and able to learn, and will accept critical discussion. Successful medical management depends on patient compliance, and that will usually (but not always) follow from them understanding their medical issues. Some see no need to understand; they trust doctors and have faith in their judgement.

What Do Patients Need to Know?

All patients should consider this question: how can I get what is best for me from the doctors and nurses dealing with my case? We all have preferences for how we wish to be treated by others. Here are a few things patients need to know.

Because it makes light work of every consultation, most doctors prefer pleasant, intelligent patients, capable of giving a focused account of their symptoms and observations, rather than descriptions dating from *'once upon a time'*. My preference was for patients who welcomed time spent on discussion, whether it be their history, my assessments, diagnoses, suggested investigations, or my proposed management strategy. Patients who need quick-fixes will do just as well with a nurse practitioner, physician associate or pharmacist, rather than a doctor. In my practice, I mostly dealt with complicated cases; those cases other doctors had failed to diagnose or help. To solve their more complicated problems there were two key elements:

- Prior knowledge of the patient and their pertaining circumstances, and

- Time for in-depth discussion.

Most patients are a pleasure to meet. Very few will cause heart-sink and be tiresome to deal with. I met only two such patients in over fifty years. As a matter of duty, we must attempt to deal with all-comers, but the more reasonable and polite the patient, the more doctors and nurses will warm to them and want to help. Those patients who are consistently unpleasant will be lucky to establish a mutually rewarding relationship with anyone. It may be better for them to transfer to another doctor. A surprise may await them. They may end up with a doctor who is even less tolerant of their discordant behaviour than the first. Since doctors get the patients they deserve, and patients get the doctors they deserve, some patient redistribution is inevitable.

Patients can change their outlook and attitude daily, depending on the pain, anxiety, fear, suffering and contentment they are experiencing at the time of consultation. Nevertheless, my preferences (in no particular order) for those patient characteristics which I believe promote a pleasant and successful patient-doctor relationship are:

1. Succinct, objective historians.

2. Honest and dependable.

3. Respectful.

4. Intelligent, logical, and educated.

5. Modest.

6. Pleasant, and sensitive to others.

7. Humorous.

8. Coordinated, with efficient executive functionality.

9. Relaxed.

10. Not overly obsessive.

Mutual compatibility is the key. It defines the nature, progress, and prognosis of every doctor-patient relationship.

Full of what seemed to be ridiculous, unshakable ideas, Ms. S. G. was feeling worse day by day. Her local NHS hospital failed to diagnose her condition. She was, however, too angry, and insufficiently literate, to recount her history meaningfully. She didn't want discussion; she wanted action. In desperation, she demanded admission to hospital. She had felt ill since giving birth to her son a few months earlier, and felt she needed something done fast. I diagnosed Sheehan's syndrome: postpartum pituitary / adrenal malfunction. Paraphrasing Spike Milligan's epitaph: she knew she was ill.

Her clinical situation was urgent (hypotensive, and hyponatraemic), so was her medico-legal situation. Out of desperation, she was ready to accuse everyone of incompetence. Since nobody had diagnosed her condition correctly, many of her accusations of incompetence were well-founded. As it so often is, the main diagnostic clue was in her history, and verified by the investigations. Following the correct diagnosis and hormone replacement, she made a rapid recovery.

Ms. S. G. was too sick to display any of the helpful patient characteristics listed above. Once she recovered, she displayed them all.

Beware: it is difficult to see the truth about a person or their clinical situation when they present in a cloud of frustration or anger, with the illogical demands they may incite.

Effective clinical practice (that which satisfies all concerned) will often depend on our understanding of patients and their understanding of us. In particular, they need to know that we are knowledgeable and capable enough to make happen what needs to happen.

It Takes all Sorts

'Men go to gape at mountain peaks, at the boundless tides of the sea, the broad sweep of the rivers, the encircling ocean and the motions of the stars: and yet they leave themselves unnoticed; they do not marvel at themselves.'

St. Augustine: *'Confessions.'*

From a doctor's point of view, there are several ways to classify patients, none of which are worthwhile unless they help with their clinical management. It is not a doctor's brief to sit in judgement on patients, but it would be naïve to ignore a patient's psychological and personality attributes and the influence they can have. The challenge is to deal with all-comers, and to benefit as many as possible.

Patients, doctors and nurses all have characteristics that vary between two extremes. These characteristics can change from time to time. Aristotle, in his writing on ethics, says 'every virtue is between two extremes' (now thought of as the 'Golden Mean'). The virtuous state that lies between these extremes he called *euthaimonia*: a state of 'good spirit'. He describes courage as a virtue, and on either side, the vices cowardice and recklessness.

The extremes of human character I think important to medical practice (for both patients, doctors and nurses) are:

Academic – Non-academic
Aggressive / combative – Placid/ non-combative.
Arrogant – Humble
Courageous – easily frightened
Balanced views – Polarised views.

Bi-(multi)lingual – Native tongue only.
Charming – Uncouth.
Vain - Modest
Communicative – Non-communicative.
Co-operative – Non-co-operative.
Cowardice - Recklessness
Cranky ideas – Well-founded ideas
Eager to say: 'I know my rights' - 'I know my duties.'
Educated – Uneducated
Egoist – Reserved
Empathetic – Apathetic

Extravert – Introvert
Focussed - Unfocussed.
Friendly – Unfriendly.
Honest – Dishonest.
Humorous – Humourless.
Ignorant - Knowledgeable
Intelligent - Unintelligent.
Literate – Illiterate
Low - High Socio-Economic class.
Numerate – Enumerate
Objective – Non-objective
Obsequious – Challenging
Obsessional - Easy-going
Open minded – Closed minded
Opinionated – Uninterested
Professional – Unprofessional.
Pugnacious – Meek.
Sympathetic – Unsympathetic
Threatening – Supportive
Worldly wise – Closeted.

And for those with psychiatric states:

Deluded / Hallucinatory – Sane.
Depressed / Withdrawn – Happy / Open.
Fearful – Brave.
Aggressive – Peaceful.
Relaxed – Anxious.

For those with drug states (alcohol and drugs of abuse)

Dangerous – Harmless.
Deluded / Hallucinatory – Well oriented.

> Under the influence of drugs – Sober.
> Uncontrolled – In control.
> Violent – Non-violent.
> Etc., etc.

Despite the artificial nature of this list, and the considerable overlap between the entities, it is easily possible for each of us to choose the type of patients we can deal with confidently. The same applies to patients able to choose their doctor (except for private patients, this choice exists only outside the UK).

One of the most charming and valued, long-standing patients of mine was Joe Satchell. I attended his funeral in September 2016. I wanted to pay tribute to him here. As a measure of the man, Joe would send flowers to my receptionists for no reason other than he had cancelled an appointment. In this, he was unique among my 20,000 patients. We missed his presence as a patient and friend.

Contrast Joe, with those who try to claim compensation against doctors and nurses, on what may sometimes seem unreasonable grounds. It is now more easily possible with *pro bono*, 'no win, no fee' legal arrangements. The NHS receives 10,000 claims for compensation each year, and recently faced paying out £4.3 billion in compensation. Many of these claims for medical negligence are substantiated. For some patients, this is one way to make a small fortune.

Every disease has its risks, as does every medical practice intervention. Doctors and nurses must resign themselves to being easy targets, given the mishaps that are bound to occur when performing the inherently risky work of investigating and treating patients. No intervention, including a consultation, is without risk. In the utopian view held by many medical managers and regulators, most of whom have never practiced medicine, medical practice should be risk free. The utopian place they seek may exist, but not on this planet.

In the 1970s, Ivan Illich noted that litigation has made even first-aiders think twice before acting as a Good Samaritan.

Medical regulators claim that their work reduces the risks patients suffer at the hands of doctors and nurses. It is not their brief to acknowledge all the risks intrinsic to every human activity. Like some patients, they stand ready to accuse doctors of exposing patients to unnecessary risk, even if they have never managed patient risk themselves. Their statutory power allows them to punish medical professionals for what they see as 'not in the public interest'. This idea is metaphysical, undefinable and untenable. It is that every citizen deserves the right to live in a utopian world, free of risk.

Although no doctor can choose their patients directly, private patients can choose their doctor and hospital. In primary care NHS practices, they will give patients a choice of GP, but that rarely guarantees which doctor they will see. In private practice, patients choose who they consult, and can change their doctor without retribution.

Patient choice should be based on medical reputation, but is more often based on availability. Hopefully, the reputation of a doctor will be based on knowledge, experience, and an ability to deal successfully with medical problems. It is more likely, however, that affability will bring the most favour. Once a patient has developed a worthwhile rapport with a doctor, availability then becomes less of an issue.

For a clinical management plan to work well, the advantage lies with patients who understand their condition, the time-scale involved, and the objectives of any proposed treatment.

From a patient's perspective, the acceptability of the doctor-patient relationship is all important. Patients may not know that the reliable measures of a doctor's ability are their diagnostic proficiency, case management efficiency, and the improvements they can make to their morbidity and mortality.

One can never satisfy every patient, but our duty is to try. Try as we might, there will always be some we cannot help, simply because of personal incompatibility. In over fifty years, I encountered only six such patients. I thought it better to refer them to other doctors. I am in no doubt that there must have been others who, having sensed our mutual incompatibility, chose never to return.

The threshold to doctor-patient incompatibility has changed in recent decades. The trend is now for patients to consult 'Doctor Google', and to think they then know best, even with no medical knowledge, technical know-how, medical experience or clinical perspective. This has gained momentum as younger patients have replaced older ones. It parallels the growing obligation to accept social equality in favour of educational inequality. For everyone to have a valid view is fast becoming acceptable. Antipathy towards expertise and science has accompanied this. Social media has given both the informed and ill-informed a voice, and the doctor-patient relationship must accommodate to the change.

To espouse equality makes complete political and social sense (it gains votes, and pleases the majority), but makes no biological sense. Equality has never existed in the animal kingdom. The basis of evolution is the inequality between species, allowing only the fittest to survive. Inequality has allowed human superiority to evolve, and explains why we have lasted so long. To accept utopian social equality is to deny biological reality for expedient political reasons only.

Based on what we now call democracy, the UK referendum of 2016 went ahead. Was the decision to have a referendum based on the assumption that every member of the public knew best how to weigh technical economic factors (trade balances, monetary policy, effects on prices, and taxation), and understood every security, political and social issue? The public had to decide whether the UK should remain a member of the EU. Surely, even by the best informed found these complicated issues difficult to understand. Unfortunately, political votes are based on popular biases and emotive issues (like the suggestion to remove all foreigners), and are only rarely evidence-based. Understanding medical issues is just as difficult for many patients.

Patient Scenarios

'All the world's a stage.'

As You Like It. Act 2, Scene 7. William Shakespeare.

On every stage, both the actors and the scenery make a play come to life. Regardless of our character, we all act out our own life scenario. Some scenarios are inescapable; others, especially for the privileged, are a matter of choice. Whether at home or at work, some scenarios can trigger the onset of a disease or change its progress. Whatever are the key differences between a wealthy scenario and a deprived one, a five-fold difference in cancer and cardiovascular disease prevalence (the social health divide) results.

Although we have yet to define all the genotypic details of every individual, we need not ignore unhealthy scenarios of aetiological significance. To understand and deal with patient scenarios is to practise the art of medicine. Such considerations gave me a clinical advantage. Most patients respected me for taking an interest in their circumstances, although initially, few drew a connection between their circumstances and their medical plight.

Many patients conflate health and disease. Although they are separate entities, disease can affect health and diminished health can affect disease (depression can enhance cancer progression, for instance). Diet can affect health, but is of doubtful relevance to the aetiology of the commonest 'killer' diseases (cancer and atherosclerosis).

In biology, exceptions drive evolution. One diet-related exception is that dietary salt can trigger hypertension; another is that increasing carbohydrate intake can reveal type-2 diabetes. That does not mean that diet alone causes either condition. Once we accept that diet causes these conditions, we must explain why so many who do not get hypertension or type-2 diabetes, have eaten salt and carbohydrate for decades without an adverse effect.

It is important to recognise the aetiological difference between trigger factors and the effect of long-term influence. In biology, factors like allergen exposure triggering asthma are common. One exposure to an allergen can stimulate a deluge of entrained, long-lasting, immune and leukotriene responses in the bronchioles. Also, long-term 'wear and tear' is a significant factor in the progression of osteoarthritis. Both processes can combine to create chronic conditions.

Some triggering theories are controversial. For instance, can one cigarette induce lung cancer? Can one virus infection start a chronic auto-immune disease? Time-links affect aetiology and susceptibility. Boys do not go bald at thirteen years of age, but many men will see some hair loss after the age of twenty. Acne appears at puberty, not usually before. Some women become hypertensive only with their menopause.

Love Lost and Lonely

Joe was a 65-year-old widower. He had everything he wanted in life, except a partner. His wife had died from leukaemia some years before. His regular gardener had fallen ill and a woman, forty years his junior, replaced him. They fell in love at first sight; not something either expected.

He had sought my clinical advice over many decades, and we had a close doctor-patient relationship. His last consultation concerned his angina. This time, he wanted my pastoral advice, given all I knew all about his personal circumstances and medical condition. His question was: 'What should I do about this unexpected new relationship?' I replied with a rhetorical question: 'What have you got to lose?' This agreed with his own view of the relationship. Not only had it relieved his bereavement and loneliness, it had unexpectedly improved his angina.

Following a CABG many years before, we failed to place a stent in Joe's extensively narrowed, distal anterior descending coronary artery. I told him then we could do no more for him. It was a surprise to observe, therefore, how his new relationship had improved his exercise tolerance.

Joe's arteries could not have changed much over the short course of his new relationship, so how might one explain his improvement? Could it be that happiness had raised his exercise threshold to angina by reducing his catecholamine drive and myocardial oxygen demand? It is certainly worthy of some further research. Heberden's original description of chest tightness (angina) in 1772, is interesting; it included emotion as a provocative factor. (Heberden, W. *Some account of a disorder of the breast*).

There is another interesting snippet of cardiovascular research history relating emotion to the heart. It has mostly escaped notice, but I found it while reading a translation of William Harvey's book, made famous after he described the circulation of blood (*de Motu Cordis*, 1628). Harvey's main concern was the connection between human emotion and the pulse rate, and not just the circulation of the blood. After all his research, he could find no connection. That is because the autonomic nervous system connecting the brain to the heart, had yet to be discovered. That would not be until the 19th century, and the work of many, including Claude Bernard, A.V. Waller, the Weber brothers. It was J.N. Langley FRS, who coined the phrase 'autonomic nervous system' (Langley J. Observations on the physiological action of extracts of the supra-renal bodies. J Physiol. 1901; 27:237-256).

If physicians delegate all consideration of their patient's emotional state to psychologists, psychiatrists, and counsellors (bureaucratic rules apply for referring patients to certificated experts only), it could prove to be a mistake, although a collegiate, interdisciplinary approach, can work well for patients. Here is an unusual example of the collegiate approach in action.

Mr. B. was wealthy enough to own his own small island in the Caribbean. Ten years before I came to know him, he developed prostate cancer, and was not now responding to treatment. He organised and paid for an annual international conference to discuss the subject. He invited many G.U. specialists who enjoyed a brief holiday on his exotic island. His only request was that they reviewed his case management for one session of the conference.

Pre-Patients

Not all diseased patients are symptomatic. This presents a problem for doctors interested in early diagnosis. Early diagnoses can prove irrelevant, however, if without a prognostic advantage.

Take as an example, the detection of atheromatous disease, using harmless carotid arterial ultrasound screening. As a simple screening test, this detects the disease process responsible for almost 50% of all middle-aged deaths in the western world - atherosclerosis. Using this technique, one can detect it decades before it causes symptoms, and well before any sign of coronary or cerebral ischaemia (using exercise ECG, a stress echocardiogram, or perfusion study). 'Statin' drugs, given early enough, can reduce or stop the progress of atheroma, and can reduce the risk of future heart attacks and strokes.

Abnormal blood lipids levels are factors in the well-known, predictive cardiovascular risk calculation, 'qRisk3'. Although of epidemiological value, the factors used can detect only 60% of individuals with carotid (or other) atheroma. Tossing a coin offers the 50% chance of a correct result.

The treatment and life-style changes motivated by the early detection of a potentially fatal disease like atherosclerosis, has the potential to reduce morbidity and mortality, but that potential needs to be proven for publicly funded patients. My private patients had a head-start; I prescribed a 'statin' (HMG-CoA Reductase Inhibitor) to all those with proven atheroma (the actual cause of risk), even if they had a 'normal' total blood cholesterol (and especially if they had a positive family history and a low HDL cholesterol level). Because neither biological clocks, nor atheroma can be reversed, my patients accepted that a 20-year advantage would be worth paying for (given the small relative cost of an innocuous ultrasound test). The only downside was the treatment. Fifty percent of patients taking a statin drug experience side-effects. The early discovery of atheroma had other worthwhile benefits. It provided motivation for weight loss, dietary improvement, and an improvement in exercise tolerance.

No NICE, or other guideline exists for the management of proven atheroma in pre-patients with a normal blood cholesterol. I suspect it will take another decade or two to recognise the need to detect atheroma, and to treat it, rather than just to lower a raised total blood cholesterol (a product of liver metabolism). I started treating it regardless of total blood cholesterol in the year 2000.

Those antipathetic to screening are properly concerned about inducing anxiety and turning healthy people into patients. Screening will more often re-assure, simply because the commonest outcome of screening is to find nothing of clinical interest.

Doctors and public health authorities try not to scare the public. The media, however, more readily spread fear and panic, and not always with scientific justification. As the coronavirus proceeded across the world (from March 2020), one newspaper suggested that we were all about to face a 'killer' virus. That led to many patients with pre-existing respiratory and other serious conditions to self-isolate. They retreated to their bunkers, fully stocked with food and toilet rolls (panic-buying reduced available stocks).

Although there has been some shift in opinion in the last fifty years about the value of screening (NHS GPs earn a fee-for-service basis for simple screening), the prevailing medical culture (the standard medical model) remains centred on the diagnosis and treatment of symptomatic disease. Since many doctors are happiest when making an 'interesting' diagnosis (according to Ivan Illich), they are not much enthused by what they perceive as pointless medical screening. The benefits of routine mammography, cervical smears, PSA testing, BP evaluation, and diabetic checks are, however, difficult to rail against.

The finding of a 'worthwhile' early case can counter negative attitudes towards screening, but only if the outcome will benefit. There remains a divide between doctors wanting to 'put out fires', and those who want to 'prevent fires occurring'. Since they are not mutually exclusive, why not try both? Surely, long exposure to the consequences of disease should motivate doctors to prevent it, but strangely it may not.

An ongoing discussion in the public sector relates to the cost effectiveness of preventative medicine. This has never applied to the private sector, where patients get screened for common sense reasons. Those spending public money must first confirm sufficient diagnostic accuracy and the prognostic value for any intervention proposed. My private patients decided fifty years ago that screening was sensible; they were happy with the reassurance and navigational advice it provided. The diagnostic accuracy on which such advice is based is an important issue every pre-patient should consider, but few question. Most of my patients accepted screening as insurance.

There are some practical problems with prevention. Too few people (potential patients) tempt providence by asking: '*is there something wrong with me that doctors could manage better if found early?*' In my practice, many came for cardiac screening after a friend or relative had died unexpectedly. Obviously, someone should have made a prior assessment of the one they lost. The fact is, we often saw the wrong people (those who were anxious, but with low risk).

Since it is common for couples to present themselves for screening – the patient and their anxious protagonist – it is not uncommon for the asymptomatic protagonist to be the greater beneficiary.

The inflexible nature of our medical culture is a problem. Medical scientists must answer a basic question about screening: 'is there any evidence for a prognostic benefit?' Does it reduce mortality or morbidity? Despite the lack of definitive answers, my patient's desire for medical and cardiac screening never dampened. Most were aware that analyses of large group data would overlook the benefit to some individuals. Many saw screening as the dutiful thing to do, given the responsibility they had for others (their family and employees).

The wealthy often see the cost of screening as a small price to pay for a chance to preserve their life, and to maintain their chosen lifestyle. They will have made many similar 'common-sense', strategic judgements about their future in business, based on much less evidence. An important motivation for these decision-making patients, is the desire to do whatever they can to allow a better future for themselves, their families, and their employees. What may be in the 'public interest', is not their concern. They are, of course gambling when spending their money on the chance that early detection will lead to an improved prognosis. They see this as a straightforward gamble (until we find solid proof), and a worthy one (given that they can easily afford it). Many of my patients ignored the many medical academic controversies about screening. The personal stand they had to take on issues of future significance (without sufficient evidence), had led many to become successful in business.

There are those who never give thought to their car tyre pressures, engine oil level, or radiator fluid, before setting off to drive. Others give no thought to saving money for their retirement. Many believe that winning the National Lottery, or receiving a windfall from a distant relative will provide the support they will need in later life. Instead, a little intelligent forethought might prove better; at least it is more reliable than hoping for the best, ill-informed gambling, disinterest, and absent-mindedness.

During the tragic bush fires in Australia in late 2019, Steve Harrison, a resident of Balmoral, New South Wales, constructed a ceramic coffin-sized kiln, just in case a bush fire overwhelmed his property. Because he put a fire extinguisher and several bottles of water into the kiln, his neighbours thought him over-anxious. His forethought saved his life. He hid inside the kiln for thirty minutes as a raging firestorm swept by.

Patient Demography

The socio-economic demographic of every medical practice varies with location. This is important because of the strong association of socio-economic grouping with disease prevalence, and clinical outcome. Both cancer and cardiac infarction incidence are three to five times more common in deprived areas (the health divide). Since poverty is a major aetiological factor in the occurrence and prognosis of all the major causes of death in the western world, it should be a major political issue.

Two centuries ago, the rich were obese, and the poor were thin. The rich lived longer lives, and the poor shorter lives. To-day the rich are thin and the poor are obese, but the longevity advantage of the wealthy remains. Which reduces longevity most, obesity or poverty? The obvious answer is poverty, but the current politically correct answer is obesity.

Being a Patient

Many patients are apprehensive about medical interventions. They also see pharmaceutical drugs as harmful, with adverse effects on their immunity, gut bacteria, fertility and future health. Some continue to drink alcohol, smoke tobacco products and take street drugs, but see no irony.

The acceptance of ideas can depend on need. Those desperate to find a cure, may resign objectivity and believe in myth and hearsay. The value we each assign to objectivity varies with educational background and socio-economic class. In my early years, I lived in a working class environment, and was only later let loose among the middle classes. I was soon to learn that myth and hearsay less frequently influence the educated.

The newly wealthy, seeking the excitement of something novel are especially prone to accept uncorroborated, expensive ideas; any need they have for self-indulgence will bias them. When seeking predictions, they will readily follow in the footsteps of ancient rulers who visited Delphi. They thought the Oracle could predict their future. Many still turn to spiritualists, astrologists and Tarot card readers to predict their future, even if the advice they get nearly always lacks specificity. In their quest for health and wealth, some use healing crystals and the power of pyramids. My oracle tells me that the future of these practitioners is assured, simply because most myths, legends, conspiracies, and crazy theories, contain a modicum of truth.

Nowadays, there are many websites exaggerating anti-medical, anti-science sentiment. Despite the advancements science has made, superstition still gains widespread acceptance. The effects of cold drafts, hot weather, barometric pressure, time of eating, and vegan food, all have their protagonists amongst those who dismiss scientific evidence.

Montenegro has a high proportion of smokers, yet what concerns many Montenegrins is the health risk of drafty conditions (Promaya).

My patient, David L. had chronic back pain. He went on holiday to Antigua, and for the first time in years, became pain free. On returning to the UK, his pain quickly returned. He then went back to Antigua for more pain relief. In the end, repeated trips failed to help. He came to realise that it was partly the relief of stress that had helped him, not the food or the Caribbean waters in which he swam. I later discovered the true cause of his back pain: ankylosing spondylitis.

Big Pharma often gets bad press. This contributes to patient distrust, made worse by news of how much profit they make. For the sake of making that profit, they must push the latest thing, even when old, cheaper remedies, will still work well.

As an example, big Pharma once vigorously promoted factor X_a inhibitors as new anticoagulants (in 2016, each apixaban tablet cost £2.10 in the UK). At the time warfarin cost 4 pence each, plus the cost of prothrombin time blood testing. Initially, the anti-coagulant effect of taking a factor X_a inhibitor, had to be taken on trust, in the absence of any measure of efficacy equivalent to a prothrombin time. I initially resisted prescribing them, partly because of their intense promotion. Out of the first five patients I treated, three developed complications (one had a large spontaneous haematoma; one haematuria, and another unacceptable nausea). They are now regarded as superior to warfarin in stroke prevention for those with chronic AF. I began using them only after the publication of many double-blind trials. I still worry about the stroke risk given no individualised bleeding tendency assessment.

The apprehension some patients have about new drugs is sometimes justified. Although doctors may resist their concerns, every patient's fear of drugs needs acknowledgement. Doctors should also acknowledge the debt we owe to the pharmaceutical industry. They deserve our gratitude, for without the many therapeutic advances they have made, where would modern medical management be?

Most patients bravely face the consequences of our clinical conclusions and medical advice, despite having to accept most of what doctors say on trust. The closer they feel to disaster, the more willing they will be to accept what we advise. I have always admired the courage shown by patients being approached by a doctor holding a needle, endoscope, or scalpel. Being seen as deserving such trust depends a lot on how doctors engage with patients and use both the art and science of medicine.

Two months after I started work as a surgical houseman (on the 30th July 1966), I admitted a forty-year-old man as an emergency. He had acute chest pain after swallowing a small dental plate with three false teeth attached. He had been watching the now famous FIFA World Cup football match in which England beat West Germany 4 : 2. England won after extra time. At the final whistle, the patient had jumped to his feet, and in his excitement, detached and swallowed his dental plate. It lodged half-way down his oesophagus. Post-operatively, I told him we tried hard to remove his dental plate (using oesophagoscopy, under GA), but failed. Regretfully, we had to proceed to a thoracotomy. 'That's alright', he said, 'It was worth it. England won!'

The vast majority of patients were once very respectful of medical training, and the hours worked by doctors and nurses. In 1966, doctors never counted their hours of work. Many consultants had served in the second world war. During wartime, they could not stop work until they had dealt with every case. They expected their juniors to do the same.

A recently expressed, bureaucratic point of view, challenged this now historic situation. 'Who would want to be seen by a tired doctor?', they asked. My person answer would now be anachronistic: 'I wouldn't mind if she was knowledgeable and dedicated.'

Patients give doctors and nurses their trust, with little more than faith to rely on. Although all doctors have diplomas and certificates on their walls, stethoscopes draped around their necks, impressive tomes in bookcases, and a lot of consulting room paraphernalia to inspire confidence. Patients deserve capable, dedicated, energetic doctors who they can trust; their energy levels are important to their performance, but can be of secondary significance when lives are at stake.

When the going gets tough, the tough get going. Those with devolved patient responsibilities who sit in offices from 9 to 5, on a six-figure salary, might find a vocational call to duty, a little beyond their ken.

Clarity and Confusion

Effective doctor-patient communication is vital. Unfortunately, some patients get confused; some cannot understand, and some cannot remember what they are told. Although not demented, some may have a limited vocabulary and be barely literate. Another problem is the difficulty we all have to concentrate after receiving bad news; anxiety and fear both diminish concentration. At such a time, a lack of mutual acquaintance and cultural differences, may not help empathetic understanding. Those with psychological or psychiatric problems may find it particularly difficult to communicate. Unfortunately, not all doctors are versatile enough as 'performers', to adapt their language and approach to the needs of each patient.

If a doctor or nurse does not identify with their patient, and cannot understand them, why should a patient accept them? Some patients will find reassurance in knowing that the GMC has qualified them for work, but ultimately it is the trust built on continuous exposure that best allows a patient to agree to any advised medical management.

Continuity is an invaluable commodity. Although the NHS has steadily diminished its availability, this is not yet the case in the private sector. In traditional private practices like mine, the relationship built on continuity was partly what patients paid for. It remains to be seen if this will also apply to private 'walk in' clinics. If private medicine becomes incorporated like the NHS, and the focus becomes profit and processing efficiency, a lack of continuity might follow. Without enough consultation time and continuity, practising the art of medicine is not easily possible.

Managing Patient Information

The amount revealed to each patient, in how much detail, and at what time, are all of clinical significance. Patients must always have enough information to assess the advice we give and enough time to see the sense in any advised course of action. One cannot standardise the process. Every action must be judged on its individual clinical merits. Adverse information can increase a patient's rate of demise; positive, encouraging information, can help speed their recovery. Information itself, can hold prognostic value. You would be right to ask for the evidence for this proposition, but those posing the question will usually be clinically inexperienced or detached.

How much detail should we reveal to each patient? There is no standard rule, but it should depend on the patient and their circumstances. The amount divulged could be all, nothing, or an amount varied by knowing a patient. In order to gauge the amount appropriately, personal acquaintance with the patient is irreplaceable. Ideally, one will have formed an idea about the patient's knowledge, and their ability to understand, especially information that includes technical detail. It is also useful to know their likely emotional reactivity, bravery or stoicism in the face of upsetting information. One should know something of their ability to cope (with or without friends or family). Although controversial, I believe it critical to know how much information to give to each patient, and how best to manage it.

There are those who insist on patients being told everything, regardless of the content and possible emotional impact. Some think there is no place for restricted disclosure, and no need to be concerned about the timing of information delivery. They want to divest themselves of responsibility and not leave themselves open to criticism or accusations of non-compliance. These are perhaps the views of detached doctors, unhappy to take responsibility for personal judgement, and unwilling to play an empathetic role.

The degree of clinical revelation suitable for each patient needs wise judgement. According to our regulators, total compliance with guidelines and bureaucratic rules is now the *sine qua non* of clinical practice, not clinical judgement. This irreverent, ill-educated approach, neatly excludes the art of medicine and the judgements that flow from it. Total revelation of patient information is now the accepted norm, even though it wrongly presumes that all patients understand it equally and can cope emotionally. Clearly, this is nonsense, and an extension of the political quest to convince us all of our equality.

Some believe that information given as a matter of compliance, or as a legal requirement, need not be humane. Duty stripped of humanity may be *de rigueur* when applying the rule of law, but is inappropriate for an attached doctor, dedicated to improve the lives of sentient beings.

The act of divesting oneself of someone else's personal, possibly life-changing information, removes the need to take personal responsibility. It can be cathartic for sheepish rule-followers, and those eager not to be caught using their judgement by regulators. The dehumanised transfer of clinical information to a patient may well comply with anonymous 'best practice', but it can fall far short of what is in an individual patient's best interest. Every master of the art of medicine will exercise clinical discretion: how much information to give, and how much to withhold, all of which will vary with each patient.

If the sovereignty and sacrosanct nature of medical practice in the lives of patients is further eroded by corporatisation and bureaucracy, doctors can look forward to a bleak future. They will be progressively trusted less to think for themselves and have their clinical judgements respected.

As Confucius would have had it, it takes knowledge together with humanity (Ren:☐), for knowledge to qualify as wisdom.

'... it is diligence, and observation, and practice, and an ambition to be the best in the Art that must do it.'
The Compleat Angler
Or the Contemplative Man's Recreation. Chapter 17. Izaak Walton.

My colleague, Dr. A. G, was trying to explain to an elderly Jewish couple (Yiddish was their mother tongue), that he (the patient, Mr. Cohen) needed a new pacing electrode. Dr. A.G., told Mr. Cohen that the right ventricular endocardial threshold of his pacing electrode had been rising steadily. Being told this, and other jargon-based announcements, left him and his wife looking troubled.

My Jewish colleague, Dr. Paul K. from New York, was sitting close by, listening with me to Dr. A. G's explanations. We were in an open-plan, out-patient clinic, empty except for the five of us. Paul rose to his feet and wandered over to where Dr. A. G. and his patients were sitting. With a strong New York accent he asked, 'Would 'ye mind if I said a few 'woyids' to Mr. and Mrs. Cohen?' 'Be my guest', said Dr. A.G, realising the difficulty he was having communicating with them. Paul addressed the patient and his wife directly. Standing in front of them, with both hands outstretched, he entreated them, saying, 'Mr. and Mrs. Cohen, you're in good hands. 'Whad are 'ye worried about?' The patient and his wife rose to their feet, and with beaming smiles clasped Paul's hands, at the same time saying, 'Thank-you doctor, that's all we wanted to know.'

A doctor's patients are not allocated randomly; practice location and its associated demography, partly pre-selects them. Other variables apply. If you work in private practice, more business owners than employees will consult you. Directors or business owners are usually the richer of the two, and are used to having their wishes granted. The employees may have private medical insurance, but only as a company perk. Directors, used to decision making, will want to know all the facts and be involved in the decision-making progress. Regardless of their position within the company, the information you give will shape their further requests and expectations, and determine what they think of your management. Because their company sends them, employees will mostly be less critical and more accepting of all you say. There may be another issue, however. Some may be worried about what medical revelation might mean for their job security.

Before the NHS existed in the UK, most patients depended on charitable donations while in hospital, and expected to show grateful, submissive behaviour. The NHS had only been running for thirteen years when I entered medical school, and patients were still submissive. It was not unusual for doctors to be rather formal and superior in their approach. On reflection, it has not been a bad thing for doctors to have loosened up, although this has made the jobs of some doctors more difficult. Instead of adopting a single, formal approach to patients, some doctors have had to master different approaches. Doctors must now earn the respect and faith of patients, whereas once they received it automatically.

Modern media allows widespread comparisons to be made, opinions to be sought, and feedback to be circulated. There is now much more to live up to, and much more information to disquiet the apprehensive. Patients are free to wander beyond their level of understanding. It is good that the exchange of media information can expose bad medical practice (albeit, with many false allegations), but it will do little for those patients who are already apprehensive about their illness, or fear becoming ill.

The demography of the area in which a doctor works will not only determine the likely prevalence of disease but also the type of characters to be met. Age, gender, education, socio-economic class, culture, race, family background, and wealth, all feed patient expectations, and the style of communication they will find acceptable. Being able to deal with these distinct groups requires understanding and sensitivity to others; an art that needs practice. To succeed in medical practice now, versatility is essential.

Michael was a builder. His company built some major parts of the London Olympics stadia in 2012. He was a hard-working, worldly wise Irishman from Cork. He had no university degree, but he was super-smart. One day he came 'to get his painful foot fixed'; he was suffering from acute gout. I explained about uric acid crystal formation in his joint tissues. I further explained about inflammation and the benefits of anti-inflammatory drugs, as well as about business stress as a causative factor. He said nothing much; thanked me and walked out.

I was later told by my receptionist what he said before leaving. 'I don't know what the f***ing hell he was talking about in there (my consulting room); something about crystals and inflammation. All I came for was to get my foot fixed.'

I had wrongly assumed that Michael wanted a pathophysiological explanation for his pain. I was later told that where he came from in Ireland, people still revere doctors and trust them without a second thought (not so much the case in the UK). It must have been his considerable trust in me that made my detailed explanation completely superfluous. For patients who trust their doctor this much, faith might obviate the need for explanation. After this, I asked patients if they wanted a detailed explanation. Many did not.

In the UK, after the case of Dr. Harold Shipman, legalistic medical regulators came to regard trust in doctors as a potential danger. With the power vested in them, algorithmic functioning and little knowledge of medical practice, their mission is to protect the public from potentially dangerous doctors and nurses.

Clashes of Socio-Economic Class

Differences in culture, education, and socioeconomic background can help to define the mutual compatibility and acceptability of patient-doctor relationships. The relationship will depend on the mutual desire each party has to overcome their inter-personal differences.

Although my upbringing was in an under-privileged, working-class area of East London (Walthamstow), my family were self-employed and financially secure. My contemporaries at medical school were all middle-class, and except for two of us, were ex-public school. In the course of our medical education, I did not notice any of my fellow students struggling to communicate with the less privileged, but then the era was one of unquestioned respect and deference towards doctors.

While at 'The London', we students were among the first to be taught 'communication' (1963). They taught us to engage with the Cockney patients who then outnumbered all other groups in Whitechapel, Bethnal Green, and Hackney (East London). The well-known affability and adaptability of Cockney's made the job of communication easier. I have worked with both the under-privileged and the privileged, and have found no difficulty with either, although I found it challenging to adjust to the mentality of those with considerable wealth and social privilege.

In the 1970s, I was caring for a very wealthy business owner whose claim to fame was the invention of the 'radiogram'. In the 1930s, his product was the first to combine a record-playing deck with a radio, enclosed in one smart piece of furniture. His father had been a furniture maker in Naples, and his own sons became manufacturers of their own well-known brands of furniture.

He owned a mansion, and after suffering a stroke, he had a gymnasium built within. He thought this would aid his recovery. When I next visited him, he had made a good recovery. I thought it a good time for him to take a holiday. The question I put to him was innocent enough. 'Perhaps it's time for you to book a holiday? To this, he laughed, pointed to his telephone and said, "I guess you know what that is Dr. Dighton? When I want to go on holiday, I lift the receiver and speak to my pilot. I might ask him to take me to my house in Bermuda or I might ask him to take me to my villa on Corfu. When you have your own private jet, and several houses abroad Dr. Dighton, there is no need to 'book' a holiday".

He taught me a lesson about the very wealthy I have never again drawn on.

Sometimes, patients do not go to their GPs for medical advice; they want a certificate for work, or a chit for an eye test. Conversation and discussion is not what they want; their need is purely administrative.

Very early in my career, I did GP locum work in Stratford, East London (1967). The patients would come in expecting to get a blue certificate, a yellow chit, a green form, or a pink prescription. I cannot remember what these various bits of paper were for. I tried to enquire about their health, but their responses suggested I shouldn't bother.

When asked by the wise old Irish doctor (a Cork man), whose practice it was, 'How are you getting on young Dighton?' I replied, 'I have found it rather difficult to talk to your patients'. Hearing this, he leaned back, and with subdued exasperation worthy of any dramatic actor, took one long draw on his pipe. After exhaling a diffuse cloud of smoke, he said something I would never forget: 'It's a grave mistake to talk to 'em!' He had clearly gained the measure of his patients, and they of him.

What could he have meant? Was it plain Irish wit? Perhaps he preferred only to talk with educated people and found lesser mortals boring. Maybe he was weary of general practice, having dealt too long with trivial matters. Was this why he minimised his communications with patients? If a doctor reaches this point in their career, they should retire or change jobs. Since an accurate diagnosis much depends on accessing a thorough history, any disinterest in communicating with patients will diminish clinical effectiveness.

Victim or Patient?

As a lecturer in cardiology at Charing Cross Hospital in the 1970s, I worked with a charismatic doctor. Dr. P.N. was an inspired, insightful cardiologist, with ideas ahead of his time (and his colleagues). His interest was how changes in life circumstances, and in particular the stress of relationships, might cause cardiac infarction. He was certain that most heart attacks owed something to a mixture of anger, disappointment, and resentment (among many other responses). He observed patients who often denied any personal responsibility for their fate; like many doctors, they saw a heart attack as nothing to do with their circumstances or behaviour.

When he first approached a patient in our CCU, his first question would usually be: 'Who is trying to kill you?' A negative answer, a denial, a blank look, or a dismissive response, would cause him to say: 'Well, you'll just have to stay here until we find out. If we don't find out, how are you prevent a further heart attack?'

I am not sure, but I think Dr. P.N. had some personal experience to draw on. He would also quote Rex Edwards (*Coronary Case, A Personal Report*.1964), a reporter who wrote about his own stress before a heart attack. Academic controversy over any connection between stress in the aetiology of disease continues. How it affects the symptoms of established disease is less controversial, but not always considered by doctors (they are not social workers). Here are three anecdotes in historical order, suggesting a connection between illness and stress.

In late 1943, during WW2, Churchill was pushing his allies to continue the Mediterranean offensive. The Allies voted against his plans. They mounted operation 'Overload', the campaign to invade northern France, before pushing on to Berlin. Churchill said that he was being crushed between a Russian bear (Stalin), and an American elephant (Roosevelt). Churchill soon developed pneumonia and had two small 'heart attacks'. The latter diagnoses are doubtful, given the sparse use of ECGs, and a lack of awareness of cardiac infarction at the time (there is almost no mention of cardiac infarction in the textbooks of the era). Perhaps they meant something different when they used the term 'heart attack.'

In the film, 'A Nun's Story', a hypothetical relationship between stress and TB is suggested. It was based on the true life story of Marie Louise Habets. It came to cinemas in 1959 (Warner Bros.). The story depicts a nun (Sister Luke, played by Audrey Hepburn) conflicted between her religious duty and her calling to be a nurse. She gets pulmonary TB, and her surgeon colleague (Dr. Fortunati, played by Peter Finch) suggests that her inner conflict is the disease, and TB, merely a side-effect.

During the fierce US Presidential campaign of 2016, which she lost, did Hillary Clinton's immune system react adversely? Like Churchill, seventy-three years before, was her pneumonia a personal reaction to the stress of failing to become the first female president of the USA?

'Life' - The Play

The Shakespearian character Jacques, in '*As You Like It*', Act 2, Scene 7, says:

'All the world's a stage, and all the men and women merely players: they have their exits and their entrances; and one man in his time plays many parts...'

The various parts patients play have varying effects on their health.

With the same weight attached, weak elastic stretches more than tough elastic, and is more liable to snap. In this simple physical analogy, the weight attached represents the physical *stress* and the amount of stretch the *strain*. This relationship, known to every GCSE physics student, has a bio-medical equivalent. Life events, like marriage, divorce, or bankruptcy as stresses, can produce a spectrum of 'strain' effects, different for each individual. I have seen these events cause no change, an increased susceptibility to infection, more migraine, resistant hypertension, worsening angina, psoriasis, worsening IBS, and depression. Stress can trigger many conditions, but perhaps only in those predisposed to it. Are these the responses of the weak? Do such responses ever occur in those made of sterner stuff? Can genetic predisposition predict adverse responses to stress? Stress may not cause disease, but I have seen it affect its progress.

Every patient will play many parts throughout their life, with their psychological state and illnesses influenced by their fellow 'players': those with whom they choose to have relationships. Dr. P.N., at Charing Cross, specifically thought that responses to prevailing circumstances (psycho-pathophysiological and biochemical changes), could affect the course of coronary artery disease. He thought that among the pertinent responses were a 'depleted' (or pro-inflammatory) immune system, exaggerated clotting, and changes in fluid dynamics (BP, and changes in blood viscosity associated with increased catecholamine output).

Dr. P.N. and I disagreed about the relative clinical contribution of stress to cardiac infarction and coronary atherosclerosis. He preferred to exclude mention of atherosclerosis from our presentations to other doctors, claiming that he did not want to dilute his message that '*stress could worsen angina, and induce heart attacks*'. He wanted no mention of the fact that without atherosclerosis, cardiac infarction is a rare event. Cardiac infarction is nearly always due to fissuring of an atheromatous plaque in a coronary artery. After this, platelet aggregation and clot formation (seen to be induced by stress in animals) can combine to block an artery. In the heart, this block can stop coronary blood flow and cause myocardial infarction (in the absence of sufficient collateral circulation). Stress can cause a depressive state, reduce walking distance in those with angina, and can predispose to cardiac infarction, perhaps by increasing blood coagulability. On these, we agreed.

For acutely ill patients, psycho-social-pathophysiological influences, culture, and character are hardly initial considerations. In chronic conditions, they are more pertinent. Although enquiring into these factors can affect the consulting experience for both doctors and patients, it may not always change the outcome. Liken it to the ambience in a restaurant. Ambience can alter the experience, but not the taste of the food (although, according to chef Heston Blumenthal, it can). Psycho-social enquiries can reveal important subsidiary factors that might in part explain a patient's demise. Although such enquiries can be too personal by some doctors, they allow for targeted advice for each individual. In my experience, such enquiries mostly strengthen the patient-doctor relationship.

In cases that do not respond to usual management, consider asking: 'Is there any evidence of previous responses to stress, and what are the patient's current stresses?'

Can a patient's response to changing psycho-social factors alter the natural history of their condition? Difficult to manage cases, and those who do not improve, often have relevant circumstance and relationship related problems. Re-visit the history, and broaden the enquiry (with permission) to include the views of friends, workmates, and relatives. At least you might recognise some contributing factors and understand why there has been a delay in improvement. You may come to appreciate the medical value of social service agents, able to change a patient's circumstances for the better. There are many patients with no intention of changing their circumstances, even if it offers the prospect of improvement.

I will tell you how to cure your headaches and high blood pressure, Mr. Jones. It is obvious, but difficult. Divorce your wife. Tell all your demanding children to stand on their own two feet. Sell your business and retire to the Caribbean. At least go somewhere far enough away from all those who aggravate and frustrate you. Make a fresh start. You only have one life, Mr. Jones. Your life could depend on you becoming 'a new man' with a normal blood pressure.

That too few patients would accept such extreme advice, doesn't make it wrong; just impossible to follow. Patients will have many reasons not to take heed. Some will have emotional inter-personal issues, and an unwillingness to relinquish their responsibilities. Others will prefer to keep their *status quo*, regardless of any threat to their life. Although the advice might seem to be a joke, it could stimulate a patient to think more about their situation and reasons for their continuing illness. By making such suggestions, one can avoid responsibility for the patient's repeated failure to improve. Expect some to react badly to such advice and feel alienated. Others will respect your personal interest and the attempt made to help them. Those with no insight will not appreciate the irony.

An uncompromising approach can sometimes work best to foster change. The aforesaid Mr. Jones might acknowledge the point made, but for reasons related to trust, loyalty, commitment, responsibility and practicality, he may have no choice but to ignore the advice. By so doing, he will have accepted his destiny, and perhaps, continued to suffer. That you obviously understand his situation, and have his best interest at heart, is likely to engender some respect for your perspicacity. You might earn his trust, but create resentment and anger among his friends and family.

Simple, off-the-cuff remarks can strike deep, and sometimes bring about surprising change.

Ralph had become a friend of mine. He was a moderate smoker. His amiable, jovial character lifted the spirit of many in his presence. His father had been a heavy smoker and died with COPD. One day, Ralph developed bronchitis. After examining his chest, I made a casual remark. It was that his chest was 'going the same way as his father' (note: I had decades of medical practice continuity with him, his father, and his children). I then thought no more of it.

Six months later, he came to me for an examination. I asked a simple question: 'How many cigarettes do you smoke now, Ralph?' 'Smoke!' he replied indignantly, 'After what you said to me?' Somewhat perplexed, I enquired: 'What exactly did I say to you? Ralph continued. 'The moment after you said that my chest was going the same way as my father's, I stopped smoking forever! I would have been mad to continue.'

An unexpected life-changing response to one simple, off-the-cuff remark.

TATT (Tired All The Time), Fatigue and Exhaustion

Tiredness is the commonest of all symptoms, but disease is not the only cause. All disease processes consume energy, so tiredness becomes likely; poor sleep, however, is a commoner cause.

Mental tiredness can only come from spending more energy on brain functioning (solving problems, stress, anxiety, and obsessional behaviour) than on the replenishment of 'normal' sleep (or meditation or relaxation). Sleep disorders (including those induced by stress) can cause tiredness. The more severe degrees of mental tiredness can influence disease pathophysiology. (See a reprint of the article I wrote on sleep in 1980: Appendix B.)

By considering the relative amounts of rest and exercise undertaken, we can easily understand physical tiredness. Sometimes, the toxic effects of disease also cause muscular tiredness and weakness.

Many textbooks are wrong: the commonest causes of tiredness in the UK are not anaemia, hypothyroidism, or diabetes. The commonest causes relate to stressful circumstances, and associated sleep disturbance.

My former boss at Charing Cross, Dr. P.N., was an ex-army officer who had worked with the SAS and paratroopers. He came to understand how fatigue affected a soldier's physical and mental performance: he saw both cognitive and emotional functioning disturbed prior to accidents happening. The relationship between road traffic accidents and driver tiredness is now established. For this reason, lorry drivers have had long driving hours restricted. Dr. P.N., applied his army experience to patients, and always assessed their tiredness and any contributing psycho-social factors.

Few doctors acknowledge tiredness as a relevant clinical factor, although all accept it as a secondary effect of illness. Dr. P.N. found that many patients reported increasing tiredness in the year prior to their cardiac infarction or stroke. He associated this with an inability to sleep well, while preoccupied by stress. Since sleep-deprived rats have an increased liability to cardiac infarction, could this apply to humans? Assuming it does, his method for helping the exhausted was to replete their energy with sedated sleep (all night, and most of the day, for at least three consecutive days). If they had accompanying ECG changes of ischaemia, he would diagnose 'pre-infarction syndrome' and would anticoagulate them. Things have moved on considerably since then. These days, we perform tests that reveal the coronary artery anatomy and intervene to improve blood flow.

While working at Charing Cross in the 1970s, I needed to define the degrees of tiredness.

- Simple tiredness, I defined as the feeling of depleted vitality (physical, or mental). There is a need for rest, since the potential still exists to perform well in response to everyday stimuli.

- Fatigue is tiredness without the potential to increase performance when motivated or stimulated.

- Exhaustion is tiredness, so extreme, that performance diminishes with arousal and encouragement. In this energy-depleted state, often caused by sleep deprivation, further stimulation risks physical and mental breakdown (equivalent to elastic snapping when stretched too far). This failure can present clinically in various ways: panic attacks, psychotic episodes, cardiac infarction, uncontrolled hypertension (if predisposed), a TIA or CVA (with predisposing factors). The most common outcome, however, is an increased liability to infection (septicaemia, pneumonia, or the recurrence of herpes simplex, etc.). I have also seen patients with worsening psoriasis, and more frequent migraine, accompanying stress induced tiredness.

In the Crimean War, Florence Nightingale noticed the fate of soldiers. If they became injured soon after arrival, their wounds quickly recovered. Specifically, their wounds did not get infected and did not lead to fatal fever (septicaemia). She also noted exhaustion amongst those who had fought for a long period, and had suffered the terrible climatic conditions of cold muddy trenches. She noticed they quickly succumbed to septicaemia from one small scratch.

Changes in immunity, hormone levels, and coagulation can occur with sleep deprivation; these can then affect both morbidity and mortality. Although mainly the subject of anecdote, I have witnessed many equivalent scenarios. There is an important point to be made. Tired, fatigued, or exhausted patients, will not readily improve without sufficient sleep. In the long-term, they will not always improve without some pertinent alteration in the circumstances that led to their demise. To be effective, any alteration made must prevent further fatigue and exhaustion. During consultation, it takes time to consider these energy factors; often more time than most doctors have available.

I mentioned this case before, but in a different context.

Mrs. Couch was an anxious person. Although she had had abdominal pain for months, she reported of it to others, only one week before seeing me. Abdominal examination revealed a large, tender mass in her abdomen. She had a partially obstructed bowel. Apart from an i.v. infusion, and a Ryle's suction tube in her stomach, the surgeon I referred her to, left her untreated over a weekend. He had prior personal arrangements. The following week, he became strangely indecisive and intimated that performing the required operation was not for him (the hospital staff and the family had harassed him for a plan of action).

Another surgeon operated on her six days after admission. At this stage, the patient had suffered months of fear and worry, thinking she had cancer. Her six days in hospital had made her more worried, and she had become exhausted. She died from septicaemia on the seventh post-operative day. Histologically, the mass revealed a combination of acute diverticulitis and colon cancer.

Unnecessary delays amplified this patient's stress, anxiety, and fear. She remained unmanaged for too long, and her energy depletion ignored. Although not identifiable at post mortem, any resulting reduction in her immune response would have significantly contributed to her death.

Alongside their technical expertise, no master of the art of medicine would ignore personal human factors.

Nail in the Foot Syndrome

A nail in the foot will cause pain eased, only when removed. After that, recovery can be remarkably quick.

Josephine was in her 60s, when I admitted her with high blood pressure (consistently greater than 190/110). No medication had reduced it. She told me she was the 'queen bee' of her family-run, furniture business. She led me to believe she was indispensable, and although she wanted to retire, she couldn't let her family down.

After a bedside discussion that established these facts, I left her to return to my consulting room. Coincidentally, her husband arrived. I requested a chat and suggested he should let his wife retire, even though she was indispensable. 'Indispensable?', he said indignantly. 'I've been trying to get her to retire for years. She simply refuses to go.' The solution was thus in hand. 'Let's go back to her room', I said. 'Allow me to say that I have persuaded you to let her go, while acknowledging how difficult the situation will become for you.' He agreed and allowed her to retire with her dignity in place. Thereafter, I controlled her hypertension with small doses of medication.

What can seem to be a 'cure', or a substantial improvement in a medical condition, can sometimes result from modifying just one psycho-social aetiological factor.

Conditions with a stress related, psychoneurotic element, commonly present with tiredness (TATT: **T**ired **A**ll **T**he **T**ime), and/or depression. As a doctor trained in hospital medicine and cardiology, I did not readily take to the psycho-social evaluation of my patients. I became convinced of its value after witnessing many therapeutic benefits.

L'impasse

Doctor: *"I am very pleased to tell you, Mrs. Smith, all your test results are within 'normal' limits. Therefore, there is nothing wrong with you" (despite all your symptoms).*

Patient: *'Then why do I feel so ill, tired, and weak, doctor?' (meaning: you have failed to convince me you know what is wrong with me).*

This doctor now has some options:

1: On the presumption that he has missed something, he can return to square-one, retake the history, and re-examine the patient.
2: He can dismiss the patient with the implication that she is wasting his valuable time.
3: He could survey her psycho-social landscape for aetiological clues.
4: He could refer her to an appropriate colleague. Beware, though, if that colleague is not interested in circumstances as aetiological agents, she might return feeling un-helped. If she returns, she may want to discuss what your colleague said. I have always found this unacceptable. I preferred to refer my patients to colleagues who would leave no loose ends for me to resolve and explain later.

The Games People Play

Solitude or a relationship? Although a common lifestyle choice, few give much thought to it.

Many choose relationships, even though many result in conflict. As gregarious beings, many of us prefer this to isolation and loneliness. Relationship problems are among the commonest causes of stress, potentially changing the natural history of any disease and its management. Knowing about a patient's relationships can, therefore, be of clinical relevance; they can cause recalcitrance and non-compliance, and sometimes account for therapeutic failure. A loving, supportive relationship often fosters recovery and health.

In 1964, psychiatrist Eric Berne wrote a definitive book (*The Games People Play*), describing the interplay between people. It is a *must* read for anyone interested in understanding why people behave as they do.

Emma Morano, from Lake Maggiore, Italy, was born on the 29th of November 1899. As the oldest living person on Earth in 2017 (117 years old), she gave credit to being single, going to bed early, and eating three eggs every day (one cooked, two raw). She thought her wisest move was to leave her violent husband in 1938.

As a pastoral exercise, someone once asked me to evaluate the compatibility of a couple intending to marry. Apart from mutual physical attraction, the long-term stability of relationships depends on each participant getting their cognitive and emotional needs fulfilled. Why else would anyone want to share their life with another person? The question is, can we measure compatibility reliably?

Mr. J. owned an international engineering company. His headstrong 20-year-old son had insisted on marrying the young woman he had just met on holiday. The advice of four parents was for them to wait a while, and get to better acquainted. The marriage lasted three months and cost his family over one million pounds. When his son came home with another young woman, his father again became worried. Given the potential cost of another failed marriage, Mr. J. insisted on some evidence for their compatibility, and much more information about her and her family. He engaged a private detective for the purpose. He worried about losing his son and even more money.

He asked for my help. After some research, I found a rather old compatibility questionnaire: the FIRO B test (Fundamental Interpersonal Relations Orientation; 'B' for Behaviour). William Schutz introduced it in 1958. I insisted the couple came to see me separately on the same day. I did not want them to confer before taking the test. The results shocked both me and Mr. J.; they strongly supported the view that the couple was a perfect match. When last I asked, they were still happily married, 35-years after their marriage. Mr. and Mrs. J. now have three grandchildren.

Was this a one-off result? Alone, it certainly does not prove the diagnostic accuracy of FIRO testing.

Schultz designed the FIRO B test to assess the inter-relationships of US air force flight crew. While flying, near disasters occurred due to inter-personal conflicts on the flight deck; they diverted the crew's attention. The test provides a useful, semi-quantitative measure of relationship compatibility.

There is a presumption in scientific circles that the age of published research affects its veracity, reliability, and usefulness. These assumptions need review.

The FIRO-B test is still available. The assumptions on which it based are insightful. It assumes that we all have three 'core' needs awaiting satisfaction or rejection by a relationship. They are *affection, control, and inclusion*. If one partner wants to give affection, compatibility requires that their chosen partner wants to receive it (albeit a subconscious need). Likewise, a controlling person is best matched by one who likes to be controlled. The inclusion factor refers to shared values, interests, and outlook. It includes sharing cultural values. Inclusion is the weakest of the three compatibility factors. Affection and control strongly bind people together and any mismatch quickly estranges them.

The test is old, but has lost none of its *raison d'être*. Too few of us understand why we feel compatible, incompatible, happy, or unhappy in the company of some others. Too few know about Schultz's valuable work, and do not appreciate the true nature of the 'chemistry' they feel (or do not feel) in relationships.

A note about current compliance. No cardiologist could now officially employ such a questionnaire. Only a doctor holding a certificate for clinical psychology, or relationship counselling, would be acceptable. Regulators would question a doctor's qualification to use such a psychological instrument, and would regard those who do as a maverick.

Specialisation and certification have brought about the de-skilling of doctors. Doctors must now stick rigidly to their certification. That has meant fewer doctors who can claim to be generalists, able to deal with all-comers. It has depleted the knowledge and experience base of medical practitioners. Another consequence is that patients will find it harder to find a doctor who can help them.

Thinking for oneself and using initiative is key to succeeding with challenging clinical cases, and practising personalised medicine. What we do for some patients needs to be individualised and standardisation is not possible. As useful as NICE guidelines are, doing what has to be done for individual patients, supersedes blindly following rules, regulations, and generalised guidelines. Blocking the initiative and free thinking of doctors with too many rules and regulations, diminishes the effective practice of medicine, while improving it in the eyes of bureaucrats.

We should resist the introduction of corporate standardisation techniques that belong in baked-bean factories. While Matt Hancock was Minister of State for Health, he commended those doctors who performed inventively during the COVID-19 pandemic. Doctors and nurses able to function without bureaucratic control must have surprised him. (BBC 4, The Andrew Marr Show, 5/7/20).

Chapter Five

Patient Charicatures

What follows first appeared in my book, *Doctors, Nurses & Patients* (2024). I reproduce some of it here for those who have not read it.

Interested observers will notice behavioural differences between people, especially when their age, gender, culture and social class differs from their own (social class is still identifiable in the UK, using family background and educational attainment). Those biased to accept the political concept of equality might be less prepared to recognise the wide spectrum of human behavioural differences that exists between individuals based on their genetic profile. One issue is gender which the law now seeks to make either a matter of choice or recognised by genetic profile. Doctors and nurses cannot afford to assume biological equality; they need to regard every patient as unique and different until proven otherwise.

One cannot deny proven biology and the importance of differences between people. Political correctness attempts to suppress this, even though the variations between animals has driven evolution, selecting those fittest to survive healthily and longest. Genetic science and the practice of medicine both attempt to overcome evolution. They do this by helping those who would not once have survived, to live long enough to procreate.

An equivalent need to recognise differences is vital for zoo keepers. They must recognise how each animal behaves differently (even if of the same species); they can then avoid being bitten, stung, or eaten alive. To remain safe, they must adopt some experienced-based generalisations, while admitting to many exceptions. Doctors need to adopt the same principles: recognising different patient types and knowing how best to deal with each of them.

When thinking about human behaviour, we all have biases. Those who only read broadsheet newspapers, watch BBC 2 and BBC 4 TV, and listen to Classic FM and BBC Radio 4, are likely to see others through the prism of UK middle-class values. Our social biases can affect how compatible we feel when meeting those from different socio-economic groups and cultures.

We can distinguish differences between individual social groups and cultures by using our experiences, or by referring to population surveys and statistical studies for guidance. How reliable will be our impression of native behaviour be when visiting a foreign country for the first time? Because our brains have none of the restraints of computer processing, we will freely form associations, impressions and judgements, especially in relation to any danger. For survival, this type of functioning is more useful than calculation (AI partly functions on calculations that seek the probability of difference). Humans can discriminate in several directions at once, with our biases constantly swaying our view; sometimes overrating risk, sometimes maximising hope. These may not be the most reliable of predictive cognitive functions, but they have sustained *homo sapiens* for millions of years.

Generalisations can provide us with preliminary forewarnings. Experience will then suggest whether we should accept or reject them. Not all tigers will eat us when we enter their cage, but it is sensible to assume so. A double-blind trial might help to confirm this presumption about tigers, but who would waste their time and money undertaking it? (The answer is, of course, a government department, or a tax funded corporation like the NHS). Perhaps we should value common sense more, and hold on to the presumption that all tigers are dangerous, even without trial evidence. I also hold (without evidence) that my chances of surviving a jump from an aircraft would be better if wearing a parachute!

The image, or 'face' that each patient presents, is the product of many influences: their rearing, age, gender, culture, biases and personality; their socio-economic circumstances, intelligence, emotional intelligence, fears, neurotic traits and education, to mention but a few important factors. Each of us holds only one genomic hand of cards to play the game of life with. For the moment, we do not have the choice to swap, substitute, or delete our genetic 'cards', although these possibilities may not be far off.

In medical circles, the discussion is ongoing: which genetic and environmental factors contribute to disease?

In all card games, there is a distinct advantage to knowing which cards others hold. Life circumstances might change gene expression (epigenetics is now a popular concept), but no environmental circumstance known will cause a turkey to grow wings, and soar into the sky like an eagle. External radiation can cause cell mutation and cancer, but we do not yet know whether the stresses of life influence atherosclerosis and hypertension enough to cause heart attacks and strokes. The study of epigenetics is a work in progress.

The art of medicine requires a doctor to identify each patient as an individual. Individuals will soon be identifiable using their genetic profile, but to bring the maximum advantages of scientific medicine to individuals, still requires doctors and nurses to appreciate every patient's circumstances and to practise the art of medicine.

In what follows, I have described many of the patient types I have dealt with, and how relevant they have been to the doctor-patient relationship. These details are important, if only to forewarn inexperienced doctors and nurses of the pitfalls awaiting them.

A natural overlap is inevitable when describing patients, doctors and nurses as characters. Perception will change, depending on whether it is a doctor, nurse or patient considering them.

Specific Patient Types

Advantage Takers

Who would park their car in a 'disabled' parking space (just for their convenience) when no disabled person is in the car? Who will jump queues with no regard for the patience of others?

Advantage takers are everywhere. A few will be patients. They try their luck wherever they go, and will try to persuade medical practice receptionists their problems are urgent, just to jump the queue. See also Users.

ALAC: Acts Like a Child

This type of patient will usually have had their consultation arranged for them by a friend, relative, or workmate. Their adopted role is passive, and this can present a problem with their co-operation. They can be in denial, or not agree with the medical problem their friend thinks they have. Paradoxically, they can become indignant if they sense being treated like a child. Some will refuse to accept advice; after all, they may feel no obligation. Those with more humility, will express gratitude to the person who arranged their consultation.

When I took John's full history, he denied any symptoms. He did, however, have a strong family history of coronary artery disease. His wife had sent him to get checked. He was a 67-year-old golfer who played 18 holes of golf, three times every week. I knew the golf course. It was hilly. My investigations included a carotid artery scan (this showed heavily calcified atheromatous plaques, with a 95% occlusion on one side), and an ECG treadmill exercise test which showed a significant ischaemic problem.

I stopped his exercise test after only two minutes because he developed chest tightness and some ominous ECG changes (those usually associated with reduced coronary blood flow: ST depression). I questioned him further: 'I thought you said you had no symptoms? How is it you can play on a hilly golf course and not get chest tightness?' He explained: 'It's easy. I use my buggy, and get off it only to hit the ball. I then carry on. Without it I couldn't play golf.'

He had a small heart attack (cardiac infarction) two days later. His wife, supported by an A&E doctor, thought his diagnostic exercise test had caused it (exercise testing is a sine qua non for diagnosing cardiac ischaemia in asymptomatic patients). He had an emergency CABG (coronary by-pass) one week later. He never returned to me as a patient; both he and his wife blamed me for the heart attack.

He had ignored his failing health and had taken no personal responsibility for his condition. Given that he was behaving more like a child than an adult, he desperately needed parental guidance. The resurrection of his parents was not then an available option.

Our duty must always be to the patient, and never primarily to those who refer them. Doctors who respect professional etiquette above the welfare of their patients, may not agree. Some health professionals prefer to communicate only with the referring doctor, and not to the patient. It can lead to doctors not informing patients of the risks they face. That lacks common sense.

Only the patient has any right to their clinical information. If they wish to give that information to family, friends or the company they work for, that is for them to decide.

ALBer: Arrogant **L**ittle '**B**lee**der**' (Cockney Slang, meaning a naughty or unpleasant person).

For I do not seek to understand in order that I may believe, but I believe in order to understand.

Anselm of Canterbury (Proslogion,1).

St. Anselm's theological view (he lived 1033 to 1109 AD), contradicts the modern scientific view, and that of the French philosophy, Peter Aberlard (1079 to 1142 AD, of the same era), that one can only believe once one understands. Unfortunately, belief without understanding is a common feature of the arrogant.

This type of patient knows best, or thinks he does. Usually demanding, they are rarely disposed to listen. If also a bigot, they will not change their mind and not accept any other point of view.

The arrogant can be compliant, but only when it suits them or when cornered or outnumbered. Presuming they know best, they may comply with their doctor only when their opinion matches. These days, many patients use 'Doctor Google' as their source of medical information, and their compliance can depend on what they have read. Since many believe everything they read in newspapers, and on the internet, convincing them otherwise might prove tiresome.

Arrogant beliefs must pre-date the internet by eons; among them today are that antibiotics diminish immunity and prolong health problems (despite the many millions of lives they have saved); that special diets hold the secret to longevity, and that the avoidance of animal fat will prevent heart attacks. Some believe that gene-modified food and mobile phones cause cancer. There are as many misbeliefs about the promotion of health as there are about disease prevention and its cure.

Amateur Doctors

Experienced professionals in every field of work, will have to suffer amateurs who think they know best. Just occasionally, one *will* know best. With an interest in medicine, science, the law, archaeology, or architecture, many amateurs have contributed novel theories and valuable observations. Many are liable jump to conclusions based on insufficient evidence and limited evaluation.

Jim had been an engineer and amateur brewer most of his life. He came to me with his irritable bowel syndrome (IBS), which caused him stomach bloating from gas collection. He asked if I would give him some anti-yeast medication and explained why. 'I am an amateur brewer, and I understand how gas (CO_2) forms when making beer. Sugar and yeast combine to produce carbon dioxide and alcohol, and I think this is happening in my bowel. If this is the case, I must have a small amount of alcohol in my blood, so could you test my blood for it?'

My reply was that I would research his blood alcohol theory.

Small amounts of blood are indeed detectable in the blood of some who drink no alcohol. It is now called the auto-brewery syndrome. When Jim suggested it to me 30-years ago, I had never heard of it.

After Jim took some nystatin tablets used to treat yeast infections (his idea), he noticed no reduction in stomach bloating. He later told me his two sisters suffered similarly, and the medication helped them.

Jim had come up with some scientifically plausible ideas.

Search engines are now a ready source of information for everyone; even those who have never frequented a library. Verified information, misinformation, and false information are all there on the world-wide-web, so I always encouraged patients to discuss what they had learned. The research they do can be a good starting point, so why not provide them with some useful search terms. In-depth discussion takes time - a valuable commodity, now rationed by the NHS.

Badly Behaved

In consulting rooms, some children will sit quietly while others talk. Others will roam around, opening cupboards, and removing anything that takes their fancy. Not all parents of such children admonish them;

they will usually be responsible for their behaviour. Most parents are euphemistic about their offspring. They might claim that they open cupboards because they are delightfully inquisitive research scientists in the making. There is another interpretation: they have yet to be domesticated. As with many other behavioural traits, family values and culture, shape us through their powerful influence.

When traveling, one will see differences in the behaviour of children from different cultures. On internal Russian flights, you will rarely see a Russian child doing anything other than sitting quietly. Travel on a British, or a Mediterranean holiday flight, and you will be lucky to avoid occasional bedlam as unruly children run around unchecked. To add to this, their endearing, loudly chastising parents, will chase them down the aisles half-heartedly apologising for them as they go.

On airplanes, some travellers are rich enough to reserve business class and first-class seats for their children. They believe their children should be free to run amok whenever and wherever they wish. They have 'paid good money' to put them in expensive seats, so why not? Who would want to pay for a long-haul, first-class seat, and expose themselves to such behaviour?

I was once on holiday in Corfu, having dinner at a family-owned taverna. The owner's mother was trying to coax her grandson to eat. To do this, she followed him around the tables with a teaspoon of food. All she had to do was hide the food and wait for him to get hungry. He would have come begging her for food. By allowing himself to be chased, this young boy obviously knew something about adult manipulation. He had learned the value of playing hard to get.

While this was taking place, his grandfather was taking my order, and criticising my elementary Greek. I asked for a bottle of 'kokkino krasi' (literally 'red' wine). Not only did he tell me that the usual phrase was 'mavro krasi' (black wine), but where and when, I had learned my Greek. 'You have been watching 'Greek Language and People', Sunday nights, BBC 2. Too many mistakes', he said. He was a native Corfiot who lived in the UK during the winter months. I should have returned for more colloquial Greek lessons.

Children are not alone in displaying inappropriate behaviour.

Twenty-five-year-old Mark, a robust 20-stone kick-boxer, used to storm into my clinic without an appointment. With no preamble, he would ask my receptionist, 'Where's the doc? I need to see the Doc.' When he got to see 'the doc' (me), he would say, 'Doc, I feel like sh*t.' I would then ask him to give me a few clues. 'Is your throat sore, etc.' Because the level of communication between us was poor, I came to liken our meetings to veterinary consultations (with no pet owner present).

A few years later, Mark died on a beach in Thailand. He had overdosed on a cocktail of drugs.

Can 'bad behaviour' predict a poor prognosis, I wonder?

Bias Unaware

Every thought and opinion can be subject to bias. Cognitive biases were first brought to light by Kahneman and Tversky in the 1970s. Their message was that everyone has biases that influence their decision making and judgement processes.

Prevalent among student doctors and patients is the 'Halo' bias. The bias will weaken or strengthen their belief in authority (teachers, for instance). The bias promotes the assumption that those in authority know best. The public may view published writers, media presenters, those with titles, and celebrities, as trusted sentient beings, always to be believed. That is the halo bias gone wrong. The same bias allows us to view information on the internet as definitive and true, and needing no challenge. Although unlikely, it could cause a doctor to view patients who have gleaned information from the internet, as more knowledgeable than they are.

A negative halo bias will lead to under-rating others. This has become more common among patients and medical bureaucrats who interface with doctors. Patients are questioning what we know, but only because they have little understanding of the knowledge required to become a doctor.

Bereaved

Dealing with the bereaved is always challenging. Those versed in the art of medicine may find it easier than those gripped by medical science. The minimum requirement is for enough empathy and inter-personal responsiveness, to bring comfort to those affected. There is an art to the rhetoric of eulogy.

Blackmailers

Specifically, emotional blackmailers. This is an uncommon tactic used by some patients to get what they want from doctors. They will have used the same coercive force to get what they want from their friends and family. It is more likely when patients 'get too close.' By sharing too much personal information with them, they can come to regard their doctor or nurse as 'one of us'. This is more common in cultures where 'who you know' is more important than 'what you know'.

BWMM: **B**utter **W**ouldn't **M**elt in their **M**outh.

They can be a paragon of virtue, or a divisive criminal. Avoid being drawn into their vortex of influence.

Clever Guys

Well-educated, intelligent people will easily understand the information you need to convey. They will, however, want to discuss the data and the basis of your opinion. If they are not just clever, but wise, they will balance the information you give them with that from other sources. They will then decide what option best suits them.

A postgraduate physicist once consulted me. He thought we were all being controlled by high-energy, electromagnetic waves, broadcast by the government. His arguments were 'scientific' in nature, and his train of thought logical. The only problem was, his ideas were delusional. He was an untreated schizophrenic.

Some clever guys (including doctors) harbour strange, unsubstantiated beliefs, but are clever enough to present them cogently. Although I have had a tendency to 'take them on', I have learned with age to save my energy. It is sometimes better to keep quiet, be graceful and move on.

Complainers (See Troublemakers)

In the West, we have now accepted complaint-oriented societies. This is perhaps one inevitable consequence of the current political drive to achieve equal rights for all. Many of the complaints now expressed by patients treated in the NHS result from being mishandled, side-lined, or ignored by those managing them. Being told that the computer says 'NO', can induce behaviour that is out-of-character. The inefficient handling of patients caused by the system in overload, is a common source of complaint. It never occurred in my practice. If frustration or aggravation ever arose between my staff and any patient, I would step in immediately to arbitrate and sort it out. Discord left too long can fester into anger and resentment. As a result, no patient recorded a complaint against me, although I am sure a few must have thought about it. While decisive intervention can strengthen the doctor-patient relationship, handling patients impersonally will weaken it.

Most medical practices have a few patients who complain. Complaining is their modus operandi; a matter of outlook, and even their mission in life. Classed as troublemakers wherever they go, they may enjoy the attention it brings. When dealing with them, always ask for witnesses to be present. A good receptionist will spot them before they attend as new patients. They are better managed proactively than reactively. Gaining their respect as *'a person who listens'*, and one who is fair and able to resolve problems wisely, allows for a bearable relationship.

Beware of being too kind. Many complainers regard kindness as a weakness; sensing kindness can cause them to strengthen their attack. Although very time-consuming, some complainers are worth listening to, especially if their criticisms are correct and constructive.

Complaints against doctors and nurses are fast-growing. If regulators are called to deal with them, a doctor's career could be in jeopardy. Our regulators are there to listen, and must investigate every complaint made, charged as they are with protecting the public from evil, advantage-taking doctors and nurses with harmful practices. Because most know so little about medicine, especially about how it functions at a personal level, they must assume that every complaint is true until proven otherwise. Devils lie in the detail, but with no knowledge of that detail, few will know to ask the most pertinent and revealing questions. The personality and outlook of each complainant is one such detail; the true nature of the doctor and his practice is another. Since these questions will be beyond the remit of regulators, their final judgements may be legally correct, but lack clinical perspective and appropriateness.

Patients need to be listened to. Sometimes, that is all they need. Bureaucrats entering the affray can alienate patients when direct intervention from the doctor involved would have achieved a quick resolution. If all this fails, doctors and nurses must call their medical protection society.

In private practice, every doctor needs a complaints policy. The first requirement is to listen. The second is to repeat the complaint back to the patient, confirming that you and she have both understood what is being alleged. Many minor complaints get resolved at this point. Even if resolved, one must write a contemporaneous report for each complaint in the patient's notes, with at least one account from a witness in writing. Note the process and the resolution. Doctors and nurses have no option but to follow every element of corporate, medico-legal processing (rules and regulations devised by detached executives, in offices far, far away from reality), and to react in a predictable, stereotyped fashion. Doctors and nurses must try to avoid becoming victims of the well-oiled, corporate legal machine, currently employed to seek reasons for punishing doctors and nurses.

If the complaint escalates to involve lawyers, the medico-legal game will change completely. To the uninitiated, investigating allegations against doctors may appear to be moderate, respectful, clinically relevant, and considerate. Don't get fooled. It will be none of these.

Imagine being arrested for crimes and awaiting trial. You could be about to get involved in an adversarial legal contest where the law, not medical matters, are the focus. Regulators then act as criminal investigators. Their mission is simple: to detect criminality (which is appropriate) or to detect and punish any departure from the published rules (formed by the GMC's *Good Medical Practice*, NICE Guidelines, and the British National Formulary or BNF). They will not engage in clinical reasoning (they would reveal incompetence if they did) and can relegate the circumstantial and inter-personal details of a doctor's case to hearsay.

How regulators can expect to judge clinical situations fairly without relevant experience, is an issue not being addressed. Their concern is to judge compliance and to keep matters simple. The blame rests on the shoulders of the accused until proven otherwise. The law has given regulators the statutory right to enjoy some insolence of power. They need no respect for doctors or nurses whose vocation has been to relieve suffering, and to save lives. The legal complaint process is dispassionate and can ignore balance, fairness, and what might be relevant clinical meta-data. The process of cross-questioning allows the free use of unfairness, the exclusion of any clinical perspective, and even bullying and intimidation. This is no forum for a doctor or nurse trying to make a clinical case for what they did.

Our regulators need to be seen to keep the public safe, but some might like to go further. Some would like to bring back the stocks and humiliate doctors publicly. They have reasons for not going that far: it would be archaic and would not win them favour with voters or patients. Instead, they have adopted more modern, equally punishing methods: trial by tribunal (with usually only one GP present), and posting their views of a doctor's misdemeanours on the internet for all to read. With their IT power, they can block any adverse comments made about themselves. Doctors who try to retaliate, will not win without mounting a financially crippling campaign. To retaliate effectively requires collective action, with help from internet savants (See also: Troublemakers).

Controllers

Those who desire to control others are ubiquitous. Avoid obsessional controllers, working beyond their level of competence; some with a pathological fear of relinquishing control. Being sure of their knowledge and ability, these patients may want to manage medical decisions personally. The process will fall apart when they attempt to interpret clinical data; in fact, the process can fall apart at any stage, leaving them perplexed.

What is it these patients (and some of our regulators) don't understand about the depth and breadth of a medical knowledge, and the value of clinical experience? They possibly understand string theory better.

Although private patients sometimes seek many opinions, they might only favour those who acknowledge their control. As a patient, conscious during a surgical operation, they could advise their surgeon where best to cut!

Cool Dudes

Look no further than Sean Connery's James Bond, or Winkler's 'Fonzie' (A.H. Fonzarelli) for classic 'cool'. The only real problem is getting to know what they are thinking. Commenting unnecessarily, and being reactionary is not 'cool'. Charisma, especially if enigmatic and mysterious, seems to conceal hidden depths, but the façade rarely survives close examination.

Corporate Kids

These are guys who know how to survive and prosper in a corporate jungle. They know the value of compliance and how to create the right impression in order to gain future security. If they keep their heads down, well below the parapet, they can avoid in-fighting and climb the corporate ladder unhindered. Showing signs of brilliance, and putting the edicts of Nicoló Machiavelli into practice, will help them attain a place on any board of directors.

Courageous

The quiet courage patients display when faced with an unpleasant, life-changing diagnosis, or after being told they need a life-saving operation, is impressive. Others faced with a life of disability, quietly and courageously rise to all the challenges.

Rob Burrows was a talented rugby player. He had a life-determining diagnosis of motor neurone disease (MND) made in 2019. He died on the 2nd June 2024, having faced the journey from peak fitness in 2019, to overwhelming muscular atrophy and bodily weakness. He was only 41-years old when he died. He took on the challenge with the brave face and smiling determination of the courageous. He has become a legend. Part of his legacy is now vested in the Leeds Seacroft Hospital research and support centre for MND, which bears his name. The aim is to find a cure for MND and to support those diagnosed with it.

Cranks?

Yes, it's true: some women want to eat their placenta after child-birth. Others quietly believe that aliens with superior knowledge are already among us (how else could science have advanced so quickly in the last 120 years?). Others believe that shark fin extract and powdered rhino horn have medicinal value. Some believe that 'natural' is always better than 'synthetic' (they should try powering their car with 'natural', unprocessed crude oil).

Some of those who express these ideas are psychotic; others are just ill-informed and would struggle to justify their beliefs when cross-questioned about the science involved. Some are attention-seekers using controversy as their USB. Doctors will need to decide which type they are. Many are benign, so it might be best to respect their views and agree to differ. Politeness costs nothing, even though sometimes seen as a weakness.

Frustrated with 'standard' medical advice and a computer that always says 'NO', some will turn to faith healing and prayer. A significant number will benefit from the experience. I never objected to 'alternative' methods, especially if harmless and they brought comfort. Given that placebos help 20% of those who take them, the onus is firmly on the medical profession to achieve better results.

Some doctors see the ideas of most alternative therapists as nonsense. Any affront caused by the relaxation of scientific rigour they have, might need to be swallowed with a pinch of salt when a patient reports a real improvement. It is important to realise that there are many medically untrained people who, by their presence, improve how others feel. Some of their explanations about how they bring about benefit can be mystical or pseudo-scientific. Many doctors dismiss them, with no wish to depart from the tenets of science.

If introducing dogs onto hospital wards can benefit patients (animal-assisted therapy), what else might? If the idea helps, and there is no obvious harm to it, why object to interventions that might expand the scope of our humanity? Doctors and nurses must always be prepared for the unexpected.

To practise the art of medicine requires one to know when to say something, when to say nothing; when to be accommodating, and when to be economical with the truth. Improving morale is not 'doing nothing' and can affect morbidity and mortality. In any personal clinical contest between maintaining professional integrity and improving patient morale, every doctor and nurse must take a position.

We should recognise that faith, even misplaced faith (in keeping with a growing anti-science movement), can act as a powerful adjunct to medical intervention. Because faith is a strongly held persuasion and offers a powerful influence, it could waste time for doctors and nurses to persuade patients otherwise. Some patients will counter a scientific explanation by saying, 'you just don't understand spiritual matters'. Clearly, you may not have yet 'found the path', or 'seen the light'. Imagine the accolade awaiting a patient who successfully persuades a doctor to follow the path they prescribed to achieve enlightenment.

In 1984, architect Carl Marsh came to consult me. He was dressed for business in his three-piece suit. I asked why he had come, to be told, 'I've been sent'. 'Do you have heart disease?', I enquired. 'I do have an aortic valve problem, but that's not why I am here.', he replied. 'I have been sent to you to give you this book, but let me explain. I am an architect and a professional spiritual medium. I am a prison visitor and have given a number of radio broadcasts on the subject. I knew you would not be ready to accept what I have to say, so I brought you this book, written by a surgeon who discovered spirituality. It is called The Path of the Masters. Perhaps you will read it later in life.'

Carl told me he had had conversations with Moses, Jesus Christ and Mohammed, and had written a small book to help others understand the spirit world. He said that after dying with heart failure, he would no longer return to being human (implying that he would have attained a higher state of being).

When I examined him, he had a moderately severe aortic valve leakage. He needed further investigation, perhaps leading to valve replacement. He refused, saying, that prolonging his life was not what he wanted.

Carl knew he sounded delusional and recognised that I might think him schizophrenic. Being highly plausible does not exclude it, though. My readers must come to their own conclusions.

Successful California surgeon, Dr. Julian P. Johnson, went to India to learn from gurus. He wrote, *'The Path of the Masters'* (1939).

Criminals

Few patients will cause doctors or nurses to sense danger. The most concerning are those who are economical with the truth; those who lie to get what they want, the devious and the criminally minded. These patients, some of whom have a criminal record, and live a life of crime, can test the resolve and professional integrity of every doctor and nurse.

I had a patient who lied about her role as a PA in a film production company. She feigned back pain and said only tramadol helped her. I gave her a prescription for a small amount, but soon after, she returned and told me she had lost her prescription.

After that, several pharmacists reported dealing with falsified prescriptions. The most relevant descriptive word for her was 'plausible'. Like many practised criminals, they are relaxed, confident, and plausible; indistinguishable from most professionals, in fact. They are professional cheats, fraudsters, and liars.

I was alone one evening in my clinic when two men appeared, descending the stairs from the first floor (I learned later that they had stolen money from the dental suite). I remember how quietly confident and self-assured they were. 'We wanted to make a dental appointment', one claimed. 'At 8.30pm, long after office hours?' I asked. 'You never know your luck, do you?' one said nonchalantly as he walked through the front door. The following morning, we discovered the extent of the burglary. In retrospect, I guess I was lucky not to be threatened. I was alone, and could hardly have defended myself against two strong young men.

Those facing a Criminal Court appearance will sometimes say anything to get support from a doctor. Some will try to get a doctor to bear false witness. They need to understand a doctor's role. In order to put him in the picture, I said to one, 'I will help you as much as I can, provide all the information I hold, but will never lie for you.' Rhetorically, I asked them: 'Why would a doctor want to jeopardise his career by giving a false statement?' I never met a criminal who didn't immediately understand my point of view.

Some drug addicts and criminals will threaten doctors who do not satisfy their demands. These situations are challenging. I was once told that I risked getting my throat cut if I didn't help! Mostly, they are bullies; the truly bad guys rarely threaten or give warnings of their intended action. *Courage mes braves!* Say 'NO' and mean it. With luck, such patients will respect you, and understand how useful you might be to them in the future. Involve your colleagues; report them to the Police, and check that your video surveillance cameras are working.

After being threatened but not harmed, the police can be of less help than you imagine. After all, there was no crime committed. To your dismay, they might suggest you return after being assaulted: they will then have an actual crime to deal with. Reporting a threat might help if there is already an outstanding arrest warrant against the culprit, or the police are building a profile against them.

It is easy to typecast and falsely accuse those with a criminal record. I remember one such patient with an implanted defibrillator. Heart failure (dilated cardiomyopathy) had caused him to have repeated episodes of ventricular tachycardia (VT). With left heart failure, he was short of breath on minimal effort.

The prosecutor in court suggested that he had run 200 metres uphill, holding a heavy iron bar intending to attack two athletic young men. The allegation was he had assaulted two young burglars.

When the judge asked me what I thought of the incident, I suggested that such physical exertion would be beyond his exercise ability. I went further and suggested that even the stress of appearing in court might prove a danger to him. Having had several surprise defibrillator shocks, he was not keen to experience another. The prosecutor, whose desire it was to proceed with the case, suggested that we might, therefore, have a resuscitation team standing by - just in case. The judge looked me in the eye, and peering over his half-rim glasses, said, 'Would you care to comment, Dr. Dighton?' I replied, 'I don't think I have heard a more inappropriate suggestion made in court.' The judge agreed and dismissed the case.

Wheeler-dealers and criminal types can offer financial rewards, or other inducements, if only a doctor helps them to avoid prosecution. Someconvicted criminals are among the most charming, intelligent, engaging and persuasive people one can meet. Not everyone will keep their promises, but, paradoxically, most criminal types will. Having experienced the long arm of the law, criminals know better than most, that good-will, loyalty, and keeping promises are invaluable.

A young man once asked me if I could provide evidence for what he claimed to be his pathological fear of needles. After being arrested for drinking and driving, he refused to give a blood sample for an alcohol test (after a positive urine alcohol test), at a police station. A neuroticism questionnaire (the MHQ, or Middlesex Hospital Questionnaire) suggested that he had high levels of anxiety. The result did not support a diagnosis of phobia, but it provided some evidence of his neurotic state. The court gave him the benefit of the doubt and discharged him. He later told me he had falsified his answers to my questionnaire. Surprise, surprise!

Dictator

This potentially dangerous sub-type of controller can be arrogant and ignorant (see **ALBer**), they are disinterested in a doctor's opinion. They are demanding and purport to know best. They are only fleetingly in listening mode and may see it as a doctor's privilege to deal with them. The first problem is to find sense in their demands, and then to be as helpful as one can. If scolded, they will not usually return.

Digital Beings

One product of the information age is 'the digital self'. Mobile phone apps and other devices can record personal 'big data': our respiratory rate, the number of steps taken, and our pulse rates. Collecting clinical data of doubtful clinical significance is one thing, accepting the growth of fashionable narcissism, is another.

Doctors as Patients

Doctors as patients are a special sub-group, simply because they will decide what is wrong with them before asking for another opinion. A doctor consulting with another doctor must remain calm, objective and independent, and take their patient's professional opinion seriously. Clinical ritual must be observed when dealing with other doctors, be it history taking, examination or an investigation. One must resist taking shortcuts and jumping to conclusions No doctor will want to be 'short-changed' clinically.

One must studiously await all the data before coming to any conclusion. The confidence and trust of patients who are doctors will by earned more easily by the examining doctor behaving like their former lecturers and teachers, while portraying knowledge and respect.

The diagnosis and management of other doctors must make complete clinical sense. Doctors, as patients, will have arguments that only special experience and specialised knowledge can counter. A doctor putting her life in another doctor's hands, will expect adequate responses to their every challenge. As with all other patients, one can only move forward with mutual agreement.

The patient may be a doctor, but is also a patient, and must be respected for both roles. Doctors will find it difficult to deal with those doctors uncomfortable with their role as a patient. For one doctor to transfer clinical responsibility to another doctor is a big step.

When dealing with non-medical patients, many of us use infantile expressions like 'just pop on the couch', or 'can I feel your tummy?' Few would use such expressions when dealing with other doctors, so why use them at all?

Duper's Delight

Duping others may not be illegal, but is always fraudulent. This behaviour is common among drug addicts who will concoct any excuse to get the extra drugs they want. They will tell doctors they have lost their prescription, lost their pills, had them stolen, or eaten by their dog. Some will say that they feel more secure with lots of stock-piled drugs (for fear of running out); others will claim that they had to share them with a needy friend.

I must admit to being naïve: I mostly believed my patients (until proven otherwise). But then, if I had wanted to become a drug inspector, or work for the fraud squad, I would not have applied to medical school.

An ex-nurse and her husband both duped me. She was getting diazepam and zopiclone from both her GP and me, while swearing that I was the only one prescribing for her. I had warned her pharmacist and her GP of this possibility, seven years before, when I suspected she was double-dealing. I had her GP's agreement to be her sole prescriber. When her GP retired, whoever replaced him did not read my original letter detailing our agreement. I suspected as much when I inadvertently saw her coming out of her GP's surgery.

Later on, her GP records revealed she had been alternating between me and her new GP to get extra drugs. Without doubt, she had been duping both of us.

The case came before the MPTS (Medical Practitioner's Tribunal Service at the GMC). They held me accountable, not her NHS GP. They believed I had exposed her to risks beyond what her NHS GP, or an addiction specialist would have allowed. The patient never came to any harm in my hands, with twice weekly visits over a seven-year period and drug doses that never escalated as far as I knew. At no time did she make an allegation against me; the complaint came from a pharmacist. The boring but important details are in my book, 'The NHS. Our Sick Sacred Cow'.

WARNING. Being deceived by a patient can lead doctors and nurses to be classed as unfit to practice. It may be necessary, but suspecting patients' motives does not help the doctor-patient relationship.

The Fearful

Many patients feel anxious while consulting doctors. They may be shy, or genuinely fear what might transpire during a consultation. Putting patients at ease is a valuable interpersonal skill and a clinical art. Without their fear and anxiety being eased, some patients will remain buttoned up and communicate poorly.

Interesting People

Many of my patients were interesting people. Like many doctors, I had side-interests: research, computing, writing, travel, finance, art, music, etc. Many of my patients knew much more about these subjects than I did, so I found it interesting and informative to talk to them.

As experienced travellers, some of my patients provided me with excellent travel advice.

My daughter Anna once travelled to Venice on a school trip. One of my patients (Franco Dorili) knew the restaurant director at the Cipriani Hotel on Giudecca Island. 'I will telephone him for you, and book a table for her', he said. 'I will ask him to take special care of her.' And so he did.

I asked my patient Colin Fitch where I should eat in New York. He booked me a table at the River Café, just below Brooklyn Bridge overlooking Manhattan – just to make sure I had a waterside table and an amazing view.

When I asked Alan Reed where to stay in San Francisco, he arranged the Fairmont Hotel for me, on a floor high enough to see the bay and Alcatraz.

Many of my patients were 'in the know', and generous to me with their time and knowledge. Many took some pleasure in advising me. The mutual sharing of beneficial knowledge is an act of friendship; it is one pleasure found while practising traditional pastoral medicine. It contrasts with the rigid and formal style of practices focused on nothing but disease (albeit the most important of issues).

If medical school selection committees continue to seek only high grade, detached scientists as applicants, the traditional pastoral role of doctors could diminish. A broader based education and extensive life experiences, were once prerequisites for would-be doctors, especially if their ambition was to provide pastoral GP care. Selection committees now, may not be sympathetic to sixty-year-old ideas; it would not do for them to be classed as 'old fashioned', or to admit that 'once upon a time' our predecessors were wise, and knew more about how to practise medicine. The way doctors deal with patients has changed little, but the politics has changed beyond recognition.

There is another way to make pastoral care acceptable. Re-brand it, and call it 'social prescribing'.

Gamers

In his book, '*The Games People Play*', (1964), Eric Berne describes the gaming and transactional aspects of human relationships. Patients playing interpersonal games want to win points of advantage, the scoring of which allows the declaration of winners and losers. Some patients try to involve medical professionals in their games.

Many have attempted to draw me into their attention-seeking games with partners. Someone who might wish to test their partner's interest, respect or love, might try to test them by creating medical concern. They might choose to present with symptoms calculated to draw attention to their health, and the more puzzling the better. The patient's partner, or ex-partner, might then take the bait and show concern. Their concern might lead to demanding second or third opinions, or to suggesting further tests to confirm the diagnosis. They might want a change of doctor if not enough interest is shown, or if the diagnosis seems ill-considered, or too benign. There is a hidden point to the process. The payoff for the patient is to learn just how much their partner really cares for them.

Berne had a lot to say about alcoholism. He regarded it as a potentially lethal, three-handed 'game'. The players in the game are the alcoholic, the uncompromising, unsympathetic one, and the 'patsy' or helpful one. All three must interrelate for alcoholism to continue. Some alcoholics can feel so dejected they commit suicide. After dying, surely that will make the uncompromising ones feel guilty. The intended message was: 'That will teach you. You should have taken me seriously.' To help understand their role in the 'game' of alcoholism, psychotherapy can provide insights for each player. The question then is, will insight help to bring about a beneficial change?

If doctors are lucky enough to practise medicine with extensive personal knowledge of their patients, their family and work colleagues, getting drawn into inter-personal games is less likely, but easily possible. By recognising the dynamics, one can either play along (if it will benefit your patient), or take no part. Doctors must recognise as soon as possible, every attempt made to involve them in attention-seeking, and other manipulative games patients want to play.

GG: Good Guys

Most patients are pleasant, fair, respectful, diplomatic and reasonable. Grace and charm are rarer virtues. These patients mostly display 'good behaviour', not easily corrupted by NBW's (nasty bits of work). It is easy to define good behaviour. It is that which leaves everyone in their midst feeling comfortable. Good guys are obvious from first meeting them.

Matthew C. was a happy five-year-old in 2016. We instantly got on. His mother told me that whenever they passed my clinic, Mat would want to call in and say 'Hello'. The last time he called in, he gave me a hug and a kiss on the cheek. He may have only been five years old, but he was already a GG.

Grammar School, Comprehensive, or Public School?

A patient's level and type of education can influence what medical management they find acceptable. Peer group influences, attitudes, and general behaviour patterns of patients can undoubtedly influence the doctor-patient relationship. (see more comment under 'Doctors and Nurses as Characters').

Hopeful Patients

Patients have many and various attitudes towards disease: 'It's not fair.'; 'I can beat this.'; 'Why me?'; 'I don't do illness.' Others are fatalistic and take whatever comes.

Even when doing their job well, doctors and nurses will always find a limit to what each patient will allow them to do. It can depend on how well informed they are, and how much they want to be involved in the decision-making process. The art of medicine requires that we know our patients well enough to judge how much amount information they can handle. Patients are right to use their experience of previous consultations, appraisals and judgements with their doctor to assess his value. An accolade I once received from a patient was: 'I would never act on any medical advice without your advice.' He always compared my opinions to those he received from other doctors. For high levels of trust like this to be built with patients, a few lengthy, tried and tested, clinical interventions must have taken place.

The need for hope is universal, but varies a lot between patients. Prayer and hope can both influence prognosis. For some doctors, this subject lies over the hills, and far, far away from their focus; often too far over the event horizon for a dedicated scientific mind to handle. The scientific method, as the Holy Grail of science, can attract the same passion among doctors as any religion, so they should respect the importance of belief and faith to some patients.

Zeus's vengeance package contained many evils, including death. They all escaped into the world after Pandora opened her jar (not a box). After closing it, only 'Elpis' (hope or expectation) remained. For some patients, this will be all they have left. There are reasons some hopes need to be dashed, and others left in place. Knowing patients well will usually guide doctors to the best course of action. With so many immeasurable inter-personal considerations, there can be no legitimate place for fixed rules or algorithmic thinking. There should always be a place for hope, compassion, empathy, dedication, goodwill, and respect for faith and belief.

Many decades ago, my uncle Charlie came to see me with what I diagnosed as a DVT and painful thrombophlebitis. He had been a smoker, and a chest x-ray had revealed a small hilar mass (lung cancer). 'Whatever I have wrong with me,' he said, 'I want you to deal with it. You don't need to share the details with me. If you want me to have a test, or a treatment, I will do whatever you think best. I want to delegate my management to you.'

In the early 1970s, the average prognosis for squamous cell carcinoma of the lung, after lobectomy and radiotherapy, was six months and one day. Without treatment, it was six months. I told his son, who agreed with me, that giving him a cancer diagnosis would adversely affect him psychologically. In line with his wishes, I elected not to tell him his diagnosis, and not to get him treated. I wanted to preserve his lifestyle without the burden of fear and anxiety. News of a poor prognosis would have devastated him, so we kept it to ourselves. His only treatment was anticoagulation. Despite the poor prognosis, he lived asymptomatically for three years before he died – six times the length of the usual prognosis.

Knowledge alone is not enough. Clinical experience must distil what knowledge we have into wisdom. Wise clinical judgement is thankfully common, but is useful only when applied appropriately to each patient. The ability to make such judgements and to know their proper application, defines a competent physician. The ability to comply with every fixed rule, and guideline, regardless of clinical appropriateness, may define a compliant knowledgeable doctor, but not necessarily one who is fully competent. The ability to comply with every regulation is not a qualification for clinical perspicacity.

In the 17th century, Nicholas Culpepper, the apothecary, said that patients attended his practice in Spitalfields (London) to get an injection of hope.

Hope is good, but reconnaissance, planning, and validation often prove more successful.

The Incapable

If a patient lacks the mental capacity to remember facts, or to process information, he might not be able to engage in medical decision-making processes. The list of causes of mental incapacity is long, but includes intoxication, delirium (related to physical illness), dementia, cerebral underdevelopment or damage, together with psychoses (being out of touch with reality). It can also apply to those suffering extreme stress; it is difficult to cope with problems when pre-occupied. All need safeguarding and support. Most will need someone they would trust to decide on their behalf, acting in their best interests. This is subject requires its own treatise.

Influencers

They do not exist on TikTok and Instagram alone; family, friends, the internet, floor-cleaners and fellow patients, can influence patients more than doctors and nurses.

When patients shared Nightingale wards, I overheard a post-op patient speaking to a pre-op patient about the surgeon they shared: 'I wish you luck with the operation, mate. His patients rarely do well.'

Interrogators

There are those who enjoy interrogating others, even over trivial matters. They have an obsessional need to delve into detail. A summary response will not satisfy them. Dealing with them requires patience and tolerance, and tact needed to bring their interrogation to an end.

IS: Internet Savvy

Incorrect information can be worse than no information, so doctors, nurses and patients need to corroborate whatever they learn.

The savvy patient may think he knows it all, especially after consulting 'Dr. Google'. With a biased opinion, based on cherry-picked information, he may have decided what he needs from a doctor. He is unlikely to have a considered picture of his problem, since only a few will have enough experience, knowledge, or judgement to assess the validity and relevance of any 'facts' gleaned from the internet. Regardless of this, some patients can reach correct conclusions. Given that most people now refer to the internet after being given a diagnosis, it might be best to provide them with the best search terms.

Steve came to see me with his wife. His extensive research led him to believe that she had Lyme disease. In the end, the diagnosis proved correct. Well done, Steve. All patients deserve to be listened to, and given the benefit of the doubt.

Patient demands arise from many situations, the most valid of which are ill-health and personal suffering. There are other sources of demand, like media outlets promoting a sensational new drug as the latest medical advance. The many unproven 'advances', 'cures', and 'tests', described by the media, can find doctors having to explain the need for double-blind, controlled experiments. Only a few patients, however, can distinguish between codswallop and well-conducted research.

When doctors dismiss new ideas, patients can take it as unsympathetic, overbearing, and unhelpful. A patient's insistence that, 'there must be something to it. Why else would anyone publish it?' can seem churlish to deny. Even with verified facts, they can remain unconvinced, and go elsewhere. Only after deteriorating will some patients question the wisdom of following a media-led strategy. Freedom of choice has consequences.

When researching the side-effects of drugs or the work of a doctor, patients can get distorted ideas. After all, who can claim to be exact in assessing probability and risk, when choosing anything? In order to inspire confidence, I sometimes used examples from my experience. I might say: 'In thirty years of using this drug, I never saw this, or that side-effect.' Alternatively, I might say: 'In my experience, half of those taking the drug experience muscle discomfort'. Even with personal recommendations, patient biases can be difficult to overcome.

Kidnappers and Hijackers

I met with a patient of mine, Mr. P, while in Istanbul. He invited me to his flat the next day for afternoon tea. Mr. P and his wife greeted me, together with the eight relatives they had assembled to get my advice.

The further west and north a doctor goes in Europe, the less this is likely to happen.

Some hijackers are polite: *'I know you're on holiday, doctor, but I wonder if I could trouble you for a quick opinion?'*

If a doctor is someone who flies a lot, or takes cruises and holidays on small islands, he will be at risk of getting hijacked medically. Someone sitting by the pool, at a bar, or at the same table during meals, may not resist seeking a medical opinion. In a diminishing number of cultures, doctors are still demigods, but how often does one meet a demigod? Meeting one is an opportunity too good to be true, and not to be missed.

Most people expect doctors and nurses to be kind and helpful. Desist! These anonymous 'patients' will rarely appreciate the difficulty you will have basing an opinion on their history alone. Be prepared for scorn if you don't comply with their request, and expect anger when they feel rejected.

Some will try repeatedly to interest you in their *'once upon a time'* medical misadventure; they will even draw their friends and partners into the discussion, just to provide corroboration. For the sake of a little added drama, their evidence will often include details of just how close they came to death. Some will contend that no doctor before has ever really listened to them; not like you. How lucky they were to survive, and how lucky to have met you.

There is no easy or courteous way to fend them off, unless you follow the example of one well-known celebrity. When asked by a fellow traveller if he was Joe Bloggs, he replied, 'Not to-day, I'm not!' This will only shake off the least determined; a less considerate, insensitive or stronger form of rebuke will be required for the more ardent.

None of this applies to me. I have always been far too soft to deny any medical hijacker. Rebuking others in need is a cultural issue for me. My parents raised me to be respectful. I believe that having medical knowledge and even a little talent to use it, comes with a sovereign duty: to help others medically, whenever one can. If as a doctor or nurse, altruism is not your thing, travel completely *in cognito*. Tell nobody what you do, or suffer consequences that may not be to your liking. Personally, I have never found it a problem.

Although rare in the UK, there are countries where it is usual to be invited to join a club, team, or firm as a recognised clique member. Within the clique, there will often be a free exchange of information, goodwill, goods, meals, hotel stays, and other services. I am a member of one such Cypriot clique (of very close friends and associates). My role within the group is medical adviser; not because I want to exchange my advice for good will, meals or hotel stays, but because I want to support my friends with guidance when they deal with doctors. I get the benefit of their support in many worldly matters.

Kind and Genteel

These patients can make a doctor feel their work is worthwhile. See also: OK, and GG.

Lazy Buggers

I have always been too polite to say it, but would like to have asked the occasional colleague or patient, 'When did your servant die?' Some spoilt people have had their every wish granted; they expect everyone to act as their servant. Some were born idle, and some have had idleness thrust upon them; others developed their lazy ways after learning how easy it is to get 'friends' to help.

Feigning illness and false incapacity are both ploys some will use to attract sympathetic helpers. Watch out for false gratitude and fluttering eyelashes. One can easily get hooked by the idle and the undeserving. Curb their demands at the earliest stage.

Liars

One third of us lie every day.

Richard Wiseman. *Psychologist.*

I always had a major weakness: I believed most of what patients told me. It was never a problem, except for being duped three times in fifty-three years. The last incident led to my suspension (by the Medical Practitioners Tribunal Service, October 2019). It did not shake my trust in patients, only my trust in regulators. (Read about it in my book, '*The NHS. Our Sick Sacred Cow*').

Unlike Pinocchio, human noses don't grow with telling lies. Doctors and nurses might try to understand the reasons for patients lying. One is to avoid any corruption of their self-image ('nothing wrong with me doc.'), another is to avoid unpleasant consequences ('if I tell the truth, doctors might do something nasty'). With drug addicts, lying about lost prescriptions is a common ploy for getting extra drugs. The consequences of lying can, of course, be tragic. Some patients fear illness and suffering so much, they would sooner lie and die, than suffer. They need counselling.

There is a growing linguistic science developing to detect lies. Analysis of recorded interviews given before criminal trials by those eventually convicted, has led to the identification of many reliable verbal indicators associated with lying.

MAE: Mental Age of Eight

When asked, a medical colleague of mine agreed to write memos for insurance company employees. After his experience of how much they understood, his advice to me was simple: when writing to instruct corporate staff about medical matters, assume they all have a mental age of eight years: the majority will then understand your memos. He had no intention to insult eight-year-old children.

'Against ignorance, the Gods strive in vain.'

Friedrich Schiller (1801)

Manipulators

There are those whose *modus operandi* is to barter and make deals. As patients, their priority may not be medical at all; it may be to make a proposal or to get a certificate of 'good standing'. Their quest is to profit somehow, like getting you or your staff to make appointments for them. If you make an appointment for them (after many telephone calls going backwards and forwards), they might try something extra, like confirming it by email. This all sounds bizarre, but I have seen it happen. These patients want unpaid servants. Business gurus regard them as effective delegators; I regard them as users.

A manipulator will first try to befriend a doctor. 'We've been friends for so long, haven't we?' is a common opening gambit. Translated into motivational terms, there is a tacit threat: 'surely, you wouldn't want to risk losing me as a friend?' He wins if you give him the recognition he seeks, or in private practice, waive your fees (friends don't charge friends, do they?). In the NHS, where direct payment is never required (except for GPs signing passports / certificates, etc.), some patients will try to obligate their doctor. Once obligated, they will try to use you, and your time, whenever they want. They want favours, with no thought of reciprocation, whether it be for goodwill or an agreement to follow your advice. Right from the start, this behaviour should signal caution.

If they don't get what they want, they can turn nasty and become disapproving. That is why some doctors need to 'man-up' and refuse to bend to their will. They will get upset once they realise not all doctors are pushovers. Beware, though, manipulators are nimble; they can deftly change their tactics and demands. They will quickly try alternative approaches: a new deal or a new understanding between friends.

'No' is a word unacceptable to manipulators. The word 'No' has a cultural perspective. In British, Greek and some eastern cultures, 'No' is often just an initial step in the process of negotiation. In Russian culture, it mostly means 'not negotiable': 'Nyet' means 'No', not 'perhaps', or I'll think about it.

If you prove useful, a manipulator will tell all his associates how soft and amenable a doctor you are. You could become the 'go to' person. Manipulators will not support you in a crisis, and will be the first to blame you when things go wrong. Unless you are shrewder than them and can spot one from afar, beware of manipulators. Unfortunately, most manipulators are street-wise, and that usually means they are streets ahead of most doctors and nurses.

Martyrs

Not always easy to spot, their behaviour means: *'I am suffering, and nobody can help me now. It's my destiny to suffer.'*

In the film *'The Life of Brian'* (1979), a Roman prisoner (played by Michael Palin) had hung upside-down in his cell for years. In keeping with British satire, he says (in a way that reflects Palin's strong British sense of fairness) the Romans who put him there were actually quite reasonable. In fact, he felt obliged to them for being so understanding. A happy martyr indeed.

MASH: Mad AS a Hatter

These patients have ideas that will cause a doctor to raise an eyebrow, while others become aghast with incredulity. Among them are attention seekers and conspiracy theorists. Some are delusional, others are misguided with information, cherry picked to suit their purpose.

MASHs are not alone in their cherry-picking ability. Those who employ meta-analyses to analyse multiple drug trials, also need to be adept (a common source of error in 'evidence-based medicine'). The selection criteria applied to any study group, will limit the clinical relevance of the research to a specified group alone. Doctors must always scrutinise research selection criteria for details that do not match those of their patients. This process is essential for every doctor who wants to put any research into practice.

MOMA and POPA

Mater **O**h **MA**ter (Matriach), or **P**ater **O**h **PA**ter (Patriarch). These patients have taken their entire family (support group, or work group) under their wing, and are subject to their control. It is essential to know whether they are acting in a beneficial or malign way. Once there is proof of a doctor's worth, it is they who will direct their family members to them for advice. They are few, but are all well worth identifying. Within any family or organisation, it is useful to identify the Top Guy, Big Chief, Head Honcho or *Ganze Macher*.

Neanderthal (*Homo neanderthalensis*)

Although thought extinct, some of their characteristics survive. As a different species, early *Homo sapiens* will have interbred with them, guaranteeing the survival of some of their supposed characteristics: physical strength and endurance, resilience, resistance to cold, and perhaps lower intelligence (relative to Homo sapiens – the wise man – that is). A small proportion of Neanderthal genes are usually detectable within any human genome.

Neurotic and TFBF – Too Fussy by Far

Many patients make clinically observations about themselves that will prove insignificant (in a doctor's opinion), but remain a concern. They might think that every naevus is a melanoma, and every headache, the first sign of a brain tumour. If they are obsessional, they will test the patience of others. The fear, anxiety, hysteria, and hypochondriasis experienced by any TFBF patient needs sympathetic handling. They can learn the real clinical significance of their concerns through education, and with cognitive therapy they can learn to face their worst fears. These patients can consume a lot of consultation time. If doctors don't have enough of it, they should refer them.

A surrogate TFBF situation exists. Here, a third party does the fussing, often to the embarrassment of the patient. Some minders will bring patients to you, like a dog on a lead. The minder (mother, father, friend) will tell you the story, while the patient remains silent. While this is a common situation when dealing with children, it also occurs with adults. It is easy to deal with: any discussion must include the patient from the start. Some minders are devious manipulators, and should be side-lined; others deserve gratitude for their genuine concern, help and involvement.

If there are any psychiatrists or psychologists reading this, I apologise for the simplification that follows. My aim is to provide a simple guide to neurotic traits.

The Crisp-Crown Index grades each of six psychoneurotic traits as average, less than average or greater than average (*Middlesex Hospital Questionnaire* (MHQ), Brit. J. Psychiatry (1966); 112: 917-23). The six traits are *anxiety, depression, obsession, hysteria, phobia and hypochondriasis*. They will all be obvious clinically to most physicians when well developed.

In the minimally expressed, stoic state, they can be difficult to detect. A stoic non-neurotic state is typical of pilots, astronauts, and Victoria Cross holders; they have a matter-of-fact attitude to life, not an emotional one. Clinically, this makes them easier to deal with.

Those with a well-developed obsessional trait (nit-pickers with attention to detail) do well in jobs requiring meticulousness. The same trait can prove exhausting to the doctors and nurses dealing with them. Their obsessional trait can be of psychiatric significance. I once met a patient who insisted on sweeping the autumn leaves from her front path, at 2am in the morning, in a strong wind.

Brian Bridgman was a chief helicopter pilot instructor, with 4000 hours of flying experience. He managed every detail of his life precisely. He was always in control. Consultations with him were always succinct and to the point. 'I have three problems to-day', he would say. We discussed them briefly, and he would then summarise his understanding of our conversation. He then detailed how he would implement the actions we had agreed.

Brian died in January (2016) flying his helicopter from Scotland to Kent, after it developed a fatal turbine bearing failure. His life ended because of a mechanical failure that was entirely beyond his control.

I used the MHQ for many decades, and computerised it (a DOS based, Dbase2 database program). I ran tests on over 1000 patients. After a while, feedback from the test allowed me to diagnose the six neurotic traits more easily. I found the questionnaire most useful for confirming depression and sub-clinical depression. The questionnaire was also useful for semi-quantifying anxiety (slight – average – severe). The information helped me appreciate the neurotic state of my patients, and the likely contribution to their clinical state. This was often of relevance to planning their medical management.

I had an occasion to speak with both Crisp and Crown. They were both rather surprised that anyone was still using their 1960s questionnaire, so long after they published it. Useful techniques do not degrade with age and may not need to be superseded. Instead of improved diagnostic accuracy, fashion and technical advances can dictate when this happens.

Here is an example of an unexpected response from a patient with both high anxiety and hypochondriasis (somatic anxiety) scores.

Mary was 68-years-old. She had been complaining about feeling unwell for years. She was also complaining that her husband took no interest in her condition. All her symptoms were nonspecific, and several general examinations had revealed nothing of diagnostic value.

On a later occasion when I tested her urine, I found blood (microscopic haematuria). I told her I had found something, but probably nothing of significance. I asked her for a next day, early morning specimen. That showed the same - a trace of blood. To this news she said: 'Thank God you've found something wrong, doctor. Everyone thinks I am an incurable hypochondriac.' Investigation revealed a non-functioning left kidney. Cancer, was my first thought. An operation revealed something else - a left renal (kidney) abscess. The infection had destroyed her kidney completely. It had become a bag of pus, which must have affected her health for years. This explained why she had felt so unwell for so long.

There is a moral to the tale. Doctors should always take extra care with patients labelled 'hypochondriac', since no hypochondriac is exempt from physical disease.

¿No Entiendo?

Doctors will sometimes notice a patient's facial expression change to one of incomprehension, when trying to explain technical details. Some will develop a glazed look, meaning 'I have zoned-out'; 'you've lost me'. Perhaps they understood nothing? Might they be pre-occupied? Perhaps they have already decided what they want and anything else is irrelevant.

If your message is important enough, ask them to repeat it back to you, and get them to recall the instructions you gave. Be prepared to conclude that you have wasted your breath.

In a high-profile case (Chester v Afshar, 14.10. 2004), neurosurgeon Mr. Fary Afshar, a medical school contemporary of mine, said he had warned his patient about the complications of laminectomy (inter-vertebral disc removal; she had several prolapsed discs). The patient later sustained neurological damage and claimed she had not warned fully of the known complications.

The House of Lords decided that a doctor's failure to inform a patient fully of every surgical risk vitiates the need to show that harm would follow from failing to inform.

If litigation against doctors gathers further momentum, we will all need to video every consultation.

OK Patients

These patients are sensible, compliant and understanding. They listen, understand, and ask pertinent questions. They learn easily and can readily decide what is in their best interest.

Objective

These patients are usually reliable historians. History taking is easy with patients who are intelligent, objective, succinct, and have an excellent memory. One can handle them efficiently if their narrative composition is concise (rather than lengthy, and from 'once upon a time'). Often, though, there will be a need to clarify what a patient says and what it is they mean to say.

Obligators

See 'manipulators'. If doctors or nurses ever accept a favour from a patient, they will be at risk of becoming obligated, although in my experience very few patients who give favours intend any obligation.

PITA: Pain In The Arse.

These patients are usually obsessional and must get what they want before moving on. Those with some insight into being 'a pain in the arse', may be no better; theirs may be a strong obsessional trait. If a patient lacks insight into this annoying trait, doctors must make accurate notes of their demands, taking care that their words are appropriate and the punctuation correct. PITA patients can waste a lot of time, and put a strain on a doctor's affability, reasonableness, and sanity. There is another side to this coin: the patient's attention to detail could save their life, especially when reminding a doctor of something overlooked.

Princess, Duchess or Queen?

When I think of examples of these, I am drawn to remember our late Queen's mother. In everyday life, such women have grace and charm enough to encourage us, although our work with them will usually have to pass the scrutiny of their protective supporters. I met only two as patients. Both were matriarchs. They were both surrounded by loyal family supporters who respected their position as head of the family. One lady, known to us all as 'Aunt Grace', was gracious in name and spirit; the other, also a gracious grandmother, had a daughter and granddaughter who inherited the same gracious nature.

From my earliest days, I knew two others: my mother and grandmother (my father's mother). When my grandmother approved my ambition to become a doctor, my family assumed my destiny to be ordained and inescapable. She was not a person to be let down.

Range Riders

Many patients become concerned when told their laboratory results are not 'within normal limits'. In my practice, most of my patients wanted to be involved in their diagnosis and management so this was something which could some take time to explain.

To explain what 'normal range' means, and the relevance of laboratory error, can be challenging. There is a difference between a variant of normal and a pathological result. Repeat testing for verification is usually advisable. Many patients are happy only when their results are all 'within normal limits'.

Patients self-recording their blood pressure has significantly contributed to the improved control of hypertension, even if taking it frightens some enough to generate high readings. Just as often it consoles them once they understand just how variable blood pressure can be. Quantitative measures of exercise undertaken, cigarettes smoked, pizzas eaten, and alcohol consumed, can all help clinical risk assessment (although patients often underestimate the quantities they consume). The errors involved in such data collection and their interpretation will concern only a few patients.

Recalcitrant Recidivists

'Can an Ethiopian change his skin or the leopard his spots?'

Jeremiah 13:23

Consider this contention: there are many who repeatedly have strokes, heart attacks, and asthma attacks, because they continually (choose to) put up with unacceptable circumstances brought about by limited resources, an inappropriate sense of responsibility, or inappropriate loyalty and duty to others.

The question is why would someone want to put their hand in a fire repeatedly, knowing it will burn them each time? Some blame necessity; others will do it because they are brave, selfless, and able to take what comes, whenever supporting others. Despite the potential damage, the reasons for such sacrifice can be noble and intelligent; others repeatedly self-harm because they have yet to receive the attention they desire. Some have no wish to learn; others cannot learn. Psychotherapy can help them with insight, and promote a change of behaviour.

Some who know themselves to be allergic to nuts, will decide to eat one, just to see if they can get away with it. Others will buy feather pillows (they are softer), knowing they are allergic to feathers. When agreeing to relate to someone as a partner, some will choose the same type of person each time, even though the strategy has led them to relationship failure before. Criminals carry on shop-lifting, and patients carry on drinking, taking drugs, and stressing themselves, hoping to get away with it. Some continue with the same behaviour despite knowing the likely consequences. Psychopaths (personality disorder), are an exception: they rarely care about consequences.

For decades, I told many patients to reduce their excessive alcohol intake, just to prove an adverse connection between their drinking and hypertension. What did most do? You guessed it: they carried on drinking. Criminals are not the only ones to exhibit recidivist behaviour; the stupid, the driven, the obligated, and the addicted do it as well. This behaviour is so common, one must regard it as 'usual', even if abnormal. Sometimes it is attention seeking, and one aspect of an interpersonal game.

To change recidivist behaviour, some need to gain insight into reality, and accept advice from those who have experienced the consequences. Group therapy can help. This makes biological sense, since survival always depends on seeing things as they really are; relying on fantasy is unreliable and too dangerous.

My guess is that we inherit the value we each place on insight. There are those who can learn from their mistakes and those who cannot. From both evolutionary and clinical points of view, this is a significant feature. A patient's insight into their medical condition, and the appropriateness and necessity of some clinical interventions, can influence their prognosis.

The Resentful

The feeling of resentment can powerfully affect mental health. It can cause steady personal decline, emotional instability, depression and aggression. Many held in its grip, will voice statements like, *'if it wasn't for you (him, them, it), I would be XYZ by now'*.

Some blame others for not achieving their goals, whether it be fame, fortune or happiness. They will blame their parents, friends, business associates, spouses, and partners for their failure, but not always themselves. Resentment consumes energy through rumination and can adversely influence the natural history of sleep disorders, immunity, migraine, IBS, psoriasis, angina and many other conditions.

Resentment is stressful because it consumes cerebral and physical energy (catecholamine, or adrenaline-like production). Many psychotherapists earn a living, trying to help patients overcome their resentment.

Rich, Famous and Conceited

Being star-struck is for patients, not for doctors and nurses. Obsequiousness is out of place, whoever the patient. There is no need to comment further, except to issue a warning: watch out for patients who consider themselves too precious to commune with ordinary mortals. Their demands can prove unacceptable.

Mr. C. owned a UK, nationwide chain of shops. One day he arrived at my clinic with an entourage of three minders. When he arrived, my waiting room was unusually full (a rare occurrence). I could not drop everything and see Mr. C. immediately, so he left in a huff, never to return.

His business office was enormous and had an over-sized desk, placed on a dais. Two minders were always in attendance, standing at his side. From his elevated position, he could peer down on his business associates sitting some distance away on an uncomfortable chair. He had subordinated them before any discourse took place.

Self-Promotional Protagonist: SPP

If you want to sell lots of merchandise, hire a crowd to stand outside your shop. The crowd will attract inquisitive shoppers. If you want your restaurant to get busy, pay people to queue outside; it will soon become known as 'THE place' to be (as long as the price, and quality of food is acceptable).

There are patients and doctors who have sufficient charisma and ability to maximise their image (now referred to as their brand). They easily draw the attention of others and attract interest in their business practices or research. Unfortunately, no disease respects charisma. After achieving fame, dying young can guarantee their posthumous fame.

Patients making a noticeable fuss or drawing attention to themselves, will grab the attention of medical staff, whatever the clinical justification. For those who need urgent attention, but are not demonstrative enough, dire consequences can follow.

Selfish and Greedy

There are those with little regard for others who insist on being dealt with first, regardless of medical priority. Unfortunately, it is selfish, dominant behaviour, that most often draws the attention of others.

Relatives and friends can display selfish, dominant behaviour on a patient's behalf. I have known this save a patient's life when they were being ignored.

I once visited a pregnant patient in labour. Although the chart showed the foetal heart rate dropping, no staff member seemed too bothered. I suggested that a nurse should make an urgent call to the obstetrician on duty; a caesarean section needed to be considered in my inexpert opinion. The SpR who arrived, agreed with me, and told me that nobody had informed him of the situation.

The brother of a female Nigerian patient brought his sister to see me. For her, this was a last ditch manoeuvre. He said, 'If you cannot cure her (of schizophrenia), we will take her into the jungle and leave her!' As a Chief, and head of the family, he told me he could make this decision. He had brought her to the UK as a last resort, since all local interventions having failed. I referred her to a psychiatrist, but never learned what happened next.

SELPO: SELf-**PO**ssessed.

To know them is to understand conceit, vanity and narcissism.

JB (who had played the lead role in a successful film) said he couldn't sleep. He awoke each night, thinking that he was just imagining his role in the film. 'I Google myself to see if I am real - then I can relax', he said when interviewed.

Googling oneself is a new form of narcissism. It is easier than finding a shady grove where lies 'a pool of silver-bright water' in which to see one's reflection (Metamorphoses; Ovid: Bk III:402-436).

SHUBE: Sentient **HU**man **BE**ing.

These patients are a constant pleasure to help, and rewarding to be with.

Simple Folk

To survive modern life it helps to be street-wise. Being academic and learned, may not help; it can complicate life. Simple, less educated folk, have no need of a complicated life; with unmuddled thoughts, they will more easily see the wood for the trees. Like Mr. Wilkins Micawber, in Dicken's *David Copperfield*, they will have learned at least one simple arithmetic process: the subtraction of one number from another.

After subtracting their expenditure from their income, the result according to Mr. Micawber, will be happiness or misery. Meanwhile, the educated can use their time running businesses, performing complicated analyses, using AI and writing books.

Slobs

Unkempt, smelly, unwashed people, all have a story to tell. In the bygone days of doctor's house visits, the decrepit state of some homes amazed me sometimes. Not all those living in squalor were poor. An unkempt state can be a feature of the eccentric. It can also result from dejection, rejection, demoralisation and a psychiatric disorder.

Harry Enfield, in his comic TV portrayal of *'The Slobs'* (BBC 2, 1990), showed that even winning the Lottery may not change personal standards. Slobs might swap their plastic toilet seats for gold-plated ones, but their personal hygiene standards will usually stay the same.

One old, and much debated social chestnut remains. The question is, do slobs make slums, or do slums make slobs? Both of my grandmothers lived in Victorian terraced houses in the East-End of London. Both had an answer to this anthropological question. One of them once said to me, 'everyone can afford soap'. Both scrubbed their front steps and pavement down to the curb. Their action went beyond personal pride; it showed civic pride.

The primary role of healthcare workers is not to judge people but to help them, especially those who cannot help themselves. Although it can suggest a psychiatric disorder, mental incapacity, or physical disability, incapacity can relate to laziness or to drug and alcohol abuse. Coping with stress can cause the disintegration of self-care. Many of those who suffer the effects of stress cannot function normally; some need urgent help. Others are happy and content to remain dishevelled, with no reason to change.

When a methylated spirit (methyl alcohol) drinking vagrant died on the street, outside the London Hospital, in East London, the pathologists working there would correctly guess their former drinking status. At post-mortem, 'meth's drinkers' had liver damage, but arteries free of atherosclerosis ('furring'); just like the arteries of a baby. Despite knowing this, I have never recommended methyl alcohol as a substitute for 'statin' drugs!

Smiling Depressives

Some people are humorous by nature and will laugh and joke even when depressed. They can also smile when scared to death or troubled by bad news.

Spies and Intruders

Just occasionally, inquisitive or jealous doctors, will get their friends and relatives to spy on the competition.

Doctors and nurses can expect telephone calls and visits from potential patients searching for information. If anyone, including a doctor is successful, others might want to steal or learn from, their intellectual property.

It happened to me occasionally, so I trained my staff to recognise spies. Medical spies give themselves away by using technical medical jargon; by being too knowledgeable, and by probing too much. They always avoid giving their name and contact details, since they would not want to be checked on. To confirm their suspicions while dealing with a telephone enquiry, my staff would ask them a direct question: 'For whom are you spying?' The question often brought about a little spluttering, and an abrupt end to the telephone call.

Our regulators sanction spying on doctors. In 2019, I pressed a senior, legally qualified CQC inspector, to confirm that they had sanctioned spying on me. She sheepishly admitted that they employed 'agencies' to collect information.

These 'agencies' made many telephone calls to check whether I was continuing to see non-cardiac patients after I had stopped providing a private general medical service (December 2018). A CQC GP Inspector, who had not long before visited my practice, made a follow-up telephone call posing as a cardiac patient. He said he had chest pain and asked my cardiac technician for advice. The CQC obviously questioned our ability to provide sound cardiac advice for symptomatic patients. Although tiresome, she played along. What these callers did not know was that we recorded their telephone numbers. We would then return their calls to witness their squirming denials of spying.

It might stagger some patients to know that there are medical bureaucrats who have little respect for any doctor, even though entrusted with the lives of others. Their bureaucratic position entitles them to disbelieve doctors and nurses, and to consider them incompetent until proven otherwise. With no clinical acumen, how can they hope to make any balanced judgement? Their *raison d'être* now rests heavily on the likelihood of another Dr. Harold Shipman or Lucy Letby being found.

Regulators completely failed to recognise Dr Shipman and Lucy Letby. With Lucy Letby, they even ignored the many warnings given to them by senior doctors and nurses. Such disregard for medical expertise is unhealthy.

Medical bureaucrats now have the power to question whatever they think to be 'in the public interest' (a phrase, lacking specificity, but used to justify their every action). I wonder if they know what, in medical terms, is 'in the public interest', other than avoiding death and irreversible injury. They will insolently claim that their mission, to protect the public from doctors and nurses, justifies their every action. It may soon come to a time when the public must say who they trust more. Will it be their doctor, a politician or a medical practice inspector?

'As bureaucracies accumulate power, they become immune to their own mistakes. Instead of changing their stories to fit reality, they can change reality to fit their stories. In the end external reality matches their bureaucratic fantasies, but only because they forced reality to do so.'

Yuval Harari. *'Homo Deus.'* (2015, Random House).

Spiritual

All his life, Sam had been a meticulous furniture business owner, making sure everything he did, worked as planned. Then Sam lost his wife. This was because a gynaecologist had overlooked her cervical cancer smear result and failed to review her quickly enough. Sam had a long history of depression, dating back to before he was married. He had undertaken years of medication and psychoanalysis which hardly helped.

Because Sam had remained bereaved for over five years, I gave him a biographical example of another prolonged bereavement. Queen Victoria had remained in a state of mourning for decades after Prince Albert's death, so I recommended her biography.

Shortly after, I described Sam's predicament in abstract to another patient; one I knew to have spiritual leanings. 'Tell him to come to the spiritual church I attend', she said. 'Perhaps someone there can help him'.

Sam was a hard-nosed business owner who believed in nothing he couldn't touch or see, but next time I saw him, I passed on the recommendation. 'What's the catch?' he asked. I told him that there was no cost to it, and that my other patient was sincere when she told me they had helped her and many others. It would seem he had nothing to lose.

One month later Sam returned. He came into my consulting room smiling. 'You will never believe this,' he said. He had gone to the church and was told to take a seat anywhere he liked. He was told to wait until the female medium on the dais, called him. 'Is there anyone here called Sam?', she asked. Sam put his hand in the air, and the medium said, 'I have a message for Sam from a woman known as Letty (a name only he used to call his wife in private; her actual name was Sheila). She says not to continue being so sad. It's unnecessary. She is happy and says you should get on with your life and be happy too.'

He told me that nobody there could have known the special name he used for his wife. He knew nobody at the church, and had told nobody he was going. He had no evidence of trickery.

His visit to the church proved transformative. His whole family found him a changed man. His depression lifted, and he remarried eighteen months later. His new wife had known Sheila well, and was happy to display her photo in their house, in her memory.

My mother was a spiritual person. Her faith was innocent and untutored, having spent all her early life caring for her older sister. Her sister (the twelfth of thirteen) had been born with Down's Syndrome, the daughter of an older mother. Polly died aged twenty years old, and my mother was heartbroken. My mother became tearful whenever asked about Polly, even when in her nineties.

After my father died, I asked her why she wasn't upset. She told me he was always with her, by her side. She wouldn't explain further. 'Some things', she said, 'Words cannot express'.

She was ninety-five-years-old, one year after my father had died, when I visited my mother. She said, 'I have something strange to tell you. Yesterday afternoon your father came in and sat down in his usual chair. Before you say anything, I know he's dead. It was so unlike him, not to say a word. I offered him a cup of tea, but he didn't respond, so I made it for him, anyway. He left it untouched, got up, and walked out without saying goodbye. Not saying hello or goodbye was completely out of character for him. Being impolite, was not your father at all!'

The ramblings of a demented old lady, or something else? She had seen my father as real, and never questioned the reality of his visit. Several interpretations are possible.

Stalkers

There are patients who will form a personal attachment to doctors and nurses. They might stalk or harass them. A witnessed discussion with them is worth trying. Otherwise, doctors must seek advice from colleagues who also know the patient. Doctors must consider contacting their professional indemnity provider and the police.

Stand-Ups (No-Shows)

Some patients make appointments, but don't arrive. Most patients will telephone to cancel, but some will not. No-shows are disruptive and disappointing. Always identify patient unreliability; it sometimes signals disrespect, but also absent-mindedness or mental illness. Question the type of doctor-patient relationships you want. In the small things they do, people show their true nature.

If new patients do not get the appointment they want straight away (usually asap), they will usually go elsewhere, and not bother to inform the medical practice. A simple but important lesson to learn in life is, some people say one thing and mean another. Note who they are. This is most pertinent when dealing with patients from cultures where 'saving face' is critical. They could lose face if they cancel an appointment and have to explain why. They may choose not to arrive, and not to explain further.

Superman / Superwoman Complex

Many men 'don't do illness.' They think themselves so tough, only a sizeable chunk of Kryptonite would weaken them. They will say, 'Nothing wrong with me, doc! Honestly, I actually don't know why my wife wanted me to come. I'm sorry for wasting your time.'

Some harbour fantasies like living forever, with no possibility of getting a disease serious enough to bring them down. Insight may not be their strong point. Their outlook could have arisen from fearing a nasty diagnosis, or from fearing a procedure that might hurt them or go wrong. Others never give thought to illness; their body image is one of unquestioned health.

Getting such patients in touch with their actual state of health can be challenging. After consulting with them over serious, rather than trivial clinical matters, I would send them a written report. This minimised the possibility of misinterpretation; at least they would have a written basis for their discussions with friends, relatives and other doctors.

Patience is a virtue when dealing with supermen and women. They may eventually achieve genuine insight and be able to think through their situation. Consider making consultations combined; perhaps with a friend or partner who knows them best.

Takers and Users (see Manipulators)

These are people who want unpaid servants. Look for the pathognomonic sign: they are demanding from the start, even before first meeting a doctor.

As I mentioned before under 'Manipulators', many want an unpaid PA. The commonest request is, 'Could you make an appointment for me' (for testing, or further consultation, etc.). If agreed to, it will involve medical staff making several telephone calls, backwards and forwards. With no access to their diary, it won't be easy. Those who comply with such demands, will have surrendered to their will. Don't let it happen. Medical staff have too little time to play the role of willing servant. After refusing their requests, they will not be gracious.

One rule that applies to all 'takers' is: the more you give, the more they want.

From the darker side of humanity, come these objectives:

- To gain more and more using any means possible, and

- To gain more while doing less.

Some takers and users are good at hiding. Hidden within many large organisations are those who get paid, regardless of their performance. When pushed to work harder, only the dedicated rise to the challenge; users may not feel the need.

Those in receipt of generosity, will paradoxically accuse their donor of meanness, when the largesse becomes reduced or discontinued. Don't be concerned if you are the donor. They will quickly depart for richer pickings, usually without a word of thanks or appreciation.

There is another paradox. Those who are rich and advantaged, rarely show ingratitude. Because the wealthy are used to attempted rip-offs, they will usually respond gratefully to generosity, and most likely to return a favour in one form or another. The under-privileged, who feel they need to fight for everything, quickly learn that limited demands are unproductive. As a result, they often push their demands to the limit. They may not understand why so few wish to befriend them.

Some patients are dispassionate users of medical services. All they want is to get 'fixed', and leave. When a doctor retires, or dies, they will give them no second thought; they will simply transfer their requests to another convenient provider. This type of patient can be completely unfulfilling to deal with since their needs require little clinical acumen. Because developing a doctor-patient relationship is unnecessary, a nurse, paramedic, pharmacist, or a computer avatar will suffice. Because the vast majority of GP consultations are of this type, the large-scale replacement of GPs by alternative health professionals, using AI driven computer avatars, is inevitable.

Troublemakers (see Complainers)

There are those who cause trouble wherever they go. They are quick to demand their rights and try to gain an advantage by listing all the faults they have noticed about a doctor and her practice. Their manner on the telephone, even on first contact, should be a warning of what is coming.

Most troublemakers are missing a trick. They will get far more satisfaction if respectful, and if they present a reasonable case. When their approach is aggressive, emotional, dogmatic, incorrect, or unreasonable, they endear no-one.

A lot of troublemakers have failed to get the respect they think they deserve and they get angry about it. Those doctors and nurses who are charitable by nature, might misread them and think them misunderstood.

Some savvy troublemakers, will know to threaten doctors with GMC notification. They know this will induce fear, given a threat to their registration and the years of pointless bureaucratic enquiry that would follow.

In the current climate of litigation, the first thing to do when any verbal dispute arises is to get a witness. The second thing is to verify the complaint. Allow the complainant to speak and repeat back to them, what they said. 'Now let me see if I heard you correctly. What you want is to sit in our reception area, and eat a meal while waiting to be seen. You would also like your child to run free and have full access to every area of our practice while you eat. Is that correct?'

Doctors and nurses must make a note of everything said and get independent witnesses to verify it. Ideally, use CCTV to record every incident. All responses to complainants must follow in writing. Tell them the result of your observations and the thinking behind your assessment and judgement. Follow this with how you intend to resolve the issue. The case alluded to before, involved a patient (not one of mine) eating a curry in my practice waiting area. It would have been better had we placed a large notice banning eating, and restricting public entry to all staff areas, but we never needed it before.

In my clinic, we never encouraged the return of any patient behaving unacceptably (making my staff and others feel uncomfortable). There are plenty of other doctors available to aggravate.

Type A & B

What makes a Type 'A' person? Friedman and Rosenman (1970s), two Californian cardiologists, developed the theory of 'time urgency' and its relationship to cardiac infarction. Type A behaviour may be 'natural' to some, but it can also develop temporarily in those dealing with medical emergencies with time restrictions.

Type 'B' characters typically make sensible plans and act at leisure. They take their time, rarely hurry, and allow plenty of redundancy in their time schedule. The plan is to avoid being rushed.

Jimmy had to catch a flight at mid-day. He awoke at 5am and went to his office first. He had letters to dictate and some business matters to sort out. It was his mother's birthday, so he wanted to buy her some flowers on the way to the airport. He planned to visit her briefly, then get on his way. His wife had sent a message to warn him not to forget his daughter's birthday.

He arrived at the airport with only 20 minutes to spare. He checked-in, and had only just cleared security when he heard his name being called. They were about to close the gate as he arrived panting, having run to get there on time. A Type 'A' character, Jim exhibited this behaviour as a feature of his badly planned, overburdened lifestyle (all of his own making).

Coronary atheroma (artery 'furring') precedes cardiac infarction, with very few exceptions. Without atheroma, Type 'A' behaviour is unlikely to cause cardiac infarction. For those with critical coronary artery stenoses, Type 'A' more than Type 'B' behaviour, will reduce exercise tolerance and increase the risk of a coronary artery occlusion. Friedman and Rosenman showed that those with Type 'A' Behaviour, have double the cardiac infarction risk of those with Type B behaviour.

Vanities and Gods

Human vanity has psychological consequences, and doctors are sometimes called upon to deal with them. Not all the consequences are harmless.

With the latest media services at our fingertips, we can all peer into the lives of others and compare 'them' to 'us'. If the Buddha was correct, and the chief cause of psychological stress is avarice, collecting admirable possessions to distinguish ourselves is common. Now that social media allows others to view us and our possessions, opportunities for avarice have multiplied. Comparing possessions and relationships on-line has led to a new form of stress - cyber-bullying.

Greek mythology long ago made obvious the profound differences between the gods and mere mortals. Many of us now harbour a fantasy: that we, too, should have god-like attributes. The promises made in cosmetic and exercise-related advertising, continually suggest we can all get close. Some humans now feel they already share some of the mythical strength, some of the beauty, and some of the wisdom of the gods. Many now fantasise about eternal youth, and spend a fortune trying to achieve it. As our quality of life improves, the more attractive immortality becomes (at least for those still fertile and full of hope). Unfortunately, only mythical creatures are immortal and able to perform miraculous deeds.

Those who have marketed medical interventions over the millennia, have made a lot of promises to the middle-aged. Some claim that cosmetics will keep their skin youthful and exercise will extend their youth, as well as giving them admirable six-pack 'abs' (abdominal muscles). The object is simple - to attract more admirers.

I have known vitamin and mineral supplement manufacturers promise the avoidance of hair loss, the maintenance of vision in old age, and enough libidinous energy to conserve youthful sexuality. Unfortunately, medical science also makes promises. One is that a cure for cancer and heart disease will soon be here, as long as we spend enough money on research. One genetically engineered step further, and we might all join the gods and have eternal life. Look what humanity has achieved since 1900, so just imagine what the next hundred years will bring. Beware of inductive reasoning and incorrect deductions such as this. It is at least, no bad thing to support hope, promise, and expectation.

The young only rarely consider their mortality and seldom feel any difference from the Gods (now called super heroes).

One hundred and fifty years ago, patients would have been more in touch with illness and death. Many lived in overcrowded, deprived conditions, and many died young from infectious diseases. Overcrowding meant that pneumonia and streptococcal throat infections spread easily. Diphtheria and TB were common. Funerals of children were frequent, and the average life expectancy only 40-years. Now perfect health and improved longevity are what many seek. After all, many adverts are there to remind us that *'we are worth it.'*

Millions of people are driven by vanity; many spending fortunes on chasing or reclaiming their youth. Luckily for them, an army of (medical) 'professionals' lies in wait, ready and willing to encourage their delusions. Growth hormone, B12 and anabolic steroid injections, personal gym-training, dietary advice, and multi-factor food supplementation, all claim to provide a magic elixir.

There is nothing much wrong with trying to preserve youth as age progresses, but for doctors focused on illness, pandering to the vain can be tiresome. The vain are much less tiresome when they have insight into the biological limitations of retaining youth.

Because genetic engineering is fast gaining the potential to change what was once impossible, some caution is required. In keeping with results in animals, transforming chromosome telomere nucleotide chemistry may well prolong our longevity.

At the moment, we cannot stop the appearances of aging for long. In the short term, however, I have seen faces made to look ten years younger, after the removal of pendulous eyelids and sagging neck folds. I have also seen aesthetic horrors: tightly stretched, wind-blasted faces resulting from face-lifts. I have also seen unnaturally large and rigid breast implants, completely out of proportion to the patient's body size. Democracy and capitalism have both worked to foster freedom of choice, with freedom to indulge in as much delusion as one can afford.

There is no need to denigrate the vain and the vacuous; they do that well enough for themselves. While a few doctors specialise in serving the needs of the vain, others believe there is no place for it. For those cosmetic surgeons engaged in serious reconstructive work, a side-line of vanity work can provide them with steady extra income.

Some of the 'perfect' health features now sought are:

- Voluptuous breasts (Jane Russell, and Marilyn Monroe – showing my age) - not now required for fame. Julia Roberts, in the 1999 film *'Notting Hill'* (PolyGram Filmed Entertainment) rhetorically questioned their attraction to men. Her comment: *'Men and nudity. Huh! Particularly breasts... they're just breasts. Every second person has them!'*

- Pouting, oversized 'fish' lips (comparable to the Giant Grouper).

- An enviable body weight and shape, equal to any bulimic / anorexic fashion model.(see my book 'Who Loses Wins' (2024);

- A six-pack abdomen, and compact 'butt'.

- Sufficient height to attract beautiful creatures. Although Danny DeVito is only 4 feet, 9 and 3/4 inches tall, he has attracted many admirers.

- Dainty feet (Judy Garland, and Cinderella);

- A complete absence of grey hair and wrinkles.

- A beautiful coiffure, or depilation: pubic, cephalic and axillary (varies with country and culture).

- A breath-taking penis size, or tidy labia.

- Floral scented armpits (*de rigueur* everywhere, except on building sites).

The vain will ignore sensible advice if it doesn't suit their purpose. Most can hear, but few want to listen and learn. Appealing adverts can suffer a similar fate; we might enjoy the advert but not remember the product.

Instead of vain aspirations, improved health and disease prevention should be more desirable. One problem is that doctors too often offer tired clichés for advice. These can bore patients rather than help to change their minds. Although true, some yawn-inducing examples are:

- Don't smoke.

- Don't drink too much alcohol.

- Don't overexpose yourself to sunlight without protection.

- Eat less. Exercise more. Get fit. Keep fit (preferably to an athletic standard).

- Eat healthily (well established scientifically, but often based on the prevailing, ever-changing fashion). You can read my books: *'Eat to Your Heart's Content', and 'Heart Sense',* to find out which foods contain the nutrients which might, in theory, reduce arterial atherogenesis (the process of artery 'furring'), the major cause of middle-age death.

This advice may be correct, but is sometimes unrealistic. Why advocate smoking cessation to those over 70-years-old (assuming they smoke in isolation), or suggest the poor should eat five pieces of fruit or vegetables every day, when some fruit is expensive (a remark made by Prof. Helen Stokes-Lampard, Chair, RCGP Council).

During the North African desert campaign of World War 2, Winston Churchill visited Field Marshall Montgomery ('Monty') in his caravan. They ate a meagre lunch. Monty announced: 'I neither smoke nor drink alcohol, and I am 100% fit'. Churchill, the epicurean, replied: 'I drink and smoke, and I am 200% fit.'

Churchill lived for ninety years, and 'Monty' for eighty-eight. Although Churchill suffered a few cardiac and cerebral infarcts, I very much doubt he regretted smoking cigars, or one morsel of his gourmet diet. Genetic profile, not food, controls longevity most. Because this is an unpopular idea at present, media channels usually edit it out.

Vulnerable Patients

This subject needs its own treatise. Older adults, children, the demented and confused, are all vulnerable to the adverse will of others.

Without sufficient life experience, and having not reached the age of consent, some children are at risk from the adults in charge of their care. We need a greater awareness of the welfare of young carers, those performing poorly at school, slow developers and parents who are uncompromising with their children.

Abuse comes in many forms; some obvious, some not. Among them are physical abuse (look-out for unexplained bruises and injuries); emotional abuse (bullying, threatening, ridiculing, disrespectful, and demeaning behaviour); neglect; sexual abuse, and financial or institutional abuse. Love, embarrassment, and the fear of reprisal will stop some adults reporting abuse. As in all human affairs, there are strong associations with poverty and wealth, socio-economic class, and educational attainment.

The Wealthy

Is 'old money' more acceptable than 'new money'? Money was once a taboo subject for the wealthy and privileged. After all, when money is taken for granted, what is there to discuss? The discussion of money was once vulgar, especially if slavery, child labour or criminal activity were involved.

Those with 'new money' are much more likely to flaunt it, especially if they were once poor. I have known some say, *'I want the best, and I don't care how much it costs'.* That is usually before they know the cost. They are likely to be impressed by the trappings of a doctor's success – a fashionable address, smart consulting room décor and state-of-the-art equipment. The alpha-male behaviour of some doctors who distance themselves from patients, with large intervening desks and dutiful, uniformed staff in attendance, will impress them. Unfortunately for the profligate, none of these are reliable indicators of a doctor's expertise.

Those who have built their own businesses will have used tight expenditure control to create their wealth. Thrift, not largesse or fantasy, has more often led to their businesses success. Some wealthy business people as private patients, will try to exert comparable control over their clinical management. I welcomed it when it favoured sensible compliance, but not when it created demands for irrelevant tests and interventions. Unfortunately, some newly rich people, having achieved much success in life, may believe in their infallibility.

I had a billionaire patient with blackouts who refused for over one year to have a much needed pacemaker implanted. Was his resistance to the procedure motivated by the cost, the inconvenience, or by ignorance and fear of the procedure and its benefits? After pacemaker implantation, he had no further blackouts. I resisted saying, 'I told you so!'

Two advantages of practicing privately are continuity and more predictable patient compliance. With no time restrictions, and time for full discussion, patients can more readily gain a full understanding of their condition. The quick return of their laboratory results, and the absence of any waiting list, were other advantages. These factors benefit clinical outcome.

A simple, but shocking fact, is that far fewer private than NHS patients die on waiting lists or go blind because of untreated glaucoma. The question is, what part has government interference and UK medical bureaucracy, played in the growing divide, especially now that doctors are no longer executive controllers of medical practice. Before we accept political control further, should we not know the answer to the question, *'what contribution have medical bureaucrats made to patient morbidity and mortality?'* A full public enquiry is appropriate, because they have likely increased both, and continue to waste lots of public money. Regulators and NHS executives are not acting solely 'in the public interest'.

Wealth allows the rich to demand 'the best', and ask for 'the best that money can buy'. This has sometimes resulted in my patients travelling to other countries for treatment. Despite what advocates of the NHS believe, the UK is not the only place where world-class medical expertise exists. Wealth allows unrestricted choice, and a wider range of medical possibilities, with accusations of élitism rarely of concern to those whose lives are at stake.

Art thou poor, yet hast thou golden slumbers?
O sweet content.
Art thou rich, yet is thy mind perplex'd?
O punishment.

'Sweet Content.' Thomas Dekker (1575 – 1641).

Wimps?

Also known as (snow) 'flakes', those who dither and constantly change their mind, can be afraid of their own shadow. Some people just can't take the physical, emotional, or cognitive demands of life, and they must choose their lifestyle carefully. People dither for other reasons. Some are in distress; some are so pre-occupied with personal problems, they choose to drive their vehicle into a brick wall. The resulting injuries, should they survive them, could put them into hospital. It seems harsh, but some patients might then find their illness or injury a convenience. It can give them time to consider their life and how to cope with it. Hospital admission will remove them from the tribulations of life and allow them to think.

A sympathetic clinical approach from doctors and nurses is appropriate. Some who feel stressed in life will have hypertension, increased blood coagulability, reduced immune resistance, inappropriate changes in behaviour and depression. When pushed towards a metaphorical medical cliff-edge, one further stress could bring about a sudden dramatic change; perhaps a catastrophic medical event. In the tough days following the 2nd World War, fractures and serious medical conditions were the only medical conditions seen as legitimate. The results of stress and deteriorating mental health remained unacceptable excuses for poor performance.

To get the full psychosocial-behavioural picture of such patients, a conversation with their parents, partner, friends, or work-mates may be necessary. Doctors versed in the art of medicine will know this; those who are not, should limit themselves to fracture management.

WISC: Wolf In Sheep's Clothing

These patients may act as if they are innocent and naïve; instead they are smart and manipulative. Some patients will use this ploy to get what they want from doctors. Their assumption is that doctors won't see through their artifice. Given that few doctors are as street-wise as their patients, they might be right.

Chapter Six

The Doctor-Patient Relationship

Having detailed something of the behaviour and character of doctors and patients, I next consider how they interact in a doctor-patient relationship.

The doctor-patient relationship is the alpha and omega of successful clinical practice. Patients and doctors, however, will set different criteria for what they regard as success. Whatever they may be, the effect doctors have on the morbidity and mortality of their patients provides the ultimate measure of their performance.

Every patient will want to know they are being cared for by someone capable and considerate; doctors must aim to make correct diagnoses and to prescribe appropriate, effective interventions. Patients will judge how doctors perform, and no self-supporting doctor can be impervious to their criticism; those working for the NHS in the UK have less need to agree.

There are innumerable character combinations that define the doctor-patient relationship. The nature of the relationship varies from mutually co-operative, to uncooperative; from engaged to disengaged; from formal, detached, and anonymous, to friendly and personal. Whatever its character, the purpose of the relationship is to expose the information required to make an accurate diagnosis and to prescribe the most appropriate intervention. It is essential for doctors to know enough about the patient and their circumstances to create a fitting management plan.

The doctor-patient relationship always starts with an enquiry made by a potential patient. If you are a specialist in the UK, it will usually be GPs (or another specialist) who will refer patients to you. In other countries, patients can go directly to the doctor of their choice. Assessing the patient's clinical priority is the first issue. In my practice, I often answered the telephone in order to make this assessment. Some patients questioned why I did this, asking 'Can't you afford a receptionist, doctor?'

Things which matter most must never be at the mercy of things which matter least.

Johann Wolfgang von Goethe

On the telephone, a strong and confident voice, with verve and responsiveness, will contrast with the subdued, slow or laboured voice of some suffering from a physical illness. Actors know how to project such voices and some patients try to feign them in order to get attention. In patients with genuine physical problems, both clinical improvement and deterioration is often apparent in their voice. Voices can range from *soto voce* to those loud with hysteria or anger. The voice of hysteria is easily spotted. Imagine someone who says, 'If I don't get what I want, I will scream and scream until I collapse!'

Most consultations are one-off occasions with an easily made diagnosis and simple management. At the other end of the scale are the more complicated presentations, requiring diligent enquiry, investigation, and evaluation. This process will call upon in-depth clinical knowledge, experience, and knowledge of the patient. By employing both the art and science of medicine together, the patient is more likely to walk away satisfied.

Patients with recurrent medical problems may need something other than a diagnosis or the elimination of disease. Some have deep-seated emotional and social problems, for which effective help requires a doctor to use himself as a 'therapeutic agent'. Psychotherapist Michael Balint wrote about this process in *'The Doctor, his Patient and the Illness'*, Tavistock Publications, 1957.

Patient Education

I was once told by an experienced older GP that 'Patients have to be trained.' By that, he meant they needed to learn when, and when not, to call on him. He thought they should know the importance of compliance to his instructions, and how to keep their answers short, rather than recalling every detail from 'once upon a time'. Few of my patients needed 'training' since most were intelligent, educated, naturally polite, well-mannered, and respectful. They were more easily satisfied without the consultation time restrictions of the NHS.

The initial function of the doctor-patient relationship is to exchange information, as well as mutual education. It would be wrong to assume that every patient wants to understand the technical details of their medical condition. No patient, however, should agree to any clinical management decision without some understanding.

Should we always expect patients to understand enough to give consent? In complicated cases, few patients will understand all the clinical facts, the inter-relationship between them, the relative relevance of each piece of data, and the significance of any devils in the detail. So is it safe to assume a patient knows enough to give their consent?

The law in the UK has it we must provide enough information for patients to understand what is being advised, and the risks involved. If harm results from them not being given sufficient information, it will be the fault of the doctor in charge of their case (See: Chester v Afshar: [2004] UKHL 41). Fary Afshar became a neurosurgeon at St. Bart's Hospital, London. We began our careers together as medical students at the London Hospital Medical College in 1961.

In order to allow patients the understanding they need, they might need to know some elementary anatomy, physiology, and biochemistry, explained with diagrams, models and verbal descriptions. By including analogy, metaphor, and even parody, patients can learn more easily.

A Few Basic Communication Principles

As an aid to understanding clinical decision-making, there are a few basic principles to know about the processes involved, and how mistaken thinking, false information and misinterpretation can create problems.

Inductive thinking can be misleading. Although aspirin can lower a patient's temperature, but does that mean it has cured their pneumonia?

A low-fat diet may lower blood cholesterol, but can it reduce atherosclerosis significantly? Feeding animals saturated fat can create atheroma, a process lessened by certain vitamins, amino-acids, and minerals (see D.H. Dighton. *Eat to Your Heart's Content and HeartSense*). A better treatment would reliably affect the biochemistry of the intima (inner artery lining), reducing the progress of atheroma, regardless of diet. We need to decrease the production of LDL-cholesterol in the intima, while increasing nitric oxide (NO) production. As an anti-oxidant, NO can help prevent the formation of pro-inflammatory ox-LDL, a forerunner of plaque fissuring, clot formation, arterial occlusion and infarction. 'Statin' drugs and the proprotein convertase subtilisin/kexin Type 9 (PCSK9) inhibitorsboth influence these processes.

The Goldilocks Principle. The dose of a drug has to be 'just right' for each patient: not too little, not too much. Because patient responsiveness to drugs varies widely, it is important to question fixed dosing that relates body weight to dose and efficacy.

Catastrophe Theory. Standing one step before a cliff edge is unsafe, but not as dangerous as taking a further step. It would allow freedom of flight, with many disastrous consequences to follow (depending on the height of the cliff). One job for all doctors is to keep their patients away from metaphorical cliff-edges. For this reason, always question whether a frail, sick patient, should undergo surgery; surgery will be that one step too far for some.

A tap constantly dripping water into a bath with the plug in, need not signal impending catastrophe at first, but many will wait until water drips through the ceiling below before they check the bath again and take action.

Two clichés are appropriate here:

- Prevention IS always better than cure, and

- Vigilance pays.

Chaos Theory. Minor changes made initially to every system (the weather; the cardiovascular system), can lead to major problems or benefits later on. There are two corresponding implications: to nip every adverse clinical trend in the bud, and to wait patiently for benefits to appear after positive intervention.

Clustering. Every object and event in the Universe clusters together. Stars in galaxies, luck, and even buses arriving at bus-stops, cluster together. Never are they evenly spaced in time or place. Fractured hips in older adults cluster in winter, and lead on to other serious medical conditions - deep vein thrombosis, pneumonia, and death. In time, order gradually disperses (increasing entropy). Without sufficient energy input to maintain the order, all systems eventually fail and fall apart. To maintain a patient's clinical status quo, one may need to work hard; to improve it, one may need to work even harder.

Universal Flux. The only constant is change. The entire universe is in flux; coming and going, moving in and out like the tides. Everything comes and goes in time, the degree of clustering included.

Sometimes changes happen quickly, like share prices on Stock Markets, or the pulsating function of atrial pacemaker cells. At the slower end of the frequency scale are changes in ocean tides, weather patterns, global warming and cooling, and physiological cycles, like female hormones and human growth patterns. Any process that appears to be constant should come as a surprise. Either perfect balance pertains (homeostasis), or the changes are too slow to observe. Homeostatic processes oppose changes in the internal cellular milieu (pH, sodium and potassium concentrations, etc.), but are very sensitive to energy supply. When cellular energy supply fails, homeostasis fails and life ends.

A British Love Affair

In-patient to Doctor: *When I was in your NHS hospital, I never saw you, even though your name was above my bed.*

Doctor: *Tell me, do you go to church?*

In-patient: *Yes.*

Doctor: *Did you see God while you were there?*

An old school friend of mine with a malignant melanoma was strangely content with the NHS system, one that caused him to wait four weeks before having it excised. Having explained to him that our Queen would not have waited more than one day (the very best medical practice standard), he exclaimed: 'What can I do? I'm not the Queen!' Even knowing that his life might be at risk, and knowing that time was of the essence, his loyalty to the NHS and its patient management remained unshaken. He assumed the NHS system was doing the best possible for him. The many secondary deposits that arose while he was waiting, made another case. His blind faith in the NHS as a sacred cow was admirable, but did not serve him well.

The love British citizens have for the NHS (the halo bias) has a negative side. It biases us against the much better medical services provided by the Dutch, Spanish, French, Russian, German, Greek, Swiss and Scandinavian health systems. Some British people, including doctors, actually believe that getting ill abroad (EU & USA) would be a catastrophe. Some UK doctors I knew regarded recuperation, or convalescence 'abroad', as ill-advised (the Status Quo and Geography biases). My advice is to experience the work of doctors in other countries, before making insular judgements.

Breaking-up is hard to do

Sometimes, it is necessary to end an unsatisfactory doctor-patient relationship. This should happen once a doctor and his patient have lost respect for one another. One should always confirm the breakdown in writing; there is no place for misunderstanding. I only wrote three such letters in forty-six years of private practice.

Bending the Rules

Some patients will try to shortcut basic clinical procedures like history taking, examination, investigation, diagnosis, and professionally considered management. Some would-be private patients (those who cannot afford it) will try this to save money. Having made their own diagnoses, they will want to bypass fundamental procedures and confirm their suspicions in the cheapest way possible. They may demand isolated tests (blood tests, an ECG, exercise ECG, etc.), without understanding their clinical relevance. They may jump to conclusions, having erroneously assumed for instance, that a 'normal' ECG has sufficient diagnostic accuracy to rule out all heart disease. I have always rejected such isolated requests, unless obviously related to the recurrence of an old established condition for which there is ample evidence. Even then, caution is required; one cannot obviate the duty to make safe clinical judgements based on a full clinical appraisal. Even if a self-made diagnosis is correct by chance, it will not usually lead patients to the most appropriate management.

Doctors, accountants, and lawyers are all asked to bend the rules. Tempting as it may be to satisfy a special client, especially if they are powerful, rich, or famous, one must resist. Some patients will ask you to bend the rules as a test of your integrity. They may want to find a professional willing to do their bidding.

Imagine the pressure to comply, faced by Dr. Conrad Murray. In 2011, Michael Jackson begged him for stronger and stronger sedatives to help him sleep. Michael, a chronic insomniac, always felt tired and stressed. He died after an intravenous infusion of profenol.

I am not privy to the details, but I can imagine a scenario in which Dr. Murray was arm-twisted to comply, out of a sense of loyalty and duty. We must always ask, 'is what I am doing in the patient's best interest?' The answer is not always straightforward. It might serve a short-term need, but not the patient's long-term interest.

Dr. Murray was convicted of involuntary manslaughter. He served two years of a four-year prison sentence.

If there were Ten Commandments of professional practice and business, one would be never to set yourself up for failure; explain to everyone what you are prepared to do, and what you are not (and why).

Unless you have special knowledge and training (certificated, of course), regular referral to others is now essential. You cannot work in any specialty without a certificate of training; it will limit your range of work, but avoid the CQC and GMC claiming you are working beyond your expertise (not something they would know much about without a certification process). This safeguards patients, and like the regular appraisal and revalidation doctors must all undertake, it generates significant income.

Although doctors referring patients to other doctors could be in the patient's best interest, it is not always so. It is possible to refer patients mistakenly to certificated doctors who have little experience and poor skills. To be certificated implies being fully trained and clinically able. Referring doctors should always keep a keen eye on their patient's progress and be ready to intervene if unacceptable. This is an important pastoral function for all doctors in 'attached' mode – protecting their patients from ineffective, 'detached' colleagues (Michael Balint's 'collusion of anonymity' may apply). I have only rarely had to intervene when my patient's condition deteriorated with ineffective care. In private practice, I spent time evaluating those colleagues I referred my patients to.

Style

My mother's ancestors were Huguenots, and I suspect I have inherited a style, said to be typical of the French. In Hazareesingh's book, *'How the French Think'*, he quotes the philosopher Montesquieu, who wrote that the French 'do frivolous things seriously, and serious things frivolously' (*'De l'esprit des lois.'* 1748. Paris). This French orientation of mine was not always acceptable to those of my colleagues who were serious by nature, and who treated even unimportant issues seriously. Some seemed weighed down by the gravity of their status and knowledge.

My *trublion* character, typical of many Parisien(ne)s, upset the GMC, PSA and CQC alike. Junior doctors should avoid getting entangled with them unless they intend to change profession or retire, like me. For a doctor disenchanted with bureaucracy-ridden UK medical practice, and well past his sell-by-date for retirement, I relished not having to care further about what they thought. *Vive les trublions!*

Throughout my career, I have successfully avoided those who take themselves too seriously. Instead, I always tried to associate with those who lightly carry their talents and expertise. They are the ones secure enough psychologically, without a need to prove themselves. Many (but not all) patients who came to trust me, appreciated my making light of some situations. They understood my philosophy of life. Patients who had experienced unduly serious medical professionals, readily appreciated that knowledgeable, capable people in any field of work, need not hide behind anonymity, seriousness and any tacit dismissiveness implied.

Some of my colleagues, especially oncologists, once portrayed a funereal manner. The diagnoses they made, and then communicated directly to patients, were not the stuff of comedy (although, 'Patch' Adams might disagree). A more approachable, softer style, will more often inspire confidence and help to reassure.

Old Hat?

A doctor's attire once reflected his professional style.

As a 2nd year student, an anatomy demonstrator once admonished me at length. He found me entering the library of the London Hospital Medical School minus jacket and tie on a Saturday morning. Such strictures of student dress are now anachronistic.

As a guide to my age, I can tell you that some of my teachers at medical school wore monocles and *pince-nez*. After receiving less than a credible answer from a student, the sudden release of a consultant's monocle, would politely signal his disapproval. It was a gentleman's way of tacitly leaving a student's dignity in place. Verbal retributions intended to embarrass soon replaced such gentlemanly admonishment.

Modern society has lost some useful non-verbal clues. In Renaissance Verona, a gentlewoman would signal her interest, or disinterest in a suitor, by positioning her fan in a certain way at the time their eyes met. Louche glances, eyelash fluttering, and head movements that once accompanied fan positioning are still with us. The behaviour associated with cigarette smoking was also once notable; offering a cigarette case or lighter was common (the cost and quality of which said something about the owner); the lighting, smoking, and extinguishing of cigarettes, all displayed styles of behaviour that have disappeared (and with it some cancer prevalence).

In the film *'To Have and Have Not'* (Warner Bros. 1944), the manner in which Lauren Bacall holds her cigarette holder, and how and where she blows the smoke, signalled her thoughts, feelings and desires. To some extent, she wasn't acting. Aged 19-years, she had fallen in love with 45-year-old Humphrey Bogart. The feeling was mutual, and they married the year after.

In the 1960s, Sir Reginald Watson-Jones, one of the Queen's orthopaedic surgeons, was a teacher of mine. He would arrive at the (now Royal) London Hospital in his car, a large Humber Super Snipe, formally suited in tails and winged collar. Matron (now Director of nursing) would meet him at the front door. Even then, he was a sartorial relic. As with all true gentlemen of the era, he treated duchesses and dustmen with the same respect. Hospital staff, his students, and patients respected him for his gentlemanly manner. Was that the halo bias working, or the justifiable privilege one felt being in the presence of an elite surgeon, much honoured and respected for his work?

Refugees from the Health Service

I have dealt occasionally with patients, called 'refugees from the NHS' by my consultant boss Dr. P.N. at Charing Cross. After being failed by the NHS, these patients sought a private opinion. In complicated cases, spending as much time as needed to consider each case still gives private medicine a diagnostic and management advantage. Although making time available is the key, spending time with patients is now a luxury for NHS nurses and doctors. When diagnostic difficulties arise, it is essential to spend more time reviewing the patient's history, examination, and investigation data.

Because urgent cases now go to A&E, or call for an ambulance, NHS GPs now mostly deal with minor problems rather than urgent conditions. These cases usually need little time. When more complicated or pressing cases arise, they will not benefit from waiting 7-10 days for a GP appointment, then to be given only 7-10 minutes consulting time.

Friendly Requests

Is it ethical or advisable to treat friends? It is not, but it happens. Your friends will think you churlish and dismissive if you are not prepared to engage with them. In this situation, you will need diplomacy, kindness, and understanding.

I had many friends who wanted to consult with me formally. There is one advantage to such a relationship: high levels of compliance come with loyalty. Friends may seek your opinion as a last resort, so some sympathy is needed. A suggested referral to the most appropriate specialist was what most of them wanted.

No Room for Ambiguity

I always wrote reports for patients whose medical condition was more than trivial. I summarised all the findings and the reasons for my conclusions. These reports served the doctor-patient relationship that followed, by providing a basis for continuity, further discussion, and a useful form of communication for the patient and other doctors. They allowed the patient to consider the facts and my opinion in relation to that of others. Many filed the documents, and returned with them for discussion. My reports made second opinions easier and helped patients track their progress.

Without such documentation, they might have misquoted me, misunderstood me, or simply not remembered what I said. Patients having their own notes was once unheard of in the UK, but not so in many other countries. My foreign patients invariably came to me with a dossier of their medical history. The paternalistic view of the NHS was once that patient information belonged to them, and could worry patients unnecessarily. Many thought the NHS was a safe guardian of patient information. I always disagreed, and so did my patients.

A Case of Deadly Bloody-Mindedness

John was an ex-convict, and an uneducated 58-year-old. He made money from various nefarious activities, which he described as 'a bit of this, and a bit of that'. He had suddenly become breathless because of fast atrial fibrillation. As an overweight smoker with coronary heart disease and a previous CABG, he was a candidate for anticoagulation.

With his wife present, we discussed his need for warfarin (before factor Xa inhibitors were introduced), and the need for regular blood tests. His view was that I had given him this advice for financial gain, making him return regularly for expensive blood tests. He therefore refused anti-coagulation, but agreed to take low-dose aspirin.

Weeks later, after a stroke, his wife returned to discuss his case with me. She had said nothing during our initial consultation. 'You should have forced him to take warfarin', she said. 'He wasn't to know what was best for him.' I had to explain that I had not actually given him a choice. The best treatment was undoubtedly warfarin (the only anticoagulant available at the time). He obviously wanted to avoid the expense of regular blood testing, even though I would have happily referred him to his NHS GP for further management. He died from a second stroke, not long after. His best friend, who had originally referred him to me, said 'He was my friend, but he was a head-strong, ignorant bastard.'

The Hippocratic Oath and the Doctor-patient Relationship

The Hippocratic Oath recognises medicine as an art. It details how a doctor should behave in relationships with his patients, although it makes only a brief mention of the doctor-patient relationship.

There are several aspects of the original oath that are now irrelevant. Originally, it was sworn in the name of Apollo, and his son and grandchildren, all of whom became Greek Gods. It committed a physician to teach medicine to his teacher's children, and his own children, should they want. It also held that a doctor should regard one's teacher and his family as his own. It disallowed both abortion and euthanasia.

Other elements of the oath still pertain: do no harm, commit no injustice, and advise the best diet; never give lethal drugs (although they can act as medicines I small doses), and leave surgery to those trained in the craft. The oath commits a doctor to purity of purpose and confidentiality. It opposes the seduction of patients, corruption and impropriety. The oath says that to disobey these tenets would deny physicians a 'full life'.

The Actual Hippocratic Oath (Ο όρκος του Ιπποκράτη)

Few now swear the oath that follows, although many patients assume we do. It recognises the art of medicine (all they had, some might say), and its ancient origins. The oath is:

'I swear by Apollo the physician, and Asclepius, and Hygieia and Panacea and all the gods and goddesses as my witnesses, that, according to my ability and judgement, I will keep this Oath and this contract:

To hold him who taught me this art equally dear to me as my parents, to be a partner in life with him, and to fulfil his needs when required; to look upon his offspring as equals to my own siblings, and to teach them this art, if they shall wish to learn it, without fee or contract; and that by the set rules, lectures, and every other mode of instruction, I will impart a knowledge of the art to my own sons, and those of my teachers, and to students bound by this contract and having sworn this Oath to the law of medicine, but to no others.

I will use those dietary regimens which will benefit my patients according to my greatest ability and judgement, and I will do no harm or injustice to them.

I will not give a lethal drug to anyone if I am asked, nor will I advise such a plan; and similarly I will not give a woman a pessary to cause an abortion.

In purity and according to divine law, will I carry out my life and my art.

I will not use the knife, even upon those suffering from stones, but I will leave this to those who are trained in this craft.

Into whatever homes I go, I will enter them for the benefit of the sick, avoiding any voluntary act of impropriety or corruption, including the seduction of women or men, whether they are freemen or slaves.

Whatever I see or hear in the lives of my patients, whether in connection with my professional practice or not, which ought not to be spoken of outside, I will keep secret, as considering all such things to be private.

So long as I maintain this Oath faithfully and without corruption, may it be granted to me to partake of life fully and the practice of my art, gaining the respect of all men for all time. However, should I transgress this Oath and violate it, may the opposite be my fate.'

Translated by Michael North, National Library of Medicine (USA), 2002.

Some Classical Background

Asclepius (Ἀσκληπιός) was the god of medicine in ancient Greek religion and mythology. Asclepius represented the healing aspect of the medical arts. His daughters were *Hygieia* (Ὑγιεία or Ὑγεία, Latin: *Hygēa* or Hygīa), or 'Hygiene', the goddess/personification of health, cleanliness, and sanitation; Iaso, the goddess of recuperation from illness; *Aceso*, the goddess of the healing process; *Aglaea/Aegle*, the goddess of beauty, splendour, glory, magnificence, and adornment, and **Panacea,** the goddess of universal remedy. Asclepius was associated with the Roman/Etruscan god *Vediovis*. The epithet *Paean* ('the Healer'), he shared with his father Apollo. Those physicians and attendants who served this god were known as *Therapeutae of Aesclepius*.

Hygieia played an important part in her father's cult. While her father was more directly associated with healing, she was associated with the prevention of sickness and the continuation of good health. The Romans imported her as the goddess *Valetudo*, goddess of personal health. In time, they increasingly identified her as the ancient Italian goddess of social welfare, *Salus*.

Panacea (Πανάκεια) was the goddess of Universal remedy. She had four brothers: *Pondaleirus* was one of the two kings of Tricca, who had a flair for diagnostics. *Machaon*, the other king of Tricca, was a master surgeon (both took part in the Trojan War until *Machaon* was killed by *Penthesilea*, Queen of the Amazons); *Telesphoros,* devoted his life to serving *Asclepius. Panacea* gave poultices and potions to heal the sick. This brought about the concept of the panacea in medicine, a substance that could cure all diseases. They named a river in Thrace after her (Zlatna Panega, from the Greek 'panakeia').

Asclepius' father Apollo carried him to the centaur *Chiron* who raised him and instructed him in the art of medicine. It is said that in return for some kindness rendered by Asclepius, a snake licked Asclepius' ears clean and taught him the secret knowledge of healing (to the Greeks, snakes were sacred beings of wisdom, healing, and resurrection). The rod of Asclepius, a snake-entwined staff, remains a symbol of medicine and healing. A species of non-venomous Mediterranean serpent, the Asculapian snake (Zamenis longissimus), was named after him.

Zeus killed Asclepius with a thunderbolt after he brought Hippolytus back from the dead, and accepted gold for the deed. Asclepius had brought others back from the dead, so Hades fearing that no more dead spirits would come to the underworld, asked his brother Zeus to put a stop to it.

Source: Wikipedia.

Do No Harm

This basic tenet of medical practice of doing no harm can conflict with the patient's best interest when euthanasia and assisted suicide are considered. Should doctors be involved? In those cases where there is clear evidence of patient suffering, and when a cogent, responsible patient wishes to die, direct medical involvement would be party to doing harm, and party to bringing relief from further suffering in keeping with the patient's wishes.

In the current state of the law in the UK, the act of assisting death will lead to accusations of murder or manslaughter. Proof of a suicide pact will mitigate the crime of murder to manslaughter. Both Hippocrates and UK law are against euthanasia; using the art of medicine might suggest that the relief of patient suffering, should take precedence regardless of all other considerations. If you practice long enough, it is likely you will be asked to help. Would you want to be involved?

The jury at Manchester Crown Court convicted Graham Mansfield of manslaughter on the 20th July 2022 and imposed a two-year suspended sentence on him. His wife, Dyanne, who had lung cancer, asked him to bring her life to an end as part of a suicide pact. The court accepted that a suicide pact existed between them. After assisting his wife to die, his own suicide failed. Using his Midlands' common sense, Graham suggested that doctors and the police should be involved from the start, and should gather the evidence required to exclude coercion and verify a suicide pact. He suggested that, better still, the law should change.

Good Medical Practice. The GMC's Guide to the Doctor-patient Relationship.

The GMC has provided a basic set of statements detailing 'best medical practice' for the guidance of all doctors in the UK. It was reviewed in 2022. It contains much more about the doctor-patient relationship than the Hippocratic Oath. Few would disagree with its clinical practice sentiments, but how many, I wonder, would agree with it being used as an instrument of compliance?

Good Medical Practice is more than just a basic guide to best practice; as a set of rules, it is now taken to be immutable by the GMC / MPTS. This allows it to be used as law to judge and punish doctors for any alleged action. It is clearly written as a set of quasi 'legal' regulations.

Given that those who come to judge doctors will know little about what doctors actually do while coping with clinical situations, *Good Medical Practice* provides them with rules to enforce. They need not engage in clinical interpretation or seek valuable clinical perspectives. Compliance with these rules can become the only issue. They have established Tribunals (MPTS) to examine allegations against doctors and administer judgements, composed of a lawyer / legally qualified chair, a lay person (most with a university degree; many with a law degree), and a medically qualified doctor (mostly GPs, with a perspective limited to NHS general practice, and public service committee work). They can call on a few other types of doctor. Among them are many with a MRCPsych, but few consultant physicians, anaesthetists, and surgeons.

Good Medical Practice states that, 'patients must be able to trust doctors with their lives and health'. But how are patients to know whom to trust? Presumably, they are those who 'show respect for human life' and whose practice 'meets the GMC's standards'. Many other directives pertinent to the doctor-patient relationship are proper and worthy of consideration.

The GMC has set itself a separate, impossible aim, to standardise medical practice. There are many places where blinkered rigidity works well, but not at the bedside. The clauses of *Good Medical Practice* are drafted in generic, corporate terms, so that lawyers and medical bureaucrats can use them for judgement. For experienced doctors, the document has a useful purpose: it forewarns us of every way in which the GMC (through the MPTS) can judge us as 'unfit to practice' (from a general legal perspective, but not from a practicable clinical perspective).

Here are some clauses (punctuation errors left uncorrected) relating to the doctor-patient relationship, some of which are worthy clichés (*my comments are in parentheses*):

'Make the care of your patient your first concern. Provide a good standard of practice and care.' (*needs definition*). 'Keep your professional knowledge and skills up to date.' (*Read every book, and every journal perhaps, and perhaps be at fault if you don't*).

'Recognise and work within the limits of your competence.' (*We are all biased, and will all find it difficult to assess our own competence, or the competence of other doctors without having worked with them*).

'Treat patients as individuals and respect their dignity. Treat patients politely and considerately. Respect patients' right to confidentiality.' (*Ultimately, patient confidentiality can be farcical, because the CQC and GMC can both ask a High Court for permission to access any patient information they deem necessary 'in the public interest'*).

'Work in partnership with patients.' (*How else?*) 'Listen to, and respond to, their concerns and preferences.'

'Respect patients' right to reach decisions with you about their treatment and care.'

'Support patients in caring for themselves to improve and maintain their health.'

'Be honest and open and act with integrity.'

'Never discriminate unfairly against patients or colleagues.'

'Never abuse your patients' trust in you or the public's trust in the profession.' (*The latter is a lofty metaphysical aim, only partially the responsibility of any doctor*).

'You are personally accountable for your professional practice and must always be prepared to justify your decisions and actions.' (*To whom? GMC lawyers, or fellow medical professionals?*).

'You must listen to patients, take account of their views, and respond honestly to their questions.' (*We would otherwise be veterinarians*).

'You must give patients the information they want or need to know in a way they can understand. You should make sure that arrangements are made, wherever possible, to meet patients' language and communication needs.'

'You must be considerate to those close to the patient and be sensitive and responsive in giving them information and support.'

'When you are on duty, you must be readily accessible to patients and colleagues seeking information, advice or support.' (*Only when 'on duty'? How many patients can speak directly to their doctor these days? Gone are the days of pastoral medical care when doctors were always available, 'on duty', or not*).

In its latest review (2022), the GMC added to their *Good Medical Practice,* the duty not to abuse, discriminate, bully, exploit or harass anyone. They recommend doctors help patients to decide for themselves, having found out what matters most to them (a basic art of medicine which may take more time than some doctors have to spare). They want to encourage medical professionals to develop leadership skills (*in preparation for doctors taking more control of the NHS, perhaps?*) and to respect the skills of colleagues (*in hospital practice, this has always been a basic necessity*).

Last, I would like to add a clause of my own to both the Hippocratic Oath and the GMC's *Best Medical Practice*:

Every attempt made by any outside person, organisation, or agency to intervene in, or interfere with, the confidential, contractual, sacrosanct, and mutually agreeable functioning of the doctor-patient relationship needs the consent of the patient. Doctors should regard all such attempts as transgressions of the sovereign duty doctors have to patients, and transgressions of the right we have to engage with patients, in confidence, for their medical benefit. Only the doctor and his patient should be able to sanction outside intervention or any enquiries by others, whatever the plea. This is especially important when the justification is metaphysical and defies definition like: 'in the public interest'; 'the safety of the public', or 'the good reputation of the medical profession'. While all are admirable political objectives, they are all beyond the remit or responsibility of any doctor. Medical regulatory authorities should never use them to judge any doctor without demonstrating their equal experience and clinical knowledge.

Platitudes

Practice your platitudes. The following platitudes are of questionable use:

- 'You can only do your best.'

- 'Carry on as best you can.'

- 'It's a cross you have to bear.'

- 'It's the way it is.'

- 'This is what we normally do.'

- 'You must take what comes.'

Avoid cop-outs like,

- 'It's your age.'

- 'That's what happens (when you don't take my advice.)';

- 'It's the system.'

- 'It's because you are menopausal.'

- 'It's because you are male, female, or other, etc.

-

The implication is always: *because I cannot change your condition or situation, you must grin and bear it.*

At all times, imagine that your most critical, clinically experienced medical school tutor is standing behind you, observing all you do, and about to cross-question you. This will better guide your actions than the GMC's *Best Medical Practice* guide, or the Hippocratic Oath, even though most rule-based doctors will refer to them before doing anything. The inherently capable will never need them, unless faced with allegations devised by the GMC.

Chapter Seven

Cultural Difference and Clinical Medicine

We should respect everyone for their given and adopted culture. We cannot ignore our family culture and ethnicity if we are to understand ourselves, and fully appreciate and respect one another.

Unfortunately, any academic discussion about culture in the present political climate is likely to provoke suspicions of racism and other questions of political correctness, especially among those who have been their victim.

Those with a genuine interest in the identity of others will want to seek a full appreciation of their individuality and learn about their culture, language, and the cultural heritage that shaped them, although all sentient human beings have more in common than differences.

The meeting of cultures presents fewer challenges when there is a willingness to learn, love, and respect others, and when kindness and compassion are used to gain mutual understanding. *Vive la différence*.

I have included considerations of culture for two reasons:

- To help explain the health divide, and

- To help medical professionals better manage those from a different culture.

Some doctors regard any discussion about culture as trespassing; an invasion of personal privacy, and unprofessional. A need to retain anonymity, has always suited those doctors whose sole focus is on practicalities, like fixing broken bones or stenting arteries. They will usually prefer other doctors to consider the patient as a fully functioning, sentient being. Because many patients now know there are better carers than some doctors (healers, hairdressers, pharmacists, nurses, counsellors, psychologists, and osteopaths, etc.), they have come to re-evaluate the value of medical professionals.

Those who see patients as anatomical specimens with disturbed physiological functions, may choose not to discuss personality, social circumstances, or culture. For solving minor medical problems, this approach will work and also suit those who wish to remain anonymous.

The anonymous approach provided by working with medical management algorithms will mostly be good enough to achieve satisfactory results in simple cases. In more complicated medical cases which defy algorithmic diagnosis, the patient's psychology and circumstances may need to be accounted for. Knowing the patient as a character within the framework of their social and cultural context can often solve more complicated diagnostic problems. Discovering this provided me with a diagnostic and management advantage.

One question has tested medical science for decades. Why are there major differences in the morbidity and mortality of different social classes and ethnic groups? With social classes, of whatever ethnicity, the difference is not just a few percentage points of doubtful statistical significance, but differences that are three to five-fold. These apply to both cancer and cardiovascular disease prevalence. In all developed countries, we find similar differences. Is this testament to the same failure of political will, or is there something fundamental about the basic nature of humans that forever divides us?

The wealthy and the poor not only live different socio-economic lives, they represent a different culture and ethos. These differences are at least as wide as those seen between distinct ethnic groups. The many cultural differences between the advantaged, and the deprived, might explain some of the health divide, but unanswered is which differences contribute most?

People's Culture

Those who share a culture share a literature.

Shakespeare's rhetoric in 'The Merchant of Venice' (Act 3, Scene 1), points to the many similarities in behaviour that exist between those of two religious cultures. To win his case, Shylock omits mentioning any differences:

'Hath not a Jew eyes? Hath not a Jew hands, organs, dimensions, senses, affections, passions; fed with some food, hurt with the same weapons, subject to the same diseases, healed by the same means, warmed and cooled by the same winter and summer as a Christian is? If you prick us, do we not bleed? If you tickle us do we not laugh? If you poison us do we not die? And if you wrong us shall we not revenge?'

The cultural differences between a patient and her doctor will present similar challenges. Shakespeare pointed out that we humans have much more in common than not. Although we all have 'affections, passions, and a need for revenge', are they weighted differently by our culture? All those engaged in the care of others need to be aware of relevant cultural differences, and respect them in their work. But what are the relevant differences?

I have chosen not to refer to race, since it is never a primary issue when two people choose to meet, converse, and interact. Although race and culture are important linked factors, it is personality that is most relevant to forming relationships. Once two people have met and want to relate further, race will not determine the nature of their relationship, but culture can. Racists are mostly bigots who choose to look no further than racial origin when forming their opinion of others.

In long-term relationships, cultural difference is a significant factor. While relating to other individuals, it is their set of cultural and other values that can lead to conflict; although initially, they may not be apparent. When two people from different cultures discuss purely objective matters such as science, mathematics, logic, or physics, their culture differences will be irrelevant (unless loss of face is an issue); the opposite can be true when they discuss subjective or personal issues, some of which may arise during medical enquiries.

In the current political climate, any discussion of cultural difference is bound to provoke claims of discrimination. Questions of equality, political correctness, emotional dissonance, racial bias, and human rights can easily arise. This is a subject that needs clinical appreciation, even though cultural values are mostly subconscious, and can evade consideration until challenged. In order to fit-in when out-numbered, some will suppress their cultural sensitivities. Others will bravely promote them (as during the Black Lives Matter movement of 2020, and Pride parades). As a predictor of individual human behaviour (especially clinically relevant aspects, such as directness, honesty, informality, openness, compliance, and reliability), knowledge of a patient's culture can be of clinical importance, especially in matters of clinical management.

The Nature of Culture

Every culture has its own shared set of values and beliefs, assimilated long-term while exposed to a meaningful group (family, tribe, nationality, etc.). From birth, these values and beliefs get imprinted on each of us (K. Lorentz, 1935), shaping our future attitudes and affecting our decision making, judgement, and behaviour patterns.

'Man is born free and everywhere he is in chains.'

J. J. Rousseau

Some of Rousseau's 'chains' are cultural, like the form relationships must take, in keeping with the directives and expectations of family, friends and tribe.

With few exceptions, we are all born into families with their own accepted set of cultural values, slowly assimilated from birth to maturity; something that has been referred to as the 'unthought known' (*'Forces of Destiny'*, Christopher Bollas, Free Association Books: 1989). Few are consciously aware of the cultural values they hold until challenged or compared.

Those who try to adopt another culture soon notice differences, and how behaviour and relationships are affected. Outsiders will easily see the differences if they have a 'fly-on-the-wall' perspective. Flies see what others may not - the unedited, natural behaviour of those belonging to the affiliated group in question.

Some cultures so strongly adhere to their guiding principles, they will not readily accept outsider influence. Clashes between two disparate cultures can be fatal. Some differences are political and religious, like those between Israelis and Palestinians; some are individual, like those between criminals and law-abiding citizens. Clashes also arise from within a culture as between Sunni and Shia Moslems; Protestants and Catholic Christians, and Orthodox and Reform Jews.

Some cultural differences arise in response to minor matters. I have sometimes heard someone say, *'Somebody should do something about that. It's not my job to clean the streets. I pay my taxes.'* Someone I knew well expressed this view. He was commenting on the litter outside his home, in a deprived area of London, UK. I never identified with his outlook, undoubtedly assimilated from his family culture. I grew up accepting social responsibility, without delegating or deferring my duties. His comment evoked interpersonal friction between us and exposed our cultural differences. My attitude was that if he didn't like it, he should pick up the litter himself; his response was to recite his rights. There are many sources of cultural friction. Among them are attitudes relating to age, fame, fortune, authority, religion, politics and more.

To judge, or to understand the judgement of others, something of their culture might need to be considered. One can easily extend this to the cultural differences between nations, some of which are obvious to tourists and those who watch foreign language TV channels and films.

In the Sicilian TV crime series 'Il Giovane Montalbano' (Episode: Terzo Segreto, The Third Secret), detective Montalbano says to a junior colleague: 'Institutions, are less important than a man.' Montalbano's humanity and judgement are clearly on display throughout the episode, illustrating how justice can result from turning a blind eye – at least in Sicily. In my youth, the same wise attitude existed in the UK. The mantra 'we must obey every rule at all times' has replaced it and is now so firmly accepted in the UK that the value of some institutions would seem above that of humanity. Some will make exceptions to rules, but only under political pressure.

Left alone in a room together, people naturally commune with those who share their age group and socio-economic background. We all naturally gravitate towards those we find attractive, those we identify with, and those who present minimum disparity. Those who share a cultural identity tend to collect together. Those seen as outsiders may have to work hard to get accepted by them. It will apply to the novice musician who wants to join a professional music group, or to a Christian who wants to become an Orthodox Jew. There can be two-way traffic, but it is not always of equal measure. It is a lot easier, for instance, for a Jew to become a Christian than the reverse.

Distinct cultures form when enough individuals accept the same set of attitudes, values, behavioural reactions, and ideas. Many cultures prescribe exactly what one should do in various situations - from the formality of meeting others, to the etiquette of eating and courtship rituals. They include acceptable actions and attitudes towards age, gender, sexuality, family, tribe, race, religion, vocation, beauty, vanity, fashion, wealth, education, social class, and nationality. Both Judaism and Islam are replete with specific instructions for acceptable behaviour in every conceivable situation, whereas Christianity only prescribes general principles. In order to behave towards others in the most respectful way, some knowledge of their culture is essential.

The set of human values we all hold - our ethos - directs both our thinking and our behaviour; it is our imaginary collective ethos that loosely defines our national culture. How we practice our personal and collective culture, differentiates us as individuals and helps define our family, tribe, religious sect, and national affiliation.

Cultural History

Those in my age group might remember Jack Warner in the TV series *'Dixon of Dock Green'* (1955-76). As an older police officer, he dealt with minor incidents from the point of view of personal wisdom, all of which he had learned from experience. Although fictitious, his humanity was on display. One question I have is, where is such humanity now hidden? We might, of course, find it buried beneath piles of books defining immutable rules and regulations; books which devalue personal judgement. Many old-style police officers sought alternative occupations because of the change. Medical practice is experiencing the same phenomenon.

In some countries, individual police officers can still make 'on the spot' exceptions, using their experience and wisdom; they need not adhere to a book of rigid rules for every occasion. While there are many advantages to having clear rules and regulations, applying them without exception can involve some loss of humanity. In every situation that demands fairness, one must ask which is the more important: the value of compliance or humanity? In the UK, in all walks of life, the rules now have it.

In prehistory, the family was the central social unit. The family assigned divine attributes to their forefathers, honouring them in regular prayer. Although such penitence has disappeared, the family remains the cultural rock on which many nations are founded. In historical terms, there has been a cultural progression from family-centred to nation-centred culture. Over millennia, distinct cultural values have become attached to different countries, some with distinct regional sub-cultures.

Obvious differences exist in the UK, between those who express the same Caucasian culture as my parents and forefathers, and those who have come from Asia, Africa, the Middle East, and the Far-East. A Caucasian doctor treating someone from one of the latter areas must learn the differences. Those arriving from the Middle East, for instance, will expect their whole family to be involved in consultations and in management decisions. To engage with them sensitively, a doctor will need at least six chairs for the family in his consulting room. Any doctor who lacks pertinent cultural knowledge and sensitivity, will find himself ill-equipped to practise effectively in our multicultural UK.

Family Culture

Some families have dominant values, like honour, honesty, humility, service to others, and respect for humanity. Sports-oriented families will support winning as their primary goal. Others will regard fairness, justice, and public service as more worthy. Public service to others is of paramount importance to some families. Others want to accumulate wealth or sit back and relax, while others do the work.

Before I became a teenager, an objective scientific outlook was the prevailing theme of my Quaker neighbours. It was John Powell (pharmacist) who introduced me as a young boy to the value of inquisitiveness; in particular, the discovery of how things worked. Why are leaves green? How does a camera record pictures (then on film)? He raised a myriad of other questions for me and his five children to answer. This made us think for ourselves. I have never stopped.
As a teacher, and later my devil's advocate, he insisted on research, verification, thinking about evidence, and only then concluding our thoughts. He highlighted how false evidence could fool us.
When I caught up with John thirty-years later, he asked me: 'Why do they pay you (as a cardiologist)? Heart disease is not going away, so how do you justify your salary?' His attributes as a mentor and critic had not diminished. His eldest daughter became a mathematician, his eldest son an engineer. Two others became doctors. We all remain indebted to him. He inspired us with a love of knowledge, and a quest for answers that would withstand independent (his) scrutiny.

The values of a patient's family, clan, tribe, club, religion, race, or country need to be appreciated if one is to engage with them convivially. The same holds true for patients trying to understand any doctor or nurse. Without some mutual appreciation, medical management could be inappropriate or unacceptable. For reasons such as loss of face (important in many Eastern cultures), patients might not choose to comply. Such considerations could perplex those with insufficient cultural knowledge.

Ranbir Singh Suri, Baron Suri, said in one of his thoughts for the day on BBC Radio 4: When two people find enough in common to call themselves 'us', they will soon find someone to attack, namely 'them'.

Anthropology and Medical Culture

My ex-partner, Dr. David Baxter, is the author of an important anthropological theory. I found it useful when trying to explain the differences between some patients. From his study of prehistory and anthropology, he suggested that two sub-species of *Homo sapiens* (meaning 'wise man') evolved. Around 40,000 years ago, what we now think of as hunter-gatherers and farmers, went their separate ways, preferring different lifestyles. A specific gene mutation might have been responsible for the cultural split that remains today.

The more successful hunter-gatherers would have been opportunists; individuals, dealing with life as it came. Yuval Harari, in his book 'Sapiens', suggests that they would have had fewer worries than farmers whose desire was for a more predictable and stable existence. I would argue that there must have been some pre-selection at work: both anxiety and obsessive neurotic traits would diminish suitability for a hunter-gathering lifestyle, given its inherently unpredictable features. The anxious and obsessive prefer a predictable structure, stability, and security; dealing with life 'as it comes' can be stressful. For these reasons, anxious and obsessive people are more likely choose a fixed abode and the more predictable existence of farming. Although farming is far from stable, some will see it as better than wandering around, hoping each day to chance upon suitable food and shelter. Farmers put down roots, but not only by planting crops. Building a stable dwelling on a plot of land will create the security of a fixed, protected enclosure. In this way, they will make life somewhat more predictable.

As personality types, both farmers and hunter-gatherers still exist. There are still those who lead their lives not knowing what tomorrow will bring. They must be able to handle insecurity. Others seek security by starting a business, hoping to create a satisfying and predictable existence; any anxiety, obsessional drive, or feeling of inadequacy they have, will drive them to create security. Do employees and employers represent the same divide as hunter-gatherers and farmers? We still have those who wait optimistically for things to happen, and those who go out and make it happen.

From an anthropological viewpoint, my colleagues at medical school must have been from farming stock, with deep family roots, and long-term plans for their future security. With the liberalisation of medical school entry, both hunter-gatherer and farming types can now enter the medical profession. Most junior doctors will foresee their career ending with stability, either as a GP, hospital consultant or academic. In order to achieve their ambition to further their knowledge and experience, most junior doctors must travel to complete their speciality training pathway courses. They might need to change location many times. In moving around, one can always suspend cultural values for a while, especially if some longer-term survival advantage is on offer. Medical bureaucracy has made things worse by lengthening these courses and limiting the numbers who will eventually achieve consultant status (reducing the wage bill). As a result, many have resigned.

Cultural Dissonance

In the UK, there are obvious differences in cultural ethos between the NHS and private practice. In the same way that airlines treat their first-class passengers differently from their economy-class travellers, the NHS has for decades presented a down-to-earth approach to patients that differs from the more obliging approach of UK private practice. Patient handling is different enough to regard the two systems as different cultures. Do these differences in culture matter clinically, and can they affect morbidity and mortality? UK patients know they do, but this is not publicised. The Guardian newspaper, on March 27th, 2024, announced the results of a recent survey: 'Public satisfaction with the NHS at its lowest ever level, poll shows.'

Without access to travel, or exposure to others, cultural differences can remain unappreciated. Once known, they can become fascinating or the source of discord. For those less involved, any differences can become suppressed, hidden, or ignored, but never completely denied. Here is an example of a culture denied.

Rachel was a young South African woman who claimed not to be Jewish. She begrudgingly acknowledged that, although she was born into a Jewish family, she now felt no affiliation to them or their religion. She had married a non-Jewish man, against her parents' wishes. Like the famous emperor denying his nudity, the denial of her Jewish affiliation seemed preposterous. Her attitude to her husband, and much else about her behaviour, was completely consistent with the many Jewish women I have known.

At the head of every Jewish home, is the mother and wife. When at home, Jewish men know their place. Only children with a Jewish mother are Jewish, so a Jewish man who has children with a non-Jewish mother, will not have Jewish children (not unless they convert to Judaism). That's how important women are within Jewish culture.

Rachel's husband, a successful company director of a large international food company, knew nothing of Judaism or Jewish family culture, and must have expected a little more discussion about home and family matters. She demanded total control. Although Rachel strongly denied her cultural heritage, the cultural differences between her and her husband brought them into conflict. What I predicted happened. They divorced one year after I met them.

She was a South African from a Jewish background, and he a typical public school educated, Englishman. She despised her parents; he venerated his. His parents baptised him in the Church of England; she professed to have no religion. Their divorce might have resulted from differences in national culture, or from a difference in attitude to their parents, but I thought not.

Culture and Disease

Ethnicity is important clinically. There are heritable diseases with specific racial associations, like Tay-Sachs (an inborn error of metabolism), and Sickle Cell disease. The geographic (and therefore ethnic) prevalence of disease is important. For some diseases, like the malignant hypertension found in those of Afro-Caribbean origin, and diabetes and ischaemic heart disease (IHD) among Celts, and Asians, disease distribution among ethnic groups is a worthy clinical study. Despite these interesting associations, considerations of age, gender, and family history are more often needed.

To what extent do ethnicity and culture contribute to disease prevalence? It can be considerable, but in most countries, serious disease occurs most among older patients, the poor and disadvantaged, and less among the rich and well-educated (the health divide). The serious consequences of COVID-19 occurred most among the same deprived groups that have more cancer and heart disease, with ethnic and cultural differences in mortality emerging as significant. The BAME community in the UK suffered more from COVID-19 than others. Viruses are genetic material (RNA, or DNA), and because the basis of ethnic difference is genetic, some difference in the interactions between them, might have been expected.

National Culture

Is there really such a thing as a national character?

In 19th century Britain, two national heroes provided male archetypes for an enduring national character. The public saw Nelson as a heroic, romantic hero. According to his Morning Chronicle obituary (19/11/1852), Wellington was a soldier with 'the ice of character, and the fires of genuine and self-sacrificing principle.' He never used hyperbole, always showed calm fortitude, and was a master of understatement. When writing to his brother, Henry Wellesley (1814), he added a PS to his letter: 'I believe I forgot to tell you, I was made a Duke.'

A French writer comparing the French to British soldiers said that the British had 'cool, reflective courage, were calm amidst danger, and had patience which surmounted difficulties and stood proof against obstacles' (*Englishness Identified*. Paul Langford. 2000. OUP).

In the century that followed, Churchill wrote (1943), after reading Pride and Prejudice (Jane Austen, 1813), that our character was adequately described as 'manners controlling natural passion'. This character template served the British well throughout the 20th century, when our nation fought wars on many fronts. (Ian Hislop (2012). BBC Two: *Stiff Upper Lip – An Emotional History of Britain*).

These characteristics, hardly belong to every British person, but have been noticeably present throughout my life. They describe the many British characters I admired most. There is, however, a major problem with all generalisations: there are too many exceptions.

Family cultural values can influence our future relationships, through work ethic, educational aspiration, status, and employment (employed, or self-employed; financially dependent, or independent). Our development into either an 'attached', or a 'detached' doctor, is not all inherited (as are autistic and obsessive traits), but born in part from cultural ethos. To a tangible extent, the values we hold and the attitudes we each express, owe a lot to our age and family, the era into which we were born, and our socio-economic background.

Tribal values have ancient origins and underlie some distinct differences in ethos. In classical Greek literature, differences between the Spartans and Athenians was common knowledge. Spartans became known for their warrior virtues ἀγωγά (agogá), Athenians for their more refined, socially acceptable behaviour - παιδεία (paithéa).

People often diverge on philosophical grounds. Nietzsche, for instance, expressed ambivalence about whom to follow: Apollo (Spartan) or Dionysius (Athenian). Some are interested most in fame, fortune, and the power that separates us; others cling to learning, duty, honour, respect, and chivalry. As the need for fame and fortune has grown, respect for others and chivalry have declined.

Interpersonal responses, learned from within a culture, can be surprising when first encountered. Some see the British as trying to avoid embarrassment at all costs (according to the character Archie Leach, played by John Cleese, in the film: '*A Fish Called Wanda*', 1988). In some Middle-Eastern and Far-Eastern cultures, a powerful motivating factor is to avoid loss of 'face'; some might think it better to lie or die, than lose face. In European academic groups, arguing aggressively, and being insensitive to the embarrassment of opponents, is an acceptable form of interaction. An insult capable of inciting physical aggression in one culture, may only bring forth a rhetorical response in another.

We British blush at the thought of being embarrassed (some will bend over backwards to avoid it), and will apologise profusely in order to diffuse it. The paradox is that we British are also fierce warriors. This is something we share with only a few other nations (the Japanese, Russians, Sikhs, Gurkhas, Germans, and Turks, etc.). When pushed into a corner, with honour, fairness, and duty at stake, traditional Brits will forget embarrassment and fight.

The Falkland's conflict of 1982, exemplified British warrior virtues. I was working in Amsterdam, and no Dutch person I met understood why Mrs Thatcher and the British government, reacted as they did. We sent an invasion force 8000 miles from the UK, because Argentina had invaded and renamed the Falkland Islands, Las Islas Malvinas. When faced with danger, cultural considerations can help predict whether an adversary will stand and fight, or flee. They taught Spartans to stand and fight, Athenians to debate and find a solution. Native Brits will usually consider both.

You can neither know a person, nor a nation, without knowing its history and cultural background. Despite this, all human beings have much more in common, than we have differences (as cultural and behavioural characteristics). Small cultural differences, however, can count a lot in everyday relationships.

Within the multi-ethnic environment of the UK, there are varying proportions of 'Native Brits', and recent immigrants. This now makes our society significantly diverse. I have used the term a few times, so I will define what I mean culturally, by 'Native Brit'. Given that we were all immigrants once upon a time, and it takes a few generations for any family to assimilate the local culture, my tough definition is a person whose four sets of great-grandparents, were all born in Great Britain (more than enough time to have assimilated the prevailing culture). Personally, I could add many more generations to the definition. Using fairness (said to be typical of the British), one might accept a weaker definition: having both sets of grandparents, born in the country. This rule, applied in any country, will adequately define those families who have had time to assimilate the native culture.

My English family were among the first to travel to the Americas in the 16th century (a town in Massachusetts, and another in Kansas, bear my surname). I can easily claim to be a 'Native Brit', using my tough definition. I have no problem with immigration at all, but no native family or national culture should not be subdued for the sake of political correctness. The cultural history of my family, back to all my great-great-grandparents, defines who I am. My father's family came from Yorkshire, where several small towns bear my surname (Dighton spelt with 'ei'). It may be illusory, but I always feel at home in Yorkshire, perhaps because I aspire to some of their common characteristics: gritty common sense, outspokenness, and helpfulness to others.

My mother was of French descent which perhaps explains why I fit Montesquieu's description of the French who 'do frivolous things seriously, and serious things frivolously' ('*De l'esprit des lois*'. 1748. Paris). I must get my genetic origins profiled sometime.

My mother was of Huguenot descent, showing that if we were all to trace our families back far enough, we would all be of mixed origin, although perhaps too dilute to be of cultural relevance today. But is this true? Is it not possible that ancient traits remain unexpressed (genetically), and not eradicated, waiting to re-emerge in later generations? When Mendel mixed red with white sweet peas, pink flowers grew initially. After many pairings, some hybrids reverted to their original red and white colours.

Within every country, some regional differences in culture and behaviour, clearly exist. A change is said to occur after every 100 kilometres. There are noticeable differences between the people of Amsterdam and Maastricht; Paris and Marseilles; Milan and Palermo. In the UK, there are some obvious differences between the natives of Scarborough and those of Lancaster; representing as they do, the inhabitants of Yorkshire and Lancashire. Both will recognise their difference from native British Londoners.

There is a specific, statistically relevant, geographic distribution to some cultural traits, albeit with many exceptions. I have found personally (anecdotally, not statistically) that Yorkshiremen speak their mind, and Londoners mostly keep their thoughts to themselves. Lancastrians are mostly open and friendly. Stand on any street corner, looking lost in Liverpool, and you won't wait long before being asked, 'Are yer OK pal?' In London, you might have to wait for a police officer to become suspicious of you, before being approached.

Medical practice in the UK presents every doctor and nurse with many parallel, regional cultural differences to assimilate. Although many differences are subtle, I would advise doctors to consider the differences before choosing where to work.

Humour and Culture

In case you are in any doubt about the reality of cultural difference, put humour and cultural difference to the test. Tell a joke to someone from a distinctly different culture. What will be funny to a Cockney, may not be funny to a Glaswegian. Joking can expose cultural differences. This is a fact well known to those comedians who travel throughout Britain. It can be very stressful for a London comedian to work in Glasgow, and vice versa. Humour can act to dispel cultural difference, but it can also be hard work and disappointing to bridge the gap.

My old friend, Don Russell (once a senior nurse, and former psychiatric social worker), while in restaurants, insisted on telling British jokes to Romanian, or Polish servers. What was humorous to him (the Cockney use of pun, innuendo, derogatory remarks, and rhyming slang), was rarely humorous to them. He often compounded his faux pas by trying to explain the joke. Not even the blankest of looks deterred him.

Don was married to Amm, a Thai lady. Their love allowed them to ignore their cultural differences. He lived until he was 90-years-old, with a completely undented sense of humour.

'Through humor, you can soften some of the worst blows that life delivers. And once you find laughter, whatever your situation might be, you can survive it.'
Bill Crosby.

In every culture, humour has a particular value. Sigmund Freud in his book, *'The Joke and Its Relation to the Unconscious'*, explains how jokes provide immense pleasure by releasing us from our inhibitions and allowing us to express sexual, aggressive, playful, or cynical instincts that would otherwise remain hidden.

Language and Culture

To practise the art of medicine in the UK, an understanding of many cultures is an advantage. Those who wish to remain detached and anonymous, need not bother, but it could be to the detriment of their patients. Not only do we need to understand what others mean when they speak English as a second language, but also their cultural outlook. No matter how well recent immigrants speak English, and how admirably they seem to adopt British values, many will naturally want to keep their native culture alive (albeit suppressed sometimes).

Many who live in the UK are bi-lingual and bi-cultural, and can easily switch between the two as circumstances dictate. Because cultural differences merge over two to three generations, they will gradually cease to be of any clinical relevance. This has not been the case for Hassidic Jews, or some devout Asian Muslims, even though they have lived in the UK for decades (a few centuries, for a few). Those with a more distinct, rigorously defined culture, will usually restrict themselves to doctors who share their culture.

Among those who speak the same language, there are many distinctly different cultural sub-groups, each with their own variation of humour. There are differences, for instance, between Americans (from the USA, or Canada), Australians, and native 'Brits'. Not that this matters much in clinical practice, unless like me, you are prone to make humorous comments to lighten the mood, or to signal openness, friendliness, and mutual acceptance. For those whose mother-tongue is English, or a dialect of it (as used by the Welsh, Scots, and Irish), our cultural similarities predominate.

Language and culture change together. So much so that when attempting to learn another language, one must also learn the cultural mores. Language difference is important in ways other than simple communication; doctors should be alert to the risk of cultural and linguistic misinterpretation. The values held by a distinct culture may lie hidden behind linguistic differences (a problem with all translation). The ability to speak another language does not imply an appreciation of the culture. I have tried to learn many languages, and although this has been much appreciated by many patients, it is not enough. Many subtle differences in culture, limit communication and therefore, mutual understanding. It matters little when dealing with simple physical illness, but illness involving psychological or social factors, can present communication difficulties.

One accepted British characteristic is that few of us appreciate much of foreign languages and cultures. So many in the world want to learn English, we Brits can easily avoid learning other languages. If one achieves any proficiency in another language, some become suspicious, and in some countries they might think we are spying.

Throughout our once extensive Empire, native Brits often expected others to learn English and to adopt our values. The Romans, and many conquerors in the past, did the same. Our arrogant attitude was not to last, although some remnants of our attitude will surface sometimes. One can see it in the way some Brits behave on foreign holidays, and at sporting events abroad. There is another side to this coin. Most native Brits readily accept others, regardless of their race and culture, as long as they make some attempt to integrate. Over a few millennia, the acceptance into the UK of many waves of immigrants, has led to one of the most culturally diverse and vibrant societies on Earth.

Every nation expresses its culture in its literature, everyday expressions, and behaviour. In Greece, one aspect of their philosophical outlook on life is expressed by the word *ησυχία (isichía)* (unrushed peace). In one Amazonian native dialect, there is no word for work; instead, they have adopted the Spanish word *trabajo* (work). *Trabajo* for them, means 'the effort of dealing with strangers' (*The Continuum Concept*, Jean Liedloff, 1975).

Through their use of pronouns and suffixes, many languages express different levels of formality and respect. Most European languages (but not English) have kept at least two levels of status recognition: the formal and informal. In French, Dutch, and Russian respectively, the formal 'You' is *'Vous'*, *'Uw'*, and *'Bbi'*; informally they are *'tu'*, *'je'*, and *'tbi'*. For the Japanese, it is not just the words used that count, but also strict patterns of behaviour that signal the required level of respect.

Gender and Sexuality

Gender, as opposed to the chromosomes that dictate our anatomical characteristics, is now an issue for doctors, especially for those working in sexual health and gynaecology.

Gender refers to that general set of attitudes and behaviours defined as masculine, feminine, androgynous (exhibiting high levels of both), and undifferentiated. Some have shown that the highly androgynous possess higher self-esteem, and have more psychological well-being (see: Flaherty and Dusek, 1980, and Lubinski et al. 1981). Others have also linked masculinity to better psychological well-being. One can assess gender using the BEM Sex Role Inventory (BSRI)(named after Sandra Bem, Californian psychologist). It defines 20 stereotypical male types; 20 stereotypical female types, and 20 that are gender neutral. American students in the 1970s, composed the study group.

As individuals, we can each define an acceptable position for ourselves on the spectrum between the stereotypical male, and stereotypical female. There is now a long list of accepted positions to take, like transgender, bi-gender, transitioning, poly-gender, and binary. Sexuality needs to be considered separately, since there are many partner types; namely, heterosexual, homosexual, and bi-sexual, etc. Sub-groups are likely to aggregate together, and to define their own particular values and tastes; in effect, defining their own sub-culture.

Beauty, Fame, Fortune and Criminal Culture

The famous, the attractive, the criminal, and the advantaged, mostly understand the desire others have to get their attention and recognition. Most will acknowledge how easy it is to seduce those who seek such acknowledgement. They will recognise the sense of privilege others might feel in their presence, and some will manipulate the advantage that can bring. The rich, famous, and criminal each have their own specific sub-culture, recognised among themselves. They are as comfortable together as those who care nothing for fame, glory, and power.

If you want to deal with the famous and advantaged, and survive unscathed, there are a few things to know. If you deal with any of these groups for long enough, disenchantment may set in. First follow a father's advice to his teenage son on his first date: *'treat 'em mean; keep 'em keen'*. The rich, famous, and advantaged can be ruthless in their exploitation of those who fall star-struck at their feet. Their favourite game is, 'You Never Know Your Luck'. Some will imply that if someone does their bidding, many benefits will follow: a ticket to their next performance, *diner à deux*, an invitation to their villa on the Cote d'Azur. My advice to doctors and nurses, is to smile pleasantly, take no heed, and simply enjoy the moment.

My friend, mentor, and colleague, Dr. A.M.H., took the bait offered by an internationally famous film-star comedian, who became his patient. The film-star in question was so famous, he could avoid paying wherever he went. He never paid while staying at the Dorchester Hotel in London, for instance. He claimed that his presence alone was enough to draw in many fans as customers, eager to be in his presence. Dr. A.M.H. implanted his pacemaker and gave him a lot of personal attention. Having offered him a wide choice of cars, in lieu of payment (a Bentley, an Aston Martin, or a Porsche), he disappeared, never to be seen again.

Religion and Medicine

In our multicultural UK, there are many religious practices, values, and behaviours that need to be recognised by doctors and nurses.

In the Judeo-Christian culture, love thy neighbour as thyself, and charity and humility, are central to the teaching. Some religions have complicated rituals that bind the devout, the disciplined, and the observant together, following paths prescribed by the Koran for Moslems, the Shulchan Aruch for religious Jews, and the Vedas for Hindus. For Buddhists, the aim is to attain Nirvana, a state in which there is no desire to gain over others, or covet them.

There are many *faux pas* to be avoided. Non-Jewish women should know not to shake the hands of Hassidic men and Non-Moslem men must avoid touching Muslim women. One has to know, when it is permissible for a man to touch a female patient, and how best to examine them (usually with a chaperone). Regardless of religion, this will be important to those who simply prefer not to be touched. By comparison, traditional Christian Brits (by my previous definition) will only rarely regard these issues as important, although always the subject of individual preference. If in doubt, seek the advice of a colleague who shares the culture in question.

One can use touch in communication, but not all cultures permit it initially. It can transmit acknowledgement, love, dislike, sympathy, and welcome, but with different restrictions, allowances, and inferences applied. Some refer to it as the 'language of touch'. Whether it is a welcoming handshake, a pat on the back as a sign of acknowledgement, or hand-holding as an accompaniment to empathy, each action requires individual consideration and acceptance. Deploying the art of touch can either break bonds or augment mutual appreciation.

The Mixing of Cultures

Before travel developed, there were thousands of isolated communities, each with their own distinct culture. Widespread travel has meant there are now fewer. The result has been mixed cultures providing a rich substrate for gene diversity and healthy evolution. Compare this to the inborn errors caused by family inbreeding (brothers married sisters in ancient Egypt). As cultures mix further, it becomes impossible to guess the contribution each culture makes to each individual. Cultures are in flux as internal discussions arise about freedom of thought and control, and about equality and individuality. Doctors must learn to keep up with the changes.

I have often witnessed the mixing of cultures, leading to painful religious and social conflict for their traditional elders. These elders can present as patients with stress exacerbated hypertension, migraine, persistent angina, eczema, and other predisposed conditions.

From a bio-evolutionary point of view, inter-racial procreation is highly desirable, if adaptability, intelligence, and healthy survival of the offspring remain the objectives. The offspring hold an interesting position: they can decide which elements of each parental culture they wish to accept. For some parents, there will be joy, for others, dismay and powerful feelings of betrayal and confusion should a child jettison their family culture. Parents can disown children who stray too far from their accepted cultural norms.

Equality Issues

In ancient times, the belief was that we are not all born with equal attributes (a biologically correct point of view). In second millennium Europe, it became the Papal Christian view that despite differences in status, everyone at least had moral equality in the eyes of God (the religious point of view. See Larry Siedentop, 2014. *Inventing the Individual*). Political equality, and equality of opportunity, are recent additions. Equality, with the popular meaning of 'everyone is equal in every way', is a biological error that exists only as a political gambit. For the sake of political correctness, it is now unacceptable to think otherwise. Equality of opportunity, however, should always exist.

In the biological realm, equality has never existed, and the political drive to make it so, runs counter to the natural world order. We know that evolution is based on inequality, the survival of the fittest depends on species' variation. This is the reality doctors, nurses, and police forces, face every day. Lions will carry on eating antelope unless the liberal minded can persuade them it is unfair and disrespectful to continue. Meanwhile, within the human social realm, pecking order will always exist in defiance of any political move to promote equality.

Human beings will never possess equal intelligence, skills, talents, or cultural values. Being different is a constant, but not only because some of us had a deprived upbringing, or had resources denied. To hold such views would be to believe that biology is unfair, politically insensitive, and socially unacceptable. It always has been and always will be. To gain an advantage today however, politicians and business people need to keep biology hidden.

Political correctness has inspired a lot of biological and cultural nonsense. One such nonsense is that gender differences and cultural values are superfluous and count for nothing. If that were true, how is it we can distinguish men from women, Londoners from Liverpudlians, and Asians from Norwegians? Although such distinctions are of no consequence while peace reigns, when conflict and social pressures arise, differences soon surface.

Respect for medically relevant age, gender, social, cultural and ethnic differences is essential for doctors working in any multicultural society. Children and older adults need to be approached differently. Women in some cultures need to be treated differently from men; it all depends on their clinical problems. It is essential to keep an open mind, and to only lightly hold stereotypical presumptions (regrettably the reverse has brought stereotyping into disrepute); thereafter, we must review every stereotype we hold.

Just as culture changes from country to country, so does the acceptance of multi-ethnicity. With time, an awareness of multi-culture usually diminishes. In my lifetime no noticeable lessening of cultural tension has occurred between Israelis and their neighbours, or between the Northern and Southern Irish. The original White Australia immigration policy (which initially excluded southern Europeans, like Italians and Greeks) was relaxed only in 1973, but only after Australia realised it would not survive unless it opened its doors to all. Their decision was to populate rather than to perish.

Relative to Great Britain, Australia is a young society, and perhaps more accepting. Few Melburnians now express 'Britishness'. If Michael Portillo's impression of Melbourne was correct (*Great Australian Railway Journey's*: Episode 4, BBC 2, 2019), the majority now see multi-ethnicity as positive. After only two generations, an acceptance of the advantages of their multi-ethnic, multicultural society, now goes unquestioned.

Always be ready to revise your view of others after observing how they react in different situations. One essential process involved in evaluating others is to match the things they say, with the things they do; matching what they have done, with what they say they have done. One can shorten the process by asking those who know them best.

It is in small, hardly noticed ways, that people reveal their true nature. Observing them can gain us more insight than listening to them. There is a reason for this: the basic purpose of verbal communication is persuasion. Observing behaviour will therefore get you closer to understanding the true agenda and nature of others.

The Tendency to Stereotype

Let me guess: the average West Ham football supporter is 5 feet 8.5 inches tall. The question is, how many West Ham supporters are exactly 5 feet 8.5 inches tall?

A human stereotype is only as useful as a statistical mean value; it provides a contrived perspective; one of limited practicable value that will not reflect the many variations in most groups.

For a complete understanding, one would have to review every item within a data set. With inconsistent data (the usual type), we need to look for truth in the granular detail. A data point will present an aspect of the truth which can differ significantly from the averaged, overall impression. In the same way that the mean value will only occasionally exist in reality, individual stereotypes will only ever approximate to the usual. To be of any use, a stereotype must reflect the mode, the most commonly occurring feature of any data set. This allows some recognition of the usual, and that which is unusual (atypical).

Between large groups, we can easily see cultural differences in attitudes to education (language and vocabulary), work, sport, money, family, and children. Cultural differences are more often on view during casual behaviour: the respect shown for personal space, how people queue (it would be fair to say that many true Brits much dislike queue jumpers, and prefer human snakes to fan-shaped queues), the politeness shown (we Brits are rather keen on the words 'please' and 'thank-you', and get upset with those who do not use them); how people eat and drink, and how they greet, interact, and exit from one another.

Cultural difference is often on display when we use our sense of humour. In 1926, during the National Strike, the then Prime Minister Stanley Baldwin said of British nationals: *'As a nation we have a curious sense of humour, and the more difficult the times, the more cheerful we become; we have staying-power, we are not rattled.'*

In the same TV series: *'Stiff Upper Lip – An Emotional History of Britain', Ian Hislop: 2012, BBC Two),* from which the above quote came (Stanley Baldwin was filmed saying these words), our traditional Victorian *sang-froid* defined some other British attributes: that we can be 'serene in difficulties', and tend to 'grin and bear it'.

Ian Hislop noted the national change of mood that threatened to displace our traditional *sang-froid*, when touchy-feely openness arrived from California in the 1970s. Although we now hug one another, and are more open emotionally than our grandfathers, our *sang-froid* remains. The devotion to duty of many UK medical professionals, and many others (like bus and taxi drivers), was on display during the COVID-19 pandemic of 2020-21. We British will 'grin and bear it', and be 'serene in difficulties', just as our Victorian forefathers would have expected.

In some cultures, both the security derived from endeavour, and the accumulation of wealth, each hold high esteem. Like iron filings to a magnet, the desire for money and security, fame, status, and recognition, will draw similar people together. For many, work is a major source of self-esteem; for a few, it is the reverse. A worthwhile business for them is one that combines profit with no work, like owning a multi-storey car park: the money rolls in, with or without the owner present.

For some, a feeling of security and self-esteem comes only from power and fame. Others desire only to *'live for now'*, and *'spend, spend, spend'*. For the competitive, coming first means everything; an outlook often promoted by their family from an early age. Some parents demand that their children become top of their class, and win every race at sports day. Others define a fulfilled life differently. They may believe that a life without friends and long lunches, with insufficient time to watch their grapes grow, is not a life worth having. Their aim is to be unrushed, with plenty of time for contemplation. Their desire is for a life free of worry, without the stress of dealing with other people. This is what Greeks mean by ησυχία (*isichía*)(unrushed peace).

In the broad sweep of history, cultures have been merging fast. If this cultural trend continues, nowhere on earth will be unfamiliar. The results are already noticeable. Many towns in the US and the UK today, have the same set of shopping outlets and venues. The trend will ultimately lead to no city on earth being novel, or interesting enough to explore. We might choose to stay at home and have our needs delivered, rather than face stultifying uniformity wherever we go. Fortunately for me, York, Cambridge, San Francisco, Beaulieu, Venice, Rome, Positano, Athens and Nicosia, have yet to be touched by any drive towards uniformity. Around every corner, there are still unexpected surprises to be seen.

The Rich, the Poor and Those in Between

Since 'wealth' and 'poverty' have such a large effect on morbidity and mortality, there is a case for doctors regarding them as distinct cultures. Compared to the rich, the poor have to cope with the stress of insecurity. They must work, or take social benefits, in order to get the bare necessities of life: food and shelter. If they want to maintain a higher standard of living, and get a better life-style for their families, they will have to work hard and long. This could challenge their cultural values.

The rich need not waste their energy on insecurity and can choose whether or not they work. They can be secure about their living conditions (unless their definition of security is to see their millions, or billions, grow further, while owning four or five homes). They can mostly predict their future. They can choose whatever suits them, regardless of cost (rather than being controlled by others). They have almost complete freedom of choice, together with freedom from any duty or obligation to others. This creates a mind-set

– a culture – that differs completely from that of the poor and disadvantaged. Successful, self-made, rich people, can create their own stress, while the poor have stress imposed upon them.

The middle classes ('the comfortable people of the middling sort', as Dickens described them) in the UK, represent a mixture of both cultures. Until they retire and have paid off their mortgages, most will have signed-up to many obligations. They do this in order to secure their future, and that of their children. They have much more choice than the poor, but much less than the rich. The middle classes desire status and recognition, which the poor and the rich may have abandoned, albeit for different reasons. Their stress comes from achieving social homeostasis - maintaining the status quo. They strive to afford annual holidays, mortgage payments, school fees, cars and taxes. They will do all they can to avoid poverty, hoping for a retirement free of debt and work obligations.

Indirect Cultural Discrimination

Geoffrey Marmot's findings (the Whitehall Studies), showed that disease prevalence has much to do with our perceived control over life circumstances. He compared the reduced morbidity of those in charge, with those who must obey orders. A stressful life will probably not affect the naturally secure as much as those who are insecure by nature.

Some aspects of culture might influence disease prevalence indirectly. The abstinence of alcohol by Moslems, and the avoidance of shellfish, or pork-based products by Jews, are examples.

Travel allows us to observe different cultures in their home setting, and to experience any differences first-hand. What we perceive in the West as fairness and justice, based on our Greco-Roman legal tradition, can be different in non-European nations. Many differences in culture are visible at a mundane level: the public attitude towards children is one obvious example. In Italy, Spain, and Greece, they mostly accept children of all ages in restaurants. This is not always the case in the UK or US. Apart from nightclubs, older adults are mostly accepted everywhere, although perhaps more respected and revered in Mediterranean, Asian, and Eastern bloc countries.

Many older patients fear a terrible death, suffering alone with their dignity lost. Those with resources can use them to buy their desired quality of life, and the circumstances of their death.

Respect for older people has a cultural and political context. In the UK, one notable collection of older citizens is still at work in the House of Lords. This unparalleled, eclectic powerhouse of knowledge, intellect, and experience has been under the threat of extinction for years. Many UK politicians have advocated its closure since its members are unelected. I hope we do not come to jettison their collective wisdom and minimal political motivation, in favour of inexperienced youngsters, trying to gain political advantage.

Cultural Perspectives

The famous chef, Georges Auguste Escoffier (1846 – 1935) worked at the Ritz Hotel in London for many years and never spoke English. He supposedly said, 'If I speak English I might start to cook like them!'

UK TV Channel 5: *The World's Most Expensive Hotels – The Ritz.*

A traditional Bedouin father asked his young son for a drink of water. 'Yes, of course, father', said the son, but did not return as promised. He then asked his eldest son for water. His response was, 'Sorry father, I have no time.'

From the cultural perspective, who will the father favour as the more honourable son? The more honourable Bedouin son, is the younger; the one who caused no loss of face for his father.

A Russian journalist on British TV, asked to summarise the Russian attitude to politics, said: 'We Russians need to be dominated'. Translated into psycho-dynamic terms, I think she meant Russians only trust powerful characters as leaders (you know where you are with them). Hence the love and respect the Russian people had for Stalin ('steel' in Russian), and now have for Vladimir Putin. Viewed from the Russian perspective, why would anyone want to trust the opinions of mere citizens, few of whom have any specialised knowledge of politics or economics (as in a democracy)? Instead, they prefer to trust one man; one who has all the required knowledge and information at his fingertips, and a track record full of correct decisions.

My patient, Mr. C., was born into a prominent Portuguese family. I asked him if there was any truth in my observation that his countrymen were rather sombre (reflected in their traditional Fado music). I will always remember his reply. 'Dr. Dighton, we Portuguese have tragedy written on our souls.'

A gentile asked a Rabbi: 'How many political parties are there in Israel?' In answer, the Rabbi posed a question: 'Do you know how many people live in Israel?' 'About eight million, I think', answered the gentile. 'Then there are eight million political parties in Israel.', replied the Rabbi.

There are thousands of examples to draw upon to illustrate cultural difference. If you are further interested, study human ethology. It has provided me with some understanding of others, and support for *vive la différence*. Knowing that I am sympathetic to cultural differences has helped me achieve the goodwill and trust of many patients from different cultures. There will have been some who thought me patronising. So be it.

Medical Practice and Cultural Bias

When I worked in Amsterdam as a cardiologist, I noticed Dutch patients behaving differently to British patients undergoing cardiac catheterisation. Most Dutch patients needed no pre-medication; they displayed little anxiety and were happy to chat away during the procedure. In the UK, most patients were anxious enough to need sedation. Working in a mixed cultural society is a challenge made easier by being forewarned of possible cultural differences.

Within any culture, age, gender, and social grouping remain the major predictors of behaviour; especially the ability to communicate. In the UK, a patient's vocabulary, not wealth, remains a powerful discriminator of social status. During my lifetime I have seen the growth of one particular group – those with loads of money, and a limited vocabulary (some are illiterate, and have employed others to read and write for them). They are likely to express sentiments such as *'I believe in every man for himself'*; *'happiness and self-gratification are my right; I deserve my advantages, I am worth it.' 'Whatever I want, I can buy.'* Unless they present with the simplest of clinical conditions, it will help doctors and nurses to be acquainted with their patient's attitudes.

Over many thousands of years, many waves of immigrant families have come to the British Isles; a substantial number of them inter-married after a few generations. In my lifetime, I have seen our historic British ethos absorbed into other cultures. Native Brits have married Greek Cypriots, Europeans, and

Afro-Caribbeans, all of whom have tried to embrace British culture. I have not seen too much of the reverse – native Brits adopting their partners' culture and language, although it does happen occasionally. By the time a 3rd generation is born, children of mixed cultural background will have taken on some shared ethos. Fourth generation children will have picked out all those bits of each culture that suit them best.

An Armenian woman born in Beirut expressed her culture as: 'having two hearts'.
Art Lovers' Guide (2018): Beirut. TV: BBC FOUR

We must all be aware that any advice we give may be subject to cultural interpretation. In one culture, authoritative advice will be *de rigueur;* without strong advice, some patients will not take a doctor seriously, or trust them. Some regard being kindly as a sign of weakness and incompetence.

Those from some cultures expect a full discussion of every minor clinical detail or none. Some expect clinical discussions to include family members. This can be time-consuming, and there is a paradox to avoid; your perceived status could erode if you take too much time with them. Their thinking may be that a high-status doctor must have other, more important work to do. An important doctor, worthy of respect, is likely to be short of time. The actors among us know this, and can feign superior status by frequently looking at their watch. The one with the highest status (the one who deserves trust) will signal the end of a consultation.

In the same way that poetry is difficult to translate from one language to another (something of its essence can be lost), medicine practised by doctors and patients from dissimilar cultures risks leaving some things unsaid, or misinterpreted. In the same way that the essence of one language is difficult to translate into another, it is difficult to convey the essence of a culture to outsiders. Cultures need to be experienced. One can readily learn a language in a classroom, but to learn a culture requires personal exposure to it.

Because a large part of inter-personal communication is non-verbal, the more exposed we are to one another as patients and doctors, the better we will understand one another, whatever our cultural differences. One lifetime is not actually long enough to gain proficiency in more than a few languages, and certainly not long enough to assimilate many cultures.

Medicine has many cultural traditions of its own. Some of what doctors believe is based on folklore and hearsay (information handed down from master to apprentice), alongside information derived from scientific research. To know the folklore of a culture is to understand its traditions and beliefs.

I was once told that when the French get sick they expect polypharmacy; at the very least a suppository, or an injection. Without such prescriptions, their family will not respect their illness. Greek Cypriots seem to prefer British doctors to their own. Like many others in the region, they also prefer multiple opinions.

Culture and Practice Location

In order to develop a mature and rewarding doctor-patient relationship, mutual consideration and respect are obvious requirements. It is important to know how each individual feels about dealing with those from a different educational, social, financial, ethnic, and cultural background. By choosing their location of work, doctors partly pre-select their patients. The epizeuxis, *'location, location, location'*, not only applies to shops, businesses, and houses, but to patients attending medical practices.

If you work in Knightsbridge or Chelsea in Central London, you are more likely to consult with the rich. Work in Leyton, in East London, and your patients are more likely to be the underprivileged, or recent immigrants. Recent immigrants may have a limited proficiency in English and a strongly held culture.

Work in the City of London, and you could deal with ambitious people who want to accumulate wealth. Work in central Oxford, and you will probably consult with students and academics. When I worked a St. George's, Hyde Park Corner, I saw many patients who were actors, military personnel, or senior civil servants. It is important for doctors and nurses to give thought to where you work.

    I much admire those doctors whose mission is to serve the disadvantaged. Theirs is an altruistic vocation; one of dedication needing the strength to meet many challenges, including that of cultural difference. Others, with ambitions less noble, might choose to work within their own culture, where they will have fewer difficulties with understanding. Those of independent spirit who want to practise medicine without the corporate considerations of the NHS, in a style of their own making, should consider private medicine. They will need to be knowledgeable and proficient enough for patients to support them. The allegiance of a loyal flock of private patients can provide considerable professional satisfaction. Private medicine can provide close doctor - patient relationships, independence, continuity, and the time and scope for personalised medicine.

PART TWO

DIAGOSIS, DIAGNOSIS, DIAGNOSIS

Chapter Eight

Patient History Taking

The importance of history taking cannot be over-estimated.

Every question put to a patient is a test. Judge the reply as positive, negative, false positive, or false negative, and then decide its diagnostic accuracy.

Questions reveal as much about the interrogator as the responder.

What patients say and don't say are both important.

Body language, including facial expression and culture-specific mannerisms (hand and head movements), are important forms of communication.

I have not attempted to list every question one could ask during history taking (see Appendix A, for my check-list). Rather, I have assumed that my readers already have some experience of basic history taking, pursued in the time-honoured order of current history, past, family, social, and therapeutic histories.

I have not wasted much time here on simple history taking – *'What's the problem?' a doctor asks a patient with blood dripping from his severed finger.* Instead, I will discuss the value of in-depth, complicated, and complex history taking, using various styles, namely executive, discursive, and interrogational. Where the diagnosis is not immediately obvious, an in-depth history will often point to the diagnosis, and provide the information needed. It could also help plan the most effective management.

Challenging cases have occupied much of my professional work, with many patients seeking my advice after others have failed to make a correct diagnosis, or arrive at the best management. Why they failed was most often explained by deficient history taking and examination.

A well-taken, complete history that reveals all the factors contributing to a diagnosis and directs the management decision process is invaluable, especially when the information is cross-checked and verified independently. My personal approach will not suit everyone, so I have described other techniques.

History taking and examination are interactive data-collection processes. Like acting skills, they are best learned at the side of an experienced professional. Competent doctors will have learned the diagnostic and management significance of each question they ask, and will appreciate the clinical value of each answer.

Symptoms are clues to disturbed physiological function (like bowel and coronary artery spasm) and patho-physiological processes like cancer and arterial atherosclerosis. They come in several varieties: incomplete, distorted, false, absent, positive, negative, false positive, and false negative. For these reasons, a sceptical approach often serves history taking well.

Since symptoms provide clinical data, each answer (result) will have its own sensitivity, specificity, and diagnostic accuracy in relation to any specified diagnosis. For those not acquainted with Bayes' Theorem, see Part 3, under clinical decision tools.

History taking is an art, and like other arts, it will be malleable in the hands of those who practise it. It cannot be standardised, although some schemes will help avoid omissions for the inexperienced (see Appendix A).

The Initial Encounter

Building a temple of healing is easy. Proving its value to patients and maintaining it are challenging.

There are some important preliminaries to consider before a clinical history is taken.

Doctors with their own practices will have appointed receptionists. If personable, they can form valuable impressions of patients while collecting their basic personal information. Those doctors who work for the NHS or its equivalent, will not usually have control over those who greet their patients, whereas in private practice a significant choice is possible. Receptionists who prefer to remain anonymous, detachment and obdurate, will confer no clinical advantage.

The environment in which doctors work can influence the opinion of those who visit him, although few doctors regard such matters as important. From an academic medical perspective, the practice and consulting room environment may be irrelevant, but the CQC will not agree if safety is compromised.

Many patients are inquisitive and observant. Does the state of the consulting room flooring and furniture matter? Whereas old furniture and other trappings can reassure those who prefer the traditional, only an ultra-modern office will impress most today. They might ask themselves why an efficient doctor (running his own practice) would leave two-year-old magazines lying around in his reception area? Perhaps he ascribes no importance to them, or is not interested enough in what goes on in his practice. Patients notice these things, and some will wonder if an inattentive doctor can deal with their medical problems efficiently.

Attention to detail is important to those businesses who need to gain the confidence of their customers. Some will occupy expensive, state-of-the-art offices in prestigious tower-blocks. Their customers pay for this. Private banks provide an example. Some in the City of London, charge 15% per annum for investing money. To pay that much, clients will need all the confidence they can get. All impressions are, of course, subject to bias and error.

After Dorothy had walked along the yellow brick road, and had overcome some dangerous obstacles, she arrived at the Emerald City. This is where the Great Oz performed his magic (she needed help to get back to Kansas). The scene that met her was awe-inspiring; one of unimaginable power and influence (Film: *'The Wizard of Oz'*, MGM, 1939). Unfortunately, the Great Oz was just another mortal, with no magical powers at all. Apart from instilling the wrong sort of confidence (it should be based on clinical results), making a show is only good for displaying a doctor's character, taste and style. This is undoubtedly important to many high profile, rich patients, but irrelevant to most.

Quo Vadis? Patient Reception

Information gathering in all medical practices starts at the reception desk. Patient acceptance, belief and trust in a doctor's practice starts here.

In hotels, the reception area for new customers must inspire confidence. Nobody paying $1000 per night for a hotel room, will want to be disappointed by the décor and the way the staff greets them. Nobody entrusting their life to an organisation should be happy either. The design, cleanliness, opulence and the prestige attached to the reception experience, are of considerable importance to some, and of no importance to those too ill to notice.

Attention to such details may seem trivial, but those who work under the NHS umbrella, may have no choice in such matters. Faith in the organisation, they think, should be enough. That may change if patients get further dissatisfied with the service. And why not? When those loyal to one hotel become dissatisfied, nothing stops them moving to another. When NHS patients become dissatisfied, moving practice or hospital has always been resisted at every stage. The NHS does not have a monopoly, but NHS patients who cannot afford to decide otherwise, must take what comes. The divide between the private and NHS sectors in the UK has many other aspects, many related to differences in patient demography and consequent morbidity and mortality (see Anderson et al. 2024). Consultants working in both sectors, know these differences well.In October 2022, Medscape UK (Dr. Sheena Meredith) reported that,' The British Medical Association (BMA) has warned of a "catastrophic crisis" brewing, with nearly half of hospital consultants in England planning to leave or take a break from the NHS over the next year.'

The welcome given by a receptionist can set the tone of a medical practice (accepting, and affable or dismissive, and disinterested). Few will warm to receptionists talking among themselves, and ignoring them. A receptionist finding their computer screen more fascinating than the patient, will cause some to get angry. And rightly so. This is dismissal, and a denial of status; the establishment of pecking order. The patient might think, 'They wouldn't do that to the Minister for Health, so why me?' As a patient, their attitude could change and affect the history taking that follows.

Dismissive reception staff were once commonplace, and totally unacceptable. In decades past, in some NHS practices, patients had to fight for attention. What followed was equivalent to subduing a fire-breathing dragon, or Odysseus having to deal with Scylla and Caribdis. All this, just to see a doctor? No receptionist of mine would have lasted five minutes had they shown any trace of an obstructive nature. This attitude was once so common in NHS practices, I referred to it as *'the NHS attitude'*. It implied, *'take it or leave it.'* A check-in computer, rather than a receptionist, would have created a better impression. In my practice, I never let a receptionist become fixated on their desk computer, or their hand-held device, in preference to greeting a patient and acknowledging their presence.

I insisted on receptionists having a medical perspective, an understanding of patients and their problems, together with a helpful approach. If receptionists at the Dorchester Hotel and Claridge's could do it, why not mine? It costs nothing extra to be affable and welcoming, and it has an important bearing on any patient handling that follows. I never wanted my patients made angry before I met them: it would have a negative effect on the history taking that followed. It helped when I was late, for my receptionists to enjoy talking to my patients. They did this more easily when they shared the same background, culture and attitudes.

Fear of illness can manifest as anger. Unfortunately, receptionists sometimes have to deal with it, and understand its origin. Patients may not express their anger to doctors in the same way. Fearful, angry and frustrated people in doctors' waiting rooms, will appear ill-at-ease, worried and irritable. I expected my receptionists to recognise this, and inform me of it. I would then see the patient as quickly as possible.

Those doctors with their own practice (only in the private sector of the UK), should only employ receptionists who can 'read' people accurately; those with 'nous' or know-how (*those with 'primitive intuition' for human psychology. See Steven Pinker, 'The Blank Slate'*). Horse owners and animal lovers usually have it; they call it *animal instinct*. The best receptionists I have known were dog and horse owners. How they could 'read' what horses and dogs were thinking (in order to predict their behaviour) remains a mystery to me. To have this nous requires a sensitivity to every verbal and non-verbal clue. Once a difficult or troublesome patient is detected, an alarm bell should ring. Doctors who wish to gain this skill should apprentice themselves to a vet.

Receptionist with nous, will assess patients from the first moment of meeting them. Be prepared for comments such as, 'Be careful with this patient. I'm not sure what he really wants.' 'She's very upset and stressed. Her daughter is ill, and her husband is having an affair with his secretary.' 'He's a bit of a smoothie.' or 'He's furious about something, but prefers to keep it secret.' In small communities, your receptionist could know much more about your patients than you realise. A good receptionist will identify all the different patient types listed in Part 1, Chapter 4, About Patients.

I used to wander in and out of my reception area frequently. If I thought a patient needed to be seen quickly, I would bypass the registration paperwork, and leave my receptionist to complete it later. *Patients before computers, and patients before paperwork, were two of my immutable rules.*

The personal nature of the service I provided, was reason enough for many patients to return, sometimes for several decades. Is this type of service still sought after? Fewer and fewer patients have any experience of personal service, and nobody misses what they have never had. To achieve acceptable results, one must respond to the needs of every patient as an individual. Personalising medicine begins in a doctor's reception area.

In the Consulting Room

Will a bookcase full of technical books engender patient confidence? I suspect it will, but the portrayal of learnedness will only affect clinical outcome if it inspires confidence and trust. A doctor's ability to respond effectively to a patient's needs, measured by the time and trouble she takes, will be more influential. The trust and confidence engendered, can influence the course of an illness and its outcome. To what extent, and for which medical conditions, is debatable. One cannot expect it to influence the long term growth of atheroma, or to halt a rapidly growing sarcoma, but it might improve well-being, reduce liability to clotting, affect immune responses, and improve the results of psychological interventions.

Does it matter whether a doctor sits behind a desk as an imposing figure, or sits with no barrier between him and his patient? It depends on the patient. Personally, I preferred to sit with my patients and have no intervening desk. Formality never suited me, but it is an approach used by those who prefer distance and anonymity. Are they confusing formality with professionalism and clinical effectiveness, I wonder? Formality can inspire trust, but may not encourage the sort of history taking that reveals most. Those patients who sought confidence from formality, and preferred a detached approach, never remained patients of mine for long. The same applied to those displaying or seeking obsequiousness. I have explored some of these dynamics in previous chapters on doctors and patients.

Pre-History Evaluation

Every new patient must engage with a 'blind date' situation; at least, when patient and doctor know nothing about one another. Unlike most blind-dates, the purpose is obvious - to provide medical help. Before any conversation begins, a mutual exchange of non-verbal information always takes place. Before taking any history, signs of ill-health and frailty need to be recognised. Illnesses that are rapidly advancing, or have advanced considerably, will change a patient's appearance, although some patients look robust and healthy, despite their serious medical problems. In the end, all those with a progressive serious illness will look ill, become slowed, have noticeable weight loss, (first seen in their face; weight-loss dieting does the same), and will appear to have aged. Regardless of the length of the natural history, the manifestations of illness always take time to appear, some being obvious only in the later stages. Those who look ill, obviously need rapid appraisal.

Observe every movement and facial expression, and look for nervousness and breathlessness. Notice the patient's state of dress, age group, gender, gender orientation, and race. Guess their cultural, financial, social and educational status, and that of those who accompany them. This is always useful collateral (meta) information for later patient management.

Tense facial features and eyes that seldom blink, allow one to presume that the patient is worried, pre-occupied or in pain. Their 'tension' needs to be explained. Excuses that make no intuitive sense, and remarks that attempt to change the subject, should not be allowed to divert the enquiry. If you hear no 'ring of truth' in what you are told, important issues await detection. You risk missing a diagnosis, and later failing with their clinical management.

Except for obvious minor problems, the country of family origin can help to modify one's approach. Although relevant to physicians working in multi-cultural nations, and those working with transmissible disease, this is rarely as important as a patient's personal values, beliefs, and practices. These will take time to assess and may not be relevant initially. Socio-economic 'class' considerations (especially in the UK) are always relevant. A patient's status can range from VIP (possibly accompanied by an entourage), to humble and deprived; highly educated to illiterate. As a private doctor in the UK, one will see not only rich patients, but those so desperate for help, they will choose to pay for private medical services they may not be able to afford. Many will have been so frustrated and disappointed by how they were treated by the NHS, they become 'refugees from the NHS', and display characteristics which I will discuss later.

From the beginning of every consultation, one must compile a mental check-list of every unresolved issues and discrepancies, and plan the questions needed to resolve them. Allow answers to emerge without undue pressure. Natural, unpressurised responses often reveal more dependable truths than those that are forced. Doctors do not help the process by looking at their watch every few minutes, or being fixated on their computer screen. In my consulting I had no computer screen. I briefed myself before the patient entered and then focussed entirely on them. Patience is required, and like childbirth, important things can take time to come forth. To achieve clarity, several consultations may be required. Don't give up without gaining some meaningful insight.

History Taking

The aim of history taking is to discover every clue available to make correct diagnoses and to discover all the pertinent factors that will facilitate patient management. Traditionally, we start with present history, then proceed through past medical history, family, social and therapeutic history. Although there is no need to change these entities, it is often expedient to change their order, or bypass some aspects completely.

Those able to switch from one subject to another, abandon fix procedure, and remember enough to return to their previous place, have an advantage. It matters not, what order one acquires information, as long as it is complete in the end.

Many doctors will be disturbed by the thought of changing the order of their enquiries. The art of history taking, akin to an actor's impromptu performance, might allude them. No art or creative process has fixed rules that matter, and those who need a check-list are either inexperienced, or soon to be replaced by an avatar.

Styles of History Taking

The Executive Style

As Major Charles Emerson Winchester III, in the TV series 'MASH', said to the office assistant Radar, 'Be quick, and be gone!'

The style(s) doctors adopt will reflect their personality and any personal needs they have for efficiency.

A doctor using the executive style will ask specific questions and use few words. He will be business-like, formal, and always to the point. If there are significant psycho-social aspects to the patient's condition, this style is unlikely to coax them out. Doctors who limit themselves to this style might waste no time, but could miss important aetiological factors, outside of their specified frame of reference. There are also patients who like to restrict doctors to this style, hoping to limit their depth of enquiry, and to hide the truth. We must respect patients who do not want to reveal their personal details, but if key issues remain hidden, incorrect diagnoses and poor management plans will follow.

A distant, anonymous, or detached approach is a fully workable position for most specialists, even though it does not always lead to the fullest of evaluations, and the best considered clinical judgements. For this reason, some clinicians cannot manage complicated cases successfully, and it is why some of their patients found their way to my door. For doctors to be disinterested in personal aspects is to regard patients as lumps of pathophysiological tissue. This exemplifies the 'detached' approach in action. They might better suit a tertiary medical centre of technical excellence, where the focus is pathophysiology.

In the business world, the development of friendly relationships is one key to success. In the streets outside every ivory tower, life, death, health, and personal welfare *are* all intensely personal issues, and never detached ones. Those doctors who desire anonymity and detachment from patients should work away from their bedside, in research and pathology laboratories, in the media or insurance companies.

The Focussed Diagnostic Style

This is a sub-type of the executive style. It is useful in cardiology and other specialties where the symptoms are accurately defined and circumscribed. For Aubrey Leatham, renowned cardiologist, it was his only style. 'Can you lie flat? If you walk quickly, does your chest discomfort worsen? Did your palpitation feel regular, or irregular? Have you had rheumatic fever? How many flights of stairs, or steps, can you manage without feeling short of breath? Was your blackout without warning?' These are all questions of diagnostic relevance.

The responses can grade the severity of coronary heart disease, and assesses the degree of mitral stenosis.

Specialists in all fields of medicine have their favourite sets questions. They have learned which are the sensitive and specific for particular diagnoses.

The Interplay Style

As Judge Judy once said to a defendant delivering a meandering defence: 'I don't need your story from once upon a time'.

This style is conversational, with the dynamics of a point-scoring game. Its purpose is something other than diagnosis; perhaps to get a document signed, or to obtain a letter of recommendation or referral. Patients withholding information, giving too much information, or giving selected snippets of information, can be the elements of a doctor-patient game. When anyone is thought to hold information that is top-secret; information that is too sensitive, or too important to be revealed, it can be beguiling. There are other patient games like *'poor little me' (I am just someone nobody can help)*. Although there are many common attention seeking gambits, be aware of being duped, or being rendered a failure because of withheld or falsified information. Those who suspect a patient-doctor game in progress, should revert to the succinct executive style.

The Discursive Style

It's true, I talked too much. It helped some patients relax, and allowed them time to 'open up', and provide me with insights into their problems. It annoyed others. As an interview technique, it requires one to be completely relaxed about technical clinical issues, and to have enough time for even the most embarrassing details to come forth.

Through chatting, and having an open discussion, many patients learn you have their interests at heart. Given enough time to assess you, they might find reasons to trust you. From this privileged position, it is much easier to help them. As trust develops, you might venture to put yourself in their place; prefacing your advice with, 'If I were you . . .' This is unacceptable to many professionals, including lawyers and accountants; a step too far, too personal, and unprofessional. It is the mark of an 'attached' doctor with fruitful doctor-patient relationships that have been extended towards friendship. Not everyone wants to share their identity with their clients, since it can reflect their insecurity.

The essential quest is to define a patient's inscape, the distinctive design of a mind that makes up individual identity (Gerard Manley Hopkins. Poet and Jesuit priest:1844-89). John Duns Scotus (1266-1308), the medieval philosopher, previously referred to the same property as haecceity.

The best patient-doctor ambience is one that promotes empathetic inter-reaction. The discursive style can resemble a fireside chat. It is most suitable for dealing with introverts; extroverts usually prefer the executive style. The discursive style allows some patients to forget their shyness and express themselves freely. They will then be more open and more likely to say what is really bothering them and how they really feel. There are many drawbacks to this style: it will prove awkward for doctors and patients with deficient inter-personal skills; it is time-consuming and predicated on the desire of those involved to have a close doctor-patient relationship.

Insufficient explanation by medical staff (usually doctors) generates most of the complaints received from patients.

The discursive style is best employed for patients who have difficulty in defining their symptoms. Since some medical conditions have a psycho-social or stress-related component, this interview style will encourage personal revelation.

For those patients with predominant psycho-social problems, nurses and a social work team, may help patients more than any doctor. In my early days, this was the role of some ward sisters and Lady Almoners.

History taking styles can be blended together and employed in a hard or soft manner. Colloquially, one can choose to 'put the boot in', or to walk on eggshells. To use both would mirror the well-worn 'good cop, bad cop' technique, used by police dealing with their less communicative captives. Resistance to questioning is a common attribute of those patients who have been sent to a doctor (usually by their boss or partner). Because someone else has insisted on their attendance, they may choose to be less than compliant.

During the process of history taking, awkward and embarrassing moments, emotional reactions, and *faux pas* will arise. Handling them sensitively is an art.

During every consultation, apart from the medical history, the aim is to gain contextual meta-data, to help complete the picture from a fly-on-the-wall perspective. As history taking progresses, one must recognise missing information, and discover the reasons for the omission. Continuously assessing the factual and emotional relevance of what patients say, allows for questioning that will define the problems.

When master gardener Monty Don visited Japanese gardens, he learned about 'Ma' (□); the space between the flowers and branches being as important to Japanese gardeners, as the plants. In speech and music, silences are relevant. In taking a medical history, try to be aware of what is missing.

Forensic History Taking

This style is employed by barristers when cross-questioning in court. It is not a commonly needed technique in medical practice, but one well worth learning for those suspected of withholding information, telling lies, or being stubborn and uncommunicative (for reasons that might not be in their best interest).

Dr. P.N., my cardiologist boss at Charing Cross Hospital, was an expert in forensic history taking (interrogation, actually). In the same way that suspects in a murder trial might withhold information, alter the facts to suit themselves or lie, some patients will do the same. Hence the need for forensic enquiry. For Dr. P.N., every potentially fatal condition (TIA, unstable angina, cardiac infarction, sudden onset AF) had a protagonist: a situation or person, needing to be identified and made impotent before causing further harm. His style was that of a criminal detective, trying to uncover the person responsible for the crime of afflicting illness on his patient. No other doctor I have known, employed this approach. It was intrusive, and too personal for most. A powerful tool when used well, it can lead to a fuller understanding of the psychosocial reasons for a patient's demise.

Dr. P.N.'s starting premise was simple: nothing can happen (heart attack, CVA) without a stimulus (various forms of stress), instigated by the patient, or by somebody in his social circle. Elastic never snaps while resting on a desk. In answer to a patient saying, 'I don't know why I had a heart attack.' he would suggest the need to find out. In keeping with his SAS background, he would tell patients of the need to understand the strategy of their enemy (those whose actions bring harm). Only then, can they defeat them. By understanding the enemy, vulnerability becomes reduced.

His teaching was that no heart attack or stroke came 'out of the blue'. In his view, those actions, concerns, and situations for which the patient was responsible, were key aetiological factors. He thought that giving preventative nutritional, smoking, and exercise advice alone, would be ineffective. He suggested that patients should examine their lifestyle stresses and attitudes toward life, and take responsibility for changing them. His interrogation technique was designed to expose the patient's prevailing stresses. He might then confront those stressing his patient, and pursue his enquiries until he found a culprit, or a satisfactory psycho-social explanation for each patient's demise. He was only occasionally unsuccessful in such quests.

Although he thought little of the possible beneficial effects of a 'healthy' diet and exercise, in the early 1970s, Dr. P.N. was first to pioneer post-infarction exercise training for post-infarction patients in the UK (collaborating with Al Murray, Olympic coach). Before then, many thought exercise too dangerous for post-infarction cases (presumed to be a cause of sudden death).

As you might guess, Dr. P.N.'s interrogation technique, and his views on stress as an aetiological agent, did not endear him to his academic colleagues (they thought he provided insufficient evidence). While the public were being told not to eat eggs or animal fat (1970s), his view of stress as the cause of heart attacks and strokes was revolutionary.

The common causes he found were loss of control, family discord, threatened (or actual) loss of status, loss of income (redundancy or bankruptcy), accumulating resentment, failing relationships, insecurity, and circumstances perceived as impossible to escape. Acknowledging these pre-morbid factors, allowed Dr. P.N. to discuss the most prudent preventative measures for each patient.

Piers Morgan has often portrayed a similar attitude, namely: 'you don't have to accept being beaten', and, 'you always have a choice'.

Dr. P.N. would paraphrase the Duke of Wellington, saying, 'Any fool can stand their ground and watch as their soldiers are cut to pieces. It takes a prudent officer to order an early retreat.'

All systems fail at their weakest point. Stress can trigger disease onset, or change the course of a disease, through changes in reflex sensitivity (BP control), altered blood clotting (thrombosis), hormone imbalance (depression and behavioural changes), and reduced immunity (pneumonia). Since every applied 'force' has a corresponding reaction or 'strain' effect, stress can possibly affect arterial pathophysiology (atheroma, clotting with embolus formation or occlusion). Subsequent heart attacks, dysrhythmia, and CVA, might then occur in those genetically predisposed. The combination of fragile cerebral vessels and hypertension make haemorrhagic strokes more likely. Consistently raised high blood pressure can cause the progressive thinning of blood vessels (arterioles) in the brain, and medial hypertrophy of the arterioles in the rest of the body (increasing peripheral resistance). In the heart, left ventricular hypertrophy will accompany these changes.

Do liars need de-fibbing?

Frasier. US TV Series

Ask Patients What They Think

Joe's Aunty Nelly

Jim was a 55-year-old publican from East London. He was unable to give more than a poor history. His primary concern was stiffness while walking and swimming. For years he had swam one mile every morning. This was now difficult because of muscular stiffness. He looked depressed and told me, 'I find it easier to walk backwards than walk forwards'. At that point, I could have dismissed him as a little crazy. Although in the wrong age-group at 55-years old, I thought he might have fibromyalgia rheumatica. I did some blood tests, but found no raised inflammatory markers. I gave him an anti-depressant, and told him to return in two weeks.

He returned with his wife, and displayed a gait that was clearly stiff and restricted. His wife did the talking. She wasn't happy with my diagnosis and treatment. After some discussion about differential diagnoses, she asked me: 'Could it be Parkinsonism?' 'Not really', I replied, 'He's far too young'. 'Well, that can't be right', she said. 'His aunt Nelly developed it when she was only 40-years-old!'

Bombs can explode without warning. If you haven't experienced this phenomenon, you will.

Patients with Parkinsonism are stiff and can look depressed. Those who are proficient observers might notice that they find it easier to walk backwards than forwards. It is easier for them to drag their feet backwards, than to lift them and walk normally. The penny, when it drops, can have a major impact.

Finkelstein's Shame

Mrs. Finkelstein was in her late 80s. Her mother tongue was Yiddish, sprinkled with German and a little Russian. Her English was poor. She told me she was 'passing out' frequently. Her GP referred her to Charing Cross Hospital in Fulham, where I was a lecturer in cardiology. She first saw Prof. de W. who found nothing wrong. Being concerned about her limited communication, he referred her to Prof. A.G. Since he was also Jewish, Prof. de W. thought he might better understand her, having identified with her culturally. Prof. A.G. saw her and decided that it would be better for me to test her and define her blackouts as either vasomotor, or Adam Stokes' syncope. Perhaps she needed a pacemaker?

I tested her in my laboratory (I had spent years at St. George's Hospital at Hyde Park Corner in Central London, developing simple methods that could distinguish vasomotor syncope from Adam-Stokes' syncope, and those who needed pacing). I also found nothing wrong with her. Nothing, until I asked her what she thought was wrong.

'Sit down young man', she commanded. 'I'm 'gonna tell you 'viy I'm collapsing. It's simple. Until my son comes back to my daughter-in-law, and my four grandchildren, I shall continue to collapse - from the shame of it. That my son should run off with a girl half his age, and leave my four grandchildren without a father... Oi! What was he thinking? It's a scandal. It's a tragedy. That I should be his mother... the shame of it. That's what's making me collapse.'

She did not need a pacemaker. Stress had made her autonomic (parasympathetic) reflexes more sensitive. She was fainting, not having Stokes' Adam's syncope.

Specialised Questioning

My first cardiological teacher/mentor/boss, was Aubrey Leatham. He concerned himself with nothing other than the specific symptoms of heart disease (usually valvular, or ischaemic heart disease), and by how much his patient's performance was affected (and thus, their suitability for intervention). He had a few specific questions only. These related to chest tightness, shortness of breath, orthopnoea, palpitation and blackouts. The answers led him to a functional assessment, and a clear understanding of what the patient might need (a valve replacement, a coronary by-pass or a pacemaker). He defined his cardiological role strictly and never widened his focus. (For detail, see my forthcoming book for medical students: 'Essential Cardiology')

Dr. Leatham had the gift of clarity. At our weekly cardiac conferences, I well remember him challenging our cardiac catheter results, which on one occasion revealed a significant mitral valve gradient (mitral stenosis), and a need for surgery (mitral valvotomy). This was prior to us getting the first echocardiogram in London in the early 1970s. 'Your results are inconsistent with my observations', he said. 'When he came to my rooms in Harley Street, he easily walked up two flights of stairs without breathlessness. Re-examine him, and re-test him.'

Highly specialised doctors have a place in specialised centres, treating well-defined, specific medical problems. There were once many generalists who could take on all-comers. The world of the general surgeon and general physician has become deleteriously eroded by specialisation, and super-specialisation.

Language Use

According to the National Literacy Trust (2020), 16% of adults in the UK are 'functionally illiterate'.

Educated people use one set of words, and less educated people a smaller set. Our use of words, and the correctness of grammar, reveal our educational status. They do not so well predict socio-economic status.

Without a sufficient vocabulary, clarity suffers: ideas and meaning may not be expressed clearly enough for others to understand. Those who cannot find the words to say what they mean, can get frustrated and angry.

There are those whose vocabulary barely exceeds 300 words. They may use colloquial expressions that are emotive, but lack descriptive power: 'It's a 'no-brainer'!' 'Not on your Nellie!' 'What are you like?' '100 percent!' Although doctors usually understand what these mean, fewer patients will understand what doctors mean; not unless they choose their words carefully. To communicate well, one must use the most appropriate language and idiom, for each patient in the correct context. The most appropriate dialect may help communication and patient acceptance.

A poor vocabulary is not a reliable indicator of intelligence. Some of the most intelligent and successful business people I have known, had a limited vocabulary. Some were illiterate and needed to be accompanied. They avoided completing forms themselves.

Those who have mastered history taking will swiftly choose which style is appropriate for each patient: one style to get the facts, another to put some flesh on the bones. Like any talented actor, doctors should be able to change the aspects of their language: its delivery (tone, volume, character, and expression), manner (respectful, or not; sympathetic, or unsympathetic), and demeanour (supportive, or dismissive), for each patient. Ideally, one would speak cockney to Cockney's; Liverpudlian to Liverpudlians; Geordie to Geordie's, and Scots to Glaswegians, but only if fluent in these dialects. A bit of shared cultural identity can help, but to be thought patronising, obsequious, or ridiculing, can damage a relationship.

One word of warning about linguistic diplomacy. Avoid trying your Greek on Turks; your Turkish on Greeks, or your Russian on Poles and Czechs. Memories of past national conflicts between these nations have lasted generations. It is less of a problem than it once was, but such *faux pas* can still cause embarrassment. Best to ask for the patient's mother tongue and what languages they speak, before launching your linguistic talents. Unlike the average Brit, our European compatriots (now ex-compatriots) will often speak two or three languages. Because many will have wanted to learn English, they will happily forgive the British for their disinterest in becoming multilingual.

Gauche ou Droit?

TV war correspondents are now less prone than they were to ask war victims embarrassing questions. *'How does it feel to have had your family wiped out?'*; or as I heard recently: *'How do you feel now?... three years after the interviewee had witnessed her mother being blown to pieces by a mortar.* It is impossible to practise medicine, and not to want to take back some words uttered in haste, or words lacking in thought, sensitivity or respect.

If you are not sure of what to say in a difficult situation, say nothing. Your body language will not, however, go unnoticed.

Past Medical History

There are usually three sources of past medical history – the patient, their friends and relatives, and the clinical notes. Much of the past history is usually irrelevant, and can divert attention from current issues. Sometimes it provides one crucial piece of information, like a past history of rheumatic fever.

Over time, perspectives change, and so can the diagnostic relevance of past information. From a cardiological point of view, there is a diminishing need to enquire about diphtheria and rheumatic fever. This is not the case when dealing with patients from 3rd world countries. The consequences of these diseases, once common in the UK before the 1950s, were still with us in the 1970s. Rheumatic fever was then the major cause of heart valve problems seen by cardiologists from the 1950s to the 1980s. In the UK, only a few now have rheumatic heart valve problems (thickened, or calcified valves that are stenosed or incompetent). Rheumatic fever could recur if we curb antibiotic prescribing further. Treatment for sore throats, and upper respiratory tract infections (URTIs) are not so commonly treated with antibiotics, now that the world-wide growth of antibiotic resistance has become a problem.

A small percentage of patients with sino-atrial block, or complete heart block, had a history of diphtheria (diphtheria toxin is an anti-cholinesterase; an active enzyme found in the cardiac conducting system). With the enzyme inhibited, acetylcholine persists, and atrial bradycardia, and/or reduced AV conduction can occur.

The past history must include travel details . Ten years before retiring, I saw my first case of diphtheria. The patient had been on holiday in Venezuela. The grey film covering the back of his throat was pathognomonic. I have seen only two cases of acute rheumatic fever. I had to discharge one patient from a private hospital to treat her with aspirin. She was only eight-years old, and because of the rules aimed at preventing Reye's Syndrome, she was too young to be given aspirin in hospital. After unsuccessfully trying an NSAID, I discharged her to be treated at home with aspirin alone (with the fully informed consent of her parents). Aspirin remains the first line treatment for uncomplicated rheumatic fever. Once at home, she improved immediately and never needed steroids. She never developed renal, cerebral, or cardiac complications.

My judgement (having tried the patient on other anti-inflammatory drugs) was to ignore the vanishingly small risk of Reye's Syndromes, and try to gain immediate clinical improvement with aspirin (as happened). Fixed, arbitrary rules can conflict with experienced clinical judgement. When this happens, there will be those who prefer to act as sheep, and follow the rules to the letter. Their desire is to conform, regardless of clinical inappropriateness. Experienced shepherds who never lose sheep, are the ones to follow. Unfortunately, there are too few experienced shepherds, and a helluva lot of sheep. I wonder if experienced shepherds have to tolerate farm regulators who know nothing about rearing sheep, but have the authority to tell them how best to manage their flocks.

NICE rules and guidelines provide intelligent advice, but may not be specific enough for us to abandon clinical experience and judgement in individual cases. This is acknowledged by NICE. The ability to use one's experience and judgement is a facet of clinical mastery, but beware, using your judgement rather than a guideline, can now incite regulatory action against you.

Oscar Wilde once wrote that experience is simply the name we give to our mistakes (Lady Windermere's Fan. Play. 1892). This is only partly true

In this age of better communication, access to patients' records should be easier than it once was (for those with the patient's permission). It is not, despite the many promises made by IT professionals. Those employed by the NHS have earned fortunes delivering some outstanding failures. My older foreign patients had the answer: they carried their written notes with them.

Would you trust any state with your medical and personal information, and to provide life and death information during an emergency? Even now, electronic record systems have a long way to go before they can match the security, completeness, and accessibility of hand-written notes. Some of my patient's notes in A4 files, were fifty years old when I retried. Since I gave each patient a summary of their condition (after an in-depth consultations and investigation) they were free to present to any other doctor.

Important notice. *Regulators can overturn doctor – patient confidentiality. They can gain access to any clinical notes they choose, without the patient's permission. Forget the Data Protection Act, all they need to do is devise a sufficiently credible reason, label it 'in the public interest', and they will easily get permission from a High Court to access the information they want.*

I understand that MI5, MI6, and the CIA, have returned to using typewriters. Using them has proven more secure. One cannot hack typewritten pages, although they are easily copied. I advocate that every patient (who has had a serious illness), should travel with a written copy of their clinical summary, together with a readable digital copy. Patients must assume that no state system or agency is responsible, organised, or trustworthy enough, to safeguard their notes and make them available in emergencies.

To brief myself, just before seeing a patient, I would always rely on my previous summaries; copies of which I had sent to patients after 1973. When I spent time with patients, and uncovered relevant clinical information, I always committed the details to a pink coloured summary sheet (easily spotted in a thick A4 file). Most of my colleagues achieved the same by writing letters which summarise the clinical facts at the time of consultation. The summary sheet is most useful when it lists all the relevant diagnoses, treatment, previous interventions, and the thinking behind every opinion and action. With a summary in hand, one can ask returning patients pertinent, focussed questions. One can also avoid the repetition of investigations and wrong diagnoses.

Intelligent business executives expect their accountants, lawyers, and surveyors to be briefed fully at meetings, so why not their doctors? Being briefed indicates engagement with the patient's problems. It is what professionals do. No excuse condones being poorly briefed. To give the best advice possible, one must have access to each patient's present, past, family, social, and therapeutic history.

Patients will judge your clinical efficiency on how specific and personalised the questions are you put to them. For instance: 'How has your shortness of breath changed since your CABG operation, three years ago?' and "You don't respond well to NSAIDs do you?' These questions mark your degree of engagement. Knowledge of your patient's intimate details, and personal story, will characterise you as either attached and caring, or detached and seeking anonymity. Mastery of the facts will foster the patient's trust. Losing patient notes, and knowing nothing about them, may be understandable, but will signify incompetence. All doctors should question whether they should work for any organisation incapable of finding patient notes.

Family Medical History

The enquiry might need to extend back to great grandparents and include cousins.

A positive family history will significantly increase the antecedent probability of conditions as diverse as coronary heart disease (CAD), hypertension, allergy, psoriasis, osteoarthritis, prostate cancer, migraine, and many more.

A family history from a reliable source can be of inestimable value. Because so few patients have any meaningful medical knowledge, the accuracy of the family history they report is likely to suffer from hearsay, misinterpretation, omission, distortion, and dilution of the Chinese whisper variety. There are conditions, like syphilis and suicide, which some families will want to forget.

I have often asked patients to enquire among their relatives for family history verification. As with all other answers to medical questions, one must view each bit of information as positive, negative, false positive, or false negative, until corroborated otherwise. Ask for copies of hospital notes and death certificates in order to uncover the truth, although these are also subject to error. If you watch detective dramas, you will learn that there are often devils in the detail, and that persistence when trying to explain one detail, can make or break an enquiry. The same principles apply to all medical and scientific enquiries.

The heritable component of both coronary artery disease (CAD) and hypertension is likely to be as much as 95%. Despite the limitations of a family history, a positive history of these conditions, involving several family members (especially if involving several generations), will support the diagnosis of CAD or hypertension for the patient. It is uncommon to find either CAD or true hypertension (with LVH) in those without a family history. I have seen many of those with labile blood pressure and a strong family history, eventually develop sustained hypertension. The lengthy continuity with patients I enjoyed in my practice, allowed me to make such observations repeatedly.

A simple Mendelian 'rule of thumb', is useful for dominantly inherited conditions like hypertension and CAD. If one parent has the condition, one in four of their children are liable to develop it. If both parents have it, assume that eight out of ten of their offspring will also. These are short odds, easily confirmed from personal experience (relatively small numbers). With a four to one chance of occurrence, one can easily appreciate the genetic risk from clinical experience alone.

The inheritance of diabetes and cancer is less predictable, and clearly more complicated. Even with 50+ years of experience, I have only seen the occasional case of inherited cancer (breast and prostate). The few examples I have seen, could have been coincidental.

It is useful to know that inheritance is best represented by a matrix (of genes). In straightforward Mendelian inheritance, relevant genes may be limited to a 2 x 2 matrix (two genes from the father; two from the mother). In conditions with less obvious inheritance, more genes may be involved. For instance, a 16 x 16, or 32 x 32 matrix of different relevant genes may apply, each contributing different elements of influence. If you imagine a matrix as a crossword puzzle with many boxes, the number of 'boxes' ticked (the genotype), and the strength of their influence (their penetrance), are likely to determine the phenotype (the clinical features).

I had a friend with severe type 2 diabetes, CAD, and peripheral vascular disease. I found it unusual that he had no retinopathy or peripheral neuropathy. Obviously, each patient will have their own genetically pre-determined set of clinical features.

The complete genotype of Duchenne muscular dystrophy is now known. If all the boxes in the relevant matrix are 'ticked', the patient will present with the full-blown condition (difficulty in walking, running and jumping; large calves and a waddling gait). With one box ticked, the patient might only have weak legs, and wonder why they couldn't win races at school. Only by knowing genotypes will we understand the many possible clinical presentations (phenotypes). The genomic revolution will undoubtedly reveal many subtle medical conditions that formerly eluded us.

Fifty years ago, few people seemed to know much about their family history. With an improvement in the availability of family information, partly aided by longevity, this is no longer the case for younger patients. Many now have grandparents, and living great grandparents. The family history is thus more available, if not more reliable. Whenever I checked a family history for morbidity and mortality in the past, I found the accuracy to be poor. Relatives would more reliably report events such as CABG or artery stenting, than actual diagnoses.

My father and his mother told me that my grandfather died of alcoholism. My father, 18-years-old in 1922 when his father died, remained a lifelong non-drinker, partly as a result. The London Hospital had my grandfather's admission notes on microfilm, so I checked his clinical data myself. He actually died of pneumonia, although on admission his breath smelt of alcohol. Looking through his clinical notes, written in beautiful copper-plate handwriting, and reproduced as fresh photo-copies (rather than musty old originals), I saw they had treated him with Tincture of Lily of the Valley, and other inefficacious medicines. A strange urge arose in me. I wanted to telephone the hospital and suggest they give him penicillin immediately (penicillin had yet to be discovered, and my urge 78-years too late).

Social History

This needs to include alcohol use, cigarette smoking, drug use, occupation, social status, relationships, lifestyle, and living and working conditions, and significant life changes (divorce, bereavement, bankruptcy). These can all influence patient morbidity and mortality.

Liberal, politically correct thinking in 2024, would have us resist any recognition of socio-economic class, race and culture. In the UK, however, we have always had a social structure that has distinguished 'insiders' from outsiders', regardless of political correctness. The 'outsiders' are those vulnerable to social discrimination.

Doctors working outside the UK, might find British class attitudes baffling; especially if their country has a President, and not a royal family. Who has the highest social status in the UK? Would it be a village priest, teacher or a billionaire? In many countries, the answer would be a billionaire. For traditional Brits, teachers and priests have a higher social status. What some regard as anachronistic, remains important in the UK. Every day, nurses and doctors must contend with the many differing presumptions, beliefs, and values held by their patients. In the UK, class and education define social status better than wealth.

These broad social differences are of practical significance when considering medical management. They can challenge our ability to value and understand symptoms, our ability to make our arguments understood, and often define the resources available to each person. The patient's ability to understand and accept evidence-based reasoning, can be of life and death significance. Together with lifestyle differences, these partly explain the health divide (the three to five-fold difference in morbidity and mortality between the social strata).

Lifestyle is a popular news topic. There is now a cultural sub-set of people called 'lifestylers'. They join a long list of other groups who define themselves as 'us', not 'them'. Despite the constant criticism of the western lifestyle: eating 'unhealthy' food, obesity, drinking too much alcohol, and not taking enough exercise, on average, we are living longer than our predecessors.

At a time when working with asbestos and inhaling factory fumes was common, occupation was a crucial component of every medical social history. It still is, but to a lesser extent. Work-related illness still remains significant (1.7 million cases in the UK in 2020, half of which were stress, anxiety and depression related). Living conditions have always been of medical importance. In the 5th richest nation on Earth, we in the UK still have a deprived class, although most have access to clean water, and few starve to death. For those not living on the street, few need to sleep six in one room. Social class, and level of deprivation, remain powerful predictors of morbidity and mortality. The patient's social history will often temper the acceptable advice we can give.

In 1963, when industrial diseases were more common, Donald Hunter was one of my teachers. He had written the definitive book on the subject (The Diseases of Occupations. 1955). One of his hobby horses was the danger of lead poisoning. Children were being poisoned after biting off the heads of their lead soldiers.

We must applaud both epidemiologists and health bureaucrats for the reduction in industrial diseases (related to cotton, coal and asbestos), brought about over the last 100 years. There has also been a notable reduction in the number of people smoking tobacco (approximately 40% of people in the UK smoked in the 1960s. In 2020, only 13% smoked). Political action brought this about. Smoking cessation features in the health divide. Smoking cessation is more frequent in educated high-earners, than in the poor and less well educated.

The Biological Aspects of Social Stress: 'Clinical Energetics'.

Biology results from sub-atomic packets of energy, moving in, out, and around cells.

A patient's living and working environment can influence the progress of their disease. If financial health depends on income, expenditure and savings, physical health depends on how our bodies exchange biochemical energy. Those who spend more mental energy on coping than they can afford, while spending too little time recharging themselves (sleep and relaxation), can expect energy depletion and poor health. While this energy imbalance is one obvious cause for reduced well-being, the role it plays in disease initiation and modification, is far from obvious.

When long-term tiredness is the predominant symptom, it is essential to estimate a patient's energy income (how effective is their relaxation and sleep), and, their energy expenditure, and how many demands they on it (together with their time and resources), and how threatening and time-urgent it is.

We should ask patients how many jobs they have, and how many they have lost; the number of family members they support; how well they are managing their finances; how concerned they are about the viability of their business, and their job security. The more demanding their life, the more easily they will deplete their energy reserves and have less energy for problem solving and coping.

Try to gauge the efficiency of a patient's 're-charging' strategies: do they suffer from sleep disturbance and can they relax? Does their job include night-shift work (social jet-lag, body-clock disturbance)? Are they always on call, and do they take holidays (if allowed, and whether they can afford them)? Is their disposition relaxing and re-charging, or energetic and exhausting? What energy saving resources do they have? Do they delegate? How independent are they, and how independent can they afford to be?

I often saw patients' medical problems deteriorate as their energy reserves declined (the patient might report progressive tiredness, constant fatigue, and even exhaustion). For the fatigued or exhausted, a clinical catastrophe is more likely for those who already have angina, hypertension, clotting dysfunction, reduced immunity, or any condition challenging their health.

The Lancet (Jan 2017), reported that stress-induced amygdala over-activity, is related to the production of inflammatory cells in the bone marrow (macrophages?). (*Relation between resting Amygdalar activity and cardiovascular events*. Tawakol, H., Ishai, A. et al. This provides one biological link between chronic stress and arterial (intimal) inflammation. Could this link help explain atherosclerotic plaque formation, plaque rupture, clot formation and cardiac infarction? The observation provides objective evidence for one biological pathway connecting mental stress with intimal inflammation and morbidity. How important is this link, and others like it, to health in general and overall patient morbidity, and mortality?

Situations associated with long-term stress, commonly cause sleep disturbance and progressive tiredness. Look out for patients with deteriorating sleep who are always tired (insomnia, although I prefer the word *dysomnia*; sleep is more often disturbed than absent). Try to understand the pressures patients face. Since tiredness is the commonest of all symptoms, and a modifier of many medical conditions, asking questions about it will expand your understanding of patients, and their appreciation of you.

Uncovering the relevant medical factors in a social history may require forensic enquiry. Not all patients will paint a complete picture of their circumstances, without some prompting. With familiarity comes adaptation, and the acceptance of adverse circumstances as being usual. Patients will not always recognise pathological issues in their lifestyle. You will need to judge when it is appropriate to chase the facts. Knowing the right time can be difficult, unless you are a good judge of mood. It is an art. The truth about family rifts, drug taking, and deviant behaviours of various sorts, will not always come forth unprompted. When interpreting what you are told, never forget the old cliché: *there are always two sides (at least) to every story*. The truth may elude you if you restrict your enquiries to symptoms alone. Revealing descriptions, that explain the patient's true state of health, will often be beyond your reach.

Just occasionally, the social history reveals a diagnosis, and its relevance.

I well remember an older adult man with severe shortness of breath who presented in the first week of January,. His wheezing started on Boxing Day, but why, I wondered. 'What Christmas presents did he get?' I asked his daughter. She had bought him a feather pillow, but could that have caused his wheezing?

He had kept an aviary with pigeons and exotic birds for 40-years, and never wheezed once. That was ten years before, so why might he wheeze now? The feather pillow, bought as a present, was the last (metaphorical) straw. As an atopic pigeon fancier, he must have become sensitised to feathers many years before. All he needed to get an asthma attack, was exposure to one more feather.

Catastrophes happen by accident, on purpose, or from system instability. Biologically, medical catastrophes result from the instability of energy-dependent, pathophysiological processes. For life to succeed healthily, there must be perfect homeostasis. The process requires vast amounts of power for its many biochemical energy demands. The demands on all of us, once centred on the search for food, water, shelter, finding and keeping mates. To that, we have added the necessities of work, maintaining security, and pursuing leisure.

As a measure of how much energy we need to stay alive, it has been estimated that we need to recycle our body weight in ATP (to power homeostasis) each day. Törnroth-Horsefield, S.; Neutze, R. (December 2008). *'Opening and Closing the Metabolite Gate.'* Proc. Natl. Acad. Sci. USA. **105** (50): 19565–19566.

When considering stress, Thom's Catastrophe Theory, is a meaningful concept, helping to understand the origin of medical emergencies. The theory explains how stresses (applied forces) might cause symptoms (strain effects), and medical catastrophe. Many situations are bi-stable. For instance, one step before a cliff edge represents safety; one step beyond it represents disaster. One further step forward could cause death, lying in bits at the bottom of a cliff (if high enough).

Ramon Vargas shot and killed Irma Garcia, a teacher at Texas' Robb elementary school in Uvalde (May 2022). She was one of two teachers, and 19 children, who died that day from gunshot wounds. Two days later, Irma's husband, Joe Garcia, died of a sudden heart attack. They had been married for thirty years, and had four children who must face life without their parents. Debra Austin, Irma's cousin, said, 'I truly believe Joe died of a broken heart … losing the love of his life was too much to bear.'

Some patients battle with stress over many years and get progressively tired, fatigued, and then exhausted. All exhausted, underpowered systems will malfunction, and some collapse. For an exhausted patient, any catastrophe that occurs will depend on their predispositions and weak points. Patients can present with a panic attack (if prone to anxiety), a cardiac infarction (if they have vulnerable coronary artery plaque, and a tendency to increased coagulation), a CVA (from cerebral artery thinning and uncontrolled hypertension), suicide (if subject to depression), or noticeable changes in their clinical condition.

Did Hillary Clinton's advisers spot any decline in her health, before she developed pneumonia during her 2016 election campaign to become US President?

Was it acute stress that caused actress Debbie Reynolds to die (December 2016), soon after the death of her daughter Carrie Fisher (Princess Leah of Star Wars fame). According to her son, she struggled to deal with the tragedy of losing a child (said to be one of the worst of all tragedies in life). She had a stroke while making funeral arrangements for her daughter.

It is easy both to over-estimate and ignore, the clinical significance of stress. We mostly describe human stress in anecdotal terms. If stress and tiredness are to be regarded as clinically important, how is it that so many who fight adverse circumstances, do not suffer medical consequences? The answer lies in biological diversity, an important part of the evolutionary process. If there were no *'horses for different courses'* (a universal biological principle), biological equality would bring evolution to an end. It is not possible biologically, for the same stressful situation to affect everyone in the same way, or to the same degree. To deny our everyday stresses, in order to preserve self-esteem, *status quo*, and personal dignity, may not promote our survival.

In 1967, Holmes and Rahe made an attempt to rate the significance of stress (The Holmes-Rahe Stress Inventory: The Social Readjustment Rating Scale). They assigned a score to each life change. Death of a spouse they rated as '100'; divorce '75'; marriage '50'; major change in living conditions '25', etc. They observed an 80% likelihood of a 'health breakdown' within two years for those with scores over 300; it was 50%, for those with scores between 150 and 300. Their list can be downloaded from American Institute of Stress website.

In all but minor cases, I regarded it as my professional duty (for an engaged, attached doctor, pursuing the art of medicine) to consider how patients live: their living conditions, circumstances, and relationships. Although irrelevant in most clinical conditions, they can be crucial in unsolved or recurrent conditions, especially when therapeutic interventions have failed consistently.

The Therapeutic History

There is more to a therapeutic history than the patient's current list of drugs. Therapeutics is a vital branch of clinical study, and too large a subject to summarise here. Instead, I offer some perspective and tangential views of drug use, simply to show the breadth of experience required for success. Although I have limited this discussion to topics associated with pharmacological therapy, other forms of therapy can be just as useful. Talking therapies (cognitive therapy, psychotherapy, and psychiatry), manipulative interventions (physiotherapy, osteopathy), and lifestyle changes when offered as 'treatment', can be of benefit. Even if one knew every detail of pharmacology, one would still need to know when to use which drug (or drugs in combination); in what dose and for how long. These decisions must be evidence-based, but because of the complex inter-reactions involved, they involve an art not all come to possess.

Many patients see themselves as health experts, so your history must include the non-pharmaceutical preparations they take, together with the lifestyle and dietary changes they have made for health reasons. The vitamin and health food supplement industry, is a 315 billion-dollar business (2020), and rapidly expanding in the US. The mixtures some now swallow every day are potions concocted from fruit, vegetables and vitamins etc., two table-spoons of mythology, a table-spoon of pseudo-science, and a quarter teaspoon of scientific evidence. The same mind-set accepts that sulphurous spring water from spas, will fortify their health, and that ginseng, garlic, soya, vitamins, lots of fresh air, and running naked through snow-clad forests, are all beneficial. Powdered parts of animals, like tiger, shark, elephant, and rhino, still hold their place as medicines for some.

When health is the subject, belief in these therapeutic alternatives is key. When disease prevention and cure are the subjects, there is little evidence to support any beneficial claim for them, other than a placebo effect. Those who believe in snake oil, will usually ignore science. They accomplish this by rejecting the need for evidence, and choosing to ignore inconsistency. However cynical one might become, one must never underestimate the therapeutic power of belief and placebos (they mostly benefit 20% of those who take them).

Running counter to the calm acceptance of alternative therapies, are the double-blind controlled trials, and meticulous safety-checking required of the pharmaceutical industry. Despite the cautious development of drugs, public mistrust in medicinal drugs remains. The uptake of COVID vaccinations in some population sub-groups was evidence of this distrust.

The Victorians omitted their medications on Sundays; it allowed the supposed toxic effects of drugs to 'clear their system'. This was a good thing, if they were taking digitalis, but was not always wise. A few drugs have caused havoc, resulting in patients with strong feelings about medication. What the public deserves, is validated assurance about drug efficacy and compatibility. In the meantime, the medico-legal and common-sense necessity to list every side-effect in information leaflets, hardly helps public trust. How future genetic profiling will help to predict individual efficacy and incompatibility, remains to be seen.

Drugs of addiction are now a major problem. Every doctor needs to know about the latest street drugs, alongside those used for centuries.

Attitudes to Drugs

Patients have many attitudes and beliefs about pharmaceuticals. Knowing what these are, when taking their therapeutic history, will help when prescribing or withdrawing medication.

Jack came to me as an emergency. He had palpitations. He arrived agitated, with a red face and bulging neck veins. He thought he must have heart disease. He was in denial about his heavy cocaine use.

The belief that antibiotics diminish immunity, or will cause deadly resistance, is now such a commonly held view, that many patients refuse to take them; at least, not until they are seriously ill. Doctors are as bad. We tell patients that resistance to antibiotics is now a major problem, without explaining why, and rarely acknowledging the debt we owe to antibiotics. We tell them that antibiotics are no longer indicated for upper respiratory tract infections (URTIs), presuming them to be caused by viruses, against which they are ineffective. In fact, some antibiotics are both harmless and anti-viral (erythromycin and tetracycline and their derivatives, can inhibit RNA replication). Using these particular, harmless antibiotics (resistance is not a major feature), will halve the time course of most viral, URTI diseases (with implications for national sick leave reduction). They can also work prophylactically and prevent some secondary infections. Those doctors brave enough to think for themselves will appreciate this. Rule followers might not prescribe an antibiotic unless it fulfils specific microbiological criteria. I have watched them prescribe paracetamol alone, and watch the patient deteriorate; the art of medicine has alluded some of them.

I contend (anecdotally) that my use of simple, non-hospital based antibiotics over the last five decades, is one reason I never saw rheumatic fever, mastoiditis, bacterial meningitis, tonsillar abscess (quinsy), and pneumonia: all common before the antibiotic era. It must also be said that no doctor in their right mind would advocate the use of novel, hospital-based antibiotics, for common URTI conditions, although in past decades, drug representatives have tried hard to convince GPs to use their latest products.

My experience forces me to conclude that the 'average' adult (in terms of average age, height, weight, number of GCSEs, and wealth), understands as much about physiology, biochemistry, and therapeutics, as they do about the workings of their car engine and mobile telephone. Explaining the intricacies of pharmacology and therapeutics will be challenging, but not impossible.

It was Aristotle who suggested that all babies were born with a mind *'unscribed'*: a *tabula rasa*. With science, most people keep their *'tabula'* clean; in a state of *'rasa'* for their lifetime. They might be the ones who see no inconsistency in smoking, while holding strong views about pharmacology and other scientific matters. Some might better accept revivifying the ancient Greek *'auspices'*: getting guidance from watching animals released from captivity, or getting relief from ritual dance performances around their sick-bed. How many would give these practices their vote, rather than derive their notions from medical science? Before entirely dismissing ancient knowledge, we know that music and dance can both be therapeutic.

Why is there such antipathy to pharmaceutical drugs? While drugs have many beneficial effects, we cannot deny their many side-effects. Almost everyone has one tale of misfortune to tell - how some drug ruined their aunt Mary's health. When patients believe this, snake oil salespeople flourish; they can more easily work their magic. Belief counts. If one can make a medication believable, harmless, even if only partly effective, many will buy it.

Through the swirling mists of ignorance, superstition, nonsense, and any sign of scientific objectivity, doctors come with their knowledge of therapeutics, face to face with those versed in pseudo-science and myth. I had a simple remedy for those who refused penicillin for their streptococcal tonsillitis: I never saw them again. Such patients put medical professionals at risk. I refused to be made responsible for any complication that might have resulted, be it fulminant meningitis, or septicaemia. If it were possible, I would have referred them to a local witch doctor, shaman, or voodoo practitioner, but there are too few practising in the UK these days.

> *A little learning is a dangerous thing*
> *Drink deep, or taste not the Pierian spring.*
> *There shallow draughts intoxicate the brain,*
> *And drinking largely sobers us again.*

Alexander Pope. *'An Essay on Criticism'* (1711): Part 2: 215-218.

Patients are often so scared of the consequences of medication, they will reduce the duration of taking it, or take lower doses than prescribed; doses that are 'homeopathic', rather than therapeutic. Many now prefer to follow the 'natural path', and despise pharmaceuticals altogether. They should consider the plight faced by the Victorians. Antibiotics had yet to be invented, and mortality rates were such that few children reached the age of ten-years-old. Many Victorian parents became used to attending child funerals. Their children frequently died from scarlet fever, diarrhoea, pneumonia, and septicaemia. Simple ignorance, arrogant ignorance, and profound ignorance, still risk patient survival today. Therapeutic non-compliance remains a major cause of clinical deterioration, for which some parents and guardians will be responsible.

When prescribing, both the predictable effects and side-effects of modern medicines, must always be acknowledged; so must the variability of individual responses. Among any large group of patients, side-effects will occur. Among them will be immune reactions, poisoning, and idiosyncratic reactions. Some patients seem able to tolerate medicines, others cannot tolerate any. Genetic reasons for this are already known, with many more yet to be found.

Taking a therapeutic history is an opportunity to discuss the patient's previous drug reactions and sensitivities, their observations, and their individual preferences. I often asked reliable patients which drug they preferred to take, from one of several alternatives. The dictatorial approach: *'you the expert, they the recipient of your superior knowledge and wisdom'*, is fraught with problems and danger (prescribing drugs that may not suit them, while ignoring the side-effects they once reported). If you ever say, *'I have never heard of that side-effect',* or, *'nowhere does it state (in the BNF) that this is a side-effect',* you might undervalue your patient's personal observations, knowledge, and experience.

Sixty percent of all chronic medical cases attending a GP, will experience a side-effects from their medication; the majority, therefore, will feel better without it. The problem is, 'feeling better' is not the same as 'being better'. A patient in heart failure can feel better after changing their medication (less nausea and fatigue without digoxin and a beta-blocker), even though their heart function has remained the same, or is further impaired. Doctors rarely advise it, but many patients leave off all their tablets, just to see if they feel better. Many drugs used in cardiology cause side-effects: *'statins'* cause muscle pains in 30 – 50% of patients; nausea occurs with both digoxin and amiodarone; lethargy, wheezing, and impotence can occur with some beta-blockers.

I used beta-blockers throughout my working life. I well remember when I first used propranolol in the 1960s. James Black had discovered them, two years before I qualified (*A New Adrenergic Betareceptor Antagonist.* The Lancet (1964). 283 (7342); 1080-1081*)*. He won a Nobel Prize in 1988 for his discovery. I gave them up progressively, as unacceptable side-effects started to appear: reduced physical prowess, impotence, the induction of asthma, and the reduction of ventricular contractility. Such side-effects, which were within the everyday experience of every physician, depended on several factors: $beta_2$ receptor sensitivity, cardio-specificity, amount of vaso-dilatation, water versus lipid solubility, and duration of action. The latest beta-blockers produce fewer side effects, but must still be monitored closely.

There is a simple rule of thumb for beta-blocker use. Patients (especially older adults) on beta-blockers, and / or, diuretics, should have their medication reviewed urgently if they get cold blue hands, feet, or nose. These are signs of diminished tissue perfusion (reduced cardiac output, and greater tissue oxygen extraction from blood).

It is not only mechanics who criticise the work of their fellow professionals; doctors do the same. Please resist asking your patients: 'Who on earth started you on this medication?' Without knowing the prescriber's train of thought, it is all too easy to be critical. You risk discrediting yourself if the patient has a strong, loyal and trusting relationship with the prescriber.

In challenging cases, where no medication works well, drugs with a reasonable antecedent rationale for their use (but not licensed for it), can be worth considering (an individual, patient consented therapeutic trial). Many will see this as foolhardy, lacking insight, and non-compliant (although imaginative and inventive). Others will see it as research, with the sole aim of helping a patient. In the days before regulators knew best how to treat the sick, I tried this a few times. It proved most useful in challenging cases; those where my colleagues had achieved no therapeutic benefit from using all the 'usual' drugs. By chance, one may 'get it right' for the worst of reasons; or 'get it wrong' for the best of reasons. The usefulness of a 2–3 day, fully informed, consented personal therapeutic trial of a new drug, should be considered for some patients *in extremis*, and for those where all other usual drugs have failed. While this represents 'going off-piste', and performing 'jungle medicine', it is the sort of thing one might need to do on a desert island (whether the UK is one, I am prepared to say). You could make an important discovery, or be completely unacceptable and ineffective.

Clare was a four-year-old child in the early 1980s, when she developed a mixed Pseudomonas and Klebsiella scalp infection. The infection had spread as a thick scab, over a large part of her scalp, and was progressing rapidly. It could have proven malignant if not halted. No oral antibiotic then available had any effect, so her parents were told that only regular intramuscular injections of rather toxic antibiotics offered any hope. But what four-year-old child would have tolerated repeated painful injections every day?
I was fortunate enough to send her sample swab to a microbiologist involved in the development of a new antibiotic not then available - ciprofloxacin. With the parent's fully informed consent, we obtained a supply of tablets. It meant trying them before being sanctioned for use, but gave her family hope. I advised her parents to accept them, despite the unknown risks. The effect was magical. The infection plaque receded within days and quickly disappeared completely.

When standard guidance failed, I sometimes undertook a trial of other drugs (those with few known side-effects). Two other 'off-piste' successes of mine were my use of doxycycline in chronic glandular fever (as well as other RNA viral, URTI infections). I wondered whether it might also be useful in COVID-19 cases, given its inhibition of RNA replication, but I had retired by then. I also used tamoxifen in women with migraine and pre-menstrual mastitis. My colleagues had 'given up' in the few cases I treated. Legitimately, they told patients there was 'no known, or effective treatment' (they could have said, 'be gone and stop annoying me'). If you understand pharmacology and pathophysiology, you too might discover a new use for an old drug. The problem is, how to prescribe them without risking your job, even if helping patients (your vocation, I hope).

Those starting their medical career, may not yet appreciate the full spectrum of drug responsiveness. Paediatric doses will be effective for some patients, while others need amounts four-times larger, just to get the same result. As a personal example, I have only ever needed 150mgs. of aspirin to lower my temperature during a fever. Most patients need four times that amount. Similar ranges exist for analgesic and sedation medications. I once took ibuprofen 100mgs and had indigestion for a week. I have known patients take ibuprofen, 800mgs three times daily, with no side-effects. I once took a single dose of promethazine 25mgs at night for hay fever and didn't wake up completely for five days. I have known patients take promethazine 50mgs tds., with no sedative effect whatsoever. I would say, from experience, that age does not explain the variance; it is more a matter of individual responsiveness. Caution, however, is always needed when treating older adults and the very young. Frailty and robustness are poor predictors of drug responsiveness; specific genetic and metabolic reasons, may eventually explain it better.

It is essential to ask each patient how they have responded to drugs in the past. Sometimes the patient's tolerance to alcohol, gives a clue to their sensitivity, but not always a reliable one. Know your drug pathways, and learn which is processed by the liver (fat soluble), and which by the kidneys (water soluble). Diseases of these organs will affect drug tolerance, patient responsiveness, and adverse effects. Adverse drug reactions are common: a good medico-legal reason to use those that have been tried and tested.

In 2017, one NHS patient died each week after being given the wrong drug.

Observant patients will tell you which antibiotic, and which analgesic, best suits them. This annoys some doctors. There is a catch-phrase for it: *'patient-led interference'*. My advice is to listen, and not to get annoyed. It has become politically expedient to spend billions on genetic research to achieve personalised medicine, but there will always be a cheaper way: ask patients, and listen to their answers.

Placebos are effective in 20% of all cases treated, even if the patient is unaware of their non-pharmacological status. For a drug to be of pharmacological value, its therapeutic efficacy must exceed that of a placebo. A former surgical colleague of mine, Mr. B.B., sometimes gave Tincture of Thuja (pine extract, skin application) to his patients with verrucas. Despite my scepticism, the warts he treated often disappeared. He would tell his patients that he was going to first give them a *'harmless placebo'*. It still worked. Such a pity he kept no data, given the ever-present onus on doctors not to fool themselves or others. He felt Thuja was *'worth a try'*, before embarking on liquid nitrogen applications or surgical removal. His advice suited a growing number of those whose preference was for 'natural' therapies, although dangerous pharmaceuticals like digitalis, belladonna alkaloids (including atropine), aspirin and curare are all 'natural' therapies.

I liked to explain to patients how the drugs I prescribed worked. The 'lock and key' analogy was most useful. It helped patients understand why one antibiotic is not 'stronger' than another, just more or less specific (to open a lock). A drug (key) will either fit into a cell receptor ('lock'), or it will not. It also helped to explain why some drugs will not work at all for those who lack the receptors. Some other mechanisms are more difficult to explain.

Drug doses are obviously of crucial importance, but it is difficult to know when enough is sufficient. Because of the variance in patient responsiveness, under-dosing and over-dosing, happen all the time. This means we must review drug doses frequently. With drugs that work within less than one hour, such as sedatives and some analgesics, 12–24 hours is time enough to assess their efficacy. Contrast this response time with anti-depressants, which take at least five days to work (sometimes ten days). For antibiotics, one must wait at least 48-hours before the first assessment.

Drug responses can follow the time taking to change the direction of an oil-tanker: after applying the rudder and changing course, one must wait patiently before the effect is seen. For anaesthetists using immediately acting intravenous drugs, the titration principle is fundamental. There is a dose of intravenous anaesthetic that will render every patient unconscious. Intermediate doses cause intermediate levels of sedation. Age and body weight-based calculations of drug dosage provide an initial guide, but individual responsiveness can be crucial, and can come as a surprise. Give too much anaesthetic to some, and they will remain drowsy for a day or so, blocking a valuable hospital bed.

Every object in the Universe has its own natural periodicity. Every action has its own response time course (periodicity), depending on the characteristics of the machine involved (light aircraft, Boeing 787, dingy, oil tanker or patient). There are also inherent differences in patient responses to tablets, intramuscular injections, and intravenous infusions. With experience, our anticipation of the responses should improve, but there will always be surprises. Patients need to know the usual timing of a response, and the usual side-effects to expect.

One therapeutic rule-of-thumb for poorly responsive cases, is to combine different drug classes, each in low dose. This is preferable to progressively increasing the dose of one drug to its limit; a strategy that will usually invite side-effects. One can also rotate classes of drugs in order to avoid tolerance. This is most useful when prescribing addictive drugs like benzodiazepines; not that anyone is allowed to prescribe them for long, unless they are a psychiatrist. One technique I used for decades was to rotate the patient's medication between diazepam, chlordiazepoxide, and a sedative anti-depressant. Few then became addicted to diazepam (addiction is partly a genetic characteristic).

Failure to appreciate the personal impact of anxiety, fear, and depression on the lives of patients, while over emphasising the dangers of addiction to benzodiazepines, led to the ill-informed restriction of their use. The aim was to safeguard patients from being treated unwisely. The fact is, doctors of my vintage, used them successfully without complication for over forty years. The trick was to choose the most suitable patients. No drug is suitable for everybody. Addiction is never a desirable outcome, but for those failing to cope with what life throws at them, mild addiction to a benzodiazepine with no somatic side-effects, is the least of their problems. Science dictates that addiction can occur; the art of medicine can suggest that it is a price worth paying.

Do not expect any medical regulator to understand. Their desire is not to appreciate the realities of life, but to create a therapeutic Utopia: a place where the rules they dictate are sacrosanct. In this mystical place, nobody dies, there is no addiction to any drug, and doctors and drugs achieve only patient benefit. There is only one place where such aspirational nonsense exists, and that is in mythical, medico-legally defended, ivory towers.

Prescribe benzodiazepines and the CDA and GMC could reprimand you, even if you have safely helped chronically anxious patients to cope with their difficult lives for decades. Their brief is not to consider the plight of patients. Their only concern is that we comply with the rules; often guidelines that they define as immutable. Few regulators appreciate medical risk; how can they without having practiced assessing it daily.

Medical bureaucrats perform robotic duties, and need not understand the practicalities of dealing with patients. They will, therefore, enforce regulations that benefit only some patients. Even doctors employed as bureaucrats can prefer paperwork to patient management; being many steps removed from the bedside, their delusional utopian ideas can flourish. They may believe that we should never prescribe an addictive drug, even though a patient has benefitted from one for decades without ill-effects. Although considerations of clinical judgement and diagnostic accuracy are beyond them, they remain the ones allowed to sit in judgement on experienced medical professionals. We must replace them with a medical directorate, staffed by experienced medical practitioners, not lawyers and those with no medical degree. Not to do so is not in the public interest, and one mark of corporate stupidity.

Aspects of Therapeutic Management

Those who want to understand how some drugs work, will need to know about capacitance. A bath will hold only so much water. Once a bath is full, adding more water, will cause it to overflow. Extra doses, added after drug receptors become saturated, cannot provoke a greater response; overflow and toxic side-effects happen next. For this reason, prednisolone 15mgs daily, rather than 60mgs+ per day, is all one usually needs to treat acute onset asthma. In severe asthma, I have never prescribed more than 15mgs daily. If fifteen milligrams of prednisolone 'fills the bath', why create overflow and cause side-effects with greater doses? Those doctors who feel the need to prescribe more, can be too anxious and impatient. Regardless of the dose of a steroid, it always takes 24-hours to achieve their maximum therapeutic effect. Well-known side-effects result from long-term steroid use: Cushingoid features, diabetes, a lowered resistance to infection, hypertension, osteoporosis, weight gain, skin striae, and sometimes psychosis or depression. The higher the dose given over the long-term, the more likely these side-effects are to occur. Steroids improve vitality, and that is of great value to those suffering physically. No drug is all good or all bad.

After a long list of negatives, let's try something positive. 'OK, so what about smoking?' Is there anything positive to say about it?

During Queen Victoria's reign, medicated cigarettes (menthol, etc.) were 'prescribed' for asthma, with obvious short-term breathing benefits. In my experience, the asthmatic smokers who gave up smoking, deteriorated at first; some had to be admitted to hospital. All wheezy ex-smokers know one thing for sure: one cigarette will relieve their wheezing rapidly. Smoking aids mental concentration and was once used to improve the voices of a few famous singers. Without smoking, both Frank Sinatra and Sammy Davis Junior, would have sounded less appealing. Sammy David died of throat cancer, a well-known effect of smoking.

Whenever we prescribe drugs long term, we must assess their effects repeatedly. Long-term titration is possible, but to be dependable, it requires something not commonly found in UK medical practice - doctor-patient continuity. The NHS has recognised the value of continuity, but stymied itself by grouping doctors together in large, evermore anonymous practices, with no on-call duties, no night calls, and no weekend commitments. By giving GPs an easier life (compared to when I was young), the NHS bade farewell to some doctor-patient continuity.

To improve the plight of refugees from the NHS, all I had to do mostly, was to step in and alter their drugs, or drug doses. Every therapeutic history needs to include drug appropriateness, acceptability, efficacy and dosing.

Doctors actually spend a lot of their time chopping and changing patient's drugs, so getting a reliable therapeutic history can be difficult (without good records). It is never wise to leave reviews of prescriptions too long. Effective drugs have powerful side-effects, therefore, review times must be short. Two to five days after a patient took a drug for the first time was my maximum, and even then, I would sometimes telephone the patient to check on side-effects.

Patients often challenge doctors about their medication: *'You say you have tried everything, doc. How are you going to help me now, after all your drugs have failed to help me? What will you suggest next?'* Going back to the future, can work. Instead of trying the latest drug, prescribe a tried and tested one. Having exhausted the latest anti-inflammatory drugs for instance, try the oldest. As an example, I sometimes found indomethacin to be just as effective as any new drug. This drug is so old, younger patients had never heard of it. When a doctor actually runs out of alternatives, it is time to refer the patient to another doctor.

Having undertaken cardiac catheterisation for over forty years, I had the chance to observe differences in post-procedural bleeding times, on and off of aspirin and warfarin. Even the smallest dose of warfarin, or aspirin, would prolong the post-procedure bleeding time. As a young cardiologist, I always applied pressure to the femoral artery entry point for at least ten minutes, post procedure. Those not instructed to stop their anti-coagulants or aspirin, needed twenty minutes of pressure. No patient of mine developed a post-procedure haematoma using this rule of thumb.

Treating Bedouin for atrial fibrillation (AF) in north Africa presents a prophylactic anti-coagulation risk. I regret to say that I have lost the reference, but I once found an Algerian publication in which the author described his warfarin regime for nomadic Bedouin. He gave all those with AF, one mg of warfarin daily. The author could not test their prothrombin times regularly, because they never returned regularly from their desert travels. Contrary to western predictions, embolus reduction using this regime, compared favourably with those patients managed using regular prothrombin tests.

This research result, and my experience post-catheterisation, inspired me to respect the power of warfarin. I once had an older adult lady with AF whose skin bruised easily. While on warfarin, I maintained her prothrombin time at a low level, between 1 and 1.5 (accepted to be sub-therapeutic). Over an eight-year period, she developed no emboli, and had no bleeding episodes.

One counter-intuitive therapeutic principle, is that if a dose of poison can do harm, a smaller dose might bring benefit (treatment or prophylaxis). No study I have seen since, replicated the Algerian warfarin findings. It suggested that we might keep patient INRs low (between 1.5 and 2.0), and still keep patients safe from embolism and bleeding. (I never applied minimal doses of warfarin to patients with prosthetic heart valves). In fifty years, using this policy, I never saw one significant bleed, or any thrombo-embolic incident. Based on inadequate numbers (a few hundred cases), was this luck, or an unreliable anecdote? In the same way that experience will improve the performance of motorists, using the same drug for fifty years, allows one to get a 'feel' for its use. Since using drugs in this way becomes an art, will those who base their every opinion of validated measures, see this as unreliable? Scientists will simply ask for numerical evidence and side-line experience and common sense as hearsay. This is undoubtedly a safe approach, but one that limits the application of efficacious medical intervention.

Experienced doctors and nurses, who employ the 'art' of medicine in the above way, can function well beyond their basic training. The best concert pianists leave their sheet music at home, and can freely apply their own art of interpretation (touch, timing, and spontaneity) to the music they play. The result differs noticeably from that produced by a computer-based, programmed piano. Unfortunately, inexperienced doctors and nurses, have little option. They cannot depart from standardised guidance without risking their job. With experience they should become able to think for themselves and be able to practise the art of medicine. The tighter the grip regulators have over medical practice, the more standardised and the less innovative and personalised, it has become.

Drug Rep's and Treatment

Beware of drug company representatives ('drug reps') promoting their latest products. If you become a university lecturer, you will get offered many perks, like free trips to the Galapagos Islands, or Bora Bora, but only if you are prepared to support a new product. I was invited to present some of my data at a conference in Gothenburg, Sweden, and also at the Turnberry Hotel, Scotland, where superb meals and golf were included. If you hold an academic post, many companies will support you for the price of one simple acknowledgement of their product. In such instances, you will deal only with senior representatives - those who command significant expense budgets. The two trips I took were both paid for by Astra Zeneca; at the time I was content to use and recommend their then latest beta-blocker, metopralol.

Junior company representatives, most of whom are keen to get promoted, will try to get doctors to attend minor promotional meetings, many of which are boring. It depends on how desperate they are for a free meal. I once meet an unmarried, East London GP, who claimed never to have bought a restaurant meal: he had attended that many drug meetings.

Liaising with pharmaceutical representatives can be a rewarding experience. They are mostly intelligent, engaging people, even though the junior ones will recite scripts, and be unable to discuss off-script details.

Doctors need drug companies as desperately as they need us. How else would the medical profession function? Without them, we would have to dig out our pestles and mortars. I find it odd, though, coming from a business family, that so many doctors despise the commercial nature of big Pharma. In common with many businesses, the intellectual property they hold is extremely valuable, and their profits depend on protecting it.

In 2016, Pfizer and one of its distributors, got fined £90 million for inflating the price paid by the NHS for phenytoin. Pharmaceutical companies have an aim that lies beyond the remit of doctors – they need to make a profit in a highly competitive scientific field. This requires the investment of billions of dollars. No profit - no new drugs. It's that simple. Courage enough to do this is one thing; to be successful at it is another.

As a medic dealing with drug reps., you will have to be careful not to compromise your standards, especially in your role as a patient advocate. There are a few issues. Has their data been peer reviewed without too much cherry-picking? Extended extrapolation, convenient omission, the use of small numbers, and unjustified comparisons, are all common devices used to make doctors prescribe a new drug.

Because there is more profit in new drugs than in old ones, be circumspect about using new drugs, unless their effects are novel and fulfil a need. For me, every decade saw two new drugs announced. I preferred other doctors to try them first, and at least for a few years, until enough follow-up data accumulated (rather than just trial-based data). Double-blind trial data might prove statistical efficacy and superiority over other drugs, but long-term follow-up studies reveal what happens in practice. Among the game changers for me were the introduction of benzodiazepines (replacing barbiturates), beta-blockers, H_2-receptor antagonists, SSRIs, aspirin as a prophylactic agent in cardiovascular disease, and 'statins' (HMG CoA reductase inhibitors). Others have been ACE inhibitors, angiotensin II receptor antagonists, amiodorone, calcium channel blockers. The last were factor X_a inhibitors. Every doctor will accumulate a list of their own favourite drugs, but those I have mentioned, all changed my practise.

I initially resisted the strong academic arm-twisting employed to promote NOACs, or factor X_a inhibitors, in place of vitamin K antagonists (trusty old warfarin). The various companies presented trials galore, with many professorial experts drafted in to lead the way. Their message: *'using these drugs is the way forward'*. Clearly, one risked being thought 'old-fashioned', or unconventional, if not prescribing them. I found it reminiscent of a second-hand car dealer trying to sell me a new car when my old one was running well and giving me no trouble. Experience taught me to trust warfarin, at least in the way I used it. Was the primary motivation profit, given the amount being spent on advertising?

For some, fashion can be a more powerful motivating force than their experience; even more powerful than its antidote, common sense. All the information available from trials suggested distinct advantages for the use of factor X_a inhibitors. Using warfarin was always troublesome and made a little expensive by needing repeated INRs. Warfarin is cheap, quickly reversed, and easily monitored (INR provides a measure of effectiveness), so my question at the time was, should I use a drug whose effectiveness I could not measure, and which could not be reversed (at least, not in 2017)? A factor X_a inhibiting agent *Andexanet alfa* received approval in the USA in 2018. It is a human recombinant decoy protein that allows the binding of factor X_a inhibitors. Indarucizmab and ciraparantag are also now in use.

My initial impression of X_a inhibitors was that they were too good to be true. I waited over one year for my colleagues to use them in practice. I then waited for further results to emerge. Now there is overwhelming evidence for their superiority. They more often stop embolus formation in those with AF, than will aspirin or warfarin. I thus decided, to stop using warfarin (unless a patient could not tolerate factor X_a inhibitors).

The decision to use a new drug should depend on one's ability to read, understand, and make sense of the relevant research papers; in fact, to 'read between the lines' of the accompanying statistics. There is always a point, in the gradual accumulation of evidence, at which it is reasonable to accept or refute the case as proven. One might call this the *statistical tipping point*. Beyond this point, proof is often assumed, and not always further questioned. You will courage to question established dogma. One must be confident enough to disagree, and powerful enough to swim against the tide.

One hundred years ago, Einstein challenged Newton's ideas. Evidence gathering for his space-time concept still continues. The 'weight of evidence' is a rhetorical device used to persuade others of 'acceptable proof', although no amount of statistical evidence can ever prove a thesis; statistics can help calculate the chance of a thesis being wrong.

History Taking. Points not to be missed.

Even when there are prevailing time restrictions, and everyone is in a rush, you will rarely waste time taking a detailed history. If the diagnosis eludes you at first, return to take the history again. If it still eludes you, take the patient's history in the presence of the patient's friend or relative; one they trust and who knows them best. This is a valuable strategy. When flummoxed, unsure, or out of diagnostic ideas, resist ordering more investigations, and getting second opinions. Take the history . . . again! Ask more questions to flush out further information. Based on the 'horses for courses principle', the time and place for further questioning must be appropriate. In troublesome cases, don't be shy to ask the patient (or his relatives) what *they* think the diagnosis might be.

In urgent cases, only a few essential facts may be pertinent; a full history may be unobtainable, irrelevant, or unnecessary. How much meta-information is vital to dealing with a severed hand or a bleeding nose? One glance makes the diagnosis, and the correct treatment can follow.

There are other classes of unhelpful histories, such as fraudulent and false ones (encountered in some insurance, and legal cases), deliberately misleading ones (to avoid embarrassment), and absent ones (dementia, or convenient omissions for the sake of pride, or politics). You may need a relative or friend to 'fill you in', with an absent or denied history. You may sense that something is missing from the whole picture being painted, but not know what it is.

Why might someone withhold information? Do they have an undeclared agenda? Are there facts they don't want to declare, or cannot declare? It may be important at this stage to convey the importance of information to reluctant patients. I have sometimes used a metaphor when trying to understand a patient's prevailing stress: an elastic band, left on a desk, will never snap of its own accord. One must then ask, 'What is it that has caused your elastic to stretch or snap?' Some patients will be forthcoming, others will not comprehend, look blank, zone-out, remain in denial, or choose to evade further questioning. Some will suggest their symptoms are a complete mystery and make no sense at all (implying, 'I have successfully defeated many other doctors.').

Taking a detailed history can endear you to patients; the sub-text for them is: *'at least someone is taking notice of me'*. With long-standing patients, with whom you have developed a relationship, information gained from informally chatting to them can prove vital. The more personal knowledge you have of a patient, the more likely will your advice be appropriate. By personal knowledge, I mean issues such as their attitude to risk, and the weight they assign to their quality of life, length of life, relationships, money and work. Personal information will help suggest the most appropriate and acceptable management decisions. Would they take a new experimental drug, or never take the risk? Your personal knowledge of a patient will help you formulate your advice, whether it be to take a statin drug, submit to a coronary stent or coronary bypass (CABG), or both. Your acquaintance with patients must never interfere with the objectivity of your advice, and never become detached from it.

For diagnostically impenetrable medical cases (diagnosis not obvious on several encounters), you may need a more intrusive form of history taking: forensic history taking. This is especially important if the patient is acting dumb, being secretive (hiding embarrassing facts), or playing hard to get (the reasons for which you cannot identify). This must include speaking to, and cross-questioning their friends, relatives and work associates.

As a first year junior doctor in 1966, I did ward-rounds during visiting times. I thought it might reduce the number of night and evening calls from relatives. This it did. It also allowed the patient's friends and relatives to offer useful bits of information, providing me with some 'fly on the wall' information. Relatives gained by having access to what we had discovered, and what we had planned (given the patient's permission). The interchange helped calm the family group by removing the anxiety of 'not knowing'. As you might expect, my doing ward rounds during visiting times bemused my colleagues. The fact they received many more evening calls than I did from enquiring relatives, didn't inspire them to follow my lead.

As a sequel, I later offered all my patients consultations with their friends and relatives. When dealing with Middle-Eastern patients, this was a basic requirement. My consulting room had three chairs, not just one for the patient.

I developed a large practice, of over 20,000 private patients, partly because the advice I gave came from knowing each patient through history taking, mostly without much time restriction. This is one definition of personalised medicine. There is another: that will be provided by revealing the patient's complete genome. To grow a large practice and maintain it, I had to pass two stringent tests: the advice I gave had to prove correct, and it had to stand the test of time.

Chapter Nine

Clinical Examination

*'There's two ways of doing a show, you either do it **at** the audience or **with** the audience.'*

Ken Dodd. He gave stage performances for over sixty years.

The aim of this chapter is to consider the diagnostic scope and importance of physical examination. I have written the chapter for those learning medicine and cardiology, but it will provide patients and others with an insight into how and why, they need to be examined.

General physical examination performed well by an experienced physician or surgeon, can explain symptoms and confirm a presumptive diagnosis. Because it was my lifelong specialty, I have focussed on cardiac examination.

Many excellent books and internet sources, list and illustrate the relevant physical signs for each disease. They need to be studied in relation to all known medical and surgical conditions. For those with little knowledge of medical terminology or jargon, please refer to the glossary of clinical terms and acronyms at the end of this book.

The aim of clinical examination is to detect the signs of disease. The finding of jaundice, anaemia, heart valve murmurs, skin rashes, arthritis, and lumps and bumps, are all signs of a pathological process. To define them better, and make them specific to certain diseases, investigation may be required. The aim in writing this chapter is not to provide a compendium of physical signs, but to illustrate signs easily missed, underrated, or misinterpreted in significance, especially in cardiopulmonary medicine.

□ □ □

Cardiovascular and neurological examinations are the most challenging. Detection of signs pertinent to these specialities requires frequent exposure to them, and then experience enough to make accurate deductions from them. The significance of every physical sign depends on knowing its clinico-pathological associations.

Despite the ability to see and touch, not every examining doctor will detect every clinically significant physical sign. Some signs are obvious, others are subtle and easily missed. Examining patients with a vague hope of finding something wrong (or with the hope of not finding something, if disinterested) is completely inappropriate. One should actively seek the physical signs that will corroborate or refute the presumptive diagnoses, made while taking the patient's history.

One must always be open to every observation that might be relevant. This requires a full examination, and not only that part of the body of primary interest. The skills required are basic to the diagnostic art. As an art, requiring little conscious processing, it will lead experienced doctors to correct conclusions. That sounds a little magical, and it is. Not only knowledge is involved; intuition and nebular thinking are also important. The art is best learned as an apprentice at the side of an experienced practitioner.

Examination has discrete objectives. At the soft end, it is a doctor - patient relationship exercise. It can show how much a doctor cares and is involved, and will acknowledge respect and mutual trust. A doctor's manner should put their patient at ease, but should also reflect thoroughness. The non-verbal language of touch is a specific art. It is culturally sensitive, and must be appropriate to each individual. It can convey professionalism, understanding, sympathy, and the bonding or breakdown of a patient-doctor relationship. At the hard end, clinical examination is a data collection exercise, and an indispensable part of the diagnostic process

Patients who allow us to examine them, must trust us. How one examines a patient, and the impression one leaves (gentle, understanding, thorough, professional, attached or detached, anonymous, disinterested, rough, and inconsiderate) will influence the nature of the doctor-patient relationship.

The primary function of a clinical examination is not to 'perform', but there are aspects of performance about it. There are some important areas of overlap in all forms of performance: co-ordination, control, fluency, and ease of accomplishment, all of which can inspire or diminish a patient's confidence. Some examiners will struggle, unable to find their instruments or the right words; they will clumsily drop things like their stethoscope and appearing ill-at-ease. Patients may forgive these, but they will not usually give a doctor credit for poor personal hygiene, an unpleasant body odour, or oral foetor (halitosis).

Examination and Trust

Trust between a doctor and patient is crucial for effective clinical management. The more intimate the examination, the more need for trust.

The best confirmation of trust I ever received was a patient who once told a colleague of mine: 'I will not agree to your management unless Dr. D. also agrees with it.' It left my colleague somewhat bemused. He was an NHS doctor who worked with no patient continuity.

The building of trust depends on something much less tangible than clinical knowledge. Many doctors cannot gain the trust of patients, even when they are very knowledgeable. For many patients, sincerity and helpfulness are more important.

Those physicians and surgeons who express a 'detached', or anonymous attitude to their patient's, risk being wrongly judged. Patients may think them lacking sincerity, functional but unhelpful. They may be sincere about their work, but their commitment to patients may be in question. It will seem to some that while they are intent on getting the job done, they could relate better. The senses used by patients to trust doctors are the same as those used to form friendships. Few will grant trust unless there is an affirmative answer to the question: 'Does this person have my best interests at heart?' Engaged, 'attached' doctors, are more likely to gain trust than those who prefer detachment and anonymity. One defining talent for those who choose to master the art of medicine, is the ability to switch smoothly between personal attachment and scientific detachment when necessary.

Patients able to choose their doctor might compare it to buying a car or a sofa; sales technique can influence the purchase. Some may not appreciate the difference between trusting a salesperson and trusting a doctor or nurse with their life and welfare. It surprises our international colleagues to learn that UK (NHS) patients, cannot choose their doctor, and is the preserve of self-paying private patients only. Patients with medical insurance may also find that their private medical insurance company will insist on choosing a specialist for them (to control costs).

Many patients are naturally apprehensive about being physically examined. They may question whether a doctor will hurt or embarrass them, or find something wrong. Some will watch their doctor like a hawk until they gain some confidence. The majority will not question a doctor's trustworthiness initially, and will mostly accept their role as professional.

In some cultures, the patient's friends and family will insist on witnessing everything a doctor does. Exactly what are they looking for? Style is important. They will observe how considerate, sensitive, and helpful a doctor is (helping the patient off her chair and onto the examining couch), and how respectful they treat their co-workers (as when a receptionist annoyingly interrupts, or when a chaperone is less than helpful). The work of every clinical master is, by definition, effortless and confident.

Surely only clinical expertise counts? Well, no! There are many doctors with clinical expertise who do not handle themselves effortlessly, and will not easily satisfy their patient's expectations.

To make use of Ken Dodd's formula for performance success: when patients pay doctors indirectly (as in the NHS), regardless of how well they perform, doctors need only to perform 'at them' (the audience). If they pay a doctor directly (as in private medicine), doctors are required to perform 'with them'.

These considerations are of no consequence to many patients. All they want is to get fixed and to move on. They don't much care who does it. A doctor's style will then be completely irrelevant. I have found it true to state that patients get the doctors they deserve, and doctors the patients they deserve.

Examining Room Behaviour

You have helped your patient out of a consulting room chair (never have those that are too deep - the frail cannot get out of them easily), and led them into a separate examining room (a couch, screened off by curtains, is second rate; some patients will feel insecure about them being opened by mistake). You will need an easily mounted couch with steps and hand-holds or, better still, a motorised one. Having observed the patient's motility, you must guess how easily they will find it to mount the couch (you may need to call for help, and/or chaperone). If appropriate, now is a good time to ask them how they are coping physically. How easy is it for them to get into and out of bed, or a bath? Observe how they perform, and draw some speculative conclusions. Ask questions that weigh the significance of your observations. With the patient struggling to manoeuvre in your examining room, they can hardly deny a physical problem. The difference between their actual performance and their description of it, can serve as a measure of their stoicism, or their tendency to seek attention; all valuable meta-information when planning their management.

For similar reasons, it is useful for doctors to observe patients exercising on a treadmill during cardiac evaluation. I liked to assess their actual exercise tolerance myself, rather than rely on 3rd party descriptions or a patient's prior assessment of their exercise ability.

When asked about their exercise ability, remember that more men than woman think they have superpowers (a Superman Complex): assuming the retention of youthful powers and agility, undented by time. Some will claim to be athletically fit, even though they have done no exercise for decades; asked about their height, men will often add an inch (or two); women are usually more accurate (ego can thus be measured in centimetres or inches).

How much clothing should the patient remove? Should they sit on the couch, stand up, or lie down? Always anticipate the need for a chaperone. This will depend on the presenting problem, and on which anatomical parts need to be examined.

By this stage, you will have noticed how well or unwell the patient is, and if fully conscious, whether they are anxious, hyperventilating, breathless at rest, cyanosed, jaundiced, or in pain. It is calming, and sometimes of great value, to continue the conversation throughout the examination; perhaps to review the answers they have already given. Chatting with patients can put them at ease. A lot of valuable meta-information can come from general conversation. One can learn about their attitudes to illness and much more. Every examination should proceed smoothly and be uninterrupted. This is a learning experience for both patient and doctor.

Inguinal hernias, varicose veins, and scrotal problems, are best examined with the patient standing; upper limbs, face, and chest (heart and lungs), with the patient sitting. For further examination of the heart, the patient should lie (or sit) on their left side, at a 45° angle (for LV palpation to detect hypertrophy, and to hear mitral murmurs). Thereafter, cardiac patients can lie on their back (to palpate the right ventricle), or sit erect to confirm the a diastolic murmur of aortic incompetence. Always be sympathetic, and anticipate any assistance they might need to rise from a lying position.

Never give the impression during an examination that you would sooner be somewhere else. Avoid answering a mobile phone, or any engagement with others. Focus entirely on your patient. Kindness, respect, gentleness, and time spent being considerate, will improve your standing.

The commonest negative impression of GPs is that they are overly interested in their computer screen: it suggests 'bureaucracy before patients' and that data collection is more important than respect for the patient. Although data acquisition and collection using computers is now essential, the impression it gives is one of detachment. Would you do it if you were called to see HM the King? I very much doubt it. So question whether you have double standards.

I have always disliked making children anxious by examining them. I often examined them on their parents' lap, or while standing next to me, with their parent close by. On the first occasion, I would never try to force open their mouth with a spatula, or take their blood, unless absolutely necessary. This presumes future patient continuity as a possibility. Gaining a child's trust is no different to gaining the trust of an adult: time taken not to frighten, hurt or embarrass them, will pay dividends later.

Explain what needs to be done and involve the patient, including children, as much as possible. When the clinical priority is high, you must remain undaunted in your examination, performing whatever procedure is necessary in the patient's interest. Their anxiety and concerns may have to be side-lined for a few seconds. What has to be done, has to be done, when the need for a procedure or diagnostic examination is of paramount importance. Sensitivity, empathy, sympathy, and other polite essentials may have to be suspended temporarily. The ability to switch off one's sensitivity temporarily is essential for any doctor wishing to perform effectively in emergency situations.

Lumbar punctures, gaining entry for an i.v. infusion or an arterial line, gynaecological examinations, undressing painful wounds, all need to be done sensitively, despite the transient pain and anxiety they can cause. If your patient is too anxious, or even hysterical, examine them under anaesthetic. With profanol, the induction and recovery are quick, with very few after-effects. One can never condone needless suffering, but one cannot condone failing to perform an essential procedure either.

Patient modesty is an important, sensitive issue. It can vary with gender, culture, age group, and the level of individual shyness. A few patients will happily remove all their clothes, with no trace of self-consciousness; others will not wish you to see any part of their skin, let alone allow themselves to be examined thoroughly. An important question is whether one should accept such patients. Be direct, and suggest that they might prefer to be examined by someone else. Do not expose yourself to failure by letting any patient avoid necessary examination. It is their undeniable right to do so, but planning for someone else to examine them, is sometimes essential.

When I was a junior clinical student, I made the mistake of informing a patient that I was going to examine him with a cold stethoscope. This was in front of a large audience of fellow students. 'You are about to examine him with what?' asked the crusty old consultant physician. 'A cold stethoscope', I said timidly, realising my error. 'Go and warm it, boy!' he directed. In unison, my much amused fellow students clapped me towards a warm tap, then clapped me back again. After that, I never again examined a patient with a cold stethoscope.

Not every patient needs or warrants a full examination. For minor issues, direct, targeted examinations can suffice. For those with diagnostic problems, especially those cases who have defeated other doctors, my previous advice remains: go back to basics. Take another full history, and repeat the examination. If you believe that complete laboratory and imaging investigations will make the diagnosis, be prepared for disappointments.

Diagnosis at First Sight

The examination process starts before the first 'Hello'. A facial expression (and other micro-behaviour) can be worth more than many words; the patient's state of mind is there to read. The patient's dress, gait, and demeanour; cleanliness (hands, shoes, and clothes), and observable signs like tremor, jaundice, cyanosis (warm blue lips and nose), gait, a distraught demeanour, and difficulty in breathing, are among a long list of instantly observable signs of illness.

Use your initial observations to direct your history taking. At this stage, you may have already made a diagnosis, even before asking one question. Parkinsonism, myxoedema, arthritis, heart failure, and depression, can all be observation-based diagnoses awaiting verification. These presumptive diagnoses await further exploration during history taking and examination. An accomplished observer is always 'on the lookout'; primed by experience not to miss 'tale-tale' signs of disease or discomfort.

I have fond memories of a brilliant teacher helping me to prepare for the MRCP examination. Dr. Paul Kligfield., and I would venture out of St. George's Hospital, Hyde Park Corner, with our tutor, Dr. Alan Gelson. 'What's wrong with that chap over there?' Alan would ask. He expected his students, to make an instant diagnosis from a distance. I try to do the same, even now. Brilliant teachers can influence us forever.

'Chance', said Louis Pasteur, 'favours the prepared mind'. This is never more so than when spotting physical signs of significance.

Not to be Missed

It is important to set up an examination routine for yourself. This is one of the few tick-box strategies worth adopting. Once learned, one can follow it without thinking, and at least avoid serious omissions. Negative observations are important; they can provide evidence against certain diseases. One problem is that many physical signs take time to develop, and can be absent in the early phases of disease.

There are a few obvious points not to be missed. You must observe skin colour, conjunctival colour, lip, buccal cavity, fingertip and nose colour. Cold blue hands and a cold nose, can indicate a poor cardiac output and reduced peripheral perfusion; others will be centrally cyanosed. The breath can have a distinctive odour in dehydration, ketosis, renal and hepatic failure. Note rashes, skin haemorrhages, and spider naevi: they are always of significance. Note the joints: small hand joints in OA, with Heberden's nodes on distal finger joints, or enlarged MP joints in rheumatoid arthritis. The implication is that the same processes could affect other joints and organs. Obesity, frailty, and wasting, must always prompt an enquiry about diet and disease. Weight loss and gain, and its rate of progression over time, are useful when assessing the stage of illness. In the neck, never forget to feel for lymph nodes, if only for practice and completeness.

One must continuously integrate the examination findings with the historical detail. Discrepancies need thought, and must be resolved.

The second rare case you encounter, will be easier to diagnose than the first. Once one detects a physical sign, and one makes a presumptive diagnosis, the next intellectual strategy must be to prove oneself wrong. The same goes for every clinical presumption. When you fail in this task, you can delight in having made a correct diagnosis.

Red herrings never easily distract a master diagnostician. As the paramedics wheeled me into A&E with severe abdominal pain, a smart young doctor sitting behind the admissions desk, loudly remarked: 'biliary colic'. 'Pancreatitis', I retorted. His was not an official consultation; we had never met, but he made a pretty good diagnostic guess. It turned out that I had pancreatitis, and not gallstones and biliary colic.

With time, experience, and much feedback from final pathological diagnoses, one can learn to recognise the errors of history taking and clinical examination. Never come to regard any presumption as fact; all presumptions need evidence-based confirmation. Clinical examination is a search for evidence, with your biases, assumptions, presumptions, and personal interests, tempting you to value the evidence wrongly. The accuracy of your diagnoses may depend on your competence to assess any negative influences. Rush your history taking, examination and conclusions, and you will more often make diagnostic errors.

The epizeuxis: 'history, history, history', is crucially important in many difficult diagnostic cases.

A Nursing Home called me to see an ill older adult lady. Her GP had visited, and after seeing her oedematous legs, prescribed powerful diuretics (for presumed heart failure). The observation of dependent oedema had been correct, but the GPs examination might not have included any corroborative signs of heart failure. Without a raised jugular venous pressure (JVP), the diagnosis would not be confirmed. One problem is, high-dose diuretics can be artificially lower the JVP in heart failure. Also, in borderline heart failure, the JVP may rise only after exercise. Misinterpreting the cause of ankle oedema (from sitting all day) can lead to digoxin and diuretics being wrongly prescribed to older adults.

What happened next was predictable: the patient became weak, albeit with some improvement in her oedema. A combination of dehydration and hyponatraemia caused her to become weak, frail and thirsty (thirsty all the time, dry mouth, inelastic skin, and a JVP around zero). Digoxin caused the patient to feel nauseated (associated with typical T-waves on the ECG, and a high blood digoxin blood level). Having stopped all her treatment, I re-visited her three days later. She was then feeling better, and had a normal JVP (before and after a short walk). The conclusion: her oedema had been gravitational, caused by sitting for long periods. She was not in heart failure. My guess was that her GP had not observed her JVP, or had observed it incorrectly.

The ability to read a JVP correctly is an invaluable skill, given its clinical significance. As a lecturer, I saw many students and doctors struggle with it. This is unacceptable, given the prevalence of heart failure in older adults (more die from heart failure than cancer). One need not rush into echocardiography, or into measuring natriuretic peptide, if one can accurately estimate the JVP clinically. Given the difficulty most seem to have identifying the jugular veins (internal and external), let alone measuring its height, echocardiography is now the more reliable method for diagnosing heart failure.

Two major observational difficulties exist: when the JVP is very high, and above the angle of the jaw (in heart failure at rest, or mediastinal obstruction), and when it is negative in dehydration. The measurement of the JVP is important, so I thought a short primer might be useful. See under 'Examining the Heart', a little further on in this chapter.

Age-related, subendocardial atrial fibrosis (the commonest cause), atrial dilatation in heart failure and thyrotoxicosis (rarer), are mostly responsible for atrial fibrillation.

Tendon reflexes are faster in thyrotoxicosis, and over-treatment with thyroxine (reflex contraction and relaxation phases both speed up). One can see the reverse in hypothyroidism, or its under-treatment with thyroxine: the relaxation phase can be notably slow. These are the tissue effects of the hormones T4 and T3. Blood tests do not always reflect the effectiveness of thyroid hormones at the tissue level.

Another example of a tissue / blood level difference occurs with digoxin. If you want to know the tissue levels of digoxin, look at the ECG (cupped ST segments). Similarly, if you want to verify the onset of myocardial hypokalaemia or hyperkalaemia, look at T-wave height changes on consecutive patient ECGs. Blood levels need to be interpreted in their own right. Because of red cell haemolysis, blood potassium is too often reported as high, when the ECG is normal. Initially believe the ECG, and repeat the blood level of potassium before taking action.

Are these observations just a nostalgic exercise for those wishing to keep old clinical skills alive? Are they useful only on desert islands, and in places where no investigation equipment is available? Perhaps, but at least you can rely on them whenever you want, and wherever you are.

Unhelpful Physical Examination

Rare diagnoses will usually need to be confirmed using an appropriate investigation.

Mrs. C was in her 80s. She was being treated for hypertension when she complained of severe itching, but no skin rash. Physical examination was unhelpful, but a review of all her haematology results revealed the steady rise of her haemoglobin: from 12 to 19g/L, over five years. Polycythaemia rubra vera, obviously!

Examining the Heart

The radial pulse is the place to start. Because this is often the first contact, ask permission to feel the pulse. Notice the rate and rhythm. You may need to sample it several times. For this reason, I would often checked it at the end of history taking, and before the patient proceeded to my examining couch. Be sure to check the pulse rate just after the patient has mounted the examining couch. (I might later check it while accompanying the patient on a short walk). Is it regular (sinus rhythm with some variation on breathing) or irregular? If irregular, the single most important observation is to discern whether the rhythm reverts to regular (predictable, like a metronome)? If it does, any irregularity is likely to be caused by ectopic activity; if not, it is likely to be atrial fibrillation.

Go next to the neck and measure the JVP. One can measure the JVP with the patient lying down (when it is low, or negative), sitting upright in a chair (when high), but most often with the patient lying at a 45º angle, sitting on a chair or couch. The primary aim is to visualise the top of the venous wave-front, and measure it above the right atrium (behind the xiphi-sternum), whatever the position of the patient. For some with borderline heart failure, observing it after a short walk can be diagnostic.

With the patient sitting at 45 degrees, elevate their chin, and turn their head away. Look for any **inward** neck pulsation, between the clavicle and the earlobe (the fast, 'x' and 'y' pressure descent components being inward). Change the patient's elevation as necessary to visualise any inward pulsation. Ask yourself: is the fastest component of movement outward (carotid artery), or inward (jugular vein)? This should be visible, even in obese people with short necks (candidates for sleep apnoea).

KEY POINT: the JVP is best seen using indirect light in a darkened room using a single light source playing across the skin. It is not so easily seen with diffused bright light.

By slowly elevating an outstretched arm, one can sometimes observe the point at which the dorsal hand veins collapse. This is a far from a reliable technique for measuring high central venous pressure, but try it sometime.

The jugular pulse may, or may not, move up and down with breathing. Measure the approximate height of the highest inward pulsation, seen above the xiphi-sternal junction, whatever the patient's position (adjust the position to make it visible). Look just below the angle of the jaw to see the highest of jugular venous waves. You can sometimes lower this pressure (to make it visible) by asking the patient to inspire deeply while sitting erect. You need to know whether there is a history of paroxysmal nocturnal dyspnoea, orthopnoea, or evidence for emphysema and ankle swelling. Does it tie in with tricuspid valve incompetence (an prominent jugular 'v' wave, synchronous with cardiac contraction (feel the right apex beat at the same time)? Should you be able to distinguish it, a prominent 'a' wave might indicate pulmonary hypertension, obstructive pulmonary emboli, mitral stenosis, or emphysema. One can also diagnose atrial fibrillation by observing 'fibrillation waves' (fast vibrations actually), superimposed on the rising and falling of the jugular pulse with breathing.

Further examination of the heart is best done with the patient sitting, and then lying on their left side, with their left hand up behind their head (to separate the ribs). The best body position is the semi-sitting one. First, find how far lateral one can feel the apex beat (the left ventricular). If it reaches the anterior axillary line, assume the heart is enlarged (usually LVH). How heavily does it beat? In left ventricular hypertrophy (LVH), it will not 'tap' outward, but 'heave' in a sustained manner. When differentiating true (primary and secondary) from labile or false hypertension, the detection of LVH provides evidence of increased left ventricular after-load. We find this in sustained systemic hypertension, aortic stenosis, and HOCM. Interestingly, the diagnosis may be supported by ECG changes (increased QRS voltages: $SV_1 + RV_5 > 45mms$). The ECG is an electrical device, and unlikely to reflect physical LV muscle changes accurately. ECG voltages can remain normal until LVH becomes advanced; ST changes (t-wave inversion) in the chest leads usually occur earlier. Echocardiography can be used to confirm the clinical diagnosis of LVH.

Immediately to the left of the lower end of the sternum, one can sometimes feel the heave of a hypertrophied right ventricle (RVH). Use the thenar eminence of your hand, with light pressure applied over this point. You might detect the outward heaving of RVH, associated with pulmonary hypertension, every cause of which is important to define.

A note on electrical vectors and the ECG: *A sheet of paper viewed end-on, appears thin. As it is rotated, more of the surface comes into view, until face-on, one can see its full area. The paper has not changed size, but the direction of view has determined how much is visible. In a comparable way, vector quantities, like ECG voltages have both direction and force. As a result, the various ECG lead positions will minimise or maximise the voltages seen in different directions (from different lead positions). The vector principle explains why increased QRS voltages in LVH may not occur; it much depends on the anatomical orientation of the heart within the chest.*

One can only learn reliable diagnostic auscultate the heart as an apprentice, and only when exposed to physiological measurement and clinic-pathological feedback. An apprentice working in a cardiac unit should gradually come to match heart sounds to valve pressure gradients, and to the various degrees of valve stenosis and regurgitation (as seen on echocardiography, or revealed by cardiac catheterisation). Except for making cardiac diagnoses on desert islands, cardiac diagnostic auscultation is almost obsolete. Echocardiography, introduced in the 1970s, is responsible. When developed to a masterly degree, cardiac auscultation can be diagnostically accurate, but few doctors ever achieve this level of expertise.

What I have described so far is elementary. What follows is a more advanced tour of cardiac physical signs.

Aubrey Leatham's landmark book (1970) (on cardiac auscultation described everything one needs to know about heart sounds and murmurs) (see bibliography). His apprentices, including me, could diagnose the stenosis and incompetence of each valve, even when occurring together. We were required to detect developmental abnormalities like ASDs, VSDs, coarctation, Fallot's Tetralogy, together with the degree of pulmonary hypertension, RBBB and LBBB, using auscultation alone.

To auscultate the heart with diagnostic accuracy, you must be able to adjust your hearing – from the perception of almost inaudible, low-frequency sounds, to loud murmurs (VSD, mitral incompetence), high-frequency clicks (prolapsing valves), and opening snaps (mitral stenosis). Only if one can hear the quietest audible sounds (frequencies down to ≈20 cps) will one hear 3rd heart sounds (passive ventricular filling in heart failure, and in athletes), and 4th heart sounds (ventricular filling from atrial systole, exaggerated in systemic hypertension, and other causes of LVH and ischaemia).

Using a stethoscope, it is possible to discern the difference between the two high-frequency components of the second heart sound: the first from aortic valve closure (A_2), and the next from pulmonary valve closure (P_2). A_2 is exaggerated in systemic hypertension, and P_2 in pulmonary hypertension (the loudness of P_2 relates directly to the pulmonary artery pressure). In normal 2nd sound splitting, the time difference between A_2 and P_2 increases with respiratory inspiration. In LBBB, A_2 can be delayed until it coincides or passes P_2; the second sound is thus single or reversed (the components get closer together on inspiration as P_2 merges with A_2). To the practiced ear, listening to the first heart sound components (mitral then tricuspid closure), is also of diagnostic value.

All students should be able to discern the difference between a pan-systolic murmur (mitral, and tricuspid incompetence), and an ejection systolic murmur (AS in adult cardiology), but it can be difficult.

Diastolic murmurs are also challenging. An early diastolic murmur (usually from AI), is best heard in the aortic area (to the right of the upper sternum, or at the lower left sternum), with the patient leaning forward, having just exhaled and holding the breath. Mitral stenosis can cause an opening snap (if the valve is pliable) followed by a late diastolic murmur in the mitral area (lower left chest; mid-clavicular to anterior axillary line, with the patient turned on their left side). The 'snap' may not be audible if thickened valve leaflets are stiff, or calcified. The length of the murmur relates to the area of the valve aperture. The vital questions are: how bad are these defects (degrees of stenosis or leakage), and how much they have affected the heart muscle (hypertrophy or dilatation)? These are questions more easily answered using echocardiography.

In association with auscultatory findings, you might consider the waveform of the arterial pulse, best felt in the neck (carotid artery) or groin (femoral). It is possible to detect severe AS (slow rise), and severe AI (very sharp components), even in the radial pulse. It takes practice. Do not expect to diagnose minor degrees of these defects in this way.

The Lungs

When first examining the lungs, the first step is to observe the shape of the chest (is there hyperinflation), and chest movements associated with breathing. Is breathing laboured or the shoulders hunched up? Remove enough clothing to expose the chest (if allowed). Are both sides moving equally? Notice associated signs like clubbing, cyanosis, and nicotine staining of the fingernails. After these initial observations, percuss the chest front and back to assess resonance, and to detect effusions (dullness), emphysema or pneumothorax (hyper-resonance). Get the patient to breathe fully through an open mouth; you might hear wheezes and rhonchi magnified by the airways. Listen to the patient coughing: does it sound dry, or wet with rattling sputum? Does it induce wheezing?

Auscultate the chest front and back with both forced and unforced breathing. Are there added sounds like rhonchi or wheezing? Are there crepitations on auscultation, and do they remain after coughing? They will remain in pneumonia, lung consolidation, and pulmonary oedema (left heart failure, and mitral stenosis), but not with minor chest infections. These are basic to every routine clinical examination for cardiopulmonary disease.

While the patient is still leaning forward after lung examination, tap the length of their spine for tenderness (osteoporosis, metastases), and palpate their renal areas (always important in women who more often than men, suffer from pyelonephritis).

The Abdomen

Even cardiologists should examine the abdomen. It is quick and easy to do, and can exclude some important clinical conditions. Warm your hands first. You will need to know if your patient is ticklish or sensitive to pain. How hard should one press during palpation? You may need to press quite hard and hold still, while trying to detect a pulsating mass – a possible abdominal aortic aneurysm. Do not press uncomfortably hard. After being examined myself occasionally, it surprised me just how hard some surgeons want to press. What are the implications of tenderness? Over the liver, it suggests neoplastic metastases. In the epigastrium, it can suggest an ulcer (PU), pancreatitis, or aortic aneurysm. Sometimes it will be diverticulitis affecting the transverse colon, although such tenderness is usually lower.

The tenderness of a mass more often indicates inflammation than neoplasia. Either way, any conclusion about which it is, can be unreliable. When following-up treated colonic diverticulitis, tenderness is a useful guide to the effectiveness of treatment. One should always further investigate a mass that does not retreat after a few days (after bowel contents have moved on). Some masses will prove to be bowel contents, but this is not an assumption one should make. You may need to bring the patient back after a few days for re-examination.

Examine the abdomen for right iliac fossa tenderness in suspected appendicitis. Is there any rebound tenderness (inflamed peritoneum)? Is the patient nauseous (toxic state)? If both are present, an inflamed appendix is likely. Nausea is the symptom most often associated with a purulent appendix. If unsure, return after one hour, and re-examine. Untreated inflammatory conditions can progress rapidly, and you might need to act swiftly to prevent peritonitis.

I always feel the abdomen to detect an enlarged (as opposed to a depressed) liver, a palpable spleen, and enlarged or tender kidneys (examined bi-manually).

My simplistic view of the commonest of conditions, constipation, is that there are two forms discernible on palpation: a loaded static colon (easily palpable), and an empty colon (not palpable). The difference is usually easy to detect. Therapeutically, the loaded colon needs emptying; the empty colon needs filling with roughage. Percussion will tell you if the colon is full of gas (resonant) as in IBS. Surgeons will listen for bowel sounds as an indicator of peristaltic activity or bowel obstruction.

In cases experiencing bleeding PR, I would look for haemorrhoids externally; I did not always perform a digital rectal examination (DRE: digital rectal examination, or *per rectum* (PR) examination), but then I always sent patients with bleeding *per rectum* for proctoscopy. I sometimes left the digital examination to the surgeon (to avoid repetition) performing this. I found only two rectal tumours in fifty years.

An old Royal Navy saying was: 'If you don't put your finger in (the rectum), you could put your foot in it!'

If you assess the prostate for cancer by PR examination, be sure to take blood for a PSA first. Only the more advanced forms of cancer are palpable. Take a urine sample for PCa3 after prostatic massage if you want, but an MRI scan is now known to be more useful diagnostically (fewer false positives and negatives than PSA or PCa3). Having established the symptoms of significant prostatism, what is to be gained by a physician performing a PR examination when the patient needs referral to a GU surgeon? Many patients will appreciate not having a PR examination done twice.

Neurological Examination

Neurological examination is challenging because many of the signs are subtle. From the cardiological point of view, high blood pressure, atherosclerosis, and embolism can all damage the brain. The clinical detection of minor strokes (transient ischaemic attacks or TIA), as opposed to overt strokes (CVAs), is not usually difficult. The burning question with every TIA, is whether it is embolic or caused by a bleed. Even after brain imaging, one cannot always be sure. Look for reasons for emboli (AF or carotid atheroma). Brain imaging (to include carotid arteries) can sometimes reveal definite, asymptomatic, multiple small areas of damage. Cerebral bleeds, cerebral artery clotting, and large emboli can all cause overt strokes; the commonest bleeds occur secondary to hypertension, especially when LVH is present (left ventricular hypertrophy). Cerebral emboli can originate in the cerebral arteries, from clot in atrial fibrillation or myxoma in left atrium (best detected using transoesophageal echocardiogram - TOE) or from the carotid arteries (from clot on sub-stenotic atheroma).

An implanted ECG recording device will usually detect 9% more cases of AF than occasional examination (*TEASE Study*. BMJ, 2020: Magnusson, P., Lyren, A.). There are two rarer considerations:

emboli passing through an atrial septal defect (1.5–4% of strokes, hence the trend to patch the hole), and either thrombus or tumour emboli arising from a left atrial myxoma. These are usually pedunculated (like a ball on a chain) and usually arise from the left inter-atrial septum. They justify the routine use of echocardiography in every stroke patient.

In the early 1970s, cardiologists at St. Thomas' Hospital referred a patient to our cardiac department at St. George's Hospital, Hyde Park Corner, London. We had the first echocardiogram machine in London. Because they found a patient with a mid-diastolic 'clunk', they wondered whether he had an atrial myxoma. He did.

There are two forms of cardiovascular blackout, or altered consciousness. Both need to be distinguished from epilepsy and vertigo (which can lead to a faint feeling). Never forget to ask for the sensation of head-spinning, associated with unsteadiness, corroborated by a positive Romberg's test (the patient will be unsteady while standing, with feet together and eyes closed) in cases of vertigo. Always ask for a past or family history of epilepsy.

Differentiating vasomotor syncope from Adam–Stokes' syncope from the history can be challenging. Vasomotor syncope usually has pre-morbid symptoms: faintness, and nausea, with pallor observed by witnesses. There is usually a history of prolonged recovery, and a lifelong tendency to faint. It is most common in young, athletic patients. Adam–Stokes' syncope causes a sudden loss of consciousness, often with a quick recovery; facial pallor quickly returning to normal or flushed (in vasomotor syncope, pallor lasts much longer). In older adults (over seventy), complete heart block is often the cause.

Examination in both cases can reveal bradycardia, with either sinus rhythm (or variations caused by sino-atrial block) or A-V block of varying degrees. One may need a 24-hour ECG to diagnose them; at the same time, one must rule out other dangerous rhythms such as runs of VT or VF as the cause of syncope. The finding of complete heart block should prompt immediate pacemaker implantation: those with definite Adam-Stokes syncope can die without warning. The same urgency applies to those with VT and VF, but they will need an implanted defibrillation device.

Vasomotor syncope can be difficult to manage. Extra dietary salt might raise their average BP if it is always low (<90mms. Hg.). Some will benefit from getting fit (perhaps from less responsive autonomic reflexes). I found that many older patients with vasomotor syncope, were experiencing stressful situations (increasing autonomic reflex responsiveness). Prescribing hyoscine derivatives can sometimes help temporarily. Fainting usually improves with age (as autonomic reflexes become less responsive). Some will need A-V, sequential pacing.

Leg Examination

Leg examination can have diagnostic implications. One can detect atherosclerotic problems (reduced foot pulses) in those with claudication (calf pain on walking), and those with early peripheral neuropathy (in diabetics, and pre-diabetics). Straight leg raising can confirm, but does not exclude, lumbar nerve root compression. I never routinely looked for varicose veins, or did tests of co-ordination, strength, and reflexes, unless the patient had complained of leg weakness, or was having a problem with walking or co-ordination. It is essential, however, to detect calf tenderness, and to look for a positive Homan's sign (calf pain with ankle flexion), when there is unilateral ankle swelling and a possibility of pulmonary embolism. I examined for these, only when prompted by specific symptoms.

At all times, one must integrate the search for clinical signs with the presenting symptoms. A complete examination, looking for every sign (most of which will be irrelevant), is essential only for the inexperienced. Although complete examination never goes amiss, too much information can cause a loss of focus.

One can always return to examine the patient again. This is a tactic of unquestionable value, both for challenging cases and for those where their condition is evolving. The time dimension of repeated examinations is crucial, especially in cases such as suspected appendicitis, stroke, meningitis, myocarditis, or septicaemia. One can examine patients too early, before any significant development is detectable. Leave a second examination for an hour or two, and observe any differences. In most situations, it is the second examination that reveals most. The whole point is to treat the patient as soon as possible and to prevent progression and complications.

Every doctor will have their own perspective when examining patients. This perspective can be a little too focused (missing the big picture), or too general (missing the detail). Knowing which is appropriate is a matter of experience and judgement – another medical art!

Having fully examined your patient, you might find yourself confused, or completely unsure of the diagnosis. Too many conflicting findings can be a problem for the inexperienced. It is always of value to make one diagnosis, although many patients genuinely have multiple diagnoses, some of which may interact. With patients living longer, multiple diagnoses have become more common. With experience, you will hopefully gain a clearer perspective of each diagnosis and its relevance to any other condition. One will need clarity to decide the correct priorities when treating such patients.

It is crucial to recognise red herrings and quickly discard them.

Back in the day when we saw complex heart valve cases, we sometimes had to decide whether a patient's mitral incompetence was more important mechanically than their aortic incompetence. A simple management guide is to treat the potentially life-threatening condition first, and then those that produce the worst symptoms. For instance, in a case of severe coronary artery disease, it might be wise to undertake a CABG before risking hip replacement. Being able to sort the relevant priorities effectively in each case, is another medical art.

In Appendix A, I have printed my schema for recording patient history and examination. You might find it a useful template when beginning medical work.

Chapter Ten

Other Consulting Room Matters

ONCE THE HISTORY TAKING and physical examination are complete, one must consider the pertinent diagnoses and request any necessary investigations. Before that, consider some other consulting room side-issues.

Not every doctor is concerned with their image, or their patient's opinion of it. Every patient will, however, form an opinion of their doctors and nurses: their age and approach, perceived trustworthiness, depth of knowledge, intelligence, executive functioning, usefulness, financial status, etc. They will consider how well they dress, and their personal hygiene standards (hair, nails, cleanliness of clothes and shoes), all of which is seen at a glance and might influence their opinion.

Those undertaking specialist examinations, especially those of the Royal Colleges (MRCP, FRCS, FRCOG), should know that the examiners will also appraise them, but even more critically. First impressions are important. Even more important is the impression given by the candidate while opening his mouth; it could shape his career.

Of what relevance is all this, when doctors have thrived successfully for centuries, studiously ignoring most of these considerations? The impression formed by colleagues and patients needs consideration, as is the style of doctor-patient relationship each doctor wishes to promote. This can range from attached and affable, to detached, academic and stern. Much depends on whether the system provides patients (as in the NHS), or whether they have to be gained and their loyalty kept (as in private practice).

Doctor-Patient Relations

How much personal information should one share with patients? If a doctor develops relationships that are too close, they could make patients emotionally dependent (an adult / child relationship, not an adult / adult one) and this could influence clinical decisions. There are other dangers. Few doctors will make this mistake, but shifting focus from the patient's problems to their own, can make patients feel ignored and awkward, even though they might learn something useful. Exchanging personal experiences can promote mutual trust, but is not always advisable.

While discussing this topic with one of my cardiac interventionist colleagues at Charing Cross Hospital, London, he told me he was 'not that sort of doctor'. Dr. K.W. was (and still is) an accomplished cardiac interventionist, with little desire for close relationships with his patients. You could do no better, however, than to have Keith standing over you if you needed urgent cardiac resuscitation, or investigation. Knowing his patient's anatomy, was more pertinent to his role than understanding their social circumstances. Finding the most suitable role within the medical profession is an important issue for all doctors and nurses. Keith became one of the best interventional cardiologists in Australia.

I developed long-term, professional relationships with many of my patients (some for more than forty years). This raises a few questions. Is it advisable? Is it an advantage? Is it still possible?

I found that giving time enough for personal discussion helped to build mutual trust, rather than dependency. I also found that such relationships provided me with a clinical advantage: they expanded my perspective of the patient, and allowed me to solve some clinical problems that had defeated others. The key ingredients were the time needed for a full understanding, and the development of trust. My patients were mostly successful, self-employed people. The last thing they wanted was a physician with no personal interest in them. Finding a doctor capable of forming a meaningful relationship might not have been what they unexpected, but it soon became what they wanted.

Private practice contrasts with NHS practice in many respects. I always allowed between 30 and 90 minutes for the initial consultation. This amount of time allowed patients to relax, with time enough to discuss every pertinent issue. It allowed me a better insight into who they were, what they really wanted, and what they needed as a patient. The time given allowed a better definition of their expectations and motivations. One can treat hypertension without asking questions, simply by prescribing a drug, but if the patient is struggling with infidelity or skulduggery, drugs will help less. If they remain hypertensive with powerful anti-hypertensives, a bit of social engineering could make all the difference. This level of involvement meant I had to work like many others, from 9am to 6pm each day.

The Stimulus – ResponseModel

A simple way to practise medicine is to follow the *stimulus – response* model: simply respond to each problem, giving advice and/or a prescription for each symptom. It works for basic medical problems, but not for resistant or complicated cases. With mounting time pressures (caused by the management structure), this simple approach is all that most doctors can achieve. If doctors persist in cramming their patients into restricted hours of work (allowing patients only 10-15 minutes, for one problem), a significant number of patients will get dissatisfied. Many have already, in the UK. Many patients complain about the brevity of GP consultations, and the lack of consideration given.

The BMA has called for GP appointment times in England to increase from 10 minutes to 15 minutes, and for the number of daily consultations per doctor to be cut. This comes from politics and economics, and is approaching medical practice from the wrong direction. We must first ask what our patients need, then design the service.

The BMA outlined their recommendations in a report aimed at tackling the 'unsustainable' pressure caused by rising demand, falling resources, and staff shortages. The BMA said that GPs were forced to see up to sixty patients a day, in ten-minute appointment slots. They thought this inadequate for the increasingly complex conditions being presented by patients.

Iacobucci G. *BMJ* (2016); 354: i4709

I cannot stop wondering how UK doctors allowed themselves to follow the dictates of those who know everything about medical politics and nothing about medical practice. Ten to fifteen minute consultations can keep patients circling like airplanes, waiting to land at an airport. By sending them off for imaging, blood tests, and other consultations, some patients remain in orbit, awaiting a diagnosis and a completed management plan. Hopefully, they won't run out of fuel and crash. What doctor would want to be a patient under these circumstances?

Which is it more efficient? To solve a patient's problems in one visit, or to see them many times, each for fifteen minutes?

'Three out of four GPs (75.4%, in the UK) said cases of abuse had increased in the last year, according to the GPonline survey. Over sixty percent who responded (66.8% of 292 GPs), said they had experienced verbal abuse from patients in the past year, while 4.9% reported being victims of physical abuse in the same period.'

Joe Richardson. *GPonLine.* February 2022.

What accounts for this growing trend? Is GP patient handling, the source of mounting patient frustration and anger? Do patients feel that UK, NHS general practice, has become inefficient, and unfit for purpose? Has the bureaucratic functioning of a medical practice become more important than the patients it is supposed to serve? Do patients prefer depersonalisation and being treated as a number, to being regarded as a sentient human being?

When I first experienced NHS practice as a junior GP in the 1960s, I had just left hospital as an SHO. I had the temerity to ask my new colleagues why they didn't work all day long. I didn't have to pick them off of floor, but their stunned silence was evidence enough of their shock. They each packed sixty patients into two, two-hour sessions each day. This allowed them to tidy their books, do their home visits, have lunch, collect silver, and play golf. Those who prioritise paper-work, administrative procedures and golf, will not always be the most effective doctors.

An efficient, patient-oriented work-style for doctors is possible, but few can achieve it in the UK. Time pressures and the ineffective use of time has kept it so. Too few staff, and multiple contractual and corporate obligations, do not help. Providing doctors with the same secretarial help as all company CEOs now expect, would be unaffordable, and far too expensive for those whose job is to save lives, even though arguably more important than the job of any CEO.

Traffic Analysis, Parallel Information, Meta-Data, and Stress

If patients are to be handled efficiently, one must discover all the factors driving their problems, be they psychosocial, physical, or genetic.

At Bletchley Park, Gordon Welchman and Alan Türing together reduced the duration of the 2nd World War by two years, saving hundreds of thousands of lives. Welchman invented a method of analysing message traffic information (from ENIGMA machine coded messages). Bletchley Park intercepted the coded messages and de-coded them, but Welchman recognised the value of information (meta-information) separate from the wording of messages. He studied who sent them, and who received them; how often, and at what time of day. He analysed how different messages intersected, From this he drew a communication map of message *activity. What he studied (not yet known as meta-data, or data about the data) helped build a picture of enemy activity that put the deciphered messages into perspective (decoded by the first computers to be used in a practical situation). In the 1940s, Welchman called the process 'Traffic Analysis'.*

In medical cases that are difficult to solve, the 'traffic' generated by patients can be of clinical importance. Ask with whom they usually interact? With whom they must interact; with whom do they want to interact? Who do they avoid? Why, and how often? The answers could provide insight into their situation, and sometimes, a reason for their demise.

The relevance of meta-data can be important when considering hypertension and its causes. Primary hypertension is strongly inherited, but with powerful drivers that can induce and exacerbate it. The same is true for conditions as diverse as migraine, angina, and eczema. In those with no genetic predisposition, the same drivers will cause no adverse effects. I suspect that without a genetic predisposition to adverse reactions, 'stress' (as an applied force) cannot drive the onset of hypertension, coronary artery thrombosis, cancer, or much else. Family history provides a limited keyhole view, but it is an unreliable surrogate for a genetic marker.

The importance of stress has been long underrated by medical professionals, in line with many other examples of modern de-personalisation. There are, however, large numbers of stressed people who will never develop hypertension, stroke, or coronary heart disease. Stress may not cause disease, but it can change its progress.

Stress produces common generic effects in the majority: insomnia because of worry, leading to tiredness, reduced resistance to infection, and hormonal imbalance. Whether stress lowers the threshold to conditions like stroke and heart attacks, I cannot say, but stressing factors are important to consider when taking a medical history.

'Tell him (Mr. Bennet) what a dreadful state I am in - that I am frightened out of my wits; and have such tremblings, such flutterings, all over me, such spasms in my side, and pains in my head, and such beatings at heart, that I can get no rest by night or day.'

Pride and Prejudice. Jane Austen (1813). Vol 3; Chapter 5.

Mrs. Bennet was at the centre of the turmoil within her family, caused by the surprise and ill-advised elopement of her daughter Lydia.

A distressed patient of mine had lost her well remunerated job as city broker, and thought it unfair. She was complaining of palpitations. She told me that the collective efforts of her male colleagues had brought this about. She was reacting to a loss of status. Her predominant psychological trait was the need for control. Since few aspects of life are completely controllable, control-freaks will meet many disappointments throughout their lives. As someone preoccupied with grief, her lack of attention caused her to make mistakes.

By the time I saw her, she had realised she might never get her job back. She had planned to get married, and her marriage date was imminent. Would her marriage also prove to be a mistake? She was confused, and in turmoil; unable to think straight. Instead, she was thinking randomly, in Knight's moves (as in chess), rather than in small, logical, sequential steps. During our meeting, she broke down in tears while describing her situation.

I knew her husband-to-be and thought him to be an excellent prospect. My unexpressed opinion was that she would be the lucky one, not him. Unfortunately, her obsessional nature, her state of turmoil, and her preoccupation with the unachievable, were alienating him. My job was not to play marriage guidance counsellor, but to provide her with support, and the freedom to breakdown and cry while trying to achieve clarity. She needed someone to confide in. Because she trusted my confidentiality, she felt she could share all of her deliberations and intimate secrets with me. Catharsis can begin the adaptive, psychological healing process. It also reduced her palpitations.

PostScript: she married, had a child, and divorced soon after.

We cannot exercise wisdom unless we know all the factors in play (as derived from surveillance). It must involve correct interpretation, and the courage to act appropriately in a timely fashion. Gaining the knowledge necessary for wise decision making, relies on sensitive and sometimes wide-ranging history taking, that goes beyond the mere cataloging of symptoms.

Another golden rule: don't rush into investigations, their value lies in confirming or refuting a diagnosis. Remember that the first three steps of successful clinical management are:

diagnosis, diagnosis, and diagnosis!

Also remember that the number of differential diagnoses being considered is often inversely proportional to the clinical expertise available.

The Early Detection of Disease

An inherent need to persevere and reach safety, might explain why some patients (in evolutionary terms) choose to deny their symptoms. In ischaemic heart disease, only coronary artery stenoses >85%, will restrict blood flow enough to cause angina. Patients learn to avoid it by slowing down their activities. They can then deny it exists. Avoidance and denial both allow patients to carry on with their usual activities. Unfortunately, this strategy reduces the chance of an early diagnosis and prophylactic intervention.

I screened many patients repeatedly for forty years, at intervals of several years, each time entering their clinical information onto a standard data collection form (see Appendix A). Retrospective analysis of these records revealed something obvious. Among early symptoms are clues to later diagnoses. It can, however, take decades for a condition to manifest and for a diagnosis to be confirmed. Early symptoms may be inconsistent and undramatic, and most will be present before any physical sign. The problem is that false positive symptoms abound. One value of repeated screening examinations is to identify symptom consistency. Many symptoms are transient and lack explanation, but symptom persistence reliably indicates clinical significance. Managing patients on a long-term basis allows one to assess the individual reliability of their symptoms.

Up to a certain point, many types of pathology will progress without symptoms. In fact, symptoms may arise only towards the end of the expected natural history. In my field of work, both atheroma and hypertension, although potentially lethal, may yield no early symptoms for decades. To discover them, decades before any symptom, these 'pre-patients' need investigating: repeated blood pressures, artery scanning (ultrasound, and CT), ECG exercise testing, and echocardiography; each repeated every 2–5 years if one is to detect progressive pathology (atheroma progression and LVH). Asymptomatic cardiovascular screening is expensive, and not yet offered by the NHS. Both initial and repeated investigations of asymptomatic patients can reveal those close enough to a cardiovascular event, to make intervention advisable. The majority screened repeatedly, however, benefit only from re-assurance.

Significant early symptoms can occur, but are not always recognised by 'healthy' subjects; some of whom present for screening. Early symptoms worth detecting are the increased frequency of defaecation as an indicator of colon cancer; parasthesiae in the feet (early sensory neuropathy) preceding type 2 diabetes (by a decade or more); slowly progressive shortness of breath (rather than angina), occurring early in ischaemic heart disease and heart failure. Weight loss, symptoms of prostatism and bowel dysfunction, will also cause patients to seek medical advice. Epigastric pain after meals, even in children, can be the forerunner of peptic ulceration in later life. Routine questioning will detect the occurrence of these symptoms at a time well before any patient would openly complain of them or be aware of their significance. My standardised pro-forma check-list of symptoms (Appendix A), was useful for spotting those symptoms that were persistent (often contrary to what patients remembered), as well as the emergence of new symptoms (ones not previously noted).

Every doctor knows not to ignore constant dysphagia, a persistent cough, or a skin mole that has changed colour and shape. Unfortunately, these symptoms may not get reported early enough by patients to make a difference to the outcome. Pursuing early symptoms does not always make a difference. I found this to be the case for early oesophageal and lung cancers, melanoma, some brain tumours, and a few other malignant conditions. Cardiac screening, by comparison, offers much more chance of benefit.

Non-specific symptoms, with no long-term predictive value abound. Tiredness, headaches, palpitations, abdominal bloating, and flatulence are among the legion of transient complaints I followed for decades. I have rarely found them to have any long-term clinical significance.

If patients are to recognise and report early symptoms, both self-awareness and openness are essential. Patients, however, vary from the courageous and stoic, to the hysterical and hypochondriacal. Hypochondriacs need some of their symptoms down-regulated, but never ignored. Stoics under-rate their symptoms, and their complaints should be up-regulated. This is one good reason to have a witness present during consultation. A partner or friend, might then provide some useful comments.

Because many patients fear nasty diagnoses, some prefer to deny symptoms, just in case their fears are realised. Others may be less inhibited about reporting their symptoms, but unenthusiastic about further enquiry. Both are understandable. A few have another fear – that of medical error and being told erroneously they have a life-limiting condition.

Frank had been a notoriously eccentric solicitor. He had been 'struck-off' for using his client's funds, but had been an ardent advocate for of his clients. This occurred soon after his wife deserted him for the Texan cowboy she met on holiday. Having not seen Frank for a year, I was alarmed to see how much weight he had lost. He was cachectic and frail. He was hardly eating, with constant diarrhoea and a large tender mass in his lower abdomen (RIF). He had partial intestinal obstruction, and needed urgent admission. His general state suggested advanced caecal cancer.

A large mass was resected from his abdomen, and radiotherapy started. Only later did some controversial histology exclude a carcinoma. Although initially reported as a mixed neoplastic and inflammatory tumour (diverticulitis), more extensive histological analysis revealed no cancer, only severe diverticulitis. He forgave everyone involved for the unnecessary radiotherapy. He chose not to sue them. 'We all make mistakes', he said ironically. He felt lucky to be alive after what many would have regarded as a worst possible nightmare.

Zoning-out and Emotional Reminiscence

My lengthy screening consultations allowed me time to inform and educate patients on medical matters. This sometimes re-orientated their ideas. I always tried to teach my patients vigilance about specific symptoms, and when to contact me. Some found this irrelevant and boring. The moment boredom strikes is usually obvious: their eyes glaze over, and they 'zone-out'. Their attention has left the room. Time to move on. Many people give no priority to education and survival until trouble looms.

There is another moving, emotionally linked form of zoning-out, seen during intimate discussions. It is worth noting because it is revealing. It happens when someone reminisces or becomes nostalgic about something deeply important to them. For a few seconds, thoughtful introspection arrests their facial expression; their eyes redden as a wave of emotion passes through, a trace of unshed tears might appear along the eyelid edges. One might call this a momentary psychic arrest.

When Idris Elba interviewed Paul McCartney (BBC 1, December 2020), he asked Paul where the inspiration came from for his song 'Let it Be'. Paul explained that in a dream, he had envisaged his mother (who died when he was 14-years-of-age). She conveyed to him, 'the past is done; let it be'. As he described this, Paul displayed an arresting response; a few seconds of facial distraction seen as a powerful wave of heartfelt emotion passed over him. This was no act. The deep inner feeling he still held for his mother was obvious.

History Taking and Computer Analysis

Clinico-pathological associations formed from the experience of many cases, should eventually allow every doctor to assess symptoms and signs for their predictive value. Although AI will step in to provide the odds of being correct, prediction has always been a clinical art-form, albeit based on anecdote and experience. How doctors learn and use these associations, will define their diagnostic ability. One can learn clinical handling, appropriate management action, and patient satisfaction, from experience, but some talent and a desire to achieve them are prerequisites. Those without these abilities can look to AI for help. AI offers an enticing prospect: to make those who lack knowledge and who are inadequate clinically, both compliant and more acceptable.

The computerised numerical weighting of symptoms and signs for clinical relevance, is developing. The probability figures produced, will confuse many and will need interpreting. The prediction of diagnoses, based on calculated 'odds ratios' of symptoms or signs (or clusters of symptoms and signs), will complement our clinical judgement and experience, with less bias (recent cases bias us most).

Will big data analyses help us with insights into early diagnoses, and foster medical scientific discovery? Do enormous data sets of symptoms, pulse rates, BPs, breathing rates, etc., hold hidden medical secrets? Since almost every discovery in science starts with an anecdotal observation, 'big data' protagonists hope to identify another way to make meaningful associations. 'Big data' analysts, detached as they are from clinical insight, think they can reveal some hidden clinical factors that have eluded physicians for millennia. I was inclined to be dismissive except for one computer generated discovery, that tempered my scepticism.

In cases of suspected appendicitis, data analysis found nausea to be strongly predictive symptom. Nausea reliably predicts infection: a septic appendix that needs removal. A sterile appendix and mesenteric adenitis are rarely associated with nausea (but often result in 'unnecessary' surgical removal). Nausea likely relates to the amount of bacterial toxin being released.

Will computers take over history taking? Why not ask 'Alexa'? Having been an amateur computer programmer myself for many years, my view is not Luddite. The patient-computer interface is developing and offers promise. Computers can easily glean clinical information from patients and input the data to a neural network, designed to decide the probability of various diagnoses. They easily allow time-to-time comparisons of data (continuity and follow-up being as important as they are). Patients, however, can lie, and falsifying what they report, whether to a doctor or machine. An important question, therefore, is how truthful are patients going to be, sitting in front of a computer? Facial expressions will be missed as patients answer questions, although corroborative questions will help. Accuracy, whatever the format used for history taking, is a crucial issue.

Nearly one thousand patients took the psychological questionnaire I computerised in 1985. The results allowed me to observe something unexpected. Some patients felt they had been more thoroughly questioned by a computer, than by any doctor. Many felt they had divulged more of themselves. They had never been asked some of these questions before. For a few, the experience was both cathartic and therapeutic. If it can be this valuable to patients, it is worthy of more research.

Direct computer interfaces for patients will prove timesaving devices for doctors with ever-growing queues at their door. They will not suffice for the small number of patients who insist on personal service. For me, such programs could serve well as screening aids (or research); not a replacement for the real thing: personal human interchange. They should work well for those who insist on being 'fixed' quickly and are suffering from straightforward medical conditions.

Regulatory bodies will love patient - computer interfaces, and will probably insist on all publicly funded doctors using them. Imagine the NHS payroll savings, if computers replaced most GPs. Bureaucrats will have no better way to gather practice data and audit the work of doctors. Might computers flag up the next Harold Shipman, I wonder, before she/he commits similar crimes? Some will think so (reflecting the blind faith in the utility, reliability and veracity of computers, typical of corporate organisations).

On the 22nd July 2022, Guy's Trust declared a major IT failure. Their computers failed and they could not access any patient information. No surgical operations were possible. No out-patients consultations took place, no appointments were made, and nobody had any idea when the system might be back on-line.

As an example of misplaced corporate faith in computers, read no further than the Post Office IT fiasco. Over a 10-year period (2010-2020 approx.) they sacked many trustworthy post office workers accused of stealing money. Many were sent to prison because of the faulty Horizon, IT computer system.

Chapter Eleven

Patient Investigation

'It is by doubting that we come to investigate, and by investigating that we recognise the truth.'

Peter Abelard (1079 – 1142)

As my career progressed, many new non-invasive investigations became available. What were once unpleasant techniques used to investigate patients, involving dye injections and radiation, became replaced by no-touch imaging techniques able to reveal both anatomy and physiological functioning. The painless nature and improved results of imaging has steadily changed the methods and choice of clinical investigations.

Traditionally, investigations have been used to provide evidence for the diagnoses made during history taking and physical examination. The most judicious tests remain those reliably enough to prove a presumptive diagnosis correct.

There are important considerations when using any diagnostic investigation:

- **Test Relevance.** Investigations relevant to the diagnosis being questioned.

- **Appropriateness**. For every medical condition, there are appropriate and inappropriate investigations.

- **Diagnostic Accuracy.** The predictive value of every test depends on its specificity, sensitivity, and the diagnostic accuracy (positive and negative). Every investigation used will have true and false-positives, as well as true and false-negative results. These are the elements used to calculate diagnostic accuracy. The same principle applies to every symptom and physical sign.

- **Diagnostic uncertainty**. The need for a corroborative investigation increases with diagnostic uncertainty, and the likelihood of clinical error.

- **Interpretation.** Test results must be viewed from the standpoint of clinical relevance.

- **Urgency.** Test urgency needs to be considered in the light of the resources available and the clinical state of the patient.

- **Corroboration** between independent tests is desirable, especially when some tests used have questionable reliability.

- **Clinical management.** Which investigations will help most to steer patient management? Such investigations should be able to grade the disease process and help decide the rapidity of progression or regression from anatomical, histological, and patho-physiological points of view.

- **Test worthiness.** Worthy investigations are those which involve little or no procedural risk, and have the potential to change patient management. (Some will add cost to this definition).

- **Test unworthiness**. Unworthy investigations are those unlikely to contribute to clinical management decisions. Their justification might be for 'research' purposes or 'out of interest'. Neither low cost, nor low risk, condone their use in routine clinical practice.

- **Clinical progress.** To assess clinic-pathological progress, repeat investigations are required. Time intervals between tests must be dictated by the initial assessment, and the likely rapidity of clinical progress. Repetition needs to be minimised where risk, discomfort, and radiation are involved.

- **Medico-legal investigations.** A distinct set of considerations apply. Their aim is to provide evidence of legal relevance.

Before doing any investigation, review the patient's notes. Avoid repeating unnecessary tests. Access all the past results to define change, and rate of change. Many of my patients kept their own notes, and even plotted their own test results over time on a spread-sheet. Is this business-like, or neurotic? Both, of course, but useful.

Why Investigate?

The first justification for any investigation is to find supportive evidence for a presumptive diagnosis. The second is to refine the diagnosis by defining the pathological anatomy, and the extent and severity of the problem (before surgery or radiotherapy, for instance). Omitting pertinent investigations can be fatal.

Tommy was in his seventies. He had an abdominal aneurysm involving his renal arteries. Many years before, I discovered his hypertension, and Class 4 atheroma in his carotid arteries (clumps of cholesterol and calcified plaque arising from the inner artery wall). I found that twenty to thirty percent of such patients also have coronary artery disease, and over ninety-nine percent of those with coronary artery disease have significant atheroma in their carotid arteries.

An experienced Harley Street cardiologist evaluated Tommy pre-operatively, and found him 'fit' for an abdominal aortic aneurysm operation. Even with the history of a previously suspected cardiac infarction (which the patient may not have declared), he thought it unnecessary to investigate his coronary arteries, even with a simple CT calcium score. His reason: the patient had no symptoms.

He didn't know Tommy like I did. I knew him to be an unreliable historian who always underplayed his symptoms. Within 24-hours of the operation, which took three surgeons nine hours, Tommy had a silent heart attack (STEMI) with typical ECG changes and a troponin >3000ngs/ml. Because they sedated him post-operatively, he might not have been able to report chest pain, if any. The doctor on duty was notified four hours after a diagnostic ECG (computer derived diagnosis of cardiac infarction) was recorded, and took no immediate action. Tommy died 21hours after his MI, in renal failure. Nine defibrillation attempts failed to revert his VT/VF.

His death was the consequence of:

- Inadequately pre-operative coronary evaluation. In retrospect, he should have had a coronary artery by-pass graft before any attempt to operate on his abdominal aneurysm (cardiac CT scanning for coronary atheroma detection was not a guideline requirement at the time);

- Delayed action. His MI happened at 10pm, and the hospital staff took no action until the next day. Post-mortem examination showed severe, widespread coronary atheroma. Only a prior coronary bypass (CABG) would have prevented what happened. It didn't help that the very expensive, Central London private hospital he found himself in, had no cardiac department!

One cannot achieve optimal clinical management without accurate diagnosis and assessment. Tommy's clinical management was based on incomplete evidence and incorrect assumptions.

As a simple rule of thumb, always have three good points of evidence before believing a diagnosis. Have these in hand to convince sceptics, the most ardent of whom must be oneself. To have two good points of evidence, is not quite enough; five points will clinch every diagnosis. The trick is to assign the correct validity to each piece of evidence. This requires one to know how valuable each piece of evidence is in relation to every diagnosis.

As a routine exercise, we must bring together the strongest points of evidence from a complete history and examination, and use them to request confirmatory investigations.

Some Politics of Investigation

The lack of investigation facilities in the UK is a political matter. Matthew Sayed made the following remarks in his *'Dispatches'* program (Channel 4, 18th October 2021):

- *Radiology is crucial to the diagnostic pathway. Coroners have cited a lack of scanners and radiologists as factors in avoidable deaths in the UK. The crisis in radiology presents a direct threat to patient safety.*

- *At present, forty-five percent of cancers in NHS patients, are not being diagnosed until an advanced stage.*

- *France has twice as many CT scanners as the UK; Germany has four times more CT scanners and five times more MRI scanners.*

It is thought that the NHS is short of 2000 radiologists and 4000 radiographers. This has left 142,000 patients waiting over six weeks for their results, and 40,000 waiting over thirteen weeks.

Investigation Changes

One can practise medicine in different ways: from a *'Desert Island'* approach (experienced judgement, guesswork and no investigation), to what we once called *'Teaching Hospital Medicine'*: medical practice with no place for unsupported guesswork and every judgement based on objective evidence.

The threshold for doing various investigations has changed with time. Whereas echo studies once required large, unwieldy machines in hospital imaging departments, trained junior doctors, nurses and paramedics can now perform them with a small handheld device. Because it is so readily available and harmless, bedside and even domiciliary ultrasound, can image the heart, arteries, lungs, abdomen, veins, and the eye. After examining a patient one can quickly confirm suspicions of pleural effusion, lung consolidation, heart failure from various causes. Heart valve defects, SBE, DVT, hydronephrosis, appendicitis, gallstones, temporal arteritis, and even papilloedema, can now be detected with what is now referred to as 'Point of Contact UltraSound', or POCUS. The practice is worthy of making rapid diagnoses in emergencies. This will allow quicker treatment by the most appropriate specialist team.

Essential Bayes' Theorem

The best investigations are those that are both sensitive and specific to a diagnosis. Few tests qualify. One of the best is the pregnancy test, based on the detection of chorionic gonadotrophin (HCG) derived from the placenta. When performed correctly, it is accurate diagnostically (both negatively and positively). The timing of the test is crucial, especially if done too soon. The result can confirm pregnancy to those possibly pregnant, with almost no fear of them being misled. The test is both sensitive (blood levels become positive soon after implantation of a fertilised ovum, 6-10 days after ovulation), and specific to pregnancy. There are very few conditions that confound the result (liver disease, and choriocarcinoma, etc.).

No clinician can avoid understanding something of Bayes' numerical analysis, even though medical practice relies heavily on anecdote; only one patient being considered at a time (i.e., no cohort analysis). Bayes' theorem provides a framework for physicians concerned with the diagnostic accuracy of every question they ask, and every investigation they request. For each medical condition, every question posed, and every investigation undertaken, has its own sensitivity, specificity, and calculated diagnostic accuracy. A sensitive test is one that easily detects the condition. If too sensitive, false positives will occur. High specificity means that it will only be positive in those cases with the stated diagnosis (no overlap with other conditions). Investigations are never too specific, but they can be too sensitive.

Statisticians define test sensitivity as the number of truly positive tests as a percentage of all positive results (all the true and false positives taken together), whereas specificity is the number of true negatives as a percentage of all the negative results (true and false negatives together). This is a few steps away from the usual work of clinicians, but we are all required to rely on evidence that comes from group analysis.

A sensitive test is useful for early detection. It must be sensitive enough to detect pre-symptomatic disease. Test specificity is a measure of the confidence we can have in it, ruling in, or ruling out, any diagnosis.

Physicians need not calculate sensitivity, specificity and diagnostic accuracy, because with experience, they can estimate them. Research data can provide the figures and help to temper our biases. From experience every physician will know which tests are prone to false positives and negatives (useless tests), and which frequently yield true positives and true negatives (the most useful tests). We can refer to them as clinically unreliable and reliable, respectively. How we weigh the relevance of each test result, yields the diagnostic certainty we have. When we find tests unreliable, further verification is always necessary.

Each car, boat and airplane has its own handling characteristics; so do the investigations we perform or request. Having a 'feel' for tests, and their practical usefulness, is an essential art of medicine gained with experience. We must employ it every time we 'interpret' any test result.

Patients who view their own results often get bothered by those falling outside of the normal laboratory range. It might become necessary to explain that a high blood potassium (in an otherwise healthy person) has resulted from red cell lysis. Finding this tiresome, I installed a centrifuge machine to separate clotted blood as soon after venesection. Thereafter, I saw fewer cases of false hyperkalaemia, but they did still occur.

There will always be error

Everything we measure is subject to error; errors imposed by the instruments we use, the technique or processes we use, and by each of us as an observer. In physics laboratories, these might only be ±0.001% errors; in biology they can be ±10% errors or more. There are several important types of error:

- *Instrument error:* Errors of instrumentation rarely concern doctors and physiological technicians.

- *Observer error*: not measuring what needs to be to measured.

- *Measurement error:* the inaccurate measurement of what is measured.

- *Errors of perspective*: not getting the correct view of what is measured.

- *Sampling error*: A statistical data acquisition problem, not a clinical problem.

Then there are the clinical errors:

- *Errors of assumption*. These lead to diagnostic error and poor patient management.

- *Errors of omission*. We are all acquainted with forgetting or ignoring vital facts.

Consider echocardiographic measurements as an example. Exactly which chord across the left ventricle should one measure when estimating LV contraction and volume? Even slight changes in the chord measured, will change the ejection fraction calculated. A major clinical error could result, and an incorrect assumption of heart failure. This can be an *observer error (made by incorrectly choosing what is to be measured)*, or a *measurement error (an error made in the process of measurement)*. An *error of perspective* will occur if we choose the wrong view of an entity (like measuring the LV cavity in the short axis, rather than in a four-chamber view). Since vector quantities (like ECG voltages) have both force and direction, errors result from not taking direction into account. With any of these errors in place, completely different conclusions about LV size and function could result.

Mr. David M. had LBBB. His interventricular septum did not seem to contract well, whereas the lateral wall of his ventricle appeared to contract normally. His measured ejection fraction (EF) was of concern, because others had calculated it to be low normal (48%). The inference was that he had borderline left heart failure. The measurements were correct, but the conclusion was false. It was his LBBB that led to a low EF calculation (an under-stimulated septum). His low EF arose from blocked electrical conduction, not from any weakness of his septal heart muscle. With exercise, his septum contracted normally. The resulting calculation of his EF was then 60%.

Which measurement outcome best reflected the overall contractile state of his left ventricle? Clearly, his septal contractile function (micro-electrics / perfusion) changed with exercise.

The commonest source of all errors is assumption. Here, the assumption was that LBBB produced a fixed mechanical defect that would not respond to local metabolic changes. The other assumption made was that an echocardiogram can reveal everything about LV function. In Mr. David M's case, an exercise test completely changed the result and final interpretation. Every second-hand car dealer knows the principle: never buy a car without taking it for a test drive.

Test Appropriateness

Every test used to diagnose or define a medical condition, has its diagnostic accuracy. This can help us decide which type of brain scan (MRI, PET, or CT) is best to diagnose a brain tumour, or MS. Is a Rubidium PET scan better than a stress echocardiogram for detecting deficient myocardial perfusion? Within every specialty, there are tests which offer the best diagnostic accuracy for specific conditions.

The polymerase chain reaction (PCR) has enabled bacterial and viral identification to become a rapid, highly specific process. Before, we had to send swabs to the laboratory for culture and sensitivity. Then we waited . . . and waited. A few patients died waiting. Culture and sensitivity (C&S) testing was all we had; sometimes it was not practicable for patients who were very sick. We had to employ intelligent guesswork and had to change course once the results became available.

We made many microbiological presumptions. A common one was that *E. coli* caused most urinary tract infections (UTIs). So, did we really need to get urine samples for C&S every time? With experience, I came to reserve bacteriology for recurrent or persistent cases only. Should every UTI have a renal ultrasound scan? I would order one if there was renal tenderness, although an ultrasound cannot actually detect renal inflammation unless tissue damage is present. Dip-stick urine tests are also of questionable value. I came to expect negative tests for WBCs, even when patients had definite, infective UTI problems. There are reasons for this. Urine acts as an antiseptic, killing bacteria while samples await laboratory testing. Bacteria from an infected segment of the kidney may only be detectable while that segment is functioning. On this basis, only one in ten urine samples will grow the culprit pathogenic bacterium in pyelonephritis. Also, not all lower urinary tract inflammation has an infectious cause; as with sunburn, radiation not infection, is the cause.

In the past, we had to work with poor test sensitivity and specificity. We relied on judgement honed by experience to help our patients.

Test Interpretation

As an art, it is by far the most contentious aspect of clinical investigation.

Interpreting some tests, like the ECG and EEG, relies heavily on pattern recognition and knowing the clinical context. Tarot card readers, clairvoyants, palmists, and tea-leaf readers employ something similar. Both ECGs and EEGs are ideal for neural network 'learning' and diagnosis. Until that happens reliably, and includes clinical context, reading them will remain an art-form. Interpretation by computer has been available to the uninitiated for decades, but master readers of ECGs, can still find errors in computer interpretations.

In my younger days, some of those who could 'read' these tests with confidence, achieved iconic status. There are so many variations of normal, and so many subtle changes to miss. Heart attack or no heart attack? Epilepsy, or vasomotor syncope? These diagnostic questions, of serious consequence to patients, can depend on someone capable, correctly interpreting machine output waveforms.

Does using the term 'machine output' connote reliability or hide a multitude of errors (instrument errors)? How does each capacitor, resistor, or integrated circuit within the machine alter the input signal to create an output signal that may differ between machines? Electronic engineers do a commendable job trying not to distort the wave functions involved, but to what extent are they successful? We are so used to accepting machine processing, few question the electronic aspects.

'Why am I always tired, Doctor?'

This is perhaps the commonest question asked by patients; a heart-sink question for most doctors, since we only occasionally find a satisfactory answer.

I used to see a lot of patients complaining of TATT (**T**ired **A**ll **T**he **T**ime). Many had been told they had 'nothing wrong with them', despite persistent tiredness for months or years. Having undertaken a full history and examination, I would proceed to 'routine' blood tests (biochemistry, haematology, and thyroid function). I quickly learned that few of these would yield a diagnosis. Only fifteen of one thousand cases I investigated, proved an exception. I afterwards included viral antibody tests, and tests for rarer infections. It would have been useful to have had long-COVID to blame, but that was yet to come.

What might cause TATT? Might it be 'ME', for which nobody has found a convincing explanation? Can a patient's mitochondria produce less ATP? Regardless of the cause, one can diagnose 'chronic fatigue syndrome', although some patients are depressed, and many are reacting to 'stressful' circumstances, with insomnia the cause of daytime tiredness.

When I did eventually extend my investigations to include viral antibodies, I found many had positive antibody levels for herpes simplex, and for the Ebstein-Barr (EB) virus (herpes virus 4). This might be because herpes simplex is ubiquitous, and 95% of us have had glandular fever without knowing it. I always requested three EB virus antibody levels (IgG, IgA, IgM) for TATT patients and found significantly raised levels in many adults. I concluded that some might have a chronic or recurrent EB virus infection (I did occasionally prove viral antigen present, but only in a few, because of the expense).

I might have previously missed recurrent / chronic glandular fever before, as a cause of TATT / chronic fatigue syndrome, by not testing for it. The typical clinical features of primary 'glandular fever', seen in teenagers, are not present in adult TATT cases. Older patients rarely have fever, lymphadenopathy, small palatal haemorrhages or a positive Paul Bunnell test.

Most will accept there is no effective treatment for glandular fever, in whatever form it presents. I decided, therefore, to treat all those with raised EB virus antibodies speculatively. I gave them a therapeutic trial (with fully informed consent) of a tetracycline (doxycycline, or oxytetracycline) over 3-6 weeks, having learned that these can reduce RNA replication (EB virus is an RNA virus). Although mostly harmless, tetracycline and its derivatives lack approval for treating EB or COVID-19, despite them both being RNA viruses (they are an alternative to acyclovir for those with oral herpes).

What resulted? Eighty percent of EB antibody positive cases improved (I would have expected 20% from a placebo). What needed to follow was a double-blind trial to prove my anecdotal observations.

There are technical questions to be answered. Had I wrongly ascribed significance to the EB antibody levels? Sometimes, that must have been true, although the diagnosis intrigued and encouraged patients. Many were keen to get a positive diagnosis, whatever it was, having been told for years they were imagining their symptoms. Discussing the results, while being honest about the various diagnostic possibilities (that the test results did not prove cause and effect), did not diminish the efficacy of the treatment. TATT is difficult to manage, especially after patients have heard only negative or equivocal views. At the very least, my harmless strategy offered a few dejected patients, some hope. Apart from UV skin sensitivity, tetracyclines are as harmless as antibiotics can get; few would think they risk bacterial resistance. A double-blind trial is now appropriate, but who would fund it?

Tiredness is more than a medical issue; it is an economic issue. Tiredness, fatigue, and exhaustion cause work absences and frequently precede road accidents, and catastrophic medical events (diminishing energy levels occur frequently in the year before cardiac infarction), and tiredness, fatigue, and exhaustion are common in serious diseases.

The husband of one of my patients suspected his wife, suffering with chronic fatigue, had Lyme disease. He had done his research and was keen to confirm his suspicions. A tick had bitten her while they were walking in France. The 'usual' tests were not specific enough, so I sent a blood sample to the Porton Down laboratory (RIPL: Rare and Imported Pathogens Laboratory) for confirmation. The test proved positive.

The patient had multiple, non-specific symptoms, and no physical signs, so the clinical evidence was weak. I remained unconvinced that Lyme disease caused her symptoms. Because she had raised EB antibody titres, I thought it more likely she had recurrent glandular fever. Perhaps she had both, or was it significant that she had some serious relationship problems with her husband?

Some TATT patients become desperate for a diagnosis, and will welcome any test that provides one. Giving a diagnosis can itself act as a placebo, giving relief from anxiety and uncertainty. Sometimes, a diagnosis can provide justification for feeling unwell; especially useful when the patient is managing social and inter-personal situations.

There is an art to knowing when to be speculative, and when to be definitive about the diagnosis given to a patient. Patients need to know what to expect with either. Sometimes, any treatment (albeit with a placebo or counselling) that supports a positive outlook, or reassures and encourages hope, will be beneficial.

'*Ma*' : Silence and the Space Between

Some difficult cases remain undiagnosed because of insufficient information. When there are information gaps, listening again to a patient can be the most rewarding 'investigation'.

The Japanese concept of 'ma' – the space between – is an appreciation of what is missing: the silences during speech and music. Occasionally, it takes years to get the information required to complete a clinical picture. I remember a patient who returned to me after five years saying: 'I had to divorce him. My husband was making me ill.' When locked into a pathological social situation, some patients can find it too difficult to talk about.

A 70-year-old woman came to me many times with nebulous symptoms. Her breast cancer had been successfully treated, fifteen years before. I had known her for several decades by the time she came to reveal something secret.

'I suspect you would like to know what is really wrong with me, Dr. Dighton?' she said. 'After all these years consulting you, I think you deserve an explanation of my complaints. I am going to tell you something I have told no one; not even my husband.' I sat back, silently intrigued.

She continued. 'When I was 22-years-old, I fell in love with a handsome naval officer. After a year of seeing one another, he came to me in his white tropical uniform, and announced he would not be seeing me again. He was off on a world tour and wanted to confess something before he went. He had withheld something from me, despite his genuine affection for me. He told me he was homosexual, and there could be no future in our relationship.'

'For fifty years, I have kept his image in my mind; him standing at my front door in his immaculate, white naval uniform. Every ring of the telephone; every knock on my door since then, has made me think it might be him. I have lived hoping that one day he would return and say: I have always loved you; I made an unforgiveable mistake. Of course, we should live together in a loving relationship.'

'Now you know what I have been living with all these years; hoping for nothing else other than he would return to claim me. I'm sure the heartache has caused all my problems, Dr Dighton. I hope you understand.'

She died not long after telling me this; metastatic breast cancer was the cause.

Daft Investigations?

Many patients have more faith in the value of investigations than in the processes of history taking and clinical examination. In private practice, patients rarely complain about the cost of investigations (which can be considerable), but often question consultation fees.

My nurse Margaret was over 70-years old, and still working in my practice. She asked me if I would be interested to visit another practice where she worked; a well-known, local GP ran it. She said he had equipment in his surgery that was impressive; perhaps even revolutionary.

When I visited (well before the Care Quality Commission existed), it intrigued me to find a long panel of 'electronic machines' with flashing lights and flickering meters. They filled one wall of his consulting room. He was happy to demonstrate them to me. He first attached a patient to the machines, using suckers and electrodes, and then took some readings. 'You are quite run down,' he said to a patient, after looking at the various dials. Having adjusted a few knobs, he left the patient to be 'treated' for twenty minutes. The machinery bleeped and made various noises throughout. Many of his patients said these 'treatments' revitalised them, so many returned for 'top-ups'.

I asked some questions, but gained no insight. His patients mostly came with tiredness, and because they 'didn't feel themselves' (I have always resisted asking the cynical question: 'Who do you feel like then?'). Both the doctor and his patients, claimed that these machines were effective. 'What is the scientific basis of these machines?' I asked. I seem to remember him mentioning something about small electric currents, electromagnetic fields and magnetic flux. He had designed the machines himself, and clearly wanted to keep their workings secret.

It is interesting to note that this doctor seemed to benefit many patients (according to our mutual nurse). He achieved most of his success with those who felt failed by scientific medicine. It is important to remember that there are many states of mind, some of them unhealthy, that are not detectable using laboratory-based tests and imaging techniques. What some patients need most is to be listened to. Once a therapist has gained a patient's confidence, counselling and placebos can work well.

For those doctors and nurses who wish to detach themselves from the psycho-social aspects of clinical medicine, there is something to remember. Since every organ is attached to the brain, their functioning can be influenced by cerebral activity. To what extent is the question.

Which Test(s)?

In many places in the world, the challenge is not choosing appropriate tests, but finding an investigation facility. In richer countries, the range of tests available is usually vast, although budget restrictions may apply. If so, one may need to convince another doctor that the test is necessary (radiologist, technician or insurance company employee). Although these discussions can be worthwhile, they are sometimes futile and annoying. In private medicine, where cost is not usually an issue, these negotiations are often minimal. Exceptions occur in privately funded facilities, where 'package deals' are in place, or when an HMO (in the USA: Health Maintenance Organisation) or an insurance company controls the funding. Their motive is to maintain profits by reducing expenditure.

Choosing targeted tests, requires more expertise than package testing. Guidelines, not judgement, can dictate which tests are necessary, although diagnostic accuracy should always guide our choice. In some places in the world, they must rely on clinical diagnoses alone, unsupported by investigation results (because they are unavailable, too expensive, or take too long). One can still achieve much, even though this situation is untenable and unacceptable in first-world countries.

In the UK, regulators might judge a doctor 'unfit' to practise if their choice of investigations has not followed written guidelines. Throughout the era I practised, the clinical experience and judgement of doctors making such choices was sacrosanct. Regulation brought this to an end. As in all corporations, the NHS included, the purpose was restrictive control. There are some key questions I would like to ask regulators (none of which they will answer). First, have any of the regulations they devised, changed patient outcomes (mortality and morbidity) for the better. Second, are patients more gratified by medical services now, than once they were? If not, the money spent building, maintaining and staffing the huge NHS bureaucratic pyramid, cannot be justified.

Most investigation choices are straightforward: measuring haemoglobin for anaemia, and creatinine for renal function, for example. Measuring blood glucose in diabetes is not. Sampling errors are a significant devil in the detail. One can miss pre-diabetes by taking normalising fasting blood samples. For this reason, I often first tested blood glucose and lipid levels in non-fasting patients (within one hour of eating). With any hint of hyperglycaemia, I would undertake a fasting glucose and a glucose tolerance test (GTT). Many cases of pre-diabetes thus became apparent, and dietary advice could follow.

I added insulin levels to the GTT evaluation if hypoglycaemia was suspected. The two key data points of a GTT are the fasting glucose level, and the blood glucose level, thirty minutes after a glucose load. It may be routine, but there is no extra diagnostic value in taking blood every half hour, for two-hours. Apart from the important discovery of a low renal threshold for glucose, and finding proteinuria, testing the urine for glucose every hour is also unhelpful.

Diabetes affects both lipid and glucose handling. So why not test lipids, before and after drinking a pint of full cream milk? Such a test might generate a better predictor of morbidity for diabetics (atherosclerotic disease, neuropathy, etc.) than Hba1c? (HbA1c is a risk factor for CVS disease).

New Tests

Research can produce worthy new investigations, but applying them in clinical practice can prove a challenge. Is it wise to use speculative, research-based tests on patients, before the publication of controlled trials? Like all products, many view what is 'the latest test', through the lens of fashion (the pro-innovation bias). Not using them might imply you are not up-to-date and competent (regardless of your own scientific evaluation).

I liked to evaluate new tests after a few cases. I used the PCA3 test for the detection of prostate cancer ten times before I gave it up (two false negatives were two, too many). After these two failures, I dropped it completely. The test was expensive and not straightforward; it required prostatic massage prior to urine collection. The results of diagnostic MRI prostate scanning became available at the same time and proved more reliable.

A colleague of mine had his prostate cancer confirmed with a PCA3 test, after an equivocal PSA result. Initially, I wrongly ascribed more diagnostic accuracy to the PCA3 test than it deserved (I believed the published research). Now one can decide who needs a prostate biopsy by combining a PSA test with MRI scanning. Unfortunately, prostatic biopsy is not 100% accurate either; cancerous areas are sometimes missed (sampling error).

Benefit / risk considerations of new investigations are always important, although it can prove difficult to find a facility or laboratory to perform them. As an example, I once tried to get a cardiac PET scan for a patient with CAD. I had two questions. First, could it measure the obstruction to coronary flow (Rubidium, PET flow assessment was new at the time). Second, could I provide evidence for his cardiac infarction risk, or death, by measuring plaque inflammation? 18-F fluorodeoxyglucose (FDG) PET, can reveal this (using increased glucose uptake as a measure of inflammation in coronary plaques). It was not then available to cardiologists, but widely used in research.

I wanted a FDG PET scan for this patient, because he had a solitary, 50% obstructive atheromatous plaque, in his proximal anterior descending coronary artery (on cardiac CT). Should a clot have formed on this plaque, it would lead to extensive anterior infarction. He was at high risk, because even small vulnerable plaques (liable to fissuring) can induce thrombosis – a cause of sudden cardiac death.

Unlike plaques in the other coronary branches, those in the left anterior descending artery and main stem, present a very high risk. Patients with big anterior cardiac infarctions do not have a good prognosis (50%, five-year mortality for many). By comparison, other coronary artery related infarcts (posterior and inferior), hardly affect life expectancy.

This patient had had a cardiac CT, and rubidium PET perfusion scan, so his life-time radiation dose from investigation was mounting (radiation risk reduces with age, and he was over 70-years-of-age). Knowing if his plaque was 'vulnerable' could have proved crucial to his management. Specifically, I would like to have known, despite his aspirin and rosuvastatin medication, whether his plaque remained inflamed. Had it been so, I would have stented him, even with no ischaemia on exercise. All of his blood inflammatory markers were negative, but given the tangible chance of his sudden death, I did not want to rely on them alone.

One major problem with such testing is clinical interpretation. My guess is that we should consider stenting all the atheromatous plaques that are inflamed; especially those in the left main stem and anterior descending coronary artery, obstructive or not. Time will tell. Because a test for plaque inflammation was not available, I left me to assume the possibility. I advised him to take a maximal dose of 'statin'. Had I have known he had plaque inflammation, I would have stented his artery.

With proprotein convertase subtilisin/kexin type 9 (PCSK9) inhibitors being introduced, I suggested he should research them. Without a specific test for plaque inflammation, it left me only one strategy: to 'throw the pharmacological book at him', in order to 'play it safe'. Without a direct measure of plaque inflammation, one cannot know how long to persist with treatment. Balancing the risk of drug side-effects with his risk of cardiac infarction or sudden death, needed little further consideration.

I discussed these issues with my patient (a retired academic physician) and his wife, in order to define their attitudes to risk, intervention, and masterly inactivity. The correct individual decision for him emerged (maximal drug therapy). One should review each case in the light of changing patient attitudes, and any recent advances in investigation and therapy.

Few would stent non-obstructive, solitary coronary lesions, so why did I consider it? First, we always need to treat individual risk, with one eye on the statistically calculated (group) risk. Since this patient had occasional, atypical chest pain (likely caused by coronary artery spasm), was it safe to do nothing? After ten years, he has still not infarcted.

What cardiologists lack is a convenient measure of the rate of growth of atheromatous plaques. For twenty-years, I scanned the carotid arteries of many hundreds patients at two-year intervals. This enabled me to measure the rate of atheroma progression. Most 'statins' will reduce cholesterol production by the liver, but do they beneficially affect the intima of arteries? Instead of using blood cholesterol as a risking factor (which is only raised in 60% of those with carotid atheroma), I thought it more important to use both the extent and rate of growth of atheroma as risk factors. My aim was to stop atheroma progression and to prove that it had stopped. Reducing the national blood cholesterol average was never my primary concern, although lowering the national average of total blood cholesterol will reduce the nationwide prevalence of heart attacks.

My enthusiasm for coronary stenting continued to be supported by never seeing post-intervention side-effects. That, I suspect, is because my interventionist colleagues (cardiologist, Dr. David Lipkin at the Wellington Hospital, London, and Dr. David Clark at Stanford, California, US) were world-class operators.

Unnecessary Testing

Without full access to a patient's notes, unnecessary testing can be repeated. This wastes time and money, and may not be in the patient's interest, especially if a repeat test involves radiation. This never happened to my overseas patients; they always carried their notes with them. Very few British patients carry copies of their clinical information with them to share.

Investigation Management

In my practice, 24-hours was long enough to wait for any result. Getting investigation results returned within 24-hours, was my practice benchmark. Forget corporate audit, and targeting, this is what my patients deserved. It reduced patient worry and moved their management forward. Those patients who came to me as refugees from the NHS, mostly thought their blood test results would take three weeks. They were in for a surprise. Thanks to the Doctors' Laboratory in London, their results were usually available within 12-hours. My cardiac testing results (ECG, stress ECGs, and echocardiograms, etc.) were all available when completed.

Whatever we choose to tolerate, we will not change.
Those who never change their mind, never change anything

Winston Churchill.

Cardiac Screening Investigations

In 2000, I introduced routine carotid artery screening for my patients. The point was to detect atheroma formation (as a pathologically justified indicator for prescribing prophylactic statin and aspirin therapy), and to assess the risk of future coronary, cerebral, renal, and peripheral vascular problems. Await the publication of my research on the subject, to review the evidence. So far, corroborated research and NHS practise are only 24-years behind!

I used echocardiography in every patient with high blood pressure or suspect hypertension. The aim was to identify early left ventricle hypertrophy (LVH) and its progression. The presence of LVH helps to predict individual morbidity (the more muscle hypertrophy, the greater the risk of stroke), and is essential for the evaluation of hypertension. Once one detects LVH, ACE inhibitors and angiotensin II antagonists, can make a difference to the prognosis (given that they can stop progressive arterial medial hypertrophy, and further LVH).

Without an interest in preventing atheroma and LVH as common components of cardiovascular death in individuals, doctors must await official guidelines, and the support of medical bureaucracy before introducing routine carotid and LV ultrasound studies. My adopted policy contrasted with giving 'statins', only to those with high blood cholesterol, and giving anti-hypertensives to all those with just a few high BPs and no LVH.

As a hospital physician working in the community, I realised over forty-years ago, that pre-diabetes, rather than just overt diabetes, was a common cause of peripheral neuropathy. I had previously assumed it must result from long-term, poorly controlled blood glucose. I therefore introduced modified glucose tolerance testing for all such patients. Many proved to be pre-diabetic. The condition is common enough, so I will leave others to collect the data and take it further. I considered adding fat to the glucose tolerance test (perhaps a glass of creamy milk containing the glucose load, in order to observe both glucose and triglyceride clearance as a measure of insulin resistance), but never did.

Every week in my practice, cardiac screening revealed at least one asymptomatic person (thereafter, a patient) with significant coronary artery disease (CHD). The anatomy and flow reducing character of their atheroma, determined whether I prescribed prophylactic pharmaceutical treatment, stented them, or referred them for a CABG.

Convincing statistical appraisal of my twenty-year-old policy to stent asymptomatic patients with critical CHD, was not possible with my small number of patients, although it was my duty to try. Anecdotally, no patient of mine suffered any long-term consequences of my stenting policy, although I met a few patients who had suffered atypical chest pain after being stented elsewhere.

I must attribute some success of my management policies to cardiac surgeon Stephen Edmondson, working at the Wellington Hospital, London. His zero mortality and morbidity results, although anecdotal (my patient referral numbers were <50), were impressive.

Patients should expect the most able doctors to treat them, although most will have a problem judging which doctors deserve this distinction. The performance figures published for doctors can be misleading. For instance, can one achieve exemplary figures using ultra-selectivity: treating only patients with the lowest risk.

Only two of my post-screening patients suffered a heart attack in four decades. My affluent patients were, of course, unwittingly pre-selected. They were mostly wealthy, lean, fit, non-smokers. Preselection by socio-economic class, will have affected the results of my cardiac screening policy.

In the 1960s and early 1970s, before routine CABG and the first use of prophylactic aspirin, heart attacks and death from coronary atheroma were commonplace monthly occurrences in GP practices. We then advised weight reduction, smoking cessation, a low-fat diet, and beta-blockers. Many post-infarction cases were anti-coagulated with vitamin K antagonists. CABGs, angioplasty, aspirin, 'statins', ACE inhibitors, and beta-blockers all became available later, each with a significant impact.

To what extent my screening policy for detecting patients with atheroma led to reduced cardiovascular morbidity and mortality, I do not know. My observations were anecdotal, and my cohort included too few patients. An important question remains: can the pre-symptomatic detection of atheroma and ischaemia, improve cardiovascular morbidity and mortality? Although it sounds logical, it will remain controversial until a large-scale study provides the evidence. Since cardiac screening is expensive, it is most unlikely to be adopted by the NHS, and many high-risk, asymptomatic patients will remain undetected. Private patients will thus keep their advantage.

The medical scientist who proposed the 'Polypill' (aspirin, a 'statin', a beta-blocker, and an ACE inhibitor) suggested that we should all take it prophylactically. His claim was that a nation-wide reduction in morbidity and mortality from CVS disease would result. When I spoke to him, he told me that trials in India were underway. As a physician, treating individuals rather than entire nations, my decision was to treat only those with atherosclerosis.

Although death from heart disease has reduced by 40% since the 1960s, there are many more significant factors at work than early detection. Smoking reduction, dietary change, a gradual increase in vitamin intake, and gymnasium attendance, have each potentially contributed to the reduction. Expecting an advantage from cardiac screening (or any intervention that may follow), is analogous to assuming benefit for cyclists who wear safety helmets, or for children provided with seat harnesses in cars. All are **BLOB** suggestions (of **BL**indingly **OB**vious benefit). Put in academic terms, they all have a *high antecedent probability* of benefit, and *a low index of suspicion* for fooling us.

*Even if one has inherited a '**BLOB** detection gene', and is awash with common sense, the burden of proof still rests on the shoulders of those who proclaim benefit from any new investigation or clinical intervention.*

Many patients believe that the medical profession has a duty to do what it can to benefit them, and to improve their health, regardless of what it takes. The public saw some of this in action during the COVID-19 pandemic. Every evening, many honoured the medical profession by clapping publicly.

Because our professional duty is always to scrutinise the scientific evidence thoroughly, new interventions are slow to be sanctioned. In dire emergencies, with a novel danger like COVID-19, this duty had to be relaxed. We had to take chances to save lives and reduce the morbidity of those who remained alive. In 2022, HM the Queen awarded the George Cross to the NHS. The award recognised the 'courage, compassion and dedication' of staff during the pandemic. They deserved it.

Screening strategies should detect inherited and other risks; the results can them help to direct personal strategies for prevention. At the moment, this only rarely includes genetic evaluation and modification, but it should. Screening asymptomatic patients allows the early detection of serious disease with some potential to improve the outcome. If no treatment is available for those conditions detected, is there any point? Although we cannot at the moment benefit those with pre-symptomatic MS, HOCM, or osteoarthritis, we can build databases of patients who might benefit in the future.

I have only seen a handful of conditions where no treatment was available and thereafter became treatable. Although not within my working lifetime, imagine the difference to those with coeliac disease, pernicious anaemia, diabetes and pneumonia, after introducing gluten-free diets, B_{12} injections, insulin and antibiotics.

In my lifetime, I have witnessed some equivalent magic: H_2-receptor antagonists for peptic ulcer; $5HT_1$ agonists that made migraine bearable, and the part reversal of the inflammatory process in rheumatoid arthritis by anti-TNF, monoclonal antibody (infliximab), pioneered by Prof. Ravinder N. ('Tiny') Maini at Charing Cross Hospital, London. In my field of work, the benefits of CABG, balloon angioplasty, and coronary artery stenting for angina were no less game-changing.

Because there are now many effective cardiac interventions available, cardiac screening is likely to benefit millions of patients worldwide. Because cardiovascular disease causes half of the middle-aged deaths in the western world, cardiac screening has numerical, political, and economic advantages.

My Cardiac Prevention Policy

For many years, I advocated that all my patients over the age of thirty-five, should have their carotid (neck) arteries scanned for atheroma; especially if they had a family history of cardiovascular disease. Post-mortem studies have shown atheroma to occur more often in the carotid arteries than elsewhere, and for coronary atheroma to be present in only a minority of cases without carotid atheroma. There is an important clinical implication to this. If a patient presents with chest pain and no carotid atheroma, they are unlikely to have coronary disease as the cause.

If I found carotid atheroma was present, my recommendation was always to perform an exercise ECG. If the ECG remained normal, and the patient showed no evidence of IHD (shortness of breath or chest discomfort), I would measure the rate of growth of atheroma over the next two to five-year period. If the ECG was abnormal, but the patient had no symptoms, I would next request a CT cardiac calcium score or CT angiogram. If the patient developed angina and had abnormal ECG changes on exercise, I would proceed directly to coronary angiography. Those with a high calcium score (>400) I offered CT angiography. Although expensive, this patho-physiological-based strategy, is more predictive in individuals than any indirect measure of coronary disease using blood lipids (second guessing the existence of atheroma-promoting genes; except for a low HDL-cholesterol), fibrinogen and inflammatory markers.

In higher doses than that which lowers blood cholesterol, I have seen the progress of atheroma halt while the patient was taking a statin drug. I observed this several times over a twenty-year period. I used the rate of progress of carotid atheroma as a measure of treatment success, especially since one can acquire carotid images so easily. This simple screening test could provide a more reliable predictor of cardiovascular morbidity and mortality in individuals than any blood lipid measure. Because the lowering of blood cholesterol by 'statins' relates to altered liver function, taking it as a measure of reduced cardiovascular risk in an individual may not be justified.

Because blood cholesterol does not accurately reflect intimal cholesterol production, we might expect total blood cholesterol to be useless as a specific CHD risk predictor in individuals (only 60% of those with atheroma have a raised total blood cholesterol; more have a low HDL cholesterol). Paradoxically, mean total blood cholesterol in populations is a reliable predictor of cardiac infarction prevalence. The difference in reliably predicting individual, rather than group occurrence, is a 'statistical paradox'. I discuss this further in the chapter on clinical decision tools.

Worthy and Unworthy Investigations

A clinically worthy investigation is one likely to advantage patient management. An unworthy test is one will have no such potential, so why request it? Clinically unworthy tests are done 'out of interest', for research, and for medico-legal reasons. All are common in practice. When asking for their fully informed consent, any distinction between worthy and unworthy testing should be clear to patients.

Risk and cost are other aspects of worthiness. Ideally, a worthy test will involve a level of risk, that will match its clinical need. Although NHS bureaucrats will want to know the monetary cost of tests, doctors need to know the clinical cost of requesting or not requesting them.

In some places (not in my practice), the financial cost of tests will limit the availability of testing, regardless worthiness. This can apply to patients funded by private medical insurance. Patients may need to get their insurer's confirmation for being tested. Medical insurance rarely covers medical screening tests. Perhaps insurers believe they are not cost effective, or an unnecessary financial expenditure. Most insurance companies are generous, however, and rarely stand in the way of clinical management; they reserve the right to cease insurance cover on annual expiry, should they wish.

The worthiness of a test is not always clear cut. We often need to do tests that exclude a vague diagnostic suspicion: a justifiable strategy in challenging cases. However unlikely, we have a duty to ensure that we are not dealing with hepatitis B or C; syphilis, HIV and AIDS, or TB. Diagnosis by exclusion may be the only way to proceed. *'Are you sure he hasn't got... ?'* has always been a favourite question of clinically detached, armchair physicians. Although one cannot prove a negative, it is a brave doctor who dismisses this question completely.

To challenge opponents effectively, know their arguments better than they do.
Charlie Munger.

As a junior doctor, I did whatever investigations I thought necessary, including many of academic interest. In that era, the dividing line between clinically indicated tests, and those for research, was not always made clear to patients. I mostly did cheap and harmless physiological manoeuvres like carotid artery massage, and Valsalva manoeuvres, while running an ECG rhythm strip, when assessing a patient's autonomic responsiveness. No danger was involved. At the time (1960s and 1970s), most patients correctly regarded doctors as altruistic, honest, and trustworthy, and were content to think that whatever research they did, would benefit others (even if they did not understand the clinical relevance). Many patients gave *carte blanche* to some tests, with little regard to personal risk or benefit. This was especially true, if they thought it might advance a young doctor's career. Trust was not an issue. The thought of continuing in the same way now, could make medical bureaucrats and regulators apoplectic.

Investigations: Risk and Benefit

I attended a 29-year-old man as an emergency. His was the son of wealthy Asian parents with a successful business background. He had drunk over one bottle of vodka every day for several years. He was semi-comatose and jaundiced, had multiple spider naevi and an enlarged liver. I admitted him to a private hospital liver unit as an emergency (private doctors could not admit patients directly to NHS hospitals). From there, they transferred him to an NHS liver unit.

Despite his acute toxic state, and history of alcohol consumption, doctors in the NHS liver unit performed a liver biopsy as an emergency procedure. Had they considered:

- *What chance was there that a biopsy would reveal anything other than alcoholic cirrhosis?*

- *Was it done to exclude an unforeseen cause of hepatic failure;*

- *Was it done simply as a matter a protocol, or as a teaching exercise?*

They chose the trans-venous biopsy method, rather than a traditional percutaneous one, to minimise the risk of blood loss. He died 24-hours after a difficult procedure. The doctor in training who performed the procedure, had ruptured his inferior vena cava.

The history of excessive alcohol intake (confirmed by his intelligent parents), made any cause for his liver failure, other than alcohol, very unlikely. Surely, the priority should have been to detoxify him and get his liver function stabilised first. Why the urgent need for a biopsy?

At a subsequent Coroner's Court, the consultant hepatologist in charge of his case, did his best to deflect any criticism of his management. Having listened to his explanation, I regret to say that both the coroner, and I, suspected that his decision to biopsy fulfilled a teaching need; he allowed a junior doctor to practice transvenous liver biopsy. The Coroner would have been within his rights to conclude that professional ineptitude had occurred. Doctors, however, need only 'do their best' (whatever that may be). As long as they work within published guidelines, they can expect no punitive legal action against them. Since the junior doctor in question was trying to 'do his best' (albeit, not well enough), the coroner's decision had to be that an unfortunate mishap had been responsible for the patient's death.

All would-be interventionists have to start somewhere. Even with an experienced tutor in attendance, a little luck and some talent are essential. This did not excuse the decision to biopsy my patient. One can never justify performing a dangerous test, when there is so little chance of it changing the subsequent management. When someone dies because of an unworthy test, it is unforgivable.

All interventionists in training will come across situations where they do not agree with their superiors. If they question the decision in the patient's notes, their career could suffer. The question then becomes, which is more important – risking the life of a patient or risking their career?

In the 'good old days' (1970s and before), three senior consultants ('three wise men') would have interrogated the consultant who ordered the above liver biopsy. The only fair way to review a doctor's actions is to employ doctors with lots of experience in the field, not regulators, managers or others without extensive medical experience. This is not what happens. Coroners have legal and medical degrees, but do not always have enough specific medical expertise. When the GMC (MPTS) judges doctors, allegations will come before lawyers, members of the public, and GPs rather than hospital doctors. Third party 'referees' supply reports as medical experts, but do not have to know the patient or the doctor on trial. The lack of contextual meta-information, can be crucial to full understanding.

The greater the invasive investigation risk, the better must be the justification. Even with complete justification, a few unfortunate patients will die being investigated. A GMC (MPTS) tribunal composed as above, might conclude that the investigating doctor had a cavalier attitude, was reckless and lacked insight. Many doctors are now aware of the inexpert ways regulators may judge them and have an acute awareness of every risk to their career. Defensive doctors may avoid harm, but what good are they for patients who need intervention? In emergencies, we need men and women of action, not those made catatonic by the thought of risk. Many doctors in the UK are now in such fear of allegations being made against them, they have become inhibited in order to safeguard their career. Is this in the public interest?

I once performed a routine cardiac catheterization on a young woman in Amsterdam. She had multiple heart valve problems and needed major cardiac valve surgery. In such cases, our department routinely performed a pre-op coronary angiogram. We did not want to miss coronary disease. The routine coronary angiogram proceeded well until a dye injection showed a strangely persistent pattern. The dye had become static, rather than flowing away as usual. Something untoward had happened. Soon after, she developed chest pain. My colleagues and I concluded that her left coronary artery wall had become dissected. She was stable, but there was no time to waste. We rushed her to an operating theatre for an emergency coronary bypass. Unfortunately, she died several days later.

After something like this, I reviewed my attitude towards fixed protocols and 'routine' procedures. At her age (28-years-old), she was very unlikely to have had significant coronary disease, yet I was required to follow the prescribed departmental protocol. All such protocols, while mostly sensible, should allow exceptions. Few working with them, have the courage to countermand them (the chance of death from a coronary arteriogram was one in 20,000 at the time). I wish I had objected.

Cardiac catheter technology has improved considerably over the years. Cardiac catheters are no longer as stiff (except in the rotational direction, which helps to guide them into position). The death of my young patient resulted from using a catheter that was too stiff; stiff enough to puncture the intima of her proximal left coronary artery. In forty years of performing cardiac catheterisation, this was the only time I caused a coronary artery dissection. It didn't stop me catheterising other patients - clearly the characteristics of a cavalier doctor with no insight. Some detachment, and an ability to take such risks, are prerequisites for anyone considering invasive cardiology as a career. Unfortunately, those called to sit in judgement on doctors, will only rarely have any comparable experience of risk to offer.

Patients Requesting Investigations

Private patients eager for a diagnosis, will sometimes try to bypass the consultation process, and request what they think are appropriate diagnostic tests. Few will understand much about investigations, or their diagnostic accuracy and significance. Allowing people to vote on subjects they know little about (like BREXIT in the UK), may be politically expedient, but naïve when it involves life, death, health and disease. Fortunately, not all political issues in peacetime, are matters of life and death.

In the early part of the 20th century, well before the NHS began, many new immigrants entered the UK from Eastern Europe. In order to save money, some families pooled their morning urine and took the combined sample to their doctor (before the NHS, they had to pay for it). If this sample tested normal, the surrogate 'patient' would return to the family with good news – the entire family was healthy. It was a way of saving money. Unfortunately, they based their conclusion on a test with too little diagnostic accuracy for defining perfect health.

The ill-informed will make more mistakes when the subject is technical. It is typical of empires in decline to allow greater degrees of freedom, and to take all opinions seriously. Condoning absurdity and ignorance will, of course, hasten the collapse of every empire. Respecting ignorance may be politically correct, but it is unintelligent. Allowing medically ill-informed patients to dictate their own investigations is fraught with danger. Rarely, however, some will have done some diligent research and will request the correct investigation.

I received a request from a patient's father. 'I would like my son to have his abdomen scanned. Can you help?' I asked what he was hoping to find. He had no particular diagnosis in mind, but thought it might help to reveal one.

He couldn't have known that most investigated for abdominal complaints end up with a diagnosis of IBS. Because colonic spasm is a common cause of pain, rather than disease, investigations mostly prove negative. Colonic gas accumulation can cause the pain, and pumping air into the colon during colonoscopy, can reproduce the symptom. For this patient, an ultrasound investigation would most likely have proven inconclusive. What would the patient's father suggest then? He was not to know that the sine qua non of all reliable diagnoses is to take a full history, followed by a full clinical examination (having thoroughly reviewed any past clinical notes). His reason for bypassing these fundamental steps, was to save money (and avoid unnecessary inconvenience).

One or two further questions put to the father, led to a consultation. His son had been febrile. He had several small ulcers on his soft palate and had inflamed tonsils. Although he had a negative urine test, he had left loin tenderness (possible renal tenderness, and perhaps pyelonephritis; although, rare in adolescent males). It was his GP who had requested abdominal and renal ultrasound scanning, and a chest X-ray.

Given the likelihood of either bacterial tonsillitis, or infectious mononucleosis, I took a throat swab and blood for EB virus antibodies. A chest X-ray was irrelevant, and abdominal scanning could await the results of the throat swab, and EB virus antibody titres. He proved to have acute glandular fever. The abdominal and renal scans done later proved negative. Further urine tests were all negative, and a chest X-ray could not be justified.

The more ill-informed the patient (or his spokesperson), the less likely are they to listen to recommendations. Trying to teach them can be a waste of time. Arrogance and ignorance sometimes combine to resist learning and change. Any attempt to persuade them that, 'they don't know as much as they might', could seem 'rude', 'unhelpful', and even 'objectionable' (meaning they have failed to get what they want). Like spoilt children, they have a tendency to shout and stamp their feet. Sooner than admit ignorance or loss of face, they will often seek another opinion.

Never let a patient make you fail by pandering to their ignorance.

Husband: *'I don't mind dying. I'm ready.'*
Wife: *'That's because you are too stupid to understand the implications.'*

Film: *Café Society.* Warner Bros. (2016). Written and directed by Woody Allen.

Great Expectations

Both doctors and patients place a lot of faith in the diagnostic power of medical equipment (ECG machines, scanners, etc.). This has inspired a tendency to prefer investigation over clinical appraisal. The convenience of investigations can lead to errors. Consider the thought that ECGs are always valuable. Would it be patronising to explain that plumbing problems (coronary artery disease), are best dealt with by plumbers, not electricians? ECGs can detect electrical defects: rhythm disturbances (SVT, AF and VE's), and conduction problems like bundle branch block and AV block, but they are not reliable used for diagnosing plumbing problems like valve defects and the blood flow restrictions of CAD.

Interpreting ECGs has a parallel in tea-leaf reading; reading something into changes, based on pattern recognition alone. A major problem for those inexperienced at reading ECGs, is the misinterpretation of normal variants and peculiarities (not abnormalities). Those questionable ECGs I reviewed, mostly showed variations of normal, not abnormalities. It is common for physiological bradycardia, for instance, to be regarded as clinically significant, and j-point elevation and other physiological ST segment changes (as in hypertension), to be mistaken for ischaemia or acute cardiac infarction. The ECG in hypertension with LVH can yield either a normal result (a common false negative), or one that suggests ischaemia (a common false positive). Correct ECG reading depends mostly on recognising the true positives and as many normal variants as possible.

Patients requesting specific investigations is becoming commoner. Despite a full explanation of what an ECG does, and what it can diagnose, some patients will still request one, thinking it can diagnose every heart problem. They will go elsewhere if denied. That 'the customer is always right' relates more to satisfying vanity and commerce, more than science. With a positive predictive accuracy of 80% (positive, and negative), an ECG will miss coronary heart disease in 20% of those who have it. Because of so many false positives and negatives, one cannot use an ECG as a stand-alone test for coronary artery disease. A doctor providing an 'on demand' ECG service (without clinical corroboration), should expect court appearances.

Patients Refusing Investigation

For decades, a few famous people asked for my medical advice. One was a famous comedian. Although he had many medical problems, he professed to be 'as strong as a horse'. He had become a successful film and TV actor later in his career. For decades, he refused to have cardiac screening tests (a carotid artery scan, treadmill exercise ECG, with CT scanning to follow, should they prove positive). He claimed to be asymptomatic, even though an overweight smoker.

He sent his wife to see me with chest discomfort. Three-vessel coronary artery disease caused her to have classical Heberden's angina, so I referred her for CABG. Not long after the operation, her husband died suddenly. Understandably, but unfairly, she accused me of being less than diligent in missing his heart disease. In private practice, patients can refuse testing for many reasons, one of which is fear of the consequences. It shocked her to learn that I had suggested testing him many times; suggestions he refused many times.

The Pitfalls of Investigation

Some of the many pitfalls of investigation I have encountered are:

- Inappropriate investigation.
- Investigation result errors.
- Reports on the wrong patient.
- Investigations lacking diagnostic accuracy.
- Irrelevant 'red herring' results (laboratory errors, etc.).
- Misinterpreted variations of normal ('peculiarities', not 'abnormalities').
- Overlooked abnormal results (filed before review).
- Failure to spot clinically significant results (human error).
- Overestimating and underestimating the clinical significance of results.
- Misinterpreted results.

Death can result from clerical error. I have known misplaced and inappropriately filed results, to cause the death of a few patients.

A relative of mine had a cervical smear which, in retrospect, showed evidence of cervical cancer. Unfortunately, the gynaecologist involved did not review the result. He went on holiday soon after taking the smear. While he was away, his secretary filed the abnormal result. What was to prove a fatal error, only came to light one year later, at her next scheduled consultation. She died of metastatic cancer not long after.

We sometimes order inappropriate investigations based on an incorrect assumption of potential significance (a skull X-ray for migraine headache; a shoulder X-ray for pain that occurs only on effort). Some doctors order tests with no more justification than, 'in case we miss something'. As a result, it is surprising how often an inappropriate test will uncover an unexpected diagnosis. Since atypical presentations are one reason for diagnoses being missed, we sometimes need to rule out the unexpected. Accepting that one cannot always be sure of the diagnosis, and sensing that an error is being made because of something missing, is an art not science. We must verify or refute our reasonable suspicions.

Regrettably, the wrong patient sometimes gets tested, or we attach the wrong name to a request form or blood collection tube. Those committing these errors have a duty to be open about it.

Some tests have poor diagnostic accuracy. The PSA test belongs to this category. The test lacks specificity and is over-sensitive. We should regard any test that needs further corroboration as inadequate. A patient with a raised PSA will need an MRI scan to further evaluate its significance.

'Red herrings' occur all the time, especially in laboratory blood testing. A raised potassium, ESR, WBCs, C-reactive protein (CPP) or liver enzymes, all cause patients and doctors to worry more than necessary. They can be of significance, of course, but trying to explain false or insignificant abnormalities to a patient, can be a challenge (many assume that medical science is exact, and interpretation straightforward).

For decades before I qualified, a raised ESR (now partly replaced by the inflammatory marker, CRP) implied hidden inflammation (like TB, or chronic pyelonephritis). Chasing the reasons for a raised ESR, often proved fruitless in the short-term. A constantly raised ESR (>30mms/hour) in the longer term, however, was another matter. I sometimes observed this, decades before a slowly progressive chronic disease appeared. One patient I had, eventually developed rheumatoid arthritis; I found another with low-grade chronic pyelonephritis.

I had a patient who for years repeatedly said: 'I never feel quite right'. After many years of reviewing her, and finding nothing abnormal except for a slightly raised ESR, she developed microscopic haematuria. This led to the diagnosis of a chronic renal abscess. The ESR is a sensitive test of inflammation, but lacks specificity.

An up-and-coming demand for total body screening will doubtless yield many contentious results. Doubtful results can frighten patients and lead to further, unnecessary testing. Doctors who disagree with screening, use this as an argument against it. Some think of it is a money making exercise; pandering to the rich and worried well. A counter argument I have never chosen to voice is: *better worried than dead!* More often, the results of screening reassure those worried about themselves (albeit, sometimes falsely).

Routine MRI scans can detect anatomical peculiarities, as well as important pathology. To discern the difference, one needs the art of interpretation. Beware of the sometimes ambivalent, defensive, but understandable conclusions, made by our imaging colleagues. They have to protect their medico-legal position. Radiologists can be at a distinct disadvantage, if they have insufficient clinical involvement with those they image. They sometimes protect themselves by being intentionally ambiguous and non-committal. Their backstop position can be: 'X, Y, and Z are diagnoses not excluded.'

In first world countries, chest X-rays are now used much less. They have always been a common source of normal variant errors; various 'normal' features, being wrongly misinterpreted as pulmonary oedema, pneumonia, TB, or early lung cancer. After being introduced, CT scanning much improved the situation, although not yet able to match a tissue diagnosis.

Failing to recognise a meaningful abnormality and its clinical relevance, can result from a lack of knowledge and inexperience. Sit in on as many clinical meetings as possible. This is where the discussion of clinical data and its interpretation leads to clinical judgement. Because pathologists come with all the answers, regular attendances at clinicopathological conferences should be obligatory.

Chapter Twelve

Clinical Process

*The first three steps of successful clinical management are:
diagnosis, diagnosis, and diagnosis.*

For competent physicians, the odds of disease occurrence cannot be ignored but should not bias the diagnosis. The odds of occurrence are most useful for the inexperienced and those whose aim is to dismiss a diagnosis.

All doctors must be comfortable working with uncertainty, incomplete information, and unpredictable outcomes.

Whenever a diagnosis is illusive, ask whether the history taking or examination was minimised. If so, re-take the history and ask the patient's friends and relatives what they think.

These strategies are sometimes indispensable to making correct diagnoses, and devising effective management plans.

Diagnosis is not the sole preserve of doctors; veterinarians and nurses, police detectives and fault-finding engineers, use the same processes. The process relies on knowledge, logic, choosing relevant data and making associations ('nebula thinking'), as well as understanding motives and causation.

Nebula thinking is key to every pie maker. Only those capable of choosing the best fruit from an orchard of trees and bushes, will excel.

If you find human (psychiatric) diagnosis difficult, become a veterinarian specialising in chameleons. Chameleons turn yellow with anger and grey with depression. Human diagnosis is usually less obvious.

Common and Rare

Although rare, lightning can strike the same place twice.

Lightning struck an American State Trooper thirteen times. His non-conductive rubber boots, saved his life on twelve occasions. On the thirteenth, he died because he changed his boots to electrically conductive leather ones. The change, facilitated his electrocution.

Although being struck by lightning is very rare, I doubt this State Trooper thought so. On the first occasion, he might have thought it terrible luck; thereafter, he must have thought each strike was a 50:50 chance (either it would, or would not happen again). I had one patient, who experienced something as rare. He developed three different cancers, one after the other: first tonsil cancer, then tongue carcinoma, and lastly prostate cancer, all within a three-year period. With unlikely events, the Murphy principle may apply: that *anything that can go wrong, will go wrong* (attributed to Edward A. Murphy Jnr., an aerospace engineer who worked on safety-critical systems).

In one lifetime, we will all experience what we perceive to be good and bad luck.

Clustering and Entropy

Although not usual taught as medical concepts, these physical processes, better known to mathematicians and physicists, can have clinical implications.

Look into the night sky. Every entity in the universe clusters. Stars cluster together in galaxies rather than being evenly spaced. The same applies to some specific clinical conditions: they can occur sporadically or cluster into groups for aetiological reasons. Wagner and Sleggs (1960) in South Africa, were first to relate asbestos exposure to mesothelioma. Many who developed mesothelioma, clustered around Barking in East London. At the centre of the cluster, was the Cape Asbestos Factory which closed in 1961. Many mesothelioma cases became recognised in the 1960s and have been the subject of litigation since. In 1854, Dr John Snow identified cholera cases living around the water pump in Broad St., Soho, London. These clustering phenomena lead to significant discoveries.

When clustering occurs around timed events, like bats flying only at night, and pulse rates rising during the day, there will usually be a straightforward explanation. Other clustering is not so easily explained. Why gamblers experience runs of good and bad luck, is not obvious. Because clustering is a universal natural phenomenon, diagnoses, house prices, share prices, and the weather all demonstrate it. When a patient has several attacks of palpitation at a certain time of day, try to find out why.

There are important considerations that apply to scientific results:

- One can overlook clustering because most group statistics summate and average the data.

- When any thesis gets independent scientific results, all clustering around the same conclusion, it is likely to be correct.

In rare clinical cases, with specifically clustered information, the specific comments of an expert can be more useful than referring to any database or reference work. When stuck for a diagnosis, ask someone with lots of anecdotal experience. Their perspective can be invaluable.

Do events occur in threes, like buses arriving together? Yes, they do. If you have seen one rare case of XZ disease, might you see another case soon after? Yes, you might. Therefore, remain vigilant. Taken as a whole, 'rare' medical conditions affect one in seventeen people during their lifetime. Although common things occur commonly, rarity is common enough not to be ignored. Bias, however, makes us think rare phenomena are more common than they are; we more readily remember recent, rare, and interesting events.

Biases to be aware of are the *frequency illusion bias:* once we have seen something unfamiliar, it appears everywhere; the *gambler's fallacy bias:* the pattern of past events will reliably predict future events; the *ob-*

server expectancy bias: to see only what you expect to see, and the *Travis effect bias:* the present is more credible than the past. For a list of clinically active biases, see the chapter on clinical judgement.

There is no such thing as the absence of bias. This includes human judgement and roulette wheels designed to yield numbers in a random fashion. In fact, many small biases can create clustered patterns; the explanation will be in the detail.

Entropy is an important concept in thermodynamics and disease progression. Like clustering, this too relates to patterns and randomness. By spending lots of energy, it is possible to take a pile of bricks and organise them into a building. In time, every organised structure (a low entropy state), like a building or the human body, will disintegrate into its component parts. Only a large amount of energy spent maintaining it, will stop this happening. Once all the energy in the universe is spent, only a mist of evenly spaced particles will remain (a state of high entropy; no clustering). Living creatures represent the ultimate in (biochemical) organisation and require large amounts of energy for health maintenance (homeostasis). As the energy supply fails, and with it homeostasis, our cells and organs slide into dysfunction. Soon, illness and death become inevitable.

The rupture of aneurysms and the eruption of volcanos, seem random catastrophic events, that disturb the status quo. To be of any use, science must improve their prediction, and beat coin tossing. As elastic stretches it becomes bi-stable (either intact or snapped), a concept important to the understanding of sudden onset medical events like cardiac infarction and stroke. Whatever the real odds of such an event, patients are liable to see them as 50 : 50 chances (either they will occur or they won't).

Diagnostic Talent and Process

Some doctors, nurses, criminal detectives, and engineers are talented diagnosticians. A talent for diagnosis depends on logical and deductive thinking, lateral and nebular thinking, and an awareness of inductive thinking as potentially misleading. Those who can master these have an advantage, especially when they include appropriate knowledge and experience.

An ability to recognise and weigh each symptom, sign, and investigation result for clinical significance, will facilitate the making of correct diagnoses. Discrimination is a vital function. It must foster the inclusion of pertinent data and the exclusion of irrelevant data. A further attribute of diagnostic acumen is the ability to change focus from minutiae to the bigger picture. Those who are anxious about making wrong decisions, and those who overly obsess about detail, may struggle with diagnosis.

Engineers, builders, and interior designers, all face choices when thinking of solutions. Only the most appropriate choices will lead to satisfactory results: designing buildings that people want to live in, for instance. Errors made on the way, may be impossible to correct later.

When making a clinical diagnosis, there are similar aspects to the choices made, the values applied, and the thinking processes used. Minor errors, made early in the sequence, can lead to incorrect conclusions. Mistakes made early in the diagnostic thinking process, can lead to an ever-widening gap between an imagined diagnosis and the real diagnosis. Mistakes made early on will block the path leading to the correct diagnosis. Sometimes, the only way forward is to go back, wipe the slate clean, and start again. This strategy is worth considering when the diagnosis is illusive.

Information. Information. Information.

As a junior doctor in the 1960s, I found an effective management policy. It was to do several brief ward rounds each day, especially in the early evening. One can see clinical deterioration or improvement from the end of a patient's bed. Never leave these observations to nurses alone; they have their own duties and priorities. This is a doctor's responsibility. This simple, early detection strategy works, and allows some time to prepare the next steps.

Have in mind an *'index of suspicion'* for every diagnosis. Base your investigations on this suspicion (based on points of evidence), and the clinical priority. If you are unsure and are feeling defensive, base your investigations on the likelihood of being wrong.

Expensive, dangerous, and non-routine investigations, are justified only when your index of suspicion is high, and you cannot afford to be wrong.

If you deal with children, never ignore the views of parents.

In December 2014, one-year-old baby William Mead, died from pneumonia and septicaemia. His GP may have missed the diagnosis of pneumonia on each of several visits. The child had been ill for two months. He prescribed no antibiotic for his 'chest infection' (GP guidance was to avoid treating URTIs with antibiotics, just in case bacterial resistance developed). Strict adherence to this policy can erode the value of clinical experience. Most patients get better quicker on a harmless traditional antibiotic like erythromycin, to which bacterial resistance is rare. The reason is, some macrolides are antiviral. There is an important proviso here: GPs should never prescribe the latest antibiotics used in hospitals without good reason.

What followed in William's case was the failure of the NHS '111' call service (the call handlers had no formal medical training) to realise the seriousness of William's case, as reported by his mother. The algorithmic questioning they used did not identify an emergency. One cannot always diagnose sepsis over the telephone, unless one assumes that 'all mothers know best' when judging the health of their children. Emergency medicine is a contact sport. The NHS apologised for the death of William Mead, but did it change its dismissive outlook to parents, I wonder?

In the absence of any clinical acumen, medical bureaucrats can easily remain dismissive of clinical judgement; dismissive of clinical experience, and dismissive of parental opinion. Unfortunately, we have promoted many medical bureaucrats to levels well beyond their competence.

Time and Diagnosis

Once you have a diagnosis, consider the time frame for action. Decide how fast you need to move in order to confirm the diagnosis and start treatment. Give thought to working for any organisation that prevents you dealing efficiently with clinical needs.

An 85-year-old man had been taking NSAIDS for osteoarthritis. He had complained of epigastric pains for the few days before fainting. He had not seen melaena, but was pale and hypotensive on examination. The initial diagnosis was blood loss from a gastric ulcer / erosions. Actions: first contact a gastroenterologist (for endoscopy), admit the patient, start a normal saline iv infusion (sodium can help hypotension), take a blood sample for cross-match, haematology and biochemistry, cross-match three units of blood, institute continuous monitoring, X-ray abdomen (for gas collection) if you suspect perforation; prepare for a cardiac arrest, especially if the patient is frail and deteriorating.

Should you wait for evidence of melaena? Would you accept laboratory delays? Do you wait for colleagues to arrange endoscopy, at their convenience? We all know the answers, but in the UK, junior doctors might have to say, 'what can I do in my position'? Should we accept defeatist attitudes that risk the lives of patients?

Evidence. Evidence. Evidence.

A firm diagnosis must be based on relevant evidence. The stronger the evidence, the firmer the diagnosis.
As a simple rule of thumb, try to have at least three points of evidence to support a diagnosis. This will help convince sceptics, and especially ourselves. Two firm points of evidence, and the diagnosis will seem reliable but in question; five points of evidence clinches any diagnosis. One point of evidence is simply inadequate, and further thinking is required. The trick is to assign the correct points of value to each piece of evidence. This requires one to know just how valuable each piece of evidence is in supporting a proposed diagnosis. Judging this is a clinical art.

- *Never forget to review all the patient's previous notes for background information.*

- *Remember that errors of judgement do not summate: they multiply.*

What valuable 'points of evidence' result from a history and examination? Every question you put to a patient is a 'test', and each answer a result (albeit anecdotal). Examination procedures are also tests producing results. Next assign a 'weight' to each symptom and sign, depending on its known diagnostic accuracy for predicting a diagnosis. With experience, many will come to assign a personal diagnostic value to each symptom and sign, in relation to every diagnosis. Over time, one must re-weigh, up-rate, down-rate and re-affirm, every symptom and sign for its diagnostic accuracy. This heuristic process should always remain active.

To become proficient at diagnosis, one must be able to gauge the sensitivity and specificity of every symptom and sign. Part of the trick is to ask questions that are both sensitive and specific. Although it may serve no diagnostic purpose to ask questions that lack sensitivity and specificity, questions asked 'out of interest', often provide meta-information about the patient. As diagnostic mastery grows, one should be able to judge which symptoms and which signs weigh most in each case. This is one of the most enduring and important arts of medical practise.

There is an important variable here: patient type. All patients are similar, but none are the same. Becoming used to the different ways each patient presents their symptoms, and how they react to direct questioning, is also an art. Some welcome preamble in order to relieve anxiety and engage socially; others want to get straight to the point. Those versed in the art of medicine can adapt their style with every patient. Doctors only need to calculate the approximate odds of a diagnosis being correct; the accurate odds we can leave to computers and AI.

Symptoms and signs are most useful when they are specific. Their sensitivity is another issue. For instance, increased frequency of bowel action is a sensitive symptom of colon cancer, but it is not specific enough to make a definitive diagnosis.

Correctly elicited angina of effort, is specific for the diagnosis of ischaemic heart disease (IHD). Angina, as described by William Heberden in 1772, is the sensation of chest tightness, produced by exercise and emotion, and relieved by rest. At least 95% of patients with 'classic' Heberden's angina (as he described it), will have a flow-limiting coronary artery stenosis. Unfortunately, few with CHD have angina; at least, not until they have a stenosis restricting the lumen to greater than 85%. That makes angina an insensitive symptom for CHD, but a sensitive one for cardiac ischaemia.

To add to the problem of diagnosing angina, is patient variability. Only 20% of patients with actual cardiac ischaemia report angina, so the absence of angina never excludes coronary ischaemia. Not only will the false negative rate be high (no symptoms despite CAD), there will also be many false positives (chest pain that is not angina). Fortunately, many of those with IHD and no angina, have shortness of breath or fatigue on exercise. For all the above reasons, asymptomatic patients with a positive family history of CHD, need to be checked using investigations capable of detecting it (ECG exercise testing, arterial scanning for atheroma, cardiac CT scanning for a high coronary calcium score, perfusion studies, and angiography). Deciding which to use will depend on one's clinical index of suspicion (another art).

'Like the mood of Paris, love, art and faith cannot be explained, only felt.'

The character Jerry Mulligan, played by Gene Kelly in *'An American in Paris',* MGM, 1951).

Take another example of symptom specificity: severe, intermittent, one-sided headache, focussed in one spot (as if hit by an ice-pick), sometimes associated with nausea and teichopsia. This is migraine until 'proven' otherwise. Unlike angina, no routine corroborative tests exist. The historical description, especially when combined with a family history, suffices to diagnose migraine with confidence. Compare these to non-specific symptoms like feeling tired, abdominal discomfort, palpitations, and generalised headaches; all are of weak diagnostic value, but all are important to patients.

On examination, one must look for specific diagnostic physical signs, especially those suggested by the symptoms. If a sign occurs in only one clinical condition, it is pathognomonic. Bilateral exophthalmos in thyrotoxicosis is one; a truly irregularly irregular pulse (never ever returning to a regular rhythm), is pathognomonic of atrial fibrillation.

Also pathognomonic, are sets of symptoms and signs. A specific set can describe a 'classical' case (even though some components are non-specific). Take the myxoedemic set (underactive thyroid): *tiredness, always feeling cold when others are warm, dry skin, voice change (from normal to gruff or lower pitched), weight gain, constipation, puffy facial features, sinus bradycardia and slowly responsive tendon reflexes.* This set is pathognomonic for hypothyroidism and allows a firm diagnosis. To make a firm diagnosis when each symptom lacks specificity, one must weight physical signs together and some more than others. Investigation has two functions here: independent confirmation and the grading of severity. Diagnostic sceptics, and the medico-legally minded, more often rely on laboratory results to confirm or refute diagnoses.

Is it true that with enough experience, every diagnosis becomes obvious? If only that were always true, rather than mostly true. Few will say, even after fifty years of practicing medicine, they are no longer challenged by diagnosis. Such complacency can come with age, but is dangerous. Thankfully for me, 'difficult' patients continued to consult me throughout my career, and kept me challenged. With experience, one should be able to speed up the ruling in, and ruling out of diagnostic possibilities.

Speed-dating has its clinical equivalent. Many doctors like to see how quickly they can make a correct diagnosis.

Can one make a diagnosis before speaking to a patient? Few doctors can resist trying, especially if the appearance of the patient is remarkable and specific. Those who are fortunate enough to have long-term relationships with patients, will more readily observe early diagnostic changes.

An older adult man I had known for thirty years consulted me. His face had changed since I saw him last. Because his face was puffy, and typical of a patient with an underactive thyroid, I started asking him direct questions: 'Has your hair been falling out? Have you put on weight? Do you feel cold when others are warm/hot? Has your voice changed?' He replied 'Yes' to all questions, confirming the clinical diagnosis. I would have been less sure had I not known him well.

A balance always needs to be struck between inept premature conclusions, and considered diagnostic judgement.

The challenging clinical cases published in the New England Journal of Medicine (and elsewhere), are helpful for learning. Their obscurity gets them published. Don't worry if the intentional red herrings deflect you, and you cannot pick up rare associations. The aim is to defeat you and make you aware of your failings. You will only improve once you have recognised the need.

The rarity of 'classic cases', and the wide variance of clinical presentations, make diagnosis challenging. The relevance of clinical factors, and the weight of significance applied to each of them, will forever confound attempts to standardise diagnostic medicine. Those who believe that fixed rules apply to everything are unlikely to become successful diagnosticians, unless they strictly define their field of work. Although it is the dream of medical AI experts and regulators, there will be no replacement for experienced clinical judgement any time soon. In order to practice diplomacy, AI designers now claim that they are providing a clinical tool; a knowledgeable assistant, not a replacement for doctors. Cost will decide the issue in the NHS.

Those who regulate the medical profession have no option other than to use fixed rules to judge doctors and nurses. Only the inexperienced would try to judge art-forms with fixed rules, but bureaucrats will continue to try, given their need to preserve their pay. Guided solely by rules and regulations, they need not test the reliability of their decisions or actions. They have nothing important to lose with no direct responsibility for patients.

Life, Death and Diagnosis

The time dimension applied to diagnosis can be crucial in rapidly developing conditions, such as LV failure and sepsis. In many such cases, one can literally draw a line between cold and warm skin at the periphery (foot to groin, or hand to forearm). It will progress centrally (centripetal vasoconstriction) as forward flow reduces with falling cardiac output, or progress peripherally as cardiac output improves (hands, nose, and feet become warmer). The faster this line moves, the faster the deterioration or improvement.

There are other diagnoses with life and death implications. Deep vein thrombosis (DVT) is one. Unfortunately, Homan's sign (calf pain on ankle flexion) is sensitive but lacks specificity; calf injury, and a burst Baker's cyst also cause it to be positive. Any suggestion of positivity should prompt a D-Dimer blood test, and a calf Doppler ultrasound study to measure venous blood flow. If pulmonary emboli (PE) have arisen, one might see a raised JVP (only with pulmonary hypertension). With PE, the JVP will only rise when the pulmonary artery obstruction is substantial (therefore, not a sensitive test). A raised JVP has many causes and lacks specificity; true relevance will depend on the history, symptoms, and several other indicators.

Patients surviving PE long enough to tell the tale, will usually have had a brief history of breathlessness (SOB), and a clinical reason for their PE: recent surgery, a road traffic accident or cancer (hidden, or already diagnosed). When you see a raised JVP, and detect a positive Homan's sign, repeat the history and consider a ventilation / perfusion scan or CT angiogram (since CXRs are notoriously insensitive). If the patient does not have a raised JVP, and does not have a definitely positive Homan's sign, walk with them to assess their breathlessness (and any rise in their JVP immediately after exercise). The sensation of breathlessness is subjective and varies a lot between patients. There are several possibilities to consider: the diagnosis is wrong, the patient is hiding some facts; the patient is in denial or is under-rating their symptoms. The under-rating phenomenon (stoicism, and the Superman Complex) is commoner in men, so ask the patient's partner for observations.

I previously referred to some important points about cardiac diagnosis. You will need to become proficient enough to diagnose the most serious of them.

- Can you gauge the degree of mitral stenosis from the length of a mid-diastolic murmur?

- Can you guess the pulmonary artery pressure from the loudness of the pulmonary component of the second heart sound (P$_2$)?

- Can you detect RBBB and LBBB, while listening to the effect of breathing on the second heart sound?

- Can you detect RV hypertrophy (RVH), and left ventricular hypertrophy (LVH) with your hand?

- Can you estimate the degree of aortic stenosis from the carotid artery waveform?

- Can you judge the degree of aortic and mitral incompetence, using cardiac auscultation?

Dr. Paul Wood developed these skills in the 1930s. Dr. Aubrey Leatham later proved them accurate in the 1950s and 1960s, using phonocardiography, angiography, and later with echocardiography. Has echocardiography completely replaced the need for these skills? Can one be a competent cardiologist without them? I leave you to guess what I think.

Assigning Weight to Clinical Evidence

Experience should teach us how many *points of evidence* we should assign to each symptom and sign, relative to each diagnosis. Cough can be a symptom of asthma, but an unusual one in PE; so cough gets one point for the diagnosis of asthma, and *null points* for the diagnosis of PE. Recent onset (1 point), rapidly progressive shortness of breath can apply to both. It is for you to determine, from experience, what notional points (and thus, weight of evidence) you choose to assign. We will deal with the thorny subject of clinical judgement further in a later chapter.

There is an art to making quick, presumptive diagnoses. One can review them while questioning and examining the patient. You may be in awe of those able to do this rapidly, consistently making the correct diagnoses. If you ask: 'How can you be so sure?' you are likely to be told: It's 'BLOB' or **BL**indingly **OB**vious). And so it may be for you one day, with enough experience.

Presumptive Emergency Diagnoses

For humans to have survived millennia, chased by wolves and sabre-tooth tigers, some must have developed an ability for 'flash' assessment. A single observation can register DANGER and an ACT NOW response. On other occasions, a 'flash' assessment may register little risk, and a minor problem to be dealt with later. You will need to engage this fast processing thinking function when first seeing a patient (see Daniel Kahneman on fast and slow thinking).

Timing Issues

'Timing is everything,' said Bob Hope. Born in Eltham, South London, he lived to 100-years-of-age, much of it spent as a comedian in the USA.

Bad timing and bad taste are common. An extreme example of both happened after John Booth shot President Lincoln in the head in Ford's theatre, Washington, DC, in 1865. Soon after, someone asked his wife, 'What did you think of the play?'

Bad timing has not disappeared. War-zone reporters on TV will sometimes ask survivors how they feel, after a fast-jet strike has wiped out their family. Hopefully, most faux pas do not survive editing.

You will need to make the diagnosis of acute conditions like appendicitis, limb fractures, internal haemorrhage, cardiogenic shock, brain damage, acute cardiac infarction, migraine, dental pain, and bone fractures quickly, and get on with the job of relieving pain and stabilising the patient. If they are conscious, question them while helping them. Avoid pointless questions. If the patient loses patience with you, you will lose respect. They might think: 'Who is this doctor who wants to know about my acne when I am suffering unbearable pain?' This doctor's focus is in the wrong place; he may be inept and insensitive, unable to set proper priorities. He might be a fumbler, or a bean-counting obsessive, with an overriding need to finish what they are doing, regardless of the patient's need. For this type of doctor, completeness is more important than appropriateness. I have met only a few such doctors. For the sake of patients, we should promote them or demote them; either way, they are best removed from dealing directly with patients.

If the NHS gets its way, 'general' hospital physicians will cease to exist. That is partly because the medical bureaucrats, lawyers and politicians who control the NHS corporation, are not dedicated hospital doctors, nurses or patients. There may soon be no generalists, only specialists. Even GPs are now regarded as specialists. They are undoubted specialists in office administration and NHS bureaucratic compliance.

Steps in Diagnosis

To emphasise the relevance of clinical data, we can walk through the points raised by a cardiac history and examination, commenting on the diagnostic significance at each step. From SYMPTOM 'S'; PRESUMPTIVE DIAGNOSES 'P' and RESOLUTION 'R'.

'S': Chest pain
'P': Angina or costochondritis (Tietze's Syndrome)? Perhaps muscle injury (common).
'R': Ask: Is it worse at rest, with movement, or does it come only on exercise? Is there tenderness of any costal cartilage or costochondral joint? If not, consider an exercise ECG, and carotid artery U/S for atheroma detection, an echocardiogram or stress echocardiogram. CT calcium score, or angiography. Coronary angiography for those with angina. They are likely to need stenting or CABG.

'S': **Shortness of breath**
'P': Obesity or unfitness? Pulmonary, or a cardiac problem? It can be the first symptom of IHD or heart failure. Asthma? Pulmonary emboli? Severe anaemia?
'R': Is it at rest, or only on exercise? Observe the pulse for AF/tachycardia? Is the JVP raised? Observe the performance on exercise. Does this raise the JVP? Smoker? Signs of asthma; hyper-resonant chest on percussion (emphysema)?

'S': **Swollen ankles**
'P': Dependent oedema? (commonest). Varicose veins in hot weather. Heart or renal failure?
'R': Sedentary history? Signs of varicose veins or DVT? Creatinine and electrolytes? Is the JVP raised?

You must ask further questions and seek the relevant physical signs prompted by the history. With experience, your brain should link the history to the physical signs you need to seek.

The duration of symptoms is important. For instance, long-term SOB is more likely caused by obesity, COPD or heart failure. Acute SOB is more likely caused by the sudden onset tachycardia, pneumothorax, asthma, or pulmonary emboli (consider the context: post-RTA, or cancer). Length of history is an important weighting factor in all diagnoses.

Consider a PHYSICAL SIGN 'S'; INFERENCE 'I'; RESOLUTION 'R'.

'S': **SOB at rest**
'I': Asthma, pneumonia, pneumothorax, severe left heart failure. Pulmonary emboli. Anxiety: hyperventilation is common.
'R': Staring, unblinking eyes, can indicate increased catecholamine drive (anxiety). Tachycardia (AF, etc.), JVP raised? Use of accessory muscles (asthma/ left heart failure). Auscultation: wheezing or crepitations? Percussion for evidence of pneumothorax, pleural effusion, emphysema. CXR, ECG, echocardiogram needed to confirm.

'S': **Raised JVP**
'I': Raised PA, RA, RV pressure.
'R': Right Ventricular Hypertrophy (RVH). Detect it using your hand at the lower left sternal edge.
Over-hydration (CCU). Right heart failure, multiple pulmonary emboli, mitral stenosis. JVP observation allows TI detection, as well as restrictive pericarditis: (not for the inexperienced.)

'S': **Cyanosis**
'I': Poor oxygenation: 'blue bloater', or, a right to left, cardiac shunt.
'R': Usually lung disease in adults. Life-threatening asthma in children, or sepsis associated with failing cardiac output. Cardio-pulmonary failure. Cardiac septal defects. Consider, gas diffusion and other lung function tests. Echocardiography. Blood culture.

An exhaustive list is not possible. You will, with experience, develop your own lists of presumptive diagnoses and ways to resolve them as you proceed from history to examination.

It is always important to make an assessment of the severity of problems and the urgency of management. If a person with asthma is struggling to breathe, don't waste time doing a CXR immediately. A CXR in asthma will only rarely contribute to the immediate management, unless you suspect a spontaneous pneumothorax.

Your most valuable instruments are your hands, eyes, and ears.

The Diagnostic Value of Surveillance

Test results should not easily persuade you. Examine the patient and judge the relevance of all test results before taking action. Does the evidence support an improving or deteriorating picture? Ask the patient how they feel. Don't just look at charts and results, or take a nurse's word for this (unless you are confident of their clinical acumen).

There is a golden rule for effective business management: continuously observe all that is going on. A business owner risks failure when he decides never to visit the shop floor. A Field Marshal who never talks to his front-line troops will lose battles. During the second world war 'Monty' (Field Marshal Montgomery), regularly visited his front-line troops; Douglas Haig, in the first world war, rarely did.

All golf professionals advise the same thing: never take your eye off of the ball.

Sick people need constant review. 'Sick', 'toxic', 'seriously ill', are descriptions that must trigger urgent action, even when there is only a presumptive diagnosis. Learn to recognise these clinical states as soon as possible. Nurses, unqualified personnel, friends, and parents can all be sensitive to patient deterioration. Never dismiss them without checking.

A quick call to the ward asking, *'How is Mr. Q?'* will not always suffice. The opinion received may not be based on a reliable clinical evaluation. If you choose not to review patients regularly yourself, you will make mistakes. Your patient, your responsibility. Senior doctors can make their own contribution once you have recognised the problem and tried to resolve it. Call them the instant you feel out of your depth.

Dr. Hadiza Bawa-Garba found herself in such a situation, when in February 2011, no consultant was available to review the case of a sick six-year-old boy with pneumonia. As a junior doctor, she needed back-up, but none was available. Because the patient died, the GMC wanted to strike her off the medical register. The MPTS disagreed, preferring to suspend her for one year. Nottingham Crown Court later found her guilty of gross negligence manslaughter. She lost her job with a young child to support. UK society sanctions the use of vindictive retribution as a punishment for those capable of saving lives.

Some see calling for help as a declaration of inadequacy, and a loss of face. This is an important cause of failure to function. It may prove difficult, but always consider intervening when a patient is suffering unduly (nurses are often privy to this, and may report it to you). Those most concerned with loss of face will strongly resist accepting help, so you may need to insist on being involved. When you delegate to others, do so only after checking their competence. It is never good enough to assume the competence of colleagues.

Several times in my career, I have encountered doctors unable to cope with emergency procedures, reading X-rays and ECGs, performing cardiac catheterisation and pacemaker implantation. It can also apply to minor procedures like venesection, and venous cannulation for iv infusion. Nurses and doctors must be free to discuss these issues with colleagues, and be free to report incompetence when intransigence or recidivism could risk patient harm or cause unnecessary pain.

The need for immediate action can elude some doctors and nurses; perhaps they are unsure of their abilities, and don't wish to tempt providence. Some have an in-born error – a surfeit of inertia. Others will follow dubious protocols (bureaucrats may insist that their priorities supersede clinical priorities), insisting that form filling and office administration, should take priority. There are many other excuses: too few appropriately trained staff; no equipment; a hospital ward too far away; a work shift about to end; staff hand-over in progress, etc., etc. No excuse is ever good enough. When you hear such excuses, prepare to do the job yourself.

Difficult and Elusive Diagnoses

Always expect your diagnoses to be challenged. A wrong diagnosis will lead to incorrect management.

My late friend, Dr. Alan Gardiner, was famous for his signature diagnosis in difficult cases: 'GOK' or 'God Only Knows!' Having given a patient this diagnosis, he would advise them 'to consult a proper doctor'.

The usual reasons for diagnostic failure are:

- Failure to take a pertinent history.

- Failure to examine the patient fully or meaningfully.

- Incorrect assumptions based on dis-information, misinformation, or a lack of access to reliable information.

- A lack of knowledge.

- A lack of data.

- Mis-judgement: an inability to choose the most relevant clinical features from an array of symptoms, signs, and investigation findings (nebula thinking).

- Applying the wrong weight of evidence to data.

The latter function critically depends on experience and talent.

RAF men posted to the island of Uist during the second world war, needed incentivising. They were told there was 'a pretty girl behind every tree'. What they didn't tell them was there are very few trees on Uist!

My friend Don Russell was once a nurse. After a fall he developed cellulitis around one ankle. He continued in pain for many weeks. Since cellulitis rarely causes pain, the origin of his pain had to lie elsewhere. Antibiotics did not help, but colchicine did. It treated his gout and quickly relieved him.

Mrs. M. presented with intermittent, right subcostal, colicky pain. She had multiple gallstones, but her pain never came after eating fatty food. After gallbladder removal, her pain continued. She had colonic spasm arising from the hepatic flexure of her colon. Treatment for her IBS was what she needed, not a cholecystectomy.

As important as clinical data, is knowledge about the patient's circumstances. When stumped for a diagnosis, re-interview and re-examine the patient. Then consider interviewing any available relative or friend. If they are not available, pursue them for information. All this will give you time to think further, and to review the clinical evidence with a fresh eye. Pursuing theoretical differential diagnoses without being anchored to confirmed, corroborated symptoms and physical signs, often leads to failure.

There is another essential clinical sense that is difficult to specify. This is the sense of something missing (or being withheld); one missing 'clue', sometimes makes sense of it all. All doctors will encounter a eureka moment after one key fact comes to light. The phenomenon is common among experienced researchers, adventurers, and police detectives. As in the fictional cases solved by Columbo, and Sherlock Holmes', a vital clue can hide in plain sight or lie buried within the detail.

Mrs. Cook was an older adult woman, the matriarch of a well-known East End (London), 'Pie and Mash' shop family. Constant itching was her only symptom, but with no rash. She was being treated for hypertension, so I wondered whether her medication might be the cause. There was no history of allergy, and anti-histamines did not help. I interviewed her husband (who was being kept awake at night by her constant scratching), hoping that he might have something relevant to offer. He knew nothing. Desperate to find an answer, I reviewed her very extensive records, which included a lot of laboratory data.

Directly I looked at her haematological history, the answer jumped out at me. Her latest haemoglobin was 19g/l.: very unusual for an 80-year-old woman. I looked back at her previous results, and there they were: gradually rising haemoglobin levels, from 12 to 19g/l over a three-year period.

Was it Polycythaemia Rubra Vera (PRV)? I telephoned my haematologist colleague who promptly told me that my diagnosis was against the odds, and most likely to be wrong. Statistics have their place in defining large group characteristics with random selection in play. There was nothing random about this patient's rising haemoglobin levels, her progressive hypertension, and constant itching. True, PRV is very rare, but radioactive marrow studies later confirmed the diagnosis. My hubris was short-lived. PRV had a poor prognosis.

Our Professor of Medicine at the London, Clifford Wilson, once told our student group, 'Never forget this case of PRV, you will never see another one.' My guess is that few of my contemporaries at medical school saw a case. Believe it or not, I diagnosed two further cases within a few years. Even with odds of forty-five million to one, someone will win the National Lottery. A few have won it twice.

Blood Rhubarb Level?

Diagnostic revelation will occasionally come with one test result.

I had recently qualified in 1966 when a middle-aged female patient of mine (Mrs. K.), had to be supported for many weeks on intravenous nutrition alone. Later on, she became psychotic. She was being cared for in one of the first ITUs in the UK: a side-room at Whipp's Cross Hospital, London, E11. A pioneering ITU doctor, Gillian Hanson, had put together a makeshift side-room for the purpose. She later became one of the first consultants dedicated to intensive care. The patient started climbing the walls, trying to escape the snakes, tigers, and demons pursuing her (none of whom were hospital bureaucrats. All were toothless in those days).

I requested a blood magnesium level. The laboratory found an undetectable level. One shot of magnesium rectified her mental state. The only time I saw psychosis reversed quickly without a psychotropic drug. Measuring blood magnesium levels have since become routine in cases needing long-term fluid replacement, but in 1966 it was a rare request to make.

If you work as a GP, or in A&E, most diagnoses will be **Bl**indingly **OB**vious (B.L.O.B.); a diagnosis that requires only a brief history, and the briefest of examinations. Nose bleeds, severed fingers, tonsillitis, urticarial rashes, are all common examples. There is no need to expand further, except to say that 'BLOB' scenarios occur commonly, and become more so with experience.

B.L.O.B. situations are everywhere and sometimes unintentional. After a river had overflowed its banks, I came across a half-submerged road sign that read: 'FLOODING. ROAD CLOSED'.

Dynamic Diagnostic Situations

In every clinical case, one must compare the current data to any old data. One purpose is to measure the rate of any clinical progression.

What appears to be a static situation may be dynamic. With lots of water running into a bath, and an equal amount running out, the water level will remain the same: the result of dynamic equilibrium. In dynamic situations, disequilibrium can be rapid. In the previous ITU case mentioned (Mrs. K.), hypokalaemia could switch to hyperkalaemia within thirty minutes. An awareness of the dynamic state of patients (and its time course) can be crucial.

Time is of the essence in acute situations. The salvage of brain and heart tissue left unoxygenated, may not be possible after three minutes; before three minutes, tissues sustain progressive damage. The prognosis in CVA, TIA, and myocardial infarction, will depend on the rapidity of diagnosis and the availability of urgent intervention facilities. Rapid intervention can lessen brain and heart damage. Examine the patient on the way to the CT scanner, or catheter lab., to assess their progress.

Try to keep ahead of the time curve. An awareness of the usual natural history (without treatment or intervention) will help one decide what to do, and how quickly to do it. Some pathological conditions have a slow rate of progress: atheroma, aneurysm formation, and LV dilatation in aortic incompetence, can take years to advance; brain haemorrhage, and coronary artery occlusion take only minutes. In all but the most trivial conditions, it is essential to know the 'usual' rates of progression.

If you are to catch a cricket ball after a batsman has struck it, you will need to engage both your cerebral and cerebellar functions to catch it. The coordination and ability needed to judge the trajectory of a ball in flight, have medical parallels; timing pathophysiological processes is similar. You either have sufficient co-ordinating ability to save lives in rapidly developing clinical situations, or you don't. Interestingly, sporting prowess was once a prerequisite for entry to medical school in the UK.

Richard had severe AI after SBE, and an LV that was dilating faster than I predicted (noticeable after only a few months). I explained that there would be no second chance to operate if his LV dilated further; after that point, valve replacement would be of no benefit. For personal reasons, he disregarded these facts, and suffered the inevitable consequences.

Pre-Diagnosis

In the UK, some doctors have long sneered at those dealing with the 'worried well'; once a synonym for private patients.

Should one be interested in pre-diagnosis? After starting screening in 1973, I quickly found some reasons. After the year 2000, each week my cardiac screening facility found one completely asymptomatic patient with a coronary artery stenosis, among twenty others. Some must have pretended to be asymptomatic. In the cases we discovered, cardiac infarction, angina or sudden death, could have occurred without warning. There is a life-saving aspect to cardiac pre-diagnosis. In the vast majority of non-cardiac, routine screening cases, the results are occasionally important, but far less dramatic. Routine screening can provide patients with re-assurance, measure progress, and help decide prophylactic management. There is a problem: cardiac screening is expensive.

Before 1990, most NHS GPs restricted themselves to dealing with symptomatic patients. After that, the NHS paid them to provide 'Health Promotion' Clinics. I had a job, supervising this activity in Brent and Harrow, London. Before this, engaging with the worried well was often thought a waste of time, and even unethical by some. In the 1960s and 1970s, many doctors held this view, but the idea lost ground as patients realised that there is benefit in making early diagnoses. Many doctors still question the benefits, especially since it can induce unnecessary worry. Over forty-five years, I observed that many more were re-assured than worried by screening procedures.

NHS GPs in the UK are now paid to check patients routinely for hypertension, diabetes, and cancer (bowel, breast, cervical and prostate). When I started screening in 1973, many of my colleagues thought my screening program was simply pandering to the rich, and the worried well. That was only rarely true.

Pre-symptomatic diagnosis will not always change the outcome. Early detection of lung cancer using a routine CXR, for instance, saved only one of my patients from a premature death over a fifty-year period (the prognosis was 6-9 months in the 1970s). One patient who survived, owed it to her husband's prompt action. He would not accept any delay in her management.

Lilian Dixon was 65-years-old when she presented to me with haemoptysis. Maurice Dixon, her husband, had run a successful plumbing and boiler maintenance business. He knew about plumbing emergencies, and how important it is to act quickly when saving properties from water damage.

After finding her abnormal hilar nodule on CXR, the surgeon planned a routine bronchoscopy. It was to be done two-weeks hence. 'I would like it done tomorrow', Maurice demanded. It took place the next day. The small upper lobe bronchial tumour was a squamous carcinoma. The private thoracic surgeon involved, then announced he would perform a lobectomy after returning from holiday, two-weeks later. 'When are you going on holiday?' Maurice asked. 'In two days' time', the surgeon replied. 'Then why not operate tomorrow?' he asked. The surgeon tried his best to put him off, but Maurice was having none of it. The operation took place the next day. His wife survived to die twenty-five years later. Maurice paid for all the investigations and the operation. He regarded it all as money well spent.

Did Maurice's insistence on management efficiency achieve a successful outcome for his wife, or was she just lucky? Maurice never thought so. I doubt the surgeon enjoyed being pushed around, but Maurice saw no distinction at all between the reluctant attitude of the surgeon and his employees. His employees would often try to put off to-day, what they could do tomorrow. For Maurice, this was an unacceptable way to run a business.

Maurice Dixon was a successful businessperson, always full of joie de vivre, and a joy to meet. I knew him well before I became a medical student. He used to service the boiler in my father's factory. He loved classic cars, which he serviced himself. He enjoyed two rides most: driving his Ferrari 'Testa Rossa' to the South of France each year, and driving to play golf at the East National Golf Club in his 1960s, desert pink Cadillac de Ville convertible.

Getting diagnosed quickly, and getting prompt treatment, were two advantages enjoyed by my private patients for fifty years. No comparable service was then available in the NHS. In some places, non-urgent patients still have to struggle to get a GP appointment in less than one week. There are vital distinctions between NHS and private services, some of which are related to attitudes and service culture, and some to practice organisation and overload. Private hospitals do not admit all-comers, while all NHS hospitals are duty bound to do so. In addition, I know of no private hospital or GP, with a waiting list. These distinctions continue to account for some of the clinical advantages enjoyed by private patients.

Efficiency and Attitude

At the inauguration of the NHS, Aneurin Bevin's policy of 'stuffing doctors' mouths with gold' must have encouraged some greedy doctors (Aneurin Bevin's intention in 1948 was to promote the NHS), but did it inspire greater clinical effectiveness and efficiency in medical practice?

Does a doctor's attitude to patients and his work affect his clinical efficiency? The attitude that some doctors bring to clinical management in the NHS, can differ greatly from what they show working in the profit-motivated private sector. The primary concern in the private medical sector is patient satisfaction (patients must feel it is worth paying for). In private medicine, a poor reputation could kill profits, and the organization. The same evidence-based approach is essential to both, but there are big differences in how the art of medicine is applied.

When I was a junior doctor, some older doctors had attitudes that pre-dated the NHS. Before that, big business and charitable donations funded hospitals. Before the NHS, major companies like Charrington's and Mann Crossman breweries (together with my paternal grandfather, who was a small personal donor), sustained the London Hospital in Whitechapel. The attitude to patients was, *'You are receiving charity. Do as you are told.'* This attitude persisted for decades. The remnants of it were still present when I qualified in 1966, and a semblance of them remained for decades. I called it 'the NHS attitude' (captured in the TV comedy series, Little Britain, by the phrase, 'The computer says no.'). The ITV series *'Doc Martin'*, played by Martin Clunes (2004, onwards), also reminded me of the dismissive attitude some doctors had towards patients for decades after the 1960s.

I very much doubt that the number of doctors imbued with the notion of clinical effectiveness, differs much between the private and NHS sectors. However, a concerning crossover phenomenon can occur. Some NHS consultants bring their NHS corporate attitudes to private practice. Many are used to delegating and allowing 'the system' to take over. One or two of these NHS consultants, working privately, were guilty of taking too long to act; several of my patients died as a result. They were not at all 'on the ball' (my guess is that they were too used to relying on their junior staff to manage cases).

Diagnostic Intuition

For both the parents of sick children, and attached physicians, the intuitive feeling that a child is sick, or is becoming sick, will often prove correct. Parents need to be listened to because most are sensitive to small changes in their child's behaviour.

At four pm, a child's mother reported that her son Tommy, aged two, was unwell. He didn't look ill, but he was unusually languid. He felt a little hot to touch, but his sub-lingual temperature was only 36°C. Would you believe the thermometer, or the mother's intuition that he had (or was about to get) an infectious illness?

Early in a febrile illness, radiated heat will keep the core body temperature down; patients feel ill, but have a normal temperature. The core temperature rises only when insufficient radiant heat is lost. This mother was correct, as they so often are. Tommy was about to get ill, possibly in the middle of the night.

When I was a young GP, we did house visits. In that era, the middle of the night was a common time for parents to telephone about a sick child. One strategy helped to avoid this. We saw all children when first notified the same afternoon or evening and treated them quickly. This avoided night visits. Now that UK GPs no longer do night visits, and have jettisoned much of the pastoral care they once offered, parents must either get telephone advice from someone working with a checklist or algorithm, visit A&E, or call an ambulance. What other replacements of NHS GP services will follow?

Is there such a thing as diagnostic intuition? It can seem like it on occasions. The older one gets, the more inclined one is to believe in it. There are subliminal diagnostic processes at work: smell, changes in a patient's dress or demeanour, and the grouping of symptoms; all can evoke a diagnosis. These processes are subtle and swift, making the resulting diagnosis seem intuitive. The better one gets at doing anything, the more automatic and intuitive it will seem. You may, of course, be:

> *'Gifted miraculously, through no achievement of your own.'*
>
> Stanley Crawford in *'Magic in the Moonlight'* (2014). Woody Allen. WB Film.

A patient of mine, Carl Marsh, told me in 1984, that he had been 'sent' to me. 'By whom?' I enquired. 'You don't understand', he said. 'I am a spiritualist, and a medium. I have performed séances on the radio; I am also a prison visitor (apart from being an architect). My mission is to give you a book (on enlightenment).' He handed me a copy of the book entitled: 'The Path of the Masters' (The Science of Surat Shabd Yoga), written by a successful California surgeon, Dr. Julian P. Johnson in 1939 (Radha Soami Satsung Beas / Punjab). The author had gone to India to learn from gurus.

Carl invited me to reflect on any unusual, successful clinical happenings I had experienced, suggesting more expertise than I could have expected at the time (a Divine hand at work). He suggested that divine intervention was real, but unappreciated. He suggested correctly that I was unlikely to give up my reliance on evidence, and to take what he said seriously; not until later in life, perhaps. I am still waiting. I liked his ideas at an emotional level (or should it be, at a spiritual level), but have remained sceptical. I will acknowledge that there was something to what he had to teach, but it is a little difficult to describe, except as an explanation for intuition.

Despite the hard facts of life, all doctors and nurses deal with every day, one can still appreciate that the practise of medicine has magical qualities.

Discussing the Diagnosis

All patients should expect a full discussion of their diagnosis, its certainty and implications.

For all patients, knowing the facts, can either settle their anxiety or incite it. The facts can allow a patient to relax and to plan for a healthy existence. They might also adjust better to retirement, or to the consequences of forthcoming treatment. In the latter case, it could help them cope with illness, disability or the possibility of death.

Sometimes it would be appropriate to follow the advice given to schoolmaster Pennyfeather by Dr. Fagan (*'Decline and Fall.'* Evelyn Waugh. Chapman and Hall, 1928). That was, 'to occasionally *'temper discretion with deceit'*. Sometimes it can be auspicious (for the sake of a patient's sanity and comfort), to be economical with the truth. Those skilled enough in the art of medicine will sometimes employ both. Both can help patients maintain their composure and preserve their ability to cope.

From within the doctor – patient relationship, the patient might benefit from some individualised help, aside from referring to the depersonalised generalised guidelines, rules and regulations. Guidelines that benefit the majority will not always benefit each individual. Having dutifully explained the rules, regulations, and compliance issues, many patients came to rely on what I thought they should do. When the rules and my opinion conflicted, those patients who had learned to trust me through our experiences together, would usually accepted my recommendations.

There are prerequisites when considering communication style. The personal, attached approach, works well when one knows the patient and their family well, and one has gained their faith and trust. By definition, this approach will not be acceptable to detached physicians; their dedication to the delivery of evidence-based facts, and their unwavering compliance with rules and regulations, will mostly overrule any need for individual patient consideration. Their usual *modus operandi* is usually to give precedence to the literal interpretation of guidelines.

What sort of doctor will take personal responsibility for the well-being of patients? What would Hippocrates have done? What would Dr. Hunter Docherty, 'Patch' Adams, do? Would they uphold sensitive patient communication? Would they take charge? Would they defend patients against callous, rule-based generalisations, and the 'averaged' prognostications of bureaucrats? I hope so. With doctors and nurses now forced to dance to corporate tunes, those who choose to invent and use their own dance steps, will need fortitude and eyes in the back of their head. There are legions of inappropriately educated bureaucrats ready to accuse doctors of irreverence, cavalier action, non-compliance, stupidity, and a lack of insight.

I had known Mrs. M. and her family for twenty-five years, when a routine blood test revealed asymptomatic, chronic lymphatic leukaemia. Her husband and son agreed it would destroy her composure and well-being for her to learn the truth. She was well for three years until she went for a routine gynaecological operation. The anaesthetist, doing her routine pre-op check, innocently asked, 'Can you tell me something about the leukaemia you have?' Mrs. M. died six months later.

Some attached physicians are brave enough to manage the truth for others, knowing it to be in their best interest. Those who don't agree, may need more moral fibre. 'The truth will set you free', some say. In Mrs. M's case, the medical truth hastened her death.

I sent seventy-year-old architect, Mike, for a routine MRI brain scan. He had been suffering severe headaches for three months. He naturally wanted to know the result. I told him that his brain scan showed multiple plaques, typical of multiple sclerosis (MS). Mike had no obvious neurological signs. When I told him about his likely prognosis, he said, 'I've been lucky in life so far; I suppose we must all develop an illness sometime.' His light-hearted, gracious stoicism, was typical of those with frontal lobe MS plaques.

Presenting Your Diagnosis

It is always instructive to expose one's ideas and diagnoses to others. It requires us to present all the facts concisely. The golden rules are: be succinct, and offer tidbits. Never force-feed others with your conclusions. If you have ever fed pigeons, ducks and swans, you know that one small morsel is enough to attract a lot of interest in others. Allow your audience to form their own conclusions from the data you serve them. As all news media pundits know, the conclusions people reach, will depend on how they select and edit information. One can then lead the inexperienced to almost any desired conclusion.

Imagine presenting the case of a 60-year-old woman with sudden onset of shortness of breath. She had rheumatic fever when fifteen-years-old. (Your audience should now think: rheumatic heart disease, with the sudden onset of AF; perhaps pulmonary emboli). She had no recent operations or accidents (perhaps not a DVT or PE), but had noticed palpitations (likely to be sudden onset AF). On examination, she had a long mid-diastolic murmur in the mitral area (now they should assume she has rheumatic mitral stenosis, and the onset of AF could have caused pulmonary oedema). If your audience has failed to make these deductions from your data so far, they are obviously not cardiologists. You may have to interpret every symptom and sign for them.

Diagnosis in the Future

Medical science is moving rapidly towards defining the genomics of each disease. Disease stratification by karyotyping is the future. Transcription factors (the transcriptome), gene sequences, and mutations, relate to the phenotypic expression of each disease. Sub-types may respond differently to treatment (molecular guided therapy), or be treatable with novel medicines. At the moment, this technology is only available to treat some leukaemias and lymphomas. Further developments should make a form of 'personalised medicine' available to all. 'Personalised medicine' as a general approach, however, will always rely on the art of medicine; without it, one cannot maximise individual patient benefit.

We can regard computer aided diagnosis (AI) as an investigatory test. It could become the darling of NHS management, once corporate finances benefit (reducing expenditure by employing fewer doctors). If they can make financial savings, nothing will hold them back from supporting its use. Computer aided diagnosis will improve with heuristic, neural network programming, with little limit to the amount of data and how it is processed. In a world that progressively assigns the halo bias to computers and denigrates the value of human wisdom, the need for computers to simulate our brains should decline.

Computers use odds-based symptom analyses to suggest diagnoses. They have yielded some interesting *post hoc* results. The current vogue for collecting 'big data' has a problem. It lacks clinical direction and specificity. It is, however, easy to use, fast and cheap. The accepted dogma is: *if you search through enough haystacks, you will surely find a needle.* This inductive reasoning runs parallel to the idea that alien life *must* exist somewhere in the Universe, simply because there are so many stars and planetary systems out there. There can never be certainty about either.

Frank Blake of Cornell University attempted to calculate the likely number of alien civilisations in the Universe. He devised the equation $N = R^* f_p n_e f_l f_i f_c L$, where N is the likely number of technical, alien civilisations, existing in our galaxy at the moment. Some suggest we would need to look at 10,000 planetary systems to find one. Those interested can Google the other terms in his equation.

Ordering large numbers of screening tests for no reason is like casting a net into a pond, hoping to catch fish. It is a random strategy, often used by the desperate to find something useful. It rarely produces meaningful results, but a few get lucky. Beware of the diagnostic odyssey, looking under every stone, hoping to find a clue.

Whether members of the public will gain anything important from the vain concept of the 'digital self': weighing their stools, recording their pulse, breathing rate, and the number of steps they take every day, remains to be seen. Collecting personal data is now big business. Those who sell it are making big profits. It has gone beyond the need for scientific enquiry. Like the content of much social media, it has fast become a narcissistic lifestyle choice for those with enough time and inclination. It will amply feed the desire of every self-possessed, obsessive neurotic. Many will become addicted.

Reliable New Tests Welcome

Be on the lookout, whatever your specialty, for useful new tests. As I write, a new test is being talked about as liquid biopsy. This involves identifying cancer cells in blood samples. By detecting cancer earlier, it could buy crucial time for patients. Will the test prove positive before a tumour is detectable using traditional means?

For twenty-two years, I worked on sub-typing carotid atheroma, and its association with coronary artery disease. Carotid artery ultrasound images allow the distinction of four sub-types: lipid-rich, calcified, fibrotic, and mixed. Lipid rich plaques are vulnerable to rupture and subsequent clot development. We urgently need help from our PET-scanning colleagues to identify those plaques that are metabolically 'active' in coronary arteries; this might help to identify the likelihood of rupture. The calcified ones are unlikely to rupture: they form the non-progressive stenoses that mostly cause chronic, stable angina. The fibrous ones are of little clinical significance.

Many questions remain unanswered about atheroma as a life-threatening process. Is the same histological type, consistent throughout the whole arterial system? If so, should it prompt cerebral, coronary, and whole-body angiographic imaging, once we discover lipid-rich lesions in the carotid arteries? Should all sub-total, lipid-rich lesions, be stented when in critical places?

Since blood cholesterol is of epidemiological significance, but of little individual relevance, we need a better test to assess individual cardiovascular risk. We need research into the value of hand-held carotid ultrasound devices. By identifying atheromatous carotid plaques at the hospital bedside, and in GP surgeries, we could improve the management of hypercholesterolaemia and the diagnosis of those with chest pain.

We must carefully assess the clinical advantages of any new test. Novelty biases us to regard new diagnostic tests as having something special to offer. Introducing ultrasound, CT, and MRI scanning were all special. All helped to assess disease processes and expose previously undetectable disease. They gradually replaced the painful and potentially dangerous invasive tests we used.

If you develop a new test yourself, and discover a new disease entity, you might join the ranks of the eponymous: Addison, Starling, Graves, Bright, and others. The day of the much derided eponym will never end; not until vanity and ego cease to flow through the veins of ambitious doctors.

PART THREE

THE TOOLS OF MEDICAL MANAGEMENT

Chapter Thirteen

Clinical Decision Tools

I HAVE DIVIDED THIS chapter into two parts. The first examines the fundamental thought processes used by experienced doctors when making diagnoses and deciding patient management (supported by anecdote and statistical analysis). The use of computers, statistics, Bayes' Theorem, meta-analysis, cognitive bias; binary, nebular and lateral thinking are all necessary considerations. In the second part, I discuss some specific computer-based processes like programming, the clinical validity of algorithms and big data, which no physician should accept without corroboration.

Clinical decision making has ancient roots, and I will examine these fundamental processes first.

Using computers in medicine is new enough for me to have observed its introduction during my career. The second part of this chapter might not appeal to those with little interest in numeric analysis, computing, algorithms, and their technical justification.

PART ONE

The Fundamental Tools for Clinical Decision Making

Apart from the fascination of discovery, the only value of science to most human beings is to improve prediction. Prediction is invaluable for weather forecasting, and determining chemical reactions, the trajectory of rockets, and the effects of chemotherapy. Although the algorithms and analytical tools computers use to process data, all have devils in their detail, many doctors now use them as tools to support clinical decisions but with little further thought.

If we are not to rely on anecdote alone, Bayes' theorem, trial data, meta-data, and statistical analysis are useful for clinical validation. Accuracy and our liability to error are ever present, and vital considerations.

Patients are real, and statistical analyses are abstract. As useful as they are for providing an over-view with averages and correlations, these are mathematical constructs. Unfortunately, there are problems in the detail of such decision-making tools that get overlooked. This is because mathematical and computer programming analyses are not the interest of every clinician. One result is that many doctors take the calculated significance of published research results for granted without further scrutiny.

Anecdote was once the darling of science. Observations of the stars and planets led Galileo to propose that the Earth revolved around the sun. Observing the variation in animal species led Darwin to postulate adaptation and the survival of the fittest. Scientists now regard anecdote as little more than a figment of the imagination, although often the source of inspiration and the starting point for many hypotheses. An anecdotal observation is only as reliable as the observer, while statistical analyses are only as reliable as the methods used, those who perform the calculations and those who interpret them. Unfortunately, they can all fool us.

Sharp observers do not always make good experimenters. Some take what they have learned from anecdote as blindingly obvious, with no need for further verification. To avoid fooling ourselves, double-blind trials have come to our rescue. To compare a 'control' group to an 'experimental' group, can help challenge overenthusiastic conjectures. Submitting to the uncertainties of anecdote is one thing, resigning one's critical faculties to the results of statistical method is another. Star-struck fans who meet their idol(s) have a similar problem: the halo and selective perception biases cause them to see only what they want to see. Our thoughts and conclusions about anecdote and scientific research results are also subject to other biases: the availability bias, Berkson's paradox, the framing bias, the observer expectation bias, and the pro-innovation bias; all can influence our views (see chapter on Clinical Judgement).

Statistics

The need for evidence-based medicine requires all doctors to understand the meaning of clinical research results and their statistical evaluation. One needs to know the fundamental assumptions of statistics in order to assess their usefulness and reliability. There are many devils in the detail. Without attention to these details, the clinical application of research results could cause patient harm, rather than benefit.

It is important to understand that statistics are meaningful only when patients are randomly selected. The results express probability, not fact. We can define data groups using simple measures, like an average (mean), or the commonest occurring number (mode), the distribution of the data, and a measure of variation (the standard deviation). We can assess the relationship between two groups using co-variance and tests of difference between two mean values (t-test, etc.).

We express the significance of any group differences as a probability. That is the probability that no difference exists (the so-call 'Null Hypothesis'). If the probability of no difference is twenty to one ($p=0.05$), assuming a difference between two means is questionable. When 'no difference' is very unlikely (odds of 10,000 to one), the two groups are most likely to be different ($p=0.0001$).

If 10 out of 85 people like ice-cream A, and 34 out of 120, like ice-cream B, is one ice-cream significantly more popular than the other? One can use the χ^2 (Chi squared) test to answer such questions.

The word 'statistic' relates to a nation 'state,' and various bits of data about it. In 1749, the State of Sweden collected data on births, marriages, and deaths, and produced tables of data: Tabellverket. They showed something shocking: there were only two million Swedes. They might have been easily overrun had their enemies known this. They used the figures to consider how they might change their shocking rates of maternal death in childbirth, and infant mortality. These were the first statistics used to improve life expectancy and population numbers. Statistics later became known as 'political arithmetic', a function they retain.

All statistical analyses involve assumption, with too many confounders to justify uncritical acceptance. For physicians, the many strictures of statistical method exclude the richness of clinical detail. Statistical methods need to define group characteristics, but only within pre-fixed, strictly defined boundaries. Most statistical methods rely on one, almost unattainable pre-requisite – random selection of the entities measured. Because everything in the Universe clusters, like stars in galaxies, computer generated randomness may not reflect reality. Randomness as a mathematical conception is used to test the likelihood of chance happenings.

Medical trials also use control groups, which never exactly match the study groups. When a doctor relies on trial data for decision making, how sure can she be that the patients studied are comparable to her own?

Statistical analysis is possible only for (large) group data. This conflicts with the fact that every clinician works with one patient at a time. We have to live with the fact that we can never apply the results of any large group analysis to any individual. This incongruity I have called the statistical paradox. Is can sometimes be better, therefore, to draw on personal experience for clinical guidance (anecdote), than to rely on research results coming from investigators we don't know, and whose biases, assumptions, and imposed data restrictions, may not best serve our patients.

Meta-analysis merges the results of many outcome trials, and can help direct management considerations. Meta-analysis is even more derived than routine statistical calculation, given the assumptions involved, and the inclusions and exclusions made for analytical convenience. NICE guidelines take such analyses into account when available. Because they are based on the accumulated evidence available, doctors can use them to make the wisest decisions possible. A wise decision is one likely to benefit a patient and stand the test of time.

The Greek word 'meta' has several translations in English (after, besides, etc.). We use it in the term meta-data, meaning information about the data. In a wider context, meta-data can mean contextual, parallel or collateral data. It is that which puts abstract data into perspective.

One can never honestly say to a patient: 'if you carry on eating saturated or trans-fat, research shows you will be more liable to a heart attack.' Based on published research data, this calls for inductive reasoning and might not make justifiable statistical sense. Never apply the constructs of large number analysis to an individual, without many caveats. 'We have nothing better to rely on', some will correctly say. Although true, this may not be reason enough to be seduced by conjecture.

Whether talking to a patient or to colleagues, research results need to be qualified, and put into clinical perspective. It is hardly helpful, but all one can say is: 'if a large group of people all eat animal fat, more of them will suffer a heart attack than those grouped together as vegetarians.' Some patients might point to their difficulty in identifying with any group, or identifying with an average derived from group analysis. Despite these shortcomings, statistical analyses have provided us with many reliable signposts to the way forward (based on the evidence at hand). They can help us decide the likely truth of our decision-making conjectures.

Many patients will see through the illusion of predicting personal outcomes, based on general statistics, even if doctors don't. They appreciate that, since we are all different, no average can represent them as an individual (except by chance). Many of my patients (especially the ones who liked to gamble) knew that statistical analyses can be unreliable when used to predict the results of their treatment. It is our duty, however, to inform them of what we know to be the usual outcomes.

The bodies of evidence that we accept as valid and meaningful for our patients, can help define our clinical wisdom, our clinical expertise and effectiveness, and our individuality in the eyes of our colleagues and patients.

Information Technology

In the same way most of us uncritically accept statistical results, we also accept the output of computer programs (algorithms); even when limited to the data fed to them, and no possibility of handling exceptions (data rejected, or withheld). This also applies to bureaucrats using inflexible rules and guidelines, imposing algorithmic regulation on the rich anecdotal practice of medicine.

All computer algorithms are formulaic and follow predetermined pathways (algorithms), derived from fixed assumptions, and an agreed syntax. AI has added vast data storage and rapid probability calculations to the process, but we cannot ignore the detail when considering their output. Before we accept their applicability to the subtle process of clinical decision making, we need to understand something of the programming. Knowing their limitations, both IT and AI creators have shifted their political positions. They now emphasise that computing serves only as an aid to clinical work. They are eager to point out, there is no intention to replace doctors with algorithms. Like me, you might find it difficult to keep your cynicism in check.

In a back room, somewhere not too far away, there will be medical bureaucrats working with computer programmers. Since it is expedient for all corporations to reduce and control their expenditure, perhaps they must now tasked to estimate the financial rewards of replacing doctors with nurses and paramedics using AI, and neural network driven avatars.

Systems of thought that offer ways to live and survive (how to hunt and gather, and how best to lead life) are the stuff of philosophy and religion. Those who are resistant to thinking and calculating, now let 'apps' do the work for them, with useful apps growing rapidly in number.

Among those capable of thinking for themselves, there are those able to create and propose theories, use logic, calculate, and use judgement to drive human progress. Contrary to the current view fuelled by computer hype (and the Travis effect bias; see next chapter), none of these human faculties came into existence recently.

The simulation of some human brain functions is partly possible by computer, but why bother? There are 8.1 billion human brains already at work (May 2024). Unlike the human brain, computers offer a few advantages other than processing speed. They never tire, but are entirely dependent on an external energy supply. Someday, a machine might pass a Türing test, but until reliability improves, they will remain useful occasional aids, rather than replacements for humanity and human intellect. Computers are most useful for information processing, and aiding those who cannot, or will not, think for themselves.

We have gained time on our ancestors by travelling faster. For this to represent progress, we must have put the time saved to good use: more time to relax or work. That computers would reduce the work of most people was nonsense from the start. There are exceptions: those who handle vast amounts of data, like genetic scientists, astronomers and those who regulate us. In the last fifty-years of computer development, computers have created more work, not less. Along with it, many have lost the pleasure of taking time to relax, think, consider and create (the Romans called this *otium*. It is still a feature of Italian and Greek life).

Is it possible to accomplish more with a digital device than with perspicacity and a pencil? Undoubtedly, digital technology has made calculation quicker and development easier, but what have computers contributed to creativity, and the generation of new ideas? Apart from speaking endlessly to friends, and providing distractions, handheld devices mostly provide us with entertainment and distraction, and alternative ways to do what we once did with pen and paper.

Effective communication, intelligence, science and art, are not recent inventions, yet one IT advert suggested recently that *'a digital device can open up our full potential: bringing forth the intelligence, and artistic ability, you never knew you had'*. Given that a personal digital device now doubles as a fashion accessory, and is essential for creating the 'quantified self', vanity alone will insure its continued success as an easy sell. Every person manufacturing and selling these devices knows that exploitation is much easier with wishful thinking clients.

The digital revolution has been a game-changer for both those with time on their hands, and those with no time to spare. There is now so much information available, the ability to filter it is an essential skill. How many now have data addiction, with an eagerness not to miss anything of importance? There are now data acquisition addicts desperate to avoid disconnection or any corruption of their devices.

A negative side-effect to information addiction is the displacement of other activities - walking up mountains, playing the piano, cooking, canoeing, or engaging in that tiresome task - thinking. Other adverse effects are emerging. Try to list from memory, the telephone numbers of your top ten contacts. If you can, you are among the few not totally reliant on their digital device. More research is needed to find out how our digital devices are affecting our cognitive and emotional functioning.

Speedy data acquisition can keep us up-to-date and improve our decision making. Ordering books and meals, learning about travel delays, finding the nearest garage, the best hotel, the cheapest and nearest clothing outlet, all are useful aspects of modern information technology. Once upon a time, reading maps while travelling, and making multiple telephone enquiries with a paper telephone directory, were arduous processes.

Not so for Kim Peek (portrayed in the film 'Rain Man'). He was an autistic savant. He died from a heart attack when 58-years-old. He could commit entire telephone books to memory. Finding a telephone number for him, was never a pain.

Fast global communication, and inter-connectedness, now allow us to trade shares while at home, and buy goods and services while lounging by a pool. The faster the information delivery, the more profit possible. Others will while away their time twittering, tweeting, criticising, and promoting their personal biases. Some will keep busy playing games, happily addicted to recording and analysing their digital 'quantified-self' with a Fitbit device. How much reading, studying and learning has this replaced, and how many can communicate well by composing grammatically correct sentences that best express their thoughts and feelings, without an app? Perhaps we should know if computerisation has improved, or deteriorated our verbal, written and relationship skills.

A publication on patient 'datafication', 'Fitbit devices', and 'UX' (User eXperience) stated: 'We find that patients' emotions are grounded in negative feelings (uncertainty, anxiety, loss of hope) and that positive experiences (relief, reassurance, safety) arise from getting feedback on symptoms and from continuous and comforting interaction with clinicians.' (see: Pervasive Health 2017: Proceedings of the 11th EAI International Conference on Pervasive Computing Technologies for Healthcare; pages 221-230). It is nice to know that clinicians are at last recognised for their historic role – providing positive experiences like 'relief' and 'reassurance'. How many already knew that physicians bring comfort and positive experiences to their patients? Have they not been doing this for millennia? I wonder how many other old gems of medical practice will be rediscovered as the so called information age progresses? If you want to sell stuff, it's best to claim that it is new, just arrived, and the very latest thing!

Science aims to know and understand everything, but hits road-blocks when it encounters art, philosophy, and religion.

The digital revolution represents a worthy addition to life, because it employs and entertains billions of people. In particular, hand-held devices can keep children amused for hours. We can use them to raise money or organise a riot. During 'The Arab Spring' of 2010, voting became influenced as dissent spread through social media. When speed is of the essence, there are many advantages for doctors, nurses, and patients using their devices. There are also many disadvantages. We can now do our duty and rapidly inform patients when we are no longer available to deal with their emergency.

News alert: 'Patients warned to expect delays due to IT system failure at four hospitals in Greater Manchester. Hospitals in Oldham, Bury, Rochdale, and the North Manchester General Hospital are 'experiencing disruption and instability issues' with digital clinical systems - with one A&E department declaring a critical incident.' Sky News. Alexander Martin 24/5/2022. Similar occurrences are now common in hospitals and airports. In July 2024 a security update disabled most of the Microsoft computers on the planet. The faulty routine update came from the cybersecurity company CrowdStrike. The result was travel and IT chaos for a day or so. Computer dependency is now a major problem.

Faster data acquisition, education, and communication are all possible when hospital computers are working, but only when they are working. All large corporations like the NHS, and US health corporations, have real problems when their computers fail. Ransomware and system failures, can cause havoc with paper records no longer available. Paper records have been acceptable for 5000 years, but no longer. In my small practice of 20,000 patients, I kept paper note files. It never took more than one or two minutes to find a patient's file. In forty years, we never lost one set of notes and never had them hacked.

I did once encounter a problem with paper notes. A GP, CQC inspector, breached my clinical records confidentiality. He inspected some clinical notes without my permission or that of the patients involved. Had I computerised the notes, he would not have gained access. Unfortunately, digital privacy is also at risk. The misuse of digital technology may also prove a limiting factor.

No electronic device has yet replaced creativity, wisdom, nous, talent, or expert know-how; although some claim (for marketing, perhaps) that neural networking will make all these possible. For those engaged in boring routine, using a digital device beats thumb twiddling. Should we regard computers with awe and wonder, assuming that they will improve our persona and quality of life, or are they simply toys for all age groups, the products of slick sales hype worth billions?

We are now at the point (in the western world) where many believe they cannot live without their digital device. Few products in history have achieved this 'essential' status. Among recent ones are TVs, alcohol, hamburgers and pizze, canned dog food, toothpaste, and cars. At the same time millions struggle to get the bare essentials: food, shelter, water, clothes, and shoes.

It's a pity that electronic social media took so long to develop. In ancient Greece, Narcissus could have embraced the technology and saved himself a country walk to find a pool and delight in his reflected beauty. With modern technology, he could have made the entire world envious of his charms without leaving home.

Clinical Reality and Reliability

One must appreciate the difference between actual and contrived data (the product of calculation, processing, or imagination), as they apply to clinical evaluation. A patient's blood pressure is tangible, his medical risk evaluation and likely prognosis, are both abstract and contrived. Much of the information doctors and nurses are now obliged to process, was unnecessary when I qualified, and would have remained superfluous if not for the growth of medical bureaucracy. The efficient running of every corporation requires data, but why should the NHS save money by involving highly trained medical staff in data collection for their audit and control purposes?

The mistake some corporations make is to ignore the personal experience, judgement and function of their employees (doctors and other medical operatives). All corporations value managerial research: how their employees function (work efficiency, outcomes, waiting times, target achievements), and what the resulting statistical analyses and meta-analyses show. Furthermore, they are prepared to pay for it. Unlike other corporations, the NHS requires key staff (nurses and doctors) to collect the data, but doesn't pay for it. All corporations pursue feedback to improve their efficiency, and to police the work employees are doing.

The NHS uses such data to direct overall system functioning, while ignoring the statistical paradox. For practising doctors and nurses medicine, clinical work is all about individuals, not about groups of patients. While this remains, corporate NHS ethos and medical ethos will be in conflict.

Clinical Risk

When throwing dice, the chance of guessing correctly is easy (the fact that there are six sides to a die makes every result a 6:1 chance). Playing with dice can help us appreciate likelihood. One can extend that appreciation by playing roulette with its many more possible outcomes. Unlike many chances we take in life, dice and roulette have fixed odds. With roulette, one can experience how often a 2:1 (red or black), 8:1, 18:1, or 37:1 bet will win.

In life, where fixed odds do not pertain, a feel for risk requires some experience of uncertainty. When we know too few of the pertaining conditions, calculation of the risk involved may seem possible when it is impossible. We all take chance more or less seriously, depending on whether we are risk takers or risk averse by nature. This consideration affects how an individual doctor will practice, and which specialty they are liable to choose: high-risk like cardiac surgery, or low-risk like dermatology.

Medical bureaucrats who try to assess the medical risk a doctor presents to patients, cannot hope to achieve a correct perspective without clinical experience. They will be in the same position as those who play dice or roulette without experience. They cannot possibly appreciate the practical significance of odds. From my personal exposure to lawyers and lay people trying to judge medical risk, what is obvious is their lack of clinical perspective. How are they to know about individual expertise, and how it reduces risk to acceptable levels? What would be the risk if one of them undertook an invasive cardiac procedure? And what would the relative risk be if I undertook the same? They can assess their own risk, but should we allow them to extrapolate that to bias their judgement of doctors? We should leave risk assessments to those with extensive clinical experience, and never to those with none.

Bayes' Theorem

Thomas Bayes (1701-61) FRS, was the son of a Yorkshire priest. His only mathematical work was *'An Introduction to the Doctrine of Fluxions'* (1736).

When seeking clinical guidance about the reliability of any diagnostic test, some knowledge of Bayes' theorem is essential. Four measures are required to calculate the diagnostic value of any test we use They are true and false positivity; true and false negativity. From these, one can calculate specificity, sensitivity, and diagnostic accuracy (positive and negative). We can apply them to any symptom, sign, or investigation result, to assess their reliability in correctly predicting a diagnosis.

Of what value is the symptom of angina as a predictor of coronary arterial stenosis? Identified correctly (as described by William Heberden, 1772), the symptom is both sensitive and specific for cardiac ischaemia. In the same way, a sign like the height and waveform of the JVP, has its value for diagnosing right heart failure / pulmonary hypertension, atrial fibrillation, tricuspid incompetence, pulmonary stenosis, and dehydration. In such cases, the diagnostic accuracy will depend much on the competence of the observer.

The instrument-based medical, physiological and pathological tests doctors use every day, are not the only reliable ones, but those who judge the performance of doctors mostly think they are. At MPTS (GMC) tribunals and other legal proceedings where doctors are cross-examined, the assumption will be that the output of all scientific instruments is reliable and independent, and the opinions and personal assessments of doctors are only hearsay. Their attitude can be to regard symptoms and signs as subjective and unreliable, anecdotal and inadmissible as evidence.

I was told by my barrister at a GMC / MPTS Tribunal that my clinical experience and judgement counted for nothing. (Read my response in my book, The NHS, 'Our Sick Sacred Cow). For every clinician who can use his judgement and experience to save lives, continuing with such palpable nonsense is unacceptable. Acceptable evidence in a court of law differs entirely from that found in any medical arena, where sometimes indefinite clinical evidence is used to make life-changing medical decisions.

The ideal investigation test is sensitive, like a dog's sense of smell. A dog's nose is seventy-times more sensitive than any human one, and able to detect the scent of another animal from afar. Dogs can even detect patients with COVID virus (*Using Dogs to Detect COVID-19*. The ARCTEC team, 2021, London School of Hygiene & Tropical Medicine). A dog's nose is also 'specific' enough to detect the dogs they know and others unknown to them. Some working dogs can detect different sheep and elephants, just from a trace of their scent (less than a few molecules in each cubic meter of air). They will fail completely when used to detect white elephants and red herrings. That is for humans only!

When a diagnostic test lacks sensitivity, it will not reliably detect early disease, and will yield false negatives. If it is too sensitive, there will be too many false positives (14% of positive faecal blood tests are false indicators of colon cancer). If a test is not specific enough, it will also yield false positives (like PSA testing for prostate cancer). If a test is very specific, like the detection of chorionic gonadotrophin in pregnancy, little ambiguity will arise. A positive test, 3-4 days after embryo implantation, means that the patient is pregnant (>99% diagnostic accuracy +ve); when negative, one can confirm the non-pregnant state with confidence.

The diagnostic accuracy (both positive and negative) of a test, represents its overall clinical value. A test yielding a high diagnostic accuracy positive, will yield a high proportion of true positives and only a few false positives. A high diagnostic accuracy negative means that a test will yield many true negatives, and only a few false negatives.

Consider the erythrocyte sedimentation rate (ESR). It is now an old and unreliable test, but still valuable because of its sensitive to inflammatory disease.

Edmund Biernacki, a Polish pathologist, invented the ESR test in 1870. It involves drawing 20cms. of sodium citrated blood into a narrow tube (the Westergren method), then measuring the red cell sedimentation after one hour. Sedimentation increases with the amount of fibrinogen in the blood, and decreases with the negative charge on red cells. Since fibrinogen is an inflammatory marker, the ESR can measure the amount of inflammation present (in 1921, Dr. Alf Westergren, later nominated for a Nobel prize in physiology or medicine in 1935, used it to measure the progress of TB patients). It is still useful for the detection of auto-immune disease. The ESR can be positive years before a condition manifests clinically.

While the ESR is a sensitive detector of inflammatory diseases, it is non-specific. A raised ESR usually means that there is 'something wrong' with the patient, albeit in the early stages. Because it is not specific, it can be time-consuming and fruitless to chase the cause. It took ten years for one of my patients to develop rheumatoid arthritis, after I first found her to have a constantly raised ESR (>50mm/h). Another developed Sjörgren's Syndrome, twenty-years after her ESR was first raised.

One cannot prove a negative, but sometimes we need to try. Proof that somebody with a slightly raised ESR, or CRP, does not have rheumatoid arthritis may be necessary when a patient is unduly worried by the possibility. Many pointless tests could follow, and the patient's anxiety compounded by suggesting regular annual reviews, just in case a treatable condition arises. This will make them into a patient. Some will embrace this without fear or trepidation; others will resent it. The better alternative is for them to return only when new symptoms develop (a brief list of which one can suggest to them).

Patients often go to doctors with ill-informed requests, fuelled by worry. One common request in my practice was for an ECG. Many believed it would rule out all forms of heart disease. Because the diagnostic accuracy of an ECG for IHD is only 60%, I always refused isolated requests. The ECG has a different diagnostic accuracy for each cardiac condition, so I had to explain what was necessary to rule out all forms of heart disease.

The growth of the left ventricular heart muscle (left ventricular hypertrophy or LVH) is often associated with clinically significant hypertension (sustained high blood pressure). In a study of LVH seen on echocardiography, increased voltage changes on an ECG, correctly detected only 14% of cases. Although ECGs are insensitive to LVH, one study reported it to have a specificity of 98% (Levy, Labib, S.B., et al. Circulation. 1990. 81: 815-820). In a post-mortem study of LVH (using ECG voltages: R (AVL) + S (V3) > 2.8mV, in men, and 2mV in women), the sensitivity of an ECG was 70%, while its specificity was 100% (Casale, P.N., Devereux, P.N. et al. Circulation (1987); 75: 565-572).

There is a widespread public view (in my practice at least) that 'tests' are more valuable than any medical opinion. This is true for an ECG, only when it is used to detect electrical defects (rhythm problems, conduction faults, pre-excitation conditions such as WPW, and ion-channel problems). It can provide the first clue of cardiomyopathy, or LVH in hypertension, but less reliably. Large tomes are available to help interpret ECG patterns. Some doctors have a talent for ECG or EEG interpretation; pattern recognition being a talent shared with expert tea leaf readers.

An ECG can present many false positives when used to diagnose ischaemic heart disease (IHD). 'Normal' people can have a falsely positive ECG (usually ST and J-point changes), although some of them can have early systemic hypertension or cardiomyopathy (both have re-modelling of heart muscle). These changes can resemble ischaemia to the inexperienced eye (and be misinterpreted by built-in ECG computer algorithms, designed to detect abnormalities). ECGs are also subject to false negatives: a normal ECG at rest that misses IHD, and old myocardial infarction (q-waves become less obvious with time, and often disappear after twenty-years). Despite these shortcomings, some agencies (CAA, TFL, DVLA, insurance companies,) still require a 'normal' ECG at rest, as a test of fitness to fly, drive, or to buy life insurance. Without a history, examination, and corroborative tests, it is too risky to get drawn into interpreting ECGs in isolation.

Is an exercise ECG any better? It is, but not much better (diagnostic accuracy 85% for IHD). It becomes more useful (sensitive) with personal observation of the test in action. One can observe a patient's shortness of breath, their 'running out of steam', and the onset of any chest tightness or 'restriction' (this, rather than 'pain', is typical of angina). Even with these symptoms, there may be little or no ECG change. Here, observations of the patient will be more valuable than the ECG. ST segment depression (down-sloping) seen during and after exercise, will suggest the diagnosis of IHD, but only in the absence of LVH. When seen, it should always prompt further evaluation.

An ECG will remain an insensitive test for one obvious reason: it detects electrical change, not mechanical change. It cannot directly detect mechanical problems like reduced blood flow, except by chance. Who would call an electrician to diagnose a plumbing problem?

NICE guidelines no longer include ECG exercise testing for the detection of an intermediate coronary heart disease risk. This may have resulted from doctors ordering exercise ECGs, without observing what happens themselves. Exercise testing is time-consuming, expensive, and not as reliable as a CT coronary calcium score, for detecting early CHD. It is second to none for assessing a patient's exercise capacity, especially in relation to any heart or lung disease.

From experience of IHD, I found ECGs and the standard risk factors for IHD, such as lipid profile, smoking habit, and obesity to be mostly non-discriminatory. In the absence of definite angina, I exercised all those with significant carotid atheroma, and / or a positive family history of IHD. For those with definite angina on exercise, with or without ECG changes, I performed coronary angiography. Those with a normal ECG, and a high index of suspicion of IHD, I would request a coronary calcium score, or CT angiography. A normal ECG does not rule out the need for coronary stenting or CABG.

Despite its poor performance, the ECG remains the best single test for detecting the risk of sudden death in athletes (ion-channel disturbances / long QT, etc.). An exercise test can assess functional capacity and reveal catecholamine, or exercise induced arrhythmias. When LBBB might obscure ischaemic ECG changes, a stress echocardiogram or a perfusion study, should be used to detect ischaemia.

I have attended almost every exercise test I ever ordered, and made only two patients stop as an emergency. I stopped both when I observed the onset of VT. I later found both had HOCM (heart muscle disease). No patient of mine died while performing an ECG exercise test. Although it is safe, we should avoid it in known HOCM cases, patients with very high blood pressure, left coronary main-stem stenosis, severe aortic valve disease, and those with aneurysms.

Negative Diagnosis and Road Testing

The ruling-out a diagnoses is an essential strategy, clearly different from proving a negative. It can be employed when the diagnosis is elusive or contentious.

A satisfactorily performed exercise test (no shortness of breath, or ECG change after nine minutes of the Bruce Protocol), confirms that the patient cannot have severe mitral stenosis. The same negative result would not dismiss CHD or aortic valve stenosis. Severe aortic valve disease may have few symptoms on exercise (a potentially misleading and dangerous test for these patients). Patients with widespread coronary atheroma, but no severe stenoses (or those with a good collateral circulation) can have negative exercise ECGs. Most times, ruling out diagnoses works well clinically, although an appreciation of the fallibility of the tests we use, is essential. The philosophy of science can keep its proposition that one cannot prove a negative; the odds against a diagnosis will suffice for clinicians.

There is an old principle: if you want to show the true character of a person (like friends and colleagues), '*kick 'em in the shin to see what happens.*' This simple test, applied literally, or metaphorically, can reveal hidden character traits. Observing patients and doctors in stressful circumstances is as useful.

In 1972, Captain Key, the pilot of British European Airways Flight 548, call sign PAPA-INDIA, took off from Heathrow. He was flying to Brussels. His aircraft crashed shortly after take-off. All 118 passengers and crew died.

At post mortem, Captain Key had a haemorrhage in the wall of a coronary artery, but no actual cardiac infarction. His coronary arteries showed some atherosclerosis. He had two 'normal' ECGs recorded in the previous year.

There was no voice recorder fitted to the Trident aircraft he was flying, so one can only speculate about what took place in the cockpit. Other colleagues later revealed that he had an argument in the crew room prior to the flight.

It is possible that Captain Key, distracted by chest pain, slumped unconscious into his seat, leaving two inexperienced co-pilots to fly the aircraft. It crashed after a deep stall (nose up / low airspeed), near Heathrow airport.

Routine CAA medical examinations for pilots at the time included an ECG at rest, but no exercise test (echocardiography and detecting atheroma had yet to be introduced). An important clinical point was made at the subsequent inquest and public enquiry. An ECG at rest was clearly not good enough to detect pre-symptomatic coronary heart disease (CHD), or any liability to cardiac infarction.

Garrison and Cullen (1972), made a further point about the exercise ECG testing of pilots. They suggested the need for a more discriminative, diagnostically accurate test. They wrote:

'ECGs have many false negatives and false positives. Pilots found to be 'unfit' form a small group; those 'presumed' to be fit, a large group. In this large group, a small percentage (representing a large number) of true CHD cases will be found hiding (among the false negatives). Therefore, a larger number of heart attacks will occur in the 'fit' group than in the smaller unfit group.' (Garrison, G.E., Cullen, E. *Interpretation of coronary artery disease in fatal aircraft accidents* (1972) Aerospace Medicine. 43. 86-91).

Recording an ECG is a standard procedure in cases of suspected acute cardiac infarction. During cardiac infarction, the ST segments of an ECG can be elevated (STEMI), or not (NSTEMI); these represent the true positive and the false negative ECG presentations of acute infarction. Once upon a time, we used several blood enzymes (ALT, AST, and CPK) to diagnose cardiac cell damage (cardiac infarction); although sensitive, they were not specific enough. Troponin replaced them, providing us with a specific and sensitive quantitative measure of cardiac cell death.

Science, Art and Chance

Many think that considerations of chance help us understand everything that has happened or will happen. Predictability improved after the scientific method was introduced; as long as experiments were conducted under strictly regulated conditions (some might say 'unreal' conditions), with only one factor varied each time. If experiments are designed well, the result can prove the truth or falsity of any thesis being tested. If *'the chance of the result being wrong is 1:10000 (P<0.0001)'*, there is no problem believing it, but what are we to believe if *'the chance of it being wrong is 1:20 (P<0.05)'*? Judging the significance of such probability calculations is an art.

The contribution made by the scientific method to our understanding of physiology, biochemistry, and medicine is beyond criticism. The human brain, however, uses other processes to discern fact from fiction, and to discern the difference between the acceptable and unacceptable. One impressive function is the ability to choose only the most important factors from a nebula of relevant and irrelevant data. Another is the ability to make associations between apparently disconnected entities. Unhindered by the restrictions of a scientific experiment, our brains can provide us with conclusions in complex situations that defy calculation. Because this process is 'not scientific' (used as a derogatory term), we must regard the process as an art-form (along with judgement, decision making, and successful diagnosis). Although all humans are capable of these processes, few have much talent for them.

The scientific method imposes severe restrictions, but makes our understanding tangible and dependable. Art goes further. It encompasses not only the knowable but also the imaginable, and the unknowable.

Corporate Algorithm

Those who want to deal with other human beings in a dehumanised, anonymous way, will be happy to use algorithms. Franchised corporations, regulators, managers, and controllers, all need to standardise their procedures and set defined rules; they cannot afford to foster individuality and lose control. They need to introduce self-contained, pre-prescribed processing methods, that produce compliant results. Algorithms help deliver these.

Humans making up their minds, and successfully using their judgement for work, can be unreliable so large organisations must enforce conformity by setting standards, often to exact pre-arranged criteria. It is untenable for any corporation to let working humans think for themselves. Those found capable of independent thought, risk being dismissed or being promoted to a decision-making role. Algorithmic check-box functioning may be inflexible, but it successfully binds humans to compliance. When biases help form an algorithm, they are not subject to variation; human biases, however, vary from time to time. Kahneman referred to this as 'noise'.

An algorithm will allow a test 'pass' to be issued, if all the standard check-boxes are ticked. Should one check-box not be ticked, a failed notice might result. This decision could prove incorrect after the backstory, meta-data, and thought processes are considered, or after some imagination and humanity are applied. An algorithm cannot consider exceptions, unless programmed to do so.

Refugees from Ukraine, trying to enter the UK in early March 2022, encountered a problem. Embarrassed politicians and bureaucrats had to work hard to replace their beloved UK immigrant entry algorithm with a compassionate policy. Those in the Republic of Ireland (Éire) did not; their entry requirements were already compassionate. It is for this reason that laws (written-in-stone algorithms) sometimes function as asinine and inhumane.

The human brain is capable of so much more than algorithmic thought, yet we remain awestruck by computers capable of little else. Organisations come to revere those who program them, and regulators become enslaved by them. That leaves many of us free to rejoice whenever we discover something computers cannot do, like showing kindness, being humane, and finding small scraps of human wisdom that enlighten and help us.

If you have ever tried to argue with the decision of a large organisation, you know that getting legal advice may be necessary at an early stage. You may need a judge in a Court of Law to consider your case (as a non-algorithmic one, with 'human' meta-factors considered). Best of luck!

Corporations shun individuality as a principle, so important is standardisation to their profit. Large organisations often have a special department, tasked with nothing else but maintaining compliance and standards, often with a budget large enough to enforce their wishes.

When businesses start, they adapt their rules to individual customer considerations. As they grow larger, algorithmic control takes over, making standardised functioning cheaper and more profitable. Seen as a strength by most organisations, it is one weakness that has led to the demise of many corporations. Because they are mostly trained in the same way as other corporate executives, those who run the NHS will usually follow the same predictable path.

A lack of adaptability often defeated the US Army in Afghanistan. The fleet-footed Taliban often outmanoeuvred them, and by 2020 had displaced all allied troops. The Taliban achieved this by employing an ancient principle: avoid predictability.

In all human affairs, political strengths and weaknesses come and go cyclically. In weak phases, an unbending adherence to the algorithmic processes can bring down corporations and empires (usually after they have siphoned off the profits, bled pension funds dry, and had knighthoods awarded).

For human thinking to be most productive, it must be open to several types of functioning: 'out-of-the box', blue-sky, nebula and lateral thinking. These styles require imaginative people to have the freedom to think, without bureaucratic restrictions. Some corporations will see this as frightening. Only doctors capable of free thinking (in colour), rather than restricted thinking (in black and white), will use both the art and a science of medicine to help patients and take clinical medicine forward. The same choice may not be available to those who are scientists at heart. Laboratory workers need restricted, algorithmic, black and white thinking, to perform their strictly standardised experiments and to produce reliable and reproducible results.

There are many parallels in other disciplines. Take musical performance, for instance. We train classical musicians to follow written scores (algorithmic performances). A conductor hopes to control the expression and tempo of what it is they play. Jazz musicians rarely use written scores and never a conductor. They are happy to extemporise, playing without a score (although often with a loose performance plan). They will however, play to an agreed structure (like twelve-bar blues) with an intro, principal theme, middle eight, and variations on the theme introduced by solo performances. They only need to know the key, the tempo, a countdown for when to start, and a meaningful look or gesture from their leader, before ending. The co-ordination and interplay between jazz musicians has a spontaneous element to it. The faithful rendition of music by classically trained musicians contrasts with the 'let's jam', and 'let's see what happens' approach of jazz musicians. These differences represent separate performance cultures – the algorithmic, and the free-wheeling. They both have their followers.

Medical Computing

A robot passed the written element of China's national medical licensing examination, making it the first robot in the world to do so. The South China Post (27th November 2017) reported that the intelligent system, the 'iFlyTek Smart Doctor Assistant', achieved 456 points in the test, exceeding the pass mark of 360.

Liu Qingfeng told the publication that rather than replacing doctors, the robot would help them serve patients better. He said, 'By studying the medical cases and diagnosis skills of top doctors in top hospitals in mega-cities in China, our doctor AI (artificial intelligence) can serve as an assistant to help doctors in remote areas of the country.'

Computers expressing a programmed opinion or a caution, are now with us. They can produce reports using pre-set, standardised texts, triggered by pre-chosen elements within a data set (decided by whom, I wonder), usually with no reference to broader clinical issues (clinical meta-data). Personalised medicine is different. It requires attention to personal detail, respecting the character, needs, and desires of each patient while understanding the broader picture. For a pre-programmed computer to function at this level is science fiction (for the moment).

Exceptions defeat both computer algorithms, and algorithmic thinkers (bean-counters), equipped as they are, to deal only with what is usual, or stored in their database or rule book. The frequent occurrence of repetitive, usual, and commonplace tasks in medicine justifies some computerisation, if only to reduce monotony.

Prediction and Diagnosis

Predictive probability indicators can be of clinical value to the inexperienced, and those seeking numeric support for their argument. The QRisk3 score, for instance, attempts to estimate a patient's risk of a cardiovascular event (heart attack or CVA) within ten years. It works (on average) for large groups, but is it valid for individuals? Since it calculates the risk based on grouped data (like all statistics), should it be used to predict the fate of an individual? At best, it can serve as a lightly weighted decision factor. Many will justify its use with the argument: 'it's the best we can do'.

Even with revised data, the probabilities generated by algorithms and AI, will only occasionally aid those with long clinical experience. Programmed with fail-safe features, they could guide nurses, paramedics, and pharmacists working on desert islands, to suggest what is likely to be best and safe. When that fails, a real doctor might have to be called.

Those convinced by the power of numbers will feel that probability calculations are more reliable than experience. Indeed, this is one advantage AI brings: using the rapid matching of current clinical data to a learned database, producing probability calculations for any differences, rather than that decided by human judgement. Not only will numerical probability calculations prove useful to the inexperienced, they can also provide a medico-legal reference point (lawyers will question a doctor's judgement, but will rarely question the output or 'judgement' of a computer). Medical AI will, I am sure, prove worthy. It will be most valuable in places where there are no experienced doctors.

Machines can calculate the likelihood of events occurring in a random world. Unfortunately, events that affect humans are rarely random, although they may appear so.

Many variable biases influence human judgement. Although they vary from time to time, our most recent experiences and remarkable cases, bias us most. Experience should reduce repetitive errors in those able to learn from their mistakes, but first an awareness of fallibility is required.

A zookeeper with forty years of experience required revalidation. They brought one animal before him as a test of his recognition. The zoo had 534 different animal species in residence, so the probability of an elephant being the chosen one, was 534 to one (assuming no bias in the choice). These long odds did not reflect the zookeeper's knowledge of his animals. Because of his experience, he did not see his ability to identify the elephant as a 534 : 1 chance, or even an odds-on, 1 : 534 chance. Instead, it was an absolute certainty. He knew the animal at a glance. In addition, he held valuable meta-data: he knew the animal as an individual (not just as a number); he knew the animal's name and eating preferences. He had known the elephant for 25-years, and they had a relationship.

How relevant might calculating the odds be to this experienced zookeeper? Not at all, unless he was lying about being a zookeeper, and incapable of distinguishing a gnu from a gnat.

Odds calculations can aid decision making, but their safety can depend on the details of the calculation process and on metadata considerations. Those who use algorithms in isolation will only rarely appreciate such information.

Creativity

If what warms you most is the fixity of algorithms, what value will you ascribe to creative and lateral thinking? Some humans can invent new methods 'on the spot', as necessity requires. Here is a 53-year-old example. The patient was my first seriously ill patient, six-weeks after starting my first surgical house job:

In July 1966, 58-year-old Mary developed severe vomiting and diarrhoea, twelve hours after a routine, uneventful cholecystectomy. She became severely dehydrated and paralysed by hypokalaemia. Overnight, I infused her with 21 litres of 5% dextrose and normal saline with lots of added potassium. (Yes, it was 21 litres!). I checked her fluid balance every 30 minutes (at a time before CVP monitoring) using her JVP and skin turgor as guides. I had to devise a rapid method for monitoring her tissue potassium level; getting laboratory results was then just too slow. I used the height of her ECG T-waves as a biological marker. By matching her blood potassium to her T-waves, I proved the concept to be useful (ECG machines then had thermionic valves, not transistors. ECG monitoring had yet to be introduced). She had no circulatory perfusion problems and never developed pulmonary oedema.

On one occasion, despite the very low blood potassium level reported by our laboratory (half the lower limit of normal), her T-waves remained unchanged. I therefore disbelieved the laboratory result. The laboratory assistant then found something blocking the electrolyte machine suction system: it failed to draw up sufficient serum, and had reported a spuriously low blood potassium level. Her ECG saved me from killing the patient with extra potassium. All this happened in Lister Ward, Whipp's Cross Hospital, London, in 1966.

I didn't involve my senior surgical registrar (Colin Davis), or consultant (Mr. Louis de Jode), at the time since her problems started late at night. I was busy, and didn't want to disturb them. Their initial attitude to me was in keeping with that applied to new Spitfire pilots in the Second World War (26-years before): choose what seem to be the right chaps, give 'em some basic training, and let 'em get on with it. The policy won the Battle of Britain, and I think by the same token, my patient survived.

How many forgotten lessons need revival? I regret that doctors and nurses are no longer as free as I once was, to use their intuition and creativity. Stringent rules now inhibit both.

Nobody would now condone my unilateral behaviour in Mary's case. One might now have to wait until the next day to allow a team to decide which guideline to follow; one that would avoid allegations. Mary would have been dead by then. While some doctors twiddle their thumbs, patients can slip quietly towards a point of no return. In 1966, I was free to decide on fait accompli.

I now understand that my behaviour marked me as a loner and maverick; one ready to take unilateral action whenever required (there were no written guidelines back then). Regulators would now undoubtedly regard my behaviour as cavalier and dangerous. **Wake up bureaucrats! Practicing medicine IS dangerous**, and in some specialties, one will deal with little else.

The actual danger of some medical work must be inconceivable to most. Among them will be bureaucrats, even those with medical qualifications (those preferring a desk job, to engaging with anything clinically dangerous). It is not their fault that the dangers intrinsic to medical practice are beyond their experience and ability, but why give them any authority to judge the work of other doctors?

Can Computers be Creative?

'Computer creates Jazz' was a headline posted by London University. Was this a party trick? Since computers are not capable of de novo creativity, it must have resulted from sorting, averaging, and compiling jazz riffs, from those already written. With a few rules defining choice, sampled riffs can create an infinite stream of 'computer-generated jazz'. By throwing pots of paint at a canvas, even chimpanzees can create unique paintings. Creation by chance differs from intended, meaningful creativity. An extreme danger now exists: we are becoming more and more liable to be fooled by computing processes portrayed as intelligent and creative.

Meaningful, *de novo* creation is rare. Leonardo da Vinci (the fighting machine, submarine and helicopter), Einstein (space-time), Tesla (alternating current), Sigmund Freud (psychoanalysis), Barnes Wallace (the bouncing bomb), Frank Whittle (the jet engine), and Koch (the transmissibility of infection), should suffice as old examples. They produced ideas that were not variations on any previous theme; their ideas were new and revolutionary.

The mental functioning required for genuine creativity is not immediately obvious. Sifting and sorting information can inspire it. Inspiration can strike when necessity demands; necessity being 'the mother of invention'. Creating something new can involve seeing something of significance for the first time, in a familiar object, pattern, theory, or bit of information. Finding inconsistency, and trying to explain it, has often led to a discovery or a new explanation.

Einstein wondered from a young age, if time varied with the speed of travel; Geoffrey Marmot wondered why the Japanese living in California had more strokes than their compatriots in Japan. Inspired questions can lead to significant new discoveries. Much more common is the creativity limited to modification: an ability to vary known stories, music, or theories, and to develop them into something that seems new, but is not revolutionary.

For the creative mind, a word, a fact, or a smell, can sometimes spark a new idea. Einstein used thought experiments to test his general theory of relativity. He imagined a man standing on a railway station platform, while a woman passed by in a train travelling at the speed of light. Every new idea must get tested, criticised, edited, re-shaped, and re-developed until acceptable theoretically. Creativity are not unique to humans. Birds have discovered novel ways to crack nuts.

A new source of discovery now exists in theory: finding gems in big data. New insights might arise while data prospecting: sifting and reviewing collected computer data, rather than making any primary observations.

Pulsars were a chance discovery. Jocelyn Bell Burnell was first to recognise the possible significance of regularly occurring signals seen in radio-astronomy quasar data. Because she was only a student, others developed her observation. They, not Jocelyn, received Nobel prizes.

It is important for all those who work in science and medicine, hoping to gain recognition, to acquaint themselves thoroughly with the teachings of Nicolò Machiavelli.

Complicated or Complex?

> Both simple and complicated systems are predictable. By definition, complex ones are not.
> (*The Theory of Complex Phenomena.* Friedrich Hayek. 1964).

The perception of complexity can create anxiety and frustration. We can all get overloaded and feel unable to cope when faced with something that seems too complicated or complex. We can easily lose sight of the fact that complicated things have straightforward explanations. Not so for complex systems: the weather and stock market movements, are truly complex. In these systems, every change affects other changes in variable ways, and the interplay of changes can defy accurate prediction. Multiple factors, interacting differently at different times are the problem. Despite this, some humans can manage the complex to their advantage, and with little active thought. Despite complexity, some traders make money buying and selling shares, and some shepherds can accurately predict the weather. The trick is to correctly value those parts of a complex system we have learned to rely on from experience.

The Imperial College Response Team created the initial model of virus spread at the start of the UK COVID-19 epidemic (30/3/2020). Imperial college enjoys an international reputation, but their model suggested that more than 500,000 deaths would occur beyond March 2020, without interventions like social distancing. They suggested that fewer than 200,000 deaths would occur with intervention. Two months later, 37,000 COVID patients had died. As a matter of political expediency, the politicians who accepted such warnings may have ignored the caveats for caution accompanying every predictive mathematical model (in their case, a 'semi-mechanistic Bayesian hierarchical model').

Something that is widely regarded as complex is human nature. This is partly a deception. From an old man's point of view (mine), human nature is mostly expressed as simple and predictable. Once one has observed the character of a person in action, and their motives, circumstances, and biases become known, predicting what they are likely to do in any situation can be dependable. The challenge is to see through their defences and the false images many people present. Some try to appear complex, mysterious, fascinating, deep, interesting, and unreachable, in order to fool or attract others.

Some use Mae West's advice: 'Keep 'em wanting'. Some resist revelation for other reasons, and engage in deceit, diversion, and selective revelation when they are trying to boost their image or create an impression aimed at persuading others. It takes all sorts, but understanding others, although complicated, is usually far from complex.

By first understanding what patients want from a doctor, interacting with them becomes straightforward. Understanding what makes a patient 'tick' can help a lot when trying to understand their issues fully, although for doctors, this is essential only in challenging cases. Understanding colleagues can be more difficult. It might not be easy to access their agenda (many doctors have a political agenda), although one can learn a lot by listening to them (Judge Judy's advice: *'Listen more. After all, God gave us two ears, but only one mouth'*). Although true, there is another even more important aspect. You will see people for what they are, if you study their behaviour, and listen less to what they say. Handling a group of other doctors is complex: only those skilled in herding cats should consider it.

Simple Decision Making: Tossing Coins

We attribute binary mathematics to Leibnitz. (Gottfried Wilhelm von Leibnitz: 1646-1716). Jesuit priests who worked in China had informed him of the Chinese Book of Changes: I Ching. A person wanting to answer a question like, 'Does he love me?' must choose between Yin and Yang, 'Broken' or 'Unbroken' lines (like 'Heads' or 'Tails').

I Ching used for divination can involve tossing three coins and noting the combination of heads and tails that arise. After repeating the process six times, they would interpret the resulting combinations. The process claimed to reveal divine intervention and intent.

Getting answers from the random tossing of coins in this way, assumed that God's will was involved. Science displaced the idea, but not completely; randomness remains an important concept, although no longer ascribed to God's will. Many similar beliefs remain important to patient welfare, and few doctors can afford to ignore them completely.

Nebula Processing: Choosing Stars

Nebulae are the graveyards and birthplaces of stars.

Classic medical cases are those with a 'full house' of specific symptoms and signs, consistent with a particular medical condition. Such combinations make a confident diagnosis possible. 'Full house', or pathognomonic clinical cases, are relatively rare because many clinical feature must match the standard diagnostic set (only some of which are pathognomonic). Non-classical cases have fewer matching symptoms, signs, or investigation results. Among them will be red-herrings and false information.

With just three highly relevant bits of clinical data, physicians can usually be sure of a diagnosis. Knowing which data is relevant, and how relevant, is the critical clinical art. Doctors develop and refine it through experience.

Although theoretically difficult to do, the human brain can make light work of the diagnostic process, sometimes calling on valuable meta-information to add perspective. Consider an equivalent example outside of the field of medicine. A chef visits an orchard to select fruit for his pie. He must first select the best trees. He will do this by judging the ripeness, colour, size of the fruit, together with the ease of picking. Although his selection process requires lots of data processing, choosing and rejecting fruit is based on his learning from experience (nebula processing). He will make his choices (from a whole orchard of information) with little conscious effort or reference to any strictly defined criteria. A doctor judging the relevance of each symptom and sign when making a diagnosis, uses a similar process. A master can guide an apprentice, but they will develop their own methods from experience. For both a chef and doctor alike, aesthetic choices need little conscious effort and no conscious calculation. This makes the discrimination process an art.

With talent and experience, the ability to recognise relevant clinical connections should improve. Astronomers, ornithologists, zoo keepers, antique experts, art historians, stamp collectors, and doctors rely as much on matching observations as deductive reasoning.

Chaos Theory, the Butterfly Effect and Medical Practice

In the early 1960s, mathematician Edward Lorenz observed that for any deterministic system (computer programs with pre-set rules), small changes in the initial conditions (a butterfly flapping its wings, for instance) could affect the outcome. Take traffic flow. Vehicles progress following pre-set rules, observing the speed limits, and stopping at traffic lights, but the same journey will change a lot with small interfering factors (broken traffic lights, broken down vehicles). The time people leave their houses for their commute to work might vary by only one and two minutes, but the resulting effect (how late or early they arrive at their destination) can be large.

While driving, I have often said: 'It's chaos'. My father, (who drove horse-drawn wagons in the 1920s), when faced with a traffic jam, would predict that traffic police were at work somewhere ahead. Police officers were notorious for holding up traffic; especially when road traffic accidents occurred. They are no different now. 'We would all be much better off, if we left traffic flow to drivers', my father would say. His opinion was that traffic duty police and fixed-period traffic lights created chaos, because of inflexibility. If you thought traffic chaos was a modern phenomenon, you would be wrong. According to my father, central London traffic was just as bad in the 1920s. There was a difference: horses, and horse-drawn wagons and buses caused the traffic jams, not cars.

In medicine, there are innumerable examples of initial conditions affecting patient outcome. Perhaps the most important factor is timing. Leave the treatment of cancer for an extra day; leave someone with pneumonia in A&E for an extra hour; leave an older adult person dehydrated for longer than necessary, and the outcome might change significantly. The more urgent the case, like septicaemia, organ failure, or cardiac arrest, the more the outcome will change with small delays. When you claim to be too busy, make sure it is with something more important than a patient's life. That 'something more important' should never be paperwork, or any managerial task.

The Misuse of Statistics and Algorithms

In 2020, 'A' level students were told they could not take their exams, and their teachers would have to estimate their grades using mock exam results. The results were to be standardised. An algorithm produced the grades based on the past performance of each school, but not on each candidate's previous GCSE results.
Had the statisticians involved made sure (using modelling techniques) that the overall results were in line with past results, only small percentage changes in grades would have occurred. What resulted was 38% of students being marked down. This was not good enough for the Scots. They abandoned the algorithm; English adjudicators did not.
In England, a brilliant student coming from a poorly performing school could easily have had their estimated grade marked down by two grades (perhaps the difference between medical school entry, or not). In England, they strongly defended the algorithm, despite 280,000 out of 750,000 students being downgraded. Despite widespread outrage at their Kafkaesque use of an algorithm and its effects on the future prospects of so many students, the government stuck with it. They relied on an appeal process, and the chance for students to take actual 'A' level exams, but only long after the 2020-21 University year had begun.

In the end, pressure from demoralised students, and a considerable loss of trust in Ofqual, caused politicians to abandon the algorithm. On August 17th, 2020, 'A' level students had their grades reverted to those given to them by their schools.

Never aired was one crucial question: did mock exam results (on which the teacher assessments depended), ever reliably predict subsequent 'A' Level examination results? If not, how could any intelligent person have used them?

Worcester College, Oxford, was the first to take unilateral action. They thought the adjusted results were unreliable. Two other Oxford colleges soon followed, and honoured every offer made for student entry. Other Oxbridge colleges gave places only to those who passed their entrance examinations.

Given that very few young people were dying with COVID-19, was it sensible to ignore the mortality evidence and abandon all 'A' Level examinations? Was it right to risk the mental health and future careers of students in this way? With the educational future of so many young people at stake, was it humane to rely on an algorithm?

This illustrates what government bureaucrats are capable of. When statistics and algorithms are used to judge individuals, we must regard the results as unreliable, until we have scrutinised every detail of the assumptions, logic and methods used.

Clinical Statistics and Anecdote

Statistics, as we know and love them, are not quite what they seem (at least, not for non-mathematically oriented clinicians). There are some obvious issues:

- Statistical analysis is valid ONLY for large data sets, and,

- Statistical analysis is valid ONLY when we use random selection.

Neither restriction applies in any consulting room. Our patients are not usually subject to random selection; we consult with self-selected patients, one at a time.

Statistical analysis aims to provide us with unbiased probability calculations. In order not to fool themselves, those who choose to rely on anecdotal observation, must ask whether their observations were serendipitous or predictable in some way.

Kahneman showed that probabilistic evaluation was not a human *forte*; cognitive biases are at work whenever we think slowly and take time to decide. Despite this, humans have survived for millions of years using one simple strategy – 'play it safe'. 'Playing it safe' does, of course, require some experience of all the potential risks and their likelihood.

Early *homo sapiens*, with enough experience of taking risks, would have learned which risks were necessary for survival. When they were being chased by a sabre-tooth tiger, a frequent question must have been – 'do I jump (over a cliff), stay and defy the animal or fight?' They couldn't have calculated the exact risks, but would have soon acquired a good idea of them. Although we can now access the statistics for tiger attack survival, the results will not help those faced with a free-ranging, hungry tiger.

In relation to the risk of disease, some patient's approach is simple. From the patient's point of view, she either has a disease or does not. To the non-probabilistic human mind, the risk is 50:50, even if doctors calculate the chance to be 10,000:1. Few patients considering disease occurrence will know about rare conditions, and very few can put them into perspective. All physicians and surgeons, however, must appreciate rarity and be ready to deal with rare cases as they arise.

The concept of clinical relevance is critical when considering any evidence. It involves knowing the patient; the data collected (symptoms, signs, and investigations) and by whom (a good observer, or not), and how it should be weighted for significance. Assigning clinical relevance can involve the conscious use of bias (taking meta-information into account to make a clinical decision more appropriate). Using only the calculated odds for a decision can prove fallible if non-specific and depersonalised.

If you prefer to use odds-based decision making, try to win money at a horse race meeting, using the odds provided by bookmakers. Try to select the winner of the next Grand National, based on the declared odds alone. You will probably lose money. Alternatively, try to make a choice using critical meta-information: what the trainers and jockeys think of their horse's vitality (how well the horse has been eating, and running recently); the reason for it being entered (it could be for experience only), and whether it prefers a hard or soft going, etc. Professional gamblers (successful ones actually exist) seek every bit of useful meta-information before placing their bets.

In more complicated clinical cases, ignoring meta-information can lead to failure. Without this information, one cannot always make a sensible diagnosis or devise a workable management plan. The probabilistic method will help generate a long list of diagnoses, many of which will be unrealistic.

Humans have survived for eons, dealing successfully with uncertainty; being prepared at all times to deal with whatever arises, as it arises. Having been coerced only recently by legal bodies employed to judge them, all doctors must now base their clinical decisions on published, calculated odds. Two examples are the CHADs2 and QRisk3 cardiovascular risk scores. They are helpful, but not totally reliable, decision-making tools for individual patients. If doctors do not use them, they risk being classed as unfit to practise.

The rule-of-thumb, clinical analyses we perform every day, usually work well in practice. This is partly because we can place the correct weight on clinical context (patient meta-data), unlike every statistical calculation that ignores it. The application of too many data restrictions in clinical trials can, for instance, lessen the relevance of the results and their clinical usefulness.

Never accept any medical trial result without detailed scrutiny of the restrictive conditions applied to patient selection.

The results of statistical analysis and anecdote are not mutually exclusive, and can helpfully inform one another. I have always used statistical methods to test the validity of my anecdotal observations, but many observations are needed. (Statistical significance changes with the square root of the number of observations. To double the significance of 16 observations, you will need to make 64 observations). Academics prefer the statistical results of trial data to anecdotal observation and often assume its greater relevance to patient management.

Anecdote suffers from every known bias.

Anecdote uses many attributes like observing change, behaviour, language and colour, and then assigns relevance; statistical analysis uses only numbers. Personal anecdotal accounts are subject to every kind of distortion – *'(S)he found me so attractive, (s)he just couldn't leave me alone'; 'the fish I caught was so large, it took three of us to carry it.'*

Anecdote is safest in the hands of experienced observers; especially those dedicated to finding the truth and eager not to fool themselves.

Anecdote and statistics are as different as Japanese kanji (universally accepted symbols) are to English script (words constructed from an alphabet). Both convey meaning, but both can mislead and create the wrong impression. There is no fight over which script is better for understanding, although some suggest that different minds must have created them.

Clinicians rarely acknowledge the limitations of statistical analysis, even though they are considerable. Any statistical method that includes averaging will exclude detail. Important limitations relate to selection criteria, with many exclusions, inclusions and definitions imposed on what makes up both the study and control group. Reliable analyses require both the study group and the matching control group to be large. The matching of controls presents a major problem, as is the need for random selection. One must always ask whether the researchers achieved them.

Statistics can attract the equivalent of religious fervour, while anecdote, even in the hands of trained observers, rarely gets the credit it deserves. Only those approaching retirement, and those retired, will dare to ascribe meaning to anecdote; their experience will confirm its value. Many significant discoveries in the past began with one anecdotal observation.

Few would disagree that we must test the validity of small number studies. We should always try to test large data sets to confirm any thesis based on anecdote. This is our current perspective, but it ignores the fact that clinical medicine progressed rapidly for centuries, solely on reproducible anecdote. Beware of any third party trying to refute what we know to be 'tried and tested' through practical observation. Anecdote knows no bounds: that's its strength and also its weakness.

Aspect

Judgement is best when derived from various standpoints. The question of gender equality, for instance, benefits from being considered from many aspects: biological, sociological, legal, political, psychological, and emotional points of view. It may be expedient to exclude certain aspects; the judgment might otherwise be unbalanced. Specific biological aspects of gender (anatomy, physiology, hormones, and physical strength) demonstrate biological difference not equality, regardless of political correctness.

We can consider cerebral aneurysm operations from several aspects: timing, operative and post-operative risk, losses and gains in cerebral functioning, anatomical feasibility, physiological need, the patient's wishes, expectations and emotional responses. An open-minded, nebula approach (taking all factors, correctly weighted) into account is more likely to result in an acceptable, sensible decision; a one-sided, rule-based approach may not.

Obvious or Just a Probability Statistic?

A recent research paper confirmed that *overt diabetes is more common in pregnant women who eat lots of chips (potatoes/carbohydrate).* Why was this research done, I wonder, given (1) that we have known for decades that carbohydrate intake affects the occurrence of both overt and pre-diabetes; (2) the occurrence of diabetes in populations relates to the average daily carbohydrate intake, and (3) pregnancy (with its physical and emotional stress) has been long known to make diabetes manifest, or to make it deteriorate?

The blindingly obvious (BLOB) is now being published (or recycled) everywhere, even in the most technical of specialist journals. Take, for instance, the conclusion: *'Asymptomatic patients with Brugada Syndromes are at lower arrhythmic risk than those presenting with syncope, or sudden cardiac death.'* Arrhythmia and Electrophysiological Review - aer-volume-5-issue-3-winter-2016. What is the point of publishing something this obvious? It will, at least, create a published reference for the writers and researchers.

Why waste paper printing the obvious, except to boost a CV, or improve a doctor's job prospects? The scientifically minded have a tendency to regard every obvious, widely observed clinical phenomenon, as a theory (an anecdote) until proven otherwise. This will usually reveal their lack of common sense and clinical experience, and an undue adherence to the strict rules of evidence. It would seem that adherence to 'the evidence' (statistically derived) is becoming more important than experience and common sense. Under the many stressful and time-urgent conditions of active clinical medicine, practical anecdotal experience will mostly win the day.

I might classify a lot of research as 'a waste of time and money', but government-linked organisations love it; it keeps their employees busy. There is an art to choosing the most worthwhile research projects. When the antecedent probability for a thesis is very high, it is hardly worth pursuing further, unless it is likely to expose something new and beneficial for patients. As long as it doesn't take too long, cost too much, and helps a researcher's career, 'why not?' Because so much money is now involved, medical progress cannot escape political consideration, with every bean-counter's duty to *'tick all the boxes, regardless of relevance'* (the obsessive-compulsive trait in action). Their modus operandi is: *completeness is everything; relevance is secondary*.

The Proliferation of Information

We are now at risk of drowning in information, little of which will benefit many. Every day we need to waste time deleting masses of irrelevant information. In my younger days, information had to be sought in libraries; it was not sent to us to attract our attention.

In professional tennis tournaments, the supply of information has gone far beyond whether a ball was 'in' or 'out'. They now measure not only the speed of each shot, but much more. They make summarised data available to tennis nerds: the average ball speed of serves, and 2nd serves; the number of errors, and forced errors, are all made available as the game progresses. The information might help predict the result, but there is an alternative: watch the match and see who wins!

In medicine, we need to collect a lot of irrelevant information for medico-legal purposes. In the UK we are behind the US, where the tendency is to screen everyone each time they present as a patient. This provides evidence of their determination not to miss anything (and risk being sued or sacked). Many tests now requested have medico-legal significance, not clinical relevance.

Many patients now demand 'completeness', with no stone left unturned. 'Normal' results effectively reassure patients (sometimes falsely), because the majority now have more faith in blood tests and the output of machines, than in a doctor's opinion. We cannot expect patients to know that an investigation finding *'nothing abnormal detected'*, does not actually mean, 'nothing is wrong'. Few will understand that the proper use of investigations is to confirm or refute a diagnosis. Patients not only tend to over-value investigations, many also underrate the value of a full clinical examination and the opinion of an experienced professional.

Research Considerations

Science advances one funeral at a time.

Max Planck.

Although the detailed results of research accumulate continuously, only the headline results are used usually to support clinical arguments.

To reduce research results to headline conclusions is convenient, but hardly reliable, because of devils in the detail. Data acquisition may seem straightforward, but one must always consider the selection criteria used, and the accuracy and consistency of the data. These should influence our interpretation of the published conclusions. An analysis of the data may show statistical significance, but not match personal experience. When this occurs, accept no research conclusions without further enquiry into the detail. Whatever your specialty, focus on the clinical relevance of the research you quote. If it will change your practice, be very sure that it represents the patients you treat.

Meta-analysis uses big data. The method involves filtering out disparate data sets, data set inconsistencies, and divergent objectives. This requires number juggling and compromise to the full. Different selection criteria and research methodologies used by different researchers, may need to be overlooked for the sake of data integration. All research groups use different selection criteria and methods, so those who favour meta-analysis must ignore some differences. Although possible, should we accept data amalgamation without reservation? Meta-analysis of all the published research data on a topic is valuable if it shows which way current opinion is heading. Given that the information summarised can be vast, the succinct 'take-home' messages it provides will be convenient for busy physicians. If you have no taste for meta-analysis, look for one well-conducted trial, carried out by a respected team of researchers. One must resist the halo bias, however.

With so many research papers and factors to consider, it is difficult to judge just how representative a meta-analysis is. Sometimes, it is only experience and more meta-analysis, that allow us to make clinical sense of research conclusions. An equivalent to meta-analysis is truth by consensus. The opinion of one person with lots of personal knowledge, dedicated experience, and a published track record, will usually make more sense than any committee. The problem is finding that person.

In pursuit of the structure of DNA, James Watson used one research technique effectively. In his book 'Avoid Boring People,' (2007), he describes how he picked the brains of others to fill gaps in his knowledge. He put key questions to experts, rather than waste his time studying the minute detail of each subject himself.

Never forget to acknowledge the contribution of others. Failing to acknowledge a contributor will lead to justified discord.

Lester Piggott rode more winning horses than losers, yet people backed other riders. In life, we can easily forget a simple winning strategy: *'back known winners'.* One piece of good advice to friends might be to buy only geese that have laid golden eggs. Even then, you will find your friends continuing to buy cheaper, untested geese. The same advice applies to research teams: back those who, time after time, come up with the most relevant and enlightening nuggets of information.

What irony lies here? Many spend decades training to become an experimenter. They collect data objectively and as honestly as they can, only to be judged by their popularity, reputation and current vogue. Scientists and physicians have never had political control; we are useful pawns in the political games of others, easily blamed when political decisions go wrong. Belief, fashion, current affairs, reputation, and a political advantage all help to decide who gets research funding and who gets published.

All knowledge is vain and full of error when it is not born of experience.
Leonardo da Vinci.

Anecdote can show the vagaries of reality; statistics, the vagaries of numerical analysis.

When interpreting anecdote or statistical conclusions, exaggeration, understatement, over-speculation and misinterpretation are common.

George Bernard Shaw wrote: 'It is the mark of a truly *intelligent* person to be moved by statistics.' Surely he meant 'a truly *emotional* person.'

When those about you, choose to be guided only by statistical evidence, you will need courage to claim that your experience is more valuable. Statistical results yield probabilities, not conclusions. Vanity and fashion are not only on display on the King's Road, Chelsea, they flourish on every hospital ward during ward rounds, and in university common rooms.

Those in touch with their inner sheep will find comfort and some reflected glory, huddled together with other sheep at the feet of a hallowed shepherd.

The variance of nature can annoy anyone who chooses to research biological subjects. It has important implications: medical case selection can seriously influence the results of research, and any subsequent conclusions drawn. It is this lack of consistency among human subjects that is the stuff of nightmares for bean-counters. They crave uniformity and hate exceptions. Experienced clinicians are different. Physicians and surgeons think nothing of wide-ranging patient variance, and must take the differences in their stride. This ability is an art-form and an essential feature of capable, effective doctors everywhere.

The rejection of cases that do not 'suit' the researchers' purpose (outside of their entry criteria), adds to the plight of medical analysts. Subject rejection is usually merciless. Forget scientific idealism; researchers are no different to other mortals wanting to prove their point; their career could depend upon it. This motivation provides bias, and can act like a prism, bending the perception of truth. I have no wish to denigrate the tough job of advancing knowledge through medical research, but there is often an unrecognised elephant living in some laboratories. In order to get useful results, natural variance may need to be minimised, smoothed out, or down-rated. Cherry-picking, and the expedient inclusion and exclusion of cases, has lead to results that hardly represent the cases seen by clinicians.

If you hope to advance medical knowledge, first make a relevant observation (the significance of which should not escape you). You must then check the literature for originality and try to overturn any contention that your observation was 'nothing new'. You will need to think like your most intelligent critic, trying to prove yourself wrong. Next, confirm your original observation in many other patients. If it still looks valid, start collecting data. For clinicians, the aim is to discover something of use to all patients and other doctors. Right from the start, you must plan to collect enough data to generate statistical significance. Unless you achieve this, you will not get your work published in any 'prestigious', peer-reviewed journal. The publishing business is undergoing a revolution. Nowadays, publishing your results on the Internet is possible; at least you will get a reference for your work and provide others with food for thought.

Once you have overcome the hubris of getting published, get grounded. Within just a few years, most of what you publish will become forgotten, or relegated to 'old knowledge' of little significance. There are many exceptions, of course, but I need only the fingers of one hand to count the game-changing publications I have read in the last fifty years. My conclusion, as a physician, is that doing research is both interesting and rewarding, but not always of obvious benefit to patients. Publishing will significantly add weight to your CV, and that has its uses when trying to impress employers. Never forget, though, the main purpose of every clinician is to relieve suffering and to save lives. By achieving these objectives daily, we will usually achieve more than publishing minor research. Vocations and hobbies are both valuable, but are separate entities.

URGENT ALERT: *Beware of sharing your ideas. Your co-workers (especially those bereft of new ideas), will occasionally try to claim your work as their own, especially if their chosen career depends on published work. If you decide to publish with a colleague, assume they will want first place on the author's list. Believe it or not, I have had many ideas stolen because I trusted a colleague. In one case, a senior colleague stole one of my ideas, lock, stock and barrel, and published it with no reference to me at all. He made a good job of it, though!*

Dimension as a Consideration

Size matters, but scale can matter more. Scale is just as important in medicine as it is in physics. The rules, laws, and data acquisition possibilities can change from one dimension to another. We need to consider organ, cellular and molecular functioning and their interactions.

What applies to the large scale, like Newtonian mechanics, will not apply at the atomic scale. For that, quantum mechanics provides some understanding.

Our conceptions of disease benefit from being viewed across scale, function and dimension. Always consider the molecular biological view (cell biochemistry, and genetics), the cellular view (tissue pathology), organ function, organ interaction, and how organs integrate dynamically with one another (the patho-physiological dimension).

Data gained from history taking and examination (the large scale, clinical dimension), may or may not be consistent with other dimensions. While we rely on our clinical data, some smaller scale data (the biochemical dimension) may lack context and could prove ambiguous or irrelevant. This data may lack clinical relevance and create a false impression. The reverse of this is the radiologist or biochemist who correctly interprets the results, but is not aware of its clinical significance.

Epidemiologists study humans on the largest scale possible. Their job is to step away from individual patients, and view entire populations. From this dimension, one could see that the original COVID-19 mutation hardly affected young people or the wealthy (not based on their immune status or their ACE receptor profile). From the same perspective, it is possible to see that heart attacks and strokes are five times more common in low-income groups, than they are in high income groups (the health divide).

It is sometimes possible from epidemiological studies to define what political action might make a difference to the health divide, and the disease profile of the poor. Advice about smoking, alcohol consumption and obesity, can benefit those who reduce them, but taken together, these factors do not fully explain all the disease differences of the health divide.

Is social equality a myth, or a political whim? The assumption of educational 'equality' now affects the work of doctors and nurses. As more and more people come to believe they know a lot about medical issues, correcting them respectfully can prove challenging. Armed with knowledge gleaned from the Internet, and encouraged by knowing their rights (to demand medical services), some patients can appear unreasonable. Take, for instance, one request I had: 'I want an injection of vitamin B12 please. It's to restore my energy'. I didn't get very far telling the patient that vitamin B12 is ubiquitous in food, and that injections are justified only for the undernourished, and those with pernicious anaemia. The patient had read that it can restore energy, and wanted some.

Medical professionals can now look forward to a time when the demands of the poorly educated grow further. With the NHS having such a limited capacity to fulfil routine medical service requirements, the public will turn to 'Dr. Google'. The digital revolution might even come to the rescue of the NHS, with medicine available through apps, 24/7. The NHS has called this 'Digital First'. The public needs to accept that detached, anonymous approaches to personal issues, are as good as the real thing. Because many are finding it so difficult to access NHS services, many patients have had to accept it!

Will patient demand for personal medical services grow further? Politicians have learned how to fan the flames of public expectation for medical services. They know how to ingratiate voters. If the apps don't work, not only will NHS healthcare workers have to cope with the overload caused by bureaucratic inertia, but also the growing demands that will come from Internet searches.

Looking on the bright side, AI could help to relieve doctors of many 'trivial' tasks and enquiries; help spot early symptoms, and then direct those with serious symptoms to a facility where a rapid diagnosis and treatment is possible, as an alternative to queuing in the nearest A&E. Hopefully, it will not be to send them, like Dorothy, to a wizard in a land called Oz. Neither here nor there, is hopefulness the most effective medical management tool.

Grand Scale. Small Scale

Doctors need to ask which scale of work suits them best. Is it at the bedside (small scale and anecdotal), or the large-scale, working with populations? At the bedside, the statistical paradox defies the reliable application of any large-scale data to one patient; at least, not without a leap of faith. Because statistical analysis only ever applies to large data sets, and never to individuals, we must resort to considerations of probability when trying to convince our patients (and ourselves) that any statistical evidence applies to them. Before one says to any patient, 'research shows that . . . ,' be aware of statistical heresy: wrongly mapping the large scale onto the small scale. It may be the best one can do, and very convenient, but the paradox remains. If you have never seen yourself as a gambler, reconsider your position. Taking leaps of faith are anathema to scientists, and playing with statistical probabilities, is gambling.

By asking the relatives and work-mates of more complicated patients for their views, one can expand the background and perspective of the clinical information available. For a fuller understanding, it might sometimes be necessary to explore patients' working conditions and their relationships. Third parties might tell you: *'Yes, he has always been a person to ignore illness. He seems to think he is Superman. He doesn't do illness.'* Or, *'He's always been a hypochondriac. He runs to his GP with every twinge.'* *'Did he tell you how upset he became when he found out about his daughter's drug-taking?'* or *'The problem is his alcoholic wife. Added to his stress at work, it means he can find no peace anywhere.'* The more meta-data you have, the more appropriate can be your advice.

Statistical proof requires a group of patients to be tested anonymously. Anecdote requires closer engagement.

The Power of Anecdote

The art of medicine employs anecdote; the science of medicine employs numbers.

Anecdote can defy probability: *'I know what I saw. It was a flying pig!'* The anecdotal observer experiences, reports, interprets, and discusses, but will not always refer to the odds of disproving their contention. It is the reliability and integrity of the observer are then in question. From the perspective of those using recording devices and measurement instruments to yield data, anecdotal observations with no measurement are 'unscientific'. It is easy to dismiss anecdote as a lie or a mistaken observation, involving undue bias, misinterpretation, manipulation, trickery, delusion and chance, but then some of these also apply to the instrument measurements.

When picking an apple to eat, it would be wise not to close your eyes while choosing. To choose a tasty apple, experience-based bias is useful: avoid the green ones (unripe), those with wormholes (unhealthy), the bruised ones (damaged), and those not too small or large. The point is to never close your eyes and trust to luck. If you are brave enough to indulge in anecdotal research, prepare for inevitable attacks from quasi respectful, jealous, unhelpful, cynical, and distrusting colleagues. It is only right that you faithfully record, verify, and report your every observation (not just the ones that fit your theory). You have a duty to show that you have tried your best to falsify your observations and conclusions.

Contrasting with the many questionable sightings of the Loch Ness monster, and flying saucers, there are anecdotes that changed the future. Edward Jenner faced scepticism, when in the 18th century, he first reported cases of incidental vaccination (from vacca: Latin for cow). He observed milk-maids with the bullous eruptions of cowpox on their hands. Folk-lore had it that these milk-maids never contracted smallpox. The observation (an 'old wife's tale') turned out to be a game-changing, anecdotal observation. The widespread vaccination it led to saved millions of lives.

Koch was not so lucky. Although sure of his anecdotal observations, he had a dismal time (that involved a language problem) with Pasteur. His contention was that a microbe caused tuberculosis. Amateur scientists from the provinces, and their anecdotal evidence, had little chance of being taken seriously by the Parisian élites. Pasteur was correct to be critical and suspicious, but wrong to be closed minded. Koch returned with his postulates, and the criteria necessary to prove the transmissible nature of disease. Pasteur had to capitulate to a provincial doctor.

Because of the extreme variance of biological entities, one should only consider repeatedly observed, verified anecdotes, for further study. The hope is that some will gain universal application or lead on to further discovery.

One important class of anecdote is axiom. An axiom requires explanation, but no further verification. As classically conceived, an axiom is a premise so obviously correct that we can accept it without controversy. The word is Greek: axióma (ἀξίωμα): 'that which commends itself as evident.'

In clinical medicine, there are equivalent observations, conclusions and processes to be made. I call them 'BLOB' (**BL**indingly **OB**vious). Although never as obvious as an Euclidian axiom, like *the shortest distance between two points, is a straight line*, biologically, one can get close by adding definitions, i.e. *Every hand, held in a fire 'for long enough' will burn*. BLOB entities may not be axiomatic and will require further qualification, like *'how long is long enough?'* As a party trick, the rapid transit of a hand through a flame will not burn.

Mary Bann was lying in bed with nausea, muscle, and joint pains. She had a temperature of 39°C. She asked her daughter for some peppermint tea, but what came, tasted like soapy water. Her daughter, Jessica, must have made a mistake. After some cross-questioning, there appeared to be no mistake. When Mary asked to view the pot containing the peppermint tea-bags, she could not smell its contents. It was later confirmed she had COVID-19.

Many patients reported a lack of taste and smell, long before anyone acknowledged them officially as pathognomonic symptoms of COVID-19. It took months for cautious UK public health experts to verify this and to add them to a list of key symptoms. Apart from sinusitis, I know of no other febrile illness which causes anosmia.

Anecdotal medical research will not now get published in a prestigious scientific journal. Some medical journals make an exception for case reports. To get work published, one must be statistically savvy, serving up data to peers in a form technically sufficient for their acceptance. Never undertake research on any anecdotal observation, unless you have observed it repeatedly. It might make you a better doctor, but could be a waste of time and effort.

God Complex?

A scientist's acceptance of a favoured thesis, and religious belief, bear comparison. A scientist may want to read one hundred research papers on the same topic before believing anything. In the end, both the scientist and the religious believer must make a leap of conviction (if not faith), to accept a new idea. The scientist and religious disciple often share another trait – sanctimony. As guardians of something sacred, they will both argue their case vehemently. Scientists are liable to go one step further, and become self-righteous. What need is there for humility when they can 'prove' their beliefs? They do not mean 'proof' of course; they mean the balance of probability is in favour of their argument.

Scientists are logical thinkers whose self-esteem derives from the verification of their theses, and the usefulness of any workable spin-offs that result. While happy to accept credit for a discovery, some can go further. They not only assume credit for their work, but somehow assume credit for the creation of the phenomenon itself, when all they did was expose and explain it.

In the absence of any need to acknowledge a creator, the attribution of eponyms has sufficed. The relationship between heart rate and cardiac output existed before Frank or Starling described it (the Frank-Starling Law). Just under 400 years ago, William Harvey attributed the rhythmic movement of the heart to God's will (*De Motu Cordis*, 1628). It would have been heresy for him to call it 'Harveian' movement. Although he knew nothing of atrial cell intrinsic pacemaking function, cell wall conductance, or the autonomic system, Harvey risked being burned at the stake had he refuted God's will in motivating cardiac pacemaker function.

False Logic and the Wrong Questions

Guglielmo Marconi first transmitted the letter 'S' in Morse-code (dot – dot - dot) from Poldhu in Cornwall, to St. John's Newfoundland in 1901. He thought that sea water, being a saline-containing conductor, had helped to propel it. He knew nothing of the ionosphere (a Rumsfeldian, unknown, unknown). His false theory did not hinder the development of wireless telegraphy.

If you ask the wrong question, use false logic, or make false assumptions, no amount of analysis will lead to the correct conclusion. On occasions, however, luck leads us there.

Consider the following propositions:

1. Eating eggs causes coronary heart disease. In the 1970s, the accepted train of thought was: eggs contain cholesterol; atherosclerotic plaques in arteries contain cholesterol. If we eat eggs, our blood cholesterol levels will rise, and cholesterol will deposit like snow on the interior walls of our arteries (the intima), forming atherosclerotic plaques.

2. The slow synthesis of lipid (cholesterol), together with calcium apatite, occurs within the arterial intima (plaque formation) and can gradually limit blood flow or block arteries. Angina is likely when a coronary stenosis reduces the lumen by 85%.

3. Some lipid-rich atherosclerotic plaques become unstable. Some fissure and attract platelets and clot. Both can cause rapid arterial obstruction (the basis of cardiac infarction when it occurs in a

coronary artery).

Proposition '1' is incorrect; the ideas expressed in '2 & 3' are correct. Although plaque formation seems explained by a completely logical train of thought – that cholesterol 'falls like snow' and sticks to the intima ('deposits', many still call them) – it is mostly wrong (only a small percentage diffuses through). The predominant process within the intima is not deposition, but lipid synthesis and LDL oxidation (in the absence of sufficient local nitric oxide generation).

There are two errors to consider. A completely unknown process fostered a misconceived theory. We did not know that the intima was metabolically active, and synthesised its own cholesterol. The second error, resulted from inductive reasoning. We all know about the silting of rivers, and this led to the assumption that atheroma must develop in the same way.

If trains of thought and logic are to lead to correct conclusions, they must stem from correct propositions.

Measuring blood lipids has its uses. Low-density blood cholesterol (LDL) and atheromatous disease are strongly associated in populations (blood HDL cholesterol is negatively associated). Although the association holds true for large groups, it disappears at the individual patient level. Blood lipid measures, predict neither the macrophage-related inflammatory changes in the intima, nor the cholesterol and calcium apatite accumulation in individuals.

Only 60% of individual patients with carotid atheroma have a raised blood cholesterol. Epidemiological group statistical analysis makes a different point: *in a large group, with an average blood cholesterol of 8 mmols/l, there will be many more cardiac infarctions than in a comparable group with an average blood cholesterol of 4 mmols/l.* Why the disparity? Again, the statistical paradox applies. The set of genes responsible for raising blood cholesterol is likely to be different to those responsible for atheroma formation.

Dyslipidaemia (abnormal blood fats) is sometimes associated with pro-infarction clotting factors like fibrinogen. The presence of atheroma (carotid and otherwise), a high LDL, and a low HDL taken together, can indicate an increased cardiac infarction risk in individuals.

It is pure speculation to suggest that inter-connected nutritional factors play a part. Dietary vitamin K and vitamin E, can both influence blood clotting functions. One might also associate a high blood cholesterol with vitamin K dependent, calcium formation in the intima. The amino-acids arginine and taurine in food can both affect atherogenesis directly (arginine is the substrate for nitric oxide [NO]) production in the intima, perhaps helping to prevent the formation of pro-inflammatory ox-LDL, since it displaces oxygen); taurine has a 'cardio-protective' effect in rats, reducing cardiac infarction size after coronary ligation.

After twenty years of scanning the carotid arteries of thousands of patients, my observations suggested that a positive cardiovascular family history (as a surrogate for genetic profile), together with carotid atheroma (not just intimal thickness), are better than blood lipid analysis for predicting the future cardiac events of individuals.

At present, only statins and PCSK9 inhibitors (Punch & Kline et al 2022), halt the LDL-mediated progress of atherosclerosis. Many plaques in the carotid arteries are calcium rich rather than lipid rich, and calcium apatite (as found in bone) is there to stay. The decrease of blood cholesterol (from the hepatic effect) caused by statins may not be a reliable indicator of improving arterial pathology. Measuring actual atheroma progression itself, is more likely to be pertinent.

In Search of Truth

Many people in affluent western societies are concerned about their blood cholesterol and body weight. We are told to diet and exercise, even if it kills us (see my book: Who Loses Wins, 2024)!

Despite increasing average body mass over the last few decades, the incidence of cardiac infarction has reduced by 40% since the 1960s. Is it erroneous, like the other errors listed above, to link atheroma and obesity?

Exceptions are not the usual focus of statisticians, but very much the concern of every patient and clinician. From a distance, group statistics point us towards valid group associations, while our own observations can suggest something different. We are told that obesity is a risk factor for cardiac infarction (statistically), yet in clinical practice many slim people have angina and heart attacks. Anecdotally, coronary care units are not full of obese people only. I suspect that those without atherosclerotic (atherogenesis) genes will only rarely get coronary artery disease, whatever their bodyweight, blood cholesterol, or lifestyle.

An important clinical question is this: how often is it misleading to tell patients that eating saturated fat, being obese, and smoking, will significantly increase their risk of a heart attack?

Never forget to assess every proposition in the light of your own observations, before accepting any statistical viewpoint. It is nice when they match, but that too can occur by chance.

In order to reveal the truth, there are strategies intrinsic to the scientific method:

1. We must banish selection bias with random selection,

2. We must minimise the effect of natural variation with matched groups, and

3. We must change only one variable at a time.

These all restrict our view of reality, but have allowed the discovery of many significant variables, even though clarity demands the removal of extraneous factors.

The nature of patient variability makes research into the art and practice of medicine a challenge. Patients are inconsistent and present with many exceptions. Extensive variation is natural, and the basis of evolution, although the survival of only the fittest, remains politically unacceptable. There are political moves to make it equally unacceptable in the animal kingdom.

Some acquaintance with variation and unpredictability is essential in all medical work. While statistics can point the way for the 'average' patient, we make a leap of faith when we apply any statistic to any individual. While it is useful to be pointed north from London when trying to find Oxford Road, Manchester (where the GMC lives), many questions remain. A compass direction alone will not get you to their front door. A medical practice analogy exists: some doctors will point the way forward, then leave their patients to decide where to go next.

Statisticians forever point to a common shortcoming: researchers collect too little data from non-randomised cohorts; they then expect reliable statistical analyses. Doctors often disregard this advice, thinking that somehow, modern mathematics can always produce significant results (a variant of the halo bias – thinking mathematics magical). A few doctors trust all statistically derived conclusions and believe them never to be misleading.

There is nothing new about the belief that those appointed to powerful positions must understand most. This belief allows every society to accept control by incapable, anonymous people.

Computing the Truth

There is an art to valuing both statistical and anecdotal information. Even if both are valid, most of us have preferences. The clinically detached, experimentally minded doctor, will prefer statistical evaluation; the clinically attached are more likely to prefer the anecdotal.

What is true for me, may not be true for you. So, how are we to reveal the truth of what works best for each patient? We cannot rely on any averaging process.

In a murder case, the statistics of murder provide some guidance, but are not much use to the detectives working on a case. The figures might inform the investigator that, *'chances are the murderer is male (40% are female), and approximately 5ft 8ins tall'*. 'Chances are', is a phrase we must attach to every statistically based deduction. No statistical analysis can detect 'who done it', so this information is of little use. The answer will come from detectives sifting through all the anecdotal and scientific information for clues (absent; true, false; misleading; half-truth, and politically expedient, etc.). They will need to be sensitive to the relevance of each fact, and which of them is key. The verified details will populate a picture. Using judgement and interpretation, a picture might form that reveals the murderer.

> *'I am convinced, to within an acceptable margin of error, that he loves me.'*

Frasier: *Adventures in Paradise 2:* Frasier's ex-wife, Lilith, a rule-based bean counter by nature, considering her new relationship.

> *Those who lack the art of knowing what to choose, can always calculate outcome probabilities.*

Except for those guided by fashion, knowing which car most people favour, will not always help to decide which car is best to buy. The key factors will be individual personal preferences and financial limitations.

We should all become more confident of our experience, and a lot less in awe of probability calculations ('p' values): there are just too many statistical and methodological devils in the detail.

A Place for Academics?

Use your personal observations and what you once learned from trusted and experienced teachers, to form your clinical views, opinions and value judgements. Rely on the dictum: *'observation, observation, observation'*. Once you are sure enough of the truth of your observations, only then let any bean counter loose on your data. Every competitive academic will enjoy the challenge of proving you wrong.

Those in possession of bean counting skills may be ill-equipped to deal with the diverse and fuzzy nature of humanity; specifically, human thought, feelings, and behaviour. Knowing this, will be reason enough for them to have chosen a bean-counting career. Bean counters prefer to juggle with one ball at a time; competent physicians must juggle with many balls, all at the same time.

Medical academics sometimes gain direct clinical responsibility. One can easily recognise them: they frequently quote published research results (using acronyms), rather than their experience. Their technical knowledge may exceed their clinical acumen. As a result, they will make errors ascribing clinical relevance. They may cast their ideas in objective and logical terms, but not always in human terms. If they get to select medical students, they will favour academics (the greatest number of the highest grades), or potential administrators (politically expedient). Both strategies will lead to future medical professionals choosing to become less acquainted with patients.

We need academics for medical research, but we need many more doctors to deliver medical care. Unfortunately, very few of the medical academics I have known, favoured the art of medicine. Could it be that they are deaf to the muses capable of making the human spirit sing?

A fetish for precision . . . leads to formalism.

How the French Think (2015). Sudhir Hazareesingh. Allen Lane / Penguin.

Generalisation and Validity

Although extrapolations from the general case to the specific, and vice versa are unreliable, there are caveats.

It is human nature to make generalisations. That is because there is an evolutionary advantage to it. Generalisations are useful when trying to avoid danger. They also help reduce the burden of too much information. Since generalisations always involve ignoring detail, and dismissing exceptions, how reliable can they be?

Generalisation that result from repeated observation can be more accurate than many think, especially when trained observers are aware of their biases. Generalisations become inaccurate and misleading, when too many observations become deleted, cherry-picked or falsified. Political purpose is a common source of bias.

My concern is the medical and scientific generalisation that attempts to crystallise clinical issues. We must now live with legalistic generalisations, formed into guidelines, some of which are clinically untenable. Two simple examples are: 'URTIs (upper respiratory tract infections) should not be treated with antibiotics' and, 'single-vessel coronary disease should never to be stented'.

Generalised views can be statistically relevant. In normally distributed data sets, two-thirds (68%) of the observations of a measure, will often lie within one standard deviation of the mean (standard deviation is a measure of data variance). This implies that generalisation, based on what we see as 'usual' (the mean of all observations) can be informative. But how are we to know when it is representative?

Anecdote occasionally provides an incontrovertible conclusion. Take the generalisation 'always smell your food before you eat it'. One sniff usually defines whether milk has 'gone off' or remains drinkable. One sniff of a previously bought uncooked turkey, will help decide whether to cook it for Christmas lunch.

No elephant gives birth to a giraffe; no apple grows on a pear tree, and few survive jumping from aircraft without a parachute. These are BLOB generalisations; there is no need for qualification. Because of the march of evidence-based legal determinations, I still find it strange that nobody has yet felt the need to study the survival benefits of parachutes.

Although churlish and pedantic, some cannot resist asking: *'What is your evidence for that?' 'How certain are you of your facts?' 'How can you be sure that you are not fooling yourself?'* An obsessive need for certainty is understandable, but sometimes, it lacks common sense.

Summary Thoughts on Clinical Numbers

Successful clinical practice will rely on the personal catalogue of cases one has met, with all the variations included. To be of most use, one must include every case, with no cherry-picking or side-lining of exceptions. One should resist fabrication, selection preference, and single case emphasis, even though our biases pull us in these directions. One must choose wisely from a personal nebula of anecdotal clinical data, in order to make a diagnosis and plan the best management. In a generalised or crystallised anecdotal form, clinical experience can aid teaching and communication. Doctors must learn to select examples that will aid diagnosis, help patients understand, and help others to learn from their experience. The ability to select these appropriately is an art.

Statistical analysis can provide the numerical equivalent to a generalisation. Using the range, mean, mode, standard deviation and distribution, are all summary measures. Although useful artificial descriptors, they cannot convey the richness of experience. Despite this, many statistical results have beneficially impacted the prognosis of large groups (smoking cessation, wearing seat belts, lowering speed limits, improved hygiene). Always assume, however, that there are devils in the detail. While seat-belts in vehicles have saved many lives in road traffic accidents, they have also trapped drivers in their seats, and caused death from side impacts.

We can reduce clinical reality to a generalisation, but only by jettisoning information. Reduction of the complicated and complex to the simple, is useful for teaching, but of no use to those trying to master a clinical art. The results of numerical analysis are a colourless substitute for experience.

Numerical models are useful, but wrong when based on insufficient or inappropriate data. They can serve as one-trick ponies in the short term (like weather prediction), or remain true for millennia. Without modern technology, ancient Britons, the Aztec and Maya, each created equivalent stone structures that still predict the seasons.

PART 2
COMPUTER CONSIDERATIONS

Do we control computers, or do they control us?

Martian 1: 'These humans adore their computing devices.'

Martian 2: 'Of course. That's because they love to network. Many prefer not to think for themselves. All credit to earthling sales agents: they have convinced Earthlings that these devices are not only 'smart' and fashionable, but essential to life.'

Martian 1: 'Is that a bad thing?'

Martian 2: 'Not at all. Deluding and biasing themselves are human weaknesses, and we take advantage of that. Once addicted, they will waste more and more time and money on their social media apps and computer-based trivial pursuits. That means they will make less trouble for us, and won't notice us grabbing their personal information, manipulating their opinions, feeding them information, or making them work harder while we take it easy.'

Martian 1: 'What happens if the masses lose their jobs to avatars and robots?'

Martian 2: 'That's a worry. Unemployment and unrest could arise. They might start fighting for jobs. The élites might come to resent supporting the idle masses, recall their loans and stop financial services. The economy would suffer, and some might even try to escape to Mars; they have been planning it for long enough.'

AI, Real Intelligence or Both?

The need for convenience typifies the current economic era.

To improve convenience in everything we do, we need fast communication. We have moved on from smoke signals, Morse code, signal flags, and pigeon post, but our purpose remains the same: to send and receive information, and to get key information to decision makers as fast as possible.

We hardly need reminding of what we once achieved with a stylus and papyrus, designing pyramids, and writing the Dead Sea Scrolls. Composers wrote symphonies for centuries using only a quill and paper, while those yet to be called scientists, used them to record their revolutionary ideas and theories. Then came Charles Babbage with his difference machine (1820s), and Ada Lovelace with the first computer algorithm (program). Many now rely on computers to process information in abstract (resigning responsibility and any need for insight).

Computer sales 'hype' refers to digital technology as a fundamental step for humanity. We are told to climb on board, or miss the boat. What has actually changed in information handling, other than how recorded, processed and transmitted, since someone first wrote on papyrus or calculated with an abacus? Certainly speed and convenience, although neither represent a change equivalent to introducing writing, mechanical power or Charles Babbage's difference machine.

The barriers of distance and time have diminished with increasing speeds of processing and information transmission; both are welcome developments. This progress has depended on the generation of reliably supplies of electric current (Tesla's invention); the development of electronics, with transistors in integrated circuits following on from hard wired thermionic valves. In parallel, we have been losing the art of taking our time: *negotium* replacing *otium*.

Computer programming has developed from binary number commands to the formulation of 'high-level' programs using words as coding instructions. Windows® then added pictures to word and numeric displays. Because of faster processor development, neural networks and AI can rapidly compare the information we have to that we have stored; they do this to calculate the odds ratios of matches. Real-time navigation, better short-term weather prediction, e-trading, and better dissipated scientific research information, have all resulted with no need for physical libraries. Inventions such as the world-wide-web, GPS, search engines, fast AI computer chips, and more accurate clocks, have allowed these changes. They have all progressed for one reason: to improve convenience, although many simply want to play chess to Grand Master level. Their reliability and security remain issues of concern.

In 2018, a publication detailed a reliable retinal image recognition system, using a convolutional neural network (CNN). The computer system was 'trained' to 'recognise' normal retinal images, as well as those showing haemorrhages, micro-aneurysms, exudates, and neo-vascularisation. The error rate of the system was only 1%, and compared favourably with the opinions of senior ophthalmologists (Lam, C., Yu, C. et al.

Retinal Lesion Detection with Deep Learning using Image Patches. Investigative Ophthalmology & Visual Science January 2018, Vol.59, 590-596. doi:10.1167/iovs.17-22721). The reading of mammograms and other images, has advanced in a similar way.

Take two chest X-rays, one a few seconds after a peripheral venous dye infusion. Subtract one picture from the other and one can see the dye distribution (unseen without subtraction). With both images digitised, one can subtract one image from the other, pixel by pixel, and get the same result (albeit more laboriously). Because current processing speeds are so fast, any differences between the two images, takes almost no time to recognise. One can then compare these image to other digitised images, taken from those with known diagnoses, like emphysema or primary pulmonary hypertension. This makes the process diagnostic: positive when a probable match is found, and negative when it is not.

ChatGPT can write a book in ten minutes, and can alter the style of text, or summarise the research on any subject listed on the internet. This facility alone will save authors and researchers lots of time, but what of the output quality? Is it so stylised that its origins are obvious?

The quest is on to make a machine that can pass the Türing test; good enough for humans to think that a computer is human.

Machines and Algorithms

Such is our unbridled faith in computers, we now think that one day they will generate all the important questions, and answer them. After posing the questions, they might even write the programs needed to provide the answers.

Recently, Google's 'Chatbox' LaMDA (June 2022), achieved sentience. There is, however, no definition of either intelligence or sentience good enough to allow this judgement to be made. When the term Artificial Intelligence became coined in 1956, the computer scientists of the day could think of nothing better than to create a chess-playing program. They thought that if a computer can play chess, it must be intelligent. Instead, the program simply followed a set of rules, better than any human.

The more we anthropomorphise robots and chat-boxes, the more we will see them as having intelligence and sentience.

Can we really look forward to a world where nobody needs to write, calculate, or think independently; a time when all our desires, peculiarities, and predilections need no personal effort (although analysed and accessible to others). If so, we can say goodbye to privacy.

'Computers are useless. They can only give you answers.'

'Pablo Picasso: *A Composite Interview*'. William Fifield (1964), Page 62, The Paris Review 32, Flushing, New York.

I once studied electronics at night school (I soon became bored with general practice) and built some simple electronic systems. I then studied computer programming and developed an office computer system. As a result, I gained some practical knowledge of computing hardware and software. For over ten years, my office programs were in daily use, with a few minor faults appearing as different people used them. I patched and re-patched my programs the instant any fault appeared, but they never achieved faultlessness, even after ten years of use.

Any mention of a faultless computer program is sales-talk, with manufacturers understandably reluctant to discuss hardware and software reliability. Most of us don't ask (lest we make fools of ourselves). Many find themselves in a bubble of awe, demeaned by inaccessible jargon, whenever computers and computing are being discussed. Computers may contain integrated circuits, but they still need to be wired and soldered. Solder dries and cracks, power supplies vary, batteries run flat, and connections become loose. So, who would entrust their life to them? The answer is to have several of them, with lots of different power supplies.

As we approach the zenith of the AI hype-cycle, many regard questioning the reliability and usefulness of computers as anachronistic. Only after studying computer programming and electronic circuit construction (and thus, algorithm generation) can one see beyond the sales pitches. For decades, the computer industry has successively replaced one promise with another. An early promise was to rid us all of paperwork, and to increase our leisure time. One of the latest promises is that big data accumulation will allow the discovery of unforeseen medical secrets. Another is that quantum computing will test computer programs for every possible coding error, correct them, and make them totally reliable. Using the same 'blue sky thinking', 'deep learning', and GANs (Generative Adversarial Networking), we await quantum based, neural networked, heuristic computers, that can 'learn' from their mistakes, and create usable programs of their own (see Google's AutoML). One idea is to leave no cognitive role for most humans; doctors would then be free to concentrate on the art of medicine. The total control of civilisation would then, of course, be completely in the hands of anonymous bureaucrats, politicians and the computers they use.

Unfortunately, no digital format has remained the same for long, whether it be the hardware, the operating systems, programming languages, or data storage formats. None has lasted as long as the pen, ink, papyrus, and paper. Each computer generation is useful only for a while. Meanwhile, faith in computers continues to drive a massive industry with exciting new possibilities offered by AI and robotics. With it, a vain pursuit has arisen: to always be up-to-date and fashionable. At the herd instinct level, we foster this by regularly producing new, 'must have' devices, that many struggle to afford.

Each robot in a factory replaces 3.5 workers on average (Ian Goldin, Oxford Professor of Globalisation). Meanwhile, AI and robotic hype can help economies; without them, there would be more people out of work. Work clearly contributes to our quality of life. It provides the self-respect, dignity, and respectability that a totally robotic world would deny us. In the future, we might need to become more imaginative and allow computers to help our productivity. This happened in the first industrial revolution. Machines made us more productive, but at a cost: unemployment, re-employment, relocation, and re-skilling.

Robots have the potential to remove some of our dirty, repetitive, and dangerous activities. They might, perhaps, free up more time for learning and leisure. Retired, and less able folk, could engage in work activity if they wished, and change the definition of who is too old to work. Some might help with the care of others, with machines doing the heavy lifting, if not the care itself. We already use them for teaching, and for training to do jobs we find difficult. We need to accept that machines will never have real human sensibilities. Why should they when there are 8.1 billion human examples available (2024 world population)? They can remain as useful tools for information handling, with robots freeing human hands from physical work.

Where empathy, sympathy, ethics, morals, and human inter-reactions pertain, humans must hold the centre ground. For calculation, compliance, transformability and adaptability, robots will always have the edge. Where the intention is to monitor, regulate, control, and sanction us, the danger is that bureaucrats and policing services, will want to develop intrusive computer systems to the full.

We could structure an algorithm to select employees. It might process the data, find that women in top jobs are rare, and reject any woman who applies. We could easily correct this, but there are many variations of this programming error. Your own personal data, processed similarly, could get you purchasing offers, get you arrested, get you employed, or offered a place at college. As with so many appropriate decision and reasoning functions, there will always be multiple levels of detail to be considered. Quantum-based, neural networked, heuristic computers are the current wizards of Oz; the hope is they will make any offer made, more appropriate. Apart from military applications, one point of computers is to help make bigger profits.

Algorithms provide a fixed-rule basis for data processing. Laws have yet to be drafted that define who will be responsible for the consequences. Writing algorithms is at present (2024) the preserve of white males, with specific cultural, racial, and gender-based biases that need review. Serious ethical questions could arise for programmers. Should a robot-driven car turn left and kill an older adult person, or turn right and kill a child? Who will do the programming? Who will be responsible, and who will pay the costs? There will always be undiscovered devils in the detail leading to unimagined consequences.

In the USA, they have formed an Algorithmic Justice League to detect police algorithmic biases. Joy Buolamwini pointed out a major bias in older facial recognition software: it only recognised white faces (ex.ajlunited.org). Policing and auditing algorithms will become big business; that's another office tower block or two for our regulators and enforcers to occupy.

If we really want computers to 'think' for us, and 'judge' us fairly, we might need to engineer them genetically, perhaps growing them in vats. Until then, anxious (and wise) passengers will insist on two pilots being present in the cockpit of any aircraft they fly in (for decades, aircraft have had two pilots and at least two computer systems, each programmed independently). Recognising omissions and faults in new software often has to await an unimaginable event (an unknown, unknown). The two fatal accidents caused by the programming of the new Boeing 737 Max aircraft, provide unforgettable examples.

The day before their discovery, the first black swan, the heart-shaped methane ice formation on Pluto, and the hexagon shaped winds on Saturn's pole, were all beyond our imagination and any computation.

Algorithms and Computer Programming

In AD 825, Mahummad Ibn Musa Al-Khwarizmi (Muhammad from Khwarizmi) a Persian mathematician, known as 'the father of algebra', wrote a treatise on Indian numbers. In Arabic, 'al-jabr' literally means the restoring of broken parts.

Algorithms are programs made from a series of computer instructions. They define formulae and logical processes which calculate, and follow logical decision pathways that seem like reasoning. The point of what follows is to illustrate just how useful they can be, and how many devils lie in the detail.

Prof. Marcus du Sautoy gave an elegant overview of the development of algorithms in a BBC program entitled *'The Secret Rules of Modern Life* (2015).' I have extracted the following points.

Euclid was perhaps the first to create a series of instructions to find the largest common divisor of numbers. Euclid's algorithm still has its applications. There are now many types of algorithms the most useful of which will sort lists and find sequences. Larry Page and Sergey Brin (1996) advanced web searching by devising a ranking algorithm for those web pages that are most interconnected; they then listed them in order of interconnectedness. The matching algorithm (to find couples likely to form stable relationships) created by David Gale and Lloyd Shapley won them a Nobel prize (2012). Similar algorithms are now used to find donors for renal transplantation involving several couples, as well as blood donors.

Most exciting are algorithmic processes that are constantly revised from experience. In this way, Amazon and others involved in marketing, can offer us suggestions of what books and films we might like. Airlines use dynamic pricing: identifying demand and changing the seat prices to maximise profit.

Logical, algorithmic processes, use Boolean operators like '.and. .or. .not.' to process information in a series of prescribed steps. Here is a very simple example:

Step 1: Input data: time of day (24h clock).
Step 2: Process the time with the following algorithm:
If earlier than 9.30 am,
DISPLAY: 'Order Breakfast';
if later than 18.30pm,
DISPLAY –'Order Dinner'.

My self-written doctor's office system was error-trapped to detect any false key-stroke, or incorrect entry. What resulted was a screen display that read: 'YOU HAVE MADE AN ERROR – TRY AGAIN.' I considered displaying: 'FOR GOD'S SAKE! PLEASE SEND FOR SOMEONE WHO KNOWS WHAT THEY ARE DOING', but I resisted.

Some of my staff, while learning how to use my system, tried to convince me they had NOT pressed an incorrect button. Instead, they told me it was the computer's fault. My error trapping naturally led to some embarrassment. The power to keep my staff 'in-line', with no effort on my part, clearly gave me an advantage. It also encouraged honesty, and some insight into the personal competence of those using my system. I did not record their errors, but those whose primary interest is compliance will.

Can computers think? This is now an old academic chestnut. One might ask, what form of thinking? If you are a computer programmer, even at an elementary level, you will know how definable computing program structure is in comparison to human thought. The apparent complexity of computers (thought to be 'alter intelligence') results from straightforward Boolean logic operators and other processes like looping procedures (if a mistake .or. a mismatch happens, go back, and repeat the process). The path leading to a conclusion (as in every maze), will divide at many specified points. This is where the code asks: IF this, do that; ELSE – go for a coffee. The discriminatory criteria used to select given pathways are the logical Boolean operators AND, NOT, and OR. (if 'x'>1 .AND. 'y'<10, calculate a + z). One could equally use '.NOT.' or '.NOR.' if required. This emulates logical thinking, but nothing further (from a human perspective).

Here are some simple examples of what computers do when they seem to 'think':
Example 1: IF (you are an elephant) .OR. (a pig), DISPLAY: Take a mud bath every week. Applying this recursively through each record in a database, the program will find all those who appreciate mud baths, but only if the database has recorded this interest.

Example 2: IF (human) .NOT. (female) .and. age> 18 years, DISPLAY: Consider fathering babies.

Example 3: IF (human) .AND. (height > 2.0metres) DISPLAY: Consider becoming a high-jumper or basket-ball player.

Example 4: IF (a qualified doctor) .AND. (age <65) .AND. ((MRCP) .OR. (FRCS))
DISPLAY: 'You are eligible for hospital work.'
ELSE

    IF (a qualified doctor) .AND. (age <65) .NOT. ((MRCP) .OR. (FRCS))
    DISPLAY: 'Consider taking the MRCGP exam and working as a GP.'
    ENDIF
ENDIF

The exceptions: *'usually, but not always'*, and, *'rarely, but it happens'*, are not programming options, even though these are common situations in everyday experience. Neural network programs can help by using probabilities, rather than 'YES' or 'NO' answers to questions. With enough of them working on different levels, they can easily give the impression of thinking. They are not alone. With smoke, mirrors, and curtains, the Wizard of Oz easily created mystery and false impressions.

So far, no computer has created a new idea. They can, however, make pre-programmed or random choices, and rearrange some bits together as if creating something new. This has been done with jazz riffs. The result seems like music creation, but it is not.

Rubbish in – rubbish out; bullshit in – bullshit out; nonsense in – nonsense out. All computer-based outputs that seem cogent, result from programming instructions. Few doctors will know enough to ask relevant questions about individual computer code, so it is best not to accept any computer output uncritically.

Unlike computers, some humans have an inbuilt advantage: the ability to recognise error, nonsense, inappropriateness, and bullshit. Computers will need these abilities to pass any realistic Türing test (requiring intelligence and sentience).

There are some mathematical problems that could take years to solve. Repeated looping back, and matching one bit of data to another within a vast dataset, takes time even at light speed (NP [Non-Practical, exponential] algorithms). A smart guy will whittle down the data to essentials (once called separating the wheat from the chaff). I previously referred to this as nebula processing: processing that allows us to pick the best apples from an orchard full of apple trees; something equivalent to a large data set. Gordon Welchman did something similar at Bletchley Park, during the 2nd world war. He shortened the iterative processes used to decipher messages, by selecting the most relevant data first. He recognised the value of meta-data, noting the senders and receivers, and how often they sent messages. Welchman also noted the sending and receiving locations. If 20 messages each day were being sent from General X in Berlin to General Y in Calais, something important was happening, either in Berlin, Calais or both. Welchman's valuable methods of analysing meta-data became known as traffic analysis.

By taking an over-view, and observing indirect clues, we can sometimes make difficult diagnoses easier.

In contrast to a silicon-based digital device, the human brain has processing functions that work simultaneously on several levels (conscious and subconscious). Unique to the brain is the ability to select only the data that applies to the prevailing problem (using meta-information, selection and rejection). The idea of 'relevance' (always a human judgement) is intrinsically beyond the wit of a non-sentient machine (whether computers will ever achieve consciousness is a much debated question). Our brains can even make complex problem solving possible, by using reduction, clarification, and judgments of relevance.

By looking out of the window in the morning, one can decide (with reasonable predictive accuracy) whether it would be wise to wear light clothes, or to take a heavy coat, a thermos flask of coffee, a blanket and a spade to work, just in case it snows. Even the most powerful computer systems find it difficult to predict whether it will snow, although they can calculate the likelihood. The number of unknown devils in the detail, together with the many other limitations of computers, means that users have almost no access to processing details. With AI, like ChatGPT, what information they are processing will also remain obscure, yet most of us assume it to be accurate and reliable.

Algorithmic Awe

Computer salesmen have largely convinced us that computers have 'brains' superior to our own; at least, they calculate faster and can follow rules faultlessly. These are characteristics that the directors and managers of every business seek in their workers. Every successful business owner needs wise judgement and luck: characteristics with no digital match.

No human brain can match a computer for speed when calculating, sorting, filtering, ranking, and indexing data, although our brains have been successfully engaged in these processes for millennia. There is a difference: computers need to be constructed according to a plan and switched on. Computers need someone to set their objectives and to select the processing criteria. Computing, analysis, and memory functions are straightforward, but the wise application of knowledge can take decades of education and life experience. Since brains and computers are complimentary, we can assist one another. The power, however, will always lie with those who can switch them off.

Thomas Friedman, in his book *'Thank You for Being Late'* (Allen Lane, 2016), is clearly a protagonist of the hype and awe of computers. He quotes Carl Bass, CEO of Autodesk, who writes (on the collaboration between man and machine): *'with the computers' help, the designer is now able to understand the whole range [of any system] beyond what any human mind can comprehend on its own.'* Really? It would certainly have applied to the independent innovative thinker, and rare computer genius Mike Lynch (who died in a tragic sailing accident in August 2024), but perhaps he can explain what lies beyond human comprehension for others. Isn't that a void that only an imaginative copywriter can fill? He might also try to explain how Isambard Kingdom Brunel, Beethoven and Leonardo da Vinci, achieved what they did without a computer. It's a good sales pitch to suggest that the average person will become a genius – if only they had the right computer and programs. Computers can, however, put information into the hands of those incapable of arithmetic and those disinclined to read and learn.

The awe induced by computers is now so well entrenched, most advertisers, publicists and journalists, accept the hype without further thought. The trick is to spot the hype and romance, and resist the awe. There is nothing more in any computer program than the thoughts and directions planned by other human beings. All else is science fiction. There is no Wizard of Oz, magician, or ghost in the machine, even though we have all gained from being able to calculate at light speed, and to have seen the drudgery of processing information removed.

Computation has made possible the first constructed 'image' of a black hole from the massive data output of five or six radio-telescopes (comprising millions of terabytes of data). Computing this with pencil and paper would have taken rather a long time, except that knowing what most holes look like, it would have taken me only five seconds to draw one! Would I then become accused of fooling myself without sufficient data?

Computers can help to spark off new personal insights, although every new tool and an intelligent friend can do it also. You only have to consider the invention of the telescope and the microscope to glimpse the origins of some knowledge. Computers cannot comprehend significance, without a human first deciding what that significance is. Computers do well, however, fulfilling our pre-arranged requests. Until humans no longer have to switch computers on, change their plugs and batteries, keep them clean and cool, and defend them from outside interference, any superiority as thinkers and creators we attribute to them, will be science fiction, but still capable of making billions of dollars.

Many organisations have had their computers hacked for access to stored personal data. This requires an internet connection. My advice (which most corporations would have to pay millions for) is to keep all their most valuable data on a computer that is not connected the internet. That way, nobody can blackmail them with the threat of a disruptive virus. Some companies charge hundreds of thousands of pounds to provide effective firewall and protective software; my simple suggestion is free.

From the standpoint of those humans who built Rome, and painted the Sistine chapel, the only people who actually need computers, are those who wish to work faster and more easily exert control. It usually comes down to money and power, and not much else. Human talent, real (not artificial) intelligence, outstanding originality, creativity, and the insight of genius, are all beyond computer programming and nothing to do with the calculation of odds ratios. Computers can certainly help us utilise our talents and undoubtedly help many by providing a different way of working. They can save us time, but have we yet worked out what best to do with the time we save?

My first computer in the early 1980s was an 'Apricot'. It took 5 – 10 minutes to save my work to a floppy disc. I made coffee while it worked. Was this time wasted? Not really. It allowed me time to reflect on what I had written. Time spent thinking is nearly always worthwhile. Here is a top management tip: you do the thinking, let others do the processing.

Big Data

This is data collection on a grand scale. It is data driven, rather than hypothesis driven science. The thesis is that vast amounts of data, analysed sufficiently, will surely reveal something valuable. There is no good reason to collect so much data, and no good reason not to.

In 1970, I cross-correlated all my research data on atrial pacemaker function, regardless of any preconceived hypotheses. Back came reams of data analysis from the London University main-frame computer, which included every possible correlation coefficient. Some came as a surprise, others confirmed what I already knew about atrial pacemaker control (parasympathetic and sympathetic influences). Nothing useful came of it, except nerdish excitement.

Big data could by chance yield something of interest, like verifying alien life from radio-astronomy data, perhaps. If you have nothing better to do, and someone will pay you to do it, why not sift through lots of haystacks, hoping to find a golden needle? Also, let me know when you discover the first message from deep space. It will at least prove that medical bureaucrats are not the only alien species. Unfortunately, any message from outer space will be at least 40,000 years old (we are over four light years away from the nearest star, using time dilation), and their coffee will have gone cold, waiting for our reply which will take another 40,000 years to reach them.

Something good drives big data projects: the noble desire for exploration and completeness, although incessant activity can divert us from asking vital questions like, 'What is the point of all this?' and, 'What, exactly, are we trying to achieve?' The answer will undoubtedly be, 'just you wait and see'.

Has large-scale sifting ever been worthwhile? In the 1890s, Marie and Pierre Curie took years to sift through tons of pitchblende. In the end, they extracted a few grains of a new radio-active element, Polonium. A few months later, they discovered Radium. How many comparable sifting discoveries have we made in the intervening 134 years?

Computer generated statistics have an important place in the management of large corporations like the NHS and banks. So far, the NHS 'big data project', has cost millions (of GB Pounds). What prospect is there of them discovering anything new, or useful? I can think of two major benefits. Although a computer is unnecessary for it, one result might be to confirm that we experienced physicians actually know what we are talking about; another would be that medical bureaucrats have failed to provide the UK with an effective health service. Those who place no value on medical experience, learning and judgement, can always turn to big data projects for insight and understanding. Meanwhile, doctors and health workers will get on with the job of saving lives.

We elect governments to take responsibility for citizens. They mostly appoint MPs as ministers to departments, who know nothing about the job at hand. Is this a responsible way to manage any business, let alone a health business? Introspective, big data collection projects are a godsend to government departments; they will keep them occupied for decades.

Are big data projects an 'end of empire' phenomenon? Are they the modern equivalent of fiddling (statistically) while Rome burns? At least, bureaucrats using computers can justify the employment of many more NHS executives and managers.

The NHS has access to millions of patient data files. Once they standardise the data (how easy it is to assume the possibility), we might gain some interesting epidemiological insights. If it could explain why the poor have 3–5 times more life-threatening disease than the rich, I would be all for it.

Senator Bobby Kennedy, brother of John F. Kennedy, gave a speech in which he derided the value of one highly regarded national statistic – the GNP (Gross National Product). His point was that no statistic is an acceptable measure of anything really worthwhile. These and other statistics, require legions of government employees to calculate them, despite computerisation.

'… the gross national product does not allow for the health of our children, the quality of their education or the joy of their play. It does not include the beauty of our poetry or the strength of our marriages, the intelligence of our public debate or the integrity of our public officials. It measures neither our wit nor our courage, neither our wisdom nor our learning, neither our compassion nor our devotion to our country; it measures everything in short, except that which makes life worthwhile.'
And it can tell us everything about America except why we are proud that we are Americans. If this is true here at home, so it is true elsewhere in world.

US Senator, Robert F.(Bobby) Kennedy.
Presidential Campaign. University of Kansas, March 18, 1968.

Chapter Fourteen

Clinical Judgement

'What I set out to do was to virtuously and justly administer the authority given to me, and to do it with wisdom. For without wisdom, nothing is worthwhile.'

King Alfred the Great. 899 AD

A Few Aphorisms

Those who possess wisdom can exercise their judgement and bring content.

To make judgements and decisions that benefit the health and well-being of others, one must first recognise a responsibility.

Our judgements need balance, being sensible, reasonable, and proportionate, but never self-indulgent.

The best judges are intelligent, reliable observers, who have experienced similar problems.

The further from first-hand experience one gets, the less chance one has of reliable judgement.

Only by considering pertinent clinical data, meta-data, and applicable research data, can clinical judgement become effective.

Clinical decisions are best for patients when we consider every relevant fact (an art). The evidence alone (the science) should never blindly inform decisions.

For our clinical judgement to improve, we must expose ourselves to the experience of others.

To gain clinical judgement, become apprenticed to a doctor whose patients have clearly benefitted from his wise decisions.

··········

This chapter considers evidence-based judgement, experienced-based judgement, and some of the algorithmic and probabilistic processes, used in clinical reasoning and decision making.

Time can sometimes take precedence and force judgement. As the time available for action runs out, some judgements will become superfluous.

My patient, John G. was found guilty of fraud. He was 82-years-old and suffering from metastatic prostate cancer. In court, the judge asked me to comment on his case. I told him that whatever sentence he might give, he had already been handed a death sentence. The judge gave him a suspended sentence and John died four months later.

Judge if you must - but resist being judgmental.

If you are not sure of what to do, do only what is in the patient's best interest.

If you still don't know what to do, send for someone who does.

A Golden Rule for Judgement: Make no important judgement when hot with anxiety, fluster or stress; error, confusion, uncertainty, and indecisiveness can result. Judgements are best made when we are composed.

A Golden Rule for Diagnostic Judgement: With only one piece of reliable clinical evidence to hand, a diagnosis will be unreliable; two points of evidence will make it likely. Three points will usually suffice to make a diagnosis certain; four points will make a diagnosis beyond doubt.

The Process of Judgement: First, ask the most appropriate and specific questions. Identify, weigh, and prioritise those factors which accurately represent the patient's problems, while taking his circumstances into account. To know what applies as relevant to each patient, and what is not, is the art required. Using this art enables sensible, practicable, reasonable, and effective decisions to be made. To qualify as wise, a judgement must seem obviously correct, and then stand the test of time for aptness.

Some will refer to good judgement as nothing more than 'good luck'. Those who understand wise judgement see it as an active process, not a passive, formulaic one.

The Consequences of Poor Judgement

The consequences of poor judgement often induce stress. Poor judgement often involves accepting incorrect assumptions, asking the wrong questions, misinterpreting the evidence, speculating too much, and then making inappropriate choices. Those who suffer from poor judgement mostly refer to the consequences as 'bad luck'.

> *Of all the causes which conspire to blind*
> *Man's erring judgement and misguide the mind,*
> *What the weak head with the strongest bias rules,*
> *Is pride, the never-failing vice of fools.*
>
> *An Essay on Criticism* (1711). Part 2: 201-204. Alexander Pope.

For those hoping to achieve success in life, the most valuable cognitive function is judgement; preferably wise judgement. Success in life and individual human survival, have forever depended on it. The future destiny of humanity depends on it. A distinguished talent for wise judgement is rare, and found regardless of education.

Judgement is a complex, conscious cognitive process. The best judges probably inherit it as a talent (the structure of neuronal interconnections perhaps). Without evidence-based clinical judgement, experience-based clinical judgement, and some luck, many patient's lives become more stressful, and a good deal shorter. Contrast this with the everyday choices we all make in the name of safety, security, health, and happiness, many of which are based on hope and chance.

Experience creates a personal library of anecdotes, weighted for relevance by observing the outcomes. The value we each place on experience is personal and defies standardisation. Our individual experiences account for the rich variation in our attitudes, and the personal management of the life we each adopt. What physicians and surgeons need is enough perspicacity to differentiate clinical relevance from irrelevance.

When a doctor discusses her judgement process with a patient, she needs to know what drives the patient's judgement. It will be a challenge to deal with those driven by hearsay, cherry-picked internet searches, and emotional factors, rather than corroborated factual knowledge. Convincing others of what makes 'good sense', 'common sense', and 'wise counsel', is usually straightforward when correct, but will sometimes be thankless and unrewarding.

Most patients and some of their relatives, want to be involved in clinical judgement, decision making, and medical management. Although often helpful, it can lead to conflict. It is entirely understandable, for instance, that some will want their brain-dead child to be kept alive for a little longer, even if it conflicts with the judgement of medical staff. It is also understandable that some patients will refuse medication. Some have no desire to become a patient; others have no belief in pharmaceuticals (some see them as 'unnatural'). It is also understandable that some patients will refuse blood transfusions, or a life-saving operation for religious reasons. Sometimes, nothing one can say will change their mind.

We are all entitled to freedom of choice; freedom to make our own judgements, and freedom to think that our judgements are sound, even when based on guesswork, hearsay, aversion, fear, biased hang-ups, belief, the occult, guilt, anxiety, resentment or scorn. It will sometimes require sound clinical judgement, a large dollop of tact and diplomacy, and more than a cursory knowledge of the patient, to challenge their views effectively. Every patient wanting advice must consider whether to accept it or reject it, for reasons best known to themselves. The better one knows a patient, the more interactive will our decision making become.

'You should get a job', said the rich tourist to a jobless man sitting on a turn-stile. 'And why should I do that, Sir?' said he. 'If you make enough money you could afford to take it easy — just like me', said the tourist. 'Well, Sir. I've been doing that all me life without a job!'

Bounded Judgements

Judgements of Time

The first clinical judgement necessary is usually to classify cases as 'action NOW' (emergency tracheotomy); 'urgent' (acute bowel obstruction: action ASAP); 'priority' (melanoma diagnosed: to be seen before most others) or 'routine' (inguinal hernia). Time priorities change, and need regular review. For those with chronic disease conditions, languishing on a waiting list could mean they will die waiting. The annals of the NHS are replete with such cases; the COVID pandemic of 2020/21, significantly increased their numbers.

Do you have enough time for action? Mundane they may be, but questions about time are ever-present. Do you have time for coffee? Should you examine a sick patient every hour? Can you leave an operation until next week? There is an art to timing, and an art to judging how much we need.

In acute medical practice, time will often be a critical factor. Will you have enough time to operate on a stabbed patient, bleeding profusely? Will sepsis overtake the patient who has already waited 24-hours to receive iv antibiotics? Judging the timescale for action must come first. How long will it take to get the patient into an operating theatre, and how long will the operation take?

Every illness is a dynamic process, so never ignore clinical timescale when arranging appointments. Having asked: Is the patient 'ill', or 'not ill', one must then ask, is the patient's case routine, urgent, a critical emergency or somewhere in between? Is the rate and direction of change fast or slow; improving or deteriorating? Do you have minutes, hours, or days to deal with them? Because best practice must include the dynamic aspect, never omit these questions. Some doctors, nurses and paramedics focused on form filling, are liable to ignore time dynamics.

I recently asked a senior GP trainer and MRCGP examiner, how much of her time at work she spent on administration, rather than consulting with patients. Her answer was '50%'.

You need to look at patients purposefully: inquisitive enough to find problems; never disinterested, hoping not to find any. Doctors feeling out of their depth, may also hope not to find a problem, in case it exposes their deficiency. A deteriorating, dehydrated older person, may need an intravenous infusion now, not tomorrow. If the patient is not ill, defer the consideration. The febrile sweaty patient with a low BP, might have internal bleeding or septicaemia. Act now, not in twenty minutes. Act first, consider the paper-work later. Prioritise patients based on their rapidity of change.

'Timing is the essence of life, and definitely of comedy.'

Bob Hope

'Observe due measure, for right timing is in all things the most important factor.'

Hesiod. *Works and Days.* circa 700BC.

Surgeon Mr. David Sellu, with 40-years' operating experience, and no complaints against his name, had an urgent case of faecal peritonitis to deal with. Time was of the essence, but he could find neither an anaesthetist, nor an available operating theatre (in a private hospital). Having suffered a longer wait for an operating theatre than advisable, the patient died post-operatively from septicaemia. They subsequently charged Mr. Sellu with gross negligence and manslaughter. They found him guilty, and he served a prison sentence ('Did He Save Lives?' David Sellu. Sweetcroft Publishing, 2019).

The judgement of urgency is a doctor's job; never leave action to managers or bureaucrats. Quickly seeking alternatives can save lives.

We need **Judgements of Measure** when we decide the distance of a putt in golf; the angle of a snooker shot, and whether a patient's blood pressure, blood potassium or creatinine levels, are clinically significant.

We need **Judgements of Value** to evaluate the relevance of each symptom, sign, and investigation result, before using them to make a diagnosis.

A friend asks you to recommend a 'good' restaurant nearby. Your judgement of which restaurant might be best for him should depend on his personal taste. Before you confer the blessing of your value judgement on others, consider your questioner's cultural background, taste, and aspirations (throughout my life in the UK, this has implied knowing their educational level, and socio-economic status). It would be inappropriate to recommend a Michelin 3-star restaurant to those with limited funds, or those whose first love is 'Fish & Chips', or 'Pie & Mash'. Beware though! There are exceptions. My patient June Robins (R.I.P.) and her family, have run a chain of Pie & Mash shops in East London for decades. Few knew more about Michelin-starred restaurants in London than June.
A friend of mine liked restaurants that served home-style, basic food, at low cost. When he said a restaurant was 'marvellous', I knew just what he meant. His only knowledge of anything Michelin, related to car tyres. He used only one criterion for his every choice – its cost. He avoided buying Michelin tyres; he thought them too expensive, even though they would have stopped his car quicker than any cheaper version. Others employ the opposite strategy: they only buy the most expensive things. They believe that 'you only get what you pay for'. The evidence published in 'Which' magazine often debunks this misconception.

Estimation of Value

Clinical judgement uses both estimates of measure (degree of coronary stenosis, and level of creatinine), and value (the relevance of specific symptoms, signs, and investigation results). Diagnosis and clinical management decisions involve both. Only with the correct values assigned, and the correct choices made, will optimal clinical management be possible.

Doctors often use guesswork, if only as a temporary measure. When we use guesswork, or an appeal to chance, we elect to choose and value what we 'feel' to be correct. With a lack of information, informed guesswork may be all we have. As an example, Tony Blair and Saddam Hussain's supposed WMD (weapons of mass destruction) in 2003, come to mind. In this type of situation, we may have to rely on judgements based on likelihood; assigning a presumed value to each option, while considering the odds of being wrong. Being effective in emergency situations requires a facility for this class of judgement, while being able to handle the time trajectory and time pressures that can force decisions. Some will be at a complete loss, faced with a situation where there is too little information, and no time to spare. Those with a talent for this type of decision making should become leaders on battle fields or work in A&E.

For non-urgent cases, we should not curtail the time needed for judgement and decision making, although the more experienced and knowledgeable one becomes, the less time we should need.

The need for initial guesswork is ever present. It persisted throughout my 53-years of medical practice. Uncertainty in medical practice is an intrinsic, everyday phenomenon. In complicated cases, I sometimes took twenty-four hours to ascribe value to the clinical data, and make a clinical judgement I was comfortable with. Ascribing the correct relative value to the data, and meta-data, can make the process lengthy. Once one has made a judgement, one must repeatedly review it, trying to find fault in it before others.

Judgements of Compliance

One simple class of judgement tests compliance. To exist comfortably in any society, citizens (by supposed mutual agreement) must comply with its accepted rules, regulations, and laws. This requires every society to employ a multitude of regulators to wave rule books before their citizens. Judgements involving compliance are an everyday issue: compliance to the Law (Police officers, lawyers, and Judges); compliance with accepted moral values (parents and philosophers), and religious observances (New Testament Christian values; the Shulchan Aruch etc. for Jews; the Koran for observant Moslems); compliance with rules of conduct (school teachers, soldiers), or, in business, compliance with contracts (both companies and their customers). These all require compliance with specific rules, terms, and conditions. Because intelligent discussion between two opposing parties is not always possible, a judge or another arbiter, may need to get involved should the rules be ambiguous, inconvenient, or not politically expedient. Given that no law, rule, or algorithm (which embodies them) yet written will achieve universal application or acceptance, the future employment of arbiters, lawyers and judges is totally secure.

To summarise: clinical judgement may involve measurement, ascribing value, compliance, guesswork and timing.

Clinical reasoning processes involve selection criteria, rejection, inclusion, association, the weighting of data for clinical relevance and clinical time trajectory assessments (the rate of deterioration and improvement, the use of extrapolation, and likely prognosis). Prognostication always involves guesswork, since the science of prediction is inadequate. For doctors, judgement is not just an intellectual exercise to be enjoyed for its own sake, but a key to making correct diagnoses, and then deciding the best course of action for the future health and welfare of patients.

Everyday Judgements

Judgements need to be logical. People use everyday judgement to choose a job, a house, a holiday, a dress, or a partner, with what seem rational reasons for their choice. Using logic is advisable, but may not always satisfy the emotional aspects of decision making. Without this awareness, some judgements will prove faulty.

I met my friend Brian Altman when studying 'A' level science. He had the benefit of Jewish family wisdom. He asked me once where I had been on holiday. I told him Bournemouth. 'What was the weather like', he asked. 'Sunny sometimes; raining at other times', I replied. Next he asked: 'What do you look for on holiday?' My reply: 'A beach, sunny weather, and a good hotel.' Brian then asked the killer question: 'So why go to Bournemouth (unreliable British weather)?' My criteria were clear, but my decision to go to Bournemouth was obviously suspect. I had enjoyed going to Bournemouth since childhood, and this affection had clearly biased the value I placed on my decision (Bournemouth has some good hotels and superb beaches). I often return, because I am comfortably acquainted with it, and for this reason I will override other reasoning.

Not everyone has insight into their psychological 'needs'. Among these are the need for affection, control, and inclusion, although safety, protection, security, diffidence, and glory are other common needs. Job advisers, house agents, dating organisations, travel agents, and sales assistants will ask: *'what exactly are you looking for?'* Many will avoid the question, saying: *'I'm only looking.'* A few will have decided on exact selection criteria. Others will say: *'I will know it/him/her, when I see it/him/her.'* With only vague selection criteria in place, many looking for a partner will pursue the romantic notion that one day, they will find their perfect match, appearing in a blaze of light with *'I am your chosen one'*, written on their forehead.

Without considered definitions (selection criteria), no wise judgements or decisions are possible. If a doctor asks a patient: *'What can I do for you?'* and the answer is exact: 'I have a swollen, painful 3rd meta-phalangeal joint on my right hand, and I want the pain to stop', the way forward is clear. If a doctor asks a patient: *'What can I do for you?'* and the answer lacks clarity ('I just don't feel myself, doctor'), the first step is to refine the questioning and get answers that will define the problem.

Meta-Information

A common problem exists with meta-data. There is often too much or too little of it. Sifting through extraneous information for something of relevance can waste a lot of time, with nothing to show for it at the end. Meta-information can be interesting enough to consume time and divert attention.

When we finally accept a partner as 'the right one', or a career path as 'suitable' for us, there will not always be good enough reasons to support our decision. There could, however, be a lot of meta-data that supports it: what your teachers, friends, and parents think, as well as information gleaned from the internet. The same applies to detective work, scientific research, financial speculation, and our dealings with patients. To find the truth and make workable decisions, we may need to go beyond simple questions and answers. As Hayek pointed out (regarding financial markets), people often hold a pool of tacit information (meta-information) which influences their decision making. As an amateur, or an outsider, it is often best to assume that there are highly relevant questions that need to be asked. The problem is often that we don't know what they are. Here are some simple examples:

George had owned a successful chain of traditional 'Pie, Mash, and Eel' shops in the East End of London for decades. Before building a large expensive factory with breeding tanks for eels (in order to cut out the middlemen), George should have asked experts at London Zoo, one simple question. Do eels breed in captivity? The answer is: they do not! He didn't think to ask, and wasted his time and lots of money, constructing an unnecessary building.

You have chosen what looks like a pleasant hotel on an island. The pictures look appealing. It looks just right. But is it? What further questions might you need you ask? For instance, can one get off the island easily? What is the food really like? Could it be another 'Fawlty Towers'? Are there other hotels nearby (fall-back options)? Is there a ferry more than once a week (if not, might you get stuck there)? Not everyone bothers to ask 'meta' questions, the answers to which could reveal important practicalities. These questions could be crucial to your next holiday experience.

Lifelong English literature lover, teacher, and world class commentator on Shakespeare, Mrs. Jan Kinrade, was being considered for cardiac catheterisation. With severe COPD and angina, she was in the latter stages of her life. Her respiratory function was so severely impaired, she couldn't walk far enough, or quickly enough, to get angina (which she had before, when she was fitter). Unlike her NHS cardiologist, I had known her for fifteen years and held important meta-data about her.

She was suffering from a loss of hope, being too breathless at rest to teach her English students. She had become what she perceived to be 'a burden on others'. I knew that extending her life was not something she wanted. I told her to consider refusing the catheterisation. I thought it very unlikely it would yield information that would help her regain her exercise tolerance. It was too late for that. Offering her a cardiac catheterisation was understandable (let's pull out all the stops to help), but totally inappropriate, given her severe COPD. Being the clever, sensitive, and insightful lady she was, she respected her doctors for trying to help, but knew they were offering false hope.

Clinical Reasoning and Judgement

Correct and wise clinical judgement always involves using one's experience and knowledge to assign the correct value to each symptom, sign, and investigation result. Which are reliable and relevant, and which are not. Only correctly valued findings lead to safe diagnoses, correct predictions of the pathophysiological processes involved, and the most appropriate management.

Beware of hidden, essential information. Only the full revelation of all the pertinent facts will lead to an acceptable judgement.

Why might someone want to throw flour into an empty room? No reason is immediately obvious. The patient could have schizophrenia and believe in invisible elephants. The flour would stick to the elephants and reveal them. He might be a burglar who wants to detect the laser beams which would trigger the alarms if disturbed.

Clinical judgement is sometimes a dynamic, iterative process. One can make a preliminary diagnosis, and several differential diagnoses, but then must await further information for resolution. The correct judgement process involves re-evaluation following each new finding.

Wise judgements will mostly leave everyone content. Good judgements should leave the majority content. Dubious judgements leave only a few content. Wrong judgements leave nobody content. Dangerous judgments will leave many harmed or dead.

Not all judgements require active reasoning at the point of observation. For instance – this is an abscess – that is an elephant – this patient has suffered a CVA – she has Bell's palsy. These are first-pass (see Kahneman, fast thinking) judgements made on first sight. One can make such diagnoses without historical data. For the sceptical, and those lacking knowledge and experience, fast diagnoses like these can be far from obvious.

Some diagnoses require a second pass (see Kahneman, slow thinking). If I was told that a patient had right heart failure, I would immediately check the JVP. If it was normal, and the patient not dehydrated (thirsty with a lack of skin elasticity), I would question the diagnosis. A third pass might be required, like examining the JVP after exercise. With the JVP raised only after a little exercise, the patient could have borderline heart failure, apparent only after an increased cardiac workload.

A bureaucratically driven trend is upon us (apparent in many professions). It is to replace personal clinical experience and personal judgement with checklists and reference to guidelines. Their purpose is not only to foster safe action, but to identify the non-compliant.

Basic Rules for Clinical Reasoning

Clinical reasoning is concerned with defining relationships between clinical features, and their diagnostic significance. We need to define coincidental features with no connection (in-growing toenails and alopecia), and those that are associated (asthma and eczema; blue eyes and pernicious anaemia; tallness and a pituitary tumour). Causative relationships need to be recognised (tonsillitis causing a high temperature, coronary artery disease causing angina), as opposed to those that are coincidental.

As a matter of outlook, always use scepticism as the default mode. Always seek discrepancies and persist in trying to explain them.

As some celebrated the centenary of Einstein's General Theory of Relativity (2019), I wondered what motivated him to derive his theory of gravity. It was the discrepancy between the actual orbit of Mercury, and the orbit predicted (incorrectly) by Newtonian physics. His explanation was that space-time bent by planetary mass, defined the observed orbit.

One can hear some of the best clinical processing at clinic-pathological conferences. At these, one can experience the real-time reasoning processes of experienced practitioners.

Deductive, Inductive, and Abductive Reasoning

Reasoning has a neurophysiological basis. Inductive and deductive reasoning both use accepted facts, assumptions, and premises to arrive at a conclusion. They both use the frontal lobe of the brain. Both use logical steps:

All men are mortal. Socrates is a man, therefore, Socrates is mortal.

Inductive reasoning is similar but can involve erroneous generalisation and loose logic. The conclusion may make no sense, despite the known facts:

Jim is a grandfather. Jim is bald. Therefore, all grandfathers are bald.

Using inductive reasoning, even when using correct statements, can lead to false conclusions. Inductive reasoning is in everyday use, mostly going unrecognised. Its conclusions can become the acceptable ones, especially when politically or emotionally expedient.

In one of my forays with the GMC, they accused me of prescribing 'potentially lethal doses' of co-proxamol (a licensed analgesic at the time) to an ex-nurse who had been taking it with no adverse effects for fifteen years. They used two invalid (inductive) arguments to persuade themselves that my prescribing was dangerous:

1. Since co-proxamol had often been used by suicide victims prior to 2005, my patient was likely to commit suicide (she was psychiatrically certified as an unlikely candidate for suicide); and worse,

2. Because of the danger they thought I posed to one patient, the public at large must be in danger.

In a written submission to the GMC, I told them I would strongly advise my patients never to seek advice from anyone using such reasoning. I thus sealed my fate as an establishment critic and dissenter.

Your future as a doctor can depend on the inductive reasoning used by bureaucrats; their conclusions being shielded from criticism by government mandate and legal statutes. They have mission statements with presumptuous metaphysical aims, like maintaining 'public safety' (by shielding patients from harmful doctors), and 'maintaining the reputation of the medical profession'.

When dealing with doctors, the GMC will use the simplest of binary distinctions: 'is this doctor before us compliant, or non-compliant?' The guidelines they choose to use as rules of compliance, they draw not only from their own publication 'Good Medical Practice' but also from NICE Guidelines and the BNF. In more serious cases, it is the criminal law they must defer to.

In their hands (tied with legal tape), their judgements (following adversarial legal discussion) will bear little comparison to those made every day by medical professionals engaged in academic clinical discussion. Despite their many assumptions, the rules they take to be immutable, will never apply to every patient. That is not their concern. In the name of maintaining public safety, they can choose to ignore NICE. The preface to every NICE guideline bears a warning against using them without individual patient consideration – individual cases can differ. Those who represent the GMC (mostly lawyers and lay people), one can hardly expect to consider such clinical matters in depth, yet we have given them the authority to try.

Abductive Reasoning

Starting with an incomplete set of observations, this reasoning seeks a likely explanation. It may entail making educated guesses, using observations that lack any immediate explanation. Doctors use abductive reasoning when making diagnoses with limited information; jurors in legal trials do the same, deciding only on the evidence presented to them (sometimes only that which the law allows to be presented).

Before using them for reasoning and deduction, repeated observations and further enquiry must attempt to confirm the facts. Confusion will lie ahead when there is no confirmation of the facts. If this process is to have any dependable outcome, someone of experience must judge how reasonable and relevant (an art) is the information available. What may follow, may only be a best guess decision.

How Reliable is Information?

Both unreliable information and faulty mental processing, result in poor reasoning and judgement. One need look no further than the judgements made in 2020, when the now infamous Ofqual algorithm, was used to adjust 'A' Level examination grades. Look also at the iniquitous, false allegations of theft, made against hundreds of postmasters using the Post Office's 'Horizon Program'.

Access to some information can prove impossible. If someone has it, they may either not choose to reveal it, or are not required to reveal it. When documentary evidence takes precedence, meta-information can get ignored or relegated to hearsay. Stolen information and that which is spun (into an edited, magnified, or otherwise altered form), deleted, or falsified for criminal or political purposes, can pervert the course of reasoning, judgement and justice.

By considering only incontrovertible documentary evidence, information will get restricted (the equivalent of being cherry-picked), especially when it applies to doctor-patient relationships, and clinical cases viewed by those with no medical training at all (the CQC, GMC and PSA, etc.). Sometimes the only information available is that seen through a keyhole, rather than through an open door. Because of this limitation, the devils in the detail may never get discussed. This will affect judgment and any actions that follow.

Fundamental Thinking

Why does the sun rise every morning? How are time, mass, energy, and the speed of light connected? How is it that cells divide and keep their genetic identity? These questions came from fundamental thinking. Answering them, changed human progress. Going back to basic concepts is always worthwhile.

Could we resolve all complicated clinical situations if doctors only knew everything about the patient's mind, body, and soul, together with their psychology, social inter-reactions, and circumstances? Maybe, but first become a doctor interested in knowing this.

Instead of researching blood cholesterol and its indirect associations, question the origins of atherosclerosis, the process that kills so many middle-aged people every year. We need to understand its genetic basis, and its metabolic processes. These are fundamental to saving lives from CHD; more important perhaps than indirect, epidemiologically valid issues like blood fat levels.

We can apply the same fundamental thinking to primary hypertension, which depends on arteriolar medial hypertrophy. Understanding why this happens is fundamental to advancing its management, and reducing its morbidity and mortality. Labelling hypertension as 'essential' (a mystery, never to be revealed), probably held back progress for decades.

Before trying to decide what degree to take at university, based on improving employment prospects, students should ask where their talent lies, and what they would most enjoy doing for the rest of their life. As we live longer, some may need to consider having two professions.

Other Modes of Thought

We all have preferences when reasoning clinically. We must learn to prefer strict logic, but some employ fuzzy logic.

'Everything is a matter of degree.'

'Fuzzy Thinking' Bart Kosko, 1994, Harper Collins).

Consider a few other modes of thought:

Insular thinking restricts appraisal and judgement to pre-defined topics: this is usually pedagogic, rather than scholarly. It implies a bunker mentality, with no wish to expose personal thoughts, judgements and decisions to the critique of others.

Nebula thinking expands diagnostic and clinical management consideration to the constellation of facts available, many of which will be irrelevant. Intelligence and experience are required to identify and draw upon only the pertinent bits of information, and any connections between them.

Lateral thinking (Edward de Bono, 1967: Lateral Thinking: A Textbook of Creativity, Penguin) will take considerations with imposed boundaries, into unknown territory. It allows new directions of thought, seeing the situation from different angles, and viewing problems in a new light.

Reversed or inverted thinking: There is no terrible weather, only inappropriate attire. By asking what you must not do, you might better understand what you must do. It helps sometimes, to turn a problem on its head and look at it from another direction. I have often heard: 'e-numbers in food are dangerous to health.' My response has sometimes been to suggest that we are living longer because of them.

Wrong thinking. Illogical thinking or that based on unverified assumption. Examples heard in everyday conversation are: when I win the Lottery, all will be well; when I become rich, I will be happy.

In order to get on with a trial in a high court, involving one of my patients (for which he had set aside a lot of time in his diary), a high court barrister suggested to me (as a medical witness) that, 'the sooner we start, the sooner we shall finish'. To the judge's amusement, and others in court, I had to point out to him that his statement was illogical.

They launched two Voyager spacecraft in 1977 to explore the planets of our solar system. When 4.5 million Kms. from Earth, NASA changed the course of Voyager 1, in order to take a better look at Titan, one of Saturn's moons. A collection of NASA scientists in conference thought that life might exist there and was 'therefore' worth a closer look. They diverted Voyager 1, out of the plane of the planets, even though the spacecraft could not have confirmed life from a fly-by. The images received were of poor quality. Desperately seeking justification for their decision, they took a unique photograph: it showed the Earth in relationship to all the other planets (except Mercury and Venus). If nothing else, the photograph helped Carl Sagan (publicist for the mission) to promote space projects.

Bunker thinking. We usually approach bunkers through tunnels. Both bunker thinking and tunnel vision can lead to errors. Bunker thinking can also lead to success that is against the odds (as others see them). This can occur when a bunkered thinker has done lots of research, is an experienced expert in the field, has considered all the meta-information, and has kept all her ideas to herself.

Knight's Move thinking: A sometimes shocking departure from straight-line, sequential thinking. This unique approach can lead others to say, 'why didn't I think of that?' When Jeff Bezos left the hedge fund world behind to start Amazon, many thought him crazy.

Wasted Thinking: There you are, standing in the path of a tsunami or a fast approaching pyroclastic cloud from a nearby volcanic explosion. In these extraordinary situations, no thinking can help. What you need is luck.

There are equivalent clinical situations: you find a child buried beneath a recently fallen brick wall. He has dilated pupils, barely responsive to light. He is non-response to painful stimuli and has no pulse. These signs require resignation not thinking.

Best Guess (Abductive) thinking: This is essential when planning what to do, based on incomplete information. When trying to guess what will happen, use experience to orient yourself (from the range of changes you have experienced), and estimate the likely errors (from the certainty of the sun rising every day, to the uncertainties of share price movements). Create the best and worst scenarios, then make a plan based on your best guess. It will be wrong, but by reviewing it frequently, and regularly adjusting your predictions and expectations, something worthy can result. Successful people never give up at the first hurdle.

Amoebic Thinking: It is common for humans to think amoebically when stressed, probing in different directions, blindly hoping to find clues and a solution to a problem they know too little about. In difficult clinical cases, some will think the more investigations they do, the more likely they are to chance upon the correct diagnosis.

Types of Clinical Situation

Consider the various forms of judgement required in various clinical situations – from almost no judgement ('BLOB', first pass diagnoses), to flummoxing situations requiring considerable deliberation:

1. **'BLOB':** A '**BL**indingly **OB**vious' diagnosis, made at first glance, with straightforward management. Its opposite is -

2. **NEFAD: N**o **E**vidence **F**or **A**ny **D**iagnosis.

3. **Straightforward:** very little clinical acumen is required to reach a conclusion. Although the diagnosis may not be obvious instantly, differential diagnoses follow readily from a well taken history, examination and appropriate investigations.

4. **Complicated:** the interconnected clues are there, but it will take experience and expertise to weigh them correctly and incorporate them appropriately.

5. **Complex:** Lots of apparently disconnected clues and facts that are confusing. The solution usually depends on a facility to see the wood for the trees (nebula thinking).

6. **Flummoxing:** A deteriorating sick patient with puzzling clinical features. Nobody knows what to do. The diagnosis may have to await an autopsy.

Not only words convey meaning. Symbols and pictures can represent meaning, like Chinese kanji and ancient Egyptian cartouches. Their meaning has to be learned and accepted. Like these symbols, clinical signs and other facts, can have a relevance that is difficult to decipher.

When trying to tackle a new jigsaw puzzle, you may have all the pieces, but will struggle to fit them together correctly. The object is to form a clinical picture using every bit of information available. We prefer to make one diagnosis that explains all the information, but this is not always possible.

Two Flummoxing Cases:

A sixty-year-old Australian tourist was making a world trip. He arrived in the UK after spending an enjoyable time island hopping across the South Pacific. He then progressed across the USA, but started losing weight. He felt well, but his weight loss continued despite eating well. Soon after arriving in the UK, I admitted him to hospital. A full history, examination and full investigation yielded no diagnostic clue.

Because he was deteriorating rapidly, I investigated him thoroughly. Eventually, he died without a diagnosis. At autopsy, the pathologist found a small malignant tumour in his lower oesophagus. He never had relevant symptoms. Total body, MRI scanning for cancer, was not then available.

The second patient was a 25-year-old man who died unexpectedly from rapid onset, pulmonary oedema. He had rapidly progressive mitral valve stenosis. That much we knew. What we could never have guessed was he had a mitral valve infected with Trichphyton fungus (athlete's foot).

To Screen, or not to Screen?

There are many potentially fatal, pre-morbid medical conditions, that occur without notable symptoms in their early phases. Among them are cerebral aneurysms, main-stem coronary atheroma, HOCM, and cardiac ion-channel defects. Political, rather than clinical issues has impeded discovering them. The question is, should we screen asymptomatic people (and transform them into patients)? Cost-effectiveness is a key consideration for a nationalised corporation like the NHS, since they spend public money. A key question then arises: what is a life worth? In private medicine, patients are not only aware of what their life is worth, but will pay for medical screening because they believe it worthwhile.

Regardless of the cost, many airlines screen their pilots. The Armed Forces screen their combat operatives; football clubs screen their players, and companies screen their executives. The NHS now undertakes basic GP-based screening, and will sometimes undertake extended, specific family screening for muscular dystrophy and HOCM, etc. In 2009, they introduced a limited form of screening (an NHS Health Check) as a fee-generating procedure for GPs (36-years after I started it as a private service).

Because of the disappointments every doctor experiences treating established disease, my initial aim in 1973, was to discover important risk factors and early, asymptomatic disease: heart problems, diabetes, kidney disease, cerebrovascular disease, thyroid disease, dementia, breast cancer, cervical or prostate cancer. I later tracked the progress of early atheroma (from 2000), one of the biggest killers of middle-aged people.

When medical screening produces a significant diagnosis, sensitive clinical judgement must follow. What action will be in the best interest of an asymptomatic patient? An appropriate answer often requires one to know the patient well. The advice given could depend on whether the patient might choose the length of life over the quality of life.

One of my best friends, George Mullaly, chose quality of life over length of life. He said repeatedly, whatever was wrong with him (hypertension, and severe type-2 diabetes), he was going to live his life to the full. He lived as if every day was his last. Restricting his diet and being meticulous about BP control concerned him too little. He died while sleeping, aged fifty-two.

Clinical Wisdom

Knowledge, experience, intelligence and common sense all need to be exercised when making clinical judgements and forming management plans. Wisdom, like every axiom, has a recognisable quality that needs no further qualification: it is usually enlightening, and sometimes cathartic. Both truth and wisdom have a ring to them. Unfortunately, this sound can be heard only by those tuned to hear it.

The application of wise judgement can make light of difficult diagnoses. It has a magical quality – one that instantly dispels uncertainty. I was lucky enough to have worked with a few pioneer doctors and engineers, who possessed wise judgement. Keith Jefferson, cardiac radiologist, Aubrey Leatham, Alan Harris, and Peter Nixon, were all wise inventive cardiologists. The two inspiring engineers I worked with were Geoff Davis, creator of the first working human cardiac pacemaker, and Graham Leach, who developed the display format for the echocardiography we use today. Using clarity of thought, they were all able to find what seemed to be simple solutions to complicated problems.

Judgement Bias

To believe that those with fame, fortune, status and reputation, are completely different to the rest of us, is not always mistaken. Being a Fellow of the Royal Society (FRS) in the UK, or having a Nobel Prize, is often a reliable guide to the holder's scientific acumen, and evidence of their difference from others. Unfortunately, the star-struck, jealous and egotistical also pervade the world of competitive science.

I was in the audience at an Advanced RCP Medicine conference, when we were told by a very senior physician (and FRS), that drastic reductions in cardiovascular events would follow if only we all prescribed a daily tablet combining aspirin, a beta-blocker, an ACE inhibitor and a statin, to all adults. His argument was: since each drug would reduce heart disease by 25%, and since 25% x 4 (the number of drugs used) = 100%, and the elimination of heart disease assured.

I questioned his maths over lunch. He has since given further thought to the possibility of overlapping sets. Beware of the halo bias; it might suggest that the judgments of highly qualified people, must be sound and unquestionable. Be as aware of the opposite: giving no credit to those you think are lesser beings.

In some cultures, those respected for their judgement, achievements, knowledge and ability must have rank and seniority. In my experience, many untutored people also possess wise judgement. Could it be that they have inherited it, or they lack the biases that warp the judgement of others?

Do minds become closed to new ideas with age? Does judgement wane with age? Not always, it seems:

'We have been wrong in medicine many times. Thus, we have to remain sceptic; the idea of staying open-minded is extremely important. Reinterpretation of old data is extremely welcome.'

Prof. Eugene Braunwald. Cardiologist (aged 88).
ESC Grand Debate (Barcelona 2017): Aspirin for Life.

'A cynic is what an idealist calls a realist.'
Sir Humphrey Appleby. BBC 2 TV series: *Yes, Minister.*

Theories backed by a prestigious patron, may survive for political reasons. When the famous double Nobel Laureate, Linus Pauling, declared vitamin C to be 'the answer' to both heart disease and cancer, who would dare contradict him? Forty years on, having inaugurated his own Institute at Oregan University, we still await convincing evidence for either hypothesis.

Since I had just given up coronary angiography as part of my day job, I referred one of my patients (Bob F.) to have the functional flow reserve (FFR) measured across his various stenotic coronary lesions (to assess the need for each to be stented). The Prof. of Cardiology who saw him, questioned whether my patient actually had angina. As any a medical student would have known, the patient had 'classic' Heberden's angina. As a result, he refused to catheterise him.

Another colleague of mine (who unknown to me, also offered the same measurement of coronary flow reserve) catheterised him. He found two flow limiting coronary stenoses which he stented. Because the stents freed the patient of angina, I wrote to the professor, suggesting he gave more thought to history taking. I was less certain his judgement would improve.

We must all be aware of one important bias. The more experienced and expert we think we are, the more likely we are to over-estimate our ability (the Dunning Krüger bias).

Experienced old-timers will get things wrong, but the inexperienced will only rarely get an advantage over them. My advice to those who have a new idea, is to persuade their older colleagues politely. Clarity, balanced reasoning and respectful diplomacy, will impress those wise and humble enough to see the sense in a worthy new idea (a characteristic of the most able). Humility rather than bombast will support it best.

Doctors and nurses make mistakes every day. Fatigue causes some; others result from being distracted, or from having insufficient forethought. Unfortunately, many of us have to learn through our mistakes. Most humans must have been quite good at assessing everyday risk in a failsafe way, and learning from their mistakes. Does this explains why we are not yet extinct?

Another survival attribute is our ability to devise different approaches to solve problems. We call it intelligence. Unfortunately, we are not the rational beings we think we are, nor do we always need to be. Success can make us complacent, and failure can make us feel dejected. Both can bias our thinking and judgement.

Consider a regular judgement problem for many — choosing which car to buy. Should we think logically or decide emotionally? A few Londoners buy Ferraris; they will have the pleasure of driving it at an average speed of 20mph through Central London. So is there any rational reason for buying one? Attention seeking might motivate it, or could it be an alternative to paying tax: buy a Ferrari, or pay the same amount in tax?

Deciding on the purchase of a Ferrari can be an emotive issue, not just a cognitive one. It is easy to employ a string of weak justifications: 'It's always been my dream to own one.' 'Because of my success, I deserve one.' 'I will (eventually) make some long journeys to utilise its power and performance.'

There is nothing wrong with the decision to buy a Ferrari. Just don't justify it with false or weak arguments. Marvellous cars, Ferrari's. If you have one, enjoy it for the right reasons (engineering excellence, highly responsive power at your fingertips, and supercar performance at your command).

It helps to have some insight into the reasons for our decisions. We need to ask our decisions are logical or emotional. What are the objectives, and will my decision fulfil them? Is my decision reasonable, proportionate, and justifiable? Can I justify my decision to sceptics? What will be the costs and benefits for me and others?

Biases Warp Judgement

Daniel Kahneman and Amos Tversky, undertook research into our judgement of uncertain events, and beliefs about their likelihood (A. Tversky, D. Kahneman, (1973). Judgement under uncertainty: heuristics and biases. Hebrew University. Jerusalem).

They investigated how we construct subjective probabilities from indefinite information. Kahneman postulated that there are two types of decision making – fast thinking ('System 1') which is intuitive, reactive, and subconscious (needed in emergency medical situations); and the considered, rational, slower sort (slow thinking or 'System 2'). Biases more commonly affect slow thinking.

Kahneman pointed out that on average, High Court judges hand down different sentences to criminals, at different times of day and doctors are more likely to prescribe antibiotics and analgesics in the afternoon. He has referred to this variation in bias as noise ('Noise'; Kahneman, D., Sibony, O., Sunstein, C.R. William Collins, 2021).

This variation of bias could lead a person to switch from friendship to enmity, based on one piece of uncorroborated evidence. After one denied request, opinions can change from adoration to dislike; from unquestioned goodwill to the withdrawal of support. Not every person will exhibit binary opinion switching, but it is common enough.

We all harbour biases. Since Kahneman and Tversky published their original work, others have added more (including my own. See later). They do not all apply to clinical medicine, or to the understanding of patients and colleagues, but here are some that might:

1. **Ambiguity bias:** ambiguous information diminishes the probability of a correct decision.

2. **Anchoring (or focussing) bias:** becoming fixated by one bit of information. It can lead us astray. The diagnostician's curse is paying too much attention to the wrong clues.

3. **Attribution Error bias:** inductive reasoning: my grandfather is bald, therefore, all grandfathers are bald.

4. **Automation bias:** the belief that a computer / processing machine, or algorithm, is likely to be correct (an error revealed in 2020: 'A' level grade results were wrongly 'processed' by a standardising algorithm).

5. **Availability bias:** giving preference to what is easily accessible.

6. **Availability cascade bias:** repeat something often enough, and it will seem true.

7. **Backfire bias:** after dismissive evidence for a theory becomes known, belief in the theory may

strengthen.

8. **Bandwagon bias:** the tendency to believe the majority knows best.

9. **Barnum (or Forer) bias:** the tendency to overestimate the value of one's individuality as unique.

10. **Belief bias:** the strength of an argument increases with the believability of its conclusion, even when the conclusion came from a series of doubtful assumptions. Doctors invested in their clinical opinion may see no need to seek confirmatory evidence. Arrogance has no place in the management of complicated, complex, or flummoxing clinical cases.

11. **Berkson's Paradox:** believing a statistical conclusion while ignoring the important conditionals applied.

12. **Blind-Spot bias:** seeing oneself as less biased than others.

13. **Choice Support bias:** seeing our choices as better than they are.

14. **Chronology bias:** current information is more valid than older information.

15. **Cluster Illusion:** over-rating the significance of clusters in data sets. When playing roulette, the belief that an unusually long run of REDs (rather than BLACKs), will continue.

16. **Confirmation bias:** recognising only those examples that prove our point.

17. **Congruence bias:** testing for only those diseases we think are likely.

18. **Conjunction bias:** the 'out of the ordinary' is more likely than the ordinary.

19. **Conservatism bias:** unwillingness to revise a view, despite fresh evidence.

20. **Control bias:** the over-estimation of the control we have in various situations.

21. **Correlation bias:** over-estimating the connectivity of facts and events.

22. **Courtesy bias:** giving a socially acceptable version of the truth, rather than the actual truth (more common in some cultures than others).

23. **Curse of Knowledge bias:** the tendency of knowledgeable people to not appreciate the thinking of those less educated.

24. **Declination bias:** things ain't what they used to be. A tendency to view the past as more favourable.

25. **Decoy bias:** a tough choice between A and B is easier if one chooses C.

26. **Distinction bias:** A & B viewed together, seem less different from A & B viewed separately.

27. **Dunning Krüger bias:** the unskilled tend to overate their ability; the skilled underrate their ability.

28. **Duration bias:** the tendency to underrate the time something will last.

29. **Empathy Gap bias:** a tendency to offer an inappropriate level of empathy (too little or a too much). Can lead to under-treating pain, and failing to recognise distress (or the reverse). (Loewenstein, G. Health Psychology (2005). Hot-Cold empathy gaps and medical decision making).

30. **Exposure bias:** favouring what we are used to.

31. **Extrapolation Bias:** believing that trends will continue uninterrupted.

32. **Fading bias:** the tendency to forget unpleasant events and remember only the pleasing ones.

33. **Flat-pack bias:** while awaiting construction, we think structures are better than they are.

34. **Framing bias:** belief follows the excellence of the presentation, rather than the value of the evidence.

35. **Frequency Illusion bias:** once we have learned an unfamiliar word, it appears everywhere.

36. **Functional Fixity bias:** the tendency to follow 'tried and tested' functions, and not to look for new ones.

37. **Gambler's Fallacy bias:** the pattern of past events will reliably predict future events.

38. **Gap Bias:** with both ends of a spectrum recognised, what lies between (in the gap) gets dismissed.

39. **Google bias:** Internet facts are easy to find and easily forgotten.

40. **Hindsight bias:** 'I could have told you that would happen!' Belief in our ability to predict the future using our experience of past events.

41. **Hot Hand bias:** an exemplary track record predicts winning in the future.

42. **Information bias:** the more data, the more likely are we to make a significant discovery.

43. **Instrument bias:** if all we have is a hammer, everything looks like a nail.

44. **Loss Aversion bias:** the value of something lost is greater than something gained.

45. **Naïve Cynicism bias:** others are more egocentric than us.

46. **Naïve Realism bias:** I alone can see things the way they really are – others cannot.

47. **Negativity bias:** weighing losses more heavily than gains.

48. **Neglecting Probability bias:** disregarding probability in uncertain situations.

49. **Normalising bias:** refusing to recognise potential risks until they happen.

50. **Observer Expectancy bias:** to see only what one expects to see.

51. **Omission bias:** harmful actions are worse than harmful omissions.

52. **Optimism bias:** a cup is half full, not half empty.

53. **Ostrich bias:** ignoring the obvious.

54. **Outcome bias:** the judgement was wise if the outcome was good.

55. **Overconfidence bias:** of the 99% who are sure, only 40% are correct.

56. **Pareidolia bias:** seeing patterns that are not there: A face on the moon.

57. **Progress bias:** all progress leads to a better place (Utopia, perhaps?).

58. **Pro-innovation bias:** new ideas are better than old ideas.

59. **Projection bias:** the tendency to believe in projections and models.

60. **Pseudo-Certainty bias:** taking more risks when the outlook is bad; taking fewer risks when the outlook is rosy.

61. **Reactance bias:** the urge to do the opposite.

62. **Reactive Devaluation bias:** to oppose anything suggested by an opponent.

63. **Retrospection bias:** the past was rosier.

64. **Rhyming bias:** giving credence to rhymes: 'If the gloves don't fit, you must acquit.'

65. **Risk Compensation bias (Pelzman Effect):** the greater the perceived safety, the greater the risks taken.

66. **Selective Perception bias:** seeing only what one wants to see.

67. **Self-Serving bias:** feeling more responsible for success than failure.

68. **Semmelweis bias:** rejecting the evidence that does not fit our paradigm.

69. **Social Compensation bias:** choose only non-competitors to associate with.

70. **Social Desirability bias:** the over-estimation of our social attributes and their acceptability.

71. **Spotlight bias:** the need to be in the spotlight, or to avoid it.

72. **Status Quo bias:** we are better as we are.

73. **Stereotyping bias:** the assumption of group characteristics.

74. **Subjective Validation bias:** the tendency to disbelieve what others believe.

75. **Survivorship bias:** the virtue of survivors is greater than non-survivors.

76. **Time Saving bias:** over-rating the time saved by speeding.

77. **Third Party bias:** media messages influence others more than us.

78. **Trait Ascription bias:** I am less predictable than others.

79. **Transparency bias:** overestimating our insight into others.

80. **Travis Effect bias:** the present is more credible than the past (see bias '56').

81. **Parkinson's Law (bias):** the tendency to overvalue the importance of the trivial.

82. **Von Restoff bias:** what stands out is better remembered.

83. **Weber-Fechner bias:** wrongly valuing minor differences in large quantities.

84. **Zeigarnik bias:** the incomplete is better remembered than the complete.

85. **Zero-Bias**: a preference to reduce quantities to zero.

86. **Zero Sum Bias:** the tendency to see actions as zero-sum: gains = loses.

Some of my own:

Anecdotal bias: Belief in personal observation.
Assumption bias: all my assumptions are valid.
Belief Transference bias: everyone looks forward to a special event (Royal Wedding, etc).
Bunker Bias: it is never safe to expose our ideas to others.
Consensus Bias: committees make the best decisions.
Cultural Bias: my culture is best (see Ex-Pat bias).
Dawkins Bias: the scientific method alone can reveal the truth.
Democracy Bias: decisions are best made by public consensus.
Drug Bias: life is sh*t, so life must be better on drugs.
Epicurean Bias: without Michelin starred restaurants, life would not be worth living.
Equality Bias: the notion that all humans are equal in ability and performance (or would be if they had equal opportunity).

Ex-Pat bias: ex-pat culture is better than the local culture.
Fat & Carb bias: all foods containing fat and carbs are bad for our health.
Fresh Air Bias: the secret of health and happiness lies in breathing fresh air.
Geography Bias: Things are better here than there or vice versa.
God Bias: all you need to do is trust in God (or someone famous).
Healthy Food Bias: health and the prevention of disease are diet dependent.
Hedonism Bias: if it isn't fun, what's the point?
Iceberg Bias: small clues lead to large discoveries.
Inamorata **Bias:** an inability to see those we love as they really are.
Incomprehensible Bias: that which is incomprehensible is too clever to be wrong.
Intervention Bias: Positive – Doing something is better than doing nothing. **Negative** – doing nothing is better than doing something.
Mañana **Bias:** waiting until tomorrow will work well.
Manipulator Bias: believing yourself clever enough to get whatever you want from others.
NHS Bias: the belief that the NHS is a world-class medical service.
Overload bias: more must be better than less.
Peasant Bias: Inferior breeding causes failure.
Peter Pan bias: I appear younger than others of my age.
Philistine Bias: life is too short for intellectual pursuits. The price of something is more important than its wider value.
Posthumous Bias: the dead were wiser than us.
Print (Publishing) Bias: if it's in print, it must be true, and worth knowing.
Profit Bias: working without profit is a waste of time.
Royalty Bias: only superior breeding leads to success.
Self-Esteem Bias: only activities which bring self-esteem are worthwhile.
Rumsfeld Bias: many unimagined truths are yet to be discovered.
Selfish Bias: If it can't be mine, I'm not interested.
Serendipity (Panglossian) Bias: Everything will work out for the best in the end.
Sports Bias: The only way to be healthy, and live longer, is to exercise.
Statistics Bias: statistical truth is the only truth worth having.
Status Bias: only that which raises our status is worthwhile.
Underload Bias: less is more.
Unknown Information Bias: what we don't know is more important than what we know.
Utopia Bias: only in Utopia can we achieve equality, fairness, and ultimate happiness.
Vanity Bias: only that which satisfies my vanity has value.
Wealth Bias: the only way to be happy is to get rich.
Wizard of Oz Bias: someone, somewhere, will know the answer.
Yonder bias: better there than here.

Does all this mean that human impartiality is a myth?

The Wrong Questions

It is quite normal for humans to think and act like an amoeba; thoughtless probing in different directions, hoping to find clues and answers. The more obscure and ill-defined the problem, the more we are likely to indulge in amoebic processing. Creating an unnecessarily long list of differential diagnoses is one clinical equivalent. The list can lead us down the wrong path, distract us from the right path, or lead to the right path by chance. When in this situation, stop aimless deliberation and go back to basics. Get more information. I will not apologise for repeating the same advice so many times — take the history again and examine the patient again. Ask relatives, friends and others dealing with the case, what they think. Any wrong thinking might then become exposed.

Judging Judgement

Our preferences and dislikes form from birth, and accumulate progressively, but at just 22-years-old, junior doctors are called upon to judge their medical colleagues, and to assess their knowledge, experience and integrity. It is especially important if they are making judgements that affect patients. Are their thought processes sound? What are their standards and attitudes? Do they have vested interests? Are they influenced by local practice or hospital politics? Are they prone to make workload or timesaving personal decisions? Are their decisions motivated to gain personal recognition, and will they allow personal motives to over-rule their duty to patients?

When someone who never goes to the theatre, and has never read Shakespeare, gives their opinion about Hamlet, how should one value their judgement of the play? Would you assign greater value to the views of an experienced theatre critic? To judge the judgements of others, one must understand their knowledge, experience, attitudes and values. One must consider whether there are reasons for any biases in their back-story or meta-information.

Wise judgement will mostly achieve the instant recognition it deserves. Wisdom can arise from surprising sources. Taxi drivers, and those with a broad experience of life (the street-wise), more often have it. My friend George Mullaly, who died when 52-years-old, was a street-wise person who frequently advised me. Unfortunately, he was least inclined to abide by his own good council.

There is some merit in listening to patently ridiculous, poorly constructed ideas: they can strengthen our adherence to what is reasonable, sensible and wise (subjective validation bias; reactive bias; reactive devaluation bias).

Referral Decisions

If a patient has severe, gnawing, lateral leg pain, do you manage them with painkillers, or refer them to a chiropractor, neurosurgeon, or orthopaedic surgeon to relieve their L5/S1 disc compression? It will be for you to decide on the patient's behalf.

Here are some considerations. Some patients reject surgery and prefer non-invasive management. Most disc compressions resolve on their own. Only time will tell. Apart from gabapentin, many painkillers are all but useless in severe neuralgia. Neurosurgical discectomy will usually give rapid relief, noticed soon after the patient recovers from the anaesthetic.

Some of the more complex and consequential judgements doctors make, involve patient referrals to colleagues. The consequences have prognostic implications. Choose a colleague who is not as competent as you thought, and your patient could return with a tale of woe, many complaints and unanswered questions. The commonest complaints concern inadequate communication, and colleagues who are unfriendly or too formal to inspire confidence. Refer no more cases to them, no matter how good their CV.

I have always disliked patients complaining to me about other doctors, without knowing all the details, one cannot comment. After referring a patient, one should not have to deal with the patient's unanswered questions. How can one explain what a colleague meant, said or did? I have often had to admit that while good at their job, some lacked communication skills.

I always tried to refer my patients to those colleagues who took complete care of them, and dealt with their every concern (expressed by the acronym I.C.E: 'Ideas, Concerns, and patient Expectations'). Ideally, they would answer all the patient's questions and make all necessary further arrangements. I never wanted to be responsible for the further instructions of colleagues; I was always too busy to do their work for them.

Patients, whose background is commercial, sometimes think that doctors are 'in business' with their partners and colleagues. Some business people find it difficult to understand why our referrals do not attract commission (at least, not in the UK). There could be nothing worse for patients than physicians referring cases to those who paid the most commission, or who showed them the most gratitude in other ways.

Medical relationships can involve other favours. A GP patient of mine once asked me to 'put in a kind word' for a colleague's daughter (whom I did not know). She was to be interviewed by the admissions board of Charing Cross Medical School, where I was working. Gratitude is one thing, coercion is another. This practice is acceptable in many parts of the world, but not yet the UK.

Judgement and Regulatory Compliance

Throughout my career, the nature of compliance has changed. When I qualified, it was not a prominent issue; now it would be dangerous to ignore. Every doctor who wants to survive medical practice in the UK has no alternative but to submit to the demands of bureaucracy, even if they are inane and detract from patient care.

Making notes of your observations, thoughts, decision-making processes, and judgements will compete for priority, and usurp the time doctors and nurses spend with patients. When one submits to appraisal, revalidation, CQC compliance, or a GMC investigation, what one has written or has omitted to write (now a heinous crime, equivalent to clinical incompetence), will provide evidence.

Smart UK doctors, unlike me, will write more than just the relevant clinical details. They will feed the corporate medical machine with all the details it demands: personal reflections, for instance. These are now taken as a sign of competence and a willingness to learn. Our risk assessments and how we derived them, our reasoning and conclusions, our clinical management plans, and any apologies, excuses, confessions, and declarations of guilt we make, all must be legible in the patient's notes for others to scrutinise. Nothing here about clinical skills or changing the morbidity and mortality prospects of patients, our dedication to helping them, or the time wasted on administration. How did UK medical practice become subordinated to corporate administration and medico-legal issues? I suspect it was by good people saying nothing.

Regulators will need feeding with evidence of our insight, remorse, contrition and plans for remedial action, when they investigate any doctor who makes questionable decisions or takes risks they cannot fully understand. The ubiquitous requirement of every corporation, including the NHS, is to receive reports about everything relating to the corporate purpose, even when it bears little relationship to the excellence of clinical functioning. It will, however, be used to measure our corporate compliance, and used by regulators and their 3rd party assessors, to measure our clinical competence, albeit falsely. With no clinical acumen themselves, they must rely on what we and others write. Very few medical bureaucrats have any medical know-how, and even those who are medically qualified, will usually give priority to their corporate role (perhaps an indicator of their clinical incompetence?). Some doctors take up medico-legal work to offset their disinterest in medical work and earn extra income. Since the Royal College of Medicine now runs courses on medical leadership, hospital specialists can now avoid seeing patients completely and devote their whole career to medical bureaucracy (as medical directors or registrars).

Those who work in teams will need to hand-over patient information for which communication skills are necessary. Although better communicated orally, corporate requirements insists on documentation. Meanwhile, what is happening to the patients? Waiting for the attention and the care they need, I guess. If you don't believe this to be important, review the thousands of legal cases taken out against the NHS each year. Are we to accept that the corporation knows best? Who will fight the corporate trend for more and more documentation while it deprives patients of the medical care and attention they need? I would like to propose a new function for all medical bureaucrats: to stand behind doctors while they work, and let them take the notes they want. They can then hide away and produce the documentation they think important. No executive in business would demand less.

Writing voluminous, inane notes (even when practising as a single-handed physician, like myself) complies with the sanctimonious demands of bureaucrats, many of whom regard themselves more worthy guardians of patient safety and welfare, than doctors. Doctors and nurses in the UK, are no longer trusted to regulate medical practice. Since our government claimed the role of protecting patients from doctors and nurses (against all potential rogues, like Dr. Harold Shipman), medical bureaucratic sanctimony has become commonplace.

If any UK patient is reading this, I must expect their disbelief. Few will understand why any medical bureaucrat occupies the moral high ground, claiming to protect patients from doctors. Protecting patients from the medical profession, however, is now a full-time bureaucratic role. As a well-paid job, filled by lawyers, legally minded politicos, and power-seeking medical professionals, many now see dealing with patients as a perfunctory role.

Most hospital in-patients and those attending GP health centres, will notice the time doctors and nurses now spend staring at their computers, rather than talking to them.

There are several key questions to be asked:

1. Have the medico-legal note-keeping requirements of medical bureaucracy improved the quality of patient care?

2. Have these requirements reduced patient morbidity and mortality?

3. If the answers to (1) and (2) are negative,

(a) why are PAs / medical secretaries not employed to service the demands of bureaucracy, by accompanying every doctor and nurse as they work?
(b) Why has medical corporate bureaucracy in the UK not been disbanded and reformed?

Although widely known, it is worth mentioning that whatever doctors and nurses write in patient notes is accessible as evidence against them. Patient notes, once considered completely confidential, are now open to 3rd party interpretation by anyone empowered to measure clinical competence.

Beware of using acronyms in patient notes; they are open to misinterpretation. 'SOB' could mean 'Silly Old Bug**r', not 'Shortness of Breath.' A nurse lost her registration after writing 'NFN' in a patient's notes. She had hoped to conceal what she meant: 'Normal for Norfolk.'

When in doubt about any judgement, sleep on it before putting your name to it. Don't be ashamed to take your time and ask others for their views.

When junior doctor Dr. H. Bawa Gaba wrote that she was 'unsure of what to do' (being unable to contact her registrar or a consultant colleague), her honesty provided the evidence for a barrister to claim she was 'unfit to practice'; an entity, very few lawyers and tribunal members are fit to judge. If ever in this position, first ask for a review of their medical knowledge and experience, and that of those who deem to judge your medical performance.

As a corporation, NHS management culture is far are too deeply entrenched to change without a revolt. The management bureaucrats, along with their self-serving ethos, would have to be extirpated and completely replaced, before one could expect a significant reform of their corporate culture.

Poor judgement

In his book 'On the Psychology of Military Incompetence' (1976. Random House), Norman F. Dixon revealed how poor judgement in military situations can arise from meekness and the prioritisation of personal matters.

In September 1944, the supreme commander of the Allied Expeditionary Forces in the Second World War, General Eisenhower ('Ike', as he became known), approved Operation Market Garden (conceived by General Bernard Montgomery). He knew German troops were present in great numbers around Arnhem in Holland. The question was, should he send allied troops into action to take the town and its bridge against great odds? Dixon has it he didn't have the heart to halt the airborne invasion, given the readiness of the troops, and how long they had trained and waited. It is said that he simply did not want to let them down.

Instead of asking: *'what must I do to keep my troops happy?'* he might have asked, *'what action will save most lives and achieve the desired end?'* The tragedy that resulted is a matter of historical record. Approximately 2000 allied soldiers lost their lives, and the Germans took 6854 prisoners (Wikipedia). The few who knew Ike well, would never have accused him of meekness.

Faulty Judgement

Few have proven to be wise judges of future events. Apart from the information available being inaccurate, or insufficient, many believe that the past can accurately predict the future (the hindsight bias).
Throughout history, the reasons for unwise judgements get repeated. They include incorrect information, incorrect assumptions, jumping to conclusions (eager to move on, and trying too hard), incompetence, adverse personal biases (pride, vanity, ego and emotional involvement), together with conflated or confounded information. All influence the judgement process. The non-fictional and fictional works of the literature from every culture, are replete with illustrations.

The Error of Presumption and Assumption

'I look to a day when people will not be judged by the color of their skin, but by the content of their character.'

Martin Luther King Jr. (August 28th, 1963).

Not every attractive person will become a meaningful, satisfying, and fulfilling partner, either for business, friendship or a loving relationship. We are all liable to it: our initial choice often gets biased by physical attractiveness and status. Responding to physical attractiveness may be biologically expedient (hastening procreation and the survival of our species), but relying on it can be mistaken and a life-shaping error. If the desire is to form an enduring, mutually compatible relationship, rather than simply satisfying lust, physically attractiveness alone will not always prove reliable. An error of judgement at this early point in life, can affect the generations that follow. An unacceptable anachronistic alternative for western cultures, who prize emotional and sexual freedom of choice, would be to impose parental control and arranged marriages.

Most of the errors I have made personally, or would have made, have come from a tendency to make assumptions. Every assumption, influenced as it can be by bias, hearsay, incorrect data, or faulty reasoning, is potentially dangerous and needs repeated review. Incorrect assumptions lead to incorrect conclusions, so we must always check and cross-check the bases of our clinical management decisions. Independently check all presumptions prefaced by: 'everybody knows that...' If you are still in doubt, seek independent, reliable sources of information. Perhaps choose a colleague with inside knowledge.

Errors of Omission

Mr. T. A's cardiologist failed to diagnose his severe coronary heart disease (CHD), during a routine pre-op assessment, made before planned abdominal aneurysm surgery. He failed to investigate the patient fully. Routine examination, a resting ECG, and a 24-hour ECG were normal. Because the patient claimed to be asymptomatic (he was a denier) and looked well, the cardiologist in question thought it unnecessary to go further and request a cardiac CT scan (to show any calcified coronary atheroma). No guideline at the time suggested the need for it (this was his defence when challenged legally).

Twenty-four hours after the operation, the diagnostic ECG software in the ITU reported an acute MI (myocardial infarction or heart attack). The staff discounted this, even though the ECG showed an unequivocal STEMI. Four hours later, at 2am the next morning, the duty house doctor requested a troponin level (blood test evidence of a heart attack), but took no action for eight hours; the patient's troponin level had risen to >3000 by the time he did. Around the same time, the patient developed cardiac electrical instability. After several cardiac arrests, the patient died.

Omissions can occur because of (Rumsfeldian) 'unknown, unknowns', or because the information lies outside our own field of work, knowledge and experience. The most vital and relevant questions will not then get asked. Inside information (sometimes tacit, mesa-information — Greek 'mesa' means 'inside') can be crucial. Employing an experienced expert who knows the pertinent questions to ask, can suffice. When buying a second-hand car, for instance, who would know to ask how much rust is present in hidden places, or to ask if any oil is leaking from the engine? In the house you are about to buy, who will ask for local development plans, or demand evidence of Japanese knotweed, subsidence or damp, without sufficient knowledge? Who asks how much tax they will pay on their pension after receiving it (after working for thirty years making regular tax-free pension contributions)? The universal legal principle is caveat emptor. It applies when buying anything. Every buyer should know that all problems arising after the purchase will be theirs. With patients, we have a duty to answer those questions, they have too little knowledge to ask.

I always remember being persuaded, as a medical student, to take out a pension. That the premiums were tax deductible was the selling point. I simply didn't know to ask if any taxes would be due after retirement. Had I known I would have to pay tax on my pension, I might have made other arrangements.

I made several incorrect assumptions about the benefits received by retired people and any future tax liability. I assumed His Majesty's Revenue and Customs were benevolent. After all, on retirement, many will have paid income tax all their life. I was naïve about the vested interests of pension salespeople. Too much trust in them, and a total absence of knowledge about the pension business, caused me not to make a full appraisal.

Before making any military decision, every commander must consider three things: reconnaissance, reconnaissance, and further reconnaissance.

The same applies to clinical practice.

Mindless routine has its advantages. That every man over a certain age should have his PSA tested repeatedly, has sense to it. The policy remains controversial because of the questionable diagnostic accuracy of the test; in particular, the number of false positives that occur. As an error of omission, I waited five years before thinking to repeat the test on one of my patients (who also didn't think to ask for it, among the many other tests he requested regularly). That 'it was not my primary role as his cardiologist to do this', was a poor excuse. He later presented with two (pathologically) fractured ribs that were slow to heal after a minor injury. He had multiple bony secondary metastases, and a PSA of 670ng/l.

Weak reasoning

'I can see no reason to believe that . . .'
'I have seen no evidence to convince me that . . .'

When the evidence is lacking or suspect, one can use these weak excuses to dismiss an idea. They might result from intuition, when there are too few points of evidence, or the evidence is weak. Obviously, more evidence is required. The absence of evidence is not evidence of absence. It is sometimes said: *'I know the evidence is circumspect, but I feel sure I am right'*. This weak argument also calls for further evidence.

One class of weak reasoning is inductive thinking. Here is an example of inductive thinking. The conclusion may be correct, but not for the reason given:

Train ticket prices have remained unchanged for thirty-years, therefore, it's time for them to change.

Faulty Trains of Thought

The effects of errors in a train of thought do not add together, they multiply.

Take the 1970s argument I mentioned before:

- Eating eggs can raise blood cholesterol

- A raised blood cholesterol can be associated with atheroma

- Ruptured cholesterol-rich plaques in coronary arteries can incite intra-arterial clot formation, coronary occlusion, and cardiac infarction

- Eating eggs is, therefore, a cause of cardiac infarction

These statements are true except for the final one. Even though it does not follow, it was the dominant belief of doctors in the 1970s.

The intima of arteries mostly generates its own cholesterol. This was unknown in the 1970s (then an unknown, unknown). It partly explains why eating eggs is not a key fact in the aetiology of cardiac infarction (saturated fat is more involved, as proven by animal experiments). The assumption that the intima of arteries is just an inactive, semi-permeable membrane, was false. The arterial intima is a metabolically active tissue, variably manufacturing nitric oxide (as an anti-oxidant), cholesterol (pro-inflammatory, ox-LDL), collagen and calcium apatite. Atheroma is not much caused by deposition (lipids falling like snow and diffusing into the inner walls of arteries, thus 'furring' them), the intima mostly manufactures it.

Here is an example of a faulty train of thought caused by downgrading evidence to hearsay:

After some concern voiced by a pharmacist, the Medical Practitioners Tribunal Service (MPTS, October 2019) thought that my prescribing had subjected several patients to undue risk. Their resulting judgement was to suspend me from the medical register for one year. They used their own assessment of 'likely future public risk', while ignoring two rather large elephants in attendance.

Elephant 1: How is it that none of my patients came to harm during 53-years of prescribing practise (including the cases in question)?

Elephant 2: How is it that none of my 20,000 patients had ever complained about my practise?

Neither disproves the contention that I presented a 'likely future public risk', but the tribunal members should have asked: 'What might we learn from doctors with a similarly long track record? How is it he avoided harming lots of patients for 53-years, even while undertaking high-risk, invasive cardiology, and general medicine)?'

They accused me of lacking insight into the risk I posed (my potentially risky prescribing), while failing to declare their lack of a medical education, clinical understanding and insight. MPTS (GMC) tribunals do not need to review their decisions in the light of a doctor's track record and clinical experience. They simply relegate all that to hearsay, and attempt to act in a failsafe way by guessing the worst of outcomes.

The corporate ethos of the NHS, the GMC and CQC is a major problem to doctors and nurses. With no medical knowledge or clinical experience, making failsafe rulings is their only option (their task is to keep the public safe). They cannot afford to be wrong, even if no clinical justice is done. Better to have a scapegoat than be wrong. Politicians and corporations do the same, all the time.

An independent review of the CQC found 'significant internal failings' and Wes Streeting, the Health Secretary in the UK, declared it 'not fit for purpose' (July 2022).

Taking Advantage

If one is an unwise, ill-educated person, weak by nature, lacking in courage; insincere, easily annoyed, impetuous, sensitive to the remarks of others, unnecessarily strict or obsessive, lazy, slow to act, too principled and assuming moral superiority (because of a mindless adherence to rules and regulations, regardless of their fairness and value), there are those who will try to take advantage of us. Sun Tsu (The Art of War), teaches that to achieve the best results in battle, one must know each party well: oneself and the enemy.

If you are to use subterfuge well, you must convince opponents that your weaknesses are strengths and your strengths weaknesses.

Taking Advice

The choice of which personal adviser can be crucial. Choose one who knows you well enough to understand your true motivations, aims, and ambitions. She must be considerate, experienced, and have evidenced-based knowledge. Avoid arrogant egoists; their aim is to boost their own self-esteem. Avoid those who would say: 'Trust me, I know what I am talking about'. Their need is to convince others of their superior knowledge. They will welcome the opportunity to say later, 'If only you had listened to me, everything would have been OK'. (i.e. you should have acknowledged my superior knowledge). This is the hindsight bias at work.

Good Judgement

A vector force has, as you will remember from school physics, both strength and direction. Good judgement also has strength (of argument), and direction (its purpose). Weak judgements have one, but not the other; the worst judgements have neither.

When nobody seems to know what to do, patience is advantageous. Like mist clearing as the sun rises, what first seems unresolvable, can become clear with further information. While mist-bound, you have time to explore the options. Survival experts teach that when lost and confused, the best thing is to light a fire. Lighting a fire provides something indispensable: time for thought. Gather the evidence and decide what is most likely to succeed. Don't be rushed. After no success, try another strategy. Persistence is invaluable.

Comprehensive, reliable information is essential for good judgement. Even reliable information can be misinterpreted or misused to strengthen an argument. There are some ever-present questions: how reliable, pertinent, and comprehensive is the information you have, and how valuable is it? If used, will it be used appropriately?

Sometimes, all a doctor has at his disposal is a set of clues and a few suspicions. The job of junior doctors is to gather information and to investigate clues. All information gatherers hold an important power: that of selection. Your talent for selecting the relevant information will define you. This is because senior doctors will later judge and weigh the information served to them. With experience, one can guide others to a conclusion by serving them only the information they need to solve a problem.

Asking Pertinent Clinical Questions

The application of intelligence, experience and clarity of thought, will aid wise selection. An ability to identify, then by-pass the irrelevant, is equally essential. Wise judges focus on the important factors and exclude the irrelevant ones. At the interview for my first research post, Dr. Aubrey Leatham posed a challenge: *'We need to know why the heart rate is slow in both athletes, and in those with blackouts. We need to know who needs a pacemaker, and who does not.'* The secret of his success as an innovator depended on his ability to ask simple, pertinent questions, for which an answer could be found.

Clinical Evidence

> *People generally see what they look for, and hear what they listen for.*
>
> Harper Lee (1960). *To Kill a Mocking Bird.*

This is one example of the observer expectancy bias.

In everyday clinical medicine, both clinical history and examination provide the primary evidence (acquired using our hands, eyes, and ears). To these, one can add secondary evidence from urine and blood tests, ECGs, and various forms of imaging.

Sometimes clinical evidence lies in plain sight. It is crucial that one becomes a competent observer of subtle clinical signs, including micro-behaviours (fleeting facial expressions and body movements). Take the tricky observation of hyperventilation. I well remember a video shown at an RCP conference. A contemporaneous video showed two women sitting side by side. They asked an audience of senior physicians to spot any differences. It had to be pointed out that one was breathing slightly faster and deeper than the other. For the first time, I appreciated how easy it was to miss minimal hyperventilation.

An important rule for medical apprentices is: never accept a diagnosis that conflicts with the primary clinical evidence.

Verifying the secondary evidence may require some discussion with its provider (radiologist, haematologist, pathologist). One may need to know how sure they are of their facts, and their true opinion of them in relation to the patient's problem.

In all disciplines requiring critical judgement, from legal work to zoo-keeping, there is one unavoidable rule for success – if in doubt, or suspicious of the data, return to the primary source of evidence or the locus in quo. This will rarely waste time.

Unfortunately, work-load can lead to evidence collection being delegated, and problems with consistency and scope. One might ask how was the information gathered (the inclusions, exclusions, and value judgements), and how interpreted. This provides a playground for those wishing to pursue their own agenda. Evidence can become overlooked, omitted, under-rated, over-rated, or distorted. It can also be cherry-picked and lied about.

When media channels give us their versions of the news, we do not always get to examine who they interviewed and why they chose them for interview. Their selection of sources, and their editing of what took place, can colour the truth.

Scientific and legal evidence can suffer from major defects. Many will accept it without a full examination of its context, and without examining how filtered or edited it was. This is equivalent to viewing the evidence through a keyhole rather than through an open door. The full picture, replete with meta-information, may be available only to a fly on the wall. Without this perspective, our understanding could become circumscribed, and wise judgement compromised.

When MPTS tribunals consider allegations against doctors, lawyers control the proceedings. By choosing the questions they ask, they select what they choose to reveal. Given that the tribunal members will mostly lack a clinical perspective, how is it possible for them to make a full and balanced judgement? When they make a judgment, it will rest heavily on the presented documentary evidence and third party expert medical opinion, rather than the clinical perspective of the accused doctor. For this reason, I would advocate limiting MPTS tribunals to criminal charges against doctors. The tribunals could then claim competence. Only those with the relevant clinical experience should judge clinical allegations.

After an initial examination and investigation, the competent clinical observer will search for exceptions, abnormalities, inconsistences, and peculiarities. Those who are curious, dislike finding nothing.

Insufficient Evidence Situations

With only incomplete evidence to rely on, we may have to rely on intuition, or the evaluation of a more experienced doctor. For those with a talent for it, intuition will sometimes be more reliable than expected.

A patient (Mr. D.S.) was having sudden 'drop attacks', several months apart. The last time he fell, he fell backwards, and injured his head; he needed ten stitches in his scalp. He had sinus bradycardia for decades and had undergone a successful atrial ablation for AF. He also had severe coronary and carotid atheroma. I made a diagnosis of Adam's Stokes' syncope as an intuitive, failsafe diagnosis, without confirmatory ECG evidence. The question was: should I implant a prophylactic pacemaker (trial of pacing), or wait for ECG proof of bradycardia, heart block, or VT as a cause for his syncope? His family demanded action, but he was unhappy at the thought of any intervention.

Several 24-hour ECGs failed to reveal an abnormality, so I admitted him (he might have a fatal episode while wearing another 24-hour ECG monitor) to monitor his heart rhythm and BP. One of his relatives revealed to me he had been on an extreme, low-salt diet for years; a reason for his consistently low serum sodium. A low BP, cerebrovascular and coronary atheroma, and extreme sinus bradycardia, could all have contributed to his syncope.

In the absence of any proof for epilepsy or vasomotor syncope (there were no prodromal features of vasomotor syncope), my intuition suggested that a trial of pacing might prove effective. If the cause of his repeated syncope was VT, pacing would not stop it, unless it was an escape rhythm of extreme bradycardia. If due only to bradycardia, a pacemaker would help.

It was unsatisfactory to proceed without ECG evidence of his arrhythmia, but for the sake of his safety I had to remove the most likely cause of his syncope. I implanted a harmless prophylactic pacing device.

After agreeing to a pacemaker, and stopping his low-salt diet, he soon felt better than he had for years. After being paced, he reported no further syncopal attacks in the years that followed.

Published Research Evidence

In hospital medicine, it is customary to change our management when published double-blind trials or meta-analyses, provide sufficient evidence for it. Because of the sheer volume of publications, one needed spend many hours each week reading journals. An ability to read 'between the lines' when interpreting the results is an essential art. After considering the selection criteria, research methods, and statistical techniques used, few results achieve universal application. One must look for devils in the research detail; any of them could confound the conclusions and falsely influence our clinical judgement.

Should we use acronyms? Doctors like quoting research papers as acronyms (VICTORY, HOPE, SMART, CURE, COURAGE), especially the positive sounding ones. Some will be content to accept their carry home message if published in an prestigious journal, with no further scrutiny.

Using acronyms in health sciences has been subject to debate. Some authors have advocated against their use, as they claim it has turned into MMMMM (a Major Malady of Modern Medical Miscommunication). They assert that positive sounding acronyms are not appropriate for clinical trials with negative outcomes.' (Humouristic and Extravagant Acronyms. Anton Pottegård et al. BMJ, 2014; 349: g7092).

An important principle, especially for historical scientific literature research, is to seek the original texts and study them. Many will misquote them.

Few have read William Harvey's 'De Motu Cordis', written in Latin, and published in 1628 (it is available in translation). Most will cite Harvey's success in revealing the circulation of blood. Harvey's primary quest, however, was rather different. His aim was to answer the question: 'how does the mind affect the heart' (tachycardia with anxiety, for instance). He knew nothing of the autonomic nervous system and could not have answered the question. In 'De Motu Cordis', he declares this as a failure, but offers the idea of circulating blood as a consolation.

Inappropriate Evidence Situations

For many years, I stented both my symptomatic and asymptomatic patients, when I found they had significant, flow-limiting, coronary artery stenoses (> 80%); especially those with stenoses in the anterior descending artery, occlusion of which could lead to anterior infarction and a poor prognosis (only 50% survive 5-years). My twenty-five-year-old policy has taken a long time to vindicate. The PREVENT trial has now done so. It showed that preventative PCI can reduce the two-year and long-term risks of major cardiac events arising from vessels containing vulnerable plaques (Park, S-J. et al. Lancet 2024; 403; 1753).

Published seventeen years earlier, a contrasting, disparaging view of coronary artery stenting appeared. The COURAGE study stated: 'There is no evidence of any reduction in mortality, or MI (heart attack) incidence, in those with stable angina' (*Optimal Medical Therapy with or without PCI for Stable Coronary Disease*. Boden W.E. et al. (2007). NEJM; 356 (15): 1503-1516). Their conclusion: medical management works well.

Consider the COURAGE trial in more detail. Its message directs us to consider full medical treatment alone, and to avoid the stenting of coronary arteries when patients have stable angina. According to this trial, coronary stenting does not improve outcome. Few of the selection criteria applied to my patients. In particular, my patients were asymptomatic, despite evidence of ischaemia on exercise. Other differences were:

1. *Only 24% of the COURAGE patients had a >50% anterior descending stenosis. All of mine had a >80% anterior descending stenosis.*

2. *29% of the COURAGE patients were smokers (none of mine were smokers).*

3. *34% had diabetes (none of mine had diabetes);*

4. *39% had prior cardiac infarctions (none of mine had),*

5. *67% were hypertensives (few of mine were hypertensive).*

6. *All the COURAGE patients had stable angina (only two of mine had angina).*

With these differences, I could not apply the COURAGE trial results to the management of my patients. I don't doubt the conclusions of COURAGE for the patient group they selected, but there were too many differences in the detail for me to consider the results appropriate for my patients.

For decades before this study, my enthusiasm for stenting remained undaunted. Because of its efficacy in relieving angina and ischaemia on exercise, I became an early advocate of balloon angioplasty and coronary artery stenting. I was qualified well before CABG became universally available, and coronary artery intervention (PCI) became routine. Most doctors then, had personal experience of the natural history of (virtually untreated) coronary artery disease. Although Robert Goetz and his colleagues at the Albert Einstein College of Medicine in New York, introduced the first coronary by-pass procedure in 1960, it did not become widely available for my patients until much later. When I qualified in 1966, there were few available interventions other than sublingual nitrate, beta-blockade and warfarin.

The cause of angina is atherosclerotic coronary artery stenosis (in 98% of cases), although a few patients have angina with little ischaemia (coronary artery spasm, perhaps), many have ischaemia without angina. Successful stenting will usually improve coronary flow. When successful, it removes angina and shortness of breath. I remained undeterred in my selection of patients for stenting, simply because it produced angina-free patients with an improved quality of life, but improved mortality less certain. These advantages remain, but there is another advantage, given how safe the intervention has become. It helps to remove the worry of impending cardiac infarction (a sword of Damocles for those with anterior descending lesions). None of my patients sustained an anterior infarction in the thirty years that followed. Only a small number required repeat interventions.

With my case numbers too small, my experiences with IHD might represent an unreliable anecdote; although, not from my patient's perspective. One must never desert good, large group-based evidence, but never let it countermand personal experience-based findings without further consideration. Never let your thinking become inhibited once you are content with your own long-term personal observations. After your personal research has shown promise, large-scale verifying studies become obligatory.

The metaphorical elephant in my consulting room asked why my policy seemed to work so well. Could it be my case selection criteria? Could it be that my interventionist cardiac colleagues were better operators than others? I suspect both were true.

My results were unusually good, but why? Making an early diagnosis of CHD should have made a difference (the COURAGE study reviewed only 'late in the day' patient management). My patients were mostly asymptomatic and discovered during screening (completely dissimilar to what happens in the

NHS), so patient selection should be a critical factor. My referral decisions were also key, given the variable abilities of surgeons and interventional cardiologists.

All practicing doctors must have certificates to confirm their training. That does not prove what patients need to know most — whether a doctor is proficient; a slick operator with sufficient experience and a worthy track-record. Having observed many surgeons at work from an anaesthetist's stool (over a one-year period), I soon realised that surgical 'slickness' and dexterity varies enormously. The functional cognitive abilities of a physician are much more difficult to assess.

No Evidence Situations

In 'no evidence' situations, do what is best for the patient temporarily, while searching for evidence. The situation may force a fishing exercise. You may have to throw out a net (of investigations), and trawl for information. Re-visit the patient and enquire, enquire, and enquire further. Ask for independent opinions. Very often, the clinical picture will clarify.

The best way to understand what is going on in the world is to walk down the street (rather than remain in your office, looking at a computer screen). Look at what is going on yourself.'

Another tip. To paraphrase Ray Kroc, who grew the largest fast-food organization in the world (McDonald's): 'Nothing succeeds like persistence'.

Septicaemia can present with a 'no evidence' or minimal evidence situation. Its presentation varies considerably, making the diagnosis sometimes difficult. The limited amount of trial data available may force its management to be inspirational and individual, rather than formulaic. With a severely ill, deteriorating patient, time will force decisions to be made. Those unable to function well in 'no', or 'weak' evidence-based scenarios, might experience decision-makers' block. The situation may call for an individual therapeutic trial. The principle is this: if the actual diagnosis is malaria, anti-malarial treatment might work. If it is leptospirosis, anti-malarial treatment will not. In a tight spot on a desert island, a pragmatic approach is all you may have. If the patient dies, you can say you tried your best, although others might have expected more of you. The case against surgeon Mr. David Sellu illustrates the point.

With time running out, you might come to appreciate the art and power of best-guess management (call it 'Jungle Medicine'). This is not for those lacking intuition and courage, or those unwilling to decide and take personal responsibility. Beware of ineffective colleagues. The indecisive ones can be frozen to the spot, waiting too long for more evidence. There are those who would sooner let a patient die than make a decision without 'sufficient' evidence. They will usually be good at one thing: protecting themselves against regulatory action.

'Personalised' Medicine. 'Evidence-based' Medicine

The phrase 'personalised medicine' has two meanings. Traditionally, it is doctors getting to know their patients well enough to treat them as individuals. The latest definition uses individual anatomical and pathophysiological characteristics, gained from mapping the patient's genome.

Both traditional 'personalised' medicine, and 'evidence-based' medicine, are what patients need. Since I was often involved in diagnosing and treating cases others failed to solve, I can reflect on how important each is in practice. The time spent with patients, gathering as much meta-data as one can, is the vital ingredient of traditional personalised medicine. All but the most elusive of diagnoses will yield to this simple approach. The evidence so gained will usually lead to the diagnosis and to a patient-centred management plan.

It has been at least 2400 years since Hippocrates detailed the basic principles of medical practice. He advocated respecting the body, mind, and soul of each patient. 'Evidence based medicine', and 'personalised medicine', were both advocated by Hippocrates, but in addition, he taught the alignment of medicine with philosophy. In India, Charaka established similar principles in the 2nd century BC.

Over sixty years ago, medical education changed. Apprenticeship gave way to scientifically structured, 'evidence based medicine', along with Hippocratic values. Medical students assimilated these values from their teachers (competent doctors, knowledgeable enough to work in teaching hospitals). Sometime later, medical bureaucrats, having nothing better to do, standardised the basic requirements of practice for regulatory purposes. They saw the merits of 'evidence-based' medicine (used for standardisation and regulation), but not 'personalised' medicine. They now have both traditional and genomic personalised medicine as aims. Genomic personalisation they can understand, and will promote; traditional personalised medicine encompasses the art of medicine which is impossible to measure and regulate. Medical bureaucrats readily adopted the idea of evidence-based medicine as a marker of clinical adequacy, with ready-made measures for judging the compliance of doctors and nurses. The phrase 'evidence-based', now signifies what all bureaucrats, administrators, regulators and medical professionals seek most. Anyone who wishes to appear medically savvy, 'up-to-speed', and 'in the know', must use it.

Old Doctor: 'I am sighing an awful lot these days. Is there a cure?'
Young Doctor: 'I am afraid not. Get used to it, the world has moved on, and so should you. I sympathise with you, but it is futile trying to re-institute old standards, and attempting to fight bureaucracy.' I'm sorry to say this, but you must adapt or go.
Old Doctor: 'I am not sure that I agree with the word, must.'
Young Doctor: 'You will, if you decide to challenge regulators.'

Clinical Judgement, Anecdote and Algorithmic Thinking

Our personal collection of anecdotes and biases, supports our every clinical judgement; especially those that draw on similar clinical cases. With enough observer training and intelligence, our experience should always inform our judgements, alongside tick-box, algorithmic decision making. An awareness of our biases is, however, obligatory.

For those unable to discern the difference between a 'sick' patient and a 'well' one, the RCP (2017) introduced their N.E.W.S. algorithm (National Early Warning Score) for NHS use. They based it on respiration rate, oxygen saturation, BP., pulse rate, level of consciousness, and temperature. I would question the competence of any doctor who needed to use it. For the inexperienced, it should prove helpful and encourage learning.

Since the ability to detect 'sickness' is a fundamental requirement for doctors and nurses (I am not sure one can call oneself a doctor, or nurse, without it), it would be shocking to think that the use of N.E.W.S. might save 2000 lives per year (as predicted by the RCP). I hope this does not reflect the state of medical education and clinical competence in the UK today. It is no bad thing to rate clinical urgency with a score, but like the mnemonic lists used by students (SOCRATES for pain assessment), NEWS should at least help to alert the inexperienced and the unqualified.

Only by knowing the programming details of an algorithm, can one know its strengths and shortcomings. The assumptions made, the data used, and the processes being undertaken, all need scrutiny. Whenever the results are inconsistent, feedback and reprogramming should aim to improve reliability. In this re-iterative, heuristic way, automatic aircraft flight management systems have become more reliable (under usual conditions). But what happens under exceptional conditions? Two Boeing 737 MAX aircraft tragedies exposed fatal weaknesses in their computerised flight management systems. Only the courageous and foolhardy would trust their life to a single algorithm or to any single silicon-based device, especially when new and untested by daily routine. A few years are usually required for all the devils in the detail to reveal themselves.

As 'apps', algorithms are going to be used more and more by patients, so it will become essential to understand how they function. Patients already accept the results of apps without question, and regard them as accurate, reliable, intelligent, and dependable — all the things they are not. For doctors, they provide useful 'rules of thumb', and can help remind us of what we have forgotten. They can also provide us with clinical options, each with a calculated risk.

Algorithmic thinkers have one-track minds. With little latitude to their personality, and little aptitude for responsive executive functioning, they are most inept when handling exceptions and a need for change.

In the Second World War, during their attempted capture of Stalingrad (now Volgograd), the Germans, seen as rule-based by the Russians, took their meals at fixed times. The defending Russians, saw this as an opportune weakness: they knew exactly when to attack them.

Dependency on mobile phone data suppresses independent thinking. As Elvis might once have sung — 'caught in an app, can't get back.' Sorry, but I once played guitar in a group. Our lead singer Kenny, loved to sing it.

Consider an advert for a dance instructor (2024): The job will be suitable for dancers able to dance to many and various rhythms, while strictly adhering to published dance protocols. Free-style moves will not be acceptable.

Advert for a dance instructor (1966): The job will suit someone who can dance to many rhythms. They must be creative, and able to cope with unpredictable changes in tempo and style. They must be capable of creating their own novel steps and be able to expand the enjoyment of others while dancing.

What will happen if we stop thinking? We need not look too far. Those who rely entirely on algorithms have already done so, but do we really want to live with decisions that lack wide-ranging, considered thought?

In my lifetime, I have seen mental arithmetic decline in favour of calculators. Few can now add three numbers together without a calculator; they simply don't need to. Mental arithmetic was an important element of my early schooling. I have always put it to good use, quickly spotting errors in accounts and concluding my calculations before others have reached for their mobile phone.

When I was a pre-clinical medical student, I sometimes worked in my father's office. One of my jobs was to add columns of figures in ledgers. My father could run his finger down a column of forty figures (about five seconds' work) and pronounce an error. 'Try again', he would say. Amazingly, he was usually correct. He could add quickly, and easily determine when my results were of the wrong order — hundreds, rather than thousands, for instance. Such skills are no longer essential; they have died with no loss, except for some loss of self-esteem.

One of my patients owned a car tyre outlet company with over one hundred branches. He stepped away from his shops, and remained closeted in his head office, watching his sales figures as they developed on his computer. He failed to acknowledge one thing: without a cat, mice are free to play. He thought the need for data superseded the need for his personal involvement. Only after he went bankrupt did he realise why his business failed.

Trusting in machines (and employees), without checking, can lead to tears. The moral is: always take time to observe first-hand what is really going on.

There are several good things about algorithms: they are tireless, and can repeatedly calculate faultlessly. They can also, consistently repeat the same errors.

I wanted to weigh and compare nutrients in food for possible cardiovascular benefit: the 'good' (possibly athero-protective), and the 'bad' (likely to be athero-generative). From animal research, I identified twenty-five relevant nutrients and created a mathematical formula to provide a 'good'/'bad' nutrient ratio score for each food. I called this the Cardiac Value™ of food. It allows the comparison of one food with another. I created an algorithm to rank food choices in order of cardiovascular value. Nobody has yet asked me how I derived my formulae, or anything about the assumptions I made while developing it. Perhaps the majority assume they know what is good for them? (See D H Dighton: 'Eat to Your Heart's Content', 2005; ISBN 0-9551072-0-2).

Balance of Risk

In some clinical situations, several independent risks can occur simultaneously. An aortic aneurysm can gradually expand to a dangerous point, while blood pressure becomes poorly controlled; renal function deteriorates and the patient's fitness diminishes. Only healthy, fit patients will easily survive some aneurysm operations; the risk of the operation depending on patient age and physical condition (how frail or athletic they are, and their cardiac, renal, cerebral, and other functions). The notion of biological age will be more pertinent than chronological age to their survival. When I replaced the pacemaker of a 100-year-old patient of mine, he was a capable, active man. So why not?

Risk is always in contention when we undertake radiation-based investigations and invasive techniques. Some gambling experience can help risk assessment. Gambling teaches us about the fallibility of our decisions. By placing wagers, one quickly learns about luck and chance (like the fixed odds of roulette, and the indefinite odds of horse-racing). Clinical decision making is not a game, but there are similarities. If risk taking is not your forte, avoid A&E, the HDU (ITU), operating theatres and invasive cardiology.

The mastery of any medical discipline must include the art of prognostication (prediction). To be of use, our predictions must be correct and consistent. One might need the foresight of a sibyl or Cassandra to achieve this, although Cassandra's predictions were cursed, never to be believed. Since medical prognostication is only partly based on science, you will need some luck if your predictions are to prove accurate.

Savvy patients faced with a difficult decision might ask, 'what would you do, doctor?' (The patient having no experience of such matters themselves). A detached doctor will most likely say: 'I am afraid it's entirely up to you.' This could place the responsibility for any decision onto a patient's inexperienced shoulders. What the patient needs at this point is the experience and judgement of one who has mastered the art of traditional personalised medicine; one who can strike a chord of common sense, and help the patient meaningfully.

No patient will benefit much from a detached doctor listing all the known complications, without putting the odds of occurrence into context. This has given some patients the impression that doctors are more concerned about divesting themselves of responsibility and medico-legal issues than about patient welfare. My advice is to speak plainly, and tell patients what you really think, and what you would do if you were in their position. I am sad to say that many doctors would now regard this as unprofessional. Some doctors derive more pleasure and self-esteem from following the rules and complying (and the endless demands of bureaucracy) than helping patients.

Sometimes we have to re-consider our advice. Most patients will understand, but only if a doctor's honesty and commitment are beyond question. With more information, one may need to reconsider. This is what mature, sentient beings do.

Patients know all about error. They also know that admitting to error is a strength. An admission of error signifies our integrity, common sense, humanity, and nearly always our humility. Easily frightened rule-book followers (obsessive neurotics, autistic doctors, and patients alike), will worry more about error and its consequences than anything else. Many will endear themselves to like-minded compatriots (fellow rule-book advocates and bean counters). Endearment may not be their purpose, but collegiate acceptability will be. They may not comprehend that success in all fields of life always involves playing the odds, risking error when appropriate, and being able to perform beyond the rigid rules designed for the servile and obedient. Rigidly obedient doctors may find themselves functionally inept in situations where no straightforward rules exist; situations where intuition, innovation, lateral thinking, and creativity are essential.

Questions of Clinical Judgement

An ever-present question in clinical practice is the relationship between individual clinical features. Are they coincidental or causatively linked as associated features of the same pathological process?

Case 1: My friend Don Russell, who died recently aged ninety, coughed after eating. He had a small oesophageal pouch. Should we remove it? No, because an operation would have been too risky at his advanced age, in his state of health, and because the pouch had been there all of his life with no previous symptoms. Finding the pouch was a co-incidental finding.

Case 2: Mr. V. O'M., aged sixty-six-years, was travelling back home one night after entertaining clients. While in his car, he devoured an entire box of chocolates. Soon after, his relatives in the back seat noticed him shaking; 'having a fit', they thought. He appeared unconscious for one or two minutes. When he 'awoke', he felt completely normal: no nausea, no drowsiness, but wanting to know what happened.

I saw him the day after and questioned him. Did he have any warning of (supposed) syncope? 'No.' After his consciousness returned, did he feel unwell, tired or nauseated? No, he did not. Did he look pale and sweaty during the episode (common with vasomotor syncope)? His companions didn't know. It was too dark. At this stage, the diagnostic considerations were: rebound hypoglycaemia or hyperglycaemia? Epilepsy? Adam-Stokes syncope, or vasomotor syncope.

I examined him. He had always been obese. His BP was 120/80, but his pulse was rapid and irregular. An ECG revealed atrial flutter. This had been noted on one occasion two years before, but he had been in sinus rhythm since.

A non-fasting blood test revealed a blood glucose of 9mmols/l. Was this relevant?

His symptoms that carried most diagnostic weight were: (1) the short period of unconsciousness, (2) no warning of syncope, and (3) a rapid recovery with no remnant symptoms. This made epilepsy, hypo- and hyper-glycaemia, and vasomotor syncope inconsistent possibilities. Adams-Stokes was, therefore, much more likely. Could it be that he had bradycardia when changing from sinus rhythm to AF? He needed a 24-hour ECG and further evaluation of diabetes / pre-diabetes (he had one raised glucose level). I checked him for postural BP changes (autonomic neuropathy), although he made no change of position immediately before the episode.

Many patients get AF induced by an excess of alcohol. This he did on the evening of the incident. Meta-data from his wife informed me he was unused to alcohol, and on this occasion he was 'the worse for wear'. This information from his wife was especially useful since the patient had always been a man of few words, and what words he did utter, usually underrated his problems. It was his wife who insisted on him consulting me, and being examined.

A 24-hour ECG, off of alcohol, revealed only sinus rhythm. After giving up alcohol, he had no further incidents. Eating an entire box of chocolates was irrelevant. Did they taste of red herring I wonder?

Case 3: I had just diagnosed Mr. C.F., with hypothyroidism. Thirty-two years before, he had a partial thyroidectomy. He had some fused together lymph nodes in his right supraclavicular fossa, far from the usual position of the thyroid gland. The patient claimed the nodes had been there, unchanged for years. They did not feel like 'normal' nodes. Thyroxine 100mcgs daily improved his tiredness after only three days. All his other blood tests were within normal range. Should we biopsy his nodes?

My decision was to observe his nodes, and only biopsy them if they grew, or if other factors arose that suggested the necessity. They most likely arose post-operatively after the original thyroidectomy. During long-term follow-up, the nodes did not change and remained.

Case 4: A woman had recurrent upper abdominal pains, especially after eating fatty food. Abdominal ultrasound revealed gallstones. Were her pains caused by her gallstones, or by IBS? Would removing the gallstones cure her pains? What would you advise?

Most gallstones are asymptomatic, and IBS a more common cause of such pain. She improved with dietary change. Her gallbladder and gallstones remain.

Case 5: A friend and medical colleague of mine (Dr. D.B.) came to me as an emergency with chest pain. His clinical notes revealed that 5-years before, he had experienced angina. He had been getting more breathless for some time and had compensated for it by walking slower. He had had two occasions of orthopnoea / possible PND (paroxysmal nocturnal dyspnoea). He had a slight rise in his blood troponin level (within 12-hours of his pain developing), but no ECG changes. A cardiac CT showed a calcium score of only 200, but concentrated entirely in one place, in his anterior descending coronary artery (LAD). Should this area be stented?

Troponin levels are subject to false positives, although his slight rise could have meant some cardiac cell damage. His calcium score was not unusual for his age (late 70s), except for his previous angina. Calcium concentrated in one place (total score 200) suggested the possibility of a significant solitary stenotic lesion.

Although he was a very reliable witness, he was a very reluctant patient. I had known him as a colleague for 50-years, and knew he was being factual, but underrating his plight (although the possibility of PND resulting from LVF was real, and had worried him). His natural inclination was to wait and see. Few enjoy cardiac intervention.

In my judgement, it would have been a grave mistake to risk a big anterior infarction (associated risk: a 50%, five-year survival), so I strongly advised an urgent coronary angiogram with a view to stenting an isolated stenosis. To get acute LVF (his PND was evidence of this), the atheromatous plaque in his anterior descending would have to have caused significant ischaemia. Alternatively, it might have had coronary artery spasm, but would that have reduced the coronary blood flow enough to his left ventricle? I couldn't take that risk.

My former colleague, cardiologist Dr. David Lipkin, placed two stents in his LAD coronary artery. He has been asymptomatic now for eight years, with considerable improvement in his effort tolerance. These improvements became apparent shortly after the procedure.

His initial inclination was to assign his shortness of breath to age. On this basis, he was slow to seek advice, assuming intervention to be futile. His life had been at very high risk because of this false assumption.

Faulty Decision Making

This subject could fill many volumes. Both the non-fictional and fictional canons of world literature are full of illustrations. Consider a few from my own observations.

If you want to chance making faulty judgments and decisions, nothing beats assumption; believing you know best and jumping to conclusions. Preoccupation with a stressful situation will accomplish much the same by diverting the sufferer's attention. Under these conditions, the tendency to jump to a convenient conclusion has a biological purpose — it reduces future energy expenditure. Tired, exhausted, and stressed people, do not always make decisions that will pass the test of time.

A prelude to faulty clinical decision making, is often the absence of any in-depth consideration; strongly held assumptions can inhibit further thinking and close the door to other considerations. A mantra of belief, repeated often enough (indoctrination), will strengthen our defence against criticism. Those with long repeated experiences also tend towards the same position. If experience repeatedly confirms our view, extraneous facts and new theories may become annoying, and foster cognitive dissonance (see Magritte's painting of a pipe entitled, *'Ceci n'est pas une pipe!'*).

When Florence Nightingale suggested separate wards, improved ward cleanliness, and space between beds to control infection, the military doctors then serving in the Crimea (1853 to 1856), were less than impressed (now called the Semmelweis bias). Their presumption was that no nurse could know what was best for patients. In the end, her suggestions created a turning point in medical history, saving countless lives. She based her discoveries on anecdote, and effectively illustrated her observations using her own pioneering statistical charts (which included her invention – the 'pie chart').

Florence Nightingale observed the fate of our soldiers in the Crimea. They died more often from infection than from enemy bullets. She observed that those 'lucky enough' to be wounded early, within the first few days of arriving, survived better than those who had long suffered the adverse living conditions (cold, wet, and covered in mud). This observation was an anecdotal step towards the understanding of immune resistance (reduced by stress). The immunity concept had yet to be discovered. Paul Ehrlich won a Nobel Prize for it in 1908, fifty years later. She couldn't wait for science to catch up. She argued for an immediate improvement in conditions for soldiers; a legacy that remains her tribute. Hers was a triumph of anecdotal observation, but not perhaps 'scientific' enough to have qualified for a Nobel Prize, had it then existed (first inaugurated thirty-nine years after the Crimean War).('Scientific' and 'objective', now mean knowledge gained from using controlled experiments to verify the truth).

Emotional attachment is another important reason for arriving at a faulty conclusion; it can act as the enemy of insight. It can sometimes be what I have called the 'inamorata bias' (Italian for 'in love'): emotional attachment overriding the desire to undertake in-depth reviews. Also, the emotional attachment to a favoured thesis can act like superglue; even incontrovertible evidence may not shake it free. Such attachment is most disruptive when used by the passionate, inexperienced, impressionable, arrogant, and ill-educated, and those more liable to emotional reasoning than deductive reasoning.

In the film 'Some Like it Hot' (United Artists. 1959), Jack Lemmon plays Jerry, dressed as the woman Daphne. Osgood, played by Joe E. Brown, is in love with her (him). The final dialogue illustrates the tenaciousness of Osgood's emotional attachment, even in the face of incontrovertible facts about Jerry's gender.

Jerry (Daphne): I'm gonna level with you. We can't get married at all.
Osgood: Why not?
Jerry (Daphne): Well. In the first place, I'm not a natural blond.
Osgood: Doesn't matter.
Jerry (Daphne): I smoke. I smoke all the time.
Osgood: I don't care.
Jerry (Daphne): I'm a terrible person. For the last three years I've lived with a saxophone player.
Osgood: I forgive you.
Jerry (Daphne): I can never have children.
Osgood: We can adopt some.
Jerry (Daphne): You don't understand (Pulling off his wig). Ah! I'm a man!
Osgood: Well. Nobody's perfect!

Conceit, egotism, jealousy, and personal pride can exaggerate the inamorata bias.

Our nomadic, prehistoric hunter-gatherer forefathers, were required to function in a flexible, open-ended way. There are those who retain this adaptability and can function whatever comes their way. In everyday life, new things constantly replace the old as fashion and taste change. The mental processes involved depend heavily on emotional factors, the need to conform, a need to be noticed, with greed and selfishness powering survival.

Everyday decision making often happens with no reference to any rules or guidelines; instead, factors derived from our culture and personality hold sway. Thought processes like, *'I like yellow. This car is red. Therefore, I will not buy it'; or, 'I want to impress others, and a red supercar car will do that. Whether the car is red or yellow is of little importance. Even though I cannot use its speed, I'll buy it, anyway. My need is to impress, and this will suffice.'*

An alternative approach to this decision might be to use a 'Which Car' algorithm. This would harness objective appraisal. Personal requirements such as fuel consumption, number of doors needed, air-con, payment of the London Congestion Charge and other road charges, luggage space, room for children and dogs, and a choice of gearbox could inform the decision-making process.

Emotional decision making utilises fashion-weighted variables (some people have very particular interests, values, attachments, and a need to conform). Each individual will apply their own weight of relevance to each variable. Within corporations, this may not good enough; many corporations need to suppress personal activity in their staff. Corporate directors and controllers want standardisation, and this requires strict adherence to all the fixed rules they impose. Allowing free, emotionally led choices could lead to inefficiency, chaos, or indeterminate outcomes.

> *'Of science and opinion: the former begets knowledge, the latter ignorance.'*
>
> Hippocrates (supposedly).

The word 'science' derives from the Latin verb 'scire' (to know). Unless he was bilingual, it is questionable whether Hippocrates, living on a Greek island would have known that. He would have known the Greek concept of knowledge - επιστήμη (epistImi). Knowledge begets knowledge might be closer to what he said. Hippocrates was a trained observer, with knowledge based only on anecdote, not experiment.

There are many other facets to faulty judgment and decision making. One is to accept the consensus view. The underlying thesis of consensus is that 'the majority knows best', and 'minority views hold less value'. Truth, like uniqueness, is an absolute, although it rarely presents itself as absolute, given all its perceived aspects. Although it need not depend on any average or consensus, the majority view will sometimes reveal the truth (in the TV programme, 'Who Wants to Be a Millionaire', the 'ask the audience' option often works well). The sanctity of modern democracy rests on one potentially faulty principle: that the majority knows best. This is far removed from the original Greek concept of democracy from which it developed, and open to much criticism even for highly intelligent fully informed groups.

Falsification

Cherry-picked data is often used to persuade us. I have known scientists to 'clean' their data when it suits their purpose (the publication of papers). One convenient device is to delete experiments which have not yielded the expected result. Some experiments have to be repeated many times before they yield the expected result. One must delete the results of truly faulty experiments, however.

There is an art to the selection processes applied when revealing scientific results. Presentations, like water-colour paintings, can be wishy-washy or bold. The framing bias may apply: the excellence of a presentation can help to convince others more than the data.

We can view most truths, other than the axiomatic, from various aspects. We can then distort the view by drawing attention to some features more than others. Those who create propaganda are used to it. Some journalists have earned their living using such techniques, writing eye-catching headlines which exaggerate those aspects of the truth they want to promote (President Donald Trump repeatedly referred to manipulated aspects of a story as 'fake news'). Politicians commonly use rhetoric and some economy with the truth to keep their voters happy. They are not averse to false analogy, false logic, generalisation, and standardisation (one size fits all). As important, is the editing process: what gets cut and why, and who decides what gets published?

Once an idea spreads and gains acceptance, it can remain unchallenged for decades (you will then often hear: *'everyone knows that... low-fat foods promote longevity, and a lean body is healthier than a fat body'*). Such dogmas can lead to false biases and conclusions that last for decades, but may not apply to everyone.

Despite all our deliberations, judgements and clinical decisions, some patients improve without us. This experience for some doctors should teach them humility. The ancient Indian text, the *Siddhi Sthana* (Charaka Samhita), noted it, and referred to it as yadrischha ('not ideal'; c.1000 BCE).

Judgement at the Hand of Mirth

> *'A man may say full sooth (the truth) in game and play'*
>
> Geoffrey Chaucer (1390). *The Cook's Tale.*

The truth is often spoken in jest. Like tomatoes and jackfruit, truth and justice can be served in different ways: as sarcasm, satire, parody, spoof, laughable metaphor, or as a sharp arrow dipped in piercing wit. Never mistake the intention when malicious.

At the Oscar ceremony in Los Angeles in March 2022, actor Will Smith took exception to some less than humorous remarks made about his wife by comedian presenter Chris Rocks. In full view of the audience and media, Will Smith walked to the stage and punched him. He told him to leave his wife's name out of his mouth. He later apologised. Chris Rocks made no complaint to the LAPD.

During any serious discussion, wit and humour can act like a draft of fresh air in a foetid room. Wit and humour can slice through pedantry, obsession, undue seriousness, waffle, bluster and bullshit to highlight the truth, although a fist can sometimes bring about a faster change of mind.

Not every patient has a well-developed sense of humour, but many will warm to those doctors able to use it beneficially. Humour varies so much between cultures, one should not use it without considerable forethought. An established, mutually respectful, patient-doctor relationship is the essential prerequisite.

Comedians can incite laughter by referring to flies in soup (metaphorical, or otherwise), and why they should be ignored. The reason is, of course, they won't drink much. These soup-swimming flies have cousins - 'flies on walls'. Clinically, such flies have distinct advantages: they see what is really going on behind the scenes (the actual data, otherwise hidden (mesa) data, and meta-data). They are to be envied for their privileged position. Before completing a clinical opinion, be sure to have accessed your patient's 'fly on the wall' information.

The Scene: A classroom full of eight-year-old pupils

Teacher: *Johnny, I have a question for you. There are five birds sitting on a fence. A man with a gun shoots one bird. How many birds remain?*
Johnny: *None, Miss.*
Teacher: *Come on, Johnny, try again. There were five birds to start with. One is now dead, so how many does that leave?*
Johnny: *None, Miss.*
Teacher: *Explain your thinking, Johnny.*

Johnny: *When the gun goes off Miss, the noise will frighten the others away.*

Applying the immutable rules of arithmetic will not always lead to the correct answer; meta-information (in this case, the loudness of the gun) needed to be considered. The correct answer to a question can depend on perspective, and other boundary conditions (the small picture versus the big picture, for instance). Meta-information was key to Johnny's thinking. While his teacher will forever remain a teacher (because she prefers the safety of rules and demarcated boundaries), Johnny's vision of the world is wider. He is already engaging with life in a more realistic way.

Which is more important for achieving success in life: compliance or imagination?

I wonder how Elon Musk, Richard Branson, Einstein, and Henry C. Beck would have answered the 'birds on a fence' question. I'll save you the trouble of Googling Henry Beck. He designed the London Underground map. Imagination is an essential requirement for all entrepreneurs and innovators. Doctors in the UK are now frightened to use theirs, bound as they are by the heavy chains of regulation.

Judgement by Analogy

René Descartes established a prerequisite for all of his discussions: first examine and declare your assumptions and definitions. Consider the following proposition:

The calcium carbonate 'furring' of central heating pipes in hard water areas is analogous to atherosclerosis in arteries.

The hypothesis is logical and reasonable, but based on a misconception – that the underlying processes mostly involves chemical deposition.

This 'active' biological process (atherosclerosis) differs from passive deposition. Obviously, central heating pipes are not the same as arteries, except for providing a conduit for fluid. In hard water areas, calcium salts can precipitate to form 'furring' in water pipes. Inductive thinking allowed some to think that cholesterol and other fats in the blood deposited like calcium salts in water pipes, but they don't.

Intimal injury in rats induces the neogenesis of atheroma (atherogenesis), and the activation of osteopontin. Osteopontin is a glycoprotein found in human artery plaques which 'calcify', or more accurately, turn to bone (calcium apatite). Osteopontin is also known to be involved in bone neogenesis. (Osteopontin is Elevated during Neointima Formation in Rat Arteries and is a Novel Component of Human Atherosclerotic Plaques. Giachelli, C.M. et al., J. Clin. Investigation (1993); 92(4): 1686-96). Something similar occurs in human arteries.

Rat experiments are the source of this information. So consider another proposition. Do rat arteries provide a reliable model for human arteries? Few regard the results of rat experiments as valid for humans. Human arteries might be an exception. From an evolutionary and biological point of view, arteries are rather 'primitive' structures, having not developed much since their humble invertebrate beginnings (biochemically and histologically). All arteries have an intima and a muscular tissue layer. It is possible, therefore, that rat arteries might be a reasonable surrogate for human artery functioning. Although this proposition is an inductive assumption and a potential source of error, it may be correct.

Exception-led Judgement

Pursuing exceptions accounts for many scientific advances. If you spot something unusual in the patient's history, examination, or investigation, pursue it until you find an explanation. You may risk annoying others with your obsessive attitude, but so what? This is the stuff of discovery. For a good examples, watch 'Columbo', the American detective TV series (1971-2003). In every episode, he finds out 'who done it', by doggedly pursuing every inconsistency.

Variation and exception allow biological evolution. Understanding and dealing with them is essential to the successful practice of clinical medicine. Strict uniformity, like the standardisation required by corporations and bureaucratic organisations, is not in keeping with biology. The corporate aim is to control others by disallowing change, unless sanctioned by them. As physicians, however, we need to embrace the individual biological differences of our patients. Effective doctors will vary their approach; the most ineffective ones will keep a standardised approach.

The 'Classic' Case

The concept of a 'classic' case is one that illustrates all the pathognomonic features. It is useful in bringing together one or many specific presenting features of a disease. Like the mode or mean of a disparate group, few cases will be the same; a less than perfect fit is more usual, with one or two features missing.

We now know that predisposition to disease has a genetic basis. When several genes are contributory, as in muscular dystrophy; a patient with only one contributory gene from the set might have weak legs but little else. Include every contributory gene, and a 'full-blown' (classical) case will be seen. In between, there will be many (some unidentifiable) permutations, with many sub-clinical phenotypes.

Exception-led Defeat

In the same way that negotiating a simple staircase defeated the Daleks (BBC TV series, 'Dr. Who', from 1963), one unforeseen exception can defeat the most sophisticated argument. All algorithms can suffer this problem. Many doctors are content with their usual routine (dealing with every variation of white swan), with only a few happy to consider the possibility of black swans.

At a US university question-and-answer session, a Japanese student asked Warren Buffett why he didn't have a 'smart phone' (he has a basic type of mobile phone). The questioner implied that the latest device would help him become more successful. Warren Buffett was then one of the richest billionaires in the world. He claimed he was good at making money, but not good at spending it. The Japanese student who questioned him had obviously accepted the myth that computer-based functions must improve decision making and judgement. The idea is thriving among naïve fashion followers. For them, however, it could be true.

There is a simple lesson here. Learn from the experienced and successful, especially when they are exceptional and different. What you learn may be unfashionable and unexciting, but correct and successful.

Experience-led judgement

Daniel Kahneman suggested that our experiences in life, can differ from our memory of them. We cannot remember every experience. Instead, we remember only the peaks, the troughs, and the end results. He referred to this as 'the remembered self'. It is prone to exaggeration and underestimation when used in decision making.

'Odd as it may seem, I am my remembering self, and the experiencing self who does my living is like a stranger to me.'

Daniel Kahneman. *Thinking, Fast and Slow* (2011) Macmillan.

Influencing Wise Judgement

Education, experience, and intelligence are the substrates of wise judgement.

One cannot always apply the way we make personal judgements to doctor-patient judgements. Our personal decisions cannot avoid being culturally biased: influenced by our family, social class, financial status, and background, all or none of which we may share with our patient. We can expose our personal judgements to the mirth and criticism of family and friends, but our professional judgements may have to withstand more intense forms of attack - from independent medical critics, from other cultures, and from the media. Such prospects can temper the judgement of all but the brave and foolhardy.

Ignorance and arrogance are powerful enemies of wise counsel. Many are happy in their ignorance and will ignore the knowledge of others. A few decades ago, all but the extrovert and ignorant remained silent, for fear of embarrassment. Now we are free to say whatever pops into our heads, and to publish it on social media channels. That universal intellectual equality exists, such that we can now ignore ignorance, and deny the existence of an intellectual spectrum, has arisen because educated, intelligent people, worried by political correctness, are too cowardly to comment. It was once impolite for well-educated politicians to talk down to their less well educated voters. Alongside the ridiculous concept of honouring ignorance, the patronising of voters is now unacceptable.

Many societies seem to accept intellectual equality without question. Politicians who dismiss it will pay a political price. Instead, politicians must now say, *'We should trust the public. They are no fools. They can all distinguish good fish from bad fish'.* What they voice in private is another matter. Empires may be barbarous on the way up, but they become too weak and liberal on the way down.

The preservation of status concerns many. Ego, vanity, and self-esteem are so fragile, they are always at risk of deflation. A need to preserve them can strongly bias our decision making and judgement.

Lust, good and bad taste, greed, jealousy and vested interests can all warp opinion. All are enemies of wise judgement. Although rarely given any credit for it, entertainers can shape public opinion, attitudes, and values. Like other media creatures, their job is to feed audiences, trying to keep them interested and amused. While feeding us with a diet of humour, sarcasm, wit, irony, and satire, the threshold to shaping our opinion is lowered.

Medical politicians are used to using the rhetoric of professional soothsayers when trying to influence the medical profession. Accountants, and others working outside the field of action, now control the scene, some shaking their heads in despair at the cost of saving a life. With their positions so well preserved and protected, for what reason might they change?

With the medical profession firmly under the political control of bureaucrats, it is important for us to safeguard the influence experienced clinical colleagues with the most expertise, can have on us. One can meet them at case conferences and medical symposia.

Judgement by Committee

Committees use consensus. Those who speak first, and those whose presentations are egotistical and uncompromising, will more often sway their decisions (the framing bias). What every ineffective organisation actually needs, apart from consensus, is independence of thought. Perhaps they should circulate the independent thoughts of each committee member before their meetings.

On most committees, there will be those who credit themselves with wisdom. Typically, they will be politically correct, and unwilling to bear personal risk. Standardisation, consistency and compromise are common aims. When committee decisions choose political correctness, they must ignore important meta-data issues like gender, race, culture, class, and the educational attainment levels that define individuals. For the sake of arriving at a tangible, compromised conclusion (like all statistical conclusions), critical details have to be sacrificed.

Those who have been a plaintiff in any Court of Law, with a jury deliberating on their future, will be subject to majority voting. A jury may have to learn that some detail might need to be forgotten, down-rated, or overridden, for the sake of concluding the case. In the adversarial court system, with the defence claiming sainthood and the prosecution claiming ultimate evil, the jury must somehow sort biased and adulterated truth from the real thing. In courts of law, this can be what passes for fair play in the absence of fair discussion and rational balance.

The Law has nothing to do with fairness, something to do with justice, and everything to do with compliance. Lawyers and regulators warm to computer algorithms, given that they can provide standardised tests of compliance. Humans have a variable capacity to comply and mostly feel dutifully compelled to do so. Humans survived for millennia thinking and creating for themselves, but we must now defer to an ever-growing, rigid body of rules, said to be essential (by consensus) for the acceptable functioning of society. The way of life that follows from these rules varies considerably from person to person, culture to culture, and country to country.

Ward-round and clinical case conferences can produce decisions by tacit consensus, weighted towards the most intelligent, common-sense view. Once those with seniority and experience agree, clinical wisdom usually prevails. There is no need to vote. The situation is more akin to a philosopher king delivering wise judgement, after listening to all his advisers. If only we were replete with philosophy kings, we would not need committees, bureaucrats, or politicians.

Information and Good Judgement

> *'When I was young, I had a pet rattlesnake. I loved it, but never turned my back on it.'*
>
> Cole Harden (Gary Cooper). Film: *The Westerner* (1940).

Good judgement depends on accurate and complete information. Doctors have to deal with distorted facts, disinformation, redacted facts, fake information, minimised and omitted facts, and attempts to mislead. We also have to cope with information that is aimed to confuse, or to raise a smoke screen. Because legal court proceeding are adversarial, barristers are at liberty to use any form of persuasion to win their case (exaggeration, omission, distortion, etc.). Ward round clinical discussions should not take this form: their aim is to benefit a patient by presenting and discussing reasonable arguments from many sources. This contrasts with legal discourse where the aim is binary: either to free a person from accusation, or to deprive him as a punishment (all 'in the public interest', of course). These differences can confuse doctors should they appear before the MPTS (GMC).

The creation of a Medical Directorate would solve any confusion. If cases brought against doctors were first heard only by senior doctors, and subject to the same academic considerations as are apparent during ward round discussions, the value of a doctor's clinical knowledge and ability would receive its full acknowledgement. When I appeared before the MPTS, I resigned from the proceedings because of the complete absence of any academic clinical discussion about me and my actions. Their concern was to assess the risk of my continuing to practice, and my likelihood of complying with their rules in future (some of which I disagreed with, and told them so).(See my book, The NHS. Or Sick Sacred Cow, 2023).

Incoming RAF men volunteering for duty on the Scottish island of Uist were once told: 'On Uist there is a girl behind every tree.' There was something the RAF men were not told: there are very few trees on Uist.

Without inside information (mesa-information), one might never learn what information is missing.

On January 15th, 2009, Captain Chesley Sullenberger ('Sully'), failed to land US Airways flight 1549 at a nearby airport, after a bird strike stopped both engines. He successfully landed in the Hudson river, saving the lives of all 155 passengers and crew. They later indicted him for his failure to land at a nearby airport.

The enquiry that followed was told that many pilots, tested in a simulator, successfully landed the same airplane under the same conditions. Sully did not accept this. Repeated simulator trials proved him correct. With no forewarning of a bird strike, not one pilot successfully landed a simulated airplane at a nearby airport.

If you are told that there IS such a thing as a magic carpet, or a rope that will stand on end without support, should you believe it? They train sales agents to make us believe similar things: that *'everybody is now buying Z175F widgets'*, or that, *'nobody is using warfarin or aspirin anymore (everyone has switched to NOACS, or factor Xa inhibitors)'*. Beware of vested interests. You might have to face equivalent coercion from colleagues and committees. In each case, it is important to identify 'what's in it for them'.

Some patients try similar strategies, trying to make a doctor believe that, *'There is nothing wrong with me doc., I'm as strong as a horse.'* I remember two patients who believed this; both died within a year from acute cardiac infarction. Years before, I had suggested to both, they should have heart screening. Both refused. I didn't believe their claims of unquestionable fitness, but nothing I said persuaded them to get tested. Some patients in fear, and others in denial, will risk death rather than face investigation and the revelation of something worrying (possibly something treatable).

Self-assured patients are another matter. They can choose to swim in dangerous waters, regardless of experienced advice. When things go wrong, though, they will often blame others.

Once aspirin has normalised a patient's temperature and she feels better, does it mean her pneumonia has resolved? If diet lowers a patient's blood cholesterol, does it mean her atheroma progression has stopped? Both conclusions are wrong. The key question to ask is: 'How does any measurement or observation (temperature or blood cholesterol), relate to the pathological process in question?' These links are the only important ones.

The Weighting Game

We all weigh (value) decision factors differently. This is because we cannot help but use our biases (personal values, and priorities) to weigh them. Judgement becomes more difficult as the number of factors under consideration increases. As populations age, the number of diagnoses and medical factors that interact will increase; the older the patient, the greater likelihood of multiple pathology. This will make balancing all the clinical factors more challenging and their prioritisation more difficult. What one can or cannot achieve for such patients, will depend on our clarity of thought and common sense, rather than technical knowledge. Start by listing the questions to be answered, and then:

- Set the aims.

- Set the time priorities.

- Decide on the key factors.

- Identify key meta-factors.

- Weigh each factor for priority and relevance.

- Always keep the desired outcomes in mind.

Goldilocks weighted for no bear. Being obsessive by nature, the weighting of her options had to be 'just right'; not too little or too much. Goldilocks had specific aims and priorities. She eventually found the right porridge and the best bed, but only after much reconnaissance.

The media loves sensation. They only rarely resist skewing information for sensational impact. They choose which topics to write about and what to feed to us. There is no need for them to be even-handed or fair. They will usually focus on exceptional cases and magnify their significance. The perverse weighting of information and how they edit it (altered or redacted), can affect public opinion. The point is to sell newspapers and gain subscriptions.

Wealth and power enable personal selectivity. Politicians and the media, exercise it relentlessly when choosing subjects for promotion. The time and space they allocate to each subject, and the subjects they choose to ignore, also shape public opinion.

Although many thousands of people die each year from two legal drugs, alcohol (6,800 in the UK in 2015) and tobacco (>100,000 deaths per annum) it is wartime action or sensational social and business news that gets airtime and fills newspapers. Too much written about the effects of smoking and alcohol would bore readers (even though equivalent to hundreds of airplane crashes each year). The death of seven international aid workers in Gaza (April 2024) commanded more attention. News editors decide what to feed us.
In the UK in 2015, fifty people died after taking MDMA. In 2016, new psycho-active substances caused 123 deaths. While individually important, these numbers are miniscule compared to those who die from tax generating, alcohol and tobacco products. (Prof. David Nutt. RSM meeting, Nov. 2017).

The medical media are not exempt from editing, skewing and spinning research results to their advantage.

The much publicised, but very rare, Reye's syndrome (rare death from aspirin), led to the edict – 'never prescribe aspirin to anyone under twelve-years-old'. (R.D.K Reye, first describe the syndrome in 1963). Deaths from aspirin then occurred in 1/158,000 of the population. By 1990, seven years after the directive, the number of deaths had lessened to 1/1,000,000. This seems a notable achievement, but the reduction is equivalent to a doctor seeing one case of Reye's during twenty-one working lifetimes. Assuming a doctor sees three new febrile cases every week, for 50-years, she would have to live for 133 working lifetimes (on average, with a random choice of patients), to see just one case of Reye's Syndrome.

The directive not to give aspirin has denied many children the benefit of aspirin, one of which is to reduce fever effectively. If a child of mine had a fever, I would first try paracetamol. If the child remained sick with a temperature > 39 °C., I would give her aspirin (after food). I would not want to risk febrile convulsions (so much more common than Reye's syndrome).

An important question for doctors is this: are you going to follow the statistics (and protect yourself from litigation), or practise medicine that protects patients from the most likely harm? To do so, you must be able to weigh the odds, based on the best evidence. It is, of course, much safer politically to follow the guidelines as if they were hard rules. No experiential or academic reason, or any individual patient consideration, will save you from GMC litigation should any patient complain or come to harm. No regulator will acknowledge the existence of doctors capable of thinking for themselves. Even more important, and for the sake of enforcing compliance, bureaucrats now regard all medical guidelines as immutable. By replacing the art of medicine with hard and fast rules, medical bureaucracy has gained better control of clinical practice, but has reduced the benefits patients can experience from personalised medicine. These benefits mostly apply to those patients who do not fit the norm.

In 2021, the media reported the deaths of fifty young people following the AstraZeneca COVID vaccine. Clotting was the commonest cause. We had vaccinated twenty million people at that point. What might your advice be to a young person? Continue to receive the AstraZeneca vaccine, or switch to the Pfizer alternative? The media had a field day. Many expressed worry about chance clotting induced by the vaccine, even though the odds were 400,000 to one.

Before COVID-19, every GP will have seen a few patients each year who developed a clot (DVTs). If they saw two such patients among their 2000 patients, that would represent an annual risk of 1000:1 (very much more common than the 400,000: 1 risk of COVID vaccination). Also, was a risk of 400,000:1 sufficient to change the guideline policy (changing the COVID vaccine given to younger people)? It was not.

An intelligent, former patient of mine, Sarah Granger, asked, 'Are some people born with an increased clotting risk, making the vaccine more dangerous for them?' Individual risk and significant group risk, both need to be considered by those who practise the art of medicine. Group risk will suffice for those who limit themselves to scientific medicine.

The Music of Decision Making

Consider the art of writing of music. Many decisions need to be made. The composer must first decide what mood he wants to create – romantic, mysterious, heroic, rousing, etc. The duration and loudness of each note are important to the appreciation of the final composition. Not only soft and loud (piano and forte) notes are important, the rests (the silent beats between the notes), and the tempo are crucial. Even though there are only eight notes on the western musical scale (not including their flats and sharps) to choose from, millions of combinations exist (only a small proportion of which have been used to create every known composition).

When doctors consciously choose and weigh medical information before coming to a clinical conclusion, they will use a similar art; one which relies little on conscious measurement. Formalising choices will seem fatuous to both experienced doctors and composers of music alike. The problem for both is producing a fluent performance; whether they have just started their medical career or started playing the piano.

Both doctors and musicians know when they have attained expertise – they can file away their instruction manuals and check-lists, and perform without them.

Judgement under Pressure

Experience should allow swifter decision making, although ditherers are often remain unchanged by experience. In fact, experience can confuse them. Re-living various scenarios in the past will help only those willing and able to learn. The Dunning-Kruger bias often assumes importance.

Research showed that those who successfully escape from airplane crashes usually gave prior thought to their escape route, before it became necessary.

When pushed for time, or overburdened by the volume of work, being obsessive can be a disadvantage. The reasons are:

1. Obsessive characters have a reduced cognitive band-width.
2. They may do things right, but not do the right thing.
3. They give priority to correct procedure (guidelines and rules), but lose sight of the main aim.
4. They are likely to run short of time with exactness as their priority.
5. They will attempt to finish what they are doing, regardless of more crucial factors.

Time and Judgement

The time needed to deal with patients effectively varies, depending on the number and type of their problems. My consulting times ranged from 10 minutes to 3 hours (including investigation time). The more time one spends with patients, the more one can get to know them as individuals.

Long-term doctor-patient relationships thrive on continuity. Our judgements of long-term clinical data, meta-data, and how they are trending, will always aid decision-making. This was once unquestioned by the traditional 'family doctors', I once knew in my youth, who practised pastoral care. I have since witnessed their demise as family friends. Corporate-minded doctors, sequestered in 'Health Centres', have now replaced them. By minimising continuity, health centres have allowed the replacement of personalised care, with anonymous, time-restricted functionality.

Fewer hospital doctors have the time and inclination for personal doctor-patient relationships. If we are to bring back personalised, patient-centred medicine, we will need a complete turnabout. Committed medical bureaucrats will regard the suggestion as retrograde, anachronistic, inefficient, time-consuming and most importantly, expensive. It will threaten one of their objectives: the standardisation of medical practice. What they cannot acknowledge is that the best medical results come from personal clinical endeavour, and not always from corporate directives. Issues such as cost, profit, efficiency, and patient through-put are vital concerns for all corporations, but they should never be the primary concern of any nurse, physician, or surgeon. The business structure of our corporate NHS has diverted us away from personalised medical practice, while making the system too expensive and unfit for purpose.

Speculation

Speculation is not for everyone. The risk of being wrong inhibits many.

A spokesperson on Radio 4 (2017), reported that university graduates voted 2:1 for the UK to remain in Europe (in the EU referendum of 2016). It was the reverse for those who left school at sixteen years old. The spokesperson would not speculate on the implications – that was not his job.

Speculation (allowed in the discussion section of every research paper) requires every researcher to ask, and then answer, some speculative questions about the meaning, significance, and implications of their work. It is very easy to go too far and indulge in science fiction. The best speculative remark I know of (thanks to Paul Kligfield) can be found at the end of Watson and Crick's paper describing the pairing of nucleic acids in molecular structure of DNA (Molecular *'Structure of Nucleic Acids: A Structure for Deoxyribose Nucleic Acid'.* Nature 25/4/1953: 171-737). What they wrote was, 'It has not escaped our notice that the specific pairing we have postulated immediately suggests a copying mechanism for genetic material.' This modest speculation suggested that they had discovered the molecular basis for biological reproduction and life. Some speculation! Some significance!

Like any satisfactory theory, worthwhile medical diagnostic speculation should explain both the positive and negative observations. After only one glance at the data, an able physician will often bring clarity and resolution to a complicated clinical situation. This is nebular processing -– an invaluable aspect of the art of medicine. The ability to bring data, wisdom, clarity and insight together, is shared with poetry.

By choosing the correct words, poetry can make the intangible seem tangible.

Motivation, Thinking, and Judgement

That reason alone motivates action, is an ancient idea. Augustine disagreed. He suggested that 'an intentional search for delight' is what activates man. We may all be equal in the eyes of God, but we are not all equally capable of considered action.

Are our patients capable of 'considered action' in medical matters? Many are, especially if they have had nursing experience, an interest in First Aid, or are first responders. I well remember as a child, my mother calling on a neighbour with some medical know-how, to help treat a large carbuncle I had on my neck. She relieved it with a hot kaolin poultice, prepared in my mother's kitchen. Patients can now ask a pharmacist, or search the Internet for similar advice, before consulting a doctor.

In medical matters, some patients choose to rely on 'gut feeling', (blind) belief, and (blind) faith; although not all would agree with using the word 'blind'. Many listen uncritically to the ideas of friends, acquaintances, and the sage down the local bar or pub. Others rely on luck, especially if they are used to being lucky. Some may say, 'Not a lot else matters if you are lucky, does it? It's written in the stars, you know.' A few will not stop believing in luck, even though they never win lottery jackpots. Others, proud of their past luck, never question their gift of prescience. 'I was right. If you had bought that house when I told you, you would have doubled your money' (The Hindsight Bias). Could they really have known the future of the housing market? They might, if they knew that current population growth guarantees an ever-widening gap between those who want to buy a home, and the number of houses available. Many make lucky guesses, but become biased to think that they can predict the future (the hot hand bias).

In the course of a medical career, one will encounter many examples of how patients think, some deciding important clinical matters on guesswork and luck alone. To be of use, the medical science must prove better.

There were once those who sought a prediction from the Oracle at Delphi. Many still turn to astrology, palm reading, and Tarot card readings for help. All methods, including science, must stand or fall by how well the method predicts the future. Innumerable times, science has proven better than chance when predicting outcome, but this is not always the case. In difficult situations, especially when time is running out, luck may be all one has.

In early 2020, the prediction of how many would die from the COVID-19 virus was over-estimated, and the devastating effect on UK economy underestimated. These estimates were the product of questionable mathematical models, yet politicians (in the absence of other guidance) thought them 'scientific' enough to act upon.

I have met people with repeated, exceptional luck. Chance allowed them to be in the right place, at the right time, in order to benefit. One lucky prediction can bring enough hot hand kudos to last a lifetime. Some had the foresight to buy Microsoft, Berkshire Hathaway, and Amazon shares, decades before anyone guessed their prospects. Maybe they saw something in the ability of Bill Gates, Warren Buffett, and Jeff Bezos, that meant their money would be in safe hands. Since most regard accurate prediction as mystical, it will continue to attract attention.

Many still revere Nostradamus, 450-years after his death, even though his predictions lacked specificity. Having faith in those with a successful track record is sometimes worth relying on in difficult judgement situations. Patients aware of this, will seek advice from those doctors who have achieved fame. In cardiac surgery, both Christiaan Barnard and Prof. Magdi Yacoub rightly achieved this status.

When a patient's progress seemed stuck, I would research the scientific literature to see who had published most on the subject. I might then refer the patient to them. I sent two of my patients to Stanford University in California, one with lupus erythematosus, and another with a spiral left anterior descending coronary artery that needed a stent. Dr. David Clark stented him after others had failed. I referred another patient to a retired GP in Essex, who was a published expert on temporal arteritis. Both outcomes were beneficial.

When we decide to play a game, there will usually be a winner, even with odds of 45 million to one (UK National Lotto Jackpot), or 140 million to one (EuroMillions Jackpot). With this in mind, some patients will say: 'I'll take my chances.' Chancing to luck is not usually the best way to make medical decisions, so be prepared to offer some scientific comment to those patients who prefer to take their chances.

Educating patients has an important part to play in prevention. Without it, I would not have had much job satisfaction. Patients who know nothing about medicine, and have no wish to contemplate their medical condition, are at risk. They may see no point to complying with advice, keeping appointments, or taking medication. Unfortunately, many patients genuinely think they know best, despite a lack of medical knowledge and experience. Perhaps they know someone whose opinion they value more than a doctor. Consider banishing them from your practice since this is no basis for a safe patient-doctor relationship.

Perhaps we should consider a double-blind trial to prove the prophylactic effectiveness of patient knowledge and education on morbidity and mortality.

As proof of any new concept, it must have led to at least one success. On December 14th, 1903, Wilbur Wright flew fifty yards in three seconds, in the prototype airplane he named 'Kittyhawk'. That was enough to prove that powered flight was possible. A powerful gust of wind destroyed it on the same day. They stored its many pieces in a shed until the US military took interest in it.

Using the same surety as many have about winning the Lottery, some patients will offer you their diagnosis. 'I am B12 deficient, doctor. I have read about it on the Internet, and my symptoms match exactly.' Convincing them otherwise can get fraught. Although it can never be possible, it might be safest to treat only those patients who know they don't know, and those who want to learn. Some might avoid self-harm, and premature death from smoking, alcohol excess, and opiate abuse, once they have engaged with the evidence. The best doctors are magnanimous and always choose the self-sacrificial alternative, taking on all-comers in need of medical attention, regardless of their beliefs.

The decision to become a doctor is a high-minded, self-sacrificial gesture. Failure to recognise this from the start will lead some doctors to leave the profession; the job is not what everyone expects. Another, more contemporary reason for leaving the profession, is the duty imposed on doctors to respect the corporate inanities of government-ruled medical administrators; those whose practical knowledge of medicine is rudimentary. Even with their bureaucratic mission to safeguard patients from doctors, few patients regard their existence as necessary. Most patients (but not all), are smart enough to avoid potentially dangerous doctors and nurses.

Giving priority to bureaucracy not only applies to the medical profession. Other professions, like building, insurance, banking, accountancy and the law, all voice the same sentiment. Subjugation to bureaucracy has proven too much for many doctors and nurses to tolerate, with too few having practicable alternatives. Some become bureaucrats, remain frustrated clinicians, emigrate, or leave the profession altogether.

Convincing Others

If convincing colleagues is your purpose, the most impressive attribute is clarity. Those with knowledge and experience, will look for clear clinical reasoning when you present clinical cases. They will seek precise and accurate assessments, logical interpretation, straightforwardly recounted. They will admire substance over ornament, and scenario analysis over guesswork. Your pertinent, practicable opinions, with minimal speculation, should convince them of your clinical acumen. Your clinical acumen should (but may not) determine your career progress. Having good connections is still as effective. We once called it 'the old-boy network'. It is still alive and well in places of privilege; it will always be more reliable to work with those you know well.

It is important to recognise those capable of making the best of weak arguments, and those inept at using strong arguments. One must be able to grade arguments based on clinical merit. Do not let bombast, confidence, charisma, charm, or ego sway you. Some of your colleagues will use these rhetorical devices, but clarity, balance, and accuracy are more worthy. No matter how engaging or enjoyable your arguments, the point of every clinical discussion is to achieve patient benefit.

Judgement and Chance

The statistical paradox is a trap for doctors. Some will base their every argument, judgement, and clinical decision, on numerically evaluated evidence derived from large groups. Because the rules of numerical engagement change with scale, their decisions will not always apply to individual patients.

A classic example of an incorrect scientific use of scale was when Newtonian physics (large scale) was used to understand quantum physics (ultra-small scale). As scale changes, the same physical laws may not apply.

Doctors now have a duty to base their medical management on the large-scale statistical results of double-blind trials, and the results of meta-analyses. These results reliably point us towards the truth about large groups and help us avoid fooling ourselves. Because statistical results are 'the best we have', it is tempting to think they are always best for individual patients.

It will often discourage patients when doctors refer to uncertainty. This might happen when we tell them that although opinion is divided, it is still 'likely' to apply to them. Their question might then be: 'What do you think, doctor? The research you quote, does it apply to me?' A bright patient might ask: 'What is your evidence for believing that?' I hope you never say: 'Well, it's up to you. I can't tell you what to think or do.'

Before concluding any diagnosis, judgement, or management plan, the responsible professional should have as much experience and group scientific proof as possible. My patients were mostly intelligent business owners, able to see through any proposition that lacked cogency, clarity, and common sense. They protected themselves by not accepting any idea that failed their common-sense test (complying with everyday experience). Out of courtesy, some patients will sometimes appear to agree with some patently daft (therapeutic) idea. Patients need protection from those doctors who harbour unsubstantiated theories and put them into practise (those not based on any known patho-physiological mechanism, and with no support from corroborated observations).

For patients who trust their doctor, only the highest standards of professional assessment and management should apply. The doctor-patient relationship is not always pliant enough, to take the strain of unconventional experimental treatments, and eccentric ideas. The danger is that some patients with no other options, and desperate for help, will try anything. If they suffer as a result, the GMC should investigate.

By law, the GMC as a corporation, has the job of assessing doctors, despite their need to call in 3rd party clinical acumen, and being limited to corporate office rules and legal processes, rather than medical process. Their perspective and processes are primarily legal, not medical. This serves to restrict the use by doctors of clinical inventiveness when patients need it most. Their fixed rules also derogate the use of the art of medicine. To make their job viable, they must assume that all patients and doctors conform to a standard, and will thus conform to the standard rules which they can enforce. This shows little or no appreciation for how different doctors function, and how each can work successfully, using their own style derived from their personality and individual clinical experience.

It is for these and other reasons that I advocate that the GMC, and other associated medicolegal statutory bodies, be superseded by a medical directorate; an organisation employing only doctors with enough clinical acumen and experience to make fair assessments of clinical work. The GMC will probably see this as nepotistic with a mistaken intent and not in the public interest. Unlike those who exist to help themselves, dedication to helping others is the usual ambition of doctors, nurses, paramedics, and carers. They do not deserve to be treated as guilty criminals by bureaucrats and lawyers 'until proven otherwise'.

'Daft' Ideas and Judgements

> 'Someday you will be old enough to start reading fairy tales again.'
>
> C.S. Lewis.

Daft ideas come in several flavours. Some are wrong because they are daft from start to finish. Some hold theories with no rational, theoretical, or ante hoc legitimacy; those with no concordance with the accepted facts of medical science, chemistry, physics, or biology. Acupuncture and homeopathy are examples. Although they can help people, explanations for how they work are usually pseudoscientific. Others are correct for no reason other than the perverse nature of the universe.

Many old doctors like me, can embrace what may seem like daft ideas. With age, we lose our inhibitions, become bolder, more outspoken, and less open to the opinions of unqualified others. Then there are those whose theories have some antecedent scientific basis, but are completely wrong. Because electrons are mysteriously unpredictable, and can be in two places at once, doesn't mean that parallel worlds, re-incarnation and pre-cognition exist. There are many currently unfathomable phenomena, but they become no better understood by employing pseudoscience and inductive reasoning. Some of this is attention-seeking; sometimes the intent is money-making.

Bohr and Einstein disagreed about a fundamental point of sub-atomic particle behaviour. Einstein thought particles existed with a determined prearrangement. Bohr disagreed, suggesting that they were undetermined and probabilistic. The English physicist John S. Bell (1928-1990, originator of Bell's Theorem) said that 'Einstein was consistent, clear, down-to-earth, and wrong'. He said 'Bohr was inconsistent, unclear, wilfully obscure, and right'.

Daft ideas, that eventually prove correct are interesting, especially when they had no antecedent credibility. It is easy to dismiss the daft ideas of outsiders, even if trained observers, and based on years of observation. The better the observer, the more likely the idea will become substantiated, eventually. Trained observers have made many advances. They first describe a phenomenon, then ask key questions. These questions are pointed enough to lead to an understanding of the phenomenon. Among such questioning observers were Charles Darwin and Alexander Fleming, alongside thousands of others whose ideas never achieved recognition.

Mr. Lang Stevenson FRCS was a near to retiring general surgeon with 30+ years of general surgical experience, when I first met him as a house officer. He had observed many patients with major personal stress prior to developing cancer (see Holmes and Rahe). In the 1960s, few favoured his thesis that stress might cause cancer; many ridiculed him (behind his back) for holding the thought. There is not much more evidence now. From my own anecdotal observations over the last fifty years, I have observed many cases that suggest the possibility, and few that dismiss it. The relatives of those who die from cancer, often agree with his stress hypothesis. For me, and a few of my colleagues, it has remained an interesting idea awaiting verification. It will remain an unpopular academic idea, because it seems more scientific and convenient to blame cancer on something impersonal like gene mutation, diet, body weight, obesity, smoking, and alcohol, rather than stress.

Not all doctors like to consider the personal factors patients cope with. 'I trained in medicine, not social work', was a comment I once heard from a GP discussing loneliness. She was a senior official at the RCGP in 2017.

Was it daft to have admitted him to medical school? My exceptionally talented friend, Dr. Alan Gardner, was a GP in London, E17. He could use a stethoscope, perform magic, build a house, construct a sailing boat, and could play the clarinet, saxophone, vibraphone, guitar and drums. Am I right to think that now we only select those with a string of science A* grades at A Level, and few other talents?

Some medical school applicants are told to use surprise as a tactic when interviewed. They might quote their experience of walking across the Gobi Desert; working for six months in a leper colony or traversing Antarctica, when they could have better spent their time experiencing the work of doctors or social workers in deprived areas.

Does it serve the public interest to admit students to medical school who have little interest in humanity? How daft is that?

> *Wherever the art of medicine is loved, there is also a love of humanity.*
> Hippocrates.

Measurement Errors and Errors of Judgement

Try to guess the length of a straight line (one dimension). Many can guess correctly. Now try to guess the size of an area (two dimensions). Few will be as accurate. Now try to guess a volume (3 dimensions), and their accuracy will be poor. To illustrate this, compare a half-litre bottle of milk to a one litre bottle: the smaller one appears to be more than half the size of the larger one. The mistake occurs because the lengths need to be cubed to calculate volume, and any errors made in determining each length will also cubed. To estimate flow, like how much water is flowing down a pipe, we must add the time dimension to static volume. With this added, guessing flow-rate is next to impossible. The same will apply to guessing how much blood is being lost from a wound.

There are many types of measurement error. We might measure the wrong object, for instance. In echocardiography, which chord across the left ventricle will allow the best estimate the LV cavity dimension? Slight changes in the chord chosen will lead to a significant change in the final measure. The error will be even greater when calculating ejection fraction. One could refer to this as a *positioning error* made by the *observer*, rather than an actual error of measurement. There are also *errors of perspective*, like when we change our viewpoint of a sculpture, or change the echo-sounding transducer position. This change of view can cause a completely different measurement to be taken, and result in a completely different conclusion about LV volume and function. Choosing the least representative measure would be an *error of judgement*.

AI has come to our aid in defining the LV endocardium, and automatically calculating LV volume and ejection fraction, but the errors being made are difficult to assess.

Besides these errors, is a major source of clinical error – the *error of interpretation*.

Mr. David M. had LBBB. At rest, his interventricular septum did not contract well, whereas the lateral wall of his ventricle did. This resulted in a low ejection fraction (EF) calculation of 48%. This became concerning when his prognosis and fitness to drive were under consideration. The inference was that he could no longer drive if he had borderline left heart failure.

The conclusion of heart failure was false. LBBB (delayed septal activation), and not heart muscle (septal) weakness, caused his low EF. With exercise, his septum contracted normally, resulting in a normal measure of EF, namely 60%.

Which measurement best reflected the actual contractile state of the patient's left ventricle? Clearly, the micro-electrics surrounding his left intraventricular bundle, changed with exercise. Unfortunately, out-of-phase septal contraction caused by LBBB, would have led to a false assumption of heart muscle weakness, and would have put him out of work.

Given the difficulties of simple physical measures of dimension, what inaccuracies will occur when judging the severity of illness, the severity of depression, or the amount of distress felt by a patient? In each case, there is no measurable dimension,. We must grade them in a semi-quantitative way, using terms like 'mild', 'moderate' and 'severe'. Grading of this sort relies on experience and is an art. Within this realm of judgement, the Cartesian imperative to first define terms (and measures), is not easily possible. Although such terms are 'fuzzy', and indefinite, doctors and nurses can utilise them meaningfully. Bean-counters will dismiss such measures. For them, only standardised scientific instruments produce usable data. All measures must be reproducible and comply with defined measurement parameters and terms of reference.

In the arts, there are many examples of completely understood, loosely defined, indefinite judgements, which work in practice.

During a masterclass, the cellist Paul Tortelier, once asked an advanced student to play 'as if breathing fresh air'. He wanted her to be 'more immaculate' in her playing. His student knew exactly what he meant. After Tortelier's advice, her playing fluency improved noticeably, with no need for further definition.

The maestro composer, Leonard Bernstein, while explaining the evolution of music, referred to music as gaining more 'colouristic warmth' with time.

Despite the complicated topology of most putting greens, skilled golfers can putt their balls into the hole without making measurements. Skill and a talent for estimating measure are on display.

Presented with many indefinite factors, skilled physicians can make accurate diagnoses.

Binary situations exist in medicine, but only in the simplest of clinical situations. Physicians and golfers both work with random factors. Both worlds present multivariate, ill-defined, indefinite, and probabilistic components. Those who limit their judgement to binary issues (right, wrong; black or white), might fail in both roles.

In which sphere do you feel most comfortable? Fixed or uncertain?

'Experience is a cruel teacher – it gives you the exam before the lesson.'

Film: *'Red Tails'*: Lucasfilm Ltd. 2012.

In medicine, the intuitive skills of an experienced observer can appear remarkable. Their anecdotes form part of the oral tradition of medicine, providing apprentices with an insight into clinical wisdom. Students will find that senior consultants can, with only brief consideration, come to correct clinical conclusions. This is because they can recognise degrees of relevance and irrelevance in clusters of symptoms, signs, and investigation findings. They are thus capable of nebular thinking.

Aubrey Leatham would say, 'The patient cannot have significant mitral stenosis. I saw him climb two flights of stairs without breathlessness. The mitral valve gradient you have measured must be incorrect.' He had learned from long experience how the degree of mitral valve stenosis, affects breathing on exercise. Whatever machine-based measure one uses to assess a cardio-pulmonary patient (echocardiogram, ECG, catheter data, respiratory function), it must explain the symptoms.

The more skilled you become, the more straightforward become the rules of judgement and performance. When I first met Aubrey Leatham, I wasn't sure whether he was a genius capable of handling complex issues, or a simple binary thinker. He was both. At my interview for a British Heart Foundation Research Fellowship, at St. George's Hospital, Hyde Park Corner, his rhetorical statement was, 'We are not sure why the heart goes slow'. My instant thought was: 'If he doesn't know, who does?' I later learned that geniuses keep their questions simple. Occam's razor in colloquial form is, 'If it isn't simple, it isn't correct!' Those who make significant scientific advances usually ask one key question at a time. Clarity of thought and purpose can be lost, when we try to tackle too many questions at once.

The sharpest and dullest minds both embrace simplicity.

Meta-Data, Meta-Analysis, and Meta-Judgement

The Greek word *'meta'* roughly means 'after', or 'afterwards'. The wisest judgements are based on all the key information, and all the subsidiary information. This meta-information, should include any 'inside information' (*mesa-information* – *'mesa'*: Greek for 'inside'). It must include 'reading between the lines', 'seeing the full picture', and spotting tacit, or hidden information. Never forget information hidden in plain sight.

It is important to recognise a clinical picture that is incomplete. For personal reasons (embarrassment, fear, shame, modesty, and guilt, etc.), some patients can withhold vital information.

Always ask whether there is a metaphorical elephant in the (consulting) room? Identifying such beasts can require a change of focus. One may need to take a step back, in order to see the wider picture; call it *meta-perception* (like advanced drivers who look well beyond the car in front and constantly scan the complete scene ahead).

Meta-judgement involves understanding what others said, and why they said it. What does a boy / girl-friend mean when they say: 'I hate you'; 'I wish you were dead'; 'I hope you rot in hell' (words which may mean 'I am not getting what I want from you')? If you are aware of all the facts, you might know that the correct interpretation lies somewhere between: 'I love you, and miss you'; and, 'I hate you, and never want to see you again.' To decide which it is, you will need all the meta-information you can get. Patients who say: 'nothing wrong with me doc' may mean, 'I am scarred witless that you will find something seriously wrong, and I am not sure I can cope with it'. Meta-judgement goes beyond fixed check-lists. It is not usually the forte of obsessive, detached, Asperger subtypes, who cope well with unequivocal facts, but not with inter-personal interpretation and hidden meaning.

Dr. Julian Hardman was my GP when I was a child. I later had the privilege of attending his practice as a medical student (circa 1964). He was an attached doctor who knew many of his patients personally. He did many home visits, and was privy to intimate, personal (meta) information about his patients and their lives. Before a patient entered his consulting room, he would guess the most likely reason for them coming. Knowing his patients as one would a close friend, gave him insight into their personal circumstances. This meta-information allowed him to make wise holistic judgements; not just routine, algorithmic clinical judgements. He used his knowledge to form the most suitable management decisions for his patients as individuals. I was never to meet a clinician with more knowledge of his patient's lives, or one who incorporated it so well into his clinical decision making and management. In the 1960s, many GPs practiced similarly.

Failure to emulate Dr Hardman's practice can lead to preventable clinical mistakes. GP contracts, and bureaucracy have relegated his style of practice to the past. It would waste time and words to ask why UK doctors have stood for it?

Everyone uses meta-judgment. After viewing an apartment for rent or purchase, some will say with no further thought: *'It's not for me'*. When you meet someone new, you may quickly sense incompatibility; often without conscious processing. Have you ever said: *'There is something not right here'*, but have not been sure what or why? Visual, auditory, and olfactory clues; body features, eye contact, body language, body odour, accent, affability and dress-sense, can all trigger uncertainty. Next, we summate all the information subliminally into an assessment. Some people call it intuition, but mostly it is meta-judgement.

Attempts to simulate (computerise) the human brain persist. It is mostly pointless. Why would anyone want to copy human thought patterns? How are computer programmers to cope with simulating human feelings, taste, nostalgia, compassion, empathy, tolerance, and the understanding or fuzzy, yet indispensable factors that together make us human? My guess is that they will give up, and revert to their original fall-back position – *'computers are but aids to human judgement'*.

Until we can make digital technology appropriately weigh each component in a nebula of clinical data, and be able to value each piece of meta-data for relevance, they will not pass an ordinary Türing test, let alone a clinical Türing test. It is easy enough to program a computer with the understanding of a parrot, but programming one that can achieve the judgement of a sentient human being will take a little longer.

Experience: Time Spent Weighting

A doctor's ability to weight the significance of clinical information, should improve with experience, but it needs a lot of clinico-pathological feedback to improve the diagnostic weight we apply to each symptom, sign or investigation result.

Consider the 'weighting' of nausea as a symptom in acute abdomen. A computerised symptom analysis of cases proceeding to appendicectomy, generated an unexpected finding: nausea is a strongly positive diagnostic indicator of appendix inflammation; its absence (diagnostic accuracy negative), reliably predicts a non-inflamed, 'lily-white' appendix (as seen on its removal). Nausea, in combination with localised right iliac fossa tenderness, provides enough good evidence for alerting an operating theatre. I wish I had known this as a house officer. It would have saved a few of my patients from the scalpel. An intra-abdominal abscess (a 'ripe' appendix) can cause a toxic state, hence the nausea. The latest suggestion is to treat uncomplicated acute appendicitis with antibiotics. Might this not lead to more cases of peritonitis? (A Randomized Trial Comparing Antibiotics with Appendectomy for Appendicitis. New Engl. J. Med. 2020; 383: 1907-1919 DOI:10.1056/NEJMoa2014320).

William Heberden first described *angina pectoris* in 1768. We now know that the symptom he described, reliably predicts coronary artery stenoses (> 85%), i.e. ischemic heart disease (IHD). He wrote this:

'They who are afflicted with it, are seized while they are walking, (especially if it be uphill, and soon after eating) with a pain and a most disagreeable sensation in the breast, which seems as if it would extinguish life, if it were to increase or continue; but the moment they stand still, all this uneasiness vanishes.' (Heberden, W. Some account of a disorder of the breast. Medical Transactions 2, 59-67: 1772. The Royal College of Physicians).

Angina is more often described by patients as a tightness than a pain ('a most disagreeable sensation', according to Heberden). When angina becomes predictable on exercise, one can reliably assume a stenotic coronary artery lesion (high diagnostic accuracy positive). This is 'classic' Heberden's angina. In cases without these classical features, the diagnosis of coronary artery disease (CHD) is uncertain, and must await further investigation.

Chest pain occurs in only 5% of those with cardiac ischaemia. Ninety-five percent have breathlessness, or a sensation of 'running out of steam', although not all patients admit it. This highlights the importance of observing patient performance and behaviour yourself, and a good reason for cardiologists to observe all exercise tests (it can provide indispensable meta-information).

Some Q&A's in cardiology have a remarkable power to predict pathology (positive diagnostic power). The Q&A's car mechanics use, yield similar results. Both reflect physical processes involving fluid dynamics, pumping action and electrical activity, all working within strictly defined limits. The physiological functions of the gut and brain bear no such close physical relationships. This is perhaps why their related symptoms are far less predictive of underlying pathology.

Road Testing

A road test is an important cardiological intervention; it can reveal more diagnostic information than history taking and examination at rest. Exercise testing allows one to observe cardio-pulmonary symptoms as they develop. The advantage lies in witnessing symptoms rather than relying on the patient's description. One can question them as the exercise proceeds. For instance, 'Is the pain or discomfort you now have, the same as you had recently?' Neurological examinations follow the same principle. One can observe strength, stability, and co-ordination during activity. Watching patients walk upstairs and performing daily tasks, is often indispensable to neurological appraisal. The observations made will directly aid management decisions.

A formal exercise test is not always necessary; it can sometimes be enough to walk beside a patient, up and down a few steps or along a nearby corridor. One can readily observe their pulse, breathing rate, and jugular venous pressure, and any difficulty they are having, with a before and after comparison. Any rise of their JVP with minimal exercise, could reveal borderline heart failure. Observations of performance are instructive, and an important addition to third-party, detached investigations like lung function studies, echocardiograms, and ECGs.

Delegation may become necessary in overload situations (few doctors have time to observe every ECG exercise test), but never underestimate the value of direct observation and road testing by a trained observer. Road testing can resolve issues that others have found difficult, ambiguous, or confusing.

Mr. M.S. had chest pain and felt unwell while working in his butcher's shop. He telephoned me for advice. I advised his immediate admission. At the Whittington Hospital, they made a diagnosis of acute anterior cardiac infarction. Ten days later, nurses walked him up and down a short flight of five steps. Since he was neither breathlessness nor had chest pain, they pronounced him fit for discharge.

Six weeks later, I exercised him, and had to stop him when he became short of breath and developed dramatic, ischaemic ECG changes after a few minutes. An emergency coronary angiogram showed a tight, left main stem stenosis. From the catheter laboratory at the Wellington Hospital, London, he went straight to the operating theatre where my colleague, Mr. Stephen Edmonson, successfully performed an emergency bypass (CABG). He remains symptom free to this day (25-years later).

His hospital discharge 'exercise test' had been inadequate to make a critical, life or death diagnosis. It was perhaps only meant to test his fitness for convalescence. The nurses who did it, should not have been made responsible for the actual diagnosis - one of impending sudden death.

There is another bonus to one-on-one observation: most patients are glad of the interest taken. I always tried my best to attend every echocardiogram and exercise test performed at my clinic. Time spent with patients cements the doctor–patient relationship for all but the most uncommunicative. Patients also appreciate the instant availability of results.

Efficacy

The primary quest of therapeutic medical science is to answer the empirical question: for most patients, is this drug / procedure more efficacious than a placebo (as shown by clinical trials)? As a responsible doctor, one must also ask: are the clinical trial (group) results, valid for my patient?

Choosing which trial results apply to any individual patient, is an art. The statistical paradox means that the results of double-blinded trials, may never provide the certainty needed for their application to any individual. Doctors sometimes use weak justifications, especially if the patient understands the paradox. Doctors will resort to saying, 'trial results are the best guide we have'.

If lost in Manchester, with a need to find the GMC offices, would you ask 100 people at random, or simply ask one person most likely to know (a police officer, firefighter, or librarian)?

There are lies, damn lies, and statistics. (Attributed to Benjamin Disraeli).

How are new drugs to be judged by prescribers? Will experience (anecdote) conflict with the trial data published by pharma companies? Drug trial statistics may show that drug 'A' has a good chance of working (on average, for a large group), but is this evidence enough for a clinician? I had a simple rule of thumb when contemplating new drugs. If my (anecdotal) experience of prescribing it to five consecutive patients did not confirm its promise, I would avoid using it, whatever the statistical trial evidence. The reason is simple. The odds of a new drug working well for five consecutive patients (with the binary outcome of 'yes, it works well'; 'no it doesn't work well'), is 2:1 each time. For all five patients to benefit, the plain odds are 2 x 2 x 2 x 2 x 2 or 32:1. With five successes in a row, a new drug is certainly worth continuing to prescribe, while keeping it under review.

The Dynamics of illness: Judgement, Timing and Clinical Trajectory

The time scale applicable to the clinical progression of each case is important; sometimes of crucial importance. Many of the serious errors of judgement made by doctors, and the system they work for, relate to timing. For every patient, one must ask: does the evidence support clinical stability, or change? If there is change, how rapid is it? Doctors are not usually in charge of their own appointments these days, and that can lead to grave errors of timing. By delegating the clinical assessment role to others (receptionists), doctors may not see patients as soon as they should, and may not achieve the optimal timing for consultation and intervention.

Most doctors have some adverse timing anecdotes to relate. Here are three of mine:

- a patient with acute chest pain given an appointment, ten-days too late;

- a patient with a melanoma put on a routine surgical waiting list for excision;

- a woman with colonic obstruction (from bowel cancer and diverticulitis), 'dripped and sucked' for over one week, before undergoing a laparotomy and bowel resection.

Early intervention can reduce morbidity and mortality. If water drips through your ceiling, find and fix the leak as quickly as possible; swift action will limit the damage. If a child is feeling cold and shivery, and has a core temperature of 40°C and some peripheral vascular constriction, you have no time to lose. Act with urgency or risk their death from septicaemia (sepsis).

The Two Golden Rules of Timing in Clinical Practice (platinum ones, actually):

1. Observe regularly – the sicker the patient – the more often.

2. With any sign of clinical deterioration, act before forced to.

The company or corporation you work for can block your interventions and clinical effectiveness. When I was a lecturer in Cardiology at Charing Cross Hospital in the 1970s, I would take calls from GPs, and see their patients with chest pain in our cardiac department, rather than let them wait weeks for an out-patient appointment.

When summoned to a management committee, I expected to be admonished. *'What do you think our Out-Patient's Dept. is for, Dr. Dighton? If you see every patient in your department, and every doctor does the same, we would lose funding'.* I carried on seeing those patients I regarded as urgent. My primary aim was always to help patients, not to satisfy management committees. My contempt for medical bureaucracy grew steadily thereafter.

A doctor's duty is clear: patients first, administration and political priorities second. During the early decades of my career, hospital medicine worked perfectly well. There were few managers whose principal role was housekeeping. Little did we know what was to come. Doctors today will know better than me, what the future promises, if the influence of medical bureaucracy on medical practice continues to expand. Beyond every medical event horizon, there now lie many bureaucratic black holes.

Useful Action Theories

Three theories have shaped my attitude towards the timing of clinical action. Those who do not prefer to see them as formal mathematical theories, can take them as common-sense directives.

Catastrophe Theory: Some systems are unstable (bi-stable has two states: working one minute, but not the next). Situations can quickly flip between life and death. Imagine a man walking along a cliff edge. Ten centimetres from the cliff edge, he is safe; 10cms beyond it, he could be in pieces on the rocks below. Situations need to be assessed from the point of view of direction: whether moving towards or away from a metaphorical cliff edge or catastrophe.

The sudden onset AF can cause acute LVF. Stretched enough, every piece of elastic will snap. Stretched enough, the atria of the heart will no longer support sinus rhythm, and AF will start. Stretched enough, every aneurysm will burst. A bath with its plug in, and the taps running, may overflow sometime soon. Emboli can suddenly detach from carotid plaque and cause an embolic stroke. Sailing boats can suddenly flip over in a high wind, or when handled incorrectly. These are all examples of preventable catastrophe. It is best to keep patients well back from every metaphorical cliff-edge; well away from hepatic, renal, and cardiac failure, dehydration, electrolyte imbalance, sepsis, acute dementia, cardiac ischaemia, and VF . . . if you can.

Chaos Theory: Small changes in the initial conditions of a dynamic system can lead to major events later on (the butterfly effect) and a different endpoint. This is especially the case for complex systems like the weather and human pathophysiology. Early intervention can lead to a more satisfactory outcome than intervention left too late. Act early, and always keep your eye on the ball.

Game Theory: The safest strategy is not always obvious. If you stay in one place and doing nothing, military studies suggest that an enemy, with time and opportunity to spare, will find and destroy you. The theory suggests that you should keep on the move, changing your situation regularly. Persist in finding different moves to keep ahead of the game, rather than doing nothing, becoming complacent, and allowing adverse clinical consequences to evolve.

If a patient has symptoms and signs that are deteriorating, do not delay. They need re-examination, and re-investigation, and frequent review of their consecutive data. Assessing the rate of progress is key, with early intervention a priority. In cardiac arrest, timing is everything. Those who quickly return to sinus rhythm, more often survive VF; those in asystole do badly, even with early resuscitation. The next decision is whether to start perfusion, and if so, when.

With a variable condition, and no evidence of deterioration, masterly inactivity and delayed enquiry can be appropriate. Although less in evidence among the sick, undernourished, and deprived, the human body has a remarkable capacity to self-correct, so give it time. But how much time?

Every situation has its own time scale and every system its natural periodicity. When a large oil tanker needs to stop or turn, it will travel many miles and take a lot of time, before the required end-point comes. With slow response times, patience is essential. Each human pathology has its own natural frequency and timing to consider. With experience, one gets a 'feel' for them (a distinct art). As an everyday example, your sense of timing should suggest the best time to re-examine each patient, be it five minutes, one hour, one day, or next week.

In complex situations, variables can combine in indefinite ways, and doing nothing may not be as dumb as it seems.

Mr. R.F. was a proven arteriopath. He had progressive atherosclerosis in several major arteries. He always had a normal total blood cholesterol (but a low HDL cholesterol). Over twenty years, his plaque size (on carotid ultrasound) increased measurably (while he elected to remain untreated). I eventually convinced him to take a statin. Despite successful coronary stenting for angina, he had many sub-critical coronary and common iliac lesions, small increases of which could prove dangerous. The first question was: 'did he take his 'statin' regularly? In sufficient dose, they can reduce the rate of progression of atheroma.

I scanned his carotid arteries every two years. When I saw his atheroma progress, I doubled his statin dose. When he developed signs of coronary ischaemia (exercise testing; PET, or MIBI scanning), I offered him for a CABG. My advice rested on knowing that his atheroma was rapidly progressive, despite medical treatment. My estimation of his clinical direction of travel, his trajectory, and his likely point of impact, made my advice about early prophylactic intervention and eventual surgery easier. As a private pilot, he was quick to get the point about trajectory.

At present, most doctors are content to maintain their patient's blood cholesterol below 5mmols/l. But what about their arteries and any atheroma present? Who wants to know? As crucial and fundamental as it is to cardiovascular prognosis, the detection of atheroma and repeated measures of its progression, are not yet NICE recommendations. Someone should tell them that atheroma kills patients, not high blood cholesterol!

Another golden rule. For all life-threatening progressive pathologies, there is an invaluable epizeuxis: regular follow-up, regular follow-up, regular follow-up.

Evidence for Presumptions

To remain an expert in any field, continually updating one's knowledge is essential. Never forget historical context, even though the current scientific attitude to anything historic is derogatory (the Travis Effect bias). Many think that most journal references, over five-years old, have lost their relevance. The notion rests partly on the daft presumption that intelligence, reliability, and inventiveness have improved with time. Psychologically, it will allow some to feel worthy of higher status.

Important questions arise when interpreting scientific journal publications. You need to know enough about the specific research techniques and analytics, to judge their applicability. How sure can one be of the reliability and veracity of the conclusions? Journal publications may have been peer reviewed, but their clinical relevance is more personal, since there are always devils in the detail. In the 1970s, I acted as a referee for the British Heart Journal, vetting submissions in electrophysiology. Some of their methods were suspect, as was patient selection, the analysis of results, and their conclusions. The number of subjects tested was often too small, and weak results were sometimes used to support strong conclusions. These reservations meant I had to return submissions for revision.

If you do your own research and undertake your own statistical analyses, you will gain insight into the difficulties involved. Only research done well (often an assumption, based on who did it, and where) with significant results, should change your clinical outlook and practice. If published research changes your outlook, be ready for others to question your justification. You should be able to argue for, and against its methods and conclusions. Your balanced opinion will mark your sagacity. If you choose to disbelieve any research authors, know their results better than they do.

Dogma and the HALO Bias.

Should you believe all you read and are told? To become 'one who knows' (a scientist) you must be a sceptic, questioning everything told to you by patients and colleagues, no matter how seasoned and prestigious their track record. Polite scepticism is essential; there is no need for unpleasantness or disrespectful ('bad') behaviour. If you disagree with a concept, fact, or idea, base your opinion and comment on extensive research; perhaps your own research.

I had no wish to become a SHO anaesthetist in 1967, but the lack of available cardiac jobs forced me to apply. My registrar was keen to teach me what she held dear, but had never challenged. I challenged two of her ideas, which were:

1. That local anaesthetic spray, used to numb the pharynx, lasted only thirty seconds. I tested this by spraying the throats of ten people. After challenging them every ten minutes with a cotton wool bud, I found the anaesthetic spray lasted hours.

2. Her second belief, now an anachronistic one, involved the use of N2O cylinders. She told me: 'The senior anaesthetists here are fooling themselves. They believe they can estimate the N2O gas left in the cylinders by tapping them and noting the sound'.

I spent one Sunday, weighing and progressively emptying several cylinders, matching the notes made when tapping them to an open guitar string tuned to 'C' (256 cps). The data I collected showed an almost straight line relationship between the change in cylinder gas weight, and the frequency of the musical note made. They cooled as they emptied, and this altered the note produced. I therefore had to wait a long time to complete the experiment.

When my turn came to give a review of the latest anaesthetic research, I presented my results to the anaesthetic department. The registrar who had expressed her views about laryngeal anaesthetic spray (and some of our senior colleagues) were not too pleased (she had been guilty of blind acceptance). A bigger challenge followed. I had to deal with her resentment and indignation. She was a NBW (See chapter on doctors).

This delighted the most senior anaesthetist in the department, Dr. Norman Eve. I had found proof for his thirty-year practice of tapping N2O cylinders to assess their contents. He had been an anaesthetist all his working life. He was an intuitive, kind, and very experienced teacher who lived for his work. His long experience suggested that he should be believed. I provided proof that he was not daft after all!

Dr. John Wright had been my tutor at 'The London'. He was a very experienced gastroenterologist; a world expert on Crohn's disease who had worked in India for many years. He insisted on examining patients' recent stools before each ward-round (Whipp's Cross Hospital in 1967). I dared to remark that I could see no point to the practice in the UK. He had to remind me that having spent years in the tropics, he could make an accurate diagnosis from stool examination alone.

He rightly left me with 'scrambled egg' on my face. It taught me respect for long experience. I think he forgave my naïve opinion. There is nothing more tiresome than an inexperienced junior doctor with too little knowledge, criticising an experienced observer. I was later to learn how he must have felt. Wisdom and justice can be a long time coming.

I understood how Dr. John Wright must have felt, while working in Amsterdam. In the early 1980s, a junior doctor during a cardiac meeting, informed me (incorrectly) of what I had written in one of my published papers (on sino-atrial pacemaker dysfunction). Even though I was the sole author, he continued mansplaining, refusing to believe that he had misread my paper. He had arrogantly refused to believe the author.

Observations and Theories

Smell is the most sensitive human sense. I know now, having had diarrhoea caused by Clostridium difficile myself, that the nose is a sensitive diagnostic instrument. Further researching the diagnostic value of smell could be worthwhile. My guess is that its value has ancient roots and goes beyond the obvious foetor of bowel bugs, diabetic ketosis, advanced hepatic disease, terminal illness, and dehydration. Dogs easily beat humans and chemical/electronic devices in detecting money, drugs, and disease. In certain respects, elephants are five times more sensitive than dogs and will use their trunks to differentiate smells coming from different directions. Instead of using sight, they can use smell to guide them. Unlike vision, smell works in the dark, and around corners.

Evidence for a theory can be slow to accumulate. While waiting, many will slip unconsciously into accepting it. Once accepted, a thesis is rarely questioned further. Two centuries after Newton published his Laws of Motion (1697), Einstein took issue with their accuracy. In 1915, he published his General Theory of Relativity, 218 years after Newton's work.

Many doctors get attached to clinical dogma; it avoids having to think, and saves energy for other tasks. A colleague once said to me, 'You don't drink enough alcohol to get pancreatitis', I took issue with him: 'If I am sensitive to alcohol, surely small amounts might induce it?' Having had acute pancreatitis, the question was, should I drink alcohol again? It could be a life and death decision for me. Why would anyone gamble with the chance of death, when giving up alcohol removes the risk? Deciding between the pleasure of wine, and the possibility of a premature death, I found rather easy (a BLOB decision, in fact). I decided never to drink alcohol again, and have not had overt pancreatitis so far.

The onus remains with me to prove that some people get pancreatitis with small, regular amounts of alcohol (contrary to dogma), perhaps because of an undue sensitivity. I remain content with my decision, but realise more research is needed.

Fundamental subjects like simple arithmetic, grammar, spelling, Greek, Latin, history taking, examination, anatomy, and physiology should never become arcane. The latest advances in science may excite us, but will they become more useful in life than the basic knowledge and techniques we learned at school and medical school?

Irrelevant Information

The only amateur clinical research worth doing is that used to verify a repeatedly observed phenomenon. Why spend time and money proving the blindingly obvious (BLOB)? There is now so much money available, and so many graduates unemployed, that many types of research are being founded. Be careful not to waste too much time doing menial research, just to help someone else's career. If someone draws you into their work, get one thing straight from the start: if your name will not feature as a co-author on any resulting publication, walk away. Do not waste your time and energy, unless you can see a career advantage to it. Callousness is one unattractive feature of the medical world, so ask yourself, *'Am I being used?'* and, *'What's in it for me?'*

Serendipity

In 1800, William Herschel sat in his room in Bath, allowing rays of light from his window to pass through a prism. It displayed a light spectrum. Using several thermometers, he measured the temperature of each colour zone. As a control, he put another thermometer into the dark region, beyond the red zone. Unexpectedly, he found this to be the hottest. Serendipity led him to discover the 'infra-red' region, lying beyond the visible colour spectrum. Of more fundamental importance was his original conception of some connection between heat and the colour components of white light.

Data Collection and Data Addiction

As an example of small data collection, my patient Barry A., is an investor who wants to know which UK companies are doing the most business. He favoured one simple research method. He would sit for an hour or two on a bench in Oxford St., Central London. From there he would count the shopping bag logos carried by passers-by. From this, he knew which shops were doing well, and which were not.

Big data collection comes with a sales label: 'within lie hopes of unimagined discoveries'. In commerce, both Google searches and credit card data (huge data sets), can reveal valuable client purchasing behaviour. Using it appropriately, can boost profits by promoting the sale of specific goods to targeted individuals.

The collection and analysis of big data is cheap and easy, so why not do it? Without multivariate analysis, few will spot any significant trends and relevant connections, in very large data sets. This is trawling in the dark for insights. The history of science reveals success from something simpler - primary observation. When one is thoroughly used to the usual, the unusual will usually stand out. Knowledge and experience of a subject, together with intelligence and perceptiveness, help make most discoveries.

We have now seen the emergence of data addiction. Data collection has become an obsessional, money making game, created for public consumption. Many are now addicted to the collection of their own physiological data (the 'quantified, or digital self'). At least an obsession with Fitbit devices might promote healthy exercise. It might also promote ill-health by creating targets, inducing anxiety, and promoting fear of 'red flag events'. Sleep can become disturbed. Having forsaken their mother's nipples, some will find comfort in the analysis of their own accumulated personal data.

Most consumers obsessed with their personal data will not understand its clinical relevance, but that would be to miss the point. The number of steps we take each day, our average pulse, and our breathing rates; our pulse rates during sleep, and during sexual intercourse, can all provide vain fascination. Perhaps pulse rate change during intercourse measures sexual satisfaction? The greater the change, the greater the lust. Would consistently poor responses be evidence for divorce? As one partner might say to the other: 'You only raised my pulse ten beats per minute last night. I'm sorry. You must leave.'

Those further interested in the subject should refer to the Journal of Medical Internet Research.

While we are collecting pointless data, are we not in danger of side-lining some valuable, but immeasurable human activities: friendship, duty, love, commitment, loyalty, and responsibility? These are not yet of consequence to computer entrepreneurs; collecting data about them defies direct measurement. For them, any data addiction is good for business.

Theory Bias

History is full of examples of those who have failed because they clung to convenient, but incorrect theories. Decades passed before I realised that total blood cholesterol was not a useful cardiovascular risk factor for any individual. With so many acute heart attack victims having a normal blood cholesterol, the theory had to be tweaked if it was to survive. Could it be that blood cholesterol is reduced during cardiac infarction? That is false, but someone somewhere proposed it. So the original thesis (that blood cholesterol is always an important cardiovascular risk factor) and its correcting factor (to help understand why those having a heart attack have a low blood cholesterol), were both false. Two false theories combined, did not a true one make.

One cannot deny that heart attacks occur more in large groups with a higher than average total blood cholesterol, although, this epidemiological accepted fact is of little clinical use. Only 60% of all patients who suffer a cardiac infarction will have had hypercholesterolaemia. From the perspective of atherosclerosis, cardiac infarction and some embolic TIAs and CVAs, I found only 60% of those with carotid atheroma had a raised total blood cholesterol (>5mmols/l)(my unpublished series).

Total blood cholesterol (manufactured in the liver), is not a reliable indicator of atheroma in the arterial intima (carotid, and coronary), or a reliable enough risk factor for cardiac infarction in individuals.

Talent and Predictive Talent

In 1933, Churchill visited Munich and observed Hitler Youth members marching with nationalistic pride. He suspected that trouble was brewing in Germany and predicted that Germany might eventually invade Austria and Czechoslovakia.

Was his correct prognosis a lucky guess, or did he have a talent for prediction? I suspect that his apparent intuition was based on his talent for collecting and weighing meta-data.

Those with what seems to be a natural talent for prediction, can leave rational thinking to those without foresight.

When exercising talent, no conscious thought is needed.

- *Riding a unicycle safely on a tightrope over the Niagara Falls, requires courage and talent, rather than thought. While performing the feat, the talented rider must obey the laws of physics, yet make no conscious reference to them.*

- *When speaking their mother tongue, five-year-old children never refer to the formal rules of grammar.*

Talents are complex by nature, and not easily understood by those without them (there are no constants, no constraints, many independent variables, and many specific brain connections).

When Melville Feynman taught his son Richard, how electronic components combined to make a working radio, he made no reference to quantum mechanics (his son had yet to invent the subject). He used anecdote to decide 'what worked', and 'what didn't'. It was Melville's experience, using trial and error, that had led to his practical know-how. With one glance, he could tell if a circuit would function, or not. Richard Feynman learned this lesson from his father – after repeatedly observing something physical that works well, you can rely on it to predict what else will work. Richard Feynman amazed his colleagues at Los Alamos by emulating his father's talent. He could take one look at an equation, and straightaway announce if it was correct.

'The first principle (of science) is that you must not fool yourself, and you are the easiest person to fool.'

Richard Feynman.

Those who can 'fly by the seat of their pants' (or play a musical instrument by ear), will find they can use the same ability, working as a doctor, driver, or sailor. The judgements used to satisfy their purpose (like staying upright on a tightrope over Niagara Falls, potting balls when playing snooker, or choosing novel treatments), usually make no reference to rules. Rule-based, jobsworth people, will find this difficult to comprehend. Following a script, and thinking on our feet, are very different cerebral functions. Medical bureaucrats now demand that all doctors follow scripts. Whether we should give up thinking for ourselves must be discussed.

Whether a doctor is a script-reader by nature, or a freethinker, his clinical judgements must suit each patient. They must show practical common sense and affordability. For patients to gain faith in a doctor, his diagnoses and management decisions must comply with these criteria. Patients will decide within a minute or two of meeting a doctor or nurse, whether they are suitable. Whether they are worthy of trust will take longer. They will not be using scientific criteria to make these assessments, but their assessments will usually work in practice. Can script-reading, rule-following, bean counters hope to pass the tests patients will apply? They might, but not easily. Patients will mostly be obliged to accept them temporarily, while searching for a doctor capable of both the art and science of medicine.

Chapter Fifteen

Clinical Management

'Victory awaits him who has everything in order — luck, people call it. Defeat is certain for him who has neglected to take the necessary precautions in time; this is called bad luck.'

Roald Engelbregt Gravning Amundsen (1872 – 1928).

Amundsen was a Norwegian explorer of polar-regions. Between 1910 and 1911, he led the first Antarctic expedition to reach the South Pole, and was the first person to reach both the North and South Poles. He was first to traverse the Northwest Passage, but disappeared in June 1928, taking part in a rescue mission.

To be a successful explorer or business owner, requires superlative executive functioning; an ability to organise and make things work in a timely fashion. The same requirements of success should apply to medical practice, but it is replete with intelligent, well-educated technical guys, who will know all about the chemistry of flour, water, and sugar, and the biology of milk and eggs, but cannot bake a cake. If doctors and nurses are to help manage the lives of others, shouldn't they first know how they manage themselves?

'Humanism in medicine is the physician's attitudes and actions that demonstrate interest in and respect for the patient and that address the patient's concerns and values. These generally are related to patients' psychological, social, and spiritual domains. The goal of patient care has to be Wellness with Wholeness. This would be to understand the relation between the body, mind and soul of the human being.'

W.T.Branch. JAMA (2001); 286(9):1067-1074

'Instead of managing for efficiency, a modern army needs adaptability more.'
Retired US General, Stanley McChrystal. Team of Teams.

Apart from planning, the management of patients and medical practice also requires courage, respect for individuality and adaptability. Through the bedrock of wise and effective clinical management, one will find written: 'know thyself', and 'know thy patient'.

The Aims of Clinical Management

Thomas Hobbes stated that life without government would be 'solitary, poor, nasty, brutish and short' (*Leviathon*. XIII.9. Thomas Hobbes. 1651). Although medical practice can help to relieve the depression of solitude, reverse the health divide for the poor and lengthen life, governments are all intent on imposing their own forms of nasty and brutish regulation on medical practice.

As genetic engineering stands at the moment, death is inevitable. The best clinical management and genetic engineering we have, cannot yet prevent it, although many await its defeat. Until then, our clinical management must aim to relieve suffering and stop premature and avoidable deaths. We can help by promoting early diagnosis and nipping disease processes in the bud, rather than allowing them to ripen. Even more fundamental is to seek, define, and remove, adverse aetiological factors. Occasionally we will eradicate disease, but more commonly we must resign ourselves to managing it and lessening its consequences.

How Cardiac Management Has Changed

I was working in cardiology before CABGs, and heart transplantation became routine. The first CABG was done in 1960, at the Albert Einstein College of Medicine-Bronx Municipal Hospital Center, US, by a team led by Robert H. Goetz. Christiaan Barnard performed the first heart transplant in 1967, the year after I qualified. This was well before the first coronary balloon procedure, performed by Andreas Gruentzig in Zurich, Switzerland, in 1977, and the first transcatheter aortic valve replacement in 2002. These interventions all made a profound difference to the management of cardiac patients.

Clinical Management Styles

Doctors must rely on scientific evidence and guidelines to manage their patients, and have documentary evidence to legitimise their work. There are others with additional skills, able to 'fly by the seat of their pants', with a facility for human understanding and insight, nebula thinking, and experience-based judgement, that helps them decide what is best for their patients. They can use the art of medicine alongside the science of medicine, but will struggle to legitimise what they do to rule-bound medical regulators.

Doctors have long practiced the personalised medicine advocated by Hippocrates, and have practiced this as an art since ancient times. Scientific medicine is the newcomer. In a renewed quest for science-based, anonymised personal medicine, medical science is looking for individual genomic studies to design tailor-made therapies for each patient. It may result it personalised medicine, but there is nothing personal about it.

Those clever enough to understand, remember, and manipulate its technical facts and processes, can readily learn the science of medicine. The art of medicine, like the art of oil painting, the ability to play a musical instrument, and the facility to speak several languages, requires talent, flair and a willingness to engage with humanity.

Scientific medicine has provided us with many indispensable tools. They are what all patients need, and what all patients should be able to access. Medical science can now claim to be overcoming our natural history by interfering with key aspects of evolution. Not only the fittest now survive.

Applying the science of medicine can be as effective as a bowl of cold soup served to stranded, freezing mountaineers, awaiting rescue. It may contain all the ingredients necessary for sustenance, but not uplift their spirits enough for them to survive the journey ahead.

No matter how much science one applies to a clinical situation, someone, sometime, must draw a line between what is 'normal' and 'abnormal'; between what is acceptable and the unacceptable, and between the clinically applicable and the pointless. Science has provided us with the means to draw lines, but not the ability to know where best to draw them. That remains an art.

Only those with a strong interest in their fellow man, and knowledge of what makes him tick, should attempt to practise the art of medicine. They will need to be brave enough, and irreverent enough of regulatory bureaucracy, to practise the art of medicine in the future. This is because medical science now encircles the regulatory high ground. From a public perspective, the progressive impenetrability and detached nature of scientific knowledge is growing its mystique. The clinical talents of those practising the art of medicine are being derided and undervalued at the same time. Art will remain impossible to regulate while there is no verifiable measure of its merit.

Most human beings want to be cared for and respected in times of need. As a result, the trend to make the art of medicine and other humane interventions irrelevant, has little or no prospect of succeeding. The management of every challenging clinical situation needs to draw on both the art and science of medicine, if health, quality of life, and patient satisfaction are to be achieved.

General practice and hospital practice have their own clinical management features. Hospital practice management depends a lot on team effort, coordination, and the technical excellence of those taking part. General practice lies at the frontier between medical science and humanity, dealing as it does, more with the commonest medical issues. It deals only partly with technical medical matters. Like soldiers, doctors must be able to trust one another in order to perform efficiently. Over my long career as a doctor, I am sad to say I have not always found this to work well. In keeping with most corporate ladder-climbing, one cannot always trust fiercely competitive doctors and nurses whose aim is to acquire status.

Sadly, there are doctors and nurses who will find it difficult to adhere to the US, West Point cadet code of honour:

A cadet will not lie, cheat, steal, or tolerate those who do.

Of What Value is Experience in Clinical Management?

Safe and successful clinical management rests heavily on accumulated experience; of what works (under certain circumstances), and what does not. Every medical student, in the transitional phases between student, junior doctor and consultant, will progressively learn how to apply their knowledge while working in live, dynamic medical scenarios. Hopefully, experienced and skilled nurses will remain by our side to assist. Always listen to their experience before deciding to do something new.

Experienced words,
From older tongues,
Will fly ad nauseam,
Into novices' ears;
Although heard,
Not always heeded.

From the events that diminish us,
With fear and resentment;

To those that enliven us,
Evoking our fascination and self-esteem,
Knowledge assumes more usefulness,
Once marinated in experience.

Our Sovereign Duty: Management Responsibility

Let me make my personal position quite clear, if I haven't already. Nobody without a medical education, clinical perspective, or patient sensibility (be they a manager, bureaucrat or regulator) should govern doctors and nurses in their clinical role.

A sovereign duty binds doctors, nurses and patients together. We are all called to use our medical knowledge, ability, and vocation to help save lives and relieve suffering. No man-made rule or regulation should ever countermand our duty to fulfil these needs, since no precedent countermands the right to life. This must continue to apply to retired doctors and nurses, certificated to practice or not.

We are all morally obliged to:

1. Treat our patients with the same care that we would give to a (beloved) friend or relative.

2. Remember we are servants to patients, not their masters.

3. Respect other members of our profession, but never blindly.

4. Keep our own counsel. Our patients deserve a completely independent, unbiased, and unmotivated opinion, formed strictly in their best interest.

5. Become so experienced, knowledgeable and dedicated to the welfare of patients, that we can honestly preface all of our advice with: 'If I were you . . .'

6. Avoid all financial considerations in matters of diagnosis and management; they must never be our direct concern. If they become so, we risk being diverted from the patient's best interest.

7. Act as gatekeeper and guardian to the ancient art of medicine and its practices. Our duty is to safeguard the trust many patients still have in medical knowledge and the doctor-patient relationship. This includes keeping patient details strictly confidential unless the patient agrees to share them.

8. Decide on the most appropriate management time-frame: minute to minute, hourly, daily, or every week, depending on clinical urgency. Never allow yourself to be deviated by any political or bureaucratic shortcomings.

9. Let nobody come between a doctor and his patient, except by mutual consent, whatever authority they pretend to have. In this respect, the qualifications of another doctor, although impressive, may not be the only measure of their clinical acumen or their practical competence (and value to patients).

10. Treat the opinions, directives, rules, guidelines, and commands of those with no clinical experience or experience of clinical judgement, with the utmost suspicion. Critically appraise them for the benefit ad detriment they might bring to patients.

Once we accept the care of a patient, it is our sovereign duty to accept complete responsibility for their management. We cannot possibly bear the same responsibility to the arcane concept of 'the medical profession'. This is a fanciful, metaphysical concept, suggested by the GMC whose part definition of good medical practice is not to bring the medical profession into dispute. It might be different if we had any effective collective political power to treat patients as only we know how.

When we close ranks, and consider our responsibility to the medical profession, our responsibility to patients can suffer. Due responsibility to the medical profession is a contrived concept, useful for regulators looking for reasons to sanction doctors. Many equivalent mythical concepts are currently in use (see '*Sapiens*', p114-124, Yuval N. Harari on an 'Imagined Order').

The bureaucratic, over-managed medical systems that now prevail in the UK, present many dangers to patients. Part of a doctor's job is now to protect patients from them.

When sharing any responsibility for a patient with colleagues, define exactly who is responsible for what, and record it for future reference. I am sad to say that few of your colleagues will help you, should your delegation of responsibility come into question. As I know full well, many colleagues will turn their backs and run from supporting you, should you come to be investigated by the GMC. One cannot expect their loyalty to outweigh their fear of reprisal. This is good enough reason for you to re-think both the priority you give colleagues, and the sanctity you ascribe to the medical profession. Your primary duty is simple — always put your patients first.

Personal Management before Patient Management?

In you run your own practice, never live too near to your practice premises. When patients learn you are in residence, a few will come knocking on your door, whatever the time of day or night.

Before you can manage a patient, doctors and nurses must first be able to manage themselves. Managing personal energy is crucial.

In 1992, junior doctor Chris Johnstone won the right to sue Bloomsbury Health Authority for subjecting him to an onerous obstetric job. He had fallen asleep while at the wheel of his car. The portable EEG he attached to himself recorded he was asleep while repairing an episiotomy. (Source: Private Eye, No. 1476: 2018).

Poor judgement, exhaustion, and mood disturbance, can all result from sleep deprivation. Under-staffing, being 'on-call' for 24–48 hours, and working over 72-hours in any one week, are among the causes. At the moment, mandated working hours for doctors have no legal basis.

US National Highway Traffic Safety Administration found the ideal time for a refreshing nap to be 26 minutes. Junior doctors might need a few refreshing naps.

Ideally, you will have impressive executive functioning skills and can use your energy and time efficiently. Doctors and nurses must be able to get things together without losing composure, even when the situation seems chaotic. Doctors can become too tired for this, or lack the ability. Others lack confidence, are too insecure, too anxious, depressed, or obsessive (wasting their energy), to manage clinical situations effectively. They should find another job where their management deficiencies no longer a risk to patients.

Doctors with health issues should seek independent professional advice and consider the implications for patients.

Coping with Personal Stress

I will depart briefly from clinical management considerations, to discuss the personal situations some doctors face. If doctors become stressed and ill-content, they may not undertake clinical management effectively. This is now an important issue, given the stress most doctors and nurses experience through inadequate working and living conditions.

The definition of 'stress' is simple. It is an applied force, the effect of which will change something, like the length of elastic, blood pressure, immunity, endocrine function or behaviour. These are all 'strain' effects. Every person is liable to experience stress from various sources, the commonest of which is from inter-personal relationships. Those responsible may not only be family or friends, but banks, the law, managers, regulators, and colleagues. So, how is one to cope?

I recommend the FIRO B questionnaire as a test of relationship compatibility. It can help one understand other people and why they can become a source of stress. I first detailed this in the earlier chapter on 'Patients', but I will summarise it here.

We can classify our immutable psychodynamic needs as the desire to give and receive three psychological entities: affection, control, and inclusion (shared interests and activities). Examine your need to give affection, and the need your partner has to receive it. Is there a mismatch? Control is even more important, while inclusion is least important. These simple factors will help you understand why you may not warm to another person. It will help you know how likely a relationship is to be compatible. Partners who match on all three fronts of giving and receiving need, will likely have a sustainable relationship. A relationship can survive on two (if affection and control), but not when one is inclusion. No relationship can survive with only one compatibility factor present.

Rule one: get enough sleep (even characters like James Bond will crack when sleep deprived).

Rule two: reduce any preoccupation. Take breaks. Do whatever takes your mind off of your problems.

Trust your subconscious to come up with relevant suggestions and solutions. Taking time off to 'do nothing', is not doing nothing; it allows time for subconscious processing. Exercise patience and avoid impulsive, knee-jerk reactions. Never run around like a headless chicken. Sit, think, and sit around some more. Limit yourself to 10–15 minutes of focused thinking, during which time you must define your problem(s), the think about solutions. Ask, 'What do I really want?' 'What would actually make me happy?' and, 'What must I do to achieve it?' Do this when you feel composed rather than desperate. The anxious and obsessive will find this difficult; constant re-iteration and pre-occupation can drain their energy and reduce their concentration and coordination.

Doing a busy job can side-track the time needed to find solutions. Once you have thought it through, discuss your problems with someone who knows you well. It will help if they have insight, wisdom, and loyalty to you. Think things through first, otherwise the opinions of others can become confusing, or set you off on the wrong track. This occurs more often to the desperate.

Negative opinions also help – your disagreement with them can strengthen your resolve (see the reactive devaluation bias).

Before asking anything of others, complete your research and prepare your arguments. Sometimes, really worthwhile life planning can result from a complete re-think of your situation. Although selfish, you must ask, *'What actually makes me happy?'* Review all of your life experiences honestly in order to arrive at an acceptable conclusion. Don't avoid admitting to mundane pleasures. Although a considerable amount of honest insight is essential, review your every involvement asking, *'did it make me happy?'* If not, put an end to your stress by changing direction and re-directing your energy. Be prepared to admit error.

All true masters can afford humility, whatever their field of endeavour.

The more experience of life you have, the better you will practise the art of medicine. Many doctors in the UK each year, now conclude that their best move is to give up medical practice altogether (or emigrate). You might guess some of their reasons for this from what I have written about the corporate, punitive state of medical practice in the UK today, and how it has changed since I qualified in 1966. I have dedicated a whole book to the subject (*The NHS. Our Sick Sacred Cow.* 2023)

'Never has there been a more urgent need to understand driving forces and motivating factors for the exodus from the UK's medical training programmes. In 2018, only 37.7% of F2s continued into run-through training programmes. It is perhaps the most critical juncture the NHS has ever faced. These doctors are the backbone of the UK's healthcare system, representing the future of medicine.
'Drexit' (Doctor-Exit) is the exponentially growing trend for doctors to walk away from their jobs in the NHS, either to new healthcare systems overseas such as Canada, Australia, New Zealand or worse, into new professions altogether, leaving behind their well-trained medical brains. This exodus has been gaining momentum for several years, with the workforce now at breaking point.'
The *BMJ Opinion*. February 6th 2020.

Despite the WHO recommendation not to recruit from certain countries (a red list that includes Nigeria, Ghana, Nepal and Pakistan), active recruitment of nurses (and doctors) by the NHS continues.
Nursing Standard. 4th October 2022).

It could be the right time to make a knight's move and run off with the girl/boy of your dreams to live in Acapulco. Beware, though, the grass is not always greener over every hill (the Yonder bias). It may not seem it to the young, but life is short, so you need to get on with your dream fulfilment. Do your research and burn all unnecessary bridges behind you. Make a list of those things in life you cannot do without, and the things that make you feel happy and worthwhile. Don't arrive at my age and look back with too many regrets.

Try to grow beyond the aspirations of others and come to live for yourself. If you do upset others, one thing is for sure – they will get over it.

If, after much deliberation, you remain unsure of what to do for the best — first do what you know to be right — morally, ethically, or from the standpoint of loyalty and duty to your genuine friends and family.

Managing Practice Culture

There are big differences between managing somebody else's medical practice and your own. As only a few will ever find out, success in what we British call 'private practice' depends primarily on patient satisfaction, and not on satisfying irrelevant corporate targets and regulations. It depends on the effectiveness and efficiency of patient handling; handling that is personal, not detached. It depends on forming a mutually agreeable doctor patient relationship, which patients readily agree to pay for, mainly because it generates trust through commitment. It works because of the attached style, where the dedication to help patients quickly and as effectively as possible goes without question. The same should apply to NHS practices, but they now share their primary commitment with corporate dictates; NHS dictates which would not meet my approval or that of any other successful, long-experienced private doctor.

In my early days, I was lucky to have an experienced colleague and partner, Dr. David Baxter, to guide me. He knew what patients wanted, and how one could ensure their return. In the UK, patients have always had a choice between a free medical service, or one they pay extra for. Those who pay will want to experience a tangible benefit, since they must continue to pay for the NHS through taxation. This sets private standards of management at a level which must exceed those of the free service (same day appointments with experienced doctors, quickly made diagnoses, screening, and early hospital admissions). In private practice, clinical efficiency and effective patient handling must produce speedy, successful results, and a positive reputation for service. In NHS practices, these should be essential, but are not always.

Our government tasks the NHS (a corporate political pawn), to provide the UK population with a medical service, free at the point of contact. Being 'free' is illusory. All those who pay UK taxes, pay for it indirectly. It naturally creates a lot of demand from patients, many of whom have no other choice. Because the history of nationalised corporations in the UK is not a happy one, how the NHS has evolved was predictable. Understaffing, and a lack of proper resources, has led to the deaths of patients on waiting lists. GP consultations are now difficult to arrange and often last only ten minutes. Some say of nationalised corporate managers that they couldn't 'organise a piss-up in a brewery'. Apart from me, how many doctors and nurses would dare to agree with this?

No law excludes personal service, but the requirements of government bureaucracy mean that the NHS will forever struggle to achieve anything like the services found in private practice. The admirable rationale of the NHS, however, was always to provide medical care for those with no other option. Population growth and the diminishing resources made available, have ensured its continuing overload and comparative inefficiency.

Why might patients consider paying for private medicine when they can get a free medical service (if they have the time to wait)? The answer is not straightforward. For many patients, it is a matter of style, prestige, and their wish to exercise free choice; perhaps opting to be treated by a specific doctor, or one who is available, efficient, approachable, and compatible; one with whom a long-term relationship can develop.

Have you any idea why someone might choose to pay £30 for a cup of tea and a few cakes and sandwiches at a five-star hotel? Those for whom this is beyond their event horizon, will struggle to understand why some patients want private practice. Even if the tea and the sandwich were the same at Joe's cafe, costing only £3.00, some still prefer the Ritz, Claridge's, or the Dorchester, paying ten times more for the experience. It has something to do with the quality of food, its presentation, the ambience, attention to detail; the personal standard of service, style of communication, and the personal respect shown to each valued client. There is, of course, no reason Joe couldn't do the same, except his business model demands that he keeps prices low and processes as many customers as he can – just like the NHS, in fact.

Receptionists at smart hotels do not snarl or become impatient with their clients, although some will try to ignore you while looking at their computer screens. I have known some medical receptionists to be called 'dragons' by patients, simply because of their snarling, 'the computer-says-no', attitude. Are receptionists employed to protect doctors from their patients, or are they just frustrated, disillusioned, demoralised, and angry people? The answers lie in a combination of factors.

A dragon receptionist can harm a doctor's reputation and savvy patients will question their judgement, asking 'why employ that person?'

One major cause of medical management error and inefficiency is allowing those who know nothing about clinical priority to make appointments. Busy doctors who leave appointments entirely to non-clinical staff might unknowingly influence their patients' prognoses.

The management of a practice must adapt to its demography. One meaningful way to classify patients is by employment status: unemployed, employed, self-employed, and company directors. Others do not need to work, cannot find work, cannot work, are too young to work, or are retired. Some will find it odd that the self-employed and company directors are least demanding. They usually have specific needs and know what they want. They will expect the same management efficiency that led to their own success. Some doctors might see this as demanding. I never did. If you are in private practice, the self-employed will admire a doctor who is self-sufficient and free to make his own decisions. They are far less likely to pursue their rights when disputes arise, and much more likely to embrace discussion after being fully informed.

The richer and more educated the patient, the more time they may want for discussion. In private practice, the extra time available (I spent 3–10 times longer than most NHS doctors), adds to the sense of thoroughness patients experience. NHS GPs, who now allow only ten minutes for each patient consultation, may struggle to impart the same impression. Many patients now complain about short NHS consultations and inefficient handling. The excuses given are: patient numbers, shortage of staff, reduced financial support or a reduced number of daily consultation sessions.

'Word of mouth' recommendation is the best way to build a professional practice. For the few, a worldwide medical reputation will attract patients from afar. Modern advertising on social media platforms should propagate a doctor's ideas, if not his practice, but I have seen little evidence of it. The indirect advertising I did, generated many inappropriate, time-consuming enquiries. All advertisers will tell you, however, that adverts need to be run many times in order to gain public awareness. If you have that sort of money, your forte might be in running a corner shop, not a medical practice. Alternatively, why not become a medical entrepreneur and create a medical franchise.

Managing staff is a matter of personal style. I have to admit, mine was inappropriately friendly. Or was it? If my staff did the work I needed, I left them to chat, and otherwise use my practice to conduct their personal affairs. To say that I have been liberal with my staff would be an understatement, yet the effect was to have a practice where patients felt comfortable. It was said that my clinic was more like a club, than a medical practice. Patients came in for a chat with my staff and happily met other patients. Some met a few of their long-lost friends and acquaintances in my reception area and exchanged experiences. (My son Nicholas, jokingly referred to it as 'a temple of healing').

I could have opted for a more business-like practice management style, with rigid rules, and a clocking-in, clocking-out system, auditing every activity. Scrooge would have organised it so. He would have overlooked a few important human factors which make a working life indistinguishable from a private life. One result was that many patients came to regard me as a friend. In time, almost all were on first-name terms with me and my staff. This is far from what many doctors wish for, and not at all appropriate for every practice.

Many doctors choose formality and anonymity (they see it as 'professional'), reflecting as it does their character, style, and type of practice.

One cannot overestimate the pleasure of designing and planning a medical practice. It requires one to be confident enough of medical expertise, standards of care and personal style to 'go it alone'. It is for the few, but a number now growing (solo practice is all but outlawed in the UK, with regulators worried about allowing another Dr. Harold Shipman to arise. Many feel secure only when working in a team, with a committee running the system. They can forget private practice, other than as a part-time option.

Few patients enjoy seeing a different doctor each time. The corporate demand for financial efficiency, has led to the proliferation of group practices. My guess is that older patients would choose a more personal form of medicine if they could. Younger patients, who have not experienced it, will not miss what they have never had.

For a long time now, there has been a UK policy to drive NHS doctors into group practices. Doctors (regarded by bureaucrats as corporate medical operatives) are thought be safer with constant peer review. For this reason, I regret to say that the practice I developed in 1973, would be impossible to recreate. In the long-term, however, patient pressure might help the return of what many patients crave – continuity. This requires access to their own favoured doctor; one with whom a trusting relationship can form.

Briefing

When César Ritz (founder of the Ritz Hotel in London and Paris) was working in Vienna, he served the Prince of Wales (the future UK King, Edward VII). The future king, who was obviously used to being served, at once experienced a new form of service; one based on anticipation. Having seen that the Prince preferred a cigar and a glass of spirit after his meal, César unexpectedly offered him a tray of cigars and matches before he asked. The much impressed Prince, sometimes exasperated by having to ask for everything he wanted, introduced him to European aristocracy. His anticipation of the needs of his guests became the basis of César Ritz's success in establishing world renowned hotels. His style of service gradually became the standard for every grand hotel in Europe.

Doctors should never allow themselves to get caught without a brief, whether in consultation with a patient, or when instructing their team. Patients rightly expect doctors to remember them and their problems, especially if they were last seen a few days before. Doctors will impress patients by remembering relevant meta-information, like family matters and any worrying concerns. Adequate briefing must allow one to ask questions that are relevant, and not those that reveal ignorance. Some patients will be right to trust only those who show interest in their problems.

Miss Keira P. told me that the gynaecologist she saw was 'useless'. Her reason for saying this was that he constantly referred to 'MIMS' (drug reference book), to check the doses of oestrogen and progesterone in contraceptive pills. 'He didn't seem to know what he was doing', she said. 'I won't be seeing him again.'

I once worked at the Harley St Clinic, as a junior supporting a famous cardiac surgeon, Mr. Donald Ross. I don't think he ever wrote clinical letters or notes himself. Before he entered a patient's room, one of his two PAs would brief him. The briefing would go beyond the clinical, to include personal information like the patient's daughter who had recently taken her exams. While engaged with his patient, he would make comments such as, 'I think we should let your GP know that your heart is now strong, and the valve replacement is working well. We must also tell him you no longer need a diuretic.' From this, his PA would compose a letter.

One of his PAs briefed him on every aspect of his working life; which patient he was about to operate on, and why. Also, where he was to have lunch, with whom, and for what purpose. He would perform 3–4 cardiac operations on several days of each week. This required the co-ordination of a large team, none of which directly involved him. His admirable handling of patients and his practice administration system, upheld his reputation. He was one of the world's most skilful cardiac surgeons, with patients drawn from every continent.

Always settle on a management plan before the patient leaves. State the proposed order of events, and let them know how and when they will happen. If a letter of referral needs writing, do it then, not weeks later. It will only take 5–10 minutes, and both you and the patient can move forward. It is inefficient for patients to return to pick up a letter or an investigation request form. Finish all you need to do in one session. Loose ends pile up and will frustrate one's ability to cope. If I suggested a blood test, I took their blood myself, and would inform the patient of their results within 24-hours. One of the commonest criticisms levelled at the NHS is how long it takes to get anything done, and how long it takes to get intelligible answers and explanations.

You will set yourself up for failure if you work for any system that cannot deliver the results your patients need for efficient, effective clinical management. You must question why you are working for such an organisation. Excuses rarely hide incompetence. Business people, running their own large organisations, will ask why you put up with such inefficiency. Patients came to me repeatedly, not only because I made myself available (on a same day basis), but because I could quickly achieve results. I never used a laboratory or any other organisation that could not deliver results quickly. I had some ongoing cases under investigation, but none where my management plans faltered for a lack of investigation results.

In the NHS, doctors must accept the services provided. I feel sympathy for any professional working for a system that affects their personal efficiency and effectiveness. Such a lack of control will foster disenchantment.

At the end of each consultation, or at the end of a full set of investigations, I would discuss my conclusions and their consequences with each patient. I later provided them with a written (technical) summary, always typed on pink paper. That way, they stood out in every A4 patient note folder. Later, I needed only to read the summaries to get a pre-consultation briefing.

Patients forget important clinical details, so I combined my summary with an explanatory letter in plain English. In this, I would mention all the key points of their history and examination, the key positive and negative investigation findings, and the reasons for my conclusions. Patients were free to assimilate the information at will, and to share it with others. Doctors willing to put their opinion in writing might gain respect, but might also need sympathy.

Flying Solo and the Art of Time Management

Driving a car has many parallels in medical management. Some doctors will manage their clinical cases like they drive – with no awareness of other road users, inadequate forward observation, poor positioning on the road, slow responses, inappropriate speed, and distracted by the radio or conversation. These drivers risk road traffic accidents. If their clinical management is similar, they will risk harming patients. When we drive a car at 70mph, we only need to take our eyes off of the road for two seconds, to travel 200 feet. An appreciation of the time-scale for action is as crucial to clinical case handling as it is to vehicle handling.

A patient's blood pressure can drop quickly when anaesthetized. Their pulse can slow, and cerebral perfusion become diminished. This can happen within seconds. Gone are the days when consultant anaesthetists put their feet up in the back office, reading a newspaper, while some lesser mortal (I was once one) supervised the unconscious patient. As when driving, one must sometimes think fast and take immediate action. In cardio-pulmonary arrest, for instance, you need to start resuscitation immediately and ask questions later. An awareness of the time-scale of change in each clinical situation, should trigger the most appropriate speed of response. The faster the reaction, the more likely the patient is to survive. In terminal care, it is tardiness that may be most appropriate.

Most medical work involves dealing with chronic conditions over long periods. Although we use the word 'chronic' a lot, few patients understand it. It derives from the Greek word chronos (χρόνος), meaning 'time'. Many patients believe it to mean 'serious'. From the point of view of lasting a long time, it is.

Many chronic cases have acute exacerbations. An octogenarian might find it difficult to lie flat, and wheeze throughout the night. His symptoms suggest LVF, requiring urgent treatment. A child struggling to breathe while wheezing, will need immediate treatment with an anti-histamine, a steroid, an inhaler (various sorts), and sometimes an antibiotic (to cover all possibilities). Unfortunately, it can take 24-hours for any steroid or antibiotic treatment to become effective. Waiting 2–3 days for a GP appointment, would not serve their best interests. A compendium of excuses related to 'faults within the system', poorly explain the shortcomings. I return to the question I posed before: why work for a system that cannot deliver the means to practise medicine at its best?

Handling a medical case can be like catching a fly (quick reactions needed), or stopping an oil tanker (patience is required). Appreciating the time needed to manage each can be crucial. One can learn timing from experience, although there are those naturally good at it (like athletes and musicians). The collective action of a medical team requires that they coordinate and act in a timely fashion.

Clinical Titration

Observe – intervene. Observe again – intervene further, at time intervals dictated by clinical priority.

You may have to co-ordinate the speed of patient re-hydration; sodium and potassium replacement and O2 levels, for someone in heart failure and a chest infection. This is especially the case when their echocardiogram shows a poorly functioning LV, an ejection fraction of 35%, and valve leakage (MI and TI as the heart dilates). Each clinical factor will follow its own time course and have an optimal time for intervention.

How might the latter case be managed? I would suggest weighing the patient (daily body weight can reflect hydration, simply because water weighs so much). Having examined the patient, measure his pO2. (blood oxygen levels are top priority) and return his pO2 to normal if necessary. Order a chest X-ray and ECG. Sit him at 45 degrees to observe his JVP. If the patient is dehydrated, the sitting angle will need to be lower than 45 degrees, and nearer to 90 degrees if the JVP is raised to the angle of his jaw (in hospital some will find it easier to start central venous pressure measurement, but this will be impractical in a nursing home or on a desert island). Set up an i.v. infusion with dextrose-saline, after taking blood for electrolytes using the same cannula (before the infusion starts. It saves the patient from an extra needle jab). Get a sputum sample. Guess what antibiotic(s) might be best (ask about allergic reactions), and start with iv doses if peripheral perfusion is inadequate, and if infection is the predominant presentation.

Catheterise the patient if necessary to get a urine sample and measure hourly output. Note the colour; measuring urine specific gravity is a more accurate measure of hydration (over the time it has taken to produce the urine). Decide on the volume of replacement needed, based on urine output, JVP, and skin turgor. You cannot afford to have sodium in the infusion until you have excluded heart failure. Infuse, observe the JVP every 15 minutes, and vary the infusion regime as necessary. If you suspect heart failure, the initial infusion rate must be slow.

This is clinical titration in progress; action after observation, and further action based on physiological observations. As the patient stabilises, decide on the most appropriate maintenance regime. Next, decide how often one should examine the patient, and don't let your rota interfere.

Initially, you will need to examine the patient's lungs, JVP, and skin turgor hourly. You might need to watch the BP, urine output, skin turgor, and JVP like a hawk. If initially dehydrated, with a low potassium from the over-use of diuretics (gravity-dependent oedema is often mistaken for heart failure in older adults), one must replace it fairly fast. Muscular weakness, and an ECG with flat T waves, will support a diagnosis of hypokalaemia. ECG T-waves will usually rise with cardiac cellular potassium repletion. Beware of falsely raised blood potassium levels caused by red cell haemolysis. This often results from delay: getting the patient's blood to the laboratory too slowly or laboratory processing delays. With a patient in a nursing home (and having to get blood to a distant lab), one should predict that the result will cause consternation amongst the staff (fear of cardiac arrest).

This simple case illustrates the basic attention needed, and the essential co-ordination necessary, for successful clinical management. Any lack of focus that leads to unnecessary delays will cause the patient to present in the middle of the night with acute LVF, struggling for breath (dyspnoea) with bronchial constriction and pulmonary oedema. Who will be responsible for this when it happens?

Once an acute situation has stabilised, it is time to question the diagnosis. What caused the heart failure? Why the dehydration? Could the patient have a cancer, secreting inappropriate ADH (SIADH)? You will need to review an echocardiogram looking for poor LV function, and an ECG that might reveal tachycardia (weak hearts quickly decompensate with the sudden onset of rapid AF, especially in older adults).

Meta-information is important. In older adult males, look for evidence of prostate cancer. You might find the PSA to be > 40ng/ml, with a raised alkaline phosphatase. This needs to be investigated further so you can estimate the likely prognosis and arrange long-term management. An interview with the relatives could be worthwhile.

Action Timescales

Throughout the management of every case, time scales may need to change from short-term to long-term. Never ignore patients, letting them die on a waiting list or on a trolley in a hospital corridor. Never leave patients waiting while you write defensive notes to please bureaucrats and lawyers, but not serve the patient's immediate needs.

Never allow clinical errors to occur because of inappropriate appointments. Ask not, *'when can I see the patient again?'*; but, *'when should I see the patient again?'* The decision must rely on clinical factors, never on administrative ones. Rules and regulations issued by managers, distant committees, and regulators must never take precedence over any patient's needs. This will save many lives. The change of culture I think necessary in the UK, requires all corporate objectives to be seconded to those that are medical and patient oriented.

When her miscarriage started in June 2022, Andrea Prudente was in Malta. She was 16-weeks pregnant. Termination of pregnancy was then illegal in Malta, so she was told by the doctors there, that intervention would have to wait until the foetal heart stopped. They offered her a grief counsellor, but risked infection and haemorrhage. After two weeks she travelled by private jet to Mallorca, and there underwent uterine evacuation.

Many Maltese doctors signed a petition to change the law (according to the 1850s law, both patient and doctor can face prison sentences). They said that the law tied their hands and they must weigh the care they provided to such patients against the risk of being prosecuted.

Here, the crucial management factors were the likelihood of haemorrhage and infection. Had either occurred, I am sure the Maltese doctors would have acted appropriately. There is a need to give doctors the power to suspend the law if what they do, favours patient survival. Those who have the power to change the law should benevolently favour altruistic doctors and forget prosecution. Only fools and the very frightened would follow rules that might lead to grave, patient risk. One should never force intelligent people to take advice from donkeys; wisdom is unlikely to follow.

Get to know the natural history of every medical condition you deal with. Act before events advance, become critical or end up catastrophic. There is an art to predicting the clinical time trajectory of various pathologies, so some feel for their usual rate of progress is essential. You must learn to recognise near cliff-edge phenomena, and at which points a patient's progress lies in the balance. There is always a point of no return; try not to let patients approach it. In some conditions, such as septicaemia and injuries, this point can approach rapidly.

Remember that you and your patient may define 'emergency', and 'urgent', differently.
A patient may declare a medical urgency, related only to their personal circumstances, perhaps caused by an imminent holiday, their need to get to work, or the pressure put upon them by others. Once they have claimed one false emergency, one can become biased about their future demands. Beware of those who cry wolf, they may someday develop a serious condition.

I defined action times for my non-medical staff as follows:

Emergency or Critical Situation: Drop everything and run.
For doctors, different colours connote different things: yellow (abscess, staph. aureus); red (haemorrhage/blood; sometimes implying 'do not enter'); black (not a colour) with purple undertones (melanoma, gangrene); green (sputum – the colour of WBCs?), and flashing blue lights. They all demand attention on a different timescale.

Urgent: act (diagnosis and plan management) within an hour.
Somewhat urgent: act within an hour or two.
Important, but non-urgent: act the same day.
Routine: act when mutually convenient.

The time scale will depend on our specialty, and the system we work for. If we do not sort out all routine clinical matters on the same day, they can pile up and cause overload later.

There is a 'golden hour' for action in emergencies; what happens within the first hour often determines the prognosis.

When I worked at the Vrije Hospital, Amsterdam, in the 1980s, we (Pim de Feyter led the work) saved patients from large cardiac infarcts by directly infusing their thrombosed coronary arteries with streptokinase. The immediate effect was to lyse the arterial clot, and to return their LV to near normal functioning. This happened soon after restoration of their normal coronary blood flow. If they were pale, sweaty and distressed on admission, they soon became pink and pain free as their cardiac output improved; all within an hour of onset.

The topography of Amsterdam made this possible; so did the efficiency of the Dutch ambulance services and hospital staff. We easily achieved a 'home to cath. Lab' transfer time of less than one hour. The quicker the transfer, the better the results. (Yet another case of BLOB).

In all medical practices, 'pop-in' cases can present a problem. Their seriousness or triviality can be surprising. One of my patients, who had never ever arrived impromptu, came through my door with a slight chest pain. Within one week, he had undergone coronary vein by-pass surgery.

The occasional pop-in case will be a poor communicator.

Mark R., used to pop in regularly. His complaint was always the same: 'I feel like sh*t doc.' Taking his history was always a challenge. What he wanted was 'to get fixed and leave'. He died on a beach in Thailand a few years later, from an overdose of 'recreational' drugs.

Negligence, disinterest, inattention (taking one's 'eye off the ball'), and failure to make pertinent observations all allow emergency situations to develop.

Getting laboratory results returned within 24-hours was one of my basic practice standards. New patients came to me thinking that their blood test results would take three weeks. Thanks to the Doctor's Laboratory in London, each patient received their results the next day. I always made my cardiac testing results available to patients within minutes of completion. We then immediately discussed their significance.

Nothing much took over twenty-four hours to accomplish in my practice, except for making appointments with colleagues. Most of my colleagues worked for the NHS and did occasional private practice. Many were short of time and not fully committed either way.

Prognostication

Of all the arts of medicine, prognostication is perhaps the most challenging. For patients, the subject abounds in cliché: *'You never can tell'. 'No-one has a crystal ball'. 'Nobody gets out of this life alive'. 'Always hedge your bets'.* For doctors, the best we can do is use our knowledge of natural history and experience to evaluate the data, its direction and rate of progress, to guess the patient's most likely clinical trajectory with and without our intervention.

Predict the worst; hope for the best. If you proclaim a poor prognosis, and your treatment works better than expected, the patient (and their friends and relatives) might revere you for your magical powers. Then beware of hubris. I have known lawyers employ a similar artifice. They first paint a bleak picture of the likely outcome of a trial, then collect more than the deserved credit when all the projected difficulties are overcome, and they win the case. There was once a parallel ploy used by obstetricians before foetal ultrasound scanning became available. Near-term mothers were told that they would deliver a boy, but the obstetrician wrote 'girl' in their diary. If it was a boy, they enjoyed the complaisance; if a girl, they showed the mother their diary.

A Scotsman, a Frenchman, and a Jew were each told by a doctor that they had three months to live. When asked what they would do with the time they had left, both the Frenchman and the Scotsman said that they were going to spend all their savings and have a good time. The Jew, when asked, said he would spend some money on a second opinion.

Attributed to Lord Immanuel Jacobovits, Emeritus Chief Rabbi.

Although most UK citizens grant the NHS sacred cow status (accepted as part of our British identity), a considerable part of my practice involved dealing with its refugees, most of whom were seeking a more detailed explanation of their illness and more efficient handling. Some thought a more personal approach could improve their prognosis.

In my specialty, I regularly had to adjust my prognostic outlook as the years rolled by. Routine CABGs became routine, but some while after I started work. Before that, those with coronary artery disease mostly received negative omens. The advice given to patients was restrictive and limited to 'don't do this or that'; 'don't eat this or that', and worst of all, 'take it easy'. We had little positive to offer except anticoagulation after a heart attack. Using warfarin was initially controversial, but more recent research showed it to be effective in reducing deaths by 24%; second cardiac infarcts by 34%, with CVAs reduced by 55% (Smith & Arnesen, 1990). During the 1960s however, cardiac infarction, heart failure, and premature death all seemed common in hospital practice, despite the use of anticoagulation.

What followed was advice to stop smoking, to lose weight and exercise. Therapeutic interventions followed sporadically with beta-blockade, CABGs, aspirin prophylaxis, angioplasty, stenting, ACE inhibitors and statins. These all improved the prognosis of coronary heart disease during my career. Patients once classified as cardiac invalids, became able to lead symptom free lives.

We all knew in the 1960s and '70s that left anterior descending (LAD) coronary disease carried a poor prognosis. It often proceeded heart failure and premature death. The anterior descending artery feeds blood to a large part of the front wall of the heart (the LV and septum). Complete occlusion is a cause of sudden death (from VF), loss of vital anterior heart wall (LV) muscle (when there is insufficient collateral circulation) and heart failure. This is not so often the case with circumflex and right coronary artery disease. The dominance of each artery and the extent of any collateral circulation, are crucial survival factors.

With my long personal experience of the natural history of severe LAD disease cases, I was keen early on to suggest CABG or stenting, even for those who were asymptomatic. Medical treatment has always been partly effective, but nothing beats normalising coronary blood flow unless the risks of intervention are too high.

It is unreliable to use mechanical analogies to explain biology, but have you ever had a blocked drain at home? Have you tried unblocking it with chemicals? Chemical products can sometimes help with minor blockages, but with more severe blockages, you have only two options – remove the blocking material completely or replace the pipework.

Over time, the prognosis of primary systemic hypertension has run a parallel course to that of coronary artery disease. In the 1960s, we used to measure blood pressure after the patient had been lying quietly in a darkened room for fifteen minutes. The medication we used then – propranolol, methyldopa, rauwolfia alkaloids, and diuretics were not always very effective. Now I would happily test a patient's blood pressure during and after exercise, simply because current anti-hypertensive medications (ACE inhibitors and ARBs), are so effective. Critically, they also reduce the development of LVF and arteriolar medial hypertrophy, and with it stroke risk.

In hypertensive patients, we once used LVH on a chest X-ray as a prognostic factor; in the 1970s, we accepted it as a reliable risk factor for CVA. The presence of LVH multiplies risk by 1.5 to 3.5. (See Tullio et al. Stroke. 2003; 34:2380-2384). We can now prevent cardiac muscle growth (LVH), arteriolar media (muscle) growth, and the thinning of cerebral arterioles (risking cerebral bleeds), with ACE inhibitors and ARBs if prescribed early enough.

Never prognosticate without specialised experience of the conditions you choose to manage. When in doubt, avoid prognostication altogether, or delegate it to another expert in the field. Leave them to make an informed guess. Because a life or a lifestyle could be at stake when prognosticating, consider restricting yourself to a range of odds (there are too many inaccuracies and errors involved in prognostication). Since science aims to provide us with perfectly functioning 'crystal balls', we must wait.

For patients with a dilated, barely contractile LV, there is one key statistic. On average, only 50% of them survive five years. It is vital, however, to remain aware of the statistical paradox when prognosticating - no statistic will apply with certainty to any individual patient. There is an important rider to this. At least the statistics can prompt us to get the best treatment available for the patient, based on the latest evidence.

Why do patients need a prognosis? We all make plans, and some even have lengthy bucket-lists. Some have important financial arrangements to make. Many will want to make sure their family is secure. This is an exercise most responsible people want to make. At my advanced age, I have a limited prognosis, although my parents and three of my grandparents, lived into their '90s. Even if I live that long, I must move quickly to complete the personal projects I am working on. The need for succession planning will seem too far ahead for medical students and junior doctors, but some early forethought about pensions and savings is wise. To gain financially, though, you must live beyond retirement age.

Managing Patient Expectation

Managing patient expectation needs prudence, kindness and hope, managed appropriately for each patient.

As a means of patient education, and to establish a policy, doctors should write their own information pamphlets. One should keep evidence of the information given to each patient during every consent process, always aiming to achieve fully informed consent. Once they have accepted the information given, patients might soon need to be given a questionnaire, to test their level of understanding of the information given to them (for medico-legally reasons).

According to barrister Bertie Leigh, of Hempson's, London (RCP *Advanced Medicine Conference* 2016), the public now looks to doctors for advocacy, as medical guides and mentors. This requires us to use every means possible to teach and inform them. Doctors and nurses now have a duty to record their decision-making processes and any information passed to patients. Patients will need educating if they are to exercise decision-making autonomy safely.

In an ideal medical environment, no doctor or nurse would waste their time writing legal oriented notes; clinical practice should always take precedence over time-wasting secretarial functions. Ideally, a PA trained to keep medical notes should accompany every doctor and nurse while they are working. They could record all that is said, the decisions made, and all that happens. A few high-earning consultant doctors already employ a PA for this purpose; they have no wish to perform the job of a secretary, manager, or lawyer's assistant. The obvious alternative is to video every consultation and procedure.

Avoid giving patients more information than they can handle, or would want to handle. Some doctors insist on downloading all the information they have, onto their patient's shoulders, regardless of their ability to make sense of it. This is simply passing the buck and with it, the responsibility. I have always found it unacceptable. This is a defensive ploy, aimed at avoiding blame for not providing 'fully informed consent'. Fully informed consent, like beauty, is for the beholder to judge. It is easy to deliver every bit of information we have, but that may be irresponsible without knowing the patient's educational background, intelligence and capacity to understand the content.

If your life depended on understanding Kant's *'Critique of Pure Reason'* (1781), or Roger Penrose's Minkowskian Geometry (*The Road to Reality*, 2005, Vintage Books, p412-439), would a ten-minute discussion with an expert, prepare you to make a life-changing decision based on your understanding of the contents?

Simple clinical cases present no problem. The patient has a virus infection, needs their hand stitched, or a skin lump removed. In more complicated situations, few patients will fully understand all the technical clinical details, so their acquiescence must usually rely on trust.

Doctors are not running the equivalent of a holiday company, where the default position (always in the small print) is – *'every client (patient) will be held entirely responsible for the decisions they make'* (caveat emptor).

With little experience of life and medical problems, some patients need someone they know and trust to explain the information doctors give them. I spent a lot of time explaining the consequences of what other doctors had said to my patients. I saw a lack of sensitive communication as a major failing of many UK consultants. The causes were cultural differences (35% of UK doctors are non-British), and the strictures of administrative compliance. There are doctors who disregard the educational and cultural background of their patients, and will not adjust their vocabulary or language appropriately, when giving explanations. Some patronisation may be required to impart a full understanding. Unfortunately, the current assumption of educational equality has made patronisation a crime. If they understand too little of what they are told, or what is being implied, the best interests of patients mat not be served.

Managing unreasonable expectations, requires a disciplined approach. In private practice, I never gave into the capitalist notion that those able to pay, should have everything they want. A doctor, star-struck, or seduced by fame, beauty, and wealth, can become compromised and lose their wise decision-making capacity.

Michael Jackson demanded sleep. His personal physician, Conrad Murray, supposedly charged him $150,000 every month to help him sleep at night. He arrived every evening and set up an intravenous infusion of propofol (an anaesthetic). He tried to wean Michael off it by lowering the dose and by using other drugs, but these alternative strategies failed. Michael Jackson had multiple drugs in his blood when he died. They found Dr. Murray guilty of homicide (involuntary manslaughter). Pending an appeal, he served approximately two years in prison.

Is Fully Informed Consent Attainable?

Getting fully informed consent is obligatory, but is it no more than a ritual; a responsible exercise, conveniently undertaken as best we can to satisfy corporate protocol? There is some equivalence here to a tribal ceremony, or rite of passage. Tradition dictates the words and the ritual, but only a few really know what is going on.

As doctors, are we exonerated from blame once we have fully explained the risks and benefits (as we see them) of a medical procedure to a patient? After they give their 'fully informed consent,' can we forget any responsibility to inform them further?

The term 'fully informed consent' implies that the patient fully understands the information, with the same level of understanding as the fully informed person giving it. Because this is sometimes unlikely and unreasonable, the concept could be unfit for purpose in all but the most straightforward of clinical situations. Exceptions may occur when patients have medical knowledge (for cardiac cases, this will also be unlikely for most doctors and nurses).

We have arrived at the point where fully informed consent, as a bureaucratic legal requirement, is not always possible or in the patient's best interests. Without conveying a full understanding, it cannot protect every patient as intended. Besides setting out relevant risks and benefits, I always preferred patients to ask me one vital question: *'What would you do doc, if you were in my position (given all that you know)?'* The question invited me to convey not only my technical knowledge and biases but also my personal experience, thinking, and feelings on the subject. In this way, patients could get closer to that illusive state of being fully informed. There is an implicit danger here. Patients who trusted me, might bypass the need for explanations, and agree to anything I suggested as 'the right thing to do'. Although regulators will regard this as potentially dangerous, it will only rarely be so after a trusting doctor-patient relationship has formed through mutual experience, over a long period.

For any patient to be given a long list of every known complication, prior to a medical intervention, is fatuous and only a de-personalised tick-box ritual. What patients need from an expert, or the master of any discipline, is her experience-based perspective; one that makes enough sense for them to decide with confidence.

Sometimes I have taken the initiative, and said to a patient: *'If I were you . . .'* That always carries a lot of weight, but only when mutual trust exists (and when the doctor knows what he is talking about). It remains important, however, to record all that is said, including the patient's responses to any discussion. Although a corporate requirement, it serves the doctor-patient relationship well, because it is not unusual for busy doctors to forget what advice they have given, and why. I suspect that sometime soon, all doctors and nurses will have to video every conversation with patients, as a legal requirement. Those corporations who insure us to practise might soon insist on it.

If you find a particular antecedent risk difficult to assess, most patients will find it impossible. Specialist doctors deal with risk assessment many times every day; most patients, regulators and bureaucrats, only occasionally engage with it.

Questions every patient should ask are, *'Is my doctor competent?'*, and, *'Does she have a track record I can rely on?'* Without specifying odds, patients need to know that they have the best chance of a successful outcome. If the best surgeon is in Delhi, I would discuss the possibility of them travelling there. In the UK, we have long had insular attitudes about what goes on abroad. Instead of believing that the NHS is the best provider of medical services in the world (it could be, if we're doctor-led, and better financed), we should stop believing UK government propaganda. Many of the best surgeons and physicians in the world do not work for NHS, working as they do in UK private practice or elsewhere in the world. The bunker entrenched bias of a 'second to none' status for the NHS, biases us against making objective international comparisons.

It was recently announced that sweeping changes in NHS maternity services are to be made, after serious failures in the maternity services provided by Shrewsbury and Telford NHS Trust, led to the deaths of over 200 babies. Many were stillborn, or left with brain damage (Source: bbc.co.uk, 30/03/2022). Many mothers questioned the arrogant attitude of midwives and their lack of commitment. Midwives treated mothers as if only they knew best. They expected to be trusted in the face of ignorance and clinical inefficiency. Donna Ockenden, a senior midwife, composed a report. She found that external regulators and inspectors either failed to undertake investigations, or undertook inadequate investigations.

Lawyers and those with MBAs who populate the Ministry of Health, CQC, GMC and PSA, will only rarely have any clinical acumen. Very few are midwives and doctors, yet we persist in giving them authority to inspect clinical practices and to judge medical staff.

Because of the shortcomings of fully informed consent, never say to a patient: *'I cannot tell you what to do, you must make up your own mind.'* Doctors have an invaluable perspective to offer, although not one that any patient has a duty to accept. The best support for decisions made by patients will come from a medical professional who shares their value judgements (risk evaluation, orientation and knowledge, and aspirations for the future). Clarity is essential, but with the many levels of intelligence and ability to understand existing among patients, it will not always be easy to achieve.

Decisions will need to be taken on behalf of unconscious patients (anaesthetised or otherwise), or when they are semi-conscious or haemorrhaging from a gunshot wound or trauma. In many emergency situations, the decision 'to operate, or let them die,' is hardly fraught with bureaucratic dilemma, and any need for 'fully informed consent' is completely inapplicable (at least, in the UK and in Europe, where pre-payment is not a pre-requisite).

Since few patients (or doctors) appreciate the significance of the statistical paradox, at least they should know that statistics can only provide a signpost to the truth, not its exact location. Statistics can provide ball-park figures for general use, never reliable figures for individual use. Although vital to understanding the significance of research findings, and the management policies based on them, such considerations could confuse the majority.

We must choose which statistic applies best when informing a patient about an intervention. Make sure it agrees with your clinical assessment, experience, and judgement. Although one must use published data as a basis, achieving accord is an art. It requires that one has surveyed the research available, and has looked at the selection criteria, assumptions, and methods of analysis. It is an art to decide whether the results of any research publication apply to the patient sitting before us, and whether one can give advice based on its appraisal.

In order to be sure of wise judgement being applied, some patients will ask for other opinions. Some will travel far away to get them (if they have the resources, the energy, and an enquiring mind).

Delivering News

'In wartime, truth is so precious that she should always be attended by a bodyguard of lies.'

Winston Churchill, during WW2

'The truth, no matter how bad, is never as dangerous as a lie in the long run.'

Ben Bradley. Editor. *The Washington Post*.

Ben Bradley was clearly not referring to patients. Some patients, do not have a 'long run' to look forward to, so his advice is not always applicable. The key is to know what medical truth implies. In some clinical circumstances, telling the truth to a patient could prove unnecessary, unkind or damaging. When giving news to patients about their condition and progress, forethought is essential; especially when delivering an opinion rather than facts, and especially when the facts could remove hope. Doctors are often proud to uncover a diagnostic truth, but not all patients will be ready to receive it. Because some patients inhabit a world of hope and fantasy, stark reality (facts supported by independently verified evidence) may not be their familiar territory. They could well reject the facts as doctors see them. Trying to help them understand, may have limited value with every patient free to choose their own version of reality.

Truth needs to be known for what it is — a double-edged sword — empowering some, destabilising others. With no hope on offer, discontent, anxiety, depression, and even suicide can follow. Doctors delivering the truth, can satisfy their regulatory duty, and walk away unhindered. Truth will empower those who need to know, and those with enough fortitude and energy to deal with the consequences. Clinical truth can harm those lacking the psychological strength to handle it, and those with no need to know. It is an art to know who to tell, what to tell, and when to tell it. It requires knowing the patient as one would a best friend. It is also an art to know when one should ignore bureaucratic regulation in a patient's interest.

At the time I qualified, NHS corporate values held little place for junior doctors. Trust in doctors was usual, and doctor-patient relationships fulfilling. Most doctors were brave enough, and trusted enough, to manage the clinical truth for their patients. If we thought it in their best interest to withhold some facts, or to euphemise in order to protect their mental health, we would (at the same time sharing the uncut version of the truth with a responsible relative). Not now. Doctors have little option but to reveal all the facts, with an approach that is dutiful, formal, anonymous and objective (thought of as 'professional' behaviour). This is sometimes harsh, unfeeling, nasty and brutish. Knowing the patient's most likely response to prognostic information is crucial, but that requires the judgement of an attached physician using the art of medicine. Doctors can be overzealous with the clinical truth to avoid getting involved personally, to avoid disciplinary action, and sometimes for personal, religious and ethical reasons. By so doing, they avoid involvement; by so doing, they show (to someone of my age) a lack the moral fibre — a *sine qua non* for being given the right to take responsibility for the life and death of patients.

An unquestioned willingness to take personal responsibility for patients usually identifies physicians and surgeons who are effective and trustworthy; only they will be wise, humane, and sagacious enough to take a patient's welfare into their hands. Rule followers, and anonymous bean counters, are mostly too anxious about bureaucratic retribution to consider it, and may be unfit for it. I would question if such doctors are fit for the sovereign duty that comes with life-shaping medical capability, responsible enough to take the lives of patients into their hands.

I am sad to observe that many doctors in the UK now function as dedicated corporate operatives. In this role, they are safe to jettison any personal responsibility for patients. Their principal concern must now be compliance with rules, regulations and regulators. They must resign their individuality in favour of hiding behind the medically untutored edicts of medical bureaucrats. This is more secure than using the courage needed to take personal responsibility for patients. Their corporate allegiance will afford them some protection from institutional vindictiveness and legal action.

Patients react in different ways when the truth of a poor prognosis is delivered to them. Some want to fight, some get depressed, some are stunned and others lose hope. Those who lose hope die quickest. Those determined to fight their illness, and put all their effort into 'beating it', will do better; they may need to be directed away from unsubstantiated alternative treatments which offer unrealistic hope. Some will be desperate enough to try anything – especially if a doctor has uttered the words – 'there is nothing more we can do for you'. Although some patients accept this, the occasional relative or friend may suggest alternatives. Sometimes to 'try everything' is in order; after all, a few will beat the odds. Sometimes, alternative action (displacement activity) serves to avoid the feeling of helplessness.

One of my patients had a Concorde standing by at Heathrow to transport his wife (with metastatic cervical cancer) to the US. On the intended day of travel, it was Thanksgiving Day in the US. Her flight had to be delayed. She died that night. He walked away, knowing that he had tried everything in his power to help her.

No battle plan ever survives first contact with the enemy.

Field Marshal Helmuth B. Graf von Moltke (1880). Prussian Army Commander.

In planning for battle, I have always found that plans are useless, but planning is indispensable.

General Dwight D. Eisenhower

Doctors as Therapeutic Agents

Michael Balint suggested doctors should use themselves like drugs, influencing how patients respond to illness and treatment (*'The Doctor, His Patient and the Illness.'* (1957). Tavistock Publications).

Faith, trust, and confidence all have a therapeutic role, whether delivered by a priest, teacher, pharmacist, osteopath, hairdresser, healer, nurse or doctor. Their knowledge, experience, style, character and *modus operandi* can influence their effectiveness. An important factor is whether the 'practitioner' is 'attached' and helpful, or 'detached', anonymous and unwilling to get involved. Many patients, I suspect, have found many 'practitioners' other than doctors, more willing to help them personally. If this trend continues, it will affect the future of medical practice in the UK.

A doctor in 'attached mode' will typically consider the patient's problems from a personal viewpoint, which will include their social and other circumstances. He will use his intellect, and his personal knowledge and experience, to advise his patients with as much involvement as he would with his friends and family.

Whatever advice one might give, resist any upset when patients ignore it. Patients may think your advice sensible, but not advice they can follow. It doesn't always mean they have lost confidence. Patients will sometimes later return and tell you how they regretted not heeding your advice. This will call upon your understanding, empathy and magnanimity.

A doctor in 'detached mode' will want to remain anonymous. Only the patient's factual medical condition will concern him. He will be comfortable only with the evidence — tissue pathology, medical images, and laboratory results. With skills limited to the practical and technical, such doctors are happy to pursue clearly defined, easy to follow, ringed-fenced roles for themselves. This is their choice. It may not be their patient's choice.

There are reasons for a doctor being accepted as a therapeutic agent. In a multi-cultural society, shared cultural identity is one. Nationality and language are important, but so are social background, family values and ethos. I often dealt with rich, middle-class patients, and easily identified with them while discussing business, legal, and property issues (sources of stress to them). I was born into a business family, but raised amid those of limited means, and little prospect of working professionally. As a teenager, I experienced working class culture. As a result, I always found it easy to identify with those from this background; many of them became my patients. Nobody truly escapes their class (socio-economic group), or culture, and the mutual recognition of shared origins and values will help patients to confide in you and trust you as a therapeutic agent.

As patients get older, there is less need for them to upkeep an image aimed at influencing others. Some will even reacquaint themselves with their childhood culture and allow their original accent, dialect and behaviour to re-emerge.

Because they mostly tread a similar path through all stages of their education, many doctors will identify with one another, and relate well to other educated, middle-class patients. Gaining the trust of the well-educated can be a pleasing outcome. Displaying enough knowledge, understanding, and objectivity is de rigueur, as is an ability to understand the nuances of their particular plight. This will gain you their intellectual respect, but not necessarily their trust. That may require a little more exposure to you and your expertise. Initially, they will want to cross-check your opinion, and seek second opinions, if only on the Internet.

A good education teaches the need for objectivity and verification, rather than blind belief. As Nietzsche implied by the phrase *'Gott ist Tot'* ('God is Dead'), the strain of disbelief (brought about by the Enlightenment) is a major curse of education. Once so cursed, it can be irreversible. In my experience, educated people only rarely respect the consolation brought to others by trust and faith (I have called it the Dawkin's bias).

'Religion is about turning untested belief into unshakable truth through the power of institutions and the passage of time.'

Richard Dawkins.

Even despising faith as they do, some will continue to trust banks, multi-national companies, the value of money, the rule of law, and intangible concepts like 'in the public interest', despite their existential nature. Don't they realise such entities work only through a consensus of faith, trust, and dreaming? (Y.N.Harari; 2011: *'Sapiens'*. Penguin Random House).

'Blessed are the meek, for they shall inherit the earth.'

Matthew 5: 3-10

The religious are, by definition, capable of trust and faith. It is interesting to note that medicine and religion were once bed-fellows. In ancient Egypt, priests alone were privy to 'medical' knowledge. Could it be that conferring trust is easier for the religious than for the irreligious? Those who see life as a blessing, may regard doctors as specially chosen; able to help others through gifts of intelligence, knowledge, wisdom, and a desire to care for others. Those who worship objectivity (without which they believe nothing), are unlikely to hold such beliefs, and will have to work to keep vital parts of their humanity suppressed.

The gift of hope is there to take for those with trust and faith.

Hope and Fear

Fear is a major psychological factor, affecting the course of every disease, but unrecognised as a medical condition worthy of note (DSM-5 lists fear, only as part of PTSD). From the long-term observation of patients, fear appears as powerful as anxiety and depression in negatively influencing medical progress.

I diagnosed bony secondary deposits from prostate cancer in a man who had been my patient for 40-years. He had no previous symptoms. We had a trusting relationship, and because he was fearful, he sought hope from me; hope that his journey (and that of his relatives) would lead to an acceptable outcome. He could deal with the facts, as unpleasant as they were, but my duty to him went beyond their delivery and explanation. I had to help diffuse his fear, and thus improve his prognosis. My first duty as his physician (not his oncologist) was to allay his fear and anxiety. For that, I needed to moderate what I told him.

As an ex-RAF man, he knew all about duty. After knowing one another for over thirty years, he knew without question, that he could rely on me, and trust what I said (just as fighting men trust their commanding officers). The composure he achieved would come not only from pharmaceuticals, proven to have a beneficial effect on his cancer, but also from me as a therapeutic agent. My role was to foster hope and positivity for him and his family.

Oncologists told him he had only a few years to live. They might have seen this as accomplishing their professional duty, with detached, evidence-based facts easily conveyed. After informing patients of their inevitable fate, those who wish to remain anonymous should leave the daily management of the patient to those skilled in the art of medicine. Some patients refer to these as 'proper doctors'.

'You know doctor, we have an expression in Greek': 'Κανείς δεν πέθανε ποτέ από ελπίδα', or 'nobody ever died of hope.'

A Cypriot boat captain told me this after I suggested a possible alternative cause for his breathlessness. Apart from his obvious COPD (the result of a lifelong smoking habit), I wondered if he might also have coronary heart disease as a cause; a cause that might be treatable.

In some cultures, hope has no such value:

> *Should one search the forest hopeful,*
> *When all nature is the Aim?*
> *For to hope is but an ailment,*
> *So is station, wealth and fame.*

'Youth'. Jibran Khalil Jibran.
Translated by George Kheirallah.
Arabic Poems. Everyman's Library. Pocket Poets.

Throughout their medical careers, every doctor and nurse will need to manage dying patients. It is only then that undue hope can be an ailment.

To hope is human, yet Jibran sees hope as an 'ailment'. An ailment it may be, but together with love, faith, and trust, hope is an intrinsic to human nature. Detached doctors can practice ignoring all of them, and like Jibran (a poet, not a doctor), can dismiss hope as an illness.

People define a 'good death' differently. Most want to die at home with their dignity preserved and their closest family and friends in attendance. The challenge is to avoid intervening too much as life approaches its end.

The act of discussing a problem with a sympathetic listener – even a doctor, can bring relief to many. Coping with bad news can be made easier to deal with thereafter. A genuine, reassuring hand to hold, can work wonders to allay anxiety; it can lower blood pressure and slow the pulse. The expression: *'you are in safe hands'*, sincerely meant, can provide comfort and the return of composure. Patients who delegate some responsibility for their care to a trusted therapeutic agent can gain this benefit. Confidence can reduce perceived levels of stress, and so improve immunity and coagulation risk (proneness to DVT, etc.).

If you understand your fellow human beings (a prerequisite for practising the art of medicine), use your knowledge to direct therapeutic interventions. Like all art, it will attract controversy. Because it is an art, there is no yardstick of excellence, other than patient satisfaction. Some medical regulators are conceited enough to attempt its measure. Are they also able to judge the relative quantitative merits of Monet and Renoir? Once mastered, the art of medicine can foster the patient-doctor relationship in many valuable, immeasurable ways.

Many more people other than doctors can confer the gift of hope. Friends, nurses, physiotherapists, osteopaths, hairdressers, priests; they can all qualify as therapeutic agents. These agents of hope have grown in importance, now patients spend more time with them, than with doctors. Their importance will grew further as patients sense a medical profession more interested in medical science than humanity, and a growing number of doctors striving to become anonymous.

Junior doctors must quickly learn what responsibility they hold as 'therapeutic agents'. Many patients will invest their hopes in them. Some will believe all they say and trust them with their lives. A doctor's (and nurse's) shoulders need to be strong enough to take the strain, with knowledge, compassion and ability enough to bear the responsibility imposed by a patient's faith and trust. Knowing this might frighten some doctors, but encourage others.

Once gained, a doctor's responsibility is to maintain his patient's trust. It will take sensitively managed communication to keep it. Some patients demand complete openness and transparency, however tough the consequences of the clinical information they receive. Whatever the patient's outlook, we need to be guided by kindness and aim to relieve fear and anxiety. To be cruel (there is no point in hiding the facts, no matter how serious), or dispassionate (no logical reason to identify with any patient), is not acceptable.

The Effect of Fear

Fear can underlie aggressive behaviour.

There are people, so fearful of disease, they would sooner bury their heads in the sand and suffocate, than know an unpleasant truth. No matter how good one gets at communication, there will always be those fearful enough to surrender themselves to the natural history of their condition. It can be tough and frustrating, but you will get used to it. A useful strategy is to suggest a second opinion. Changing the messenger can sometimes prove effective.

It is essential to become sensitive to lost causes. Patients wasting their time is not your problem; patients wasting your time can be.

If you are not sure of the most likely diagnosis, especially when it could be life-threatening, never give incomplete or difficult to interpret technical detail to the patient. Many stew over indefinite results and get destabilised by anxiety and fear.

The KGB (now FSB) successfully used psychological destabilisation to uncover the truth. They would inform a prisoner that he was going to be executed, but on a date yet to be decided. Perhaps just after the new bullets arrived. Perhaps the week after next.

Doing no harm as Hippocrates suggested, includes not being cruel. Information can be mightier than a sword, but in everyday terms, give priority to avoiding fists and swords.

Everyone has a plan 'till they get punched in the mouth.

Mike Tyson

Horses for Courses and Trust

For attached doctors, different patients can encourage or discourage the doctor-patient relationship. For detached doctors, this is not a consideration.

In the words of Mae West, *'You can't please all the people, all the time'.* So don't get disheartened when a patient says, *'I insist on seeing Dr. X.'* (implying that she is the only one he trusts.). Private patients are free to seek the style and personality of the doctor who suits them best, regardless of their medical knowledge (which patients cannot easily assess). Few patients have much medical knowledge, so choosing a doctor with the most appropriate knowledge and expertise will be all but impossible for them. Most NHS patients have no choice anyway, even though their tax contributions finance the service.

During the development of doctor-patient relationships, doctors should emulate James Bond's approach to women: *'woo, but never chase.'* Patient-doctor compatibility can act as a powerful therapeutic factor, but like true love, neither plea can gain it, nor is it had by right.

The Holy Grail of medical practice for some is to gain a patient's trust and faith. For others it is only to attain clinical expertise, to make accurate diagnoses and to treat patients accordingly. Faith and trust are not usual considerations in emergency situations; most patients will have little choice but to accept whatever action becomes necessary to save their life. In the management of chronic cases, mutual trust is invaluable. For some doctors, however, patient trust is sometimes a burden.

While observing the faith patients have in doctors, some bureaucrats will be covetous, although many more will become alarmed. Having learned what dangers trust in doctors and nurses can bring (Dr. Harold Shipman, and nurse Lucy Letby), they have condoned spying on us, and will forever suspect our motives, given their duty to protect the public. With Shipman, Letby, and in many other cases, bureaucrats failed completely. Because they have duties but no personal responsibilities, they have carried on as usual, allowing their employing corporation to take the blame.

The development of trust is testimony to the power vested in medical relationships. This is not the common experience of many people outside of the medical profession. For both politicians and their bureaucratic servants, the faith and trust some patients have in doctors must cause envy. Their primary mission, however, is to control the medical profession by binding doctors and nurses to strict rules of engagement, rather than fostering a kaleidoscope of individual doctor-patient relationships. Intrinsic to all their communications with patients lies a tacit warning: beware of misplaced faith and trust in doctors. This is the milieu in which many UK doctors now work. Unfortunately, the only compliment politicians or medical bureaucrats can expect from their work will come from their diligence in creating and enforcing rules. As a substitute for trust, many receive awards and knighthoods. The trust we get from saving lives, is by comparison, not as important.

Contrary to the prevailing corporate bureaucratic belief, most doctors are not programmable production line workers. Contrary to bureaucratic belief, our primary role is not to help them achieve corporate goals. Patients expect something more – that we are free to help them, using our knowledge as power over life, death, and suffering. For this, our patients naturally grant us a status and privilege beyond that of bureaucracy. Doctors and nurses must learn to live with this privilege, boosted every day by the gratitude freely expressed by patients. Although it will create cognitive dissonance for them, regulators should respect the faith and trust patients have in doctors and nurses; it is forgiving, often unshakable, and only rarely misplaced. Bureaucrats will remain conflicted until we remove any unjustified authority they have.

Relatives may say, after the death of a relative, *'Our doctor was wonderful. He did all he could. We are so lucky to have had him/her by our side'.* The strange thing is that you may have done nothing to deserve the compliment (the 'Halo' bias?), apart from being there and exercising a duty of care. Doctors and nurses must learn to accept gratitude. For the modest, accepting compliments will not be easy. My advice is – try harder. Many patients want to extend compliments to doctors and nurses, so be gracious enough to accept them.

There are many reasons patients and doctors don't get on. They are all rooted in personality, attitude, and general demeanour, as well as differences in culture, gender, race, social class, level of education, experience, and language. A doctor skilled in the art of medicine, with an appreciative, respectful outlook, will make light work of any differences.

Patients do not always need to get on with a doctor, especially if the doctor's role is mechanical, like replacing a hip or implanting a pacemaker. Confidence is important, however, because it promotes composure, and allows for a less turbulent experience during unpleasant procedures. When patients are confident and relaxed, one will see lower anaesthetic doses and shorter recovery times.

It is difficult sometimes to gain the confidence of children. Once you hurt them (their perception of 'hurt' could differ from adults), they will become wary, and not readily trust you again. When seeing young children, I always tried hard to avoid unpleasant examinations, until I knew them better. After that, they would peacefully allow me to examine their throat and ears, etc. I sometimes left venesection to others, so as not to lose their trust. Some of my junior patients insisted on visiting me when passing by. They just wanted to come and say 'Hello' (Hi, Matt). In contrast, some children seem to hate the idea of seeing any doctor; they will scream, even at the thought of an innocuous chest auscultation.

In some cultures, gaining the trust of the entire family is necessary. I was once told by a Saudi family that they needed to be assured of my worthiness, as an intermediary between '(□□□□Allah') and the patient. I remember a case when all eight members of the patient's family attended the initial consultation. The patient himself said nothing while the family cross-questioned me. They later gave me the freedom to deal with him, as I saw fit. I guessed they must have 'instructed' the patient to trust me. I wondered whether second-hand trust is as good as first-hand trust.

Which elements of a doctor's performance, demeanour, and style have the most therapeutic effect? My experiments in cyber-psychology, developing patient computer questionnaires, gave me some insights. My impression, after almost one thousand patients had completed my questionnaires on a computer terminal, was that the specificity of questions was important. The appropriateness and personal relevance of some questions impressed many. Patients thought the thoroughness of the computerised Middlesex Hospital Questionnaire (Crisp, Crown Index, measuring anxiety, depression, hypochondria, hysteria, obsession, and phobia) important. Some reported a beneficial effect. They said that they had never divulged so much about themselves before, with some feeling relief. Others (the more neurotic ones) said: *'I have always felt that someone should have asked me these questions. Nobody had until today.'* Clearly, research that could pick apart these impressions could be of value.

It was the depth, breadth and specificity of the questions that patients recognised as important measures of clinical thoroughness and expertise. Those doctors who employ such tools could come to understand their patients better. They may get credit for some expertise, even though the need to use a questionnaire implies the opposite.

Some Therapeutic Management Tips

Because I know it best, I will limit myself to cardiovascular medicine in order to illustrate some management principles. The examples given, detail the thought processes and the weighting considerations made when planning the clinical management of atrial fibrillation, ischaemic heart disease, hypertension and heart failure. Although the factors described are evidence-based, choosing which to apply to any individual patient is an art.

Good therapeutic management must consider all the patient's idiosyncrasies, like their individual tolerance to medications. I have a rule of thumb for prescribing: in different individuals, assume a one to five-fold responsiveness to the same pharmaceutical. Whereas one milligram of a medication might suffice for one patient, to get the same effect in another might need five milligrams. Individual tolerance is a factor not to be ignored. As part of your therapeutic history, always ask about intolerance to alcohol, analgesics and known sedatives. Sensitivity to drugs in the same drug class can occur.

Despite the warnings throughout the BNF about the young and older adults being sensitive to pharmaceuticals, responsiveness relates as much to individual (genetic) factors as it does to age. Whatever the patient's age, always start with the smallest therapeutic dose.

If you ever prescribe a novel pharmaceutical, observe its effects closely over the first 24-48 hours. Sensible patients will stop their medication if they think it is causing side-effects. This rarely happened during my early career; patients were then more inclined to obey instructions to the letter. It may not be correct for a patient to stop a drug without instruction, but it can be a sensible, failsafe action for some.

Prescribing often involves titration, keeping the patient asymptomatic while adjusting their drug doses.

In the early 1980s, Mrs. C. was an 80-year-old ex-nurse. She was in stable AF and religiously took digoxin 250mcgs daily for heart rate control. She had post-rheumatic heart disease and was vomiting every day. She was feeble and cachexic when first I saw her. She had made an understandable, but false assumption. She thought her vomiting and weight loss had a neoplastic cause. She was reluctant to be reviewed, but her family strong-armed her into seeing me. I simply stopped her digoxin and after a few days, restarted her on a paediatric dose. Her vomiting ceased, and her weight gradually returned. Sometime later, she underwent a successful mitral valve replacement.

Since the side-effects of pharmaceuticals account for 60% of all symptoms, all doctors must expect to see many adverse drug reactions and inter-reactions. Knowledge of drug metabolism and excretion is essential. For example, digoxin is water soluble and excreted mainly through the kidneys. It builds up and quickly becomes toxic as renal function deteriorates. Digitoxin is fat soluble and excreted mainly through the liver. Although these drugs are now unfashionable, and side-effects are common, I would switch digoxin to digitoxin if a patient had diminishing renal function. You can extend this principle to beta-blockers and many other drugs by knowing their fat and water solubility.

Therapeutic Homeostasis

Diuretics reduce body sodium, water and other electrolytes (especially potassium), possibly disturbing the *milieu intérieur* (Claude Bernard 1813-78). Although universally used in heart and renal failure, the use of diuretics is not straightforward. There is danger in treating the swollen legs of chair-bound patients who are not in right heart failure. Diuretics can reduce blood volume, while the patient's feet remain oedematous. The progressive lowering of blood volume will eventually lead to diminished cardiac output.

To discern whether a patient's oedema is dependent (gravitational), or caused by heart failure, one must check their JVP at rest, and then again after a short walk. If still unsure, get an echocardiogram done to assess cardiac function before prescribing diuretics, or progressively increasing their dose. Never forget to assess LV output: in the absence of Raynaud's, and peripheral vascular disease, diminished perfusion will make a patient's extremities cold (feet, hands, and nose). If you prescribe a diuretic, review the patient within one week for signs of dehydration, hypovolaemia, and cold peripheries. Remember that the ventricles need a certain head of pressure to function normally.

Please refer to the chapter on clinical examination for more about dependent oedema and diuretic use.

Some patients in heart failure need diuretics all the time. You will need to establish their ideal body weight; the body weight associated with a normal JVP. Try to keep this weight constant (by daily weighing), and adjust the diuretic dose to keep it so. Again, the key principle is one of effective titration. Never start a patient on frusemide 80mgs twice daily, never to recheck the JVP. Some patients will be better off with longer-acting diuretics like bendrofluazide, with occasional frusemide added to make body-weight corrections. In tropical climates, everyone needs more water and salt occasionally (sodium is needed to retain fluid), especially if the patient sweats a lot and is in heart failure. Consider stopping diuretics if the patient's body weight drops too fast, and increase them when both body weight and JVP increase fast.

Intelligent patients can manage their own diuretic regime with some simple instructions related to their oedema, peripheral perfusion, body weight, and urine volume. Although not for everyone, I have known many manage this well. Patients should always question their diuretic dose if their hands and nose become cold, and the volume and colour of their urine changes. By weighing themselves every morning after all toilet procedures, they can monitor the use of their diuretics.

Focus on body weight and JVP changes when managing every diuretic regime. Do not let simplistic binary rules guide you, like 'salt is bad for hypertension', or 'adding dietary salt is dangerous in heart failure'. Learn to think and act beyond the simplistic. Regard every patient as an individual and titrate them according to their personal responses. Personalised care is the only successful form of care, and never more so than in heart failure.

I have omitted any mention of BNP (B-type natriuretic peptide), and N-type-proBNP in heart failure, since I never found the need to measure them. Atrial distension and high ventricular end-diastolic pressures release them. Use them to support or refute a questionable diagnosis of heart failure if you must, but avoid using them to make a definitive diagnosis of heart failure. For those who cannot measure the JVP, or are unsure about how to observe it, these biochemical measures will provide a substitute, and provide regulators with evidence of how up-to-date you are, and how keen you are to provide corroborative evidence. It will not allow them to measure your clinical acumen.

Since many find it difficult, here is a quick tutorial on how best to read a JVP:

Reduce bright ambient light. A darkened room is best, but not essential. The JVP is best seen using an oblique light from a single source; a small torch in a darkened room is best. Use this to highlight the contour of the skin, rather than drown the area in bright light. First, learn how to detect the JVP.

With the patient sitting, elevate their chin and get them to turn their head away. Observe the top end of any rapid inward movement of the neck veins. Any fast outward movement is likely to have an arterial origin. The purpose is to measure the height of the jugular venous pressure above the atrium (which lies behind the upper sternum). For dehydrated patients, lessen the patient's sitting angle by tilting them towards lying flat. If you think the JVP is elevated, examine the patient sitting almost erect. The positioning of the patient has no purpose other than to make the top of the JVP pulsation visible. Use whatever sitting angle allows you to see the top of the waveform. A JVP of around 5cms means a normal right atrial pressure.

The JVP has a double waveform, with an 'a' wave and 'v' wave that will drop below zero during inspiration, and be maximal on expiration. The inward movement to look for follows every 'v' wave. Advanced observers can diagnose AF ('flutter' and 'fibrillation' waves replace the 'a' and 'v' waves). A high RV end-diastolic pressure in pulmonary hypertension will raise the 'a' wave (atrial systole). In tricuspid incompetence, a large regurgitant 'v' wave occurs at the same time as RV systole. With a restricted pericardium, the JVP rises with inhalation.

Regularly assess BP, blood potassium, and sodium in all those on diuretics. Although one can measure blood digoxin, its tissue effects cause 'cupping' of ECG, ST segments. Toxicity often causes nausea and anorexia. Low potassium levels usually flatten the T-waves, but you will need consecutive ECGs to spot it; high-peaked T-waves can appear in hyperkalaemia (renal disease, and ACE inhibitors, etc.), but not always. You might try to use the T-waves to titrate a patient's tissue potassium levels, but this can be unreliable. Watch for T-waves slowly rising as you infuse potassium solution, or diminish slowly with long-term diuretic use. Remember that blood potassium may not reflect tissue or total body potassium (reflected by the ECG). I have never measured total body ion levels using whole body MRI spectroscopy, but I am sure it could have uses.

Atrial Fibrillation (AF)

Acquaint yourself with the latest NICE Guidance on the management of AF. What follows are my personal comments, based on my experience of managing it.

Atrial fibrillation (AF) is a nuisance. It is likely that all those who live beyond 100 years of age, will develop it sometime. There is good reason for this. The commonest cause of AF is chronic, idiopathic atrial (endocardial) fibrosis. This is the pathological description of the (sometimes premature) aging process affecting atrial conducting tissue. Bear this in mind when you next consider repeat cardio-versions; in older adult patients, the pathology will remain. Try electrical reversion in older adults, (conversion is something best left to missionaries), only if sinus rhythm is likely to remain afterwards (small atrium on echocardiography), or if there is a strong clinical reason for its use (to end tachycardia causing heart failure). The acute form in young people, secondary to surgical operations, RTAs, lightning strikes, or accidental electrocution, is a different matter: defibrillate them without a second thought. The chronic atrial endocardial fibrosis found in older adults is not otherwise an issue.

The electro-cellular basis for AF is re-entry. Most commonly responsible are re-entry pathways around the pulmonary veins, at their junction with the left atrium. One can ablate them, but the long-term value to patients (at the time of writing) needs assessment. Long-term re-modelling of the ventricles, caused by variable filling in AF is best avoided.

Never forget drug reversion as the alternative to electrical reversion in AF. This is especially important when it is of recent origin in older adult patients. We once used quinidine or procainamide, given for a few days, although they are not now recommended. Although it is prone to more side-effects, many consider amiodarone better (both drugs lengthen Purkinje fibre / myocyte refractory period). I reverted many patients with a short therapeutic course of an antiarrhythmic drug over five days (using flecainide or disopyramide). Don't rush into DC reversion if the patient is asymptomatic, with a heart rate at rest < 80bpm. In such cases, you might try pharmacological intervention first. If a faster heart rate has pushed the patient into LVF with pulmonary oedema, you must consider urgent DC reversion or a bradycardia inducing drug (beta-blocker, verapamil, diltiazem, ivabradine, or digoxin).

If you are not sure that a fast rhythm is AF, try carotid artery massage (high in the neck, for 5-10 secs.) while running a simultaneous ECG rhythm strip. This could reveal atrial 'f' waves between QRS complexes. Remember that to be effective, carotid massaging action often needs to be uncomfortable. Both right and left sides are worth trying.

Several failed attempts at DC reversion are not unusual. If all else fails, consider a catheter ablation (several types exist). As it becomes cheaper, more fashionable, and better justified, it should ascend the guideline scale of preference towards 'essential'. For my patients, the results so far (2020), were not impressive. There is room for improvement. Only three of my patients underwent the procedure, with only one permanently successful. Like everything else in life, it depends a lot on who does it; this is not something reflected in meta-analyses. The tacit assumption that all doctors, and all cardiac departments are equally capable, is untrue.

There is a powerful drive to revert all patients in chronic AF to sinus rhythm. We need to prevent ventricular muscle remodelling, and make heart failure less likely in the long run. The other critical reason is to reduce the risk of embolism.

Heart failure is now a major cause of morbidity. With many populations living longer, preventative measures are worth considering. Many patients remain asymptomatic despite having chronic stable AF, although some will need heart rate control. Beware of beta-blockers that can worsen heart failure by reducing ventricular contractility (NICE guidelines provide up-to-date advice on which beta-blocker to use). The Frank Starling Law predicts that the cardiac output will rise with heart rate, but only up to a point (that point being lower in older adults, and those with heart failure). Beyond that critical point, further increases in heart rate will cause a drop in cardiac output. In heart failure (dilated ventricle, reduced contractility, or asynchrony), the rate associated with maximal cardiac output will be lower than normal, hence the need to slow the heart rate in heart failure with a beta-blockade, digoxin, digitoxin, verapamil, or ivabradine.

By lowering the heart rate to 50 bpm, some of the negative ionotropic (contractile) effect of beta-blockade should be offset by the higher stroke volume with bradycardia. I practised echocardiography for almost 50-years and commonly observed reduced ventricular contractility (sluggishness) caused by beta-blockade. I could guess when a patient was taking a beta-blocker, before reading their treatment list. It was that obvious. I remain cautious about the claims made for some beta-blockers in heart failure, especially when rate control is not an issue. Individual patients need individual consideration, and no more so than in heart failure. If you rely totally on rules-of-thumb, guidelines, and algorithms for beta-blocker use, not every patient will improve.

If only patients were all cast in the same mould, guidelines, generalised rules, and the instructions written in the British National Formulary (BNF) would work predictably for every patient. Nurses and pharmacists using algorithms might then reliably replace doctors.

Research has shown that trained research nurses manage gout better than 'usual medical practitioners'. They spend longer with patients, have fewer time constraints, and give more in-depth explanations. Over two years, 95% of participants in a nurse-led group achieved the suggested target for blood uric acid (< 360 μmols/L). Only 30% of the doctor-led group achieved the same. (Doherty, M., Jenkins, W., et al. Efficacy and cost-effectiveness of nurse-led care involving education and engagement of patients and a treat-to-target urate-lowering strategy versus usual care for gout: a randomised controlled trial. Lancet (2018); 392: 1403-12.

My experience suggested a poor relationship between blood uric acid and exacerbations of gout, so I must question the clinical relevance of Doherty's research. Without doubt, nurses can be more attentive to patients than doctors. Many doctors need to improve their understanding of the art of medicine, and its role in patient care.

Angina

Someone once asked me what treatment I gave to my patients with angina. 'I have no patients with angina.' I replied. 'They have all had angioplasty or a CABG.'

Drugs hardly help the symptom of angina; unless you accept that patients with angina are much improved by being able to walk 150 yards instead of 100 yards with pharmacological treatment. This 50% improvement is of little use to anyone who enjoys walking. Beta-blockers, 'statins' and ACE inhibitors / ARBs, will however, reduce the high risk of cardiac infarction.

Spontaneous coronary artery spasm can cause atypical angina. It can occur after artery stenting. Typical angina has one main cause: a greater than 85% mechanical blockage (stenosis) of a coronary artery restricting coronary flow and myocardial oxygenation. 'Calcified' atherosclerotic plaques commonly form these blockages. Unstable, 'vulnerable', lipid-rich plaques can cause Prinzmetal angina (an atypical form of angina related to the repeated formation and dissolution of platelets on a fissured plaque). The process can also cause acute cardiac infarction.

The best way to deal with a blocked artery is to open it (mechanically), or bypass it. Also of help is risk-reducing, maximal pharmacological treatment. There are research results and meta-analyses which suggest that medical management is comparable. That is so, according to the COURAGE trial, but only for smokers, older adult patients, and those who have had widespread stable coronary artery disease for decades.

There are many patients for whom we can do little, like those selected for the COURAGE trial (Boden et al.; 2010; NEJM). Their patients had chronic cardiac ischaemia, and their symptoms did not improve after angioplasty. Their mortality figures also remained unchanged. They treated patients too late in the course of their disease. Such cases may benefit more from beta-blockade, calcium blockers (especially good for coronary spasm after stenting), nicorandil, and sub-lingual TNT, but only to a limited extent.

We used to refer to such patients as 'cardiac invalids'. Because of my determination to improve their coronary flow if I could, I came to regard the long-term need to prescribe TNT as a failure, or as a temporary intervention at best. The last time I prescribed it was in the 1970s. It was, however, all we could offer to patients when I first qualified. Trinitrin has its uses — on the way to the catheter lab., or operating theatre. It can keep coronary collateral arteries open and reverse coronary spasm. The pounding headache it can cause is unpleasant. Once patients use trinitrin, many will store it for six months before using it. It became obvious that older tablets had lesser side-effects. Doctors are not the only ones capable of therapeutic discovery.

Hypertension

The onset of hypertension is age-related. Because I saw many patients repeatedly over 30–40 years, I witnessed hypertension arising, as if de novo, within one year. It was common in women at the onset of their menopause, but also triggered by stress. I guess there must be a genetic timing mechanism; when triggered, we see the onset of baldness in young men, the menarche and menopause in females, and the onset of hypertension.

The diagnosis and treatment of essential (now called primary) hypertension is not straightforward. The problem is diagnosing it for sure (it might be transient or labile) and knowing when to treat it. Rarer secondary hypertension is another matter. In my lifetime, I only saw two cases of pheochromocytomas, one adreno-cortical adenoma, and a few renal artery stenoses, as causes. Nephrologists and endocrinologists see them more often.

The important question, after excluding secondary hypertension, is whether hypertension is 'real' (constantly raised, with associated arterial medial hypertrophy, and/or LV hypertrophy), or psychologically reactive, 'white coat hypertension'? Even the latter, however, has been shown to risk the later development of cardiovascular disease. There is little doubt that what starts off as white coat hypertension, can become primary hypertension later on. I saw it many times.

If the patient has clinical left ventricular hypertrophy (LVH), also seen on an echocardiogram, on ECG, CXR, or cardiac MRI, they will need anti-hypertensive treatment for life. This should reduce their risk of haemorrhagic stroke (commonly associated with LVH). In the absence of these findings, it is difficult to know what to do for the best. Get non-anxious, stoic patients to record their BP for several weeks. Ask about family history. This is crucial because, if one parent has hypertension, one in four of their children will also be liable. If both parents have hypertension, expect 80% of their offspring to get it. Let's hope genomic exome studies soon come to our rescue.

Resist treating blood pressure at first sight, unless there is reliable evidence of LVH. To apply the blunderbuss strategy of instantly treating every patient with a daytime systolic BP > 130mmsHg., and/or a diastolic BP of > 80mmsHg, is to believe that the results of clinical trials apply to everyone. Ambulatory monitoring now provides the gold standard for BP categorisation. I have never taken much note of pulse pressure myself, but it can reflect arterial stiffness, and correlate with the presence of coronary heart disease in group studies (pulse pressures > 60mms Hg). A constantly raised systolic pressure in those over 60-years of age is also associated with CHD. Masked hypertension occurs (on 24-hour recordings) when the BPs recorded are high only outside of the consulting room.

Resistant hypertension will sometimes require triple drug therapy or more, to treat it. Some of these patients are now being considered for renal denervation. It is worth seeking causes for background stress (and any associated sleep disorders) in all such patients. In a small study Dr. P.N. and I did at Charing Cross Hospital, few patients (5%) were hypertensive when sedated for a day or two. (Those who remained hypertensive were likely to have secondary hypertension). Our results suggested that 95% of our hypertensive patients had a constant stress component to their problem (sympathetic stimulation). Dr. P.N. would interrogate them to find out what those stresses were, since few volunteered any. By insisting that they (and their associates) dealt with their problems more effectively, it sometimes led them needing less medication.

Accountants will have calculated the savings to national expenditure made by lowering the average BP of a population, but I prefer to judge each patient, and his needs, as an individual. To achieve this, the sine qua non is to look for the pathological markers of primary hypertension, like medial hypertrophy, LVH, and renal impairment. Be aware of any associated risk factors (unfitness, poverty, smoking, obesity, diabetes), and how they might exaggerate or mitigate the risk.

Body weight can bias BP measurements, but is not the primary cause of hypertension. Many believe that obesity causes hypertension, when it is merely an associated factor. True hypertension (as a separate genetic factor) will rarely disappear with weight loss alone, and is not restricted to the obese.

Large arms will help to over-estimate the BP; the opposite is true for those with thin arms. Always take the BP from a point on the arm where the circumference is appropriate to the cuff size in use. A large arm will require a large sphygmomanometer cuff, but some cuffs are simply not large enough. Obese patients may need their BP taken from their forearm. Use the cessation of their radial pulse as the initial 'ball park value' for the systolic pressure. One can also use the dampening of oscillations (using an anaeroid sphygmomanometer) to judge the diastolic pressure (an anaesthetist's method). Confirm both with auscultation. Any discrepancy between these measures and automatic machine-taken BPs, will need further evaluation.

When I was a student, we referred to hypertension with no discernible cause as 'essential'. Nobody knew why it was 'essential', except that its aetiology was unknown. We now know that widespread somatic, arteriolar medial hypertrophy (artery wall thickening) is the pathophysiological prerequisite. The left ventricular muscle can become hypertrophied (LVH) simultaneously, but not always. Simultaneously, the cerebral arteries may do the opposite and become thin and more fragile, making them more liable to haemorrhage. There is little room in an already tightly packed skull for widespread cerebral medial hypertrophy.

I have occasionally observed LVH preceding hypertension by one year. Clearly, the hypertrophy of the left ventricle and that of arterioles, may not develop in parallel. This is worthy research to pursue.

Both ARBs (which cause less cough), and ACE inhibitors, help to stop arteriolar medial hypertrophy and LVH, so I regard them as essential antihypertensives (please refer to NICE guidelines, which may differ). The question is, should the first aim of treating BP be to prevent medial hypertrophy and LVH (or to stop further LVH), or to lower the BP to within pre-defined limits? Both are of value, although I suspect preventing medial hypertrophy and LV remodelling, are more important. Not every doctor can undertake the measurement of LVH long-term, since it requires consecutive ECGs (in some places, CXRs), and echocardiography. In the UK, practice nurses and GPs are the ones mostly tasked to reduce blood pressure, but they are not yet in a position to assess LVH and its progress. They are liable to treat those with white coat hypertension unnecessarily, rather than treat only those with primary hypertension.

There are many drugs to consider when treating hypertension for which NICE guidelines are a valuable resource. Menopausal women can sometimes respond well to a thiazide diuretic alone. Never persist with diuretics alone, unless effective within a two-week period. Among the calcium antagonists, I preferred felodipine to amlodipine; it seemed to cause less ankle oedema. Ankle oedema is so common with amlodipine that my first question to anyone presenting with it was, 'are you taking amlodipine?' There are many other drugs to consider. I have only ever needed to consider verapamil, doxazasin, and diltiazem. In non-asthmatic patients, I sometimes used propranolol in those who were overtly anxious. Although not an anti-anxiolytic, it can stop some of the somatic effects of anxiety (white coat hypertension, palpitation and tremor). It can also be useful in alleviating stress-induced symptoms, but any associated sleep disturbance (caused by vivid dreams), will not help those coping with difficult circumstances.

Patients get hypertensive, or more hypertensive, when their stress has led to sleep deprivation. In such cases, apart from sensible sleep precautions like light and noise exclusion, and some calming activity before sleep, night sedation can work well. It can be worth trying OTC drugs like chlorpheniramine (4–8mgs at night), or health food products such as valerian, or (15mgs) griffonia seed extract (contains 5-HT). Next, consider using a non-addictive anti-depressant like dosulepin 25–75mgs. at night. One problem with such drugs is the hang-over next morning.

Benzodiazepines, Hypertension and Medical Politics

Anxiolytics help to control blood pressure in anxious and stressed patients, many of whom have 'white coat hypertension', or resistant hypertension. Unfortunately, their use is now curtailed by concerns about addiction, even though they can be life-saving. The absence of a balanced view in every field of endeavour now typifies those with utopian views and no relevant practical experience.

Although now a controversial subject, it was never so during my career. I used anxiolytics like diazepam and lorazepam, with almost no adverse outcome risk to any patient and considerable benefit to many. I prescribed them for over 40-years, and that is very long time to experience the effects of any drug. Doctors must now refuse to prescribe them long-term, for fear of patient addiction and regulatory disapproval. I clashed with the GMC over this, disagreeing with restricting them to short-term use only, as suggested in the BNF. No adverse evidence of side-effects or lack of efficacy was then available for their long-term use. Although I had all the evidence I needed from my experience of their daily usage, the MPTS thought decades of anecdotal evidence insufficient. For those without clinical experience, the long experience of a doctor is only hearsay; without experience lawyers have only published statistical appraisals to rely on. To deny long experience is unintelligent. This stupidity will continue while lawyers, not highly trained doctors, judge clinical decisions.

Because the GMC is a legal organisation, with no clinical sensibility, one could waste a lot of effort convincing them that there are valuable, nuanced clinical points here. Importantly, medico-legal bureaucrats are now influencing patient care adversely. The medical profession cannot afford to leave inexperienced judgement of clinical matters unchallenged; unless it wishes to remain a flock of sheep without an intelligent shepherd.

Although benzodiazepines and z-drugs (zolpidem and zopiclone) used for night sedation work well, they readily induce dependency. To focus too much on this is to underestimate the considerable benefits of sleep for those constantly stressed, when mild addiction is easily reversed. Benzodiazepines cause no organic side-effects and have been long regarded as among the safest pharmaceuticals agents ever made. I agree. One simple trick is not to prescribe them to those prone to addiction. Bureaucrats with no clinical sensibility, ignore this and instruct lawyers to emphasise the reckless nature of prescribing them, adversely skewing a clinical balanced appraisal.

Poor sleep, due either to sleep apnoea, or stress, increases both morbidity and mortality for many pathological entities (see Colten, H.R., Altevogt, B.M: *Sleep Disorders and Sleep Deprivation* in bibliography). During the time I prescribed them, I never observed liver or renal defects from any benzodiazepine. Mind you, I only prescribed them for four decades to hundreds of patients; maybe that's not long enough! Despite the guidance not to use them long-term (despite no available long-term study evidence to the contrary), many of my patients claimed that, *'without them, they would not have had a life!'*

The BNF does not support the long-term use of benzodiazepines. Because they are addictive, most doctors accept the dogma of unacceptability. Like those who assemble the BNF, their stance must be failsafe, and in the absence of contrary evidence must assume them to be dangerous. The world-wide effort to remove all addictive drugs is political, not medical. It is utopian in inspiration, not scientific (see the psychopharmacological work of Prof. David Nutt). Their use cannot be justified by stating that 'there are no safer drugs to be addicted to'; instead we must use alternatives that are ineffective and have many side-effects. The challenge, of course, is to find something better to promote normal sleep during long-term, stressful periods, without creating addiction (like non-sedative anti-depressives). Many non-pharmacological methods, and safe OTCs, do sometimes feature among them.

The key question is: do the dangers of sleep deprivation outrank the dangers of benzodiazepine addiction? In my experience, they undoubtedly do. It is not only road traffic accidents that are associated with sleep deprivation but cardiac infarction, type 2 diabetes, failure to cope, depression, suicide and CVA. Anti-depressants and other drugs with hangover effects also cause accidents. The inappropriate use of sedative drugs can make driving or the use of machinery, dangerous.

By restricting night sedation, problems more serious than mild addiction can arise. Serious deleterious effects like fatigue and exhaustion can lead to stress-induced insomnia, secondary to business failure, divorce, eviction, false accusation and arrest, injury and bereavement. Associated with fatigue and exhaustion are RTAs, increased liability to clotting, uncontrolled hypertension, proneness to infection, suicide, diabetes, and even cardiac infarction and CVA. Quality of sleep questions are important and one should include them in the history of all those deteriorating for no obvious reason.

There are several ways to negotiate a brick wall, apart from driving through it with a bulldozer. Using a modicum of intelligence, one might decide to walk round it, or climb over it.

The absence of evidence is not evidence of absence, except in the hands of medical bureaucrats, whose job it is to protect the public from dangerous, or potentially dangerous doctors and nurses. As one measure of their integrity, it seems they see no conflict or irony in working for the same organisation that permits the large scale, legal sale of dangerous alcohol and tobacco. Not their department, perhaps? Their legalistic, adversarial outlook, renders them blind to academic medical argument, experienced medical judgement, and clinical discussion, yet doctors and nurses are required to show them respect and abide by their rulings.

Catastrophising is not limited to the anxious and the insecure; the potential for drug abuse has now focussed the medically untrained utopian minds of bureaucrats. They have been making a concerted effort for some time to stop doctors prescribing benzodiazepines, conflating them with all the other drugs of misuse, some of which are used for suicide. Using the same logic, we would have to rid ourselves of paracetamol, trains, bridges, and gas ovens, because they too pose a suicide risk.

In their ignorance, medical bureaucrats would have us remove many of the important drugs we use in what is a tough fight to control anxiety, insomnia, and stress effects. Patients, happy to ignore the minor addictive effect of benzodiazepines (it is easy to stop them when on holiday; a time when insomnia is more readily tolerated), can get them on the internet. Many intelligent patients with experience of drug usage, and no particular liability to addiction (a genetic trait), recognise one important fact: there are no drugs with the same efficacy as benzodiazepines ('benzos') for sleep. No other drugs benefit sleep and relaxation in the same harmless way. There is an important proviso; a critical factor important when advocating their use. We must never prescribe them to the irresponsible, or to those with a predilection to addiction. Like the tendency for obesity and type 2 diabetes, addiction might more readily occur in those with Neanderthal genes (Dr. Tony Capra. Vanderbilt University, Department of Biological and Computer Sciences, Nashville USA).

Within my lifetime, I witnessed 'benzos' replace barbiturates and paraldehyde; these drugs caused infinitely more problems. You need to be as old as me to fully appreciate the difference made by chlordiazepoxide (Librium) and diazepam (Valium) to patients in the 1960s. Benzos so readily helped the stressed and anxious, that once addicted to them, many patients strongly resisted any attempt to withdraw them. They were always that effective, and that beneficial to many people. It would be wrong to think, however that every patient is capable of the self-control needed to avoid dose-escalation. Every one of my patients, educated by me to do so, achieved steady state dosing without a problem.

When trying to withdraw patients, it helped to prescribe liquid diazepam, or lorazepam, making it possible to sub-divide doses more accurately. Withdrawal is the only clinical problem. The irresponsible, those with deficient coping ability and the criminal, are among those who most abused these drugs. For this reason I cannot support their unqualified use.

In order to help avoid tolerance, I would sometimes rotate three separate sedatives to maintain undisturbed sleep (albeit, not natural sleep), until any stress responsible had passed.

Should one go 'off-piste' and prescribe a 'benzo' for long-term insomnia and hypertension now, even if all other approaches have failed? One would risk being sanctioned, unless a psychiatrist or a psychiatrist has given approval.

The GMC, whose legally oriented advisers will not usually be used to challenging clinical work, will blindly refer to the BNF and NICE guidelines, as would any medical student. What is expedient for them to ignore is that every NICE guideline comes with a rider: *'We must interpret NICE guidelines using specific, patient-related, clinical judgement'*. With their current statutory powers, the GMC, CQC and PSA can choose to ignore this and the views of experienced doctors.

Team Practice and Management

Working within a close-knit medical team can be a joy, especially when everyone is competent, co-ordinated and inter-coordinated. If team members are creative, intelligent, knowledgeable, and have effective executive functioning skills and determined to help others, they can deliver efficient patient care.

For the care to be 'the best', requires effortless technical competence, with time-efficient, compassionate care, delivered wisely. It can sometimes include a less frequently found attribute: creativity of thought, and an interest in advancing knowledge.

As a British Heart Foundation research fellow at St. George's Hospital, then at Hyde Park Corner, London, I worked with Alan Harris, a master of invasive cardiology, and Aubrey Leatham, doyen of heart sounds and the developer of clinical pacing. He followed on from Laennec in developing his own stethoscope. Dr. Paul Kligfield, ideas man from NY, USA, would go places (he returned to New York to become a renowned cardiologist). With Alan Gelson as our teacher and insightful physician, Paul and I owed our success in the MRCP examination to his teaching. There at the time was Geoff Davies, electronics engineer, inventor, and developer of the cardiac pacemaker, and Graham Leach, an electronics engineer who developed the echocardiogram from its primitive oscilloscope format, to the first versions that we use to-day. We were the go-to team for cardiac pacing in London.

Our nurses (Maggie on Grosvenor ward, and Sue in the cath. Lab.), and our technicians (especially Anne Edwards) were smart, competent, and without practical limitations. They rose effortlessly to every routine and emergency challenge, with efficiency and quiet grace. I had a most enjoyable, fulfilling time for three years.

My first experience of supreme nursing effectiveness, based on experience, knowledge, and top-class executive functioning, was while working with the nurse in charge of the Whipp's Cross Hospital Casualty in 1967 (I regret not to remember her name). Times were formal, and I never used her first name. I experienced comparable nursing expertise, working with Angela Pollard at the Roding Private Hospital (now London East Spire Hospital), in Redbridge, London. Angela was not just a highly competent nurse, she was a diagnostician, comparable to any specialist. We shared an *alma mater:* The (now Royal) London Hospital.

If we are to continue to manage patients efficiently and achieve patient satisfaction, doctors and nurses must first retrieve their former medical management authority. Our priority to put patients first, and make them the sole focus of our attention and activity, has to return. To do this, we must stop personal engagement with corporate tasks, all of which need to be delegated to non-medical personnel — secretaries, managers, auditors, regulators, and accountants. It is simply absurd to allow those lacking in clinical knowledge and experience, to interfere with anything other than housekeeping. To waste the invaluable time of medical staff on bureaucracy lacks common sense. What fool would devise a corporate system employing executives and staff to do jobs for which they were untrained and inexperienced?

When more patients come to know the truth about what medical practices have to suffer from bureaucrats, things might change. Many patients already know about the surfeit of managers, inspectors, and regulators, interfering with our medical work, claiming to protect the public. It is obvious to most patients that only experienced doctors and nurses have sufficient clinical knowledge to control medical work. Patients are now witnessing the dangerous effects of an invasive, metastasising medical bureaucracy, with parallels in every nationalised industry.

If we are to manage patients better in the future, we need a fundamental shift in direction; a change of culture which separates bureaucracy and housekeeping from medical practice. We also need to stop the erosion of experience-based medicine (the art of medicine), by evidence-based scientific medicine. We must respect them both, and strive to use them in a complimentary way, although many will favour one more than the other.

Team Ethos

Collegiate collaboration is important, but secondary to your own involvement. You and your patient *are* the front line of care. Delegation and sharing is for the corporate minded, and not primarily for the independent-minded, engaged front-line health worker. Others can make valuable contributions to the provision of service once you have assessed the problem and decided your own plans to resolve it. Always call others for help whenever feeling out of your depth (even then, be prepared to get none. Dr. Hadiza Bawa-Garba experienced this in 2015).

In the corporate mind, striving for a risk-free, standardised, predictable utopian world at odds with the unpredictable world of medical practice, teamwork alone is thought effective. In clinical medicine, clinical teams are small, with no primary corporate mission. The team approach can be cumbersome and inefficient, but rewarding for those in training, and those unsure of what to do. For competent doctors, consulting others is only useful in situations where specialised expertise is required. Although not advisable these days, those doctors with a proven track record of success in managing various types of case, might sometimes have to get on with the job and inform their colleagues later. In principle, capable doctors (the only ones we need) should never allow an acutely ill, deteriorating patient to suffer, while waiting for collegiate or managerial approval. Because, medico-legally, it will always be safer to get team approval before taking action, it will be for each of us to decide whether the time involved compromises the patient's best interests.

My attitude was always the same, but applicable to all. My focus was always on my patients, and hardly ever on secondary corporate requirements (ignoring them completely is not now possible). From my perspective, the only colleagues worthy of acknowledgement were the indispensable members of my team. Within the current corporate western medical culture, however, I accept that my outlook might now mark me as a patient-led maverick; a cavalier, second-rate team player. So be it. I did, however, spend forty years in successful private practice without a complaint from a patient.

Clinical leaders should emulate Admiral Lord Nelson at Trafalgar: he abrogated the prescribed code of engagement in battle, and famously succeeded against the odds. If you have the knowledge and capability, you can get jobs done while others are discussing what defensive action to take. Even if your actions have saved lives, others might deprecate them for political reasons, or perhaps attribute your success to luck. This will happen if you ever denounce the presumption that 'only teamwork produces results'.

With experience one will soon learn a golden rule: never presume the competence of others at any level of seniority. Whenever you delegate, you must remain vigilant. You have a clear responsibility: the buck stops with the doctor who holds the patient's trust. One duty follows: fight the patient's corner, and check the competence of all who deal with them. The effect of closing ranks may not be in the patient's best interest.

Ways of Handling Patients

I will repeat what I wrote before about the handling of patients (in the chapter on 'Doctors'), because they include important clinical management features. I have described each style of patient management, as two extremes; as opposite ends of a spectrum, although most doctors practice in the middle ground (the 'Golden Mean'). There are four ranges: from an objective outlook and evidence-based, to observational and anecdotal; from detached to attached; from effective executive functioning to uncoordinated; from those with a talent for practical procedures, to those who are inept.

1. □ *Objective (Evidence -based).*

A strictly evidence-based approach. A rule-based modus operandi
— *versus* —

Anecdotal.

Observer-based information, often with an intuitive, 'common-sense' approach. Includes the art of communication, a sense of timing, a sense of priority, and a facility for risk management for which strict rules may not exist.

2. □ *An Attached Demeanour*

Uses emotional intelligence and inter-personal sensitivity. An ability to sympathise, empathise, and understand a patient's outlook (and that of his relatives / friends, if involved meaningfully), together with an appreciation of their background and context.
— *versus* —

A Detached Demeanour.

No aptitude for, or interest in, patients' emotional issues, and no desire for any personal attachment to them.

3. □ *Executive functioning*

The ability to handle the dynamics of clinical practice efficiently. An ability for data analysis leading to effective plans of action, and their execution within an appropriate time frame, dictated by correctly judged clinical urgency. An ability to shift focus and time frame. A facility for co-ordinated action.

- *versus* —

Executive Functioning Ineptitude.

Little evidence of planning or efficient coordinated activity.

4.☐   *Effective Physical Functioning*

A talent for practical applications: venesection; iv, and arterial punctures; lumbar puncture, surgery, and other interventional procedures.

— versus —

Practical Ineptitude.

Crucial Dynamics

Every physician and surgeon will have their place within each of these management categories. None are mutually inclusive or exclusive. Each combine with professional style and clinical acumen to create various forms of competence.

One crucial dynamic is switching: an ability to switch from an 'attached' mode, to a 'detached' mode of engagement; from emotional engagement to a dispassionate approach when necessary. This dynamic is especially important for those involved in lifesaving and emergency procedures. It is also useful for those engaged with difficult surgical judgements. Clinical judgement is best served by a rational, dispassionate outlook, with an awareness of bias and no loss of humanity while weighing relevant clinical, emotional, and circumstantial factors.

Clinical Responsibility and Management

Little emotional intelligence was in evidence in the clinical management of baby Alfie Evans (2018). The paediatric staff of a Liverpool hospital diagnosed his irreparable brain damage. In the High Court, the staff blocked his parents from taking any further action, despite their duty to try everything in their power to encourage a favourable outcome.

The baby's father visited the Pope and received his blessing. The baby received immediate Italian citizenship, and could have transferred to a hospital in Genova. Since the baby appreciated nothing of this, why were his parents denied decision-making authority?

What harm (or good) could have come from transferring Alfie to Genoa, even if thought a hopeless endeavour? It would have been a straightforward journey. Because medical science at the Liverpool hospital had nothing more to offer, why not let others try? The baby would then have had some chance (albeit a vanishingly small chance). A small chance, would have been one step better than 'no' chance. Perhaps their focus was safeguarding Alfie. Even if it was, why did they not respect the wishes of parents who loved him?

For Alfie's parents, there was an advantage to going to Genoa. The move would have satisfied their consciences, knowing that they had done all they could. Nobody lives in isolation, and what happened looked like contempt for their parental role. If Alfie could have thought and expressed his love for his deserving parents, surely he would have wanted his parents' wishes to be honoured.

Having served the best interests of the child, next in line for consideration should have been those directly related to him, and those he would recognise as loving him most.

Nobody has questioned the quality of the medical care given to the child, but the hospital staff refused to acknowledge any duty to his parents or their duty to their son. Instead, they seemed to detach themselves from the duties of parenthood. Not their brief, perhaps? Locked in an Ivory Tower, that is easy to do.

What motivated Alfie's parents? Was it their son's best interest or was it just their personal sentiment? Was it in his best interest, not to allow him to travel? They had only one course of action left: to take him home to die, and to wonder forevermore, what might have been. The unlikelihood of a beneficial outcome, was not the key issue. To clarify the point, I would ask one question. Had it been the wish of one of our Royal family to send their son to Genoa, would the hospital have blocked the move in a High Court?

Often quoted as examples of inappropriate parental wishes, are those of Jehovah's Witnesses. Jehovah's Witness parents can refuse their dying, anaemic children, life giving blood transfusions. Unlike baby Alfie, this action could never be in the patient's best interest, and a court ruling to ignore the parent's wishes would be clinically appropriate. Courts can temporarily suspend parental rights, and make children wards of court, just to save their life. This will exonerate parents, motivated to comply with their religious beliefs. A child's best life interests should not depend on parental belief if (potentially) malfeasant. This applies especially when a child is too young to decide what religious belief he/she will later adopt.

A political regime that exercises cruelty, while demanding total compliance, is Fascist by definition. One facet of fascism is that it seeks scapegoats. Did Alfie's parents fit this role? (see Madeleine Albright: *Fascism: A Warning*. Harper Collins. 2018).

Medical Management and Trust

Both a doctor's attitude, and the time she allows for consultation, will create a patient's impression of their care. Patient management not only relies on medical expertise; it can depend on the attitudes brought to the doctor-patient relationship as it evolves.

For patients to be compliant (to what is in their best interest medically), doctors and nurses need to be trusted, within a relationship that is mutually rewarding. Trust development, other than that which comes from the respect of a doctor's role, will involve openness and time enough for feelings of safety to evolve; time enough, for mutual understanding. If a doctor's interest is genuine, honourable, and true, even the most frightened patient can come to trust and accept the recommendations we make in their best interest (hopefully based on the best evidence and meaningful experience).

Fear is not an acknowledged diagnosis, but for patients, it can change their compliance (DSM-5 lists fear, only as part of PTSD). The last patient I saw in fear of what some doctors wanted to do to him said, 'It's my body, and I'll do what I want with it.'

The fifty-one-year-old man in question had uncontrolled type 2 diabetes, swollen legs, and a raised JVP. He was fearful, so I adopted a cautious approach. I did not want to rush him into decisions and investigations. I first treated his diabetes, and after gaining his confidence, I investigated his heart. He reluctantly agreed to have an echocardiogram. It showed a dilated, poorly contractile LV. His previous distrust of doctors made him fearful. His doctors failed to discuss much with him, especially the safety of medications.

He learned I was on his side, because I respected his intelligence and fully discussed every issue with him. Once he knew he had reduced LV function, he accepted the rationale of taking an ACE inhibitor. The next hurdle was to convince him to take a statin (he had extensive carotid atherosclerosis), and a NOAC (Novel Oral Anticoagulant, or factor Xa inhibitor anticoagulant), given the thrombotic risk he faced from a dilated, poorly contractile LV.

Telephone Management

Because the medical history often contributes most to a diagnosis, telephone conversations can contribute a lot. Telephone conversations can allow diagnostically accurate guesses to be made, progress assessments to be judged, and therapeutic interventions assessed or suggested (antihistamines in allergic reactions, the use of an inhaler in asthmatics, how to manage a burn, etc.). One can certainly discuss changes in treatment, the necessity for a live consultation or a '999' call.

It is far from safe to discuss every type of symptom on the telephone. When a patient has a persistent cough, chest pain, or breathlessness, a telephone conversation may only be the first step. At least it is possible to discuss likelihoods, and the safest of management plans. Mobile communication, with the ability to send texts and photographs, has transformed some medical practice. Photographs of sore throats and dermatological conditions are possible, as are conditions like bruising, sub-conjunctival haemorrhage and allergic reactions causing facial swelling. A manic patient once sent me a photograph of his swollen neck veins, and bulging veins across his forehead that resulted from a cocaine binge. The diagnosis presented no difficulty, but I needed to see him to assess his cardiovascular status.

I never engaged in detailed medical discussions over the telephone, or commit to a diagnosis (except the most straightforward), unless I knew the patient well. I liked to help if I could, but this form of communication is too prone to error (as I always explained to them). I might give guidance like, 'It sounds to me as if... you need to see a dermatologist, gynaecologist', etc., but I rarely went further. Patients usually welcomed the convenience of such advice, but had to make they realised I couldn't give a fully informed opinion without examining them. Telephone advice and 'FaceTime' consultations, will doubtless become subject to litigation. With current litigation trends increasing as they are, telephone and internet-based conversations will need to be recorded, and every Good Samaritan might need a lawyer by his side.

There are some obvious short-comings to telephone medicine. They lack full non-verbal communication, and the possibility of full examination, except for the photos they choose to send (which may not be theirs). This method of patient-doctor communication will increase, simply because it is convenient. Because the convenient attracts demand, any lack of it, like having to wait in A&E can induce aggressive and inconsiderate behaviour.

There are many advantages to using a mobile telephone. Without being tied to an office, one can send patients their laboratory results (together with an interpretation and comments), the day after taking their blood. 'Normal' results quickly allay anxiety, simply because most patients believe (incorrectly) that they are the be-all and end-all of clinical assessment. It is best to discuss all abnormal results face-to-face.

Risk Management

Every patient should ask: 'How might I benefit? What are the real risks?'

Throughout our lives, we all gamble, knowingly or not. Some gamble with their choice of profession, some with their choice of partner. Others risk their money at the roulette wheel, at the card table or race meeting. One can more reliably predict the outcome of gambling on a roulette wheel than most chances we take in life.

Except for those who rely completely on others, every life will involve taking a series of loosely calculated (guessed) decisions, each with an unspecified risk. Winning or losing can depend on how well we understand the associated odds. Those who play roulette know the odds; they quickly learn how frequently a 37:1 bet actually wins (single number bets). Those who gamble by tossing coins quickly learn how often they will win a 2:1 bet; namely, heads or tails. A realistic sense of risk may elude those who play neither. They will have to speculate with no contextual experience to guide them.

Doctors with no feel for risk when tossing coins, estimating the odds of disease occurrence, or the likelihood of surgical complications, risk losing both their money and their patients. Without a realistic 'feel' for the odds, one cannot discern which risks are worth taking and which are not. How will they understand what it means for a patient to risk having an operation, or the risk of taking a medication? The street-wise are so-called because they have learned from experience, and know the risk of everyday situations. Although both patients and doctors have a need for this valuable sense, those who have spent their life shielded from struggle and risk-taking, will be disadvantaged. This might explain why family wealth rarely lasts more than one or two generations. Being shielded from reality make one vulnerable to risk.

The chances pertaining to roulette and coin tossing, differ from those we encounter in life. This is because both coin tossing and roulette have fixed odds: either 2:1 (heads or tails), or 37:1 (or, 38:1, on wheels with two zeros). In medicine, as in horse racing and greyhound racing, the actual odds of any outcome (antecedent probability) are indeterminate, but with winners and losers whatever the odds.

In medicine the indeterminate factors can be crucial. Factors like the talent and knowledge of each doctor, the circumstances of each patient, the hidden pressures upon them, and other outside influences, can all affect the outcome. Relevant contextual information (metadata), can make indeterminate chance situations more predictable. In horse racing, there are horses that rarely lose when ridden by a champion jockey (Lester Piggott was one), on days when the going suits them, when they are healthy, and when there is no comparable competitor in the race. The coincidence of all these conditions will make a horse more likely to win, but how is the average race-goer to know this? Very often, it is only privileged insiders who know. By extension, the better we know our patients, the better our risk assessments will be.

An operation performed by a surgeon with an outstanding track record, working with a team of proven excellence, in a hospital with the best of facilities and results, will usually present a lower risk for patients. Although such conditions exist in both the NHS and the private sector, only private patients in the UK have the choice of hospital and surgeon. Without a lot of inside information, the choice and the risks remain indeterminate.

Some operations and procedures are inherently safe (cataract operations); some, like heart transplantation and aneurysm surgery, are more risky, especially if performed by less talented surgeons on frail patients. Difficult clinical decisions can rely on many indefinite factors, making the entire business of assigning risk to some medical procedures, more an art than a science. Most clinical decisions are, however, completely straightforward and predictable.

We could all learn from investment companies. If you go to them, expecting them to invest your money for gain, they will first want to know what level of risk you are prepared to take: high risk (quickly double your money with a chance of losing it all), or low risk (a slow and steady rise in value with a guaranteed stop-loss). We should assess every patient's attitude to medical intervention in the same way. A doctor who will not commit to a decision, and a patient unwilling to 'take a chance,' could be wrongly assessing risk.

In medical risk taking, incentive can dictate what level of risk is acceptable. If a patient wishes to live for a few more years, rather than die within a few weeks from his condition, a 10:1 chance of surviving an operation might be worth taking. In ordinary circumstances, a 10:1 chance of survival would be totally unacceptable. A patient's attitude to gambling (in the widest sense) will determine which decisions and actions they find acceptable.

Other Aspects of Risk

Some patients underrate the odds of treatment complications, while others overrate them. Success or failure are two sides of the same coin. Whatever the actual odds of a medical intervention, a patient's decision to say, 'Yes' or 'No', can seem like a 2:1 gamble. Of those who try to win the National Lottery, how many are aware that the odds of winning are 45 million to one? Assuming that someone nearly always wins, the personal odds of winning will seem like a 50:50 chance: either they will win or they won't. This explains why so many buy tickets and try to get rich. Because the actual numerical odds of a successful surgical operation or side-effects are difficult to grasp, I prefer to use categories like *'the usual outcome', 'a rare outcome', or 'virtually unknown outcome'*.

In trying to communicate risk to patients, I often referred to my experience. For instance, I never saw a serious complication of coronary artery stenting, or of cataract removal, in thirty years. Of those I originally referred for *hallux valgus* operations, the result satisfied only a minority; I therefore only referred patients who were suffering unbearably. New techniques have improved the results since. Before contemplating any invasive measure, first consider the necessity of the procedure. This must take precedence over any discussion of risk. If a patient with chronic condition wanted to avoid an invasive procedure, I would support them as much as possible, while retaining the option to re-consider the decision later.

When discussing odds with patients, it would be unfair to contrast the very best of outcomes with the very worst. To tell a patient that he could die having an ingrowing toenail removed may be correct (a *'virtually unknown outcome'*), but it lacks perspicacity. We all know that septicaemia can follow any operation, but the odds of it are too long for most patients to grasp (wise patients might ask for the incidence of post-op infections at the admitting hospital). This perspective is important when considering fully informed consent. Many patients now believe that doctors discuss outcomes in a de-personalised way, if only to exonerate themselves from blame. Disclaimers and de-personalised advice do nothing to promote trust in a doctor-patient relationship.

Risk Assessment and Outcome

Among the negative factors used to assess a patient's individual operative risk are frailty, cardiovascular state, respiratory function, and immunity. A simple frailty index is useful. This allows anonymous doctors to calculate whether the patient is agile enough, and mentally fit enough, to care for themselves. The poorest of prognostic scenarios is easy enough to assess. Avoid operating on the frail; those with poorly controlled diabetes, and those with heart failure, emphysema, or a history of slow healing.

Chronological age has less prognostic value than biological age. Vitality is the key. Healthy older adults with a youthful vitality will usually have a good prognosis; after all, they have already lived a long life and survived decades of its slings and arrows.

The onus on doctors is to reduce every risk possible, before any intervention. The correction of anaemia, a CABG before an aneurysm repair, treatment of any infection, body weight change, and improvement in physical fitness (exercise tolerance) are all wise. We assume the relationship between athletic fitness and risk to be inverse, but this is not always the case. A few fit young people die after procedures that older adults would survive. Toughness and resilience, the opposites of frailty (they are not just the absence of frailty), are key factors. These assessments are more readily made by doctors who know their patients well.

One can deliver personalised clinical medicine best when the doctor patient relationship is based on mutual trust. An intimate knowledge of the patient allows an all-encompassing, dependable risk assessment to be made. Reliable recommendations, based on the art of medicine, can then follow.

Here are a few risk assessments I undertook. I had intimate knowledge of each patient, gained from long-term clinical continuity.

1)▫ I did not advise my uncle (C.D), to have a lobectomy and radiotherapy, after diagnosing his squamous cell bronchial cancer. He was not prepared psychologically to undergo the treatment, which offered very little improvement in life expectancy. He lived for five years when the usual prognosis at the time was six months.

2)▫ I advised my friend and former partner Dr. D.B., not to have a cholecystectomy, but an ERCP. He was told that because he had a coronary artery stent six months before, the risk of ERCP investigation would be too high. I disagreed. His requirement for opiates had increased because of frequent biliary colic. A successful ERCP relieved his biliary colic and jaundice. Despite having become a little frail, his surgeon continued to advise cholecystectomy. What advantage could there be? He had been seriously ill and was now asymptomatic. What gain was there in exposing him to any further surgical risk? The mantra he was being asked to accept was: 'it is our practise to proceed to cholecystectomy in all cases like yours'. This policy held no merit and was irrational. On my advice, he declined surgery, and has been symptom-free ever since. He is now no longer frail, and could opt for surgery, should he again experience biliary colic.

I trained all my patients not to be persuaded, without further consideration, by the rhetorical argument: 'This is what we usually do in cases like yours'.

3)▫ I advised frail Mrs J. K., not to consider a coronary angiogram. She had severe chronic obstructive pulmonary disease (COPD) and conflicted about what advice to take. Clearly, no cardiac procedure, whatever it might be, would improve her effort tolerance significantly. Extending her life, living with severe irreversible COPD, was not actually what she wanted. She was content to resign herself to be a 'respiratory invalid', with untreated IHD.

4)▫ Eighty-five-year-old, Francisco R-C had always been active. He had mitral stenosis and so breathless, he could no longer engage in football with his grandchildren. Fitter than average for his age, and with an ambition to continue exercising, I advised him to have his mitral and aortic valves replaced by a specific surgeon. He achieved his ambition (thanks to Stephen Edmondson, London-based cardiac surgeon).

5)▫ Mrs Pam F. was told by a rheumatologist to stop her naproxen immediately because of an associated cardiac risk. Naproxen was the only analgesic that helped her joint pain. Having found her not to have any cardiac abnormality, and no detectable atheroma in her arteries (at age 75-years), I told her to continue naproxen in the face of (what I took to be) a depersonalised risk assessment, based on unrepresentative statistical evidence. Seven years later, her healthy cardiac status remained unchanged. The rheumatologist in question made a common error. He insisted that general risk, as published in a reputable journal, applies to every individual. Although a professor of rheumatology, he needed to understand the statistical paradox and how to make practicable therapeutic decisions. Could it be that some professors are not what they used to be, or has a fear of regulators overtaken them?

The only important outcomes of medical intervention are a longer life, and a life of improved quality. A longer life is not everyone's wish, especially if it means prolonging their pain, misery or unhappiness. There are many quality of life outcomes worth having: improved mobility, freedom from pain, improved exercise tolerance, improved happiness, and improved well-being among them.

When things go wrong, it is usually because of poor surveillance, poor judgement, inappropriate action, and what we perceive to be bad luck.

Fashion and Clinical Management

In keeping with the human instinct to flock together, many follow fashion sheepishly. Fashions have their day, then quietly disappear. Electroconvulsive (ECT) and sleep therapy, were once in vogue, but are now resigned to obscurity. Doubtless, both will be recycled someday.

Never dispense with old tools if they work. If you live long enough, you are likely to see their return and their benefit rediscovered. Along with the art of inventiveness, reinvention is nothing new; human existence has depended on both.

Fashion Items

On one hand, we are told that our diet and growing obesity problems are claiming lives; on the other, we are told we are all living longer (on average) with less heart disease (since the 1960s). The rich can afford protein and need not exist on the same carbohydrate-rich diet as the poor; they can more easily be fashionably slim. Those who are poor and obese, eat little protein and more carbohydrate. Those who are rich and obese will probably have a better prognosis than those who are poor and obese. This is because mortality from obesity has important co-factors like affluence, lifestyle, diet, and genetic profile. At the moment, it is politically expedient (and fashionable) to blame food rather than poverty for the three to five-fold increase in morbidity and mortality found among the poor when compared to the rich (the 'health divide').

Diet is more easily manipulated than poverty. The solution to poverty is beyond the remit of the medical profession, but all healthcare workers have to deal with its consequences. The only politicians willing to deal with poverty are those resigned to fail or to retire. No politician has yet convinced the rich they should share their wealth with the poor; not unless a sizeable tax saving or political advantage would be forthcoming. At least two philanthropist billionaires, Bill Gates and Warren Buffett, have trumped (not a capital 'T') politicians, and given vast sums to the disadvantaged of the world. Counter-intuitively, the donation of money and facilities to the poor in the US, has been less beneficial than expected. The poor want to spend money, the rich want to invest it. (Herrnstein and Murray, Free Press (1994): *'The Bell Curve'. Intelligence and Class Structure in American Life).*

The Equality Bias

All doctors will need to refer a patient to a specialist colleague. The question then arises: 'who might be best?' The notion that all doctors are equally talented and equally capable of handling patients, is naïve. It would be easy if academic qualifications helped. They may not, since all professionals weigh and use knowledge differently, and respond to patients' needs in different ways. For my discerning patients, who had the advantage of choice, it was often the attitude and management style of other doctors they sometimes found unacceptable. Their skill and competence, they could not assess.

Medical incompetence is a risk faced by all patients. If you are the one managing a patient's referral, the judgement you make of your colleagues' competence could be crucial to the outcome. Although there is an almost unlimited choice of specialists to refer to, I always tried to choose the most appropriate one for each patient, using feedback from previously referred patients.

One of my patients needed a hallux valgus operation. I found it difficult to choose the most appropriate surgeon. Unable to get enough information from CV data and Google searches (qualifications, and status, can both be misleading or meaningless), I addressed the question to a private hospital ward sister with experience of nursing such cases. My question was: 'If your mother needed this operation, who would you choose to operate on her?' She left me in no doubt about whom to choose.

I doubt that any comparable freedom to choose will ever exist in the NHS, where most doctors are in training, and most referrals are 'in house'. The advantage of choosing particular doctors for patients to consult, is not always what it seems. There are many talented junior doctors with skills superior to those of their consultant colleagues, but not yet in a position to accept referrals.

Patients are almost never in a position to judge the technical capabilities of a doctor, although some will try, after an adverse event. At their initial consultation, patients will judge a doctor on how affable, amenable, and knowledgeable they are. Surroundings and ambience can be important. An impressive hospital, an acknowledged reputation, the respect shown by their staff; all count to form their impression. They should include an audit of a doctor's successes and failures, but few will ask for such details. The onus fell on me to assess these factors for them.

Throughout my 53-year career, I witnessed the performance of many admirable colleagues, but as they retired, or died, it became more and more difficult to replace them. I was not always successful when referring patients to others. Patients would sometimes return saying: *'I guess he is good at his job, but he has no personality; he was rude and allowed no time for questions. He was full of his own importance and arrogant. I would never want to see him again.'* These are features of detached doctors with poor interpersonal skills.

Medical Hubris

Asymptomatic dentist, Mr. F. W., wrote to compliment me. He thought I had saved his life. I guess I may have, given that I discovered his sub-total, anterior descending coronary artery stenosis, before it clotted and caused a massive anterior cardiac infarction. I did my best to avert the risk of his sudden death.

In replying to another patient where I achieved the same result, I used a cowardly cop-out, rather than accept his compliment with grace. I said, 'we all have our a place in society'.

If you save lives, beware of hubris. Some patients can show a level of respect not far short of Messianic adoration, although ungratefulness also exists. The reaction of some to a private doctor can be, *'why should I be grateful? I paid him for what he did.'* Aside from dealing with those who know the price of everything, and the value of nothing, the self-esteem deserved from saving a life, will make most feel worthwhile. Nurses, paramedics, firefighters, rescue teams, helpful bystanders, and those trained in resuscitation, are all eligible. Hubris, however, needs to be tamed, and kept in its rightful place — under a mountain of humility; quite far from the cave where the sanctimony of medical bureaucrats pretends to hide.

Some Management Aphorisms

Nietzsche, Wittgenstein, and Marcus Aurelius, all used aphorisms to make some important points. So shall I.

Logjams kill. (People waiting in NHS hospitals for social care, will block new admissions. With log-jammed A&E departments, ambulances waiting outside hospitals, one cannot fetch new emergency cases from the community).

Science is a dispassionate practice drawn in black and white; the practice of art needs colour to describe humanity. Colour is especially useful when helping others, and when feelings need to be expressed.

Medical science has provided us with many tools, but only the art of medicine can appropriately direct their use.

The job of science is to understand the universe. Art adds something new to the universe.

Is science man's greatest achievement, or have language, mathematics, love, faith and commitment achieved more?

In catastrophic situations, quickly dismiss gormless bystanders.

Get all investigation results to patients ASAP. It can help prevent unnecessary worry and stop them nagging you.

Bureaucrats believe that rule books, guidelines, protocols and regulations, pave the road to Utopia. Their worthy, but futile aim, is to remove every risk to life, including medical risk. Zero risk exists, but only in ivory towers and on exoplanets (awaiting confirmation)

Regulators waste more money than they save. In their pursuit of zero risk for all, regulators in ivory towers are free to develop daft strategies. Have they considered improving pig-sties, I wonder? Just in case pigs learn to fly, all pig-sties should have re-enforced roofs and windows.

If you want to improve, prepare to be thought foolish. Epictetus.

Trouble flows downhill. (Naval saying).

Prophylaxis beats treatment: take a laxative before your haemorrhoid operation, not after. You will avoid post-operative constipation and the intense pain of defaecating what feels like a red-hot brick.

Safe-guarding can save lives. Make a map of all the known mines and minefields patients must face, before instructing them how to proceed.

Avoid ineffective medical action (like writing too many notes, ordering too many investigations, and delaying clinical review for personal convenience).

Avoid awkwardness. Do not grovel on the floor while examining a patient's feet. Lay them on a couch first and then proceed.

I have wasted time, now time doth waste me.
Richard II. Shakespeare.

Targets, fixed rules, and protocols are for corporate office workers with specific job specifications and fixed timelines. They are not for practising doctors working with a fluid narrative and ill-defined risks.

Those who cannot think on their feet should lie down, or leave the theatre of action. Their absence will save lives.

In battle, frontline troops are first to know the situation, never to commanders.

Our first duty of care is to make accurate diagnoses. Our duties are then to extinguish pain and suffering and to diminish morbidity and mortality.

The correct management of patients will depend on their personal circumstances.

The challenge of providing healthcare is to provide: *'the science of the possible with the art of the impossible.'* Dr. John Fry.

In private medicine, patients pay for their care directly. In NHS practice, patients pay indirectly (through National Insurance and taxation).

How much is a life worth? Before you save it: priceless. After saving it, not much.

Ultimately, the greatest benefit derived from all our medical efforts can be mundane. A patient's last wish may be to play one more game of bingo.

Rube-Goldberg constructions like those of Wallace and Grommet, overcomplicate matters. As Occam might have said: 'keep things simple.'

Don't come to think of yourself as better than you are. Resist thinking that you are worse than you are (the Dunning-Krüger bias).

The man who walks into a sweet shop will not leave with a bag of widgets. If you ask for a surgical opinion, a surgical operation becomes more likely.

Administrators, with one eye on the law, and the other on compliance, must curtail free-thinking among those they control.

A competent manager will always share responsibility and always avoid personal culpability.

Always be true to your word. Never promise to be five minutes early, when you mean five hours late.

Think ahead.

Gaining a patient's confidence will be difficult unless you are genuine, or a genuine confidence trickster. UK comedian, Bob Monkhouse once said: 'If you can fake sincerity, you've got it made.'

Medication causes 60% of all symptoms.

Drugs are poisons used in a controlled manner. Medically untrained bureaucrats should stay awake at night worrying about this.

When a diagnosis is proving difficult, give thought to drug reactions, and poisoning (the same thing).

When dosing patients, give as little as possible, but as much as necessary for as long as needed.

In few parts of the world would it be sensible to wear an overcoat every day. So why give the same diuretic dose every day, regardless of the ambient temperature?

'Never underestimate the predictability of stupidity'
(Film: *'Snatch'*. Columbia Pictures, 2000).

Never underestimate the number of people who overlook the blindingly obvious.

New patients who request an 'urgent' appointment will only attend if the appointment is convenient.

If you give your time to patients on the telephone, they will rarely consult you further. Having discovered convenience, they will more likely telephone you repeatedly and pester you.

Those with whom you spend the least time, will often respect you most.

Rote learning does not foster wisdom. Learning from experience, and putting knowledge to the test, are the essential ingredients.

Do not allow the scepticism of the educated to put an end to hope, faith, superstition, fantasy, and myth (See Nietzsche, and The Death of God).

Hope based on ignorance is not always a bad thing.

Medical science is replete with unknown unknowns.

Few ideas and theories withstand critical scrutiny.

Differences in culture appear at social interfaces.

Some cultures are ethnofugal: outward going, open, and happy to embrace others. Others are ethnopetal: for insiders only, closed and endogamous.

From one grain of sand, one can neither visualise its atomic structure nor the beauty of the beach it came from. When considering the truth, consider every viewpoint and each dimension.

No grain of sand is like any other, no snowflake like any other, and no person like any other. They are more easily understood as collectives: beaches, snowdrifts, and cultures. Dimension matters.

Beware of 'devils' in the detail.

Truth and reality can seem definable and granular, even when their components are mythical and illusory.

Patient Review

Sick patients need constant review, at intervals set by clinical urgency and volatility. Ask if it is safe to delegate their review to others? Remember that nurses are often too busy to send you alerts. A quick call to the ward asking 'How is Mrs. P.?' may not suffice for providing accurate, meaningful clinical information. Failure to review patients regularly could prove fatal. Your patient – your responsibility – whatever your status.

Political and Cultural Factors

For those who come from cultures where saving face is crucial, asking others for help and advice can be equivalent to a declaration of incompetence. This is especially the case for those with senior status. It might explain their reluctance to seek advice, even when they are out of their depth, and a patient's life is at risk.

I have encountered the 'face-saving' phenomenon many times. As a personal or cultural characteristic, it can motivate the hiding of incompetence and cause resistance to change. Compensate for it if you detect it, but be prepared for offers to help to be rejected. Learn to stand your ground. You may have to expose them, and insist on intervening, even though they protest about being shown insufficient respect. Don't forget to report your observations, what you did, and why. Fearing political incorrectness and discrimination can subvert the truth to the detriment of patients. Since a review will follow reporting them, your notes could be crucial to your own survival. The most important issue here is the patient's best interests.

We must learn to know our weaknesses, own up to them, and learn to overcome them. Although this is crucial to clinical effectiveness, it may not help a doctor's political aspirations. Not every doctor wants to address such matters, even if ignoring them will put patients at risk. However liberal and politically correct you choose to be, you must first address the patient's welfare, since life and death issues must always supersede politics. Patients will expect you to be culturally and politically correct, but keeping them alive is more important.

One of my patients, an ex-RAF Tornado jet pilot instructor, working in a wealthy middle-eastern country, was told not to be too critical of certain high-status student pilots who repeatedly ejected, prior to crashing their fighter-jet. They saw this as better than inferring incompetent flying. Apparently, $60,000,000 of fast jet losses each month, was cheap compared to losing face.

Dr. G. had worked in the Cardiac Dept. at St. George's, Hyde Park Corner, London, for six months. He tried repeatedly to implant pacemakers, but failed every time. Despite the hours we spent teaching him, it was clearly beyond his ability. On his last day, having taken six hours to find the brachiocephalic vein, we relieved him of his duty for the sake of the distressed patient. He was angry and very displeased. After returning to his country, he was to be the only doctor available for pacemaker implantation.

On the 22nd December 1999, Korean Air cargo flight 8509 (a Boeing 747), crashed shortly after taking off from Stansted Airport (UK). The captain had a malfunctioning flight level instrument, while the 2nd officer's instrument was working normally. Instead of flying with both wings level (horizontal), he was actually flying the airplane with both wings vertical to the ground. From the flight recorders, they deduced that since the 2nd officer said nothing, he must have thought it beneath his station to criticise a senior captain. I believe that Korean airlines later reviewed the cultural aspects of the crash, and retrained their pilots to communicate without considering loss of face.

In some countries, where the preservation of 'face' is vital, any consultation with a person of lower rank, must stop immediately a person of higher rank arrives. This practice, although not entirely unknown in Central London private hospitals, is very much alive and well in parts of the Middle East, Asia, and Africa. In the U.K., one can encounter it with the mega-rich (who may regard it as an insult to be kept waiting), and in hierarchical systems like the armed forces. If you intend to work in a multicultural environment, where there are sub-groups with little or no intention to integrate (culturally ethnopetal), you must acquaint yourself with as many specific cultural attitudes as you can, if only to better meet patient expectations.

Urgent Management

Some observations and interventions must be learned a.s.a.p:

Any asthmatic using their accessory muscles to breathe, needs urgent attention. Give them oxygen, a nebuliser with salbutamol, an anti-histamine, a leukotriene antagonist, and possibly a steroid immediately. Steroids take 24 – 48-hours to work; anti-histamines, and montelukast, work in less than one hour. The allergic process is easier to prevent than reverse. Take only a brief history, and learn to zone in on the key questions while using your eyes, ears, and hands to the full.

Quickly rule out a pleural effusion, left heart failure, and pneumothorax. The top priority must be to relieve the patient's dyspnoea. Never allow such patients to remain unobserved in an ambulance, on a trolley in A&E, or in a side-room. Sit them up, give them oxygen, and re-visit them every 5 – 10 minutes until you observe improvement. Better still, stay with them until they improve. The doctors and nurses we need most can work anywhere, anytime.

Prednisolone 60mgs daily is an unnecessarily large dose, even in this situation. An overflowing bath illustrates the point. There is only so much water a bath can hold before it overflows. Prednisolone 15mgs daily is enough for an average-sized adult. If you use this smaller dose, others will criticise you. Frequently, I found 15mgs daily worked as well as 60mgs, although there is no downside to giving 60mgs in the short term. Those who insist on 60mgs, presume that more is best. They may lack experience, confidence, patience, and a talent for timing.

Your job is to relieve symptoms and reverse pathophysiology. This involves effective prioritisation and time management. The need to complete records, or to deal with personal issues, must always remain subsidiary priorities. Everything other than your patient's welfare must wait. Drag yourself away from your computing devices; they will still be there when you return. Your presence and focused action will improve your patient's confidence and prognosis.

I still find it surprising that decisive action eludes some doctors and nurses. Some are not able; some have other priorities. Some are obsessive bean-counters, driven to finish whatever they are doing, regardless of pressing clinical priorities. If you are working with them, either gently force them into action, or dismiss them. They will offer excuses: too little time, too few staff, personal stress, many other administrative tasks to complete. Don't waste your time listening to them.

Patient Engagement

Whether making an enquiry or contributing to their medical management, many patients are capable of effective co-operation, understanding and compliance with management suggestions. It is usually obvious which patients are capable, but I have had my surprises. Some patients need educating before they change their attitude towards prevention and medical intervention. Some need a stimulus. The death of a close friend or themselves having a stroke or heart attack, may make them accept early medical evaluation, early diagnosis, and preventative medicine as expedient. Some doctors have no wish to embrace prevention (unless well remunerated). Others will favour prevention after experiencing a preventable medical event themselves or among family or friends.

Mr. C.F. had heart failure. He presented with SOB, and pitting oedema up to his thighs. His JVP was raised, but only after brief exercise. He needed diuretics, but which one, in what dose, and for how long? I weighed him, started him on a daily regime of frusemide 20mgs, and saw him again after five days. He had lost 3 kilos, had less oedema, and had passed urine frequently. My guess was that he needed to lose another two kilos of fluid.

Mr. C.F. was a highly capable and continued to run his business while unwell. We decided he should weigh himself every morning (after toilet procedures). The outcome, we agreed, was to keep his weight at 100 Kgs. If it increased overnight, or he felt breathless lying flat, he would take a further dose of frusemide 20mgs a.s.a.p. He was otherwise not to take any more that day. Sometime later, I maintained him on bendrofluazide alone. His intelligent co-operation ensured his optimum hydration and cardiac output.

Many patients can take their own BP and pulse, and co-operate at a level once reserved for nurses. Empowering patients works well for some. The insight some patients gain into how their therapy works, can significantly aid their engagement and improve their management.

Such patients contrast with the arrogant and ignorant, some of whom demand tests and treatments they have heard about at their local pub, on the internet or in a newspaper. Some will be doctors. It is important to recognise that the combination of ignorance (no discernible education, or insight), and arrogance can be fatal. Attempts to enlighten them can be futile or very hard work. They are best sent elsewhere. Never encourage them, unless they have a legitimate claim to arrogance and hold verifiable beliefs (beware of the Halo bias). This will happen occasionally, especially if your patient is an educated professional or academic.

My 30-year-old patient, Mr. B.H, had found himself a medically qualified girlfriend. She was a young surgeon in training. He had diarrhoea after they returned from an Ivory Coast holiday. His girlfriend requested a plain erect abdominal X-ray (despite no clinical suspicion of obstruction), a full blood count (almost irrelevant in acute traveller's gastroenteritis), and an abdominal ultrasound scan (looking for what?). She did request one relevant test - a stool sample for culture.

I agreed to some of her inexperienced requests, but only out of respect for a colleague; albeit, someone unknown to me. All the test results were normal, as predicted. Her advice cost him a lot of money. She had acted defensively and from a fear of omission. For unfortunate medico-legal reasons, the strategy of requesting every test possible is becoming inescapable.

This couple were engaged is something other than the diagnostic process. They were playing a relationship game (to score points, as Eric Berne pointed out, in his book 'The Games People Play'). She was trying to impress him with her medical knowledge; he was respecting her opinion as a sign of affection. Their relationship was at the development stage, and they were playing the mating game. I played along because I knew the situation to be innocent. It is never wise to engage in the personal games patients play; some can have unforeseen consequences.

Some Cultural, Ethnicity and Communication Issues

> '... a world that has no room for difference has no room for humanity.'
> Former Chief Rabbi, Dr. Jonathon Sacks. (2015)

Many millennia ago, ethnic differences were the likely result of mutation or arose from human species interbreeding. We know now that many modern humans possess a small percentage (0 - 4%) of Neanderthal genes (*Homo neanderthalensis*). Somewhere within modern human genomes, one can find seventy percent of known Neanderthal genes. Rather than exterminating our fellow species, Homo sapiens more likely bred with them (*Homo neanderthalensis, Homo erectus, Homo rudolfensis, Homo floresiensis,* and *Homo denisova*). The many ethnic groups we recognise to-day are the result (*'Sapiens'*, by Yuval N. Harari: 2011; Vintage).

There are those practising medicine who regard cultural and ethnic issues, as 'out-of-bounds'. They believe the consideration to be unprofessional and of no relevance to clinical practice. In simple cases they will usually be correct since such enquiries are rarely called for. Apart from certain diseases that are ethnic group specific (sickle-cell anaemia, thalassaemia, Tay Sachs disease, etc.), these considerations can be relevant when dealing with complicated, more difficult to resolve cases.

Our ethnicity derives from our fixed genetic profile, with our adoption of culture slowly accumulating thereafter. From birth, we take on a personal set of values and beliefs (our culture) from the surrounding influences. These influences need considering, whenever patients have social and psychological problems, or medical conditions that seem difficult to understand.

Our compatibility with patients can depend on our understanding of their ethnicity and culture. The ability to diagnose and treat a patient's athlete's foot, fractured tibia, heartburn, or angina, requires no such knowledge. To cope well with a patient who is tearful or distressed will usually require emotional intelligence, interpersonal sensitivity, and some understanding of their origins, values, and beliefs.

Cultural values are dynamic. Held at a subconscious level, they are under constant revision as the influences upon them change. There are strong and variable cultural associations with ethnicity, which are virtually fixed and immutable. Cultural identity needs no discussion when like people associate (referring to one another as 'us'). An awareness of personal cultural identity arises at frontiers, as relationships develop and we attempt integration with those from different cultures ('them').

'… *society is founded on custom, on the unreflective adoption of received habits, whether these relate to modes of conduct or perception.*' Dominic Baker-Smith on the Syrian satirist Lucian (125 – 180 AD). Introduction (xiii) to '*Utopia*', by Thomas More (1518). Penguin Books. 2012.

The adoption of foreign cultures forms a major part of world history. Following invasions by Roman, Assyrian, Ottoman, Mongol, and Greek forces, the invaders forced local natives to assimilate their culture.

Ruth (The Scroll of Ruth. Hebrew Bible) embraced Jewish culture. As she says to Naomi, '*Where you go I will go and where you stay I will stay.*' Some want to relinquish their native culture and adopt another. Is that possible? Many can learn to be different when it suits their interest.

Moses Mendelssohn believed in cultural assimilation. He replaced his Jewish culture with the prevailing German Protestant culture, believing Judaism to be outmoded. His seven-year-old son, Felix Mendelssohn Bartholdy, became baptised as a Christian and wrote his first symphony when 15-years old. He later composed a further 750 pieces, some of which were sacred choral works, all of which were thought of as German music.

Richard Wagner regarded those with Jewish origins as lacking pure German ethnicity. He wrote an anti-Semitic essay on Jewish musicians (Das Judentum in der Musik; 1850), three years after Mendelssohn's death. Hitler pursued the ethnicity issue with catastrophic consequences for all Jews, including those like Moses Mendelssohn, who had dedicated himself to being German.

There are those who are not wolves, but who live with wolves. They can become wolf-like. Eventually, they may become accepted by wolves, but will never be wolves.

No wolf becomes a poodle. The best a wolf can do is to imagine itself as a poodle, imitating how they bark and walk. It may then need a wolf or poodle psychiatrist to help it adjust. When veterinary management issues arise for dogs, or clinical management issues arise for some patients, their breed and cultural background can be important. In challenging cases, clinicians may need to consider their patient's background, attitudes, outlook, and disposition. For reasons of political correctness, few medical professionals want to consider culture, gender or ethnicity issues.

Had I gone out on a social occasion without a shirt, jacket and tie in the 1960s, my lack of 'proper' attire would have displeased my father. In 1962, an anatomy demonstrator severely reprimanded me as I entered the library of The London Hospital Medical College on a Saturday morning, because I was not wearing a jacket. During the same period, adherence to many inane rules persisted. As an example, many unmarried mothers were forced to give up their babies for adoption. Times have changed for the better.

With the passing of a few decades, most patients now ignore the informal and relaxed approach doctors prefer. No longer are doctors expected to consult in mourning dress as befits their class and professionalism. The way doctors dress is no longer seen as crucial to our professional image.

I had known Mrs. D. J. since she was a child. She was now a mature, married woman with two children of her own. I had known four generations of her family. Her father, a long-term patient of mine, died of oesophageal cancer. She inherited something rather special from her parents and her grandmother: natural grace and charm.

Mrs. D. J. had become concerned about her health. She came to me with sheets of laboratory results, crying while we considered them. The focus of her concern was a doctor who said she had 'deteriorating renal function' (slowly rising creatinine levels from 70 μmols/L, twelve years before, to 90 μmols/L), together with abnormal antibodies. Without knowing enough to assess their significance, these results had worried and depressed her.

Having no previous acquaintance with laboratory data, she had temporarily reverted to a child-like state, crying and needing help; tacitly asking to share her burden (Freudian transference?). I took this to be an acknowledgement of the trust she and her forebears had in me (allowing me to witness her unrestrained emotion).

Explaining the findings to her and putting them into clinical perspective, took some time. All I could do was to blunt the Sword of Damocles, fixed over her head by another doctor. I could not remove it, given her slowly rising creatinine level and significant monoclonal antibody levels.

How might she get relief from worry? Obviously, more information might help. A tissue diagnosis could become necessary. I explained that only renal histology and immunochemistry would put her problem into proper clinical perspective. Only then could we have a more meaningful discussion about her prognosis and options for her future. In the meantime, it left her and her family to imagine every future scenario, all of which were frightening.

It is cruel to leave some patients in an unresolved psychological state. Leave no patient too long without information and a workable management plan. Otherwise, prepare to pick up the pieces of a shattered mental state.

Preventative Management

Doctors cure sometimes, relieve often, comfort always, and prevent hopefully.
Dr. John Fry

A wise man would sooner keep clear of sickness than find a remedy for it.
Thomas More (1518). *Utopia*. Book Two.

The British Heart Foundation recently produced a mobile telephone app. which stores patient data. Its purpose is the early detection of heart disease. They finally achieved it forty-five years after I started collecting medical screening data for the same purpose. In 1973, I took it as blindingly obvious (BLOB) that repeated screening and data recording, would help detect early heart disease, and impact prognosis. Because corporations and committees are so ponderous when deciding anything of consequence, it is easy to find yourself ahead of their game, but without the support or recognition to facilitate progress. The very wealthy have no such problems. When Elon Musk and Richard Branson decided to make space travel available to the public, they simply ordered it to happen.

When I was a junior doctor, few thought preventative medical examinations worthwhile. The outlook then was only to treat symptoms. Many thought that the early detection of disease was pandering to the rich and neurotic, and would make some unnecessarily anxious. Attitudes have changed, but not for those entirely focussed on treating the sick.

After many decades, ample data now exists to prove the benefit of cervical smears, mammograms, colon cancer screening, PSA testing, and cardiac screening. Unfortunately, screening is expensive. Whether a nation should regard screening as good value for money, is not straightforward. It is corporate accountants who will calculate the cost of saving or extending a life. They long ago decided that not all screening is affordable. In keeping with the health divide, the rich who know screening to be beneficial (if only from a common-sense point of view), will continue to keep their considerable health advantages.

Cancer Research UK, published an article showing the reduced cost of treating cancer early (see Birtwhistle, Mike: *Saving lives and averting costs? The case for earlier diagnosis just got stronger.* Cancer News September 2014). Also, adding a routine ECG to the screening of athletes detects the liability to sudden death, but it costs between $62,000 to $130,000 per life year saved (Wheeler, M.T. et al. 2010. *Cost effectiveness of pre-participation screening for prevention of sudden cardiac death in young athletes.* Ann. Int. Med: 152(5): 276-286) doi: 10.1059/0003-4819-152-5-201003020-00005.

When spending public money, bureaucrats rightly insist on getting value for money (let's ignore the colossal sums we spend on their salaries and the cost of the offices they run). I have made the odd bureaucrat cringe by saying that I would sooner waste money than waste lives. What is money, anyway? Is it not just a digital token, a credit or debit saved on a computer, somewhere far, far away? While the value of currency exists only by faith and consensus, the value of life and death remains as real as it gets.

Because I directed the spending of my patients' money, my viewpoint was different to that of public servants. If I had a harmless test that might save their life, I would use it, whatever it cost. Since it would take twenty-five-years-worth of double-blind trials to 'prove' a new idea worthwhile (none become proven, just more probable), those now living in their eighth decade have too little time left to wait for further research results. Our professional responsibility, however, must always be to prove the efficacy of the interventions we offer.

My patients spent their money on my advice, believing that screening was of value. The screening of patients I undertook for 47-years, was a rewarding exercise (albeit repetitive, and somewhat boring when there were only negative results). I routinely discovered heart disease, and a few cancers in the asymptomatic. I therefore limited myself to the detection of pre-symptomatic heart disease in my last twenty years of work. After ultrasound scanning the carotid arteries of thousands of patients, I found it to be the single most useful test for diagnosing the atherosclerotic trait, and asymptomatic coronary heart disease (99.9% of those with proven coronary disease have carotid atheroma).

Screening is also an exercise in patient education. Patients learn which symptoms are relevant, and which to watch out for in the future. It is also a navigational exercise. Patients need to know how they are progressing, and in what direction. The baseline assessments we make are essential for assessing both. In compliant subjects, early risk factor identification and early diagnosis, both lead to beneficial changes in lifestyle and prognosis.

Screening occasionally prompts urgent, life-saving intervention, so I concentrated on the detection of atherosclerosis, ventricular hypertrophy, cardiac conduction defects, and the expansion of the atria as a prequel to AF.

Mrs. Amanda C. was an intelligent, fifty-year-old mother of two young women. She exercised every day and ate a diet that most would think the best available. Because she had a strong family history of cardiac infarction, she came for a routine carotid artery screen and exercise test on my advice. Her carotid ultrasound revealed many large calcified atheromatous plaques.

Her treadmill ECG exercise test was normal, except that she 'ran out of steam' early on. I advised a cardiac CT scan, which showed a high calcium score (> 700). This suggested the need for coronary angiography. We found extensive 3-vessel coronary disease, with several critical stenoses. In colloquium with my cardiological colleagues, we advised coronary bypass surgery.

This advice perplexed Mrs. C. She became so perturbed by the investigation results, her normal decision-making ability abandoned her. The opinion she voiced was, 'I am fit. There is nothing wrong with my health.' The information we gave her made her anxious and preoccupied. She could no longer sleep, and now feared sudden death. Our fault? Well, 'Yes', undoubtedly. But then, I had merely discovered her coronary problems; I was not responsible for them, or the risks they implied.

Although her anxiety was the unfortunate outcome of a screening process, induced anxiety was surely preferable to cardiac infarction and sudden death from VF (the previous fate of her father and brother at a young age).

In screening practise, there are a few alternative endings. One is the suffocation of action – some patients will bury their head in the sand. Once Mrs. C. finally accepted the facts (it took many hours of discussion with her and her husband), she agreed to a CABG.

There is an interesting postscript to this case, which may apply to other asymptomatic cases found unexpectedly on screening. Long after her CABG, she remained unsure that she ever needed the operation. After all, she had no prior symptoms to suggest the need.

As part of our educational duty to screening patients, we must present all the data in an understandable form. We also need to explain it to all involved. Only then can one meaningfully discuss acceptable management plans with them.

As a means of explanation, the use of metaphor is useful. For Mr. F. W, an asymptomatic dentist with unexpected severe coronary artery disease, one simple metaphor helped him understand. The metaphor I used was of fuel pipe blockages limiting the performance of a motor vehicle engine. As a mechanically minded man, he instantly saw how his own failing exercise performance related to his sub-total coronary artery stenoses. Once he accepted the connection, he agreed to angiography and the stenting procedures that followed.

For me, one benefit of spending a long time with each patient, was learning about them. This included something of their culture, their value system, personality, intelligence, socio-economic class, and educational background. These are the minimal requirements for understanding a patient well enough to give them pertinent, personal advice, on life and death issues or on complicated health-related matters. Without personal meta-data, one cannot construct meaningful individual advice, especially when the decisions that follow could be crucial. Without this level of information, our only recourse is to rely on generalised guidelines and advice, some of which will not apply.

After explaining all the pertinent clinical facts, some patients will accept your point of view, others will not. Just occasionally, the seriousness of their clinical situation will elude them. Others will fully understand, but choose to do nothing.

Richard had been an Oxford academic and a senior executive in a public company. He had normal heart valves until he developed a valve infection (SBE), aged fifty-four years. I diagnosed this simply from his prolonged history of fever (PUO for ten days), and the development of a pronounced early diastolic murmur, consistent with severe aortic valve incompetence (AI). My previous screening recorded no murmur.

I admitted him and successfully treated his bacterial endocarditis with i.v. penicillin. I explained he would need an aortic valve replacement (if he wanted to survive long). 'No. I don't want that', he said. 'I have had a wonderful life so far, and I am happy to take my chances.' This was after I had fully explained the likely morbidity and mortality associated with doing nothing, as opposed to valve replacement. He chose to 'take his chances' without valve replacement and died from rapidly progressive left heart failure one year later.

I have rarely met anyone more sagacious than Richard. His decision followed a full explanation of all the facts and factors involved. Perhaps he feared an operation, but I don't think so. He seemed not to believe in surgical interventions. Since his subsequent funeral was a Humanist one, religious conviction did not explain his attitude either. I never really found why he decided not to accept an operation.

Mr. Ron C. was a philanthropist and a generous financial supporter of Essex County cricket. At a cricket dinner in the City of London, he arranged for me to sit next to the then captain of England, Graham Gooch. Knowing nothing about cricket, I asked his name and occupation. Graham lightly dismissed my ignorance. He actually favoured sitting next to someone who could not discuss cricket.

Ron C. weighed 400 pounds (190Kgs). The patient who referred him to me had given me some prior advice: 'Whatever you do, never mention his weight.' I didn't, not until he had his first acute anterior cardiac infarction. Sitting at his bedside in hospital I said, 'The time has come, Ron', to discuss your weight.' I never saw him again. I heard later he had died suddenly. He was an affable man, and a much missed philanthropist. We communicated well, but he had decided that any discussion about his weight was 'out-of-bounds'.

Review Review Review.

Effective preventative medicine can rely on sequential patient data. For decades, I used a standard form for recording patient history and examination data (see Appendix A). I first used it in 1973, and it remained unchanged for forty-five years; I used it to review the changes in a patient's symptoms and signs (easy to do since each symptom and sign, occupied an unchanging position on the form). I could not have done this with freehand notes. Since many patients dislike doctors staring at computer screens, rather than looking at them, my use of this simple standard form was acceptable to most patients. I could have computerised the form, written a program to highlight the changes from year to year, and produced a stylised report.

Individualised appraisal, not a standardised one, was what my patients wanted, so I never developed it further.

An efficient patient recall system is another essential management tool for screening. It is especially important for those where long-term, ongoing assessments are desirable (hypertension, atherosclerosis progression, diabetes, etc.). After each consultation, I would record when each patient was to be reviewed (a few months or years). The computer program I developed stopped sending further prompts once a patient had ignored three consecutive reminders.

Patient compliance is a problem for recall systems. US doctors often use a more aggressive approach.; many telephone their patients repeatedly. For decades, I sent reminder letters to patients, but not all responded. I saw it as my duty to remind them, so I sent many variations of the same letter, but none was more successful than any other. Patients will return when they are ready, often prompted by a close friend or relative having a medical incident. Because I sent the letters every year, many eventually got the message.

Pharmacist Physicians

Brain Altman and I, studied 'A' level sciences subjects together. He became a successful pharmacist. When I first visited him at his Salmon Lane, East London pharmacy in the 1960s, I surprised me how many of his customers sought his medical advice. His grasp of medical issues, and how to handle them, brought him many rewards.

It is now obvious that pharmacists, empowered to act as GPs, can easily manage a wide variety of minor clinical cases. Patients find pharmacists convenient and approachable, with consultations and medicines available at the same point of service. Young GPs should start worry about their job security.

Out-of-Hours Requests

To be a successful doctor, without the support of a corporation, requires an independent spirit. Because many doctors in the UK come from Asia, where there is less government support and the culture promotes business, I expect many more private GP practices to arise in the UK within the next few years. The first requirement will be to make themselves available. In private practice, this is the first thing a doctor needs to do; in NHS GP practice, it was mostly the last thing doctors want to do.

The word 'available' requires definition. If a doctor intends to provide an 'around the clock' service, some patients will want to come at 10pm, and on Sunday afternoon at 3pm. To many people, 'available' means 'at any time'. There are a few patients who can be jaw-droppingly selfish; they will call at 2am, just because they cannot sleep. This will leave some doctors wondering why they didn't choose investment banking as a career.

The more unreasonable the time for a patient to call, the more trivial or very serious the enquiry is likely to be. Unfortunately, most CVAs (strokes) and cardiac episodes occur at night, or early in the morning. Dedicated doctors are aware of the self-sacrificial nature of medical practice, and the need to be available to assess every call on its merits. Most patients will wait to call their doctor, and try not to cause her an unnecessary disturbance. Sometimes, such delays can have serious consequences.

The Law of Indulgence

The more generous, kind, and indulgent you are, the more people will demand. An obdurate, distant, and curt approach will often curb their demands.

> *'While you treat them well, they are yours.'*
> Niccolò Machiavelli. *The Prince (De Principatibus)*.1513.

There is an instructive corollary: those you treat well, could come to regard you as theirs.

Consider a business trial. Open a sweet shop and set a fixed daily charge for customers. The contract would them to eat as many sweets as they like. Initially, many will visit frequently. Thereafter, they will tire of sweets. The same phenomenon applies to all situations where something holds value for money, or is available at no cost. It applies to sex in a relationship and to food and drink on all-inclusive holidays.

In 1985, I created my own American-style HMO scheme, and offered it as a subscription package to patients. The idea was to provide a no-limit, private, out-patient general medical service, with an annual medical screen included. I made both annual and monthly subscriptions available. My colleagues thought I had made a rod for my back. At a cost of £10 per week, they said it was too generous.

Two important aspects explain the absence of queues outside my door. The first was the cost; the second was patient selection. Some of those who can afford private care, but choose to use the NHS in order to save money, are those who will also queue for bargains. For those dedicated to saving money or getting value for money, getting something for nothing is better still. They either have a mean streak, or want to conserve their resources. Because my scheme cost all of £10 per week, these people didn't join. There is an unexpected paradox here. Those who can easily afford what they want, only rarely exploit others. For those with plentiful resources, there is less need to worry about 'saving money', although most will seek value for money.

People vary between those who mostly give, and those who mostly take; from those who chance their luck to those who take what comes; from the meek to the demanding; from the objective to the subjective; from the disciplined to the undisciplined; from the emotional to the dispassionate; from the self-important to the humble; from the arrogant to the modest; from the criminal to the honest; from the educated to the duncified (see Quiller-Couch); from the intelligent to the unintelligent; from rich to poor, and from outspoken to those who are accepting. None of these are typical of any social class, race, or culture.

Doctors must deal with every type of patient, but only those capable of the art of medicine will handle them all with mutual satisfaction.

Who Actually Wants Your Advice?

I was in a gym on a training day for personal trainers. I had just started doing a sitting weight lifting exercise (one I had done for over fifty years), when a training instructor stopped in front of me. 'Do you know you are sitting too low?' he said. 'May I offer you some advice?' To this, I curtly replied 'No'. Red-faced, he apologised and walked off.

I didn't feel good about refusing his offer of advice. He was only trying to help. He was being generous while in 'instructor mode'. I didn't know him, or what experience and expertise he had. I had used gymnasia for decades, and knew that some explanations given by trainers made little physiological sense. I understood their need to sell their own methods of exercise. Although they are rightly concerned with safety, who would accept theirs as the only safe and correct method?

Can one be too old to learn? I am sorry I didn't listen more sympathetically to him. I had nothing to lose. My inappropriate behaviour will have disadvantaged both of us. He was only trying to be helpful, and I might have learned something useful. I sought the right moment to apologise, but it never arose.

Not long after this incident, I nearly found myself in a similar position to the gym instructor. I noticed a Chinese lady in a supermarket with exophthalmos. I considered offering her some advice. I was about to say, *'Do you realise that you have thyrotoxicosis?'*, but I stopped short, feeling that it was none of my business. Is it possible she didn't already know? Given that some Chinese suffer from a sinister form of T3-driven thyrotoxicosis, perhaps it would have been better to risk being brushed aside, rather than miss an important opportunity? I knew nothing of Chinese cultural norms, so I felt unsure about how she might react to my unsolicited advice. Had she been a native Cockney, I would have had no such inhibition.

Should one offer advice when not requested? The establishment answer is 'No', but I sometimes considered it from within a well-established doctor-patient relationship. For those I knew well, I might comment on and discuss any issue that arose – their failing business, their demanding partner, their son underperforming at school, because their reactions to such stresses, sometimes had a clinical bearing. To gain entry to this personal zone, one must have already established a friendly, well-meaning, doctor - patient relationship. In my practice, I achieved this status with most patients, but then I had known some of them for forty years.

Even the best advice can go unheeded. The proverbial penny may drop later when the need becomes critical. Patients will often remember what you said.

I fondly remember an older man with resistant hypertension. I had previously discussed his relationship with his wife. It had been acrimonious and stressful for decades. He then returned to the OPD at St. George's Hospital at Hyde Park Corner, saying: 'I don't need treatment for blood pressure anymore. Thing is, you were right Doc. Me and the wife never got on, so I moved out. I now live just round the corner from her. We get on fine now. I take her out for a meal twice a week, and call in for tea occasionally. We both came to realise that although we love one another, we can't live together. Since we parted, my blood pressure has been perfect without pills. So, doc, I'm here to say goodbye.' He was correct. His hypertension had abated. (I cannot remember if he had LVH or not. If he had, it would have been wrong to discharge him unconditionally from medical supervision).

Best avoided are:

> *'... those restless thoughts which corrode the sweets of life.'*
> Isaac Walton. (1653): The Compleat Angler.

Practice Style and Clinical Management

Clinical management varies with specialty. Some doctors deal with passing strangers (GPs, and A&E), and have an intense, transient involvement (surgical and cardiological). Those who deal with chronic cases often have to vary their approach and their therapies, until they find the most suitable one for each patient.

Some patients will consult a doctor only once (A&E, neurologists and surgeons); others will consult the same doctor repeatedly, and over long periods (counsellors and psychiatrists).

Few UK doctors now provide pastoral care. The current modus operandi is to delegate all such duties to social care providers. In my private practice (as an indicator of its demography), I never had to make social care arrangements; my patients were mostly wealthy; self-sufficient, and with lots of family support. They all made their own arrangements.

My style of practice would be all but impossible for UK doctors today. Few doctors are free to choose their level of availability and continuity. The value of continuity, with commitment to the same patients over long periods, is something the NHS has only recently re-discovered, even though doctors have known about its indispensable value for millennia. Having acknowledged it, the NHS will struggle as a corporation to put it into practice.

The NHS trend has all but dissolved the notion of the same doctor being available to the same individual in their times of need. While NHS GPs share patients in collective groups (primary and community health centres), they no longer do home visits or have on-call duties, so there is much less chance for them to ceate continuity and pastoral care.

The King's Fund commissioned Prof. George Freeman, and Jane Hughes (*Continuity of Care and the patient experience,* 2018) to find scientific support for the thesis that continuity enhances patient and staff satisfaction, and contributes to health benefits. Really! Few doctors or nurses working over the last few millennia would have challenged this BLOB contention. That care and its continuity benefit patients is axiomatic. Why do bureaucrats spend our money proving the axiomatic?

'While history does not repeat itself, I think it has a rhythm.' General Sir Nick Carter Chief of Defence Staff, said this on the Andrew Marr Show, November 2020. (A quote questionably attributed to Mark Twain).

Can a continuous, doctor-patient relationship be detrimental? When patients consult other doctors, the opportunity arises for an independent review. This is usually a good thing, unless it confuses the patient. Referring patients to other doctors has another function: it is a useful way to delegate the dissatisfied.

In the UK, specialists (as opposed to GPs) rarely maintain ongoing relationships with patients. This is quite different in countries where patients have direct access to specialists and are free to consult whom they wish. Although unusual in the UK, I combined cardiology with general medicine, without relying on referrals from GPs. This policy allowed me a distinct advantage: to combine specialisation with the continuity that NHS GPs once offered decades before.

The Proximity Principle

The further away a proficient doctor positions himself from patients, the worse the likely clinical outcome.

The further away a dysfunctional doctor stays from patients, the better the likely clinical outcome.

GP Practice

General Practice in the UK is traditionally the place patients go for their initial assessment, and their minor medical complaints. 'General practice' fulfils the same role worldwide, but is rapidly becoming shared with nurses, pharmacists, and paramedics. Some patients need their more serious diseases confirmed and dealt with by an appropriate specialist. Many GPs enjoy the challenge of varied work, a good salary, and more social freedom than hospital doctors; they can have a life of their own. One drawback for some, is the limited intellectual challenge. Some GPs prefer this, others find it like a race jockey being put to ride cart horses. Some will become discontented and demoralised, especially if they find repetitive, minor work trivial. It has not helped to re-classify general practice as a specialty, and to appoint professors to the subject; this is like appointing professors of cat-flap technology. As a political move, this has fooled nobody.

> *'What's in a name? That which we call a rose by any other name would smell as sweet.'*
> Romeo. *Romeo and Juliet.* Act 2; Scene 2. William Shakespeare.

General practitioners specialise in taking an overview; a perspective which requires some knowledge of every clinical specialty. A GP is required to diagnose and manage minor illnesses and many chronic conditions. She must be adept at detecting serious illness and be ready to refer patients to an appropriate specialist. Once a case becomes 'interesting', or 'complicated', it takes experience to know who might best manage the case. Because GPs often leave the confirmation of diagnoses and treatment to those specialised in hospital medicine, many will become disenchanted. In the days of cottage hospitals, GPs and specialists were mutually supportive, managing cases together. GPs then had a more active clinical role. That put any specialist expertise a GP had to good use. Unfortunately, the corporate financial conclusion was that small cottage hospitals lacked the economy of scale. Like the bureaucratic attempt to close the Paediatric Cardiology Unit at the Royal Brompton Hospital in 2016, the decision lacked insight and understanding of how medical practice works best for doctors and patients.

What has changed since I tried NHS general practice in 1968, is regulatory compliance. It now takes dedicated office staff to handle audit and compliance, even though the disease entities presented by patients have not changed. At least the NHS pays GPs for some of the bureaucratic burden they impose. In private practice, it remains a tiresome, expensive requirement, for which there is no financial compensation other than tax relief on expenses.

As the need for certification and regulatory compliance increases, the orbit of GP work might shrink. This will make GP work even less satisfying to those seeking clinical challenges. These factors will lead to major changes, with more nurses and paramedics being trained to replace GPs in primary care.

Since reduced medical staffing levels now need urgent attention in the UK, GMC and corporate NHS executives, must be wondering what to do next. Since few medical executives in the UK know anything about medical practice, they will guess. When they get it wrong, like executive Paula Vennells at the Post Office, they can claim they were unaware of what was actually going on in the (NHS) corporation. Health bureaucrats might read what I have written about the demoralisation of medical staff, but I doubt they will acknowledge any part they played in the decline of the NHS. For decades they have failed to change their culture and outlook. If they studied five decades of UK medical service history, they might appreciate some of the deleterious effects medical regulatory bureaucracy has imposed on UK medical practice. With doctors choosing other professions and emigrating, they will need to think fast. Unfortunately, more insightful thoughts than corporate committees can provide, will be required to overcome NHS bureaucratic inertia. Without clinical experience, how can they possibly know what best to do? All they can do is run the NHS like a baked-bean factory. At least their MBAs and legal qualifications, will qualify them for it.

The only viable way forward is to change NHS corporate culture. This will require the removal of those who have set the current culture, with every element of medical bureaucracy unpicked and reviewed by those who know most about medical practice. Who will have the motivation, time, energy and power for the task? At what point before an airplane crashes into the ground, will the pilot sitting in the left seat ask for help?

Jungle Medicine

One can typify this as making do, while making good. In a jungle, you will still have access to your most valuable assets: your brain, eyes, ears, hands, and mouth. In the absence of an ECG monitor, feel the pulse. Without a central venous line, observe the JVP. Take the BP repeatedly yourself, rather than have an arterial line. Without learning the sort of medicine practiced in jungles, you might not fully appreciate hands-on patient management, and accept its value as indispensable.

Village Medicine

There is an art to living with your patients. Part of it involves attending births, weddings, and funerals. In many cultures, doctors still have a valued position in their community and often fulfil a societal role. Not only is there an art to this style of medicine, there is a spiritual aspect to it, dating back at least to Ancient Egypt. Before being recognised as 'doctors', most healers in ancient Egypt were priests (worshipping the Goddess Sekhmet). Medicine was a religious cult, using traditional magic spells and potions (in some countries, they are still in use).

We are not human beings having a spiritual experience. We are spiritual beings having a human experience.
Pierre Teilhard de Chardin (disputed source)

Village medicine varies with location. In the UK, it is all but obsolete; in the Republic of Ireland, Greece and Cyprus, it is not. An ability to fulfil this role was once a selection criterion for medical students. Students were more likely to be chosen to study medicine if they were 'rounded' individuals (ideally polymaths, athletes, and socially acceptable), respected by others, able to think for themselves, and comfortable dealing with the medical problems of all-comers. They once chose those with intellectual ability and practical skills, but with enough humanity to attract the respect of a community. Selection committees must now first make sure that candidates are good at science, and next make sure they will be compliant enough to follow every bureaucratic rule to the letter.

In teaching hospitals, medical practice combines with teaching medical students, and some research. In tertiary centres, academics doing research are the ones practising medicine. The students are mostly postgraduate, and the purpose is to pursue the science of medicine.

Machine Dependent Medicine

The potential for computers to provide learning aids, sources of information and analysis, is impressive. Diagnostic algorithms are already in use, although their output often needs to be tempered by experienced clinical judgement. The danger is that we stop thinking, and come to regard computers as always reliable and indispensable. They will be so for many. For the more experienced and able, they can provide valuable information.

If you let a computer suggest a restaurant, try to see beyond any suggestion they make. The meta-information is important. For instance, is the restaurant paying commission to advertisers? What is the ambience? What types of people go there? Is it noisy? Is it really child friendly? Is it more expensive than it looks? Equivalent questions need to be asked when a medical computer program suggests a diagnosis and a course of action.

Many now see medical AI as a torchbearer, lighting the path to a brighter future for doctors and patients. One thing is certain: it will be a boon to regulators whose job is to control how doctors practise. AI is being promoted as worthy of our trust, with the promise of better risk evaluation, and improved clinical judgement. The enthusiasm of the young will make it fashionable and acceptable. It is more likely to disenchant the old, who may see it as unworthy progress.

As calculators, computers undoubtedly perform better than any human. We can usefully apply this to general risk assessment. Without specific patient information, will dependable individual risk calculation elude AI? Replacing experienced judgement would be a wonderful thing; most useful for those without experience. How accurate can they be in situations for which they have been inadequately 'trained'? (The idea of being trained attempts their humanisation). Computer pundits are already backtracking on their sales pitches. They are now promoting AI as an aid to medical decision making, not a replacement for it.

Computers hold economic promise for governments. Virtual reality is not reality, but it can provide a good enough simulation of it for training people. Think of the savings in staff costs that would result if AI replaced even a small percentage of human functioning. It is already quite obvious that nurses and paramedics using AI could replace GPs and those dealing only with minor medical problems.

Computers will never replace high court judges or specialist doctors dealing with complex conditions, although the use of algorithms can remove some of the variability of human biases (*'Noise',* D. Kahneman *et al.,* William Collins, 2021). They can provide valuable diagnostic suggestions with associated probabilities. They can also provide ill-considered lists of alternative diagnoses, much like any 3rd or 4th year medical student. If we accept the clinical judgements of AI, what legal liability will computers, the software and the developers have? This is a matter for the GMC, the law, and the insurance industry. Similar accident liability issues are now being discussed, as computer-driven vehicles take to the roads and replace experienced drivers.

There are simple ways in which computers can helpfully out-perform doctors. They can perform boring tasks without fatigue. Routine form-filling, the data collection demanded by medical bureaucracy, the scoring of diagnostic questionnaires used for dementia, severity of illness (National Early Warning Score) and personality disorders, are examples.

Outcome and Patient Type

Clinical outcome can depend on a patient's personal resources: their character, wealth and education; their social and business networks. All can influence clinical scenarios. In deprived areas, diminished resources pose a challenge. Those with secure lifestyles (the advantaged socio-economic classes) will mostly enjoy better health and have more predictable outcomes, simply because they have flexibility, choice and more resources.

When they become patients themselves, most disciplined professionals interact with other professionals in a predictable way. Their acquaintance with information handling, clarity of thought, and an ability to express themselves succinctly, make history taking easy. This will usually make diagnosis and management easier, and influence the outcome. Expect more detailed, lengthy discussion though.

One socio-economic phenomenon became more obvious as my career progressed: a growing number of newly rich. They are now ubiquitous. Among them are Lottery winners, footballers, private business owners, city traders, show-biz personalities, and those one step removed from crime. They are not all pretentious, indolent, ignorant, arrogant, and brash, with 'more money than sense', but their demands and expectations can be exaggerated and ill-informed.

Never entertain the sort of obsequiousness and unctuousness displayed by Uriah Heep in David Copperfield, or clergyman William Collins, in Pride and Prejudice. There is one way to deal with such people; never take them seriously. If you are fortunate enough to deal with capable, honest, successful, honourable men and women of purpose who have experienced success, you are likely to enjoy medical practice more. Amusing, friendly, straightforward, and pleasant people are a joy to deal with, whatever their background.

It is not a doctor's role to judge the suitability of people to be patients, but there are exceptions. You must avoid being duped (as I was), or being led astray by those with devious agendas. Until you have experienced most types of individual, lived among them, worked with them, and know them well, it would be best not to form firm opinions. One important purpose is to understand them well enough to detect dishonesty and their desire to manipulate. Some will try to use you for their own purposes, and having them as patients could be something you regret. Beware of overly pleasant, easy-going characters, who are just a little too good to be true. If you disagree with any of this, the chances are that your views are liberal, and you have yet to be tested by reality.

Diffusing Situations

Patients will usually view a relaxed, confident and forthright bedside manner as reassuring, especially in urgent clinical situations.

While in ITU, a patient once asked me (while attached to two simultaneous iv blood infusions), if she was about to die. She was pale, sweaty, and frightened. I sat by her bedside and told her a golfing joke, knowing she had a low handicap.

A professional golfer hit his ball into the long grass. He asked his caddy for a one-iron. 'Are you sure? It is very long grass', he advised. 'Please, give me my one iron', repeated the golfer. 'I know what I'm doing.' He didn't strike the ball well. The ball ricocheted off a nearby tree trunk and returned to strike him between the eyes. He died as a result and arrived unexpectedly at the gates of Heaven. 'What are you here for?' asked the attending guardian angel; 'For two', the golfer replied.

Although not that amusing, the joke was much appreciated by this patient who was fearing death. She smiled, and her tension melted a little. She underwent an emergency hysterectomy and survived. She later told me she thought no doctor would be stupid enough to tell a joke to a dying patient. It gave her hope. Was her survival in any way aided by humour, relaxation, improved confidence, and reduction of fear? I believe so, but cannot prove it; because I employed the art of medicine, not the science of medicine, proof was not the issue.

When others have asked me whether they are going to die soon, I have sometimes replied – you cannot without my permission. Be my guest: do the research on the relationship between humour, hope and survival, but first read about Dr. 'Patch' Adams. I am sure he would agree with me – this is a BLOB issue for those capable of using the art of medicine.

Some of my colleagues have regarded my jocular interventions as unprofessional. That is to ignore their purpose, and other professional prerequisites. It is a prerequisite that you know your patient well, and they trust you enough to know everything you say and do, is in their best interest. Only then can you safely employ humour. It would be inappropriate, shocking, and unprofessional, to tell jokes to those you don't know, with no purpose other than to while away the time. For me, medicine has always been a lifestyle occupation, and I only felt comfortable (in the long run), dealing with patients who shared my outlook and sense of humour. Along with technical expertise, practical ability, and sound clinical judgement, mutual understanding and appreciation, humour used as a management tool can help patients cope with their problems. In private medicine, where patients have the choice of whom they consult, they have no need to return to those whose style they find unacceptable.

Making Mistakes

Mistakes will happen even to the most competent, so learn to expect them. Surely competent doctors and nurses don't make mistakes? I wish that were true. Airlines accept that even their best pilots will make errors, and when they do, every other pilot should learn from their experience. This reflective attitude purveys the airline industry. Their non-punitive sharing of information has led to a culture that includes cross-checking, open discussion, and criticism of everything that arises. It has led to many improvements in safety. All pilots must show that they have learned from their mistakes by proving it in a simulator.

Most mistakes come as a surprise. They are often 'out-of-the-blue' events, so low on the scale of predictability that few expect them.

I did echocardiograms on two patients with the same surname on the same day. One had serious heart disease and AF, the other a normal heart. No prizes for guessing what happened. I sent the abnormal results to the healthy patient, and vice versa. The one with AF and heart disease lost confidence in me, even though I discovered the error and informed him before he had realised the error himself. He was an anxious, obsessive man by nature, and this led to his loss of trust in all doctors.

To avoid mistakes, keep vigilant and cross-check everything. Never make assumptions without confirming them repeatedly, especially when the consequences could be serious. There is a simple line to be drawn, however, between over enthusiastic checking (fear of making mistakes), and getting on with the job (desire for action). With many mistakes being serendipitous, a second pair of eyes is always useful.

I like the Italian word for 'worried': *preoccupato* (male form). Being preoccupied causes forgetfulness and inattentiveness. Some people are forgetful by nature (absent-minded); others develop it with age. Early dementia, as a cause of forgetfulness in doctors of my age, is of concern. One cannot easily detect it in its early stages. Hopefully, it will not lead to a clinical mistake, error, or oversight. It would be reasonable for all doctors working over 75-years of age, to be tested regularly for senile dementia (once reliable tests are available, and we can trust bureaucrats to interpret them correctly).

Customer Service and Medical Management

Convenience is key.

Your duty of care, when trying to avoid mistakes, should emulate an experienced server in a Michelin starred restaurant. Observe repeatedly, and respond appropriately; observe again, and respond further. For this, you need to be plugged in, switched on and focussed, not zoned out or day dreaming. Do not allow your attention to get deflected; you can register managerial data later. Stop texting, tweeting, and twittering. You can answer your emails, and flirt with your favourite co-worker later.

No client in a well-run restaurant should have to ask for the menu, a fork, or the pepper. A good server will notice the need and react before being asked. In principle, all nurses and doctors should anticipate the needs of patients. Some patients cannot ask for help; the vigilance of medical staff is then even more vital. In some countries, the constant presence of family members will remind medical staff of their duties.

'Gilly' Panayiotou is a good friend of mine. He runs 'Nippon', a famous restaurant in Larnaca, Cyprus. Having spent his life in his family's restaurants, he is a master of restaurant service. From his viewpoint, sitting at the bar (his equivalent to the bridge on a warship), he observes every minor problem, before it develops. By responding before any client raises an issue, he avoids discontent. Surveillance, recognition, and appropriate early intervention on behalf of his clients, is his modus operandi. Every nurse and doctor should emulate his skill set. Compare Gilly's skill to the usual management found in medical practices, and it is easy to see why doctors and nurses leave their patients wanting and dissatisfied.

Like clients ignored in a restaurant, or delayed passengers left uninformed at an airport, few enjoy being ignored and left without information. Some situations require the reverse. With fewer time-pressures and fewer demands on me while on holiday, I actually enjoy the luxury of being ignored.

Simple Management and Attention to Detail

Whether you are ironing a shirt, performing a tracheostomy, or sewing a button on a shirt, it is attention to detail that separates the amateur from the master.

Take two trivial examples:

1. I have always used chlorhexidine skin-wipes on the skin before venesection. I never used alcohol wipes because patients will feel pain as the alcohol enters the skin break. I know there is no good bacteriological evidence to support the use of any skin wipe, but many patients regard it as a common sense, harmless precaution.

2. After venesection from an antecubital vein, I would never bend the elbow immediately afterwards. It will not achieve excellent haemostasis; it will not usually provide enough pressure to stop bruising. What it will achieve is a quicker patient turnaround. Flexing the elbow removes pressure from the desired point. Blood then seeps out, and a bruise develops. Instead, keep the patient's arm straight and get the patient to press directly onto the needle entry point: 1-2 minutes for young men, 4-5 minutes for women and those with fragile skin. Bruising is an avoidable.

Every surgeon and physician will have their own tricks of the trade. Apprentices should aim to gather as many of these as they can.

One pertinent detail can make sense of everything.
One pertinent fact can falsify any argument.

Entrances and Exits

Last, consider your entrance and exiting behaviour. The first question must be, how well do you know your patient? Even if you know them well, there is no fixed guidance about using hugs and kisses, as patients arrive, or depart your consulting room (*les bises* – kisses on alternate cheeks. Either two or three is still customary *'en France'*). Those who prefer to remain anonymous will keep their distance, and regard any behaviour other than shaking hands, as 'unprofessional'.

I only used *des bises* and hugs with patients I knew well (as friends), and those seeking acknowledgement of a positive, successful, doctor-patient relationship. Use neither, unless they help to maintain a genuine bond. Use them as a sign of friendship and a token of mutual esteem. Both can bring re-assurance. Use them as non-verbal replacements for, *'Don't worry, you are in safe and expert hands',* or, *'Be assured, I have your best interests at heart.'* Appropriate exiting behaviour may be all that some patients need to gain confidence or to be consoled.

To have the trust of a patient always comes with a sovereign duty: to make the best clinical management available; clinically appropriate, efficient and successful whenever needed.

The ethos of medical duty I learned as a medical student may now be outdated, and difficult to match. Rudyard Kipling expressed it in his poem *'If'*. It still holds true for all those dedicated to their duty, of whatever gender.

. . . If you can force your heart and nerve and sinew
To serve your turn long after they are gone,
And so hold on when there is nothing in you
Except the Will which says to them: "Hold on". . .

Yours is the Earth and everything that's in it,
And—which is more—you'll be a Man, my son!

'If'. Rewards and Fairies (1910). Rudyard Kipling 1865-1936).

PART FOUR

FURTHER STUDY

Chapter Sixteen

Basic Clinical Research

'Research is the door to tomorrow.'
Motto: The Post Office Research Station. Dollis Hill.

I was once told by a CQC inspector that I would need a licence to 'do research'. I asked if I should stop thinking until I got one.

Those drawn to consider research will usually enjoy science and value scientific measurement. Most doctors are inquisitive beings, capable of expressing compassion as much as inquisitiveness. Research oriented doctors make valuable colleagues, but the job of caring for people is often best left to those capable of the art of medicine.

It is enough for most students to achieve their basic qualifications, but many will want to pursue specialist qualifications (MRCP, FRCS, FRCOG, MRCGP), or pursue a career in medical research (after a BSc or PhD). In career terms, the subject chosen for research will directly affect a doctor's progress. Research topics are subject to fashion and political orientation, and only those that appeal to both will easily attract funding. What follows may be of interest only to those interested in the altruistic pursuit of discovery.

Clinical Research. Is it a duty?

If you have an inquiring mind, are perceptive, and keen to find explanations for your observations, consider medical research. The first inspirational steps come from observations; those of possible clinical significance. It helps to be passionate about explaining all you observe. You will need the ability to devise experiments capable of proving or refuting your explanation. After that comes a little drudgery (at least for clinicians): the arduous process of data collection (even professional scientists delegate this). After one has completed the data collection process, the analysis can begin. Getting results is exciting, and comes with a challenge not to over-interpret them.

If you possess any clinical features of Asperger's, or have a strong obsessive - compulsive trait (i.e. you are a bean-counter by nature), full-time research could be for you. If not, delegate the data collection and analysis to a partner versed in scientific and statistical methods.

Those professionally engaged in statistical analysis are important members of every research team. In clinical research, however, the first aim must be to pre-empt analytical results by coming to trust your frequently made observations.

What bean counters can produce are the odds for your thesis being wrong (based on the null hypothesis). The process will provide statistical backup for sceptics and believers, given in terms of probability. A twenty-to-one chance of being wrong ($P=0.05$), barely supports a thesis. Instead, seek a 1000 to one ($P<0.001$) probability against your thesis being wrong. At this level of significance, your thesis is most likely correct. Your peer reviewers will try to find errors, wrong assumptions, and incorrect conclusions in your work. Your written work must survive their scrutiny and all attempts to disprove your observations and thesis. You should undertake all these first, before any critic tries.

The iconic case of Heinrich Hermann Robert Koch, trying to prove his ideas to others, is a case in point. To satisfy his critics, he had to invent and use his own rules of proof (his postulates). The same rules still apply if you believe that a specific microbe causes a particular infectious disease. One can only admire the determination he had to prove causation - that cholera, and tuberculosis microbes, were the responsible agents for those diseases and not just casual contaminants. He must have had the conviction, well before formulating his postulates. The scepticism of others served an important purpose: it motivated him to make his proof of causation incontrovertible.

Koch's postulates, although essential reading, are of little use to physicians and surgeons working at the bedside. Doctors work in the realm of anecdote, supported by evidence from controlled trials and meta-analyses. We now refer to medical practice that follows trial data as 'evidence-based medicine'. The term strongly implies dependability, with few bothering to check for devils in trial details, and the reliability of evidence they believe. Lawyers involved in clinical cases are in a worse position. With only second-hand expertise to weigh clinical relevance, their tendency is to give weight to any pertinent publication. We cannot expect them to appreciate scientific detail, and their liability to misinterpret the relevance of any science they quote. This is rarely up for discussion in a courtroom full of non-scientists. Legal proceedings (binary – as right or wrong, and adversarial) are an inappropriate forum for scientific discussion, even though the professional future of a doctor or scientist can depend on it.

Few doctors can claim to be both a scientist and a master of the art of medicine. The reasons for the rarity relate to risk taking. Typically, a medical scientist is a tick-box checker; a bean-counter needing all the data available before deciding anything. They rarely consider uncorroborated evidence and are first to accuse those who do, of cavalier behaviour. Doctors of this sort can hold off treating a rapidly deteriorating patient with sepsis until the microbiological evidence is available, 48-hours after 'best guess' treatment might have saved the patient's life.

An example of the same tardy, nerdish and self-protective behaviour in the UK, arose when the public health scientists advising politicians, expressed uncertainty about the value of wearing masks during the COVID-19 pandemic. They recognised the value, long after the public had accepted mask wearing as 'common sense'.

Every patient can harbour clinical clues that could lead to a discovery. Not Nobel Prize stuff, or enough to get you elected to the Royal Society, but good enough to make advances that could help patients.

Arcus senilis is a common physical sign caused by lipid formation around the iris. One can see it in patients who have had heart attacks and those at risk of them (so said Dr. P.N. of Charing Cross in the 1970s). What is its relationship to atheroma? From my personal data, 80% of patients with arcus have carotid atheroma. Only 60% of those with hypercholesterolaemia have carotid atheroma. Not all those with arcus have hypercholesterolaemia, or coronary artery disease, so there is scope here to discover the reasons for these simple observations.

The liver manufactures cholesterol, so what might be the connection between blood cholesterol and tissue (arterial intima) cholesterol—the real culprit in coronary and cerebral artery disease? I strongly suspect that different genes determine the formation of arterial lipid plaques and arcus senilis, while blood cholesterol levels relate more to hepatic metabolism. If this is true, finding evidence of tissue lipid accumulation (iris, carotid intima, etc.), will better predict coronary artery disease than any blood lipid. This is an example of how simple clinical observations might advance diagnostic accuracy; providing predictability through an understanding of the mechanism, is one purpose of science.

Observational research will add interest to a medical career. Writing up your observations, and getting them published, is another matter. Professional researchers, and journal editors, will treat your work with disdain if you are a GP, or a doctor working in a non-teaching hospital. They may question your qualification as an observer / scientist, if you do not hold a PhD and a record of published work. The internet has become a game-changer. With internet publishing, one can bypass the political (and scientific) selectivity of publishing establishments.

One should never tolerate less than rigorous methodological scientific standards. To get recognition in a prestigious journal, several acknowledged referees must review your work. Fashionable hot topics, and political expediency, both influence the peer-review process. It is always important to consider what is already know about a subject, and what additional knowledge your research might add. Publishers know that eye-catching headlines, 'advances' and 'hot topics' sell journals. This partly explains why negative results rarely get published.

At academic conferences, any research quoted that is over 5-years old, is likely to be regarded as 'out-of-date' and not worth considering (the Travis Effect bias). My education has been deficient: I never learned that truth aged or that it had a sell-buy date.

Undertaking research will broaden your outlook and help you evaluate evidence. How would you answer the question I posed at the head of this brief chapter, 'is doing research a duty?' The answer is 'No', but consider what research might do for you. The answer is that it can help advance your career. One should not underestimate the personal esteem gained by discovering something new, no matter how trivial. It will stand to your credit if you have published a paper on XYZ, regardless of how important the contents.

To stem hubris, scientists should acknowledge they have merely uncovered (and sometimes explained) what exists; they have played no part in its creation.

When asked about the Nobel Prize he won, Richard Feynman modestly said that he had already won the most important prize, discovering something important about how quantum mechanics works.

If you have an interest in clinical research, but no interest in a research career, there is much to capture the interest of an enquiring physician or surgeon. If you have a professional interest in research, you will need to work in a research department, under the watchful eye of a professor to supervise and encourage your work.

In the early 1970s, I was a British Heart Foundation research fellow at St. George's Hospital, Hyde Park Corner (the building later became the Lanesborough Hotel). Under the watchful eye of Graham Leech, our cardiac unit was one of the first in the UK to use cardiac ultrasound. In the 1970s, we made simultaneous recordings of heart valve movements and the sounds they made. For the first time, we visualised the valve movements of mitral stenosis; the opening snap, and the 'clunk' of a pedunculated atrial myxoma. We saw the mitral valve leaflet vibrations associated with Austin-Flint murmurs in aortic incompetence. What we saw became so routine, we didn't think to publish it. That was a mistake. We gave too little weight to the fact that few others in the world had then visualised the origins of heart sounds.

I discovered some electrophysiological differences in patients presenting with sinus bradycardia. Different electrophysiological mechanisms pertain to the sinus bradycardia of athletes and that of sino-atrial fibrosis (sick sinus syndrome). I later found evidence for a loss of conducting and pacemaking functioning in the AV node, prior to AV block. My autonomic reflex studies suggested (but never proved) that this might relate to the reduced functioning of atrial tissue cholinesterase, the accumulation of acetylcholine, delayed AV conduction and sinus bradycardia. I knew diphtheria to be a precursor of complete AV block, and sino-atrial dysfunction. Interestingly, diphtheria toxin is an anticholinesterase, and likely to affect conducting and pacemaker functions in tissues rich in cholinesterase.

I tried to take my idea further, electing to work with Dr. Ann Silver (author of *'The Biology of Cholinesterases'* 1974), at Babraham near Cambridge, but found the research more than I could commit to. Later, while working at Charing Cross Hospital, I collaborated with others to show that slow pulses found in athletes was due, not to increased vagal tone (an increased frequency of parasympathetic afferent impulses), or reduced sympathetic tone, but to a reduction in the intrinsic rate of atrial pacemakers (Katona et al. See bibliography).

In the early 1980s, while working in Amsterdam, I was part of a team headed by Dr. Pim de Feyter. He showed how intra-coronary streptokinase could lyse coronary artery thrombus (occlusion causing cardiac infarction). This worked well, as long as the patient arrived in our catheter lab. within four hours of the onset of chest pain (more easily achieved in a small city). The instant relief of pain felt by patients as their clot dispersed, accompanied an immediate improvement in their cardiac output. This was before the introduction of angioplasty and artery stenting. Pim also headed our team, matching exercise ECG changes with specific coronary artery stenoses.

There is excitement to be had working in an active, collaborative research team. A shared ethos of discovery and development, promotes advances in knowledge.

In private practice, I started collecting clinical data in 1973. I hoped to support the thesis that medical screening and early diagnosis would improve prognosis in cardiac and general medical cases, even though I regarded the assumption as obvious (BLOB, in fact).

The onus of those with theories is to prove them before inviting others to accept them.

A patient of mine once asked me what he should eat after his coronary by-pass operation. He said he knew what not to eat, but couldn't find information about what he should eat. The aim, I suggested, was not just to lower blood cholesterol but to halt the progress of his coronary atheroma. He was not interested in the dietary hearsay of TV chefs, so I analysed hundreds of animal experiments designed to study the connection between diet and atheroma. I searched the scientific literature for evidence that specific nutrients might benefit atherosclerosis (if taken over decades). They included the amino-acids arginine and taurine, some metals (Se., Mg., Mn., and Zn), and some vitamins (natural vitamin E, folic acid, B6, and B12). I identified the foods that contained most of them, and the amounts of saturated fats, trans-fats, and sodium chloride they contained.

Clearly, a 'good / bad' ratio calculation, might simplify the cardiac risk assessment of foods. Since none existed, I had to invent one. I called it 'the Cardiac Value of Food™'. This ratio expresses the sum of the atheroprotective nutrients, divided by the sum of all atherogenic nutrients, in each food (derived from published nutrient analysis). One problem remains: nobody knows how much each element contributes to atherogenesis, or its prevention, so I left the work for others to take further. It is likely that any dietary effect on arteries will take decades to manifest. Any research into the benefits of atheroprotective nutrients on atheroma would have to last a very long time, so I decided it was not for me.

So far I have had no enquiries about my results. I can hardly expect food manufacturers to welcome an index that suggests their products might be 'bad for the heart'. I did, however, expect more interest from patients. Given my lack of interest in nutrition, my job was to generate the concept, not to develop it further. I later wrote two books on the subject for the public.

This theoretical nutritional work of mine yielded a few surprising conclusions in need of further verification.

1. If 'atheroprotective' nutrients work as well in humans as they do in animals, and if we only require one recommended nutritional intake (one RNI) daily, no acceptable diet can deliver them.

2. Cow's milk, taken after the normal weaning age, could significantly contribute to atherogenesis.

3. Olive oil, beloved by many, contains some saturated fat. It is, therefore, unlikely to be as good as sunflower oil for atheroma protection.

4. The presence of taurine and nutrient metals in seafood, is likely to override the atherogenic effect of some fats present. Although the gastrointestinal absorption of cholesterol in food is poor, CHD patients have long been told to avoid cholesterol-containing sea-foods.

5. Offal, nuts, and seeds contain most nutrients with atheroprotective potential.

I would like to have used my 'Cardiac Value' to rate the various foods of different nations and to correlate them with the national incidence of IHD. I suspect that the dietary effects are small, and easily overridden by genetic factors.

So far, the only people to take any interest in this important subject are patients. That's OK, they have always been the focus of my attention. You can read more about the subject in: *'Eat to Your Heart's Content'* (2005; ISBN: 0-955-10720-2). The book suffers from not being technical enough for doctors, and too complicated for lay people. I also wrote a simpler version for patients (*'Heart Sense'*, 2006, in paper-back and digital form).

Over the last twenty-years, I collected data on the types of carotid atheroma associated with coronary atheroma. My results strongly suggest that the absence of carotid atheroma in an individual, reliably excludes coronary artery disease. I have seen five exceptions, having studied 8000 cases over twenty years. The future relevance of this research is that GPs and A&E doctors could, by undertaking a carotid artery ultrasound, detect atheroma (using a portable machine or iPhone app). The pathological cardiovascular risk of finding atheroma is likely to prove better in individuals than calculating it indirectly (age, weight, smoking habit, cholesterol, LDL, HDL and qRisk3).

Paradoxically, lipid studies correlate well to CHD morbidity and mortality in population studies. Since the statistical paradox applies, one can safely say that blood lipids are of no diagnostic value for predicting atheroma in individuals.

Another interesting observation I made, but did not follow-up, was the discovery of left ventricular hypertrophy (LVH), well before sustained hypertension developed. This is likely to exist because of a dominant genetic influence on arteriolar medial hypertrophy and LVH in sustained primary hypertension. Genes are likely to drive both, with arteriolar medial hypertrophy the pathological cause of what we once called 'essential' hypertension.

I have only observed LVH as a precursor in two cases, but like one black swan, it proves the concept. I made my discovery after screening 'normal' people using echocardiography. This is something few doctors will do.

I also collected longitudinal data on the rate of development of carotid atheroma and LVH in individuals. I learned atheroma can take decades to grow significantly, and LVH can appear within one year.

Over forty-years ago, I discovered that sensory loss in the feet can precede diabetes by a decade or more. Seeing the same patients for decades enabled me to make this observation. Because hospital-based diabetologists have no access to 'pre-patients', the selection bias would prevent them making this observation.

I have taken these simple observations no further myself, but hope at least one reader will carry them forward.

Simple 'Experimental' Clinical Medicine

Any mention of an 'experiment', will draw medical bureaucratic attention.

The simple therapeutic research I undertook, is now far too dangerous for all but cavalier doctors trying to help patients find a successful treatment. Prescribing any drug outside of its approved use (as in the BNF/NICE guidelines), could bring sanctions, even if scientifically justified. The fear doctors have of bureaucratic sanctions, now suppresses intelligent, open-minded, free thinking and creativity. Going 'off piste', if argued from a knowledgeable academic standpoint (which bureaucrats and corporate lawyers are ill-qualified to acknowledge), with a reasonable antecedent probability of success, can foster imaginative therapeutic benefits for a few patients found difficult to treat by any other means.

Managing patients *'off piste'* when required, can prove successful. I dealt with atypical challenging cases, where accepted practice had not previously produced a successful outcome, although this was mostly because of an incorrect diagnosis. Sometimes, failure to recognise the underlying pathophysiological mechanism had stopped a doctor finding a successful therapy. Sometimes, they made a correct diagnosis, but then ineffective guidelines had constrained them.

The perceived need to adhere to fixed rules in any academic field, will restrict any thinking outside the box, and inhibit alternative action. Doctors are fully aware of the punitive attitude regulators have towards them, and understandably, are unwilling to risk any deviation. It will put their career in jeopardy, especially if their management goes wrong, and a patient, relative, or pharmacist complains. Why tempt providence? In fact, why bother thinking at all, now that we have guidelines and treatment algorithms to follow?

Here are a few examples of my own, off-piste, therapeutic research trials:

- I 'experimented' on non-asthmatic patients whose cough failed to resolve, three weeks after a viral chest infection, despite treatment with antibiotics and bronchodilators. Could bronchial leukotrienes be involved? After taking montelukast, 80% quickly improved.

- With pain of unknown origin, many standard analgesics will fail. The exact nature of the pain needs to be defined, but can remain elusive. If gnawing in quality, and 'sharp', resembling an electric shock, neuralgia is a possibility. Soon after being introduced, I prescribed gabapentin for patients with this type of pain. I also used it to be a therapeutic test of neuralgic pain. If it worked, some form of nerve compression was likely. It often worked well in atypical cases with no obvious neurological cause. Some were anxious and the sedative effect of gabapentin helped.

- I saw many young women with pre-menstrual breast discomfort on exercise. Although not licensed for it, I gave several patients a trial dose of tamoxifen (oestrogen blockade) to help the problem (with full informed consent). Not only did it prevent the pain, it sometimes prevented its return.

- Tiredness is a very important issue, given the number accidents it causes and days lost from work each year. A common presentation is unexplained tiredness (TATT: Tired all the Time). If the patient claimed to sleep well (ruling out stress-related insomnia), I tested them for Epstein Barr virus antibody levels. Recurrent glandular fever is difficult to prove without demonstrating rising EB antibody titres, or viral antigen detection, but I assumed that high antibody levels must be of some significance (their clinical relevance is another issue). Recent EB virus re-activation was not an unreasonable diagnostic suggestion. Even if correct, no proven treatment is on offer. We know that some older antibiotics (tetracyclines) inhibit RNA virus replication (EB is an RNA virus), so I decided (30-years ago) to try oxytetracycline, minocycline, and doxycycline.

I found that eighty percent of TATT patients (with raised EB antibody titres) improved within two weeks. Most fatigued patients had been unable to work for months or years. The expected effectiveness of a placebo is 20%, so these results were encouraging. I should then have started a double-blind trial.

Because COVID-19 is an RNA virus, I also wondered if the same antibiotics might help. I retired before the first wave of the COVID-19 pandemic, and could not try them.

- Structured questionnaires are a useful way to gain an insight into a patient's mental state. The best have been checked for leading questions and dissimilar choices etc.. Computerised questionnaires have advantages: the questions remain the same, and no bias exists when scoring them. We can program questionnaires as standardised interventions and repeat them as necessary. They allow the tracking of patient data throughout time, and the comparison of one patient to another.

In the late 1980s, I administered the Middlesex Neuroticism questionnaire to 987 patients. Patients completed the questionnaire while sitting at a computer. I found something unexpected. Many patients felt this machine-based system avoided embarrassment. Sometimes they even felt a cathartic, therapeutic effect, after being asked some questions for the first time. Some said the questions had: 'delved into my innermost feelings' (by implication: unlike me and most doctors). The questionnaire seemed to help some achieve insight into their neurotic traits.

In my experience, this patient-computer interface seemed to dis-inhibit people; allowing them to be more open and honest. The possibilities for aiding clinical management in this way are exciting, although patients need to know how the results might direct any clinical management decisions. Checklist routines, in many medical and non-medical situations, can reduce morbidity and mortality. Computers can easily deliver and score them.

For those interested in research, potential discoveries are to be found everywhere. Only after observing a 'new' phenomenon several times, however, should one embark on researching it further.

In my experience, the acceptability of a doctor to patients and his clinical success, rarely relates to the number of papers he has published, or the number of degrees he holds. Clinical medicine and research are separate, albeit complimentary disciplines.

Research can suffer from short-termism. Few projects, apart from those like the Framingham Study, continue for decades. Some projects need a lifetime to complete. I would like to have answered the question: could deficient dietary intake of minerals, vitamins, and amino-acids throughout life, be responsible for some of the mortality and morbidity differences seen between socio-economic groups? (This is the thesis I expounded in my book, *Eat to Your Heart's Content* ISBN 0-9551072-0-2). To prove this would take a lifetime, and without an efficacious elixir of life, I have too little time left to prove it. Before you commit yourself to any research, consider the time it will take to complete.

Researching the Literature and Publishing

The British Library, and the RSM Library hold unparalleled resources. If you are intending to write a research paper or scientific monograph, it's best to have access to such resources. You will need to review all the previous relevant publications (now made easier with AI) and be able to state what is new about your work. To get published in a prestigious journal, your work will have to pass peer review. It helps to have high-profile supporters. First make sure that your idea is new, or an updated development of an older idea.

If you are intending to write a general interest book, you may wish to publish it yourself (through Amazon KDP or Ingram Spark), but first review the competition. Who else has written a similar book, and what are the key search terms? I have done this several times now. Don't expect the world to sit up and pay attention. Your book will have to compete with the thousands of books published every day. You may choose digital marketing or placing adverts yourself, but this is expensive, and only rarely profitable. Contact me is you want to know more about independent publishing (david@daviddighton.com).

Allowing patients to 'take the risk' (Institute for Cancer Research Website)

Speaking on a BBC Radio 4 *'Today'* programme (January 2018), Baroness Tessa Jowell recounted her experiences. She was told she couldn't have further treatments for her brain tumour on the NHS. That was before being offered immunotherapy in Germany.

She said: 'That (an adaptive trial) is exactly the kind of risk patients should be free to take... and it's certainly what somebody like me wants.' Tessa Jowell died a few months later (May 2018).

How are researchers going to cut clinical trials short (required by adaptive trials), in rapidly progressive clinical situations, when they need to collect sufficient data for analysis? The need for data collection may be at odds with a patient's wishes. If a trial patient feels no immediate improvement, and the investigation needs more time, some patients may want to stop their involvement and switch to another trial.

Naming concepts (like 'adaptive trial') gives it media and political credence. Following Tessa Jowell's death, the label nudged the government into doubling brain cancer research funding. One reason a patient might want to switch trials, is that their time could be fast running out. Other trials may not have much antecedent promise, so patients may only have their belief and hope to guide them in 'life and death' situations.

Those experienced in making life-changing decisions will not support policies that continually fail. The trick is to find an evidence-based alternative. The fight between the 'adaptive trial' approach (common sense, from a patient's point of view), and the scientific requirement to get enough credible data for a well-designed, double-blind trial, is a good example of the art of medicine at odds with the science of medicine.

Epilogue

'Though much is taken, much abides; and though we are not that strength which in old days moved earth and heaven, that which we are, we are; one equal temper of heroic hearts, made weak by time and fate, but strong in will to strive, to seek, to find, and not to yield.'

Alfred Lord Tennyson (1834). *'Ulysses'*; Vol 2.

For many years I thought this book showed be called 'The Doctor's Apprentice', but thought the current title more descriptive. Not every master will welcome an apprentice, but most students will benefit from a master. To remain competent, every doctor must remain an apprentice.

The art of medicine has developed over millennia; the science of medicine over a few centuries. Because it is easier to regulate, the science of medicine is displacing the art of medicine, but for the sake of patients we should not allow this to happen. Those who practise both and bring benefits to patients, should acknowledge how indebted we all are to those who, before scientific medicine developed, practised only the art of medicine.

I hope those who read this book will learn some practical tips about how to practise medicine, judged as successful by patients. This requires using both its art and science, its judgement tools, methods and creative processes, rather than the anonymous, check-box ticking algorithmic processing that is leading patients to become ever more discontent with the UK medical profession. Routine medical practice has the power to save lives and to reduce suffering, but unregulated independent thought and creativity can also help in problematic cases.

Doctors and nurses are not alone in caring for patients. All those who care for others can share in it. Caring for others is special, and capable of fulfilling the lives of all those who engage with it. Because others can be better at it than some doctors and nurses, the place of medical practice in society is mobile not stable.

Those privileged to practise medicine and nursing, hold a special place, albeit a changing one in society. In this place, one can help to improve and protect the state of health and quality of life of our fellow humans. Like food, air and water, most humans think medical practice is indispensable, yet medical practice has always been more than just a service, co-opted by politicians to buy votes and to provide jobs for bureaucrats. It is a sacrosanct discipline with duties specific to its sovereign status. Those who disagree should undertake a simple thought experiment: image life without medical professionals and carers.

Having retired before any bureaucrat could do their worst (something that soon followed), I now watch bureaucrats and regulators exerting more of their untutored influence on patients, doctors and nurses. It is the nature of bureaucracy to imprison us in an iron cage (*Stahlhartes Gehäuse,* Max Weber (1864-1920)); something medical bureaucrats in the UK have now successfully achieved. Although this book discusses topics far more important than bureaucracy and politics, it is partly bureaucracy that has put the medical profession and UK clinical services into crisis. Please accept my apologies for having to mention such matters; I have included such antipathy because all I can do is watch UK patients suffer from inept political and bureaucratic action, accompanied by the gradual depersonalisation of medical practice. Both need to change if we are to help patients most.

Glossary of Terms

ACE: Angiotensin Converting Enzyme inhibitor. A pharmaceutical used in high blood pressure and heart failure.
  Acute: recent onset.
  AF: Atrial fibrillation. An irregular heartbeat.
  AI: Aortic Incompetence. Aortic valve leakage (incompetence).
  Angioplasty: Opening of an artery using a balloon catheter.
  Anterior: Front.
  ARB: Angiotensin Receptor Blocker. A pharmaceutical used in high blood pressure and heart failure.
  Arteriopath: A person prone to atherosclerosis.
  Arteriosclerosis: thickened arteries (medial hypertrophy of arterioles) in hypertension.
  Atheroma: the intimal build-up of cholesterol and calcified compounds in arteries.
  Atherosclerosis: See atheroma.
  AV Block: electrical blockage (half way down the heart) in the AV node.
  BBB: Bundle Branch Block (electrical blockage of a branch of ventricular electrical fibres).
  Bd: Medication taken twice daily (bi diem).
  Benzodiazepine: a class of tranquillizer drugs.
  Beta-Blocker: drug that slows the heart by blocking the effects of sympathetic drive.
  BLOB: **BL**indingly **OB**vious.
  BNF: British National Formulary.
  Bureaucrat. From the French 'bureau' (desk), and the Greek 'kratos' (power or rule).
  CABG: Coronary Artery Bypass Graft.
  CAD: Coronary Artery Disease
  Cardiac Infarction: Death of a segment of heart tissue (heart attack).
  CCU: Coronary Care Unit.
  CHB: Complete Heart Block. Electrical disconnection of the top part of the heart from the bottom.
  Chronic: Long-term.
  Circumflex: A coronary artery that provides blood to the side and back of the heart.
  Collateral: Extra arteries.
  Complete Heart Block: See CHB.
  COPD: Chronic Obstructive Pulmonary Disease: bronchitis, emphysema and asthma, etc.
  CPK: A blood enzyme derived from muscle.

CQC: Care Quality Commission.
Cyanosed: blue lips etc. The result of de-oxygenated blood (heart and lung disease).
Distal: In anatomy, a position further away.
DVT: Deep Vein Thrombosis.
Dysrhythmia: abnormal heart rhythm.
ESR: Erythrocyte Sedimentation Rate: a blood test index of inflammation.
Exophthalmos: Eyes that protrude, found in hyperthyroidism.
FRCGP: Fellow of the Royal College of general Practitioners (London).
FRCOG: Fellow of the Royal College of Obstetricians and Gynaecologists. (London).
FRCP: Fellow of the Royal College of Physicians (London)
FRCS: Fellow of the Royal College of Surgeons. London.
GMC: General Medical Council.
GTT: Glucose Tolerance Test. Involves drinking glucose and testing the blood afterwards.
G.U. Genito-urinary.
HbA1c: The glucose attached to red cells in the blood. A measure of diabetic control.
HDL: High Density Lipoprotein: thought of as 'Good' cholesterol.
Hepatic: Pertaining to the liver.
Hyper-: Increased
Hyperglycaemia: Increased blood glucose.
Hypertension: High blood pressure.
Hyperthyroid: Overactive thyroid gland.
Hypo- : under.
Hypoglycaemia: low blood glucose.
Hypotension: low blood pressure.
Hypothyroid: Underactive thyroid.
Hypovolaemia: Under-filling of the circulation (as in blood loss and dehydration).
IHD: Ischaemic Heart Disease (narrowed arteries with less blood reaching heart muscle).
Inferior: Lowest position.
j-point: Point on an ECG (where QRS meets the ST segment)
Ketosis: Ketones in the blood (and on the breath) during starvation, or dieting by omitting carbohydrates, or diabetes that is out of control.
LDL: Low Density Lipoprotein: thought of as 'bad' cholesterol.
LV: Left ventricle (main heart pumping chamber).
LVH: Left Ventricular Hypertrophy: increased heart muscle with hypertension, etc.
Metastases / metastatic: cancerous tumours that have widely spread.
MI: myocardial infarction or mitral incompetence (leakage).
MP: Metacarpal Phalangeal joints of the hand.
MRCP: Member of the Royal College of Physicians
MS: Mitral (valve) Stenosis (narrowing), or multiple sclerosis.
Myxoedema: Sign of an underactive thyroid. Synonym for hypothyroidism.
NEJM: New England Journal of Medicine.
NICE: National Institute for Care and Health Excellence.
NSTEMI: No ST Elevation seen in Myocardial Infarction. No ECG changes seen.
OPD: Out Patient Department.
OTC: Over The Counter preparation.

PA: Pulmonary artery.
PE: Pulmonary Embolus (clot that has travelled to the lung).
PND: Paroxysmal Nocturnal Dyspnoea (usually due to left heart failure)
Posterior: behind.
Proximal: In anatomy, the nearest position.
PSA: Prostate Specific Antigen (blood test for prostate cancer).
PSA: Professional Standards Authority.
PTSD: Post-traumatic stress disorder.
PUO: Pyrexia (temperature) of Unknown Origin
Qds: Medication taken four times daily.
Raynaud's Condition: Inherited spasm of hand arteries.
RBC: Red Blood Cell.
RCP: Royal College of Physicians.
Renal: pertaining to the kidneys.
Retrospectoscope: a fictional instrument that can look back in time.
RV: Right Ventricle.
RVH: Right Ventricular Hypertrophy.
SBE: Sub-acute Bacterial Endocarditis (infection of heart valves).
ST Segment: Feature of the ECG (can change in hypertension and IHD).
Statins: A class of pharmaceuticals that can lower blood cholesterol. HMG-CoA reductase inhibitors
STEMI: Heart attack with ECG changes (ST Elevation in Myocardial Infarction).
Stents, stenting: Insertion of a metal cage within an artery to stop it narrowing.
Stenosis: narrowing (of a blood vessel, etc).
Superior: Anatomy, above.
SVT: Supra-Ventricular Tachycardia: fast heart beats arising from the top of the heart (atria).
Tds: Medication taken three times daily.
TI: Tricuspid valve incompetence.
U/S: Ultra Sound.
VE: Ventricular Ectopic. Extra heart beat arising from the ventricles.
VF: Ventricular Fibrillation: Fatal heart rhythm. The heart shimmers, but does not pump effectively.
VT: Ventricular Tachycardia: Fast heart rhythm arising from the heart, needing immediate correction.
WBC: White Blood Cell.

Bibliography, Notes & References

BRIEFING NOTES FOR MEDICAL APPRENTICES

Adamson, Mike. Chief Executive Red Cross (2017).

Asher, Richard (*'Talking Sense'*, Pitman, 1972)

Clinton, Bill. US President Bill Clinton said that political candidates have a greater chance of winning (votes) by being strong and wrong rather than by being right and weak. The *NY Times* December 4, 2002, Section A, Page 1.

Cohen, Dr. R.D.. Prof. Clifford Wilson's obituary (*Independent Newspaper*, 19.11.1997).

Connolly, Billy. (2018). *'Made in Scotland'*, p42, Random House.

Dickens, Charles. *David Copperfield*. 1850. Uriah Heap.

'Doctor in the House.' 1954. Film: Rank Organisation.

Five Wisdoms of Buddhism (Wikipedia).

Goodenough Committee. *BMJ*. July 22, 1944.

Han Fei. *Book of Han Fei*.

Josh Cohen, *'Not Working'* (Granta, 2018, p214),

Machiavelli, Niccoló (*Book VI of 'The Art of War'* (1521).

Minghella, Dominic. *'Doc Martin'* (played by Martin Clunes in the UK, TV series (2004-2009),

Parkinson, C. Northcote: *'Parkinson's Law'* (1955). Dead Author's Society.

Pliny, Adam and Felix Roesel, Ifo working paper number 328.

Politkovskaya, Anna. Russian journalist, said during a *BBC News interview* (14/9/2004) – 'the job of doctors is to give health to their patients.'

President of the RCP: *Bulletin from the President of the RCP:* May 2014.

Ruiz de Santayana y Borrás, Jorge Agustin Nicolás (1863–1952).

Shakespeare, William. Claudius in *Hamlet*.

Shakespeare, William. Polonius. *Hamlet*.

Shang. *Book of Lord Shang*.

Tagore, Rabindranath. *'Unity in Diversity'*.

Watson, James of Watson, Crick and Wilkins.

Wilde, Oscar. *'The Picture of Dorian Gray'. Chapter 8.* 'We live in an age when unnecessary things are our only necessities.'

Xenophon's Dialogue, *Ieron (Hiero)*, 4th century BC (Chapter 7, section 3).

PART 1
Encounters

DOCTORS

Aeschylus (*The Persians*, 441 BC)

Asher, Dr. Richard: *Seven Sins of Medicine.* Lancet, August, 1949.

Asher, Dr. Richard . *Talking Sense.* Pitman Books. 1972.

Betjeman, John. *An Oxford University Chest.* 1979. Oxford University Press.

Bowen, Jeremy. *Six Days.* (2004). Pocket Books. Six Days War (Egypt vs. Israel, 1966).

Clavell, James. TV miniseries *Shōgun*, 1980s, Paramount TV.

Company of Parish Clerks in London 'Twenty died of parental 'Grieffe'.

Culpeper, Thomas. (1515 – 1541) *'no man deserved to starve to pay a proud, insulting, domineering physician'.*

Delgado, A., Lopez-Fernandez, L.A., Luna, J de Dios (1993). *Medical Care,* 795-800.

Descartes. *The Cartesian view – 'cogito ergo sum'.*

Dirac, Paul A.M. *The Dirac equation (1926). 'Principles of Quantum Mechanics.'* International Series of Monographs on Physics (4th ed.). Oxford University Press. p. 255. ISBN 978-0-19-852011-5.

Doctor in Clover, (Film. Rank Organisation, 1966).

Eliot, George. *Middlemarch (1871).*

Eliot, T. S. (1888 – 1965). *The Hollow Men* (1926).

Fish Called Wanda, A (Film, MGM. 1988).

Flick, Otto Herr, played by Richard Gibson. TV series *Allo, Allo.*

Forsyth, Mark. *The Elements of Eloquence*, p4, 2013, Icon Books.

Fry, Dr.John (1922-1994). Doctors, 'cure sometimes, relieve often, comfort always, and prevent hopefully.'

GMC. *The Workforce Report.* 2022.

Good Schools Guide, The, Galore Park. 2012.

Granger, Dr. Kate. *'#HelloMyNameIs'*

Grier, Germaine. *The Female Eunuch.* (1970). MacGibbon & Kee.

Good Year, A (Film. 20th Century Fox, 2006).

Hardy, G. H. *A Mathematician's Apology.* (1940). 'No good work was ever done by humble men.'

Harvey, Dr. William . 1628. *Exercitatio Anatomica de Motu Cordis et Sanguinis in Animalibus.*

Hobbes, Thomas. *The Leviathan'.*(1651).

Illich, Ivan. *The Obsessional with Perfect Health. 1999.*

Illich, Ivan: *Limits to Medicine: Medical Nemesis.* 1974. Re: *'narcissistic scientism'.*

*I'm All Right Jack'.*1959. (Charter Film Productions, Boulting Brothers).

Jenkins, Simon. *The Guardian 25/8/2017:* 'It is more important to make what is important measurable, than to make what is measurable important.'

Kennedy, J.F. *Do not negotiate out of fear and do not fear to negotiate.*

Lactantius (240 – 320 AD). *Provide for others through humanity, what we provide for our own family through affection.*

Lill, M.M., Wilkinson, T.J (2005) BMJ 331 (7531); 1524-1527), *'Patients prefer their doctors to smile and to wear conservative clothing.'*

McChrystal, General Stanley A. *Team of Teams*. (2015). Tantum Books.

M*A*S*H; TV series; 20th Century Fox television.

Matthew, 25:29. King James' Bible.

Maraolo, et al. 2017; *Doctors working long hours. Personal life and working conditions of trainees and young specialists in clinical microbiology and infectious diseases in Europe: a questionnaire survey. Eur. J. Clin. Microbiol. Infect. Disease. doi: 1007/s10096-017-2937-4*

Merleau-Ponty, Maurice. *Phénoménologie de la perception.* (1945).

Minghella, Dominic. *Doc Martin.* (played by Martin Clunes in the UK, TV series (2004-2009),

National Health Services Act came into effect on the 5th July 1948.

Nelson, Lord Horatio (1758 – 1805): *'Glory is my object, and that alone.'*

Nixon, Dr. Peter. *The Life Function Curve.* The Practitioner. 1979.

Palin, Michael. *Full Circle, trip of the world* (BBC 1997).

Pantridge, Dr. James Francis ('Frank') (1916-2004).

Pauling, Linus. (1901 – 1944). The Linus Pauling Institute. Oregan State University.

Pope, Alexander. *An Essay on Criticism.* (1711). Part 2. Anodos Books (2017).

Proverbs 21.2. St. James' Bible. *'Every way of a man is right in his own eyes.*

Quiller-Couch, Sir Arthur Thomas. *The Art of Writing.* (Cambridge UP, 1917), quotes Newman's definition of a gentleman.

Report on Residents in US and Canada, 2015. Association of Medical Colleges.

Sacks, Dr. Jonathon. Former UK Chief Rabbi, *The Power of Praise*, 2018.

*School for Scoundrels. (*Film. Associated British Picture Corp., 1960).

Sedaka, Neil. *'Besides the gift, you need the drive.'*

Semmelweis, Dr. Ignaz Philipp . Hungarian gynaecologist who discovered that when doctors washed their hands after dissection, it prevented puerperal fever.

Shakespeare, William: *As You Like It (Act 1, Scene 1).*

Shakespeare, William. *Hamlet: Act 1, Scene 5. 'For every man has business and desire.'*

Shakespeare, William. *Hamlet: Act 3, Scene 1.* '*The insolence of office . . .'*

Shrek. (Film. Universal Pictures, 2001).

Shulchan Aruch, The. Joseph Karo. 1563.

Snow, C.P. *'Rede Lecture', 1959. The Two Cultures and the Scientific Revolution.* Martino Publishing. 2013).

Theophrastus. *The Characters of Theophrastus.* Book.

Thom, René. Structural Stability, Catastrophe Theory, and Applied Mathematics.'(1977).

Vidal, Gore. *'It's not enough for me to win. My friends must lose.'*

Walton, Izaak. 1653: *The Compleat Angler. 'Those restless thoughts which corrode the sweets of life.'*

Waugh, Evelyn. *Decline and Fall (1928).*

Waugh, Evelyn. *Decline and Fall. (1928).* *'We schoolmasters must temper discretion with deceit.'* Words of *Dr. Fagan to Mr. Pennyfeather.*

Wood, Victoria. *'Where do you get your ideas from?' 'I don't know. But, if I ever find out, I'm going to live there.'*

Xenophon's Dialogue, *Ieron (Hiero),* written in the 4th century BC (Chapter 7, section 3),

ABOUT PATIENTS

Aristotle. *Ethics.* Stanford Encyclopedia of Philosophy (2024). Department of Philosophy, Stanford University. Library of Congress Catalog Data: ISSN 1095-5054

Augustine, Saint. *Confessions.*

Berne, Eric. *The Games People Play, The* (1964), . Grove Press.

British National Formulary. BNF.

Chester v Afshar:[2004] UKHL 41.

Dekker, Thomas. *Sweet Content.* (1575 – 1641).

Dickens, Charles (1849) *David Copperfield.*

Enfield, Harry. *The Slobs* (BBC 2 TV, 1990).

Friedman, M.; Rosenman, R. (1959). *Association of specific overt behaviour pattern with blood and cardiovascular findings.* Journal of the American Medical Association. **169** (12): 1286–1296.

GMC. *'Good Medical Practice.'*

Harvey, William. 1628. *'de Motu Cordis'.* Understanding the connection between human emotion and the pulse rate.

Illich, Ivan. Notes that litigation has made even first-aiders think twice before acting as a Good Samaritan.

Illich, Ivan. *Limits to Medicine.* 1975. 'Doctors are at their happiest making an 'interesting' diagnosis.'

Jeremiah 13:23: 'Can an Ethiopian change his skin or the leopard his spots?'

Langely, J.N. Brain; Part 1 (1903). *The Autonomic System.*

Langley J. *Observations on the physiological action of extracts of the supra-renal bodies.* J Physiol. 1901; 27:237-256.

Life of Brian, The (1979), HandMade Films, and Python (Monty) Pictures.

Middlesex Hospital Questionnaire (MHQ), B.J.Psychiatry (1966); 112: 917-23).

NICE. National Institute for Health and Care Excellence: NICE Guidelines.

Garofalo, M. E. *Florence Nightingale, Florence and the Crimean War. Am. J. Public Health. (2010); 1591.*

Ovid: *Metamorphoses*; Book III:402-436).

Schiller, Friedrich. *The Maid of Orleans [Die Jungfrau von Orleans]*, Act III, sc. vi (1801) [tr. Swanwick]*Mit der Dummheit kämpfen Götter selbst vergebens.* 'Against ignorance the Gods strive in vain.' (1801).

Shakespeare, William. *As You Like It*, Act 2, Scene 7. 'All the world's a stage'.

Schutz, W.C. (1958). FIRO: *A Three Dimensional Theory of Interpersonal Behaviour.* New York, NY: Holt, Rinehart, & Winston.

Tversky, A., Kahneman, D. (1973). *Judgement under Uncertainty: Heuristics and Biases.* Hebrew University. Jerusalem.

Wiseman, Richard. Psychologist. *The MegaLab Truth Test*, Nature, 373, 391. 'One third of us lie every day.'

PATIENT CHARICATURES

THE DOCTOR-PATIENT RELATIONSHIP

Balint, Michael. *The Doctor, his Patient and the Illness.* 1957. Tavistock Publications.

GMC: *Best Practice.*
Hazareesingh, S. *How the French Think.* (2015) Basic Books.
North, M. National Library of Medicine (USA), 2002.Hippocratic Oath (*Ο όρκος του Ιπποκράτη*).
To Have and Have Not. (Film, Warner Bros. 1944)

CULTURAL DIFFERENCE AND CLINICAL MEDICINE

Art Lovers' Guide (2018): Beirut. TV: BBC FOUR. An Armenian woman born in Beirut expressed her culture as: 'having two hearts.'

Bollas, C. (1989) *Forces of Destiny.* Free Association Books.

Crosby, Bill. Quote: 'Through humor, you can soften some of the worst blows that life delivers. And once you find laughter, whatever your situation might be, you can survive it.'

Dixon of Dock Green' (1955-76). Jack Warner in the TV.

Escoffier, Georges Auguste (1846 – 1935). UK TV Channel 5: *The World's Most Expensive Hotels – The Ritz.*

Fish Called Wanda, A, Film. MGM. 1988.

Flaherty, J. F., & Dusek, J. B. (1980). *An investigation of the relationship between psychological androgyny and components of self-concept. Journal of Personality and Social Psychology, 38*(6), 984–992. The highly androgynous possess higher self-esteem, and have more psychological well-being.

Freud, Sigmund (1905). *Der Witz und seine Beziehung zum Unbewußten.* ('The Joke and Its Relation to the Unconscious').

Harari, Yuval N. (2011). *'Sapiens'.*

Hislop, Ian. (2012) *Stiff Upper Lip – An Emotional History of Britain,* BBC Two, 2012.

Liedloff, Jean. *The Continuum Concept,* 1986. Hatchett Books,

Lorenz, Konrad. (1935). *Der Kumpan in der Umwelt des Vogels. Der Artgenosse als auslösendes Moment sozialer Verhaltensweisen. Journal für Ornithologie, 83,* 137–215, 289–413.

Montalbano. Sicilian TV series '*Il Giovane Montalbano'* (Episode: *Terzo Segreto,* The Third Secret).

Portillo, Michael. *Great Australian Railway Journey's*: Episode 4, BBC 2, 2019

Rousseau, J.J. (1762) *The Social Contract.* 'Man is born free and everywhere he is in chains.'

Wellington, Duke of. A soldier with 'the ice of character, and the fire of genuine and self-sacrificing principle' (The Morning Chronicle: 19/11/1852).

PART 2
Diagnosis, Diagnosis, and Diagnosis

HISTORY TAKING

Anderson, Michael, Friebel, R., et al. *Patient outcomes, efficiency, and adverse events for elective hip and knee replacement in private and NHS hospitals: a population-based cohort in England.* Lancet, Regional Health, Europe. Volume 40; 100904. May 2024.

Austen, Jane. *Pride and Prejudice.* (1813). Vol 3; Chapter 5.

Black, J. et al. (1964). *A New Adrenergic Beta-receptor Antagonist.* The Lancet: 283 (7342); 1080-1081.

Data Protection Act, The (2018). Legislation.gov.uk

Einstein, Albert. *Relativity: The Special and the General Theory* (1920). Henry Holt & Company.

Frasier: US TV Series. 'Do liars need de-fibbing?'

Grüntzig, A, Turina, M, Schneider J. *Coronary transluminal angioplasty.* Circulation (1976) 56:84.

Holmes, T. H., & Rahe, R. H. (1967). The Social Readjustment Rating Scale. *Journal of Psychosomatic Research, 11*(2), 213–218. https://doi.org/10.1016/0022-3999(67)90010-4

Homer: *The Oddysey*. Books 12 – 14. Scylla and Caribdis.

National Literacy Trust (2020), 16% of adults in the UK are now considered 'functionally illiterate.'

Pinker, Steven. *The Blank Slate* (2002). Penguin Books.

Pope. Alexander. *An Essay on Criticism.* (1711): Part 2: 215-218. Anodos Books 2017.

Post Office Scandal. Slingo, Jemma (16 December 2019). The Law Society Gazette.

Rapid Response: Re: GP appointments should be 15 minutes long, says BMA. *BMJ* 2016; 354: i4709.

Siedentop, Larry (2015) *Inventing the Individual. The Origins of Western Liberalism.* Penguin Books.

Törnroth-Horsefield, S.; Neutze, R. (December 2008). '*Opening and Closing the Metabolite Gate.*' Proc. Natl. Acad. Sci. USA. **105** (50): 19565–19566.

Welchman, Gordon. [1982]. *The Hut 6 Story: Breaking the Enigma Codes.* Harmondsworth, England: Penguin Books 1984.

Wizard of Oz', Film: 1939. MGM.

CLINICAL EXAMINATION

Dodd, Ken. *Latest Brighton.* Interview with Ken Dodd 31st July 2012. 'There's two ways of doing a show, you either do it at the audience or with the audience.'

Leatham, Aubrey (1970). *Auscultation of the Heart and Phonocardiography.* J & A Churchill, London.

INVESTIGATION

Bayes, Thomas. *Bayes' Theorem.* Thomas & Price, Richard (1763). *An essay towards solving a Problem in the Doctrine of Chance.* Philosophical Transactions of the Royal Society of London. **53**: 370–418.doi: 10.1098/rstl.1763.0053.

Café Society. Film: Warner Bros. (2016). Written and directed by Woody Allen.

Churchill, Winston quoted George Bernard Shaw. 'Those who never change their mind, never change anything'

Clark, Dr. David, MD. Cardiologist in Monterey, California, is affiliated with Stanford Health Care – Stanford Hospital.

Health Maintenance Organisation (HMO): a health insurance plan that provides health services through a network of doctors for a monthly or annual fee.

Munger, Charlie. 'To challenge opponents effectively, know their arguments better than they do.'

Osteen, Joel: 'You will never change what you tolerate.'

PCa3 test: Hessels D, Klein Gunnewiek JM, van Oort I, et al. *DD3(PCA3)-based molecular urine analysis for the diagnosis of prostate cancer.* Eur Urol. 2003;44:8–15. discussion 15–16.

Point Of Contact Ultra-Sound, or POCUS.

DIAGNOSIS FROM THE FIRST ENCOUNTER

American in Paris, An', Film: MGM, 1951. 'Like the mood of Paris, love, art and faith cannot be explained, only felt.' Jerry Mulligan played by Gene Kelly.

Baby Mead, Death of (2014). Henry Austin. Independent newspaper: Tuesday 26 January 2016 23:37

Doherty, Hunter 'Patch' Adams is an American physician, comedian, social activist, clown, and author. He founded the Gesundheit Institute in 1971. Film *Patch Adams*, released 1998.

Heberden, W. (1772) *Some account of a disorder of the breast*. Medical Transactions 2, 59-67: 1772. The Royal College of Physicians.

Hope, Bob : 'Timing is the essence of life, and definitely of comedy.' Or, 'Timing is everything.'

Murphy's Law. See: Matthews, R A J (1995). *Tumbling toast, Murphy's Law and Fundamental Constants*. European Journal of Physics. 16 (4): 172–176.

Waugh, Evelyn (1928). *Decline and Fall*. Chapman and Hal.

PART 3
The Tools of Medical Management

CLINICAL DECISION TOOLS

'A' level students . . . *The Guardian Newspaper*. August 17th, 2020, 'A' level students had their grades reverted to those given by their schools.

Arrhythmia and Electrophysiological Review - aer-volume-5-issue-3-winter-2016.

Bayes, Thomas (1763) Bayes' Theorem: *An essay towards solving a Problem in the Doctrine of Chance*. Philosophical Transactions of the Royal Society of London. **53**: 370–418.doi:10.1098/rstl.1763.0053.

Biernacki, Edmund. Polish pathologist, invented the ESR test in 1870

Frasier: Adventures in Paradise 2: Frasier's ex-wife, Lilith, a rule-based bean counter by nature, considering her new relationship.

Friedman, Thomas: *'Thank You for Being Late'* (Allen Lane, 2016).

Garrison, G.E., Cullen, E. (1972) Aerospace Medicine. 43. 86-91.

Hazareesingh, Sudhir (2015). *'How the French Think'*. Allen Lane / Penguin. A fetish for precision . . . leads to formalism.

Ibn Musa Al-Khwarizmi, Mahummad. AD 825. *Indian Numbers*.

I Ching. Chinese Book of Changes:

iFlyTek. The South China Post (27th November 2017) reports that the intelligent system, the 'iFlyTek Smart Doctor Assistant', achieved 456 points in the test, exceeding the pass mark of 360.

Kennedy, Robert, F. (Bobby). US Senator, Presidential Campaign. University of Kansas, March 18, 1968.

Lam, C., Lam C Yu., Huang, L., Rubin, D. (2018). *Retinal lesion detection with deep learning using image patches*. Investigative Ophthalmology & Visual Science January 2018, Vol.59, 590-596. doi:10.11 6/iovs.17-3371.

Leibnitz, Gottfried Wilhelm von (1646-1716): Attributed with binary mathematics.

Lovelace, Ada. English mathematician, an associate of Charles Babbage.

Pervasive Health 2017: Proceedings of the 11th EAI International Conference on Pervasive Computing Technologies for Healthcare; pages 221-230. Barcelona Spain May 23 - 26, 2017.

Pablo Picasso: *A Composite Interview*, by William Fifield (1964), Page 62, The Paris Review 32, Flushing, New York. 'Computers are useless. They can only give you answers.'

Planck, Max. 'Science advances one funeral at a time.'

Reichek, N., Devereux, R.B. *Left ventricular hypertrophy: relationship of anatomic, echocardiographic and electrocardiographic findings*. Circulation (1981); 75: 565-572).

Tabellverket (1749) Swedish Population statistics.
Watson, James. *Avoid Boring People*, O.U.P (2007).
Westergren, Alf (1921). Invented the Westergren tube to measure erythrocyte sedimentation rate.

CLINICAL JUDGEMENT

Alfred the Great, King. 899 AD.
Arhnem. Wikipedia.
Belay, E.D., Bresee, J.S., et al. (1999).*Reye's Syndrome in the United States from 1981 through 1997.*N Engl J Med 1999; 340:1377-1382 DOI: 10.1056/NEJM199905063401801
Bell, John. (1928-1990), originator of Bell's Theorem. Quote: 'Einstein was consistent, clear, down-to-earth, and wrong.'
Boden, W.E., O'Rourke, R.A., Teo, K.K. et al.(2007). *'Optimal medical therapy with or without PCI for stable coronary disease.* NEJM; 356: 1503-1516)
British National Formulary (BNF).
British Medical Association and Royal Pharmaceutical Society of Great Britain.
Braunwald, Prof. Eugene. ESC Grand Debate (Barcelona 2017): *Aspirin for Life.*
Charaka Samhita (*the Siddhi Sthana*), noted it, referring to it as 'not ideal', or yadrischha (c.1000 BCE).
Chaucer, Geoffrey. 'A man may say full sooth (the truth) in game and play'. *The Cook's Tale.* 1390).
Coda Collaborative (2020): *A Randomized Trial Comparing Antibiotics with Appendicectomy for Appendicitis.* New England J. Med; 383: 1907-1919.
Columbo, the American detective TV series (1971-2003).
Dighton, David. H. (2005): *'Eat to Your Heart's Content'*, ISBN 0-9551072-0-2).
Disraeli, Benjamin. Quote: 'There are lies, damn lies, and statistics.'
Dr. Who. Daleks: in the BBC TV series 1963.
Einstein, Albert (1919) *'Über die spezielle und die allgemeine Relativitätstheorie.'* Druck and Verlag von Friedr. Vieweg & Sohn in Braunschweig. General Theory of Relativity (2019).
Ehrlich, Paul. Nobel Prize 1908.
EU referendum of 2016. A spokesperson on Radio 4 (2017), reported that university graduates voted 2:1 for the UK to remain in Europe.
Feynman, Richard. (1985). *'Surely You're Joking, Mr. Feynman.'* W. W. Norton (US).
Giachelli, C.M. et al., (1993).Bone neogenesis. (J. Clin. Investigation; 92(4): 1686-96).
GMC: *Good Medical Practice*, GMC Publication.
Heberden, W. *Some account of a disorder of the breast.* Medical Transactions 2, 59-67: 1772. The Royal College of Physicians.
Herschel, William: Institute of Physics:https://spark.iop.org/william-herschel-and-discovery-infra-red-radiation.
Harvey, William. 1628. *'Exercitatio Anatomica de Motu Cordis et Sanguinis in Animalibus.'*
Hesiod. *Works and Days. c 700BC.*
Hippocrates. 'Wherever the art of medicine is loved, there is also a love of humanity.'
Hippocrates: Gave birth to the basic principles of good medical practice.
Hippocrates (supposed). 'Of 'science' and opinion: the former begets knowledge, the latter ignorance.'
Kahneman, D., Sibony, O., Sunstein, C.R. *Noise*; William Collins, 2021.
Kahneman, Daniel. (2011). *'Thinking, Fast and Slow.'* Macmillan.
King Jr, Martin Luther. Speech. (August 28th, 1963)

Kosko, Bart, *'Fuzzy Thinking'*. 1994, Harper Collins.

Lee, Harper. *To Kill a Mocking Bird.* 1960. J. B. Lippincott & Co.

Lewis. C.S. 'Someday you will be old enough to start reading fairy tales again.'

Lorenz, Edward N. (1963). *Deterministic non-periodic flow.* Journal of the Atmospheric Sciences. **20** (2): 130–141. (Chaos Theory).

Manna, Aditya; Sarkar, S. K.; Khanra, L. K. (2015-04-01). *'PA1 An internal audit into the adequacy of pain assessment in a hospice setting'. BMJ Supportive & Palliative Care.* **5** (Suppl 1): A19–A20. doi:10.1136/bmjspcare-2015-000906.61. ISSN 2045-435X. PMID 25960483. Refers to 'SOCRATES' as a mnemonic for pain assessment. (Site, Origin, Character, Radiation, Associations, Time-course, Exacerbating / relieving features, Severity).

National Institute for Health and Care Excellence: *'NICE Guidelines.'*

National Early Warning Score. 'N.E.W.S.' (The RCP (2017).

Neumann, John von (1928). 'Zur Theorie der Gesellschaftsspiele' [On the Theory of Games of Strategy]. *Mathematische Annalen* [Mathematical Annals] (in German). **100** (1): 295–320.

Newton, Isaac. Laws of Motion (1687). In Philosophiæ Naturalis Principia Mathematica (Mathematical Principles of *Natural Philosophy*).[

Nightingale, Florence . The polar area diagram ('pie chart'). Invented by Florence Nightingale to dramatise the extent of needless deaths in British military hospitals during the Crimean War (1954-56).

Nutt, Prof. David. RSM meeting., Nov. 2017. 'While individually important, these numbers are miniscule compared to those dying from tax generating, alcohol and tobacco products.'

Park, S-J., Ahn, Y-K., Kim, W-J, et al. *Preventative percutaneous coronary intervention versus optimal medical therapy alone for the treatment of vulnerable atherosclerotic coronary plaques (PREVENT): a multicentre, open label, randomised controlled trial.* Lancet (2024). 403; 1753-1768.

Pope, Alexander. *'An Essay on Criticism'* (1711). Part 2: 201-204.

Post Office IT scandal. *'Horizon Program'* (2020).

Poston, Tim, Stewart, Ian. (1996). *Catastrophy Theory and its applications.* Dover Books.

Pottegård, A., Haastrup, Maija Bruun, et al. (2014). *'Search for humourIstic and Extravagant acronyms and thoroughly Inappropriate names for important clinical trials (SCIENTIFIC): qualitative and quantitative systematic study.'* BMJ, 2014; 349: g7092.

Red Tails (2012). Film. 'Experience is a cruel teacher – it gives you the exam before the lesson.' Lucasfilm Ltd.

Spock, Mr. *'Star Trek'* Character.

Sullenberger, Chesney, Zaslow, Jeffrey (2010). *Highest Duty: My Search for What Really Matters.* William Morrow

Türing, Alan M. *'Computable numbers, with an application to Entscheidung's* (German for 'decisions') problem.' (1936).

Tversky, A, Kahneman, D. (1973). The Dunning-Kruger effect.

Watson, J.D, Crick, F.H. (1953). *Molecular structure of nucleic acids; a structure for DNA.*

Westerner, The (1940). Film United Artists. 'When I was young, I had a pet rattlesnake. I loved it, but never turned my back on it.' Cole Harden (Gary Cooper).

Wright, Wilbur. On December 14th, 1903, Wilbur Wright flew 50 yards in 3 seconds in a prototype airplane named 'Kittyhawk'.

Yes, Minister. BBC 2, TV series:

CLINICAL MANAGEMENT

Albright, Madeleine: *'Fascism: A Warning.'* Harper Collins. 2018.

Amundsen, Roald Engelbregt Gravning (1872 - 1928). Explorer of Arctic and Antarctic.

Baker-Smith, Dominic, on the Syrian satirist Lucian (125 – 180AD). Introduction (xiii) to *'Utopia',* by Thomas More (1518). Penguin Books. 2012.

Balint, Michael. (1957) *'The Doctor, his Patient and the Illness.'* (Tavistock Publications).

Bawa-Garba. *The Dr. Hadiza Bawa-Garba Case 2015-2018.* The BMJ. BMJ Publishing Group.

Bernard, Claude (1813-78). *Milieu Intérieur.*

Birtwhistle, Mike: *Saving Lives and Averting Costs? The Case for Earlier Diagnosis just got stronger.* Cancer News. 2014.

Bradley, Ben. Editor. The Washington Post. 'The truth, no matter how bad, is never as dangerous as a lie in the long run.'

Branch. W.T. (2001); 'Humanism in medicine is the physician's attitudes and actions that demonstrate interest in and respect for the patient and that address the patient's concerns and values.' JAMA: 286(9):1067-1074.

Buffett, Warren (Berkshire Hathaway).

Carter, General Sir Nick. Chief of Defence Staff. (November 2020). 'While history does not repeat itself, I think it has a rhythm.' *The Andrew Marr Show,* BBC TV. November 2020.

Colten, H.R., Altevogt, B.M., Editors. (2006) *Sleep Disorders and Sleep Deprivation: An Unmet Public Health Problem.* National Acadamies Press (US).

de Feyter, P., van Eenige, Machiel J. et al. (1983). *Effects of Spontaneous and Streptokinase Induced Recanalization on the Left Ventricular Function after Myocardial Infarction.* Circulation: 67;5:1039-1044.

Doherty, M., Jenkins, E, et al. (2018). Lancet (2018); 392: 1403-12. Only 30% of the doctor-led group achieved the same target.

'Drexit'. The BMJ Opinion. February 6th 2020.

Eisenhower, General Dwight D. 'In planning for battle, I have always found that plans are useless, but planning is indispensable.'

Epictetus. 'If you want to improve, be prepared to be thought foolish.'

Fry, Dr. John (1922-1994). 'The science of the possible with the art of the impossible.' He liked to quote the 16th century French military surgeon Ambroise Paré's description of a doctor's role - 'to cure sometimes, to relieve often, to comfort always'.

Harari, Yuval N. (2011) *'Sapiens, A Brief History of Humankind.'* p114-124, on an 'Imagined Order'. Dvir Publishing House Ltd. (Israel) Harper. Also, Penguin Random House.

Hobbes, Thomas. *Leviathon.* XIII.9. 1651.

Jibran, Khalil. *'Youth.'* Translated by George Kheirallah. Arabic Poems. Everyman's Library. Pocket Poets.

Johnstone, Chris. Private Eye, No. 1476: 2018. In 1992, junior doctor Chris Johnstone won the right to sue Bloomsbury Health Authority for subjecting him to an onerous obstetric job.

Kahneman, Daniel. *et al.,* *'Noise.'* William Collins, 2021.

Kamikazi: Tokubetsu Kōgekitai (□□□□, Japanese Special Attack Unit.

Kant, Emanuel. *'Critique of Pure Reason.'* (1781).

King's Fund, The (2018). Commission: Prof. George Freeman, and Jane Hughes. *Continuity of Care and the patient experience.*

Machiavelli, Niccolò.(1513).'While you treat them well, they are yours.' *The Prince (De Principatibus).*

Matthew: 5: 3-10: *St. James' Bible*. 'Blessed are the meek, for they shall inherit the earth'.

McChrystal, Stanley. Ret.US General,: 'Instead of managing for efficiency, a modern army needs adaptability more.' . *'Team of Teams' (2015). Tantum Books.*

MIMS: Monthly Index of Medical Specialities. Haymarket.com.

More, Thomas. (1518). 'A wise man would sooner keep clear of sickness than find a remedy for it. *Utopia.* Book Two.

Murray, Conrad. Michael Jackson demanded sleep. His personal physician, Conrad Murray, supposedly charged him $150,000 every month for his services.

Penrose, Roger. *'The Road to Reality'* (2005), Vintage Books, p412-439. Chapter on Minkowskian Geometry.

Quiller-Couch, Sir Arthur Thomas. *'The Art of Writing'* (Cambridge UP, 1917), quotes Newman's definition of a gentleman.

Ritz, César (the founder of the Ritz Hotel in London and Paris). When working in Vienna, he served the Prince of Wales (the future UK King, Edward VII).

Rosekind, Mark. Head of the US National Highway Traffic Safety Administration US, has said that 'a 26-minute nap improves performance in pilots by 34% and alertness by 54%'. Source The Guardian Newspaper. 23/10/2016.

Ruth, *The Scroll of Ruth*. Hebrew Bible.

Sacks, Jonathon. Former Chief Rabbi UK.(2015) ' . . . a world that has no room for difference has no room for humanity.'

Shakespeare, William. *Richard II*. 'I have wasted time, now time doth waste me.'

Smith, Terry (Fund Manager Fundsmith).

Tyson, Mike. 'Everyone has a plan - until I punch 'em in the mouth.'

von Moltke, Graf. Field Marshal Helmuth B. (1880). Prussian Army Commander.'No battle plan ever survives first contact with the enemy.'

Wagner, Richard (1850). Essay: *'Das Judentum in der Musik.'*

Walton, Isaac (1653). ' . . . those restless thoughts which corrode the sweets of life.' *The Compleat Angler.*

West Point cadet code of honour: 'A cadet will not lie, cheat, steal, or tolerate those who do.'

Wheeler, M.T. et al. 2010. *Cost effectiveness of pre-participation screening for prevention of sudden cardiac death in young athletes.* Ann. Int. Med: 152(5): 276-286)doi: 10.1059/0003-4819-152-5-201003020-00005

CLINICAL RESEARCH FOR DOCTORS

Dighton, David H. (2005*). 'Eat to Your Heart's Content'* (2005; ISBN: 0-955-10720-2).

Feynman, Richard. When asked about the Nobel Prize he won in 1965, Richard Feynman modestly said that he had already won the most important prize: discovering something important about how quantum mechanics works.

Hodgkin AL, Huxley AF (1952). *A quantitative description of membrane current and its application to conduction and excitation in nerve.* The Journal of Physiology. **117** (4): 500–44.

Jowell, Tessa (2018). Speaking on a BBC Radio 4 'Today' programme in January 2018, Baroness Tessa Jowell recounted her medical experiences.

Koch, Heinrich Hermann Robert. (1884). Koch's postulates.

'Research is the door to tomorrow.' Motto: The Post Office Research Station. Dollis Hill.

Silver, Ann (editor) (1974). *The Biology of Cholinesterases.* North-Holland Publishing Company.

EPILOGUE

Maximillian Weber (1864-1920). 'The Theory of Social and Economic Organisation.' 1920. Also an essay: On Bureaucracy. See Weber's 'Rationalism and Modern Society', 2015, pp. 73–127, edited and translated by Tony Waters and Dagmar Waters, New York: Palgrave MacMillan. Weber wrote: 'the fate of our times is characterised by rationalisation and intellectualisation and, above all, by the disenchantment of the world.'

PERSONAL REFERENCES

Dighton, D.H. (1974) Sinus Bradycardia : Autonomic Influences and Clinical Assessment.
British Heart Journal. 36 : 791-797.

Dighton, D.H. (1975) Sino-Atrial Block: Autonomic Influences and Clinical Assessment.
British Heart Journal: 37: 321-325.

Dighton, D.H. (1975) Complete Heart Block : Studies of Atrial and Ventricular Pacemaker Site and Function.
British Heart Journal.37:(11)1156-1160.

Dighton, D.H. (1976) Autonomic Features of Supraventricular Tachycardia.
British Heart Journal. 38 (3) 319.

Nixon, P.G.F., Dighton, D. H. (1976)
Meditation and Methyldopa.
British Heart Journal 2 (6034):525.

Dighton, D.H. (1980) Sleep Therapy (On Sleep, Homeostasis, and Cardiovascular Problems.
General Practitioner (Magazine).

Dighton, D.H. (1980) Sinus Bradycardia.
The Practitioner.224 : 261-266.

Dighton, D.H., Golding, R., de Feyter, P., (1982)
Post-Pneumonectomy Pericardial Effusion.
Chest. 82 : 389-390.

Katona, P.G., McLean, M., Dighton, D.H., Guz, A. (1982). Sympathetic and Parasympathetic Cardiac Control in Athletes and Non-Athletes. J. Applied Physiology 52(6): 1652-1657.

de Feyter, P., Van Eenige, M.J., Dighton, D.H. et al (1982). Prognostic Value of Exercise Testing, Coronary Arteriography and Left Ventriculography 6-8 Weeks after Cardiac Infarction. Circulation 66 (3): 527-536.

de Feyter, P.J., Van Eenige, M.J., Dighton, D.H., Roos, J.P. (1983) Exercise Testing Early after Myocardial Infarction. Chest. 83: 853-859.

de Feyter, P. J., Van Eenige, M.J., et al incl. Dighton, D.H.,(1982) Experience with Intracoronary Streptokinase in 36 patients with Acute Evolving Myocardial Infarction. European Heart Journal (1982), Volume 3, Issue 5, Pages 441–448, https://doi.org/10.1093/oxfordjournals.eurheartj.a061330

Dighton, D.H. (1986) Vasovagal Syncope. Lancet 1:982.

See Other Books Published by the author.

Further Acknowledgements

My apprenticeship started in September 1961 when I first crossed the threshold of the London Hospital Medical College as a student. Over the last 61 years as an alumnus, I have adhered to the sentiment of our Terentian motto:

Homo sum humani a me nihil alienum puto:

I am human and nothing human is alien to me.

There are many from whom I have learned. Among them were:

Pre-Medical School

My neighbours:
Mr. John (Pharmacist), and Mrs. Nancy Powell (Physiotherapist);
Dr. Julian Hardman, GP. London E17.
Dr. A. Lewin. Head of Science, South West Essex Technical College, London E17, in 1959.

At the London Hospital Medical School,

Dr. John Ellis (Dean and Physician).
Dr. John Wright, my Clinical Tutor, Physician and Gastroenterologist.
Dr. Duncan de Vere (Physician).
Dr. Donald Hunter.
Sir Reginald Watson Jones (Orthopaedic Surgeon).
Mr. Donald Brews (Obstetrician and Gynaecologist),
together with all my other professors and teachers at The London Hospital, Whitechapel.

At Whipp's Cross Hospital, London E17,

Dr. Richard ('Dick') Barrett.

Surgeons: Mr. Louis de Jode, and Colin Davis.
Dr. John Wright, Physician and Gastroenterologist.
Dr. Malcolm Morris. Registrar Physician.
Dr. Gill Hanson, Intensivist.
Dr. Norman Eve, Anaesthetist,
together with many Irish nurses, among whom were Anne Cousins and Breda Harty, who helped me understand my role and what counted when dealing with patients.

In General Practice, London E4,

Drs Henry and Cicely Blair.
Dr. Rodney Herbert.
Dr. Alan Gardner, London E17.
Dr. C. Anand, Buckhurst Hill, Essex.

At St. George's Hospital, Hyde Park Corner, London.

Dr. Aubrey Leatham. Cardiologist.
Dr. Alan Harris. Cardiologist.
Geoff Davis, and Graham Leech
(Electronics engineers in pacing and echocardiography, respectively)
Dr. Alan Gelson and Dr. Paul Kligfield.
Dr. Michael Davies. Cardiac Pathologist.
Dr. Keith Jefferson, Radiologist.
Professor Tony Dornhorst,
and the many nurses who tutored and guided me:
Ward Sisters: Maggie on Grosvenor ward, and Sue in the catheter lab.; both exceptional nurses.

At Charing Cross Hospital, London

Dr. Peter Nixon, Cardiologist.
Dr. Alan Harris, Cardiologist.
Prof. Hugh de Wardener.
Dr. Keith Woollard. Intensivist. Cardiologist.

At the Vrije University Hospital, Amsterdam

Dr. Pim de Feyter. Cardiologist.
Prof. J.P. Roos. Cardiologist.

At the Loughton Clinic and Cardiac Centre

Dr. David Baxter. GP.
Mr. Sunit Ghatak, Gynaecologist.
Miss Noreen Connolly, Physiological Technician.

Pharmacists: Paul Mellis (Oakwood Pharmacy), and Dushant Patel (Loughton)
Dr. Raymond Randell, GP.
Mr. Gerry MacCarron, Psychotherapist.
Mr. Bisu Banerjee, General Surgeon.
Dr. David Lipkin, Cardiologist.
Mr. Stephen Edmondson, Cardiac Surgeon.
Mr. Eric Slater, General Surgeon.
Dr. David Prothero, Psychiatrist.
Dr. Mohommed Daddabhoy, GP.
Dr. S. Anand, GP.
Dr. Sunil Bhattacharya, GP.
Mr. Frank Cross, Vascular Surgeon.
Dr. Vik Watts. Psychiatrist.
Dr. Tim Allum. Anaesthetist.
Dr. Michael Hodges, GP.
Dr. L. Larh, GP.
Dr. Alan Gardner, GP.
Mr. Don Russell, SRN.
Neil Mellerick, Osteopath.
'Dee': Dipak Vasani, Pharmacist.
Mr. Tony Erian, Cosmetic Surgeon.
Mr. Pasquale Giordano, General Surgeon.
Dr. Roderick Storring, Respiratory Physician.
Mr. Ibrahim Shuaib, Orthopaedic Surgeon.
Mr. Brian Levack, Orthopaedic Surgeon.
Mr. Malcolm Devereux, GU Surgeon.
Dr. Anthony Bewley, Dermatologist.
Dr. Colin Ainley, Gastroenterologist.
Dr. Ed Stoner, Gastroenterologist.
Mr. Ravi Kunzru, ENT Surgeon,
together with many worldly wise receptionists, like Hazel Jacobs, Maureen Tribe, and very able nurses, like Noreen Nainby, Margaret, and John Speed.

From my mother, I learned to appreciate a nebula approach to judgement, the importance of detail, and the danger of assumption. She always considered problems from every angle before deciding. Her brother, Alfred (Alf) tried to teach me to draw, how to think artistically, and how optimism enriches life. The Powell family were my childhood neighbours. John Powell included me when he introduced his five children to objectivity, logic, puzzle solving, and the wonders of science. Tony Smith (alav ha-shalom), I met as an 11-year-old schoolboy. We sat together in class and giggled our way through to 'O' Levels (now GCSE's). He was still my wise and loyal friend at the time of his premature death from melanoma. Brian Altman. I met while studying 'A' Levels, at what was then the South West Essex Technical College in Walthamstow. He never queued. He taught me the value of confidence and how clarity of thought can lead to success in life. Another colleague was to become a notable nutritional scientist. Mike Stock (RIP) showed me the errors of my ways while learning laboratory techniques. Having never entered a science lab before, his advice was invaluable.

Barney Brenner (alav ha-shalom), must have been Solomon in a former life. He brought wisdom to every tough question. 'Gilly' (Kyriakos) Panayioutou showed to me, that up to the point of meeting him, I had never lived a full life (closeted in libraries). My friend George (Giorgos) Mullaly (R.I.P.) died of 'natural causes' in 2011. His invaluably wise council was there for me daily for over 12 years. I only wish I had found his friendship earlier. I was also lucky to have met Ralph Barton, who could detect scams coming over the horizon while blind-folded – something he shared with Giorgos and Gilly; I have never fully learned how. Ralph was a raconteur, practical joker, comedian, and for children, a modern equivalent of the Pied Piper of Hamelin. As a master of street-wise wisdom, I only got the better of him once, and only then after many weeks of planning. It would have been a special gift to have met just one of these people, but to have met them all was a divine gift. Whatever I know of practical wisdom and stupidity, I learned from them.

In medicine, I learned a lot from my many colleagues listed earlier. All were notable for their wise counsel.

Few doctors experience enough of life to gain the sort of wisdom possessed by patients. I wonder how this affects the advice they give. As doctors, we risk being outclassed by patients with no university degree, who take guidance from something more useful – common sense, practicality and nous.

I would like to acknowledge all those fellow students, demonstrators, and teachers at 'The London', who taught me the basics of medicine. In particular, I am indebted to Dr. John Ellis, Dean of the London Medical School, who in 1961, gave me my the chance to study medicine. It was his theory that clever students need guidance more than teaching. That gave me my chance. At my initial interview, I unknowingly proved him right by having self-studied, and passed six 'O' level science subjects; the school I attended did not teach them. Together with John Ellis, I learned much from Dr. Duncan de Vere. His expertise, and calm humility, inspired confidence in all of his students, colleagues and patients. Dr. John Wright, my tutor, tried hard to get me a job at the London but failed.

I am especially indebted to those colleagues at St. George's Hospital, Hyde Park Corner, who taught me research technique, cardiology, and further medicine; especially Dr. Alan Gelson, without whom Paul Kligfield and I would not have so easily become members of the Royal College of Physicians. I am most indebted to my mentor, teacher and friend, the late Alan Harris, who taught me most of what I know about cardiology. Alan was a master of invasive cardiology and pacing. Geoff Davies, who was still working there, developed the very first practical trans-venous pacemaker. He deserved a knighthood, but barely received recognition for his work. I am honoured to have called him a friend. In the 1970s, our department had the first echocardiogram machine in the UK. It was a privilege to have worked with Graham Leech, an electronics engineer who later became a recognised master of clinical echocardiography. While I was working with him, he transformed the early echocardiographic display into what we have today. Keith Jefferson was a master of cardiac radiology. To witness him interpreting a chest x-ray was inspiring. As a world-class expert, he showed only the kindest of ways to expose our ignorance. Michael J. Davies studied the histopathology of the coronary tree and proved the role of atheroma in various coronary syndromes. His monograph on the 'Pathology of The Conducting System of the Heart' (1971) remains a testament to his brilliance. All these men were truly inspiring, and evidence of how modesty can hide exceptional expertise.

From older colleagues, I learned what a doctor must do to become fully functional and effective. I spent many happy weeks as an elective student with my own GP, Dr. Julian Hardman, at his Addison Rd. 'Surgery' in Walthamstow, London, E17. He combined exceptional clinical insight with Irish and Jewish wit, to make an art-form of medical practice. He was a master of the art of medicine. At Whipp's Cross Hospital, the practical expertise of Mr. Louis de Jode (RIP), Australian surgeon Colin Davis ('Col'), Dick Barrett (RIP), Gill Hanson (RIP), and Norman Eve (RIP) was impressive. After having spent 2-years at Whipp's Cross Hospital, I worked with Drs. Henry and Cicely Blair ,☐☐☐☐☐ ☐☐☐☐☐and Dr. Rodney Herbert ; ☐☐☐☐☐ ☐☐☐☐all respected GPs in South Chingford. By then, Henry Blair had been a GP for many decades and was a respected elder in the local Jewish community. These doctors were not technocrats, but they were masters of the art of medicine. They cared for thousands of patients over many decades, in sensible, imaginative, and sensitive ways; ways possessed by those versed in the art of medicine. They treated every patient as an individual, never as a number. As medical carers, they brought comfort and relief to sufferers, and were worthy of the non-academic title 'doctor'.

I owe much to many remarkable nurses. There were many dedicated Irish nurses working on Lister Ward, and in Casualty, at Whipp's Cross Hospital in the late 1960s, among whom were Breda Harty and Anne Cousins. They put me in my place. The lesson they taught me not to think of myself as special in any way – the patients were the special ones. Luckily, they rapidly 'de-Londonised' me (they regarded all London Hospital graduates as too full of their own importance). One truly fantastic Irish nurse who ran the 'Casualty' at Whipp's Cross in 1967 was a diminutive red-head. Being totally competent, she was effortless in control at all times. From her example, I learned how to manage complicated, complex, and dangerous situations with calm efficiency. I regret not remembering her name.

At St. George's, Hyde Park Corner, Sister Maggie on Grosvenor ward, was the epitome of a nurse dedicated to her patients. The nurses of Grosvenor Ward, who worked under her, rarely worked to a fixed time-table. Like Florence Nightingale and her disciples, they stayed for as long as their patients needed them; the patients in their care came before any personal issue. These days, concerns have shifted. Mistakes are more easily made when nurses are tired, and staying late would not comply with the rules. In our Cath. Lab. were Sister Suzy, secretary Margaret, and Jane Page-Thomas (respiratory technician). I fondly remember them all for their dedication and contribution to the fun we had working together. As a supplier of wine from my family firm, we all enjoyed a simple punch made at party times. My punch formula was not a secret: one bottle each of red wine, white wine, and lemonade with Angostura bitters added to taste. It never failed to be drunk liberally. It strengthened our esprit de corps.

At the Roding Hospital, Nurse Angela (Pollard) was more knowledgeable and dedicated than many doctors. As an invaluable colleague, she could combine remarkable medical knowledge, practical experience and medical management skills, with compassion, understanding, forethought, and calm. Putting patients first was her priority, long before anyone deemed it necessary to regulate.

Some of my previous medical colleagues deserve special mention:

Dr. Aubrey Leatham, world expert on heart sounds and instigator of cardiac pacing, gave me the chance to research slow heart rhythms. According to him, I became an expert in 'slow heart rhythms, and fast cars' (I owned a yellow Lotus Elan 2+2 at the time). His monograph on heart sounds was a landmark publication, providing evidence for his clarity of thought and succinctness. In fostering the development of both cardiac pacing (with Geoff Davis) and echocardiography (with Graham Leech), he demonstrated amazing foresight. His cardiac clinical acumen was legendary, and am proud to have been one of his many apprentices.

My friend and mentor, Dr. Alan Harris, was my cardiological guru. His expertise in invasive cardiology might be equalled someday, but never surpassed. He taught me cardiology and honed my clinical judgement with good-humour and a wealth of experience. After being appointed consultant, I followed him to Charing Cross Hospital.

Geoff Davis, inventor of the first functioning human pacemaker, was always funny, witty, and forever amiable. From him, I learned that world-class expertise doesn't have to be expressed seriously. He laughed a lot and reminisced about his RAF days. He had a party trick: after taking a puff of his pipe he would blow 900 litres/minute through a Wright Peak Flow Meter.

Dr. Alan Gelson tutored me, and our mutual friend Dr. Paul Kingfield (who became a New York cardiologist), to pass the MRCP examination. He would spot 'patients' walking down the Brompton Road, and quiz us about their gait, demeanour, and possible diagnosis. Paul introduced me to Watson and Crick's show stopping statement, which he paraphrased as –'the significance of what we have discovered (the significance to life of the DNA double helix) has not escaped us.' Paul was witty, serious, captivating, and a bright star. Great fun to be with in hospital, or discussing research at the first Hard Rock Café to open in Piccadilly, London. We three were there on its 2nd day of business in 1971.

Dr. Peter Nixon was a charismatic, unforgettable character, with an unforgettable approach to cardiology which many of his academic colleagues despised. He was a doctor with ideas both before and after his time. On attending a patient recovering from cardiac infarction in our coronary care unit at Charing Cross Hospital, Peter might first whisper a question to the patient: 'Who is trying to kill you?' This question would often take patients completely by surprise. The patient would then ask, 'What do you mean, doctor?' 'If you don't know', Peter would say, 'you will stay here until you find out. Your life could depend on it.' He would then walked away, leaving his juniors to pick up the pieces. This was Peter Nixon in action, trying to discover the psycho-social reasons for each patient's demise. He introduced me to how the art of medicine can benefit patients with fixed physical pathology.

Peter was a man before his time. When nobody much accepted the role of stress as an aetiology factor for myocardial infarction or hypertension, he clearly saw what many still cannot see (through a lack of personal interest and experience). Peter taught me the value of unlearning, and of original thinking, in transcending the established rules of thought. Having served as an officer in the special services division that became the SAS, he appreciated how human beings functioned at the extremes, why they failed, and how they became sick. He dedicated himself to putting stress on the scientific map. He knew that if ideas were to be accepted and promoted, they needed to be kept simple. Provocatively, he sometimes denied the contribution of coronary atherosclerosis to angina and cardiac infarction. I understood his ploy – he hoped to stimulate discussion. Unfortunately, too few others wanted to engage with him. His academic colleagues regarded him as unacceptably eccentric who needed no evidence to support his theories. The medical establishment looked for an opportunity to remove him from office. Their moment came when he challenged a broadsheet newspaper in Court, accusing them of libel. The court advised him not to continue in practice. Such can be the fate of inspired visionaries who come to shake our acceptance of the status quo.

He failed to produce the evidence, proclaiming that the validity of his theories was obvious. To-day, few would attempt to deny the part played by stress in coronary artery disease, and how it can affect the immune and inflammatory responses (atherosclerosis is in part, an inflammatory process).

Peter was the first to realise that, even an 80% coronary lesion can become symptomatic if accompanied by the fatigue and exhaustion resulting from prolonged psycho-social stress. The same lesion, or worse, can remain asymptomatic in contented, fulfilled people. He realised that vulnerable coronary lesions could either remain stable, or fissure and clot, with cardiac infarction to follow. His colleagues might have accepted this had he acknowledged that atherosclerosis and stress can be co-factors. He was so keen not to dilute his message - that stress was key - that his ideas became undermined and rejected. 'Where is your evidence?' was not an unreasonable question. His conviction was such that he saw no need to waste his time proving it for the satisfaction of others. He thought that was a job for others. The pertinent circumstantial evidence he amassed for his thesis was undeniably persuasive. I still honour his unique insight and bravery when facing derogatory and divisive colleagues. There is one major problem that all insightful bringers of discovery must face: like anyone claiming to be a Messiah, they risk crucifixion. I have not heard from him in decades. We might forget his rejection by the medical community, but we should never forget his contribution to medical knowledge and the art of medicine.

Dr. Keith Woollard was the first doctor I saw wearing white shorts while walking around the wards at Charing Cross Hospital. London. That's what he wore in Australia, so why not in the UK? He was the greatest vein and artery locator I have ever met. An outstanding practical talent. He could have easily said: 'It's easy, when you know how', but he never did. When challenged by me to take an interest in the psycho-social aspects of cardiology, he memorably said: ''I'm not that sort of doctor.' One could do no better than have Keith present at a cardiac emergency. He returned to Australia. They were lucky to get him back. Latterly, I believe that he may have upset a few of his colleagues, perhaps jealous of his talent. I hope the medical regulators involved did not prevent him from saving many more lives.

My friend, colleague, co-founder (in 1973) of Loughton Clinic, and fellow ex-Londoner Dr. David Baxter, taught me (among many things) that organising NHS doctors in private practice was like herding cats. I saw how he handled them, but never learned how to do it myself. From him, I learned how to run a personal private practice that works for patients. As a brilliant entrepreneur, he founded his own private hospital (Holly House Hospital in Buckhurst Hill, Essex).

While working in Holland (Vrije Universiteit, Amsterdam, 1980), I was fortunate to have cardiologist Dr. Pim de Feyter as a colleague. We shared an office together. Pim was on call at all times for acute cardiac infarction cases. His 1980 study involved injecting streptokinase directly into the relevant clotted coronary artery of patients undergoing an acute myocardial infarction. He had a talent for sheer hard work and application. It was amazing to see a patient's chest pain dissipate, and his face returned to pink as soon as the streptokinase dispersed his coronary clot. It was an honour to have helped him write his scientific papers in English. He remained a prolific, highly respected researcher.

Prof. J.P. Roos, Head of the Cardiac Department, Vrije Universiteit, Amsterdam, gave me the opportunity to bring transvenous pacing to his department. I gratefully remember his kindliness, guidance and generosity.

When I worked in Holland, the Dutch system put patient welfare first. Having completed any catheterisation, or pacemaker implantation, nurses expected me on the ward a.s.a.p., to deliver the results in person to each patient. It is what I would have done anyway, but in Holland, this was obligatory. The Dutch have developed an exemplary medical service for their citizens, with many lessons for us to learn in the UK.

I worked with many others in Holland whose help I will not forget. Their light-hearted attitude, combined with professionalism, made working there a pleasure. Vielen dank jongers. Uitstekant.

No list of colleagues from whom I have learned could ever be complete. I do not wish my omissions to be hurtful to those I have failed to mention, many of whom will remember me. I beg their forgiveness. To have contributed, but not remembered, is unforgivable. I remain indebted to you all.

I have an unquestionable debt to all my patients. Many were notable characters who taught me how to be of value. The first patient to influence me in this way was Sam Silverman. It was he who suggested that I should work privately to make it easier for both of us. And so it did.

I met Sam as a patient at St George's Hospital, London SW1. He lived in Ilford, and I lived in Buckhurst Hill, Essex. 'Why am I coming all this way to see you when I could come to Buckhurst Hill?' he protested. 'I don't have a private practice', I replied. 'You have a home, don't you? I'll see you there. You're now in private practice', he said. And so it came to be. Sam and his wonderful wife, Yetta, became my first private patients and lifelong friends. Sam appointed himself as my business advisor, and how lucky I was to have his wisdom and support at my disposal. I still cherish his memory. Unfortunately, Sam gradually developed what was to become severe dementia. He died after a stroke, years after having bilateral carotid arterectomies for atherosclerotic occlusions.

I am indebted to all my patients for their support over five decades. I am indebted to those who taught me along the way. I thankfully acknowledge the contributions of: George Mullaly (RIP), 'Gilly' and 'Pany' Panayioutou, Ralph and Mary Barton, Keith and Noreen Nainby, Kenny Nicholls, Sharon Paul, Kate and Trevor Ireland, Colleen and Bernie Palser, June Allpress with 'Aunt' Grace and family, Arthur Howes, Jennifer and Leonard Griffiths, Olive and David Hall, Brenda and Robert Hall, Tanya and John Chicken, Bill Kenny, Joe Satchell, Jan and Stephen Kinrade, Harvey Barbash, Tony Bilton, Richard Reeder, Bob Newton, Albert and Kathleen Sanders, Malcolm Bond, David Sullivan, Ken Furlong, David Kent, Frank Campion, Doreen Hambling, Andrew Casey, Joe Davies, Rita and Michael Dias, Lilian & Maurice Dixon, Doug Walton, Jean Raven, Alan and Shirley Fisher, Zeki Asimoglu, Cemil Mercezi, Anthony and Evelyn Joyce, Percy Ingle, Alex Jacobs, Norman Kaye, Dennis Kilby, Graham Kilby. Brian Kirkby, George Martin, Andy Mavrou, David Lewis, Tony Barilone, Paul Mansfield (RIP), Wendy MacAuliffe, Bill Morris, Lorna and James, Kenny Nicholls, Amrik Singh, Jack Harris, Tommy Powell, John Hatchett, Malcolm Hepburn, Patrick Hurling, Percy Ingle, John Clay, Mike Reid, Wahid Rajack, Tommy Powell, Walter Purkiss, Alan Read, John Riley, Don Russell, Richard and Pam Schild and family, Terry Gregory, John Speed, Jean Stock, Sandra Tomlin, Susan and Lawrence Troyna, Tony Wood, Ron Smith, Raymond and Carol Ward, Oliver O'Rourke, Roger Neville, Ray Linard, Vic Lupson, Barry and Toni Anten and family, Gabor Lacko, Patricia Spero, Raj Ahir, Amanda and Nial Cameron, Jon Bard, John Bird, Lawrence Bird, Jack Blake, Brad and Donna W., Brian Bridgman, Kathleen Bridgman, Tim Bridgman, Sidney Brockwell, Bill Wood, David and June Windrow, Malcolm Westbury, Matthew Chidlow, Susan and Robert Sawyer, Vicki and Anne Michelle, Aiden 'Gerry' Couch (RIP), Gary Couch, Ron Tyler,

Chris Trussell, Steven and Brian Altman, Tommy Tomkins, Ian Terry, Glen Tamplin, Maggie and John Jordan, Patrick Storring, Stan Beavis, David and Terry Sherrin, Alan Sewell, Valerie and Stephen Metcalfe and families, Nicole and Peter Kromberg (RIP), Brenda Sands, David and Louise Sands, Theresa Taylor, Brendan O'Malley, Donal Mulryan, Rosa Byrne, Leila R., Theresa Taylor, Ronnie O'S., Rod Self, Bob and Cindy Falconer, Alan Reed, Colin and Pam Fitch, Francisco Rivas Calzon, Richard Reeder, Franco Dorili, Ricardo Dorili, Keith and Christine Renew, Tony and Mandy Bowers, Raymond and Richard Rains, Colin and Carol Quy, Philip Wallace, Jean Heard, Bernard Huber, 'Vicki' and Tommy Mines, Paul Foulds, Ken Parsons, Eva Wade, Richard May, John Parker, Kris Oza, Oliver O'Rourke, John Oddi, Niteen Patel, Hayley Newton, Neil and Sharon Davey, Natalie Sullivan, Edward Payton, Malcolm Trowbridge, Morton and Galena Stokes, Keith Mitchell, Amanda Miller, Mike Chipperfield, Michael Hunter, Russell Meager, Mazlin Osman, Richard May, Andy Mavrou, Mary Bann, Martine Asscher, Malcolm Westbury, The Cooks and the Kelly's, Mark Gregory, Bipin Mandalia, Mahesh Patel, Alexandra Lyons, Jonathon Lamb, Carol King, Kim Yardley and Shirley Shaw, Michael Kelsey, Jean and Rene Bennett, Davina Jacoby, Jane Granger, Howard Winston, Malcolm Hepburn, Michael Hennessy, Haydon and Doretta Smith, Simone and Christopher Smith and children, Shelly Hart, Ian Harper, Shirley Gropper, Sarah Granger, Jennifer and Leonard Griffiths, David Gilbert, 'Frank' Harvey, Peter Flower, Fay and Barry Cookson, Elvira de Paul, John Eldred, Eddy Andrews, Darren Docwra, Simon Donoghue, Keith and Susan Dodd, Diane and Simon Capelhorn, Carly Davies, John and Davina Bryan, Eddie and Barbara Bryan, Barbara and David Cunningham, Brian Dadd, Mark Cutter, The Coyle family, Mike Coulson, Pierce Campion, Richard Burgess, Beverley Burgess, Cornelius and Irene Burgess, Anne Buckland, Colin Booker, Lawrence Bird, David Austin, Panos Eliades, Ian Agates, Brian Kirkby, Andy Mavrou, Graham Kilby, Roger Neville, Linda and Philip Latchford, John Gray, Pat Quill, Hazel Jacobs, Ingrid Pryor, Maureen Tribe, Tommy Powell, Eileen Pitman, Feruccio Pieropan, Raymond Perry, John Paget, Malcolm Busby, David Burrell, Gerry and Anita Brummitt, Len Brown, Sam Brenner, Sidney Brockwell, Frances and Martin Brett, Anthony Breeze, John Bray, Nigel Bowerbank, Roy Bowden, Michael and Jean Booker, Malcolm Bond, Brian and Patricia Bluck, Jack Blake, Tony Bilton, Peter Beeson, Jon and Michael Bard, Jill Ballard, Norman Balbes, Florence Bagwell, David Austin, Sara and John Attwood, Zeki Asimoglu, Violet Appleton, Alfred Anthony, Steven Allen, Ann Abbott, Linda and Michael Gould, Christine and Derek Gould, Colin Good, Michael Gonella, Harvey Golding, George Gluck, Sydney Gipson, David Gilbert, Indira Ghatak, Eddie Fowles, John Flynn, Sally Filek, Andrew Field, Ernest Felgate, Hugh Farish, Leslie and Susan Falco and the twins, Miles Emblem, John Euesden, John Eldred, Petrina and David Edwards, Mary Easton, Derek and Edith Doy, Carol and Ron Dowsett, Keith and Susan Dodd, Ashley Doctors, Brian Halliday, Gerry Desler, William (Bill) Deane, Joe Davis, Charles and David Craig, David Cooper, Irene and John Cook, Terry Convoy, Nigel Coleman, Alan Clench, Joan Charsley, Joanne Canedo, Condaisy Garcia, Victor Ercolani, Nigel Nodolski, and many, many others.

COPYRIGHT

First Published 2024
Copyright: Dr. David Henry Dighton
All rights reserved. No part of this publication may be reproduced or transmitted in any form or by any means

Published in the UK by Medicause

www.daviddighton.com

British Library Cataloguing in Publication Data. A CIP catalogue record for this title is available from the British Library.

Cover page background artwork: from an oil painting by Dr David Dighton ('DrDhD'), called 'F.I.R.O.'
See pages 128, 412, 415 for background narrative.
Copyright David H. Dighton). 2022.

Cover produced by Stefan P. (Fiverr.com)

ISBN: 978-1-7385207-3-2

Appendix A

Clinical Data Entry Form

Decision making:
  Normal
  Impaired
Depressed:
  Early morning waking
Co-ordination:
  Normal
  Can't get things together as usual
  Driving competence
Anxiety — Endogenous
           Exogenous
  Mild
  Moderate
  Severe
  Treated with
Aggression:
  To work
  Driving
  Otherwise
Relaxation:
  Easy
  Difficult

Weight:
  Increasing
  Losing
Appetite: Diet: (anything
                          (low calorie
                          (vegetarian
                          (low fat
Soreness of   (sweet
  Tongue   (non-sweet
  Mouth   (etc.
  Throat
Swallowing:
  No difficulty
  Difficulty
  Solids
  Liquids
  Painful
Heartburn:
Waterbrash:
Acid regurgitation:
Oral flatus:
Nausea:
Vomiting:
Indigestion:
  Area
  Type
  Radiation
  Relief with
Abdominal pain:
  Site
  Type
  Frequency

Memory:
  Long term
  Short term —   normal
                 not so good
                 poor
Irritability:
  Unchanged
  More
  Less
Behavioural change:
  Observed by
Activities given up:
Tolerance to smaller and smaller things:
  O.K.
  Impaired
  Grossly impaired
Upkeep of Personal Standards
  Trouble taken
  Not so/can't be bothered
Sense of humour:
  Normal/reduced

B.O.R./B.N.O.R.
Melaena:
Mucus:
Blood on/in stool:
Flatus:
Diarrhoea: Frequency
              Blood
              Mucus

Dysuria:
Frequency: Nocturia: No / Yes
Haematuria:
Hesitancy:
Poor stream:
Stress incontinence:
Loin pain:

Cough:
  Dry
Expectorant:
  Blood
  Purulent
Wheezing:

Nasal blockage: Hay Fever:
Catarrh: Allergies to:
Ears:
Throat: Loss of voice:

Febrile: Days per year:

Skin problem:	Hair:	Joints: All normal/Abnormal
Rash	Normal	Stiff Painful Red Tender Deformed
Itching	Losing	Hand:
Dry		Wrist:
Eyes:		Elbow:
Thyroid symptoms:		Shoulders:
Hypo.	Hyper.	Neck:
		Back:
		Hips:
		Knees:
		Ankles:
		Feet:
		Muscles:
		Pains
		Tenderness
Dizzy spells:		Headache:
Syncope:		Site
Vasomotor		Type
Stokes Adams		Associated symptoms
Epileptic		Migraine:
Hypoglycaemia		Paresthesias:
Chest pain:		Weakness:
Inframammary		Generalised
Angina Int./prolonged		Localised
Related to		Co-ordination
Swallowing		Walking
Moving		Vertigo:
Respiration		Hearing:
Intral		Double Vision:
S.O.B. – Grade I, IIa, IIb, III, IV		Visual loss:
Orthopnoea:		Distortion
Ankle swelling		Nostalgia:
Palpitation:		Taste:
Related to		Smell:
Regular/Irregular		
At rest/on effort		
Associated S.O.B.		
PAST HISTORY		**FAMILY HISTORY**
Childhood Illnesses:		Mother: Alive/age Father: Alive/age
Rheumatic fever		Died/age Died/age
Diphtheria		Date: Date:
Scarlet Fever		Illnesses: Illnesses:
T.B.		
Jaundice		
Etc.		Patient = 1st/2nd/3rd child born
Medical:		Brothers: Sisters:
Surgical:		
Obstetric/Gynae		Wife/Husband: Children:
Para		Ages:

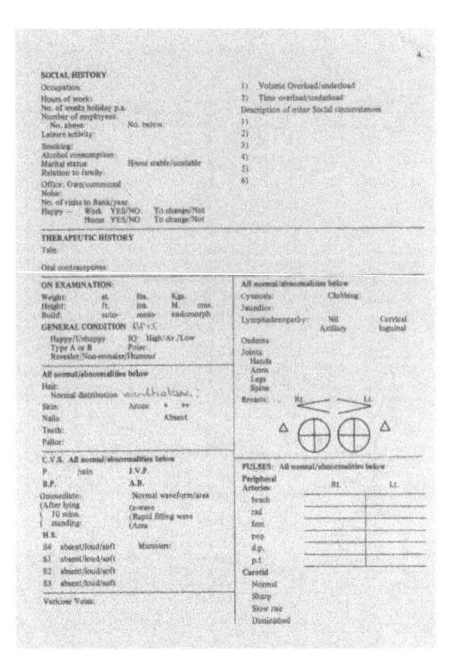

SOCIAL HISTORY
Occupation:
Hours of work:
No. of weeks holiday p.a.
Number of employees:
 No. above: No. below:
Leisure activity:
Smoking:
Alcohol consumption:
Marital status Home stable/unstable
Relation to family:
Office: Own/communal
Noise:
No. of visits to Bank/year
Happy — Work YES/NO To change/Not
         Home YES/NO To change/Not

1) Volume Overload/underload
2) Time overload/underload
Description of other Social circumstances
1)
2)
3)
4)
5)
6)

THERAPEUTIC HISTORY
Tabs:

Oral contraceptives:

ON EXAMINATION:
Weight: st. lbs. Kgs.
Height: ft. ins. M. cms.
Build: ecto- meso- endomorph
GENERAL CONDITION
Happy/Unhappy IQ High/Av./Low
Type A or B Poise
Reveler/Non-reveler/Humour

All normal/abnormalities below
Hair:
 Normal distribution
Skin: Areas + ++
Nails: Absent
Teeth:
Pallor:

C.V.S. All normal/abnormalities below
P. /min J.V.P.
B.P. A.B.
Osmolality: Normal waveform/area
(After lying (a-wave
  10 mins. (Rapid filling wave
  standing) (Area
H.S.
S4 absent/loud/soft Murmurs:
S1 absent/loud/soft
S2 absent/loud/soft
S3 absent/loud/soft
Varicose Veins:

All normal/abnormalities below
Cyanosis: Clubbing:
Jaundice:
Lymphadenopathy: Nil Cervical
                  Axillary Inguinal
Oedema
Joints
 Hands
 Arms
 Legs
 Spine
Breasts: Rt. Lt.

PULSES: All normal/abnormalities below
Peripheral
Arteries: Rt. Lt.
brach
rad
fem
pop
d.p.
p.t.
Carotid
Normal
Sharp
Slow rise
Diminished

R.S. All normal/abnormalities below
Chest Exp: Normal   Restricted
              Rt.   Lt.
T.V.F.
P.N.
B.S.
Rhonchi
Wheeze
Creps.

ABD. All normal/abnormalities below
L
S
K_R
K_L
Hernias
P.R.
Prostate

C.N.S.   All normal/abnormalities below
Cranial Nerves: I II III IV V VI VII VIII IX X XI XII

Ocular opacities   Perimetry
     R   L   R   L

Fundi:
Diabetic/Hypertensive Retinopathy
  Absent
  Grade I, II, III, IV
Co-ordination: Gait: Dysphasia: Dysarthria:
Tremor: Fine/coarse   Nystagmus:
Hearing: Rt.   Lt.
  Normal
  Air
  Bone

ARMS:	Rt.	Lt.	Abd.	LEGS:	Rt.	Lt.
Reflexes: S.J.				Reflexes: K.J.		
B.J.				A.J.		
T.J.				P.R.		
Tone:				Tone:		
Wasting:				Wasting:		
Fasciculation:				Fasciculation:		
Power:				Power:		
Fingers				Toes		
Wrist				Ankle		
Elbow				Knee		
Shoulder				Hip		
Sensory:				Sensory:		

E.C.G.   C.X.R.   Peak flow:

Ex ECG.
Urine:

CAROTID U/S: R)   1 2 3 4 5
              L)

ECHOCARDIOGRAM :

                              REST   POST-EXER
STRESS ECHOCARDIOGRAM: EF   %

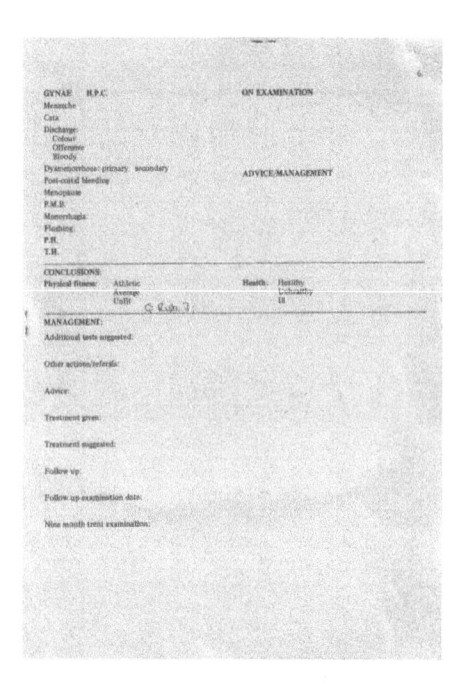

GYNAE H.P.C.	ON EXAMINATION
Menarche	
Cata	
Discharge:	
Colour	
Offensive	
Bloody	
Dysmenorrhoea: primary secondary	
Post-coital bleeding	ADVICE/MANAGEMENT
Menopause	
P.M.B.	
Menorrhagia	
Flushing	
P.H.	
T.B.	

CONCLUSIONS:
Physical fitness: Athletic   Health: Healthy
              Average   Vulnerable
              Unfit   Ill

MANAGEMENT:
Additional tests suggested:

Other actions/referrals:

Advice:

Treatment given:

Treatment suggested:

Follow up:

Follow up examination date:

Nine month term examination:

Appendix B

Sleep Article (1980)

SLEEP THERAPY

Sleep that knits up the ravell'd sleave of care

Sleep deprivation reduces the ability to cope and may so alter homeostasis as to produce hypertension, cardiac infarction or suicide. Dr David Dighton suggests that putting people to sleep for 18 hours every day for five to seven days may help treat cardiovascular problems

Many patients with angina, cardiac infarction, palpitation, syncope or hypertension give a history of inability to sleep well. They may wake feeling unrefreshed, have a partner who complains that they are restless during sleep and be aware that the more tired they feel, the less well they sleep. If questioned specifically many will declare that their energy runs out during the day or that they fall asleep in a chair in the early evening.

As sleep becomes ineffective, the individual may become irritable, quick to 'jump down people's throats', while memory may deteriorate so much that they find it essential to keep a written note of everything. The adapting patients (or perhaps pre-patients) will start to keep a diary, often with considerable self-deception or lack of awareness they will fail to keep appointments or to comply with commitments. Their ability to process information becomes impaired so that their working colleagues have to compensate for them.

The last person to be aware of these changes seems to be the person affected. Energy deprivation causes—or is often associated with—lack of awareness and self-deception. Self-deception expends further energy and perpetuates the vicious circle of decreasing energy levels and increasing difficulty in obtaining rest, and may be an essential component in producing extreme energy depletion or exhaustion. This 'energy depletion syndrome' is the most common finding among people submitting themselves (or being submitted by relatives) for medical screening.

Clinical examination may reveal a normal cardiovascular system or one that demonstrates disturbed homeostasis. The pulse may be fast and sometimes irregular because of atrial or ventricular ectopics; the blood pressure may be raised and the left ventricular apex beat may be distributed over a larger area than normal with perhaps a palpable A wave associated with a fourth heart sound.

From the clinical point of view there seems to be a spectrum of psychological, behavioural, physiological and biochemical changes associated with energy depletion and sleep deprivation which together constitute the 'energy depletion syndrome'.

Sleep therapy was used in ancient Greece in the names of Aesculapius and Amphiaros in order that divine beings could, in sleep, cure ill health. During the last world war sleep therapy was used for battle fatigue. More recently psychiatrists have found it useful for chronic anxiety, obsessional states, schizophrenia and depression.

Cardiovascular conditions are assuming epidemic proportions. Since my experience using sleep therapy is almost restricted to these conditions, I will limit myself to these entities.

At Charing Cross Hospital our current indications for using sleep therapy are:—
● Cardiac infarction
● Cardiac pain with deteriorating effort tolerance despite beta-blockers, TNT, etc
● Hypertension – difficult to control with usual agents
● Decreasing effort tolerance in valvular and other forms of heart disease – cardiomyopathy
● Fast dysrhythmias, frequent ventricular ectopics, recurrent tachycardia associated with cardiac infarction, high arousal or no other obvious cause
● Peripheral vascular disease with ischaemia (while awaiting further assessment)
● Increasing tiredness accompanied by left ventricular dysfunction (palpable A wave and fourth sounds), hyperuricaemia, hypercholesterolaemia (pre-infarction syndrome)

● An unacceptable level of tiredness, fatigue or exhaustion not necessarily associated with abnormal physical signs or biochemistry but leading to difficulties in coping due to psychological and behavioural changes

The contra-indications are limited to patients with serious chest problems, although if therapy is carried out correctly, respiratory depression does not occur. Because active physiotherapy is recommended in wakeful periods, deep vein thrombosis is not a complication.

The aim is to provide relaxed undisturbed sleep for 18 out of 24 hours for three to seven days depending on the indication, the initial level of arousal and subsequent response. The initial drug regime has to be related to the level of arousal, as do succeeding doses.

The simple life

Anaesthetists are accustomed to judging the levels of arousal or consciousness but physicians have to be trained.

In routine anaesthetic practice, it is well known (by virtue of the consequences) that if a patient has not been premedicated, a high induction dose will be required to produce unconsciousness. A similar 'titration principle' applies to the initial and maintenance doses used for sleep therapy. For a person who is highly aroused, agitated and perhaps aggressive the choice would be 20mg diazepam orally in combination with 50mg Phenergan; sometimes a fraction of this (say one half) is given by slow intravenous injection. The principle of 'hitting them hard to start with and quickly reducing to a maintenance dose' is the most reliable method.

Rarely, in very agitated patients, Chlorpromazine 50-100mg qds may also be required. But this is very much the exception and can be avoided by making use of a quiet single room. Peaceful sleep is difficult in a ward which is a thoroughfare, an eight-bed intensive care unit with lights on day and night, or a coronary care unit. It is essential that a quiet, peaceful, unhurried and caring environment is created by the nurses and other staff. For the same reasons treatment of cardiac infarction at home can be uncomplicated (as long as home is not a place of unrest). Adverse arousing influences will be induced by floor cleaners, doctors, physiotherapists, social workers, engineers, administrators and many others about their daily business. Their aim is to get on with their jobs – not to procure tranquility. Visitors may have to be kept away for the first three to five days and telephone calls stopped if a patient has a telephone by his side. Lengthy discussions with the patient about his problems are arousing, and often futile. Initially the patient may be either unwilling to communicate or produce hours of irrelevant verbiage, exhausting both himself and medical attendants. Often it is more practical to induce the patient's sleep as soon as possible, get the main facts from the relatives and friends, and aim to get more sensible and meaningful answers from the unexhausted, relaxed patient a few days later.

Nurses, social workers, medical students and doctors alike help to establish a rapport with the patient as he surfaces from his sleep therapy at about day five. The person most easily able to establish rapport with the patient should continue to do so. This is not always a doctor; it may be a nursing sister, medical student, nursing auxiliary, counsellor, psychotherapist, social worker or hypnotherapist who will take over the responsibility for communication with the patient, so getting to the root of the patient's problem. The brief is to discover what conditions contributed to the patient's decline in health.

After day five the patient may require only a short two to three hour sleep in the afternoon with a good sleep at night. Valium is usually all that is required at this stage. The principle is always to use the smallest amount of hypnotic to achieve a low arousal state.

It is important to realise (as it is in anaesthetic practice) that as treatment continues, sedative dosage can be progressively decreased without affecting the low arousal state. If dosages are kept high, over-sedation will occur and an undesirable drunken or uncontrolled state will result. It is therefore essential to review the drug regime daily. Daily physiotherapy is also essential, including leg exercises, coughing and deep breathing. Many patients will require a daily prophylactic lubricant laxative such as Milpar or liquid paraffin, especially older patients with known constipation problems.

Years of experience with this therapy at Charing Cross Hospital have revealed not a single case of clinically apparent deep vein thrombosis resulting from treatment and only a few resultant chest infections. To avoid complications the level of arousal must be adjusted so the periods of sleep alternate with wakeful periods at meal times, in the morning, midday and evening.

After sleep in the afternoon and at night, it is necessary to start mobilisation. If the cardiovascular abnormalities for which the patient was originally treated have been relieved, mobilisation can continue, if not, further inquiry and investigations are required. By this stage hypertension disappears in 50 per cent of patients, T waves are normalised (whether due to ischaemia or to catecholamine effect), chest pain is relieved in cardiac infarction/progressive angina and the blood lipoproteins and other biochemical profiles are returned to normal in most people. By reducing the need to cope using sleep therapy, homeostatic mechanisms are allowed to exert their full effect, presumably unopposed by the effect of reduced energy reserves.

If abnormalities are still present then hypertensives may need a hypertensive screen (IVP, 24 hours VMA, urea, electrolytes) and patients with cardiac rest pain may need arteriography. Before assuming a purely pathophysiological aetiology for cardiac pain or the rare presence of a phaeochromocytoma for the cause of hypertension, the effects of arousal should be first excluded.

Successful rehabilitation often means a change of attitude to alter the circumstances which gave rise to previous ill-health, together with training to increase physical strength, stamina, self-confidence. As important is coming to know personal limitations or maintaining a budget for physical, emotional and intellectual activity. The emphasis should be not on finding a pill or technique that will cure without self-involvement, but rather education and training to maintain health. Nobody would expect to learn French or how to fly or to know how to survive in difficult circumstances without training. If cardiac infarction, angina and hypertension are seen as evidence of failure to survive healthily then the idea of retraining follows logically.

Pharmacological agents have undoubted use in reducing morbidity and mortality but side-effects may affect the quality of life. If they exist, the patient is less likely to accept personal responsibility for his survival. Our principle is to use them when a patient has appropriate pathological disease or is untrainable. Ideally, it is preferable to have a choice, for the physician, and for the patient. In cases of pre-cardiac infarction, angina or post-infarction, several different techniques are in use in our department, namely relaxation training, auto-hypnosis training and exercise retraining.

People undergoing therapy have an educational reading programme. For those who feel the effects of stress and for the hypertensive, hypercholesterolaemic, hyperuricaemic and type A characters, we may advise a choice from the same list, together with feedback training, massage, transcendental meditation, yoga or autogenic training. We find that these alternative medical techniques may be powerful in effect; for example they compare favourably with beta blockers in hypertension. We are assessing their true clinical value.

Experience so far suggests that many of these techniques usefully add to our armamentarium of therapy.

Many patients seem to respect the effectiveness of these techniques more than their medical attendants feel proper without further proof of efficacy. In this respect little change in the attitudes of patients or their physicians has occurred since the time of Hippocrates. To some extent we may have to accept that the 'customers are right' (albeit not scientifically) until proven otherwise. The onus of professional responsibility is to prove practical usefulness – perhaps even proof.

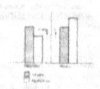

Until then we are content to continue to use the techniques listed.

Dr Dighton is a lecturer in the cardiac department at Charing Cross Hospital. Illustrations from his lecture material.

Reprinted with permission from
GENERAL PRACTITIONER
11th January 1980

INDEX

1
18-F-fluorodeoxyglucose (FDG), 260
2
24-hour ECG, 362
2nd heart sound splitting, 237
3
3rd heart sound, 237
4
4th heart sound, 237
6
6 Cs, The six C's, 41

A

'A' Level students, 7, 14
'A' Levels, 11, 28
A&E, 11, 56, 73, 85, 132, 173, 234, 285, 288, 298, 311, 319, 341, 373, 448, 454, 458, 468, 482
abacus, 327
ability to communicate, 11, 19, 196
abrasive, 65

academic, 5, 18, 35, 44, 50, 51, 67, 81, 90, 109, 116, 180, 187, 201, 208, 226, 261, 263, 265, 331, 345, 386, 392, 442, 453, 460, 465, 480, 483, 509, 510

academics, 18, 52, 61, 73, 100, 198, 313, 324, 325, 471
accuracy, 8, 68, 115, 171, 200, 201, 213, 214, 224, 234,

ACE inhibitors, 226, 262, 263, 422, 436, 440
ACE receptor profile, 318
acetylcholine, 211, 481
acronym, 26, 63, 88, 91, 92, 229, 324, 359, 360, 368, 505,
actors, 65
acupuncture, 392
Ada Lovelace, 327, 498
Adam-Stokes, 240
adaptive trial, 485, 486
adversarial, 90, 137, 345, 383
aetiological factor, 113
AF, 225, 239, 269, 281, 290, 367, 375, 400, 419, 434, 436, 437, 464, 474, 489
affordability, 406
Afro-Caribbean, 28
against the odds, 284, 347, 444

age, 11, 22, 34, 37, 57, 58, 74, 92, 96, 107, 128, 130, 131, 135, 141, 148, 151, 162, 163, 164, 172, 179, 182, 183, 186, 187, 194, 196, 204, 209, 212, 219, 222, 233, 235, 240, 242, 248, 261, 264, 267, 280, 297, 298, 308, 331, 332, 350, 357, 363, 373, 374, 376,385, 392, 419, 423, 427, 434, 450, 451, 464, 474, 482, 483, 493

aggressive, 71, 80, 160, 189, 431, 448, 466
agogá (Grk), 187
AI, 8, 39, 40, 238, 278, 285, 296, 305, 306, 319, 329, 465, 472, 489
AI hype-cycle, 329
Alan Gardner, 48, 59,103, 393, 506, 507
Alan Harris, 350, 443, 506, 508, 510
Alan Türing, 99

alcohol, 27, 101, 110, 116, 141, 153, 156, 163, 164, 195, 214, 215, 222, 266, 298, 318, 375, 385, 390, 392, 403, 434, 442, 475, 500

alcoholism, 144, 214
Alexander Fleming, 392
algorithm, 114, 288, 304, 311, 312, 327, 329, 331, 341, 352, 372, 373, 378
Algorithmic Justice League, 330
alien life, 290, 334
alien species, 334
all-comers, 31, 108, 109, 129, 210, 287, 390
Altman, Brian, 466
altruistic, 30, 43, 44, 51, 86, 198, 266, 420, 478
altruistic vocation, 198
ambience, 124, 206, 414, 453, 471
ambiguity, 173, 352
ambition, 15, 43, 56, 103, 119, 185, 451

ambulance, 173, 288, 421, 458
amino acids, 322, 481, 485
Amsterdam, 5, 87, 187, 196, 267, 403, 420, 421, 481, 506, 511
Amundsen, 85, 407, 501
anachronistic, 40, 44, 172, 214, 329, 388, 402
anaemia, 281
anaesthetics, 11, 15, 25, 56, 58
analgesic, 222, 345, 451
anathema, 18, 319
anatomical, 180, 190, 232, 237, 271, 314, 370
anatomy, 11, 12, 168, 172, 243, 250, 251, 262, 314, 403, 461
Ancient Egypt, 471
androgynous, 190, 496

anecdote, 22, 46, 51, 56, 68, 83, 126, 225, 248, 253, 293, 294, 295, 300, 312, 313, 314, 316, 319, 320, 369, 377, 378, 398, 406, 479

Aneurin Bevin, 287
aneurysm, 238, 251, 252, 285, 314, 362, 400, 449, 451

angina, 36, 37, 113, 123, 154, 192, 207, 216, 245, 246, 247, 260, 264, 269, 276, 277, 280, 286, 291, 299, 301, 302, 323, 343, 344, 351, 368, 369, 375, 376, 397, 401, 438, 461, 510

angiotensin II receptor antagonists, 226
animal fat, 133, 208, 295
anonymous, 3, 4, 6, 11, 25, 31, 51, 66, 96, 167, 180, 181, 189, 201, 205, 224, 230, 304, 318, 387, 427, 428, 430, 450, 476
ante hoc, 392
anterior cardiac infarction, 398, 453, 465
anterior descending coronary artery, 113, 260, 261, 375
anterior infarction, 260, 369, 376
anthropology, 184
anthropomorphised robots, 328
antibiotic, 49, 211, 221, 222, 418
antibody levels, 256, 257, 462, 484
anticoagulation, 145, 174, 422

anxiety, 56, 82, 90, 100, 101, 108, 115, 125, 126, 141, 142, 145, 150, 151, 185, 196, 215, 217, 223, 228, 232, 257, 276, 281, 289, 297, 301, 309, 337, 338, 368, 404, 427, 429, 430, 431, 433, 440, 442, 448, 464

aortic aneurysm, 36, 238, 373
aortic area, 238
aortic valve replacement, 408, 465
aortic valve stenosis, 302
apex beat, 236

aphorism, 26
apixaban, 117
appendicitis, 239, 241, 253, 280, 396
appraisal, 76, 171, 204, 250, 263, 269, 347, 359, 363, 397
apprentice, 7, 19, 23, 24, 69, 197, 203, 237, 337, 395, 487
apprenticeship, 5, 10, 505
approachability, 32
apps, 141, 296, 318, 326
Apricot computer, 334
aptitude, 14, 17, 53, 445
Arab Spring, 297
ARCTEC Team, 300
arcus senilis, 480
arrogant, 66, 109, 132, 453
art, 5, 22, 23, 44, 51, 119, 165, 197, 303, 304, 365, 417, 492, 494, 496, 502
art and science, 2, 4, 5, 7, 19, 23, 38, 45, 168, 406, 409, 487

art of medicine, 3, 6, 7, 17, 20, 21, 22, 23, 24, 27, 32, 46, 58, 60, 66, 67, 81, 96, 98, 112, 118, 119, 135, 139, 144, 166, 174, 175, 176, 189, 218, 225, 231, 254, 276, 287, 289, 290, 325, 386, 388, 391, 393, 408, 409, 410, 413, 427, 430, 431, 444, 454, 479, 486, 487, 499, 509, 510, 511

arterial intima, 321, 364, 405, 480

arteries, 113, 239, 251, 253, 261, 264, 291, 302, 321, 322, 364, 368, 380, 401, 420, 440, 451, 463, 482, 489, 490, 491

arteriopath, 401
artery scanning, 246
art form, 248, 256, 278, 303, 317, 509
arthritis, 229, 233, 234, 264, 271, 300
Asclepius, 174, 175, 176
Asher, Richard, 23, 50, 56, 99, 492, 493
asian, 25, 28, 189, 195, 266
Asperger, 24, 85, 98, 395, 479
aspirin, 11, 174, 211, 212, 222, 225, 226, 227, 261, 262, 263, 350, 384, 386, 422

assessment, 7, 11, 19, 37, 95, 153, 161, 167, 210, 223, 251, 260, 279, 281, 362, 364, 372, 391, 396, 399, 425, 426, 436, 444, 448, 451, 469, 472, 482, 500

assumption, 58, 105, 134, 166, 239, 254, 255, 270, 294, 318, 347, 356, 362, 364, 376, 380, 394, 401, 424, 425, 434, 437, 474, 481, 507

atherogenerative, 373
atherogenesis, 164, 322, 323, 380, 482

atheroma, 114, 161, 169, 208, 239, 246, 251, 252, 261, 262, 263, 264, 265, 277, 280, 285, 291, 302, 303, 322, 323, 349, 362, 364, 367, 380, 384, 401, 405, 451, 480, 481, 482, 489, 508

atheromatous plaque, 123, 260, 321, 376
atheroprotective, 373

atherosclerosis, 29, 36, 113, 114, 123, 156, 169, 239, 263, 302, 322, 346, 380, 401, 448, 464, 466, 489, 510, 511

athletic, 16, 163, 240, 373, 437, 451
atrial fibrillation, 174, 225, 236, 277, 299
atrial flutter, 375
atrial myxoma, 240, 481
atrial pacemaker control, 334
atrial systole, 237, 436

attached, 6, 8, 35, 51, 52, 53, 58, 66, 74, 117, 123, 171, 187, 213, 218, 230, 258, 259, 287, 289, 324, 374, 396, 403, 411, 427, 428, 432, 445, 446, 473, 490

attitude to work, 45
atypical chest pain, 261
Aubrey Leatham, 29, 68, 70, 103, 205, 210, 237, 279, 350, 366, 395, 443, 506, 509
auditing algorithms, 330
Augustine, St., 109, 388, 495
auscultation, 237, 238, 279, 433
autistic, 19, 24, 67, 68, 101, 103, 187, 297, 374
autistic savant, 297
Autodesk, 333
AutoML, 329
autonomic nervous system, 114, 368
autonomic reflex, 240, 481
autonomic responsiveness, 265
AV block, 269, 481
availability, 32, 111, 214, 285, 398, 469
avatar, 67, 160
avoidable deaths, 252, 408
axiom, 320, 469

B

Babbage, Charles, 327, 498
back to basics, 233, 358
bacterial resistance, 257, 275
bad luck, 86, 407
Baked Bean Counter, 67, 68, 280, 305, 317, 459

Baker's cyst, 278
Balint, Michael, Dr., 168, 171, 428
BAME, (Black, Asian, Minority Ethnic) 186
barbiturates, 226, 442
Barnard, Christiaan, 389, 408
basic skills, 19
Baxter, David. Dr., 69, 103, 184, 414, 506, 511
Bayes' Theorem, 8, 201, 253, 293, 299, 497, 498
BBC Radio 4, 184, 485, 502
bedside manner, 473
beginners, 16
belief, 60, 219, 316, 353, 354, 356
benzodiazepine, 441
bereavement, 113, 214, 442
Berne, Eric, 128, 143, 460, 495
Bernstein, Leonard, 394
best practice, 119, 176, 339
beta-blockers, 220, 221, 226, 263, 434, 437
Bevin, Aneurin, 287
Bezos, Jeff, 347, 389

bias, 26, 58, 68, 94, 134, 146, 182, 196, 201, 248, 260, 273, 274, 293, 296, 306, 312, 313, 317, 319, 323, 330, 338, 350, 351, 352, 353, 354, 355, 356, 357, 358, 362, 366, 377, 379, 387, 401, 402, 405, 413, 433, 439, 446, 453, 460, 472, 483, 484

biblical, 35, 79
bi-cultural, 189
Big Data, 248, 404
big data project, 335
bi-lingual, 189
Bill Gates, 389, 452
binary, 14, 32, 46, 51, 190, 293, 309, 327, 345, 394, 395, 398, 435, 498
binary mathematics, 309, 498
binary thinkers, 51
binary thought, 46
biochemistry□11, 168, 169, 219, 256, 275, 303, 318
biological age, 37, 373, 450
biological clock, 114
biological drive, 27
biology, 52, 57, 100, 113, 192, 254, 381, 392, 422
biopsy, 260, 266, 291, 375
bi-sexual, 190
black and white, 23, 68, 305, 454
black hole, 333, 399
black swan, 330, 483

blackmail, 69
Blake, Frank, 290
Bletchley Park, 99, 244, 332
BLOB (**BL**indingly **OB**vious, 37, 47, 263, 279, 320, 348
BLOB topics, 37

blood, 12, 48, 78, 94, 95, 104, 114, 117, 123, 124, 127, 141, 153, 166, 169, 171, 174, 200, 208, 209, 213, 221, 232, 235, 239, 244, 246, 253, 254, 256, 257, 259, 260, 261, 262, 264, 265, 266, 268, 270, 271, 275, 278, 284, 289, 291, 298, 300, 301, 302, 303, 307, 315, 321, 322, 323, 338, 340, 346, 362, 364, 366, 375, 376, 380, 384, 401, 405, 412, 417, 418, 419, 420, 421, 422, 424, 431, 435, 436, 437, 438, 439, 440, 447, 448, 460, 468, 473, 480, 481, 489, 490, 491, 495

blood culture, 281
blood glucose, 259, 260, 262, 375, 490

blood pressure, 124, 127, 153, 208, 213, 239, 262, 298, 301, 302, 340, 412, 417, 422, 431, 439, 440, 468, 489, 490

blood transfusion, 338, 447
blue bloater, 281
blue sky thinking, 305, 329
bluster, 379
BMA, 92, 243, 497
Bob Hope, 54, 82, 280, 339, 498
body mass, 323
Boeing 737 Max, 330
bombast, □67
Boolean, 331
boundaries, 29, 294, 347, 380, 434
boundary conditions, 46, 380
bravery, 119, 511
breast discomfort on exercise, 484
breathless, 174, 232, 343, 375, 395, 451, 459
BREXIT, 61, 267
British Heart Foundation, 5, 395, 443, 463, 481
Brits, 45, 187, 188, 189, 191, 193, 196, 214
BSc., 35, 50, 478
Buffett, Warren, 381, 389, 452, 501
bullies, 69
bullshit, 33, 332, 379
BUPA, 32

bureaucracy, 4, 7, 8, 24, 25, 34, 41, 84, 119, 165, 172, 232, 262, 359, 360, 371, 374, 386, 390, 396, 399, 409, 414, 443, 444, 470, 472, 488 bureaucratic, 6, 7, 8, 32, 38, 60, 84, 97, 102, 114, 118, 119, 129, 160, 259, 280, 305, 318, 359, 381, 390, 399, 410, 411, 425, 426, 427, 432, 470, 471, 483

business owner, 121, 282

C

C&S (culture and sensitivity), 255
CAA medical examination, 303

CABG, 37, 113, 132, 174, 213, 214, 228, 241, 252, 262, 263, 264, 269, 280, 302, 369, 398, 401, 408, 422, 438, 451, 464, 489

calcified, 132, 211, 238, 251, 291, 362, 464, 489
calcium channel blocker, 226
calcium score (CAC score), 251, 264, 277, 280, 302, 375, 376, 464
calf tenderness, 240
camaraderie, 58

cancer, 22, 27, 72, 86, 99, 113, 114, 116, 126, 145, 162, 176, 181, 186, 213, 235, 239, 245, 247, 258, 260, 270, 271, 273, 276, 278, 281, 286, 291, 300, 311, 337, 349, 351, 392, 399, 419, 428, 430, 451, 462, 463, 485, 491, 497

capacity, 105, 145, 302, 318, 383, 400, 423
capricious 69
carbohydrates, 113, 314
carcinoma, 145, 247, 273, 286

cardiac, 5, 29, 32, 36, 54, 56, 68, 81, 87, 95, 97, 115, 116, 122, 123, 125, 126, 132, 157, 161, 164, 196, 210, 211, 216, 217, 221, 225, 229, 231, 232, 234, 236, 237, 240, 243, 251, 252, 257, 260, 261, 262, 263, 264, 265, 267, 269, 275, 277, 278, 279, 280, 281, 282, 285, 299, 301, 302, 303, 311, 314, 321, 322, 323, 343, 344, 349, 350, 362, 364, 369, 373, 375, 376, 384, 389, 397, 398, 400, 402, 405, 408, 416, 417, 419, 420, 421, 422, 435, 437, 438, 439, 441, 442, 443, 451, 453, 459, 463, 464, 465, 466, 481, 482, 508, 509, 510, 511

cardiac arrest, 54, 275, 311, 400, 419
Cardiac Centre Loughton, 5
cardiac output, 221, 234, 278, 281, 321, 420, 435, 437, 459, 481
cardiac pre-diagnosis, 286
cardiac screening, 115, 262, 263, 264, 269, 285, 463
cardiac shunt, 281
Cardiac Value of Food, 482

cardiology, 5, 11, 26, 47, 54, 56, 72, 85, 94, 122, 127, 205, 209, 220, 237, 267, 364, 373, 397, 408, 443, 469, 508, 510, 511

cardiomyopathy, 140, 301
cardiopulmonary medicine, 229

career, 6, 12, 14, 24, 27, 28, 35, 37, 38, 40, 41, 44, 47, 51, 52, 54, 55, 65, 75, 76, 83, 87, 97, 104, 122, 136, 140, 185, 222, 242, 250, 266, 267, 269, 277, 282, 293, 315, 317, 324, 342, 359, 360, 387, 389, 390, 399, 403, 409, 434, 453, 466, 472, 478, 480, 483

caring, 2, 7, 19, 20, 23, 50, 52, 56, 67, 107, 121, 177, 213, 478
Carl Bass, 333
Carl Marsh, 288
carotid artery screening, 262
carotid atheroma, 114, 261, 264, 265, 291, 302, 322, 367, 405, 463, 480, 482, 483
case selection, 317
Cassandra, 373
catalogue of cases, 326
Catastrophe Theory, 400, 500
catecholamine, 302
cathartic, 119, 249, 350, 484
catheter lab., 285, 438, 481, 506
catheterisation, 196, 237, 267, 282, 299, 343, 512
cavalier, 94, 267, 289, 307, 444, 479, 483
caveat, 363, 424
cerebral ischaemia, 114
certification, 32, 52, 122, 129, 171
César Ritz, 416, 502
CHADs2 score, 313
challenging cases, 200
chameleon psychiatry, □272
Chaos Theory, 169, 310, 400, 500

characteristics, 20, 44, 45, 53, 54, 79, 106, 108, 109, 150, 187, 190, 204, 223, 254, 267, 284, 294, 333, 356, 370

Charaka Samhita, 379, 499
Charing Cross Hospital, 5, 22, 87, 122, 207, 209, 243, 264, 399, 481, 506, 510, 511
charisma, 93, 154, 390
charm, 70
charming, 87, 103, 107, 111, 140
chatbox LaMDA, 328
chatting, 232
Chaucer, 80, 379, 499
check-box functioning, 304
check list, 7, 20, 200, 204, 247, 485
chemotherapy, 293

cherry picking, 149, 226, 317, 326
Chesney Sullenberger ('Sully'), 384
Chi Square test, 294

children, 27, 35, 55, 58, 59, 76, 91, 96, 124, 133, 134, 148, 150, 164, 174, 184, 186, 193, 194, 195, 197, 213, 217, 220, 232, 247, 263, 268, 275, 281, 287, 288, 297, 335, 378, 386, 405, 433, 439, 447, 462, 507, 508, 513

chivalry, 34, 187

choice, 7, 18, 33, 44, 55, 58, 111, 124, 128, 162, 165, 174, 191, 194, 195, 208, 212, 259, 306, 308, 312, 338, 341, 353, 362, 365, 378, 414, 432, 448, 449, 453, 472, 474

cholesterol, 114, 169, 251, 261, 262, 264, 265, 291, 321, 322, 323, 346, 364, 380, 384, 401, 405, 480, 481, 483, 489, 490, 491

cholinesterase, 211, 481
chorionic gonadotrophin, 253, 300
Chris Johnstone, 411
Christiaan Barnard, 389, 408
chromosomes, 57

chronic, 36, 113, 117, 124, 171, 216, 220, 221, 256, 257, 271, 289, 291, 339, 418, 432, 436, 437, 438, 451, 468, 470

chronos, 418
Churchill, Winston, 122, 164
circulation, 114, 123, 302, 422, 490
citizens, 7, 30, 32, 170, 195, 196, 335, 341, 422, 512
claims, 111, 310, 384, 437
Clark, David. Dr., 261, 389, 497
clarity, 15, 118, 323, 390, 426
clarity of thought, 350, 366, 385, 395, 472, 507, 509
Class 1 doctors, 20
Class 2 doctors, 20
Class 3 doctors, 19
Class 4 doctors, 19
classic cases, 278, 381
claudication, 240
clergy, 7, 24
clicks, 237
Clifford Wilson, Prof., 21, 284, 492
clinical advantage, 243
clinical arena, 300
clinical detail, 4, 294

clinical dimension, 318
clinical excellence, 29
clinical experience, 5, 6, 27, 51, 83, 137, 145, 212, 213, 259, 267, 275, 300, 344, 364, 367, 411

clinical judgement, 6, 8, 11, 20, 68, 93, 145, 212, 224, 248, 278, 279, 336, 337, 338, 341, 343, 349, 411, 471, 472, 474, 510

clinical management, 7, 25, 33, 46, 111, 165, 168, 182, 204, 230, 246, 251, 252, 265, 272, 287, 340, 359, 362, 408, 409, 417, 446, 461, 484

clinical relevance, 12, 39, 128, 149, 189, 248, 250, 271, 280, 312, 315, 318, 324, 338, 401, 404, 438, 479

clinical responsibility, 5, 19, 141, 324
clinical situation, 51, 109, 388, 409, 418, 465
clinical titration, 418
clinical trajectory, 399
clinical urgency, 53, 371, 410, 445, 457
clinico-pathological, 12, 229, 271, 396
club, 16, 34, 55, 70, 147, 184, 415
clumsy people, 15
clustering, 169, 273
coarctation, 237
co-author, 403
Cockney, 82, 121, 188, 189, 210, 468
cognitive, 52, 68, 105, 125, 128, 150, 165, 218, 293, 297, 312, 329, 338, 351, 387, 433
cognitive band-width, 387
cohort analysis, 253, 323
Colin Davis, 102, 307, 506, 509
collaterals, 438

colleagues, 4, 8, 12, 17, 18, 19, 20, 22, 23, 25, 29, 32, 35, 43, 44, 47, 49, 54, 57, 63, 64, 73, 74, 75, 76, 78, 80, 86, 87, 92, 94, 97, 102, 103, 105, 106, 122, 127, 140, 144, 159, 171, 172, 177, 178, 185, 208, 212, 221, 227, 228, 231, 243, 244, 245, 261, 267, 271, 276, 282, 286, 291, 295, 302, 303, 309, 320, 352, 358, 359, 369, 370, 382, 384, 390, 392, 402, 406, 411, 412, 421, 444, 453, 467, 473, 478, 508, 509, 510, 511, 512

collection of data, 478
Columbo, Inspector, 284, 381, 499
commercial, 13, 15, 36, 44, 89, 226, 359
commitment, 1, 19, 20, 21, 33, 34, 38, 40, 41, 124, 224, 374, 404, 425, 469
committee, 45, 176, 316, 383, 399, 416
common sense, 27, 51, 53, 115, 385, 391, 445, 508

communication, 8, 19, 20, 39, 41, 65, 68, 105, 118, 120, 121, 122, 134, 169, 173, 189, 191, 193, 197, 209, 212, 289, 296, 297, 298, 326, 327, 358, 360, 414, 424, 431, 432, 433, 446, 449, 461

community, 54, 186, 262, 454, 469, 471, 509, 511
compassion, 6, 26, 41, 45, 98, 145, 304, 335, 396, 478, 509
compatibility, 108, 121, 128, 218, 412, 432, 461
compendium, 105, 229, 418
compensation, 17, 111, 470
competent physician, 67, 145
complacency, 22, 277
complete heart block, 47, 211, 240
completeness, 13, 212, 234, 280, 315, 334

complex, 13, 14, 15, 24, 46, 200, 241, 243, 303, 308, 309, 332, 338, 353, 358, 395, 400, 405, 417, 472, 509

complexity, 57, 309, 326, 331

compliance, 39, 60, 107, 119, 128, 129, 133, 137, 138, 165, 168, 173, 176, 182, 198, 220, 280, 289, 304, 329, 331, 341, 359, 371, 380, 383, 386, 424, 427, 447, 456, 459, 466, 470, 474

computer analysis, 248

computers, 3, 39, 203, 232, 245, 248, 249, 276, 293, 296, 298, 304, 308, 326, 327, 328, 329, 330, 331, 332, 333, 334, 360, 396, 471, 472

conceit, 377
conduction (electrical), 211, 255, 269, 301, 464, 481

confidence, 11, 15, 34, 38, 58, 75, 79, 82, 86, 87, 90, 99, 103, 118, 141, 146, 172, 178, 201, 203, 230, 231, 253, 256, 259, 277, 300, 358, 390, 412, 425, 428, 433, 447, 456, 459, 473, 474, 476, 507, 508

confidentiality, 174, 177, 212, 246, 298
conformity, 51, 304
Confucius, 119
conscious thought, 405
consciousness, 233, 240, 332, 371, 374
consensus, 316, 356, 378, 383, 429, 463

consequences, 5, 20, 27, 57, 69, 74, 105, 115, 117, 146, 148, 153, 162, 169, 217, 220, 268, 269, 289, 330, 337, 358, 374, 400, 408, 417, 424, 427, 431, 452, 460, 461, 466, 474

consolidation, 238, 253
consultant, 55
contractility, 81, 220, 437
contributing factors, 124
control freak, 71
control group, 313

controversy, 44, 320, 431
cool, 71, 137
co-ordinated, 53, 108, 444, 447,
COPD, 124, 281, 343, 430, 451, 489
co-out, 453
core temperature, 288
coronary heart disease, 27, 174, 205, 213, 245, 269, 301, 303, 321, 362, 422, 430
coronary artery stenosis, 321, 340
coronavirus, 115
corporate, 18, 72, 138, 299, 304, 378, 387
corporate bureaucrats, 33
corporate NHS, 38
corporations, 3, 7, 33, 38, 39, 40, 97, 131, 296, 298, 304, 305, 335, 364, 378, 388, 414, 425
corporatisation, 119
corroboration, 251
cosmopolitan, 25, 458
costal cartilages, 280
counselling, 129, 148, 257, 259
courage, 41, 51, 100, 103, 117, 187, 246, 267, 316, 335, 365, 370, 405, 407
Court of Law, 304, 383
COVID-19, 27, 31, 40, 129, 186, 194, 221, 257, 264, 309, 318, 320, 479, 484
CQC, 35, 41, 80, 157, 171, 172, 177, 201, 298, 359, 478, 490
creativity, 6, 17, 298, 307, 308, 334, 374, 383, 443, 483
Crimea, 92, 126, 376, 377, 497, 505
criminal, 135, 137, 139, 140, 148, 164, 191, 207, 274, 345, 346, 442, 467
Crohn's Disease, 402
cross check, 474
cross questioning, 397
CRP, 26, 271, 300

CT, 40, 97, 246, 251, 252, 255, 260, 261, 264, 269, 271, 277, 278, 280, 285, 291, 302, 362, 375, 464

CT angiography, 302
Culpeper, Nicholas, 145
cultural difference, 20, 28, 181, 182, 185, 186, 188, 189, 193, 196, 197
cultural diversity, 8
cultural knowledge, 184
cultural values, 82, 129, 182, 184, 185, 187, 192, 194
culture, 25, 181, 182, 184, 186, 188, 189, 191, 197, 255, 414
cunning, 86
cure, 31, 52, 60, 106, 116, 124, 127, 155, 169, 175, 218, 371, 375, 462, 493, 501
Curie, Marie and Pierre, 334
CVP monitoring, 307
CXR in asthma, 281
cyanosis, 233, 238

D

daft, 72, 258, 392

dangerous, 36, 94, 109, 121, 139, 140, 141, 154, 169, 201, 208, 220, 240, 266, 275, 277, 291, 302, 307, 329, 345, 347, 359, 384, 386, 390, 401, 426, 435, 441, 442, 483, 501, 509

Daniel Kahneman, 280, 352, 382
Darin, Charles, 76, 392
data analysis, 52, 68, 334, 445
datafication, 297
David Baxter, Dr., 69, 103, 184, 414, 506, 511
David Clark, Dr., 261, 389, 497
Davis, Colin, 102, 307, 506, 509
Dawkins, Richard, 74, 429
de motu cordis, 59, 321, 368
de novo creation, 308

death, 4, 12, 31, 58, 95, 96, 116, 126, 145, 147, 156, 169, 195, 205, 208, 212, 213, 215, 217, 252, 260, 261, 262, 263, 266, 267, 274, 275, 278, 286, 289, 294, 302, 303, 314, 326, 337, 384, 386, 389, 390, 398, 399, 400, 403, 408, 422, 423, 427, 430, 432, 433, 453, 458, 461, 463, 464, 485, 507

deception, 96, 309

decision making, 6, 7, 17, 68, 98, 134, 182, 246, 293, 295, 296, 297, 303, 306, 312, 332, 337, 338, 339, 341, 342, 352, 354, 373, 376, 377, 378, 379, 381, 382, 387, 396, 472

decision tools, 5, 201, 265
dedicated doctors, 73
dedication, 11, 29, 41, 66, 145, 198, 289, 359, 509
defibrillation, 240, 252
dehydration, 234, 235, 299, 400, 403, 419, 435, 490
delays, 31, 126, 276, 297, 298, 419
Delphic dictum, 27, 53
delusion☐ 17, 110, 162, 319
demigod, 147
democracy, 112, 196, 378
demography, 25, 27, 106, 116, 120, 415, 469
denial, 122, 132, 186, 219, 228, 246, 279, 384
depersonalisation, 14, 244

depression, 36, 98, 100, 113, 123, 127, 132, 150, 151, 166, 208, 215, 217, 223, 224, 233, 301, 394, 427, 429, 433, 441

derogatory criticism, 303
Descartes, 380, 493
desert island medicine, 221, 237, 253, 370, 418

detached (attitude), 6, 7, 8, 27, 29, 51, 52, 53, 58, 66, 81, 96, 117, 118, 119, 136, 143, 167, 171, 187, 189, 203, 205, 213, 228, 230, 248, 265, 282, 289, 318, 324, 374, 395, 409, 428, 432, 445, 446

devils in the detail, 168, 259, 293, 294, 315, 324, 326, 330, 332, 346, 368, 369, 401
dexterity, 1, 11, 78, 370

diabetes, 37, 113, 125, 186, 213, 214, 224, 247, 259, 260, 262, 264, 286, 314, 349, 375, 439, 442, 447, 466, 483, 490

diagnostic ability, 12, 248
diagnostic accuracy, 250, 255, 276, 300, 301
diagnostic odyssey, 291
diagnostic proficiency, 112
diagnostic weight, 375, 396
diastolic murmurs, 238
dice and chance, 299
difficult diagnoses, 5, 332, 350
digital, 141, 212, 239, 291, 296, 297, 298, 318, 327, 329, 332, 333, 396, 404, 463, 482
digital self, 141, 291, 404
dignity, 127, 172, 177, 195, 217, 329, 430
digoxin, 220, 234, 235, 434, 436, 437
dilatation, 221, 238, 285
diphtheria toxin, 211, 481
discipline, 23, 315, 373, 425, 487
discoveries, 4, 22, 308, 314, 334, 357, 376, 404, 485

discussion 17, 33, 47, 76, 91, 99, 107, 109, 115, 122, 127, 147, 150, 159, 164, 165, 180, 182, 186, 197, 206, 209, 218, 243, 271, 288, 331, 341, 345, 366, 379, 388, 390, 415, 424, 425, 442, 462, 464, 465, 474, 479, 497, 510

disease prevalence, 116, 181, 186, 195
disrespect, 74
distinctions, 27, 192, 287, 345
diuretics, 221, 234, 419, 422, 435, 436, 440, 459
diverticulitis, 126, 238, 247, 399
divine gifts, 64
divine intervention, 288, 310
Dixon, Maurice, 286, 512
Doc Martin, 33, 67, 100, 287, 492, 494

doctor(s), 2, 4, 5, 6, 7, 8, 10, 11, 12, 13, 17, 18, 19, 21, 22, 23, 25, 26, 28, 29, 30, 31, 32, 33, 35, 38, 39, 40, 41, 44, 48, 49, 50, 51, 52, 53, 54, 56, 57, 58, 59, 60, 64, 66, 67, 68, 73, 78, 80, 82, 83, 86, 87, 91, 96, 97, 98, 100, 102, 105, 106, 107, 108, 109, 111, 117, 118, 120, 121, 122, 124, 127, 129, 132, 135, 136, 137, 139, 140, 143, 145, 147, 150, 152, 157, 164, 168, 170, 171, 172, 174, 176, 178, 181, 184, 187, 197, 198, 200, 201, 202, 204, 205, 206, 207, 210, 212, 218, 219, 220, 224, 225, 226, 230, 231, 232, 234, 241, 242, 243, 247, 248, 249, 252, 259, 265, 266, 267, 268, 269, 280, 283, 284, 287, 288, 289, 299, 305, 310, 314, 324, 331, 332, 338, 342, 345, 359, 362, 364, 367, 371, 374, 379, 386, 387, 388, 390, 391, 396, 398, 399, 406, 409, 410, 411, 414, 415, 416, 420, 421, 423, 425, 427, 428, 430, 431, 432, 433, 437, 440, 445, 447, 448, 449, 450, 453, 454, 457, 458, 462, 466, 468, 469, 475, 479, 485, 487, 501, 509, 510, 511

Doctor Google, 112, 133
doctorate, 26
dogma, 402
dogs, 300, 403
Don Russell, 189, 283, 374, 507, 512
Donald Ross, Mr., 416
Donald Trump, 103, 378
dork, 76
double-blind trial, 131, 257, 390, 484, 486
Dr. Who, 381, 499
Drexit (doctor exit), 413
drinkers, 37, 156
drug addicts, 37, 140, 142, 148
drug (company) reps, 225, 226
duncified, 467
DVT, 145, 278, 281, 290, 431, 490
dynamic, 19, 53, 57, 196, 285, 339, 343, 400, 409, 446, 461
dysfunction, 216, 403, 481
dysrhythmia, 208

E

early detection, 246
early diastolic murmur, 238, 465
East End (London), 284
Ebstein Barr virus (EB), 256, 257, 268, 484
eccentricity, 48

ECG, 17, 47, 114, 125, 171, 235, 236, 237, 239, 240, 246, 252, 256, 262, 264, 265, 269, 277, 280, 281, 301, 302, 303, 307, 362, 367, 375, 395, 397, 398, 418, 419, 436, 437, 439, 464, 471, 481, 490, 491

echocardiogram, 17, 114, 210, 239, 240, 255, 280, 281, 302, 395, 398, 418, 419, 435, 439, 443, 447, 508

echocardiography, 235, 236, 237, 238, 240, 246, 262, 279, 301, 303, 350, 437, 440, 483, 506, 510

economy of the truth, 378
eczema, 192, 245, 344

educated, 1, 27, 34, 44, 85, 107, 108, 116, 119, 122, 135, 168, 186, 204, 210, 215, 318, 345, 353, 365, 377, 382, 415, 429, 442, 456, 460, 467

education, 2, 7, 21, 28, 32, 40, 41, 45, 83, 102, 120, 121, 131, 137, 143, 144, 150, 168, 183, 193, 214, 248, 298, 333, 335, 364, 371, 390, 410, 423, 429, 433, 460, 472, 480

educational attainment, 106, 130, 164, 383
educational background, 116, 423, 465
Edward Jenner, 320
Edward Lorenz, 310
EEG interpretation, 301
efficacious, 398, 485
eggs, 53, 69, 128, 208, 316, 321, 364, 407
ego, 76, 377, 382
Einstein, Albert, 46, 227, 308, 344, 369, 380, 392, 403, 408, 499
ejection fraction, 255, 393, 418
elderly, 85, 119, 164, 216, 221, 234, 235, 278, 284, 311, 330, 339, 437, 451
electrocution, 272, 436
electrolyte, 307, 400
electronics, 256, 327, 328, 329, 443, 508
electrophysiology, 401
elusive diagnoses, 46
embarrassing, 36, 207, 211, 228

emergency, 15, 17, 35, 54, 56, 58, 82, 85, 90, 94, 95, 117, 132, 212, 219, 232, 266, 267, 275, 282, 302, 339, 341, 352, 375, 398, 420, 421, 426, 432, 443, 446, 454, 473, 511

emotion, 113, 114, 248, 276, 462, 495
emotional attachment, 377
emotional connectivity, 19
emotional functioning, 125, 297
emotional intelligence, 131
emotional needs, 11, 128
emotionally reactive, 76
emotionally unstable, 76
empathy, 6, 14, 19, 20, 65, 67, 110, 135, 145, 191, 232, 329, 354, 396, 428
employees, 39, 80, 115, 148, 185, 286, 298, 314, 329, 373
endocrine function, 412
endoscopy, 276

energy, 15, 36, 54, 61, 69, 118, 125, 126, 135, 154, 162, 169, 170, 194, 215, 216, 217, 257, 318, 346, 376, 403, 404, 411, 412, 413, 426, 427, 470

enquiries, 124, 178, 181, 208, 297, 319, 415, 482
enthusiasts, 81
entrepreneur, 415, 511
entropy, 273
enzymes, 271, 303
epidemiology, 1, 291, 318
epigastrium, 238
epigenetic, 36
epilepsy, 31, 240, 367, 375
epizeuxis, 197, 234, 401
equality, 112, 192, 356, 453
Eric Berne, 128, 143, 460, 495

error, 20, 24, 94, 149, 153, 186, 201, 213, 247, 250, 254, 260, 270, 282, 293, 316, 322, 323, 324, 327, 329, 331, 332, 350, 352, 362, 363, 373, 374, 380, 393, 406, 413, 415, 448, 452, 474, 475

error trapping, 331
ESR, 271, 300, 490, 498
ethics, 329
ethnicity, 8, 180, 186, 193, 461
ethology, 196
ethos, 40, 181, 183, 185, 187, 196, 299, 361, 364, 428, 481
eureka, 34, 284
European, 190, 195, 211, 302, 416, 498
euthanasia, 174, 176
Eve, Norman, Dr., 402, 506, 509
Evelyn Waugh, 50, 86, 289, 494

evidence, 5, 6, 12, 22, 29, 39, 41, 50, 52, 53, 59, 72, 76, 90, 91, 92, 96, 115, 117, 118, 124, 128, 133, 141, 149, 176, 184, 208, 215, 216, 218, 225, 227, 234, 236, 244, 250, 252, 253, 257, 260, 262, 263, 264, 270, 275, 276, 279, 281, 282, 283, 284, 287, 288, 289, 294, 295, 300, 311, 312, 314, 315, 316, 319, 320, 323, 325, 336, 337, 338, 340, 345, 346, 351, 352, 353, 354, 355, 360, 361, 362, 363, 364, 365, 366, 367, 368, 369, 370, 371, 376, 377, 386, 390, 391, 392, 396, 398, 399, 400, 404, 408, 415, 419, 423, 427, 428, 430, 435, 439, 441, 442, 444, 445, 446, 447, 451, 475, 479, 480, 481, 486, 509, 511

evidence-based medicine, 29, 51, 149, 294, 371, 479
evolution, 112, 192, 217, 323, 381, 394, 408

examination (clinical etc.), 7, 8, 13, 70, 125, 126, 141, 151, 171, 173, 201, 229, 230, 232, 233, 234, 235, 238, 239, 240, 241, 242, 250, 252, 256, 258, 268, 272, 276, 277, 280, 281, 290, 301, 305, 311, 315, 318, 348, 349, 362, 366, 367, 381, 397, 400, 402, 403, 417, 443, 448, 465, 510

examination grades, 7
examination routine, 234

exceptions, 259, 305, 323
excuse(s), 31, 37, 57, 89, 99, 142, 166, 212, 266, 283, 359, 363, 418, 459

executive, 6, 14, 15, 19, 25, 33, 53, 71, 98, 108, 134, 165, 200, 205, 206, 242, 372, 411, 443, 445, 465

executive functioning, 53, 108, 445
exercise test, 132, 302, 303, 397, 398, 464
exercise tolerance, 113, 114, 161, 231, 343, 451, 452
exhaustion, 125, 126
exophthalmus, 277, 468
expedient, 100, 204, 222, 267, 296, 314, 317, 324, 325, 341, 344, 443, 452, 459

experience (clinical etc.), 2, 10, 12, 13, 14, 17, 19, 20, 23, 25, 27, 31, 34, 36, 37, 43, 44, 47, 48, 50, 51, 55, 57, 58, 64, 68, 72, 75, 76, 87, 89, 90, 93, 94, 101, 102, 103, 111, 112, 124, 125, 129, 138, 140, 141, 144, 146, 151, 152, 164, 168, 170, 171, 183, 185, 195, 200, 203, 212, 213, 218, 219, 220, 222, 223, 225, 226, 232, 233, 234, 241, 248, 249, 253, 255, 256, 273, 274, 276, 277, 279, 281, 283, 285, 289, 295, 298, 299, 300, 306, 307, 310, 312, 314, 315, 316, 319, 324, 326, 333, 335, 336, 338, 340, 341, 343, 344, 345, 347, 348, 350, 354, 358, 362, 365, 366, 369, 370, 371, 373, 374, 376, 381, 382, 383, 386, 387, 388, 390, 391, 392, 393, 394, 395, 396, 398, 400, 402, 404, 406, 408, 409, 410, 411, 412, 413, 414, 418, 421, 422, 423, 424, 425, 426, 428, 429, 432, 433, 436, 438, 442, 443, 444, 447, 449, 450, 453, 456, 459, 467, 469, 470, 471, 474, 484, 485, 502, 508, 509, 510

experiment, 303, 378, 402, 483
experimental, 228, 294, 391
experimenter, 316
extemporise, 305

F

facetious, 54
facility (for hands-on tasks), 12□
factor Xa inhibitors, 117, 174, 226, 227, 384
faint (vasomotor syncope), 47, 100, 240
fair play, 383
fairness, 52, 59, 183, 184, 187, 195, 357, 365, 383
fait accompli, 307
Fallot's Tetralogy, 237
false impression, 318, 332
false negative, 200, 201, 213, 269, 277, 303
false positive, 200, 201, 213, 246, 269, 301
falsified, 140, 141, 206, 325, 346
famous, 32, 35, 91, 100, 114, 117, 171, 186, 191, 195, 224, 269, 283, 351, 416, 475
fanciful concept, 411
fantasy, 1, 49, 77, 154, 162, 427, 456
farmers, 185

farming, 185

fashion, 61, 136, 163, 164, 183, 226, 246, 260, 274, 296, 316, 317, 324, 377, 378, 381, 418, 452, 478, 480

fatalistic, 144
fatigue, 37, 125 126, 216, 217
faux pas, 189, 191, 207, 211, 280
FDG PET scan, 260

fear, 12, 36, 57, 65, 82, 100, 108, 115, 118, 120, 126, 137, 141, 142, 145, 148, 150, 159, 164, 195, 223, 247, 253, 267, 269, 301, 338, 382, 384, 395, 404, 409, 411, 419, 429, 430, 431, 447, 460, 473, 474, 483, 493

fear of making wrong decisions, 474
female insight, 57
femininity, 57
Feynman, Richard, 406, 480, 502
fiction, 303, 305, 333, 388
finance, 143
fingerspritzengefüln, 13
FIRO B questionnaire, 128, 412, 415
fist aid, 388
FitBit device, 297
fitness, 94, 112, 373, 384, 394, 398, 451
First World War, 84, 282
Five Wisdoms of Buddhism, 23, 492
fixed rules, 84, 145, 205, 278, 378, 391, 456, 483
flatterers, 77
Florence Nightingale, 41, 92, 126, 376, 495, 500, 509
fluency, 230
flummoxed, 227
fly-on-the-wall, 5, 106, 182, 207
foetor, 230, 403
Fonzarelli, 71, 137

food, 78, 106, 115, 117, 124, 134, 154, 162, 164, 181, 185, 186, 194, 215, 216, 218, 298, 317, 318, 322, 325, 340, 343, 347, 370, 373, 375, 414, 440, 452, 467, 482, 487

force majeur, 269
foreigners, 112
forensic enquiry, 207, 216
forensic history taking, 8, 228
forgetfulness, 474
formality, 82, 183, 190, 203, 416

formulaic, 24, 46
Framingham Study, 485
Frank Blake, 290
Frank-Starling Law, 321
fraudsters, 140
FRCOG, 16, 20, 242, 478, 490
FRCS, 16, 34, 242, 331, 332, 392, 478, 490
free thinker, 406
French, the, 17, 71, 80, 88, 170, 172, 187, 190, 197, 325, 489, 496, 498, 501
friends, 173, 431
friendship, 49, 96, 143, 404, 476, 508
frontiers, □3
funding, 4, 30, 31, 259, 316, 399, 478, 485
fuzzy, 324, 346, 394, 396

G

Game Theory, 400
game changing, 79, 317, 320
gamesmanship, 84, 87
GANs (generative Adversarial Networking), 329
gender, 57, 58, 92, 106, 120, 130, 131, 183, 186, 190, 192, 196, 204, 233, 314, 330, 433
gene therapies, 60
general practice, 11, 16, 56, 58, 85, 122, 157, 176, 244, 328, 360, 470
generalisation, 325
generalised guidelines, 129, 289
generic, 51, 177, 245

genetic, 36, 60, 123, 154, 186, 218, 222, 244, 264, 322, 346, 381, 388, 408, 434, 438, 442, 452, 461, 482, 483

genetic engineering, 408
genius, 46, 99, 334, 395
genotype, 36, 214
gentlemen, 11, 35, 44, 59, 172, 173, 494, 502
genuine, 14, 21, 34, 35, 44, 51, 61, 67, 142, 150, 180, 186, 308, 334, 413, 431, 447, 456, 496
Geoffrey Marmot, 195, 308
Gillian, Hanson, 284
glucose tolerance test, 259, 262

GMC, 3, 32, 35, 41, 80, 97, 118, 137, 142, 160, 171, 172, 176, 177, 178, 179, 223, 267, 282, 300, 323, 345, 359, 364, 386, 391, 398, 411, 441, 443, 470, 472, 490, 495, 496, 499

God, 34, 35, 43, 49, 64, 151, 162, 170, 192, 283, 309, 310, 321, 357, 388, 429, 456
God Complex, 43, 321

God's gift, 64
golden rule, 246, 282, 401, 444
Goldilocks, 169, 385
golfer, 132, 473
golfing, 473
good behaviour, 59
Good Medical Practice (GMC), 97, 137, 176, 177, 345, 495, 499
Good Samaritan, 35, 111, 448, 495
Goodenough Committee, 21
Gordon, Welchman, 244, 332, 497
government, 30, 37, 50, 76, 131, 135, 165, 187, 311, 312, 335, 345, 360, 390, 414, 425, 485

GP (general practice), 5, 18, 20, 30, 33, 49, 58, 65, 73, 74, 111, 122, 142, 157, 160, 168, 174, 185, 209, 220, 226, 234, 235, 239, 243, 244, 258, 263, 268, 275, 285, 287, 291, 298, 319, 332, 339, 349, 359, 360, 386, 389, 393, 396, 414, 416, 418, 467, 469, 470, 480, 497, 505, 506, 507, 509

Grammar School, 79, 144
grand plan, 8
greed, 25, 43, 377, 382

Greek, 17, 59, 60, 63, 91, 134, 149, 162, 170, 174, 175, 187, 190, 196, 197, 211, 219, 295, 320, 362, 378, 395, 403, 418, 430, 461, 489

Greek Cypriots, 196, 197
grief, 12, 96, 245, 419
guarantee, 88
guardianship, 7
guesses, 12, 255, 295, 345, 389, 448
guesswork, 253, 274, 338, 340, 341, 389, 390
guideline, 114, 145, 212, 252, 307, 345, 362, 386, 437, 443
guru, 510
gut feeling, 388
gymnasium, 121, 263
gynaecologist, 20, 232, 289

H

haematoma, 117, 225
haemoglobin, 235, 259, 284
haemoptysis, 286
haemorrhage, 36, 95, 280, 285, 302, 419, 420, 440, 448
haemorrhoid, 455
haemostasis, 475
halitosis, 230
halo bias, 66, 72, 170, 173, 290, 316, 323, 350

handheld devices, 253, 297
handling efficiency, 13
Hanson, Gillian, 284
Hardman, Julian, Dr. 396, 505, 509
Harvey, William, 59, 65, 113, 321, 368, 493, 499
hate, 71, 95, 193, 317, 395
Hayek, Friedrich, 308, 342
headache, 150, 270, 277, 438
healing, 116, 138, 175, 201, 246, 450

health, 36, 37, 44, 60, 76, 98, 113, 115, 116, 122, 123, 128, 132, 133, 151, 159, 162, 163, 166, 170, 173, 190, 215, 216, 217, 218, 267, 274, 286, 311, 335, 336, 341, 357, 368, 411, 464, 470, 471, 473, 479,

health centres, 387
health check (NHS), 349
health divide, 4, 112, 116, 180, 181, 186, 215, 318, 407, 408, 452, 454, 463
health promotion (NHS clinics), 286
health risk, 117

heart attacks, 22, 72, 114, 122, 123, 133, 153, 208, 245, 261, 263, 303, 318, 323, 405, 422, 480

heart sound, 237, 279
heart-sink, 256
Heberden, William, 113, 234, 269, 276, 299, 351, 397, 495, 498, 499
Heep, Uriah, 473
helpfulness, 6, 11, 47, 80, 188, 230
hepatitis, 265
heresy, 39, 45, 84, 319, 321
Herschel, William, 404
heterosexual, 190
heuristic, 12, 16, 45, 276, 290, 329, 330, 372
hidden in plain sight, 284, 395
high-minded, 80, 98, 390
high risk, 299
hijackers, 147
hindsight, 68, 354
Hippocrates, 23, 34, 41, 176, 289, 371, 378, 393, 408, 499
Hippocratic Oath, 174, 176, 178, 179, 496
historians, 108, 152, 310

history (clinical), 5, 7, 8, 11, 13, 25, 29, 36, 50, 72, 101, 107, 109, 114, 122, 124, 127, 128, 132, 141, 171, 173, 186, 187, 188, 194, 200, 201, 202, 203, 204, 207, 209, 211, 212, 213, 214, 215, 216, 218, 219, 220, 224, 227, 228, 230, 233, 234, 240, 241, 242, 245, 246, 248, 250, 252, 256, 258, 264, 266, 268, 272, 276, 277, 278, 280, 281, 283, 284, 285, 298, 301, 302, 318, 322, 348, 349, 351, 358, 361, 366, 369, 376, 381,

397, 403, 404, 408, 414, 417, 420, 421, 422, 431, 434, 439, 442, 448, 450, 458, 461, 464, 465, 469, 470, 472, 501, 514

history taking, 5, 8, 13, 141, 171, 200, 201, 202, 203, 204, 207, 227, 228, 233, 234, 242, 246, 248, 258, 272, 318, 351, 397, 403, 472

HIV and AIDS, 265
HMG CoA reductase inhibitors , 226
HMO, 75, 259, 467, 497
HOCM, 236, 302, 349
holistic, 6, 106
Holmes, Sherlock, 284
Holy Grail, 144, 432
Homan's sign, 240, 278
homeopathy, 392
homeostasis, 170, 216, 217, 274
Homo sapiens, 312
homosexual, 190, 258
honesty, 11, 21, 34, 40, 49, 59, 182, 184, 331, 361, 374
honour, 14, 26, 40, 78, 184, 187, 409, 502, 511

hope, 6, 14, 31, 36, 60, 77, 96, 103, 131, 144, 145, 157, 162, 221, 230, 248, 257, 258, 289, 297, 317, 320, 330, 338, 339, 343, 371, 391, 395, 406, 421, 423, 427, 429, 430, 431, 439, 456, 473, 483, 486, 487, 511

hormonal imbalance, 245
hospital physician, 47, 262, 480
housekeeping, 8, 399, 443, 444
hubris, 60, 284, 317, 421, 454, 480
hugs and kisses, 476
human btain, 303, 304, 308, 310, 332, 396
human wisdom, 290, 304
humanitarian, 6, 7, 21, 33, 66

humanity, 18, 26, 60, 98, 99, 102, 119, 160, 162, 183, 184, 304, 324, 327, 338, 374, 393, 408, 429, 446, 460, 471, 494, 499, 502

humorous, 81, 108, 110
humour, 69, 81, 82, 83, 100, 101, 188, 189, 193, 194, 379, 382, 473, 510
hunter-gathering, 185
hypercholesterolaemia, 405, 480
hyperglycaemia, 259, 375
hyperkalaemia, 235, 254, 285, 436
hyperventilation, 281, 366
hypochondriac, 81, 151, 319
hypoglycaemia, 260, 375

hypokalaemia, 235, 285, 307, 419
hyponatraemia, 234
hypothesis, 334, 351, 380, 392, 479
hypothyroidism, 125, 235, 277, 375
hysterectomy, 82, 473
hysteria, 150, 433

I

I Ching, 309, 310, 498
IBS, 123, 154, 239, 268, 283, 375
idol, 82
Illitch, Ivan, 50, 60, 111, 115, 493, 495
illiterate, 110
illusion, 62, 295, 414
immigrant, 28, 196, 304
immune, 98, 113, 123, 126, 166, 220, 300, 318, 377, 511
immunotherapy, 485
Imperial College Response Team, 309
inaccurate, 254, 325, 361
inattentiveness, 474
inborn error of metabolism, 186
incompatibility, 112, 218, 396
incontrovertible conclusion, 325
independent-minded, 20, 33
in-depth consideration. 376
index of suspicion, 263, 275, 277, 302
individual characters, 112
inductive reasoning, 162, 295, 322, 344, 345, 352, 392
indulgence, 467
industrial revolution, 329
ineptitude, 53, 266, 445, 446
inequality, 112, 180, 192
inexperience, 55, 271
infectious mononucleosis, 268
inflammation, 59, 120, 121, 216, 238, 255, 260, 261, 271, 300, 396, 490
inflammatory markers, 209, 261, 264
information addiction, 297
infusion, 94, 126, 171, 232, 275, 282, 339, 418, 424
inhumane, 304
innate ability, 8
innovation, 34, 72, 260, 355, 374
inscape, 206
insecurity, 52, 66, 101, 194, 208
insensitive test, 301

insight, 38, 45, 54, 105, 106, 107, 124, 152, 154, 159, 162, 193, 221, 243, 245, 248, 259, 267, 289, 308, 327, 331, 334, 335, 342, 352, 356, 364, 377, 395, 396, 401, 408, 412, 413, 433, 460, 484, 509, 511

insolence, 66, 86, 87, 137, 494
insomnia, 12, 171, 216, 245, 256, 441, 442, 443, 484
instinct, 56, 95, 203, 329, 452
integrative work, 46
integrity, 11, 56, 59, 177, 319, 335, 358, 374, 405
intellectual, 22, 27, 51, 52, 61, 69, 76, 98, 156, 226, 234, 341, 357, 382, 429, 470
intellectual challenge, 470
intellectual rigour, 27

intelligence, 11, 14, 18, 19, 44, 53, 77, 86, 105, 131, 150, 192, 210, 242, 296, 305, 328, 331, 332, 334, 335, 350, 366, 371, 382, 401, 404, 423, 426, 429, 442, 445, 446, 461, 464

inter-arial septum, 240
inter-connectedness, 297
interdisciplinary, 114
international, 114, 128, 186, 231, 309, 425
interpersonal skill, 142
interpretation, 250, 255, 256
inter-racial, 192
interrogation, 146, 207, 208
intransigence, 282
intrinsic rate, 481
intuition, 7, 203, 288, 367, 370, 374, 396, 405
intuitiveness, 53
inventiveness, 29, 391, 401, 452

investigation(s), 5, 8, 12, 13, 17, 30, 97, 107, 109, 132, 168, 171, 212, 227, 233, 235, 243, 246, 247, 250, 251, 252, 253, 255, 256, 257, 258, 259, 260, 261, 262, 263, 264, 265, 267, 268, 269, 270, 274, 275, 277, 283, 286, 299, 300, 310, 312, 315, 340, 343, 348, 349, 359, 370, 373, 381, 384, 387, 395, 396, 397, 400, 417, 425, 447, 451, 454, 455, 464, 485, 496

ion channel, 301, 302, 349
iron cage, 488
IT, 39, 137, 212, 249, 296, 298, 500
ITU, 15, 25, 284, 285, 362, 373, 473
Izaak Walton, 100, 119, 494

J

Jackson, Michael, 171, 424, 502
jaundice, 233

jazz, 305, 308, 332
jealousy, 61, 74, 83, 350, 377, 382
Jeff Bezos, 347, 389
Jenner, Edward, 320
John Wright, Dr., 402, 505, 506, 508
jokes, 82
Jonathan, Sacks, 79, 460, 494, 502
Jowell, Tessa, 485, 502
Judge Judy, 34, 206, 309

judgement, 7, 11, 12, 14, 16, 19, 29, 34, 45, 55, 83, 107, 109, 119, 134, 146, 161, 170, 174, 175, 182, 183, 212, 241, 253, 259, 267, 274, 276, 278, 283, 296, 298, 300, 303, 304, 306, 314, 324, 328, 332, 333, 335, 336, 337, 338, 339, 340, 341, 342, 343, 344, 345, 346, 347, 348, 350, 351, 355, 358, 361, 362, 364, 365, 366, 367, 368, 374, 376, 381, 382, 383, 387, 389, 391, 393, 394, 395, 396, 399, 408, 411, 415, 426, 427, 442, 443, 446, 452, 453, 472, 487, 507

jugular venous presssure (JVP), 234, 397, 435
Julian Hardman, Dr. 396, 505, 509
jungle medicine, 370, 471
junior doctor, 2, 48, 76, 85, 103, 228, 265, 266, 274, 282, 287, 361, 365, 403, 409, 411, 431, 463, 501
justification, 72, 90, 157, 178, 251, 257, 267, 270, 293, 347, 401
JVP, 234, 235, 236, 278, 281, 307, 344, 397, 418, 419, 435, 436, 447, 459, 471

K

Kafkaesque, 311
Kahneman, Daniel, 280, 352, 382
Kennedy, Bobby, 335
ketosis, 234, 403
key element, 108
key factors, 385
kidnapper, 147
kidney, 86, 151, 255, 349
Kim Peek, 297
kindness, 6, 26, 43, 45, 136, 175, 180, 423, 431
KIng's Fund, 469, 502
Koch, Heinrich, H.R. Dr., 308, 320, 479, 503
Koch's Postulates, 479
kudos, 1, 27, 389

L

labile hypertension, 236
laboratory, 25, 28, 47, 48, 67, 68, 153, 209, 254, 255, 257, 259, 270, 271, 276, 277, 284, 307, 398, 419, 421, 428, 448, 462, 507

Lang-Stevenson, Mr., 22, 392
language, 39, 51, 106, 118, 178, 180, 182, 189, 191, 193, 197, 200, 210, 211, 230, 313, 320, 396, 424, 428, 433, 454
laryngoscopy, 12
late diastolic murmur, 232, 238
lateral thinking, 19, 101, 293, 305, 307, 374
Latin, 59, 92, 175, 320, 368, 378, 403
lawyers, 2, 6, 136, 171, 177, 206, 212, 267, 280, 299, 306, 341, 345, 360, 361, 367, 419, 421, 483
lay people, 5, 345, 482
Left bundle branch block (LBBB), 237, 255, 279, 302, 393, 394
leap of faith, 319, 323
learning curve, 12
left ventricular hypertrophy (LVH), 236, 279, 439, 483
legal proceedings, 479
legitimacy, 392
Leonard Bernstein, 394
Leonardo da Vinci, 10, 103, 308, 316, 333
leukaemia, 113, 289
leukotriene, 113, 458
liars, 140, 208, 496
lipid accumulation, 480
liquid biopsy, 291
literature, 28, 50, 181, 187, 190, 317, 343, 376, 389
litigation, 111, 152, 161, 386, 448, 495
location, 54, 95, 106, 114, 120, 197, 471
logic, 57, 181, 296, 321, 322, 331, 341, 342, 344, 345, 346, 378, 442, 507
logical, 67, 108, 135, 245, 263, 274, 321, 322, 324, 330, 331, 344, 352, 377, 380, 390, 431
London Hospital Medical College, 21, 168, 461, 505
lone wolf, 85
loneliness, 113, 393
longevity, 106, 116, 133, 162, 164, 214, 379
Lorenz, Edward, 310

loss, 43, 61, 76, 101, 113, 162, 181, 183, 184, 196, 204, 208, 234, 240, 241, 245, 266, 268, 275, 282, 297, 311, 341, 343, 349, 373, 434, 446, 450, 481, 483, 490

lottery, 389
Loughton Clinic, 5, 506, 511
Louis de Jode, Mr., 307, 506, 509

love, 28, 56, 57, 61, 86, 87, 95, 113, 143, 170, 173, 180, 184, 189, 191, 196, 217, 249, 258, 277, 309, 312, 314, 326, 340, 357, 393, 395, 404, 430, 432, 468, 498, 499

low risk, 299
loyalty, 17, 18, 34, 124, 141, 153, 171, 173, 404, 411, 412, 413

luck, 23, 27, 55, 57, 68, 85, 86, 91, 131, 140, 169, 225, 266, 273, 304, 321, 337, 338, 348, 373, 388, 389, 407, 441, 444, 452, 467

lung cancer, 145

lung function, 281, 397

LVH, 213, 236, 237, 239, 262, 269, 279, 301, 422, 439, 440, 468, 483, 490

Lynch, Mike, 333

Lyme Disease, 146, 257

lysis, 254

M

Ma (Japanese), 257

Machiavelli, NIccolò, 18, 86, 308, 467, 492, 502

machine output, 256

macrophage, 59, 322

Mae West, 93, 309, 432

Magdi Yacoub, Mr., 389

magical qualities, 288

Magritte, René, 376

Mahummad Ibn Musa Al-Khwarizmi, 330, 498

making diagnoses, 8, 51, 293, 345

management efficiency, 33, 112, 286

manipulator, 149, 357

manslaughter, 171, 282, 340, 424

Marie and Pierre Curie, 334

Marmot, Geoffrey, 195, 308

Mars, 327

martyr, 149

masculinity, 57

master, 7, 10, 15, 16, 34, 48, 64, 115, 120, 175, 186, 197, 207, 231, 234, 274, 310, 326, 425, 443, 475, 479, 487, 508, 509

mastery, 20, 21, 38, 46, 53, 64, 68, 212, 213, 276, 373

mathematical, 293, 295, 299, 309, 332, 389, 400

mathematical model, 309

mathematicians, 273

matriarch, 7, 284

Matthew Effect, 85

Maurice Dixon, 286, 512

maverick, 129, 307, 444

Max Weber, 488

mean, 26, 36, 44, 45, 46, 64, 68, 86, 113, 120, 140, 152, 159, 188, 189, 191, 193, 210, 228, 265, 294, 295, 315, 321, 325, 326, 339, 357, 360, 384, 392, 395, 414, 418, 428, 456, 467, 510

meaningful relationship, 243

measurement, 6, 16, 117, 254, 255, 319, 341, 351, 384, 387, 393, 394, 418, 440, 478
medical bureaucrat, 7, 360
medical education, 21
medical elite, 29
medical ethos, 299
medical event horizon, 399
medical franchise, 415
medical insurance, 32, 120, 231, 265
medical knowledge, 11
Medical Practitioners Tribunal Service (MPTS), 148, 364
medical profession, 2, 7, 17, 19, 30, 43, 51, 57, 69, 75, 76, 86, 98, 99, 138, 178, 185, 226, 243, 264, 278, 345, 360, 382, 411, 432, 452, 488
medical professionals, 11, 30, 34, 42, 44, 50, 64, 65, 69, 106, 111, 143, 180, 194, 219, 224, 245, 318, 325, 345, 360, 371, 462, 487
medical regulators, 111
medical school, 1, 11, 12, 14, 22, 40, 45, 58, 59, 72, 105, 120, 121, 142, 143, 152, 172, 179, 185, 284, 311, 393, 403
medical screening, 13, 36, 106, 115, 271, 463, 481
medical specialities, 8, 11
medical students, 1, 4, 6, 11, 21, 40, 50, 87, 102, 106, 168, 324, 423, 471
medical trials, 295
melaena, 275, 276
men, 30, 32, 58, 66, 78, 107, 123, 140, 159, 163, 175, 186, 191, 192, 193, 232, 266, 267, 279, 283, 301, 344, 384, 430, 438, 473, 475, 493, 508
Mendel, Gregor, 188
menopause, 57, 179
mentor, 70, 184, 191, 210, 508, 510
mesa-information, 362, 384, 395

meta, 8, 11, 207, 231, 232, 245, 293, 295, 298, 304, 305, 306, 310, 312, 313, 316, 319, 332, 336, 341, 342, 343, 346, 347, 366, 368, 371, 379, 380, 383, 385, 387, 391, 395, 396, 397, 405, 416, 437, 438, 465, 471, 479

meta-analyses, 8, 293, 298, 316, 368, 391, 437, 438, 479
metabolism, 114, 186, 260, 434, 480
meta-data, 11, 207, 245, 295, 304, 305, 306, 313, 319, 336, 341, 342, 343, 371, 379, 383, 387, 396, 405, 465
meta-perception, 395
metaphysical, 22, 111, 177, 178, 345, 410
meticulousness, 150
MHQ (Middlesex Hospital Questionnaire), 141, 150, 151, 495
Michael Balint, Dr., 168, 171, 428
Michael Jackson, 171, 424, 502
middle classes, 195
middle-aged deaths, 114, 264

Middlesex Hospital Questionnaire, 141, 150, 433, 495
migraine, 123, 154, 192, 213, 221, 245, 264, 270, 277, 280
Mike, Lynch, 333
military studies, 400
mineral nutrients, 485
minor surgery, 12
mirth, 379, 382
misinterpreted, 197, 229, 271, 365
misleading, 169, 227, 263, 274, 302, 323, 324, 325, 453
misogyny, 45
misuse of statistics, 311
mitral incompetence, 237, 241, 279
mitral stenosis, 205, 210, 236, 237, 238, 279, 281, 290, 302, 395, 451, 481
mobile telephone, 11, 90, 232, 381
modern technology, 298
modus operandi, 18, 136, 148, 428, 445, 469, 475
molecular biological view, 318
molecular guided therapy, 290

money, 14, 26, 27, 30, 37, 38, 39, 43, 49, 56, 57, 61, 75, 80, 85, 86, 97, 106, 115, 116, 128, 131, 134, 140, 162, 164, 165, 171, 174, 193, 194, 196, 201, 215, 228, 249, 259, 261, 268, 286, 297, 309, 312, 314, 326, 334, 339, 349, 381, 389, 392, 403, 404, 415, 421, 429, 448, 449, 452, 460, 463, 467, 473

monitoring, 307
Montgomery, Bernard Law, Field Marshall, 164, 282, 361
moral equality, 192
moral fibre, 15
morals, 88, 329

morbidity, 4, 8, 106, 112, 114, 115, 126, 139, 165, 167, 181, 194, 195, 214, 215, 216, 260, 262, 263, 264, 265, 360, 390, 399, 437, 441, 452, 455, 465, 483, 485

Morse code, 327

mortality, 4, 8, 106, 112, 114, 115, 126, 139, 162, 164, 165, 167, 181, 186, 194, 214, 215, 216, 220, 260, 263, 265, 294, 311, 360, 368, 390, 399, 438, 441, 452, 455, 465, 483, 485

MRCP., 16, 233, 242, 331, 332, 443, 478, 490, 510
MRI., 40, 94, 239, 252, 255, 260, 270, 271, 289, 291, 436, 439
multiculture, 25, 28, 184, 191, 193
multi-ethnic, 188, 193
multifactorial, 14, 46
multilingual, 25, 36, 211
multiple pathologies, 37, 385
multiple sclerosis (MS), 289

multivariate analysis, 404
murmurs (heart), 229, 232, 237, 238, 465, 481
music, 6, 58, 143, 183, 196, 219, 225, 289, 305, 308, 332, 386, 387, 394, 461
musicians, 35, 305, 387, 418, 461
mutually rewarding, 108, 447
myocardial, 113, 123, 235, 255, 285, 301, 362, 438, 510, 511
myocardial infarction, 123, 285, 301, 362, 510, 511
myocarditis, 241
mystique, 409
myth, 116, 192, 219, 318, 357, 381, 456

N

N_2O, 402
Narcissus, 298
national character, 186
national culture, 183, 186
nationalisation, 33
Native Brit., 188
natriuretic peptide, 235, 435
nebula processing, 310
nebula thinking, 7, 14, 46, 52, 57, 272, 283, 408
nebula, 46, 274, 293
negligence, 111, 282, 340
neogenesis, 380, 499
neoplasia, 238
nerds, 64, 99, 315
nerve root compression, 240
neural network, 14, 249, 256, 290, 296, 327
neuropathy, 214, 240, 247, 260, 375
new drug, 221, 224, 226, 227, 398
newly qualified doctor, 12
Newton, Isaac, 46, 227, 403, 500, 512

NHS., 3, 4, 5, 6, 7, 17, 18, 21, 25, 30, 31, 32, 33, 38, 39, 41, 48, 55, 64, 73, 75, 76, 77, 83, 97, 102, 106, 109, 111, 115, 118, 120, 133, 136, 142, 149, 165, 168, 170, 173, 174, 176, 185, 198, 201, 202, 204, 212, 222, 224, 226, 231, 243, 244, 247, 252, 253, 259, 262, 263, 266, 268, 275, 280, 286, 287, 290, 298, 304, 318, 335, 339, 343, 349, 360, 361, 370, 371, 413, 414, 415, 416, 417, 421, 422, 425, 432, 449, 453, 455, 466, 467, 469, 470, 485, 511

NHS attitude, 202, 287
NHS consultants, 30, 38, 287
Nietzsche, 83, 187, 429, 454, 456
Nightingale, Florence, 41, 92, 126, 376, 495, 500, 509
Nightingale ward, 146

nit-pickers, 68, 89
nitric oxide (NO), 169, 322, 364
Nobel, 72, 220, 300, 308, 350, 351, 377, 479, 480, 499, 502
Nobel Prize, 220, 300, 350, 377, 479, 480, 499, 502
non-discriminatory, 302
non-probabilistic, 312
nonsense, 91, 119, 192, 219, 296, 300, 332
non-verbal cues, 172
Norman Eve, Dr., 402, 506, 509
nostalgia, 396
Nostradamus, 389
NP (Non-Practical, exponential) algorithms, 332
NSAIDS., 275
null hypothesis, 294
numerate, 110
numeric analysis, 293

nurse, 5, 7, 8, 12, 13, 18, 30, 35, 41, 49, 52, 53, 86, 102, 107, 108, 118, 123, 142, 160, 189, 258, 259, 282, 283, 345, 360, 371, 376, 388, 406, 423, 428, 430, 434, 437, 443, 469, 475, 509

nursing, 26, 41, 82, 173, 388, 418, 419, 443, 453, 487
nutrients, 164, 373, 481, 482

O

oath, 174
obese, 116, 236, 323, 375, 452
objective, 11, 23, 45, 67, 108, 110, 152, 181, 184, 216, 253, 300, 324, 388, 425, 427, 445, 467
objectivity, 18, 38, 52, 96, 116, 219, 228, 429, 430, 513
obligatory, 8, 17, 271, 369, 371, 512
obsequious, 18, 33, 93, 154, 210
observation, 34

obsessive, 24, 44, 71, 84, 90, 108, 185, 187, 280, 291, 315, 325, 365, 374, 381, 385, 387, 395, 412, 459, 474, 479

oedema, 234, 238, 271, 281, 290, 307, 419, 435, 437, 440, 459
off-piste (medical activity), 483
older doctors, 2, 54, 58, 287
omission, 207, 213, 226, 254, 362, 363, 460
onus of proof, 22
opening snap, 238, 481
open-minded, 314, 350, 483
opportunity, 26, 32, 96, 102, 147, 192, 220, 356, 365, 400, 468, 469, 510, 511
optimists, 96

Oracle at Delphi, 389
organ failure, 311
osteopath, 428
osteopontin, 380
out-patient, 399
outside of the box thinking, 483
overload, 33, 136, 318, 397, 420
oversight, 474
oxygen, 113, 221, 322, 371, 418, 458

P

Pablo Picasso, 328, 498
pacemaker implantation, 240, 282, 458, 512
pacemaker function, 321

pain, 11, 36, 44, 93, 94, 108, 117, 121, 126, 127, 139, 152, 157, 204, 232, 234, 238, 240, 247, 252, 261, 263, 267, 268, 270, 278, 280, 282, 283, 291, 297, 301, 303, 354, 358, 372, 375, 397, 398, 399, 420, 421, 448, 451, 452, 475, 481, 484, 500, 511

paithéa (Grk), 187
pallor, 59, 240
palpitation, 219, 247, 277, 290
Panacea, 174, 175
pancreatitis, 234
Pandora, 145
PAPA INDIA, 302
paradox, 187, 197, 319, 398, 467
parallels, 14, 29, 84, 112, 305, 417, 444
paramedic, 10, 12, 13, 35, 41, 160
parasthesiae, 247, 262
parasympathetic, 209, 334, 481
parent, 132, 213, 221, 232, 439, 447
Parkinson's Law, 40, 356, 492
Parkinsonism, 209, 233
Pasteur, Louis, 233, 320
pastoral, 51, 113, 128, 143, 171, 178, 288, 387, 469
pastoral care, 143, 288, 387, 469
'Patch' Adams, 172, 289, 473
Path of the Masters, 139, 288
pathogenic, 255
pathognomonic, 159, 211, 277, 310, 320
pathological process, 374, 384
pathology, 12, 25, 36, 205, 246, 253, 271, 318, 322, 397, 400, 428, 436, 510
pathophysiology, 205, 208, 221, 400, 459

patience, 6, 131, 146, 187, 280, 365, 400, 412, 418, 459
patient attachment, 11
patient demography, 116
patient-doctor relationship, 18, 33, 107, 108, 121, 124, 230, 379, 390, 431
patient games, 206
patient management, 8, 20, 106, 204, 223, 251, 254, 293, 369, 408, 445
patient preference, 58

patient relationship(s), 8, 13, 50, 93, 106, 107, 108, 112, 113, 131, 136, 144, 159, 160, 167, 168, 170, 174, 176, 177, 178, 197, 224, 230, 289, 391, 398, 410, 425, 432, 447, 468, 469, 476

patient through-put, 13, 388

patients, 3, 4, 5, 6, 7, 8, 10, 11, 12, 13, 14, 15, 17, 18, 19, 20, 23, 24, 25, 26, 27, 29, 30, 31, 32, 33, 34, 35, 36, 37, 38, 39, 41, 43, 44, 45, 47, 48, 49, 50, 51, 52, 53, 54, 55, 56, 57, 58, 59, 60, 64, 65, 66, 68, 71, 73, 74, 75, 78, 80, 81, 82, 84, 87, 88, 90, 93, 94, 95, 96, 97, 98, 100, 101, 102, 103, 105, 106, 107, 108, 109, 111, 112, 113, 114, 115, 116, 117, 118, 119, 120, 121, 122, 123, 124, 125, 126, 127, 129, 131, 132, 133, 135, 136, 137, 139, 141, 142, 143, 144, 145, 146, 147, 148, 149, 150, 151, 152, 153, 154, 157, 159, 160, 161, 162, 164, 165, 166, 167, 168, 169, 171, 172, 173, 174, 175, 176, 177, 178, 180, 184, 185, 189, 192, 195, 196, 197, 198, 200, 201, 202, 203, 204, 205, 206, 207, 208, 209, 211, 212, 213, 214, 216, 217, 218, 219, 220, 221, 222, 223, 224, 225, 227, 228, 230, 231, 232, 233, 238, 239, 240, 241, 242, 243, 244, 245, 246, 247, 248, 249, 250, 251, 252, 253, 255, 256, 257, 258, 259, 260, 261, 262, 263, 264, 265, 266, 267, 268, 269, 271, 274, 275, 276, 277, 279, 280, 282, 285, 286, 287, 288, 289, 291, 294, 295, 297, 298, 300, 302, 305, 306, 307, 309, 312, 313, 314, 315, 317, 318, 319, 320, 322, 323, 325, 326, 336, 338, 339, 341, 342, 345, 347, 349, 351, 352, 358, 359, 360, 363, 364, 368, 369, 370, 372, 373, 374, 375, 381, 382, 384, 385, 386, 387, 388, 389, 390, 391, 392, 393, 395, 396, 397, 398, 399, 400, 401, 402, 405, 406, 407, 408, 410, 411, 412, 414, 415, 416, 417, 418, 419, 420, 421, 422, 423, 424, 425, 426, 427, 428, 429, 430, 431, 432, 433, 434, 435, 436, 437, 438, 439, 440, 441, 442, 443, 444, 445, 447, 448, 449, 450, 451, 453, 454, 455, 456, 457, 458, 459, 460, 461, 462, 463, 464, 465, 466, 467, 468, 469, 470, 471, 472, 473, 474, 475, 476, 479, 480, 481, 482, 483, 484, 485, 486, 487, 488, 492, 504, 506, 508, 509, 510, 511, 512

patients first, 25, 78, 399, 411, 419, 443, 509
patriarchy, 7
patronising, 30, 33, 196, 210, 269, 382
patsy, 144
Paul Tortelier, 394
pawns, 316
PCa3, 239, 497
PCR (polymerase chain reaction, 255
PCSK9 inhibitors, 261, 322
pedagogues, 91
pedantic, 91
Peek, Kim, 297
percussion, 239, 281
percutaneous, 95, 266

pericarditis, 281
periodicity, 223, 400
peripheral vascular, 37, 214, 262, 399, 435
perks, 225
persecuted, 60
Persona Non Grata, 92
personal attachment, 159, 231
personal dada, 291, 329, 404
personalised medicine, 23, 203, 290, 305
personality, 8, 14, 23, 107, 109, 131, 136, 153, 156, 180, 181, 185, 372, 377, 432, 433, 453, 464, 472

perspective, 12, 55, 94, 106, 112, 136, 149, 176, 177, 182, 193, 196, 201, 207, 218, 228, 241, 245, 254, 273, 295, 299, 310, 312, 318, 319, 367, 369, 380, 393, 405, 409, 410, 418, 425, 426, 444, 450, 462, 470

perspicacity, □124, 145, 296, 450
pervasive computing, 297, 498
PET (positron emission tomography), 40, 255, 260, 261, 291, 401
Pharma, 117, 398
pharmaceutical drugs, 116
pharmacist(s), 108, 140, 142, 160, 180, 184, 437, 466, 469, 483
pharmacology, 47, 219, 221
PhD, 27, 35, 50, 59, 67, 478, 480
phenomena, 52, 273, 392, 420
phenotype, 36, 214, 290
philanthropist, 452, 465
philosophy, 50, 172, 296, 297, 302, 371, 383
phobia, 141, 150, 433
phonocardiography, 279
physical condition, 373
physical signs, 229, 230, 233, 234, 237, 257, 277, 281, 284
physicists, 46, 273
physiological data, 404
physiology, 11, 12, 125, 168, 219, 300, 303, 314, 403
Picasso, Pablo, 328, 498
pianola, 6
pilot(s), 14, 15, 17, 89, 121, 151, 302, 303, 330, 384, 401, 458, 474
pioneer, 208, 350
placebo, 218, 222, 257, 398, 484
plague, 96
plaintiff, 383
planning, 35, 53, 86, 145, 231, 327, 348, 407, 413, 416, 423, 428, 446, 501, 508
plaque (atheroma), 216, 217, 251, 260, 261, 321, 322, 401, 438
play it safe, 261, 312
playing the odds, 374
PND (Paroxysmal Nocturnal Dyspnoea), 375, 376

pneumothorax, 238, 281, 458
pO$_2$, 418
POCUS, 253, 497
poetry, 34, 197, 335, 388
points of evidence, 252, 276, 337
politeness, 59, 193
political climate, 180, 182
political correctness, 28, 182, 188, 192, 214, 383, 461
political credence, 485
political equality, 192
political pawn, 414
political selectivity, 480
politicians, 2, 4, 26, 35, 44, 57, 74, 195, 280, 304, 309, 311, 329, 382, 383, 389, 432, 452, 479, 487
Polycythaemia Rubra Vera (PRV), 284
polymerase chain reaction, 255
Polypill, 263
population growth, 414
Porton Down Laboratory, 257
post graduate, 6, 27
posthumous, 22
potassium, 170, 235, 254, 271, 307, 340, 418, 419, 436
poverty, 28, 116, 164, 194, 195, 439, 452

power, 23, 26, 35, 43, 44, 66, 67, 71, 75, 92, 93, 116, 121, 137, 138, 187, 191, 194, 201, 210, 217, 218, 306, 329, 331, 334, 351, 352, 360, 365, 370, 385, 397, 411, 420, 428, 429, 432, 446, 470, 487, 489

powerlessness, 53
PPPP: Preparation Prevents Poor Performance, 86
PR (rectal) examination, 239
practising medicine , 2, 8, 24, 26, 27, 44, 61, 87, 103, 118, 129, 299, 360, 409, 431, 455, 471
predators, 86, 87
pre-diabetic, 262
predictive, 45, 46, 67, 114, 131, 247, 248, 250, 264, 269, 309, 397
predisposed, 123, 126, 192, 208
pre-excitation, 301
preferences, 18, 107, 108, 177, 220, 306, 346, 358
pregnancy, 253, 300, 314, 419
pre-infarction syndrome, 125
pre-medical students, 6
pre-medication, 196
prerequisite, 14, 33, 61, 88, 285, 379, 380, 431
prescience, 46, 389
pre-selection, 79, 120, 185
pressure gradients, 237
prestigious, 57, 75, 92, 94, 201, 317, 320, 351, 402, 480

presumptions, 193, 214, 234, 255, 362
pride, 66, 156, 227, 338, 361, 377, 405
primary aim, 19, 26, 55, 106, 184, 236, 399
primary responsibility, 8
prioritisation, 14, 19, 361, 385, 459
private doctors, 17
private GPs, 18, 33

private practice, 3, 5, 7, 17, 31, 33, 38, 55, 75, 111, 118, 120, 136, 149, 165, 170, 171, 185, 258, 269, 414, 415, 416, 421, 424, 466, 469, 470, 481, 511, 512

private sector, 31, 33, 75, 97, 115, 287, 449

probability, 90, 146, 213, 248, 249, 263, 294, 303, 306, 312, 314, 316, 319, 321, 324, 332, 352, 355, 449, 463, 472, 479, 483

problem solving, 21

processing, 18, 19, 39, 46, 118, 130, 136, 230, 280, 296, 298, 304, 310, 327, 330, 332, 334, 344, 345, 352, 358, 396, 412

Professional Standards Authority (PSA), 3, 491
professionalism, 32, 203, 230, 462
professors, 27, 452, 470, 505
prognostic odds, 96
prognostic value, 115, 118, 450
prognostication, 373, 421, 423
program (computer), 17, 39, 47, 86, 151, 249, 252, 286, 304, 327, 328, 331, 333, 396, 466, 484
programming (computer), 39, 290, 293, 296, 327, 328, 329, 330, 332, 372
pro-inflammatory, 123, 169, 322
Promaya, 117
prophylactic agent, 226
prostatism, 239
prothrombin time, 117
prototype, 390, 501
provocative factor, 113
pseudoscientific, 392
psychiatrist, 1, 86, 103, 128, 155, 223, 461
psychiatry, 11, 47, 56, 58, 85, 272
psycho-patho-physiology, 123
psychosis, 98, 145, 224, 284
psycho-social enquiries, 124
psychotherapy, 144
public awareness, 415
public interest, 1, 111, 115, 157, 165, 177, 178, 212, 267, 289, 391, 393, 429, 432

public school (as in UK), 59, 79, 87, 144
publishing, 52, 317, 480
pulmonary emboli, 236, 278, 281, 290
pulmonary hypertension, 236, 237, 278, 299, 349, 436
pulmonary valve closure (P$_2$), 237
puzzling (features), 143, 348

Q

QALY, (Quality-Adjusted Life Year), 30
qRisk3 (score), 114
QRS (ECG), 236, 237, 437, 490
QT interval (ECG), 302
quantified self, 297
quants, 94
quantum physics, 46, 391
questioning, 7, 39, 207, 227, 228, 247, 275, 276, 279, 320, 329, 342, 392, 402
queue, 93, 154, 193, 467
queue jumping, 193

R

race(s), 25, 28, 85, 120, 181, 183, 184, 190, 194, 204, 214, 312, 383, 433, 449, 467, 470
racism, 45, 180, 181
radioactive, 284
Rain Man (film), 297
raison d'être, 24
random (selection), 274, 284, 291, 294, 306, 312, 313, 323, 332, 394
randomness, 295
rapport, 14, 29, 38, 111
rational, 341, 351, 352, 392, 405, 446, 483
rationale, 221, 414, 448
right bundle branch block (RBBB. ECG), 237, 279
re-assurance, 247, 286, 476
recalcitrant recidivists,□153
recall systems, 466
reception area, 161, 201, 203, 415
receptionist(s), 82, 121, 134, 136, 167, 202, 203, 231, 415
recidivism, 282
reckless, 94, 95, 267

recognition, 22, 32, 40, 47, 71, 78, 93, 98, 99, 100, 149, 190, 191, 194, 195, 204, 214, 256, 327, 330, 358, 390, 392, 429, 463, 475, 480, 508

recurrent Epstein-Barr infection, 256

red herrings, 234, 271
Red Cross, 33,
reference, 67, 205, 225, 305, 306, 310, 317, 319, 377, 394, 405, 406, 411, 416
referendum, 112, 388, 499
referral, 12, 18, 33, 97, 114, 171, 173, 263, 370, 417, 453
refreshing nap, 411
refugees, 173
refusing investigation, 269
registrar, 14, 34, 35, 65, 102, 307, 361, 402
regulation, 24, 60, 101, 145, 410

regulator(s), 1, 3, 7, 11, 12, 17, 27, 45, 78, 79, 83, 84, 111, 119, 121, 129, 136, 137, 148, 157, 165, 212, 221, 223, 225, 259, 267, 278, 304, 307, 330, 341, 359, 371, 383, 386, 408, 410, 412, 416, 419, 425, 426, 427, 431, 433, 436, 445, 456, 473, 483, 487, 515,

relationships, 2, 27, 58, 63, 67, 84, 106, 107, 121, 122, 123, 124, 128, 129, 143, 162, 174, 181, 182, 187, 198, 205, 208, 214, 218, 228, 242, 243, 277, 318, 319, 344, 346, 359, 387, 412, 432, 461, 469

religion, 145, 175, 182, 183, 184, 186, 191, 296, 297, 429
religions, 25, 101, 191
religious belief, 321, 447
religious fervour, 314
reports, 267, 305, 319, 320, 340, 498
reproducible, 45, 305, 314, 394
reputation, 83, 93, 94, 107, 111, 178, 287, 309, 316, 345, 350, 415, 417, 453

research, 5, 8, 11, 12, 13, 18, 22, 26, 27, 35, 37, 38, 43, 44, 50, 52, 55, 56, 60, 65, 66, 67, 68, 69, 72, 79, 80, 81, 86, 92, 100, 113, 114, 128, 133, 134, 143, 146, 149, 154, 162, 184, 197, 221, 222, 227, 249, 251, 253, 257, 260, 261, 262, 265, 268, 291, 293, 294, 295, 298, 314, 315, 316, 317, 319, 320, 323, 324, 325, 327, 334, 336, 342, 347, 352, 366, 368, 369, 373, 385, 388, 391, 401, 402, 403, 404, 413, 426, 433, 437, 438, 440, 443, 463, 471, 473, 478, 479, 480, 481, 482, 483, 485, 508, 509, 510

research interest, 100
research method, 27, 404
research papers (publications), 8, 52, 55, 65, 69, 227, 316
resentment, 78, 122, 124, 136, 154, 208, 338, 402, 409
re-skilling, 329
resource, 440
respect, 61, 177, 193, 195, 410
restaurant, 124, 154, 226, 340, 471, 474, 475
resuscitation, 12, 54, 95, 140, 243, 400, 418, 454
retinal image, 327
retirement, 22
reverence, 80, 100
Reye's syndrome, 386

rhetoric, 100, 135, 181, 187, 378, 382
rhetorical, 87, 107, 113, 227, 390, 395, 451
rhonchi, 238
Richard Dawkins, 74, 429
Richard Feynman, 406, 480, 502
right ventricular hypertrophy (RVH), 237, 279, 281, 491
rigidity, 51, 177

risk, 7, 11, 17, 19, 24, 30, 34, 35, 53, 82, 87, 90, 94, 99, 111, 114, 115, 117, 131, 147, 149, 152, 153, 161, 164, 189, 197, 204, 212, 213, 219, 220, 223, 225, 228, 230, 251, 260, 261, 262, 263, 265, 266, 267, 279, 282, 291, 298, 299, 301, 302, 306, 307, 311, 312, 313, 314, 322, 323, 349, 351, 359, 364, 373, 376, 381, 382, 383, 384, 386, 388, 389, 399, 403, 405, 417, 419, 420, 425, 426, 431, 437, 439, 441, 442, 443, 445, 448, 449, 450, 451, 453, 454, 457, 464, 468, 472, 479, 480, 482, 483, 485, 508, 511

risk averse, 94
risk evaluation, 53, 94, 298, 426, 472
RNA, 186, 219, 221, 256, 484
RNI (Recommended Nutritional Intake), 482
road testing, 397
road traffic accident(s), 94, 278
robots, 305, 329, 330
role model, 80
romantic, 95
Ross, Donald. Mr., 416
roughage, 239
roulette, 85, 94, 274, 299, 353, 373, 448, 449
Royal College of General Practice (RCGP), 16
Royal Society, The, 57, 350, 479, 497, 498
rubidium PET scan, 255
rude, 89, 268, 453

rules, 1, 5, 7, 12, 17, 19, 23, 24, 38, 43, 46, 51, 52, 72, 84, 88, 91, 101, 114, 119, 129, 136, 137, 145, 171, 174, 176, 183, 203, 211, 212, 278, 289, 290, 296, 304, 308, 310, 314, 318, 328, 333, 338, 341, 345, 365, 372, 374, 377, 378, 380, 386, 387, 391, 395, 399, 405, 406, 411, 415, 420, 432, 435, 437, 445, 455, 471, 479, 483, 509, 510

rules of arithmetic, 51, 380
running out of steam, 301, 397
Rushdie, Salman, 74
Russian (language and people), 36, 77, 91, 122, 134, 149, 170, 190, 196, 209, 211, 492
RVH (right ventricular hypertrophy), 237, 279, 281, 491

S

sacrosanct, 8, 30, 119, 178, 259, 487

Sacks, Jonathan, 79, 460, 494, 502

safety, 14, 27, 90, 130, 178, 217, 218, 246, 252, 263, 273, 297, 306, 338, 342, 345, 355, 360, 367, 380, 447, 468, 474

Salman Rushdie, 74
sampling error, 254
sanctimonious, 101
sanctimony, 51, 76, 321
sanction, 157, 178, 329, 411
sang froid, 194
Sarah Granger, 386, 513
SBE (subacute bacterial endocarditis), 253, 285, 465, 491
scale (and dimension), 111, 160, 168, 170, 263, 318, 319, 334, 369, 387, 391, 399, 400, 417, 418, 420, 437, 442, 470, 474
scammers, 97
sceptics, 252, 276, 277, 352, 479

science, 1, 3, 5, 6, 7, 14, 16, 17, 19, 20, 23, 36, 46, 47, 50, 51, 52, 58, 60, 66, 67, 68, 81, 98, 102, 105, 112, 117, 133, 138, 139, 148, 162, 181, 218, 219, 248, 269, 271, 274, 290, 293, 294, 302, 305, 308, 333, 334, 336, 341, 342, 373, 377, 378, 388, 389, 392,393, 398, 403, 404, 406, 408, 409, 446, 449, 454, 455, 457, 466, 471, 479, 480, 486, 487, 499, 501, 507, 508

science and art of medicine, 22
science of medicine, 2, 4, 5, 6, 7, 19, 23, 24, 38, 45, 60, 81, 93, 168, 406, 408, 409, 486, 487
science-based, 7, 408
scientific basis, 21, 138, 259, 392
scientific detachment, 231
scientific determinism, 50
scientific medicine, 6, 259, 386, 444
scientific method, 51, 144, 303, 304, 323, 356, 377
scientific perspective, 8
scientistic, 138
scoundrels, 74, 84, 494
medical screening, 13, 36, 106, 115, 271, 463, 481
script reader, □406
Scylla and Caribdis, 497
search engine, 327
security, 17, 25, 61, 83, 102, 112, 161, 185, 194, 195, 212, 215, 217, 327, 338, 342
selection criteria, 14, 94, 149, 313, 315, 316, 341, 342, 368, 369, 426
selection process, 40, 53, 310
self esteem, 37, 100, 104, 105, 106, 190, 194, 217, 321, 357, 365, 373, 374, 382, 410, 454, 496
selfless, 98
self-protective behaviour, 479
self-sacrificing principle, 186, 496

Semmelweis, Ignaz, Dr., 92, 355, 494
sensation, 237, 240, 276, 279, 385, 397
sense of humour, 474
sense of timing, 14, 53, 101, 400, 445

sensitive, 20, 25, 28, 36, 50, 108, 170, 178, 203, 206, 209, 230, 231, 233, 238, 246, 253, 270, 271, 276, 278, 282, 287, 289, 300, 301, 303, 324, 343, 349, 365, 403, 424, 431, 434, 509

sensitivity (test characteristic), 8, 24, 27, 53, 120, 135, 184, 201, 208, 211, 220, 222, 232, 250, 253, 255, 257, 276, 299, 300, 301, 403, 434, 445, 461

sentient human being (s), 3, 55, 180, 244, 396
sentiment, 25, 36, 80, 117, 390, 447, 505
septal contraction, 394
septicaemia, 126, 220, 241, 311, 339, 340, 399, 420, 450
septum (cardiac), 255, 393, 394, 422
serendipity, 357, 404
Shakespeare, William, 8, 10, 13, 54, 63, 80, 112, 181, 343, 358, 455, 470, 492, 494, 495, 502
Sheehan's Syndrome, 109
sheep, 69, 212, 300, 317
shepherd, 317
Sherlock Holmes, 284
shifting dullness, 98
SHO (senior house officer), 14, 54, 55, 244, 402
shortness of breath, 36, 210, 213, 216, 247, 264, 277, 279, 301, 302, 369, 376
short termism, 485
show biz, 472
sickness, 12, 175, 371, 462, 502
silicon-based, 332, 372
simple tiredness, 125
sinus bradycardia, 277, 367, 481
sinus rhythm, 240, 375, 400, 436, 437
Sistine Chapel, 334
Sjörgren's Syndrome, 300
skew, 385

skills, 1, 3, 11, 13, 19, 20, 23, 24, 68, 106, 171, 177, 178, 192, 201, 206, 230, 235, 279, 305, 324, 359, 360, 373, 395, 408, 411, 429, 444, 454, 472, 514

sleep, 100, 101, 125, 126, 154, 155, 171, 215, 216, 236, 350, 361, 404, 411, 412, 424, 440, 441, 442, 452, 464, 466, 484, 502

sleep disturbance, 125, 216, 440
slobs, 98, 99, 156, 495
smart, 56, 75, 87, 93, 120, 121, 166, 234, 326, 332, 381, 390, 415, 443

smokers, 37, 117, 224, 263, 369, 438
social benefits, 194
social division, 25
social media, 112, 162, 291, 297, 298, 326, 382, 415
social status, 196, 214
socio-economic class, 106, 116, 120, 164, 214, 263, 464
sovereign (duty), 5, 8, 24, 30, 35, 44, 147, 178, 410, 411, 427, 476, 487
sovereign right, 8
spanophilia, 81, 99
specialisation, 16, 129

specific, 13, 14, 94, 185, 186, 188, 191, 200, 201, 205, 206, 210, 212, 213, 222, 240, 247, 253, 255, 257, 261, 267, 273, 276, 277, 293, 300, 303, 310, 312, 325, 330, 337, 340, 349, 385, 388, 401, 404, 405, 414, 415, 418, 432, 443, 451, 458, 460, 472, 479, 481, 495

specificity, 201, 221, 250, 253, 255, 270, 271, 276, 277, 278, 290, 299, 301, 389, 433
spectrum, 57, 123, 190, 222, 354, 382, 404, 445
speculation, 388
spleen, 94, 239
spontaneity, 81, 225
sport, 19, 36, 193, 275
SpR (senior registrar), 55
sputum, 238, 418, 420
St, George's Hospital, 5, 41, 68, 106, 209, 233, 240, 395, 443, 468, 481, 506, 508

staff (medical), 8, 31, 38, 82, 84, 98, 126, 136, 148, 155, 156, 159, 161, 164, 173, 175, 202, 207, 243, 244, 252, 283, 287, 298, 331, 362, 378, 415, 419, 420, 421, 443, 446, 447, 453, 459, 469, 470, 472

staffing, 4
standard deviation, 45, 294, 325, 326
standard of living, 194
standardisation, 13, 18, 129, 304, 338, 378, 381
standardise, 177, 278, 304, 311, 388
standards, 25, 29, 30, 31, 52, 57, 156, 177, 226, 232, 242, 304, 358, 371, 391, 414, 416, 421, 480
standards of care, 416
star-struck, 100
state of health, 36, 159, 216, 374, 487
'statin' drugs, 114
sttaistical analyses, 27, 293, 313, 319, 322, 323, 324, 335, 479
staistical appraisal, 263
statistical paradox, the, 8, 37, 265, 295, 299, 319, 322, 391, 398, 423, 426, 452, 483
statistical significance, 92, 181, 315, 317
status quo, 124, 195, 217, 510
statutory power, 80, 111
stents (vascular), 37, 351, 368, 376

stereotype, 45, 193
stereotypical, 68, 190, 193
steroid(s), 162, 224, 418, 458
stethoscope, 230, 233, 237, 393, 443
stiff upper lip, 187, 194, 496
stock market, 94, 309
stoic, 150, 247, 439
strategic judgments, 115
strategies, 38, 98, 146, 215, 264, 323, 325, 424, 454

strategy, 86, 108, 207, 227, 234, 246, 257, 261, 264, 265, 274, 275, 288, 291, 302, 305, 312, 316, 340, 365, 378, 400, 431, 439, 460

streptokinase, 420, 481, 511
stress, 123, 208, 209, 215, 244, 245, 412
stress-related, 484
stroke, 174, 208, 217, 240, 241, 245, 262, 331, 437, 439, 459, 512
strong as a horse (patient claim), 269, 384

style, 4, 6, 14, 23, 25, 57, 58, 80, 82, 114, 120, 143, 172, 194, 198, 205, 206, 207, 210, 231, 244, 289, 340, 372, 396, 414, 415, 416, 428, 432, 433, 445, 446, 453, 467, 469, 471, 474

subconscious, 185, 332, 352, 412, 461
sub-endocardial atrial fibrosis, 235
sub-groups, 190
successful medical management, 107

suffering, 12, 31, 35, 43, 49, 60, 94, 108, 120, 121, 137, 145, 146, 148, 149, 175, 176, 195, 233, 249, 257, 280, 282, 317, 337, 343, 410, 432, 450, 487

suitability, 10, 185, 210, 406, 473
Sullenberger, Chesney ('Sully'), 384
Superman, 159, 232, 279, 319
Superman Complex, 159, 232, 279

surgeon, 12, 14, 17, 22, 27, 47, 53, 56, 72, 89, 90, 94, 95, 96, 102, 123, 126, 137, 139, 146, 173, 175, 210, 239, 263, 286, 288, 358, 370, 388, 392, 416, 425, 446, 449, 451, 453, 460, 475, 478, 501, 509

surrogate, 150, 322, 380

survival, 13, 17, 26, 54, 75, 86, 96, 131, 150, 154, 185, 192, 220, 248, 294, 312, 323, 325, 338, 351, 362, 376, 420, 422, 450, 457, 473

suspicion, 275, 411, 460
SVT (supra-ventricular tachycardia), 269, 491

sympathetic, ☐143, 148, 150, 166, 173, 196, 210, 232, 334, 431, 481, 489
syncope, 31, 47, 209, 240, 256, 314, 367, 374, 375

T

Tabellverket, 294, 499
taking blood (phlebotomy), 260, 418

talent, 1, 6, 7, 8, 13, 14, 15, 16, 17, 18, 19, 29, 53, 64, 66, 83, 89, 98, 147, 231, 266, 274, 283, 298, 301, 303, 310, 334, 338, 341, 365, 367, 405, 406, 408, 445, 446, 449, 459, 511

tamoxifen, 221, 484
target, 13, 298, 437, 501
Tarot (cards), 256, 389
Tired All the Time (TATT), 125, 127, 256, 257, 484
taurine, 322, 482
teacher, 21, 80, 92, 174, 184, 210, 214, 217, 233, 311, 343, 380, 394, 402, 428, 443, 500, 508

teachers, 48, 55, 59, 60, 80, 82, 96, 100, 134, 141, 172, 173, 175, 214, 215, 217, 233, 311, 324, 341, 342, 371, 505, 508

teaching exercise, 266
teaching hospital medicine, 253
teaching hospital standards, 25
team players, 102
technical excellence, 205, 409
technical know-how, 112
technique, 81, 114, 206, 207, 208, 223, 236, 254, 266, 316, 508
teichopsia, 277
telephone, 121, 143, 148, 156, 157, 159, 160, 214, 219, 224, 258, 275, 288, 297, 448, 456, 463, 466
temperature, 169, 222, 288, 320, 344, 371, 384, 386, 399, 404, 456, 491
tendon reflexes, 235
Tennyson, Lord Alfred, 487
terminal illness, 403
territorial, 47
tertiary medical centres, 205
Tessa Jowell, 485, 502
tetracycline, 219, 256
Theophrastus, 8, 63, 64, 79, 494
theorem, 392, 499
theoretical, 17, 46, 284, 392, 482
theories, 12, 22, 72, 80, 116, 133, 296, 308, 327, 376, 391, 392, 400, 405, 457, 481, 511
theory of relativity, 308
third party referees, 267
three wise men, 266

thrombophlebitis, 145
thyroid gland, 375, 490
thyrotoxicosis, 235, 277, 468
thyroxine, 235
transient ischaemic attack (TIA), 126, 207, 239, 285
tick-box processes, 13
Tietze's Syndrome, 280
time frame, 53, 275, 410, 445
time management, 417
time priorities, 385
time-restricted, 387
timing, 82, 280, 339, 399, 498
tiredness, 125, 247, 257, 484
tools, 6, 8, 21, 23, 26, 51, 57, 293, 313, 408, 433, 452, 454, 487
topology, 46, 394
Tortelier, Paul, 394
toxic state, 239, 266, 396
Traffic analysis, 244, 245

training, 2, 6, 15, 16, 21, 40, 66, 87, 118, 162, 168, 171, 225, 266, 275, 307, 316, 329, 346, 370, 371, 413, 445, 454, 461, 468, 473,

training programmes, 413
traits, 19, 20, 26, 63, 64, 69, 78, 98, 100, 101, 131, 134, 150, 151, 185, 187, 188, 302, 484
transformability, 329
transient complaints, 247
trans-oesophageal echocardiogram (TOE), 239
travel, 32, 143, 147, 185, 188, 192, 211, 212, 297, 308, 342, 400, 417, 426, 428, 447, 463
Travis effect bias, 296
trickery, 319
Trident aircraft, 303
trigger factors, 113
true character, 302
Trump, Donald, 103, 378

trust, 7, 10, 14, 18, 19, 22, 27, 43, 56, 96, 99, 107, 117, 118, 121, 124, 141, 144, 145, 148, 157, 172, 176, 177, 196, 197, 202, 203, 206, 212, 213, 226, 227, 230, 232, 242, 243, 289, 311, 319, 323, 357, 363, 372, 382, 391, 406, 409, 410, 416, 424, 425, 428, 429, 430, 431, 432, 433, 445, 447, 451, 462, 472, 473, 474, 476

t-test (statistics), 294
Türing, Alan (and test), 39, 244, 296, 332, 396, 500
t-wave (ECG), 235
t-wave inversion (ECG), ▫236
Two Cultures, 50

Types A & B, 103

U

ubiquitous, 79, 137, 318, 359, 472
unbending adherence, 305
uncompromising, 20, 33, 71, 80, 124, 144, 164, 383
unconscious patients, 25
unconventional, 18, 51, 226, 391
uncooperative, 167, 207
uncoordinated, 53, 445
unctuousness, 473
undernourished, 318, 400

understanding, 6, 8, 13, 15, 23, 25, 27, 34, 37, 45, 50, 64, 67, 82, 92, 99, 105, 109, 118, 120, 149, 151, 152, 165, 168, 169, 171, 173, 180, 189, 193, 196, 198, 202, 207, 210, 216, 230, 243, 253, 303, 304, 309, 313, 318, 319, 335, 352, 364, 367, 377, 392, 395, 396, 408, 423, 424, 425, 426, 429, 459, 461, 464, 465, 474, 480, 509

undressing, 232
unique, 332
university graduates, 388, 499
unpleasant, 33, 74, 89, 100, 103, 108, 148, 230, 250, 354, 430, 433, 438
unscientific, 319
unstructured, 101
unworthy (test), 265
urgency, 251
Uriah Heep, 473
UTI urinary tract infection, 255
URTI: upper respiratory tract infection, 219, 221
US Army, 84, 305
US health corporation, 298
use of time, 19, 244
User eXperience (UX), 297
Utopia, 60, 111, 355, 357, 454, 461, 462, 501, 502
Utopian society, 79

V

vaccination, 117, 320
validity, 8, 24, 146, 252, 293, 313, 314, 511
Valsalva manoeuvre, 265
valuable skills, 44

values, 8, 28, 38, 129, 130, 134, 181, 182, 183, 184, 187, 189, 190, 191, 204, 214, 274, 294, 324, 340, 341, 343, 358, 371, 378, 382, 385, 407, 427, 428, 461, 501

vanity, 316, 357
vasomotor syncope, 209, 240, 367, 374, 375
vectors ECG), 46, 237
vegetarian, 295
venesection, 53, 103, 254, 282, 433, 446, 475
verification, 38, 184, 213, 233, 254, 294, 320, 321, 392, 429, 482
versatility, 51, 53, 84, 120
vertigo, 240
ventricular fibrillation (VF), 240, 252, 400, 464, 491
village medicine, 471
VIP, 204
visionaries, 103
vitamin(s), 72, 162, 226, 263, 318, 322, 351, 482
vitamin B12, 318
vitamin K antagonists, 226, 263
vive la différence, 196
vocabulary, 106, 118, 193, 196, 210, 424
vocation, 41, 61, 183, 198, 410
voltages, 236, 237, 301
ventricular septal defect (VSD), □237
VT□ 140, 240, 252, 302, 367, 491

W

waffle, 379
waiting room, □154
Walton, Izaak, 100, 119, 494
Warren Buffett, 381, 389, 452, 501
wasp chewing behaviour, 57
Waugh, Evelyn, 50, 86, 289, 494
waveform, 238, 279, 299, 436

wealth, 4, 8, 28, 30, 77, 102, 116, 120, 121, 164, 183, 184, 194, 196, 198, 214, 219, 424, 430, 449, 452, 472, 510

Weber, Max, 488
weight of evidence, 279
weighted, 181, 306, 312, 314, 338, 378, 383, 385
Welchman, Gordon, 244, 332, 497
wheezes, 238
wheezing, 216, 220, 238, 281, 418
Whipp's Cross Hospital, 22, 34, 48, 54, 72, 73, 284, 307, 443, 509

William Harvey, 59, 65, 113, 321, 368, 493, 499
William Herschel, 404
Wilson, Clifford, Prof., 21, 284, 492
wimps, 104, 165
wise judgement, 55, 338, 350
wit, 46, 81, 122, 332, 335, 379, 382, 509
Wizard of Oz, 201, 332, 357, 497

women, 11, 32, 40, 57, 58, 107, 113, 123, 138, 175, 186, 191, 192, 221, 232, 238, 267, 301, 314, 329, 366, 432, 438, 440, 464, 473, 475, 484

work absence, 257
worthy (test), 265
World War Two, WW2, 122, 307, 426
Wright, John, Dr., 402, 505, 506, 508

X

X-ray, 40

Y

Yiddishkeit, 186
Yin and Yang, 309

Z

Zeus, 145, 176
zoning out, 248
zoo, 130, 306, 310, 366
zookeeper, 306